ENGINEERING ECONOMICS

McGraw-Hill Series in Industrial Engineering and Management Science

Consulting Editor

James L. Riggs, *Department of Industrial Engineering, Oregon State University*

Barish and Kaplan: *Economic Analysis: For Engineering and Managerial Decision Making*

Blank: *Statistical Procedures for Engineering, Management, and Science*

Cleland and Kocaoglu: *Engineering Management*

Denton: *Safety Management: Improving Performance*

Dervitsiotis: *Operations Management*

Gillet: *Introduction to Operations Research: A Computer-oriented Algorithmic Approach*

Hicks: *Introduction to Industrial Engineering and Management Science*

Huchingson: *New Horizons for Human Factors in Design*

Law and Kelton: *Simulation Modeling and Analysis*

Lehrer: *White-Collar Productivity*

Love: *Inventory Control*

Polk: *Methods Analysis and Work Measurement*

Riggs and West: *Engineering Economics*

Riggs and West: *Essentials of Engineering Economics*

Wu and Coppins: *Linear Programming and Extensions*

ENGINEERING ECONOMICS

Third Edition

James L. Riggs
Professor and Department Head
Industrial and General Engineering
Director, Oregon Productivity Center
Oregon State University

Thomas M. West
Associate Professor of Industrial and General Engineering
Oregon State University

McGraw-Hill, Inc.
New York   St. Louis   San Francisco   Auckland   Bogotá
Caracas   Lisbon   London   Madrid   Mexico   Milan
Montreal   New Delhi   Paris   San Juan   Singapore
Sydney   Tokyo   Toronto

ENGINEERING ECONOMICS

6 7 8 9 10 11   BRBBRB   92 9 8 7 6 5 4 3 2 1 0

ISBN 0-07-052873-X

This book was set in Times Roman by Monotype Composition Company, Inc.
The editor was Anne Murphy;
the production supervisor was Leroy A. Young;
the cover designer was Joe Gillians;
the designer was Charles A. Carson.
Project supervision was done by The Total Book.

Library of Congress Cataloging-in-Publication Data

Riggs, James L.
  Engineering economics.

  (McGraw-Hill series in industrial engineering and management science)
  Bibliography: p.
  Includes index.
  1. Engineering economics. 2. Managerial economics.
I. West, Thomas M.   II. Title.   III. Series.
TA177.4.R53   1986   658.1′5   85-18018
ISBN 0-07-052873-X

ABOUT THE AUTHORS

JAMES L. RIGGS is the founder and head of the Oregon Productivity Center and is concurrently department head and professor of industrial engineering at Oregon State University. He holds degrees in forest engineering, mechanical engineering, and industrial engineering from Oregon State University. He was a Fulbright Scholar to Yugoslavia in 1975 and has since been a visiting professor at universities in Mexico, Costa Rica, Hawaii, Israel, Australia, and Greece. He is also a faculty affiliate at the Japan-American Institute of Management Science. He is the author or coauthor of 18 books in the areas of engineering economics, production, operations research, and management. He has also authored over 45 technical papers and 120 articles. He is a registered professional engineer and has served as vice-president of the American Institute of Industrial Engineers.

THOMAS M. WEST is an associate professor and assistant department head of industrial engineering at Oregon State University. Prior to joining the faculty at OSU, he taught at the University of Tennessee and the Georgia Institute of Technology. He holds B.S. and M.S. degrees from the University of Tennessee-Knoxville and the Ph.D. from Oregon State University. He is a member of Alpha Pi Mu, Sigma Xi, and Phi Kappa Phi honoraries. Professional activities include previous positions with the Chemstrand Division of Monsanto Chemicals and IBM Corporation. He has also served as a consultant to over two dozen companies and government organizations and regularly teaches short courses on a consulting basis throughout the Northwest. He is a registered professional engineer and has served as vice-president of the American Institute of Industrial Engineers.

CONTENTS

PREFACE

The curriculums of most professional schools include a course in applied economics under such titles as *engineering economy, financial management,* and *managerial economics.* Practicing professionals usually rate their "econ" course as one of the most useful subjects taken in college. This book is written to make that econ experience as rewarding as possible and to provide a comprehensive reference for future applications.

Engineering economics is a fascinating subject. Its core is decision making based on comparisons of the worth of alternative courses of action with respect to their costs. Decisions vary from those that are concerned with personal investments to those that have to do with corporate capital budgeting. They must be made at all organizational levels in both public and private sectors of the economy. Tools for decision making range from standardized worksheets for discounted cash-flow evaluations to refinements necessary for sensitivity and risk analysis. The practices followed are grounded in classical economic theory, operations research, and other disciplines. Most of the applications are intuitively logical and computationally simple, but the underlying principles are conceptually demanding.

Text Contents

The following brief tour through the third edition of *Engineering Economics* indicates how the diverse characteristics described above are coordinated into a logical flow of subjects that is both comprehensive and comprehensible.

- An introductory chapter traces the history of engineering economic thought to show how the generic title *engineering economics* represents a blend of subjects that builds on the traditional engineering concern for operating economies to include *financial considerations, management concepts,* and *decision-analysis techniques.*

- The conventional mathematics of money is presented in Section One. Five chapters are devoted to the mechanics of *time-value calculations* and comparisons of alternatives based on their *equivalent annual worth, present worth,* and *rate of return.* Interest factors follow ASEE-suggested functional notations and interest calculations are displayed on cash flow diagrams. A wide variety of examples and numerous worked-out review exercises are provided.

- Applications of discounted cash flow comparisons are detailed in Section Two. Chapters 7 and 8 present ways to structure evaluations to determine a *preferred investment alternative* or *replacement policy. Financial considerations* for private investments and *benefit-cost analyses* for government projects are discussed in Chapters 9 and 10. This section thus examines economic evaluation models and delineates operating features.

- Practicalities of *money management* is the subject of Section Three. A chapter highlighting current *depreciation* and *corporate income tax* considerations precedes a chapter about *industrial practices* that explores estimating techniques, proposal justification procedures, and financing methods. These subjects are followed by a pair of timely chapters, 13 and 14, on the *effects of inflation* and *sensitivity analysis.*

- Readily applicable *decision models* are introduced in Section Four. Chapter 15 shows how and why *breakeven analyes* are used to examine current conditions. *Risk analysis* in Chapter 16 starts with probability concepts and proceeds to the fundamental *expected-value criterion* that guides decisions in which future conditions are uncertain. The last chapter extends the expected-value concept to *discounted decision trees* and addresses the controversial subject of "intangibles," advocating a tabular technique that accommodates both convenient and difficult-to-quantify data for *economic decision making.*

- The economics of production, as characterized by the *money value of time,* is featured in Section Five. Fundamentals of *accounting* and ways to use financial data to *forecast* future economic conditions are discussed in Chapter 19. Methods for evaluating and improving the *economic efficiency of production operations* are given in Chapter 20, followed by a chapter on the cost-effective *management of time and materials.* Chapters 21 and 22 treat the vital topic of *productivity;* the former comprehensively relates *productivity factors to profitability* and the latter concentrates on *productivity measurement.*

- Section Six returns to consideration of risk and uncertainty. Chapter 23 expands upon the concepts introduced in the previous two sections to provide more sophisticated *risk analyses* of the factors involved in *production operations.* Comparable analysis methods for evaluating the effects of *risk on cash flows* are given in Chapter 24. The book concludes with a final chapter on basic criteria for making decisions under conditions of uncertainty, an explanation of *utility theory,* and a discussion of *game theory,* exotic but enlightening tools to equip the complete engineering economist.

Third Edition Additions

Everyone familiar with the previous edition of *Engineering Economics* will find many unfamiliar features in this third edition. Since it is an uncommon privilege to have three opportunities to enrich and refine a book, it has been revised solicitously. The more significant amendments include

- *Margin notes.* Most conspicuous are notes added in margins to explain or comment on adjacent subjects without interrupting the flow of the text. They are anecdotes, amplifications, and annotations.

- *Updated material.* Recent changes in tax laws required a complete revision of the chapter on depreciation and corporate income taxes. Changing economic conditions are reflected in revisions to such other subjects as inflation analysis and industrial practices.

- *Expanded coverage.* Growing interest in productivity and risk analysis led to an additional chapter for both subjects. These expansions recognize topical areas that are receiving ever-increasing attention from engineering economists.

- *New appendixes.* A glossary of frequently encountered words and terms is located in an appendix. Answers to over half the end-of-chapter problems are included in another appendix.

- *More extensions.* Additional thought-provoking and skill-development topics make the popular *chapter extensions* even more useful. Included are such current-interest subjects as the use of microcomputer spreadsheets, adjustable rate loans, advanced-technology replacements, increment graphing, potential tax reforms, stock market workings, and many more.

Besides these obvious amendments and supplements, numerous other revisions have been made to make the contents more understandable and easier to assimilate. Based on input from users, examples were modified to better illustrate key points, whole sections were moved or combined to improve the flow, and many new problems were added to provide greater variety.

Continued Features

A text on engineering economy is expected to have plenty of detailed examples, comprehensive interest tables, and an ample assortment of practice problems. This book has them. Realistic examples are interspersed throughout each chapter, and a collection of worked-out review exercises at the end of each chapter further expands on solution techniques. There are over 1050 questions and problems. A complete instructor's manual provides solutions to all of them.

The format is designed to provide a livelier and more student-oriented text. Informal prose is intended to appeal to practitioners without insulting academicians. Overly elaborate theoretical developments are avoided. Concepts are reinforced with examples and are related to actual practices wherever possible. An overview at the beginning of each chapter serves as an introduction to key topics and as a summary of the contents for review purposes.

Unique insights and general approaches favored by users of the second edition, and retained in this revision, include

- Examining financial realities from the business world—both opportunities and restrictions—that influence economic decisions.

- Providing special attention to the analysis of public projects and service programs, while recognizing the often conflicting issues and responsibilities involved.

- Bringing out current concerns confronting today's engineering economists as they face tomorrow's challenges, such as measures for protecting the environment, utilizing resources more effectively, and improving productivity.

- Presenting a number of amplifications of conventional economic comparisons which use probability-based models, thought-provoking concepts for decision making, and historical developments that provide a perspective for current practices.

There is ample material in the text for a two-term course sequence of basic and advanced study. In addition to the 25 chapters that examine all the conventional subjects—and many more—the *extensions* that appear at the end of each chapter provide further flexibility. These are supplementary topics that can be treated as study cases, because questions are included for each one; or they may be considered as readings, because they are outside the regular flow of subject matter. Some of them deal with personal finances and socioeconomic situations (creativity, pollution control, noncorporate taxation, causes and consequences of inflation, value of a human life, etc.), whereas others introduce analysis methods that complement those in the chapters (life cycle costing, analysis of financial statements, fault trees, cause-and-effect analysis, learning curves, queuing theory, PERT, productivity audits, etc.).

Tribute

Money is the modern equivalent of the long-sought "philosopher's stone." For centuries, alchemists vainly sought the stone that could transform one type of metal into another. Now we have that capacity, indirectly. With money as the medium of exchange, we can convert one type of resource into other types easily and rapidly, but not always wisely. The worthiness of this transformation is a subtle aspect of the mission of engineering economists. Telltale analyses of alternatives reveal the innermost workings of a project and burden the analyst with ethical responsibilities atop fiscal obligations. A prerequisite to bearing the burden is a thorough knowledge of accepted economic principles and practices. This knowledge facilitates putting a legitimate monetary value on the transformation of each resource and allows an accurate and conscientious appraisal of worthiness when combined with technical expertise about the subject. Resource commitments monetized as cash flows set a quantitative framework for ensuing qualitative value considerations. This book is dedicated to the economic analysts who contribute to resource-allocation decisions. May you do so wisely.

It can be said that those who manage people manage people who manage works, but those who manage money manage all. We hope that you accept the challenge and enjoy a satisfying and profitable experience from *Engineering Economics*.

James L. Riggs
Thomas M. West

ENGINEERING ECONOMICS

CHAPTER 1

INTRODUCTION TO ENGINEERING ECONOMICS

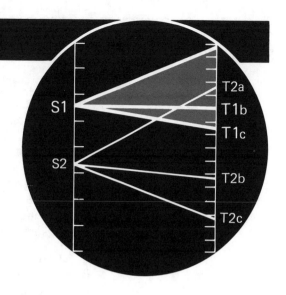

OVERVIEW

Engineers are planners and builders. They are also problem solvers, managers, and decision makers. Engineering economics touches each of these activities. Plans and production must be financed. Problems are eventually defined by dollar dimensions, and decisions are evaluated by their monetary consequences. Much of the management function is directed toward economic objectives and monitored by economic measures.

Engineering economics is closely aligned with conventional microeconomics, but it has a history and a special flavor of its own. It is devoted to problem solving and decision making at the operations level. It is subject to suboptimization—a condition in which a solution satisfies tactical objectives at the expense of strategic effectiveness—but careful attention to the collection and analysis of data minimizes the danger.

An engineering economist draws upon the accumulated knowledge of engineering and economics to identify alternative uses of limited resources and to select the preferred course of action. Evaluations rely mainly on mathematical models, but judgment and experience are pivotal inputs. Many accepted models are available for analyses of short-range projects when the time value of money is not relevant, and of long-range proposals when discounting is required for input data assumed to be known or subject to risk. Familiarity with these models, gained from studying subsequent chapters, should guide your passage through the engineering economic decision maze.

ENGINEERING DECISION MAKERS

Which one of several competing designs should be selected?

Should the machine now in use be replaced with a new one?

With limited capital available, which investment alternative should be funded?

Would it be preferable to pursue a safer conservative course of action or follow a riskier one that offers higher potential returns?

Among several proposals for funding that yield substantially equivalent worthwhile results but have different cash flow patterns, which is preferable?

Are the benefits expected from a public-service project large enough to make its implementation costs acceptable?

Two characteristics of the questions above should be apparent. The first is that each deals with a choice among alternatives, and the second is that all involve economic considerations. Less obvious are the requirements of adequate data and an awareness of technological constraints, to define the problem and to identify legitimate solutions. These considerations are embodied in the decision-making role of engineering economists to

1 Identify alternative uses for limited resources and obtain appropriate data
2 Analyze the data to determine the preferred alternative

The breadth of problems, depth of analysis, and scope of application that a practicing engineer encounters vary widely. Newly graduated engineers are regularly assigned to cost-reduction projects and are expected to be cost conscious in all their operations. As they gain more experience, they may become specialists in certain application areas or undertake more general responsibilities as managers. Beginners are usually restricted to short-range decisions for low-budget operations, whereas engineering managers are confronted with policy decisions that involve large sums and are influenced by many factors with long-range consequences. Both situations are served by the principles and practices of engineering economics.

Engineering economic decisions often occur at the interface of technology and management, where engineered efficiencies meet financial practicalities.

A decision is simply a selection from two or more courses of action, whether it takes place in construction or production operations, service or manufacturing industries, private or public agencies. Some choices are trivial or largely automatic, but other decisions can be challenging and exciting experiences. Most major decisions, even personal ones, have economic overtones. This consistent usage makes the subject of engineering economics especially challenging and rewarding.

ENGINEERING AND ECONOMICS

Before 1940, engineers were mainly concerned with the design, construction, and operation of machines, structures, and processes. They gave less attention to the resources, human and physical, that produced the final products. Many factors have since contributed to an expansion of engineering responsibilities and concerns.

Besides the traditional work with scientists to develop new discoveries about nature into useful products, engineers are now expected not only to generate novel technological solutions but also to make skillful financial analyses of the effects of implementation. In today's close and tangled relations among industry, the public, and government, cost and value analyses are expected to be more detailed and inclusive (e.g., worker safety, environmental effects, consumer protection, resource conservation) than ever before. Without these analyses, an entire project can easily become more of a burden than a benefit.

Most definitions of engineering suggest that the mission of engineers is to transform the resources of nature for the benefit of the human race. The types of resources susceptible to engineering enrichment include everything from ores and crops to information and energy. A growing awareness of finite limits for earth's resources has added a pressing dimension to engineering evaluations. The focus on scarce resources welds engineering to economics.

Dr. Paul S. Samuelson, Nobel laureate in economics, says that

Neglected in most definitions of engineering is the sales function by which engineers convince users to implement their suggestions.

> Economists today agree on a general definition something like the following: Economics is the study of how men and society end up *choosing,* with or without the use of money, to employ *scarce* productive resources that would have alternative uses, to produce various commodities and distribute them for consumption, now or in the future, among various people and groups in society. It analyzes the costs and benefits of improving patterns of resource allocation.*

The relation of engineering to economics can be likened to that of engineering to physics. Scientists are devoted to the discovery and explanation of nature's laws. Engineers work with the scientists and translate the revelations to practical applications. The "laws" of economics are not as precise as are those of physics, but their obvious application to production and the utilization of scarce resources ensures increasing attention from engineers.

ECONOMICS: A CAPSULE VIEW

Economics, like engineering, has informal roots deep in history. The construction of the pyramids is considered to be an engineering marvel. It was also a significant economic accomplishment in that it funneled all the necessary resources into monuments rather than consuming them in commerce. The formal roots of economics stretch back two centuries to the publication (in 1776) of Adam Smith's *The Wealth of Nations.*

Early writings deplored government intervention in commerce and promoted a laissez-faire policy. In *An Essay on the Principles of Population*

* P. A. Samuelson, *Economics,* 9th ed., McGraw-Hill, New York, 1973, p. 3.

(1798), Thomas Malthus conjectured about the causes of economic crises, saying that population tends to increase geometrically and the means of subsistence only arithmetically; his forecasts of misery for most of the population predisposed the "dismal science" nickname for economics. Later John Stuart Mill, in *Treatise on Political Economy* (1800), argued against Malthus' pessimism by suggesting that the laws of distribution are not as immutable as are the laws of production. Modern doomsday scenarios indicate that the issue is still in doubt.

Economic theories are plentiful. Among them are beliefs that control of the money supply is the key to control over the economy, or that adjustments should be left entirely to free-market factors.

In *Das Kapital* (1867), Karl Marx argued that capitalism would be superseded by socialism, which would then develop into communism. According to his view, workers produce more value than they receive in wages. The surplus takes the form of profit and allows capital accumulation. He argued that the capitalist system will eventually fail, owing to cyclic depressions and other inherent weaknesses. About one-third of the world's population agrees with Marx.

"New Economics" evolved from the work of John Maynard Keynes in the 1930s. In his *General Theory of Employment, Interest, and Money,** Keynes clashed with classical economic theory by proclaiming, for example, that interest rates and price-wage adjustments are not adequate mechanisms for controlling unemployment in capitalistic economies. Refinements and extensions of the original work are collectively called Keynesian economics, which is one of many current schools of economic thought.

The Keynes and Marx theories deal with the entire economic system with respect to national income, flow of money, consumption, investment, wages, and general prices. This level of analysis concerned with the economy as a whole is called *macroeconomics*. It produces economywide statistical measures such as the national cost of living index and total employment figures.

Microeconomics is the study of economic behavior in very small segments of the economy, such as a particular firm or household. It is generally assumed that the objective of a firm is to maximize profit and the objective of a household is to maximize satisfaction. Measurement statistics for a small economic unit might be the number of workers employed by a firm and income or expenditures of a given firm or family.

Engineering economics, with its focus on economic decision making in an individual organizational unit, is closely aligned with microeconomics.

ENGINEERING ECONOMY: A SHORT HISTORY

The Economic Theory of the Location of Railways, written by Arthur M. Wellington in 1887, pioneered an engineering interest in economic evaluations. Wellington, a civil engineer, reasoned that the capitalized cost method of

---

* J. M. Keynes, *General Theory of Employment, Interest, and Money*, Harcourt, Brace & World, New York, 1936.

analysis should be utilized in selecting the preferred lengths of rail lines or curvatures of the lines. He delightfully captured the thrust of engineering economy:

It would be well if engineering were less generally thought of, and even defined, as the art of constructing. In a certain important sense it is rather the art of not constructing; or, to define it rudely but not inaptly, it is the art of doing that well with one dollar which any bungler can do with two after a fashion.*

In the 1920s, J. C. L. Fish and O. B. Goldman looked at investments in engineered structures from the perspective of actuarial mathematics. Fish† formulated an investment model related to the bond market. In *Financial Engineering,* Goldman proposed a compound-interest procedure for determining comparative values and said that

It seems peculiar and is indeed very unfortunate that so many authors in their engineering books give no, or very little, consideration to costs, in spite of the fact that the primary duty of the engineer is to consider costs in order to obtain real economy—to get the most possible number of dollars and cents: to get the best financial efficiency.‡

The confines of classical engineering economy were staked out in 1930 by Eugene L. Grant in *Principles of Engineering Economy.*§ Grant discussed the importance of judgment factors and short-term investment evaluation as well as conventional comparisons of long-run investments in capital goods based on compound-interest calculations. His many contributions resulted in the recognition that "Eugene L. Grant can truthfully be called the father of engineering economy."¶

Modern approaches to discounted cash flow and capital rationing were influenced by the work of Joel Dean.** He incorporated the theories of Keynes and other economists to develop ways to analyze the effects of supply and demand for investment funds in allocating resources.

Current developments are pushing the frontiers of engineering economics to encompass new methods of risk, sensitivity, and intangible analysis. Traditional methods are being refined to reflect today's concerns for resource conservation and effective utilization of public funds.

Increasingly sophisticated cash flow analyses are accommodated by increasingly powerful computers and evaluation methods associated with management science, industrial engineering, and operations research.

Example 1.1
Economics of Energy

The divergent missions of classical economics and engineering economics are apparent when a specific application is examined. Consider the energy problem. How serious is it? What are its dimensions? What can be done about it?

* A. M. Wellington, *The Economic Theory of the Location of Railways,* Wiley, New York, 1887.
† C. L. Fish, *Engineering Economics,* 2d ed., McGraw-Hill, New York, 1923.
‡ O. B. Goldman, *Financial Engineering,* Wiley, New York, 1920.
§ E. L. Grant, *Principles of Engineering Economy,* Ronald, New York, 1930.
¶ A. Lesser, Jr., "Engineering Economy in the United States in Retrospect—An Analysis," *The Engineering Economist,* vol. 14, no. 2, 1969.
** J. Dean, *Capital Budgeting,* Columbia, New York, 1951.

Energy supply and demand relations fit familiar economic concepts. Collection of data about who uses how much energy falls into the province of economic demographics, a growing branch of economics that deals with the consequence of changes in the characteristics of a nation's population. An optimal energy mix that best utilizes the national energy supply can be determined according to market prices and market risks, economic principles that give priority to energy sources that are cheapest and carry the least risk, other things being equal.

From an understanding of supply-demand relations, engineering effort can be unleashed to overcome technological constraints. If solar or geothermal energy appears to be the most promising source, engineers must provide the means to convert the promise into reality. Engineering economics can be applied to evaluate alternative solutions.

A proposal from the National Academy of Sciences is that of easing the energy crisis by reducing waste. Amazing advances have been made since the first oil shortage in 1972, but more are needed. In 1984 American automobile drivers used the equivalent of 25 million barrels of oil each day. Experts say gasoline consumption can be cut by 40 percent by the year 2000 by requiring carmakers to meet established fuel economy standards. Other actions to plug energy leaks in the national economic machine include correcting heat losses in offices and homes, creating better building designs, building more energy-efficient machines and automobiles, and developing co-generation of electricity and steam. Estimating cash flows, comparing investment proposals, and testing the sensitivity of specific energy-conservation actions are tasks for engineering economists.

PROBLEM SOLVING AND DECISION MAKING

An engineering economist draws upon the accumulated knowledge of engineering and economics to fashion and employ tools to identify a preferred course of action. The tools developed so far are not perfect. There is still considerable debate about their theoretical bases and how they should be used. This concern is wholesome because it promises improved procedures, but the variety of analysis techniques can frustrate practitioners, especially inexperienced ones: There are many aspects to consider, and many ways to consider them.

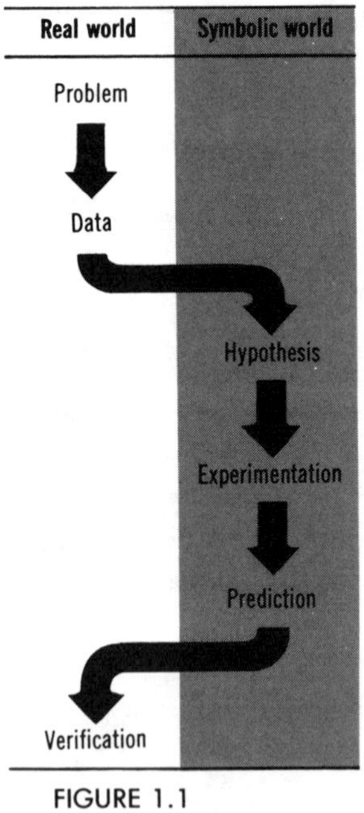

The fundamental approach to economic problem solving is that of elaborating the time-honored "scientific method." The method is anchored in two worlds: the real everyday working world and the abstract scientifically oriented world. As depicted in Figure 1.1, *problems* in engineering and managerial economy originate in the real world of economic planning, management, and control. The problem is confined and clarified by *data* from the real world. This information is combined with scientific principles supplied by the analyst to formulate a *hypothesis* in symbolic terms. The symbolic language aids the digestion of data. By manipulating and *experimenting* with the abstractions of the real world, the analyst can simulate multiple configurations of reality which otherwise would be too costly or too inconvenient to investigate. From this activity a *prediction* emerges, usually.

The predicted behavior is converted back to reality for testing in the form of hardware, designs, or commands. If it is valid, the problem is solved. If not, the cycle is repeated with the added information that the previous approach was unsuccessful. Fortunately, a host of successful approaches have been discovered and validated for economic analyses; the challenge now is to use them wisely.

Real world | **Symbolic world**

Problem

Data

Hypothesis

Experimentation

Prediction

Verification

FIGURE 1.1
Problem-solving process.

Intuition and Analysis

Because engineers generally attack practical problems with solution deadlines instead of engaging esoteric issues for long-term enlightenment, their mission might appear relatively simple. Engineering economic evaluations could even seem mundane, since they usually rely on data from the marketplace and technology from the shelf: Simply grab prices from a catalog, plug them into a handy formula, and grind out an answer. Occasionally, such a routine works. Spectacular workbench discoveries and overnight fortunes attest to the fact that plungers sometimes win. There are also innumerable instances where rule-of-thumb, skin-deep evaluations are absolutely unsatisfactory.

Methods to improve creativity are featured in Extension 1A.

As represented in Figure 1.2, a decision made now is based on data from past performances and establishes a course of action that will result in some future outcome. When the decision is shallow and the outcomes are unnotable, a reflex response based upon intuition is feasible. Instinctive judgments are often formalized by *standard operating procedures* (SOPs). In economic analyses, SOPs often take the form of worksheets for the justification of investments. Such short-form justifications are typically limited to smaller investments, say $2000, which can be recaptured from savings generated by the investment within 6 months or 1 year. These forms or similar SOPs represent collective intuition derived from experience. They have a secure place in economic evaluations, but their use should be tempered by economic principles and a continuing audit to verify that previous judgments are appropriate for current decisions.

Most significant problems require both analysis and personal judgment. Initially the analyst settles on which evaluation technique to utilize and how to apply it. As the solution procedures progress, factors that are difficult to quantify often arise. These are called *intangibles;* they represent aspects of a problem that cannot be translated readily into monetary values. Intuitive ratings are frequently assigned to intangibles to allow them to be included in the decision process. Judgment also enters the process in determining whether a solution is well enough founded to be accepted. Thus intuition and judgment complement analysis methods by contributing to better decisions.

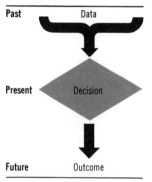

FIGURE 1.2
Decision-making process.

Example 1.2
To Intuit or to Analyze

Most decision makers informally set boundaries for routine responses to noncritical problems of a personal and professional nature. Three possible parameters to identify routine responses are shown in Figure 1.3. The level that separates an automatic decision from a problem that requires more investigation varies among decision makers.

Since there are limits to a decision maker's time and energy whereas the reservoir of problems often seems infinite, guidelines are necessary to confine involvement. SOPs do save time. An intuitive response is quick. Both draw upon experience to yield a reasonable solution. However, handy answers may mask better solutions that could have been exposed by analysis. What was good for yesterday's operations may not be adequate for tomorrow's.

FIGURE 1.3
Criteria for routine responses to an economic problem.

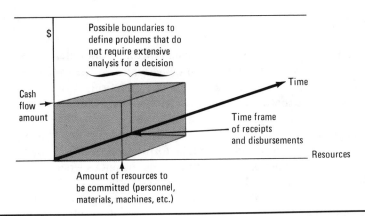

Tactics and Strategy

About the only thing more frustrating than a wrong decision for an important problem is the right decision for the wrong problem. Some problems are virtually handed to an analyst on a platter, complete with data trimmings. More commonly, a problem is ill-defined and the analyst is forced to seek the intent of a solution before applying analytical tools. Recognizing the difference between tactical and strategic considerations may clarify the purpose.

Strategy and *tactics* historically are military terms associated with broad plans from the high command and specific schedules from lower echelons, respectively. Strategy sets ultimate objectives, and the associated tactics define the multiple maneuvers required to achieve the objectives. Strategic and tactical considerations have essentially the same meaning for economic studies.

There are usually several strategies available to an organization. A strategic decision ideally selects the overall plan that makes the best use of the organization's resources in accordance with its long-range objectives. A strategic industrial decision could be a choice from among several different designs to develop or products to promote. In government, strategic evaluations could take the form of benefit-cost analyses to select the preferred method of flood control or development of recreational sites. The measure of merit for strategic alternatives is *effectiveness*—the degree to which a plan meets economic targets.

A strategic plan can normally be implemented in a number of ways. For example, each industrial design or product has tactical alternatives, such as which kind of machine to employ or materials to use; tactics for flood control might involve choices among dams, levees, dredging, etc. The relative values of tactical choices are rated according to their *efficiency*—the degree to which an operation accomplishes a mission within economic expectations.

The relationship between strategies and tactics offers some constructive insights. The effectiveness of each strategy is initially estimated from the effect it will have on system objectives. It thus serves as a guide to the area in which

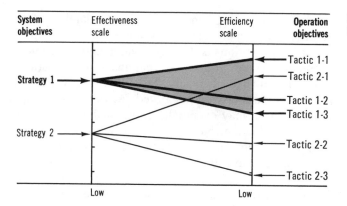

FIGURE 1.4
Relation of tactics and strategies.

tactics will produce the highest efficiency. The actual efficiency of each tactic is determined from a study of the activities required to conduct the tactical operation.

Two strategies, each with three apparent means of accomplishment, are depicted in Figure 1.4. The average efficiency for the tactics associated with strategy 1 (tactics 1-1, 1-2, and 1-3) has a higher value than that for strategy 2. However, it could happen that a strategy with a lower effectiveness possesses the tactic with the highest efficiency. Tactic 2-1 is close to the most efficient tactic of strategy 1. If it were the highest on the efficiency scale, it would be the leading candidate for selection, regardless of its strategic origin.

Sensitivity and Suboptimization

The decision situation related by Figure 1.4 has high sensitivity; that is, it is vulnerable to small changes in the controlling conditions. With tactics 1-1 and 2-1 so close on the efficiency scale, a slight change in operating conditions or external influencing factors could switch the positions of the top tactics, or even the strategies. An insensitive situation occurs when all the tactics for a given strategy have a higher efficiency than the best tactic of any other strategy. The consequence of high sensitivity is to force a complete investigation to ensure the validity of the data being evaluated.

A *sensitivity analysis* can be conducted on any problem to explore the effects of deviations from the original problem conditions. Since most engineering economic problems extend over a period of years, future cash flows are necessarily estimated. These estimations may be quite reliable, but it is often enlightening to observe how the attractiveness of alternatives varies as the initial estimates are altered.

Whenever multiple objectives are present in a decision situation, it is probable that there is no single course of action that will optimize all the objectives simultaneously. In general, suboptimization occurs when there is a larger problem than the analyst has visualized. It is always tempting to employ intact a classical textbook solution to a real-world problem, whether or not it

"You don't understand the big picture" is a common complaint that represents tactical suboptimization—a solution that promises tactical efficiency with little regard for strategic effectiveness.

truly fits the actual conditions. The availability of "canned" computerized solutions to complex problems increases the temptation. Another cause of suboptimal solutions is the legitimate analysis technique of partitioning a large problem into parts during a preliminary investigation to avoid being bogged down in a deluge of details. Trouble enters when tentative solutions to the problem's parts are not integrated. Advances in computer science and operations research may eventually allow analysis of an entire complex system in a single evaluation, but until then it helps to be aware of the areas in which suboptimization is most likely to occur. Three regularly encountered perspectives that lead to suboptimization are described below.

1 Cross-Eyed View Both organizations and individuals can be confused by opposing objectives. An example of the danger inherent in focusing on only one parameter while blurring others is what would happen to a company that redeployed its resources to save its ailing flagship product at the expense of the rest of the product line. The rescue could boost sales for the previously eminent product while total sales declined owing to the drain on resources suffered by the rest of the company's products; thus, the battle could be won but the war lost.

Individuals seeking "the good life" also get caught by conflicting goals. If "good" is interpreted as "long and full," then unlimited pleasure seeking for a full life would undoubtedly jeopardize the health needed for a long life. Moderation, however, should produce a temperate plan to satisfy both goals, resulting in a life less full but longer. Of course, there are also irreconcilable objectives such as those pictured in Figure 1.5.

2 Shortsightedness Tactics based on a planning horizon of 1 or 2 years may not have the same efficiency as do those based on a longer span of years. Suppose that a

FIGURE 1.5
Symbolic world strategy and real world tactics.

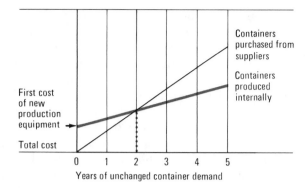

FIGURE 1.6
Pattern for potential suboptimization owing to shortsightedness.

manufacturer anticipates using a fixed number of containers each year. The containers can be purchased or the manufacturer can make them by acquiring new production equipment. Costs for the choices are displayed by the breakeven chart in Figure 1.6. A planning horizon under 2 years would indicate that purchasing is the preferable alternative; beyond 2 years it is more attractive to make the containers. Individuals face the same danger of suboptimization in lease-or-buy decisions for housing and transportation.

Organizations are very susceptible to situations in which departments understand the common goal but individually go about working toward the goal in ways that hurt each other. A typical example is the goal to reduce material and inventory costs, as viewed by

3 Tunnel Vision

- *Purchasing*   "Buy in large quantities to get quantity discounts."
- *Comptroller*   "Buy in smaller quantities to avoid paying interest on the capital required for purchases."
- *Production*   "Larger inventories allow longer production runs which reduce manufacturing costs."
- *Warehousing*   "Larger inventories cost more to store and increase the cost of material handling."

If each of the involved departments acts independently, inventory levels will behave like a yo-yo. A workable plan will obviously be a compromise, probably satisfying no one completely but still producing lower total material costs for the organization as a whole.

THE ENGINEERING ECONOMIC DECISION MAZE

Most important decisions in engineering economics entail consideration of future events. A focus on the future has always had a special and irresistible appeal, but it also encumbers the mission of engineering economists. Not only must they search the past to understand the present and survey the present for hints about the future, but they must consolidate the accumulated results into a pattern that is susceptible to analysis and then select a decision rule to yield a verdict. An indication of the complexities involved is apparent from the maze shown in Figure 1.7.

See Extension 1B for a something-for-nothing exposé.

It would take a much larger maze to portray all the pitfalls and challenges of economic analyses, but enough are included to expose the anatomy of engineering economics and to map the contents of this book. All the channels in the maze represent subjects treated in the chapters that follow. As is apparent in the decision labyrinth, there are many paths by which to progress from a problem to a solution. Which path is utilized depends on the nature of the problem and the type of analysis that is most appropriate. Because problems come in such profuse variety and there are so many ways to evaluate them, engineering economics is rich in application opportunities and offers rewarding challenges to its practitioners.

Review Exercises and Discussions

Exercise 1 The past few years have witnessed a remarkable number of events and activities that will undoubtedly have far-reaching effects on engineering practices. Some of the most prominent areas of activity include:

1 The passage of national, state, and local legislation that regulates industrial operations and developments
2 The formation of public pressure groups whose efforts are directed toward improving the quality of life.
3 Concern about low productivity in United States industries.

Select a specific example that illustrates activity in each of the areas above and relate it to the work of engineers with particular reference to engineering economic considerations.

Solution 1 Numerous examples can be cited to show how changing societal interests direct the practices of engineers.

1 *Laws.* The Occupational Safety and Health Act of 1971 (OSHA) established stricter standards for worker safety. Many states and communities have passed zoning laws that set tighter requirements for new plants. Most legislative actions require added investments, and engineers are called upon to devise ways to meet new standards with minimal expenditures.
2 *Pressure groups.* Environmental groups influenced the design of the Alaska pipeline. Consumer-protection groups successfully campaigned for automobile modifications and control of waste discharges. New technologies must be developed to protect the quality of air, water, and land; the relation between control costs and quality levels is largely an engineering function.
3 *Productive investments.* From the late 1960s to the early 1980s, productivity gains in the United States lagged behind most other developed nations. Engineers contributed to a United States resurgence by developing advanced technologies and methodologies to improve quality and operating efficiencies. Investments required for these advances

FIGURE 1.7
Engineering economic decision maze

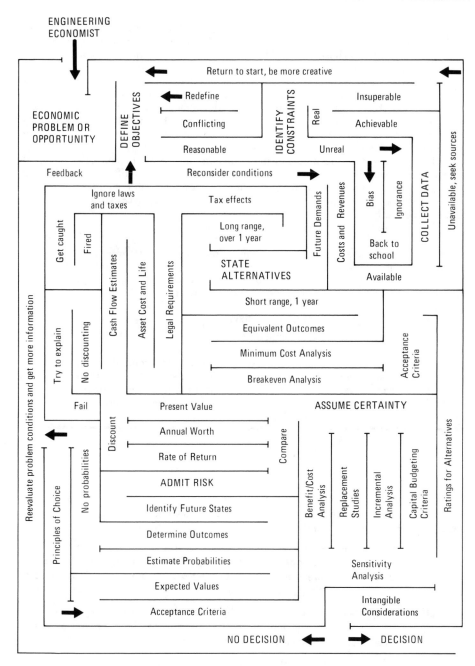

were justified by engineering economic evaluations, which in retrospect may appear obvious but at the time were often controversial.

A subassembly line has been giving the production manager nightmares for months. All **Exercise 2** kinds of minor modifications have been tried and all failed to improve output. The current per-unit cost is $4.20, which seems reasonable, but output has failed to reach the required 10,000 units per year. A check with the purchasing department reveals that supplementary units are now being purchased for $4.75 each; but one vendor agrees to provide them at $4.50 each if the entire annual demand is ordered.

The subassembly-line supervisor suggests acquiring three new machines to mechanize successive stages of the production process. Engineers calculate that the machines' total purchase price of $100,000 would make their discounted annual cost over a 10-year machine life, coupled with yearly operating costs, amount to $27,000. If needed, the machines have the capacity to double the present output. At the present output level, the remaining subassembly costs using the new machines will annually total $18,000.

While investigating the problem, the engineers uncovered another alternative: All three successive operations could be combined and handled by a single machine. This one combination machine would have the same capacity, speed, life, and remaining production costs as the three-machine alternative, but ownership and operating costs would be reduced by $3000 per year.

Which alternative should be accepted, and why?

Solution 2 Apparently it has been decided that something must be done to improve subassembly production, so the do-nothing alternative is eliminated. The single machine is obviously more attractive than is the three-machine alternative because the combined operation costs $3000 less. The unit cost for the one-machine alternative is

$$\frac{\$27,000 + \$18,000 - \$3000}{10,000 \text{ subassemblies}} = \$4.20/\text{subassembly}$$

This unit cost is the same as the current cost but promises more reliability. It is also $0.30 per unit less expensive ($4.50 − $4.20) than is purchasing all the subassemblies from a supplier. However, more information is needed about the long-range (10-year) expected demand for the subassemblies. Without this information, the decision is subject to the make-or-buy pattern of suboptimization shown in Figure 1.6.

PROBLEMS

1.1 There are many general definitions of engineering, and specific ones for different branches of engineering. Look up one of them and comment on the explicit and/or implied attention paid to economic considerations.

1.2 It has been said that economists are very busy people because they have to spend full time telling what's going to happen and full time explaining why it didn't. Engineering economists also work with forecasts of the future but are not usually subjected to such joking comments. Why?

1.3 Efficiency is defined as output divided by input (times 100 percent). Engineering efficiency is commendable when it approaches 100 percent, but financial efficiency must exceed 100 percent before it is considered adequate. Explain.

1.4 "In the 1930s many economists maintained that the United States economy had reached the apogee of its growth; in the period just after World War II, many predicted an immediate depression; and in late 1969, some economists predicted without qualification that there would be no recession in 1970; but all these assertions proved to be erroneous. This is not to deny, of course, that the predictions made by economists often are accurate; indeed, at a given moment reputable economists make so many conflicting forecasts that one is almost certain to be correct. (To give one illustration, in December of 1969 Milton Friedman said that a recession on the order of that in 1960 seems to be in the cards for 1970; Raymond J. Sulnier said there was a 50-50 chance of a recession; and Pierre A. Rinfret said that 'There ain't gonna be no recession in 1970, period.')"*

    (a)   Why do economic projections for a future event tend to vary more widely than do engineering estimates of performance for a new design?

    (b)   Why might there be less confidence in the economic performance than in the operating performance of a new machine?

The following four cases illustrate everyday situations in which individuals are required to decide the most advantageous use of limited resources. Identify the strategic or tactical nature of each decision situation and discuss the factors that should be considered (including sensitivity, if appropriate).

1.5   A fisherman stands in his crowded cabin and studies a map. Three fishing grounds are circled on the map. He mentally compares recent reports and gossip concerning the grounds. He also considers the market demands, the weather forecasts, the condition of his ship, and the supplies on board. He has to decide where to fish.

1.6   An engineer contemplates the bulging walls of a large concrete culvert under an interstate highway. The collapsing culvert is the result of thousands of tons of rock recently stockpiled on the roadbed above in preparation for new construction. In a few days the spring thaw will soak the ground and send torrents of water through the culvert. The engineer speculates on possible designs in terms of the conditions and the restrictions imposed by available equipment, materials, and time.

1.7   The owner of a wholesale distribution center seeks to improve his delivery service in order to meet competition. To do so he can buy or rent more trucks, subcontract his deliveries, open additional outlets, and/or improve his handling facilities. His capital is limited, and the outlook for increased volume is uncertain. First he must decide if any action is needed. If it is, he must select the most suitable alternative.

1.8   The manager of a large manufacturing company surveys a collection of proposals laid out neatly on the long table in the board room. Each proposal represents many hours of staff work. Each is a detailed plan for the development of a new product. She must select the proposal which best serves the interests of the company within the constraints of restricted physical and human resources, competition, legal requirements, available capital, and corporate objectives. Her decision will affect the activities of hundreds of people.

1.9   Since engineering economics has been compared with a maze in Figure 1.7, it is consistent to use a word puzzle to review some of the terminology associated with the discipline.

Across

  4   Science concerned with the economy as a whole

 12   Selection from two or more courses of action

 13   Pioneer in engineering economic evaluations

 14   Rating scale for tactical operations

* R. Handy and E. C. Harwood, *A Current Appraisal of the Behavioral Sciences*, rev. ed., Behavioral Research Council, Great Barrington, Mass., 1973.

17   Fifth step in the problem-solving process
18   _____ analysis explores the effect of deviations from original conditions
19   Perfection of a part at the expense of the whole

Down

1   "Father of engineering economy"
2   Key contributor to economics in the 1930s
3   Critical resource that fuels economic concern
4   Made his mark with *Das Kapital*
5   Science concerned with specific economic units

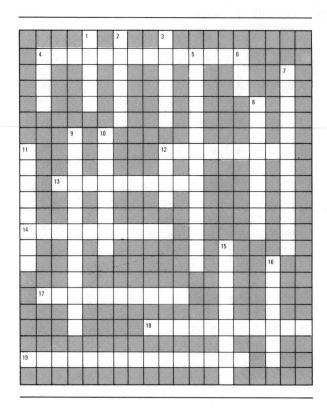

6   Standard operating procedure
7   View that could lead to suboptimization
8   Limited _____ attract attention of engineering economists
9   Measure of merit for strategic alternatives
10   Examination of the components of a problem
11   Overall plan for a major accomplishment
12   Information
15   Immediate knowledge without conscious reasoning
16   Mode of operation to achieve an operational objective

EXTENSIONS

1A Creativity

Creativity is essentially the ability to produce new and interesting results. Pure basic research seeks new enlightenment from nature. Innovation discovers a novel relation. The ability to relate things, sometimes in odd yet striking fashion, is the heart of creativity, no matter what the field or discipline. An innovative engineering economist has the ability to identify more productive uses of capital within the problem-solving process.

Originality enters the problem-solving sequence of Figure 1.1 more as an attitude than as a procedure. The first five steps could be replaced by the following personal outlooks: *What's wanted? → How can it be done? → By whom? → What if? → Maybe . . . → Aha! Eureka!* Many techniques have been proposed to expedite this chain. Some involve groups of participants, to take advantage of shared experiences and the snowball effect of one person's idea triggering many more ideas from others in the group. "Brainstorming" is the most famous group approach.*

There are also ways to improve solo ideation. Since economic analyses are often individual exercises, methods to enhance individual inventiveness are emphasized below.

Analogies. A structured approach that attempts to make the strange familiar and the familiar strange is called *synectics* (a Greek word for combining diverse elements).† It achieves an "out-of-focus" look at some aspect of the known world by employing analogical mechanisms to tap the subconscious mind. A *direct* analogy can be made between objects or systems which are not usually associated (e.g., an ant hill and an urban development). By identifying oneself with a nonhuman entity, a *personal* analogy is formed (e.g., "if I were a rudder" or "how would I behave if I were a rat in a maze?"). Images may be formed by *symbolic* or *fantasy* analogies (e.g., a rope stiff enough to climb, or if we could train bees . . .).

Triggers. Several techniques rely on words or phrases to stir innovation. The *modifier* approach fuels imagination by progressing through a list of modifiers that pertain to a key characteristic of a problem. For instance, if size were a key factor, the list could include bigger, shorter, thinner, flatter,

etc. *Checklists* can be developed for regularly encountered types of problems. They can be categorized by actions (e.g., what is done, where, when, and by whom?), changes (e.g., combine, substitute, rearrange, use differently, alter quantity), and functions (e.g., bottlenecks? waste? delays? duplication?).

Matrices. Morphological synthesis‡ involves listing the major variables of a problem in a two- or three-dimensional matrix so that interrelations can be systematically examined (e.g., a 3×3 matrix for the solution to a project-management problem could have rows labeled organization, facilities, and operations, and columns labeled personnel, money, and time. The operations-time cell might have an entry such as "speed up services"). *Trigger cards* are strips of paper on which are vertically listed possible factors that pertain to the parameters of a problem. For instance, parameters for a design could be locomotion, structure, power, and material. Factors listed under material could include wood, plastic, metal, fiberglass, paper, rubber, and foam. Ideas are generated by placing all the cards side by side and slipping them vertically, one at a time, to reveal different combinations of factors across each row. Five parameter cards, each with seven factors, would suggest 16,807 combinations, many of which would be infeasible or even ridiculous, but which nonetheless might spark an inspiration.

The relation between decision making and creativity is clarified by classifying people who deal with problems. *Problem recognizers* can detect when a problem exists, and this certainly requires intelligence. *Problem solvers* need a higher order of intelligence to develop and implement the best solution. *Problem anticipators* combine extraordinary intelligence with creativity to sense an approaching obstacle and devise means to avoid trouble before it happens.

QUESTIONS
1A.1 It has been said that the universal idea killer is silence. But vocal putdowns are effective too. Negative phrases include: It's against company policy. Has anyone else ever tried it? Write it up on one sheet of paper. It's not new; it reminds me of. . . .

---

* Alex Osborn, *Applied Imagination*, Scribner, New York, 1957.
† W. J. J. Gordon, *Synectics*, Harper & Row, New York, 1961.
‡ M. S. Allen, *Morphological Creativity*, Prentice-Hall, Englewood Cliffs, N.J., 1962.

a What other idea killers have you heard?

b What idea-boosting phrases can you suggest?

1A.2 The challenge is to suggest labor-intensive products suitable for production in cottage industries but which require no outstanding talent. They must also be salable. The products will be manufactured in homes and therefore must be produced with hand tools, and these tools cannot be expensive. Materials must also be inexpensive and readily available. Since finished products will be sold at booths in supermarkets, fairs, and craft shows, the units must be easy to transport. They should be designed to sell for less than $25 apiece, and materials should not cost over 25 percent of the selling price.

Develop a set of trigger cards to stimulate ideas. Some of the parameters might be material, use, construction, and market. Use the trigger cards to generate a list of possible products. Evaluate the list according to the economic criteria, to select the five most promising products.

1B Perpetual-Motion Mystique

The idea of perpetual motion—something for nothing—seems to be a persistent, insidious dream of humans. It has been around a long time. Over 2000 years ago the Chinese searched for an unpowered "everlasting going." Archimedes tried to find it through hydraulics, and da Vinci experimented with gravity-powered mechanisms. In 1670, John Wilkens, the Bishop of Chester, designed a ramp leading to a pedestal where a magnet was mounted. The magnet was supposed to attract an iron ball up the ramp until it fell off onto a chute that returned it to the bottom of the ramp, again and again and. . . . (What was wrong with the bishop's reasoning?)

A perpetual-motion machine was exhibited in New York in 1813. People paid to see little carriers ceaselessly moving up and down inclined planes to drive a wheel which offered free energy. Robert Fulton, of steamboat fame, exposed the hoax by showing that the contraption was connected by a hidden strand of catgut to a hand-powered crank in an adjacent room.

The infamous John E. W. Keely perpetual-motion machine, unveiled in 1875, enriched its inventor for years without disclosing any practicality. "Whatever other laws he may have violated in his long career," wrote Stanley W. Angrist in *Scientific American,* "Keely had left the first and second laws of thermodynamics inviolate."

The mystique of perpetual-motion machines has a counterpart in economic ventures—rewards without inputs.

QUESTIONS

1B.1 Cite an example of a get-rich-quick scheme or a similar hoax that promises a lot for a little.

1B.2 How could a trip through the engineering economic decision maze expose a financial hoax or at least make a potential investor wary of a scheme such as that cited in answer to Question 1B.1?

SECTION ONE

DISCOUNTED CASH FLOW MECHANICS

*I*nterest is the cost of using capital. Its history extends as far back as the recorded transactions of humanity. In earliest times, before money was coined, capital was represented by wealth in the form of personal possessions, and interest was paid in kind. For example, a loan of seed to a neighbor before planting was returned after harvest with an additional increment. We can surmise that the concept of interest in its modern sense arose from such loans for productive purposes.

Capital and credit have been important to human progress since about 5000 B.C. At that time Neolithic man was engaged in agriculture and animal culture to provide his own food. Capital was counted by seeds, tools, and herds of animals. Cattle were probably the first true productive assets and are the origin of many financial terms. *Pecuniary* stems from *pecus*, meaning "flock" in Latin, and the Egyptian term *ms*, meaning "interest," is derived from the verb *msj*, which means "to give birth." Early Greeks measured wealth in terms of cattle; in the Odyssey, Ulysses was promised a contribution "of bronze and gold to the value of twenty oxen."

By the time the Greek and Roman empires were in their ascendancies, interest rates were somewhat standardized and occasionally legislated. The amount charged for loans to the most reliable borrowers was around 10 percent, with the range from 4 percent in first-century Rome to about 50 percent for grain loans in Egypt during the same period.*

* S. Homer, *History of Interest Rates*, Rutgers, New Brunswick, N.J., 1963.

Along with the development of money and credit came abuses. Aristotle pointed out that money was "barren," and it was unfair to charge interest for loans. Early Israelites did not permit lending at interest. Romans permitted credit but limited the rate of interest to about 5 to 12 percent. Greeks encouraged credit without limit but forbade personal bondage for debt. Biblical utterances against usury were aimed at loans for consumption rather than production, yet interest was forbidden by canon law throughout the Middle Ages.

The concept of interest has not changed much through the centuries, but the modern credit structure differs markedly from that of antiquity. Lending or investing was relatively inconvenient in ancient days because transactions were made directly between individuals. There were no banking organizations to act as intermediaries and no credit instruments in the money market. Governments were not often able to float loans since they could not pledge the private resources of their people. And they had not discovered the practice of deficit financing.

Today there are many credit instruments, and most people use them. Business and government are the biggest borrowers. Businesses seek the use of capital goods to increase productivity. Governments borrow against future tax revenues to finance highways, welfare programs, and public services. Households also borrow to make purchases in excess of their current cash resources. Such borrowers, and the corresponding lenders, must acknowledge the time value of their commitments.

The following examples reveal the significance of interest in economic transactions and confirm the importance of understanding how it operates, whether it pertains to personal finances or to professional practices.

- The purchase of a home is the largest investment most people make. The two tables below vividly portray the impact of interest rates and loan periods: A shorter repayment period at a given interest rate or a lower interest rate for a given loan period begets a conspicuous saving.

Repayment Period, Years	Monthly Payment	Total Interest
15	$295.50	$23,190
20	260.50	32,520
25	241.75	42,525
30	230.75	53,069

A $30,000 loan at $8\frac{1}{2}$ percent interest for four repayment periods.

Interst Rate, %	Monthly Payment	Total Interest
$7\frac{1}{2}$	$210.00	$45,600
$8\frac{1}{2}$	230.75	53,069
$9\frac{1}{2}$	252.50	60,899
$10\frac{1}{2}$	274.50	68,820

A $30,000 loan for 30 years at four interest rates.

- During the War of 1812, New York City loaned $1 million to the nation's capital. During the period of New York's financial crises in 1975, it was suggested that Washington might be billed for the original loan and the accumulated interest. At 6 percent interest compounded annually, the $1 million loan would have increased to an $11.2 billion debt.

Laws have been passed to encourage truth in lending by requiring standardized statements of interest charges. The need for such laws arises from the many ways interest can be calculated and the unique vocabulary associated with the subject. The profusion breeds confusion.

In this section we will explore interest vocabulary and discounted cash flow calculations (Chapters 2 and 3) and then consider three methods for comparing the economic time value of alternatives: present worth (Chapter 4), equivalent annual worth (Chapter 5), and rate of return (Chapter 6). From the economic decision maze in Chapter 1 we observed the prominent position of these comparison methods in the problem-solving process; they are the foundation for refinements that ration capital to the most rewarding projects. The mechanics of discounted cash flow are straightforward, relatively simple mathematical operations, but they are absorbing because they have so many practical applications.

CHAPTER 2

TIME VALUE OF MONEY

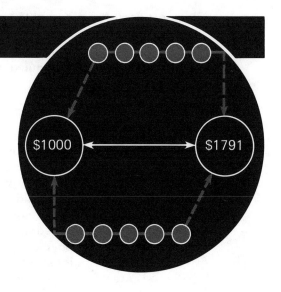

OVERVIEW

Nearly everyone is directly exposed to interest transactions occasionally and is indirectly affected regularly. Credit cards are a mainstay of commerce; they have an interest load for delayed payments. Key parts of a contract for purchasing an automobile or home are the interest stipulations. The rate of interest paid on municipal bonds directly affects tax rates for property in the affected area. Businesses borrow to expand or simply maintain operations, and the cost of their borrowing must be repaid from more profitable operations allowed by loans. All this borrowing taken together adds up to an enormous debt, and it all has interest charges.

To fully appreciate interest charges, one must comprehend the reasons for the charges, understand how they are calculated, and realize their effect on cash flows. Interest represents the earning power of money. It is the premium paid to compensate a lender for the administrative cost of making a loan, the risk of nonrepayment, and the loss of use of the loaned money. A borrower pays interest charges for the opportunity to do something now that would otherwise have to be delayed or never done. *Simple interest I* is a charge directly proportional to the capital (principal P) loaned at rate i for N periods, so that $I = PiN$. *Compound interest* includes charges for the accumulated interest as well as the amount of unpaid principal.

A *nominal interest rate r* of 8 percent compounded quarterly, for example, indicates an interest charge of 2 percent per period compounded four times a year. If m is the number of compounding periods per year, the equivalent *effective interest rate* i of a nominal rate is

$$i = \left(1 + \frac{r}{m}\right)^{m} - 1$$

Continuous interest is the nominal interest rate as m approaches infinity, and its equivalent effective interest rate is $i = e^{r} - 1$.

Time-value mechanics involves the use of compound-interest factors to translate payments of various amounts occurring at various times to a single equivalent payment. Interest factors are symbolized by notations based on i, N, P = present worth, F = future worth, and A = annuity payment. An ordinary annuity is a series of equal payments, at equal intervals, with the first payment at the end of the first period. When payments in an annuity increase by a constant increment G each period, an equivalent ordinary annuity is determined through use of the arithmetic-gradient factor.

Seven discrete interest factors are commonly used to evaluate cash flows and convert them into summary statements that define alternative uses for capital. These factors are represented by functional symbols that assist calculations; the values of the factors for different interest rates and numbers of compounding periods are tabulated in Appendix D.

The fundamental concepts of interest and the basic constructs of interest calculation introduced in this chapter are the foundations for discounted cash flow applications developed in the next four chapters.

REASONS FOR INTEREST

Example 2.1
All You Have to Do to Be Rich Is Live Long Enough

The famous purchase of Manhattan Island from the Indians for $24 is often referred to as an exceptional bargain. This incident reputedly occurred in 1626, when Peter Minuit of the Dutch West India Company bought the rights to the island from local residents. Was it a bargain? For the sake of argument, suppose that the Indians could have invested the money at a reasonable interest rate of 6 percent compounded annually. Over the years since then, the original $24 investment would have grown by the proportions shown in the table on the right.

Whether Minuit or the Indians got the better deal depends on the perspective taken. If they could have invested their wampum at a higher rate, say the 18 percent return earned

Year	Value of the Original $24 Investment
1626	$24.00
1676	442.08
1726	8,143.25
1776	149,999.92
1826	2,763,021.69
1876	50,895,285.76
1926	937,499,015.11
1976	17,268,876,484.38

on money market funds in 1981, the Indians' theoretical investment would now be worth

$$\$790,000,000,000,000,000,000,000,000$$

or \$790 septillion. On the other hand, at the 6 percent rate, its value would have increased to slightly more than \$27 billion by 1984, when the assessed value of taxable real estate in Manhattan was about \$24 billion. Furthermore, Minuit bought Manhattan from the Canarsee tribe, who were native to Brooklyn. Most of Manhattan belonged to the Weckquaesqeeks, who warred with the Dutch after being left out of the deal. Finally, the Dutch paid them for Manhattan too.

The significance of interest is obvious in the example above. The reasons for this effect become more apparent when we examine the uses of capital. In our economic environment, capital is the basic resource. It can be converted into production goods, consumer goods, or services. It has the power to earn and to satisfy wants.

From a lender's viewpoint, capital is a fluid resource. Capital can be spent on goods expected to produce a profit or on personal satisfaction. It can be hoarded or given away. It can also be loaned. If it is loaned, the lender will normally expect some type of compensation. The common compensation is interest. Interest compensates for the administrative expense of making the loan, for the risk that the loan will not be repaid, and for the loss of earnings which would have been obtained if the money had been invested for productive purposes.

From a borrower's viewpoint, a loan is both an obligation and an opportunity. A borrower must expect to repay the loan. Failure to repay leads to a damaged reputation, loss of possessions, and other consequences. The loan offers an opportunity to do something immediately that would otherwise have to be delayed. In some cases an objective would no longer exist after a delay. In order to take advantage of an existing course of action or to fulfill a current need, the borrower agrees to pay a certain amount in addition to the sum immediately received. This premium is the interest paid to avoid waiting for the money.

Implied in both the lender's and borrower's viewpoints is the earning power of money. For money to earn something, the owner or user must wait (*waiting-earning* is obviously opposed to the *spending-owing* use of money to gratify immediate desires). Interest payments have been likened to the reward for waiting, but it is more appropriate for engineering economists to view interest as the productive gain from efficient use of the money resource. The prevailing interest rate is essentially a measure of the productivity to expect from the resource. An owner of money can lend it at the prevailing rate and wait to be repaid the original amount plus an extra increment. Equivalently, the borrower could reloan the money at a higher rate to acquire a gain larger than the amount to be repaid, or the money could be converted to productive goods that would be expected to earn more than the amount needed to repay the loan. In both cases the prevailing interest rate sets the minimum level of expected productivity, and both cases involve time between receipt and return of the loan to secure the earnings: the *time value of money*.

"If you would know the value of money, go and try to borrow some," advised B. Franklin in *Poor Richard's Almanac.*

Example 2.2
Borrowers Be Aware

"One/ten-a-week" loans are offered by shady characters to tide borrowers over until the next paycheck. The lenders charge $1 for each $10 borrowed for a week. Thus, $140 would have to be repaid for the use of $100 for 4 weeks. The simple interest rate for this arrangement, based on a 52-week year, is an exorbitant 520 percent: A $10 debt held for a year would accumulate $1/week $\times$ 52 weeks = $52 interest expense, so that the simple interest rate is

$$\$52/\$10 \times 100 \text{ percent} = 520 \text{ percent}$$

SIMPLE INTEREST

When a *simple interest rate* is quoted, the interest earned is directly proportional to the capital involved in the loan. Expressed as a formula, the interest earned I is calculated by

where P = present amount or principal
i = interest rate/period
N = number of interest periods

$$I = PiN$$

Since the principal, or amount borrowed, P is a fixed value, the annual interest charged is constant. Therefore, the total amount a borrower is obligated to pay a lender is

where F is a future sum of money

$$F = P + I = P + PiN = P(1 + iN)$$

When N is not a full year, there are also two ways to calculate the simple interest earned during the period of the loan. Using *ordinary simple interest,* the year is divided into twelve 30-day periods, or a year is considered to have 360 days. In *exact simple interest* a year has exactly the calendar number of days, and N is the fraction of the number of days the loan is in effect that year.

An example of simple interest as the rental cost of money is a loan of $1000 for 2 months at 10 percent. With ordinary simple interest, the amount to be repaid is

where N is 1 year and $(2/12)N$ is 2 months

$$F = P\left(1 + i\left(\frac{2}{12}\right)N\right)$$
$$= \$1000(1 + 0.01667) = \$1016.67$$

With exact simple interest when the 2 months are January and February in a non-leap year, the future sum is

$$F = P\left(1 + i\left(\frac{31 + 28}{365}\right)N\right)$$
$$= \$1000(1 + 0.01616) = \$1016.16$$

COMPOUND INTEREST

Again assume a loan of $1000, this time for 2 years at an interest rate of 10 percent compounded annually; the pattern of interest compounding is shown in Table 2.1.

TABLE 2.1

Future value of a $1000 loan when interest is due on both the principal and unpaid interest

Year	Amount Owed at Beginning of Year	Interest on Amount Owed	Amount Owed at End of Year
1	$1000	$1000 × 0.10 = $100	$1000 + $100 = $1100
2	1100	$1100 × 0.10 = $110	$1100 + $110 = $1210

The amount to be repaid for the given loan is thus $1210 − $1200 = $10 greater for compound than for simple interest. The $10 difference accrues from the interest charge on the $100 earned during the first year that was not accounted for in the simple-interest calculation. The formula approach for the calculations in Table 2.1, using previously defined symbols, is

$$\frac{\text{Compound amount}}{\text{due in 2 years}} = \frac{\text{amount}}{\text{borrowed}} + \frac{\text{year-1}}{\text{interest}} + \left(\begin{array}{c}\text{amount bor-}\\ \text{rowed plus}\\ \text{interest due}\end{array}\right)\left(\begin{array}{c}\text{interest}\\ \text{rate}\end{array}\right)$$

$$F_2 = P + Pi + (P + Pi)i$$
$$= P(1 + i + i + i^2)$$
$$= P(1 + i)^2$$
$$= \$1000(1 + 0.10)^2$$
$$= \$1000(1.21) = \$1210$$

The key equation in the development above is $F_2 = P(1 + i)^2$. Generalized for any number of interest periods N, this expression becomes $F = P(1 + i)^N$, and $(1 + i)^N$ is known as the *compound-amount factor*. It is one of several interest factors derived in this chapter for which numerical values are tabulated in Appendix D.

Example 2.3
The Timing Tells the Value

Let the interest on a $1000 loan for 2 years at an interest rate of 10 percent compounded annually be paid when it comes due. That is, at the end of the first year an interest payment of $1000 × 0.10 = $100 is paid, and at the end

of the second year the principal ($1000) plus the interest earned during the second year ($100) is paid. The total interest charge is thus $100 + $100 = $200 at 10 percent compounded annually, which is the same as the interest charge at 10 percent simple interest ($I = PiN = \$1000 \times 0.10 \times 2 = \$200$). How come?

The mirage that the two interest payment plans are equal is dispelled by noting the difference in the timing of the payments. When compounded annually, interest is calculated and charged to the account each year. If the interest so calculated is not physically withdrawn from the account, it is added to the balance and it earns interest during the next period.

The cited payment plans appear to produce the same interest because no mention is made of what was done with the $100 received after the first year. If the $100 interest payment were reinvested at the given 10 percent rate, it would earn $100 \times 0.10 = \$10$, making the total amount due at the end of 2 years $200 + \$10 = \$210$. The sum of the principal ($P = \$1000$) and interest is then $1210, which is the expected compound amount due in 2 years.

Nominal Interest Rates

For a discussion of interest rate statements, see Extension 2A.

Interest rates are normally quoted on an annual basis. However, agreements may specify that interest will be compounded several times per year: monthly, quarterly, semiannually, etc. For example, a year divided into four quarters with interest at 2 percent per quarter is typically quoted as "8 percent compounded quarterly." Stated in this fashion, it is called a *nominal interest rate*. The future value at the end of 1 year for $200 earning interest at 8 percent compounded quarterly is developed as

$$F_{3\,mo} = P + Pi = \$200 + (\$200)(0.02)$$
$$= \$200 + \$4 = \$204$$

$$F_{6\,mo} = \$204 + (\$204)(0.02)$$
$$= \$204 + \$4.08 = \$208.08$$

$$F_{9\,mo} = \$208.08 + (\$208.08)(0.02)$$
$$= \$208.08 + \$4.16 = \$212.24$$

$$F_{12\,mo} = \$212.24 + (\$212.24)(0.02)$$
$$= \$212.24 + \$4.24 = \$216.48$$

The result of the nominal interest rate is to produce a higher value than might be expected from the 8 percent figure stated in its expression. At 8 percent compounded annually, the $200 mentioned above would earn in 1 year $F_{12\,mo} = \$200 + \$200(0.08) = \$216$, which is 48 cents less than the amount accrued from the nominal rate of 8 percent compounded quarterly. An interest of $1\frac{1}{2}$ percent per month is also a nominal interest rate that could appear to the uninitiated as being quite reasonable. Using the compound-amount factor to calculate how much would have to be repaid on a 1-year loan of $1000 at a nominal interest rate of 18 percent compounded monthly ($1\frac{1}{2}$ percent per period with 12 interest periods per year) gives

$$F_{12} = \$1000(1 + 1\tfrac{1}{2}\%)^{12} = \$1000(1.015)^{12}$$
$$= \$1000(1.1956) = \$1196$$

This can be compared with the future value of the same loan at 18 percent compounded semiannually (9 percent per period with 2 interest periods per year):

$$F_{12} = \$1000(1 + 9\%)^2 = \$1000(1.09)^2$$
$$= \$1000(1.1881) = \$1188$$

Thus, more frequent compounding within a nominally stated annual rate does indeed increase the future worth.

Effective Interest Rates

Confusion about the actual interest earned is eliminated by stating the charge as an *effective interest rate*. Efforts to protect borrowers from exotic statements of interest charges was the thrust behind the national truth-in-lending law passed in 1973. The effective interest rate is simply the ratio of the interest charge for 1 year to the principal (amount loaned or borrowed). For the $1000 1-year loan at a nominal interest rate of 18 percent compounded monthly,

$$\text{Effective interest rate} = \frac{F - P}{P} = \frac{\$1196 - \$1000}{\$1000}$$
$$= \frac{\$196}{\$1000} 100\% = 19.6\%$$

For the same loan at 18 percent compounded semiannually,

$$\text{Effective interest rate} = \frac{\$1188 - \$1000}{\$1000}$$
$$= \frac{\$188}{\$1000} 100\% = 18.8\%$$

The effective interest rate can be obtained without reference to the principal. Based on the same reasoning utilized previously, the effective interest rate for a nominal interest rate of 18 percent compounded semiannually is

$$i = \left(1 + \frac{r}{m}\right)^m - 1 = \left(1 + \frac{0.18}{2}\right)^2 - 1$$
$$= (1 + 0.09)^2 - 1 = 1.188 - 1$$
$$= 0.188 \quad \text{or} \quad 18.8\%$$

where i = effective interest rate
r = nominal interest rate
m = numbers of com-
pounding periods per year

which means that a nominal interest rate of 18 percent compounded semi-annually is equivalent to a compound interest rate of 18.8 percent on an annual basis.

The ultimate limit for the number of compounding periods in a year is called *continuous compounding*. Under this accrual pattern, m approaches infinity as interest compounds continuously, moment by moment. The effective interest rate for continuous compounding is developed as follows:

The interest periods are made infinitesimally small:

$$i = \lim_{m \to \infty} \left(1 + \frac{r}{m}\right)^m - 1$$

The right side of the equality is rearranged to include r in the exponent:

$$\left(1 + \frac{r}{m}\right)^m - 1 = \left[\left(1 + \frac{r}{m}\right)^{m/r}\right]^r - 1$$

The bracketed term is recognized as the value of the mathematical symbol e, [$e = 2.718$ is the value of $(1 + 1/n)^n$ as n approaches infinity]:

$$\lim_{m \to \infty} \left(1 + \frac{r}{m}\right)^{m/r} = e$$

By substitution,

$$i = \lim_{m \to \infty} \left[\left(1 + \frac{r}{m}\right)^{m/r}\right]^r - 1 = e^r - 1$$

As an example of continuous compounding, when the interest rate is $r = 18.232$ percent,

$$i = e^r - 1 = e^{0.18232} - 1 = 0.20 \quad \text{or} \quad 20\%$$

and, correspondingly, when the effective interest rate is $i = 22.1$ percent,

$$0.221 = e^r - 1$$
$$1.221 = e^r$$
$$r = 20\%$$

Example 2.4
Relative Effects of Nominal Interest Rates

A loan can be arranged at a nominal rate of 12 percent compounded monthly, or 13 percent compounded semiannually. Which arrangement provides the lower debt at the end of the loan period?

The more attractive arrangement is the one with the lowest effective interest rate. At 12 percent compounded monthly,

$$r = 0.12 \quad \text{and} \quad m = 12$$

and

$$\text{Effective interest rate} = \left(1 + \frac{0.12}{12}\right)^{12} - 1$$
$$= (1.01)^{12} - 1 = 1.127 - 1$$
$$= 0.127$$

At 13 percent compounded semiannually, $r = 0.13$ and $m = 2$, and

$$\text{Effective interest rate} = \left(1 + \frac{0.13}{2}\right)^2 - 1$$
$$= (1.065)^2 - 1 = 1.134 - 1$$
$$= 0.134$$

The loan at 12 percent compounded monthly is thus seen to produce a lower-cost loan.

Continuous Compounding

Economic studies are occasionally conducted with continuous compounding rather than conventional discrete interest rates. The most obvious computational effect of using continuous interest is that it produces a larger future amount than does the same rate compounded discretely; as demonstrated above, a continuously compounded r of 20 percent compared with $i = 20$ percent yields an annual return greater by

$$\frac{0.221 - 0.20}{0.20} 100\% = 10.5\%$$

The rationale for using continuous interest in economic analyses is that the cash flow in certain situations is best approximated by a continuous pattern; that is, cash transactions tend to be spread out over a year in a more or less even distribution rather than being concentrated at particular dates. Some mathematical models are also facilitated by the assumption of continuous compounding rather than periodic compounding.

In actual practice, however, interest rates are seldom quoted on a continuous basis, and the vast majority of organizations use discrete compounding periods in their economic studies. The reason for this is probably custom or the familiarity that makes it easier to understand periodic interest charges. Accounting practices that categorize receipts and disbursements as end-of-year values and financial experiences with annual tax, insurance, or mortgage payments contribute to thinking in terms of discrete periods. Yet continuous and discrete compounding are both only approximations of true cash flow, because cash neither flows like a free stream of water nor gushes like a geyser at given intervals. Receipts and disbursements are often irregular in amount and in timing.

In the following discussion of the time value of money and in subsequent chapters on economic comparison methods, end-of-year compounding is utilized. Tables of interest factors are provided in Appendix D for discrete compounding and in Appendix E for continuous compounding.

Discrete compounding is implied in all examples and exercises in this text unless specified otherwise.

TIME-VALUE EQUIVALENCE

Two things are equivalent when they produce the same effect. The effective interest rate computed for a nominally stated interest rate is an equivalent

Several price indexes are published regularly to indicate the equivalent purchasing power of money. See Chapter 13.

expression of the interest charge. Both interest charges produce the same effect on an investment. In considering time-value conversion, the equivalent numerical values of money are determined, *not* values with equivalent *purchasing power*. The amount of goods that can be purchased with a given sum of money varies up and down (more often down) as a function of special localized circumstances and nationwide or worldwide economic conditions. Ways to include inflation effects are discussed in Chapter 13. In this chapter about time-value mechanics, attention is directed toward calculations based on the *earning power* of money, which relates time and earnings to locate time-equivalent money amounts.

If $1000 were sealed and buried today, it would have a cash value of $1000 when it was dug up 2 years from now. Regardless of changes in buying power, the value remains constant because the earning power of the money was forfeited. It was observed earlier that $1000 deposited at 10 percent interest compounded annually has a value of $1000(1 + 0.10)^2 = 1210 after 2 years. Therefore, $1000 today is equivalent to $1210 in 2 years from now if it earns at a prevailing rate of 10 percent compounded yearly. Similarly, to have $1000 in 2 years from now, one need only deposit

$$\$1000 \, \frac{1}{(1 + 0.10)^2} = \$826.44$$

today. In theory then, if 10 percent is an acceptable rate of return, an investor would be indifferent between having $826.44 in hand or having a trusted promise to receive $1000 in 2 years.

The $1000 could also be used to pay two equal annual $500 installments. The buried $1000 could be retrieved after 1 year, an installment paid, and the remaining $500 interred again until the second payment became due. If the $1000 is instead deposited at 10 percent, there would be $1100 available at the end of the first year. After the first $500 installment was paid, the remaining $600 would draw interest until the next payment. Paying the second $500 installment would leave

$$\$600(1.10) - \$500 = \$660 - \$500 = \$160$$

in the account. Because of the earning power of money, the initial deposit could have been reduced to $868 to pay out $500 at the end of each of the 2 years:

First year: $868(1.10) - $500 \doteq $955 - $500 = 455
Second year: $455(1.10) \doteq $500 = $ second installment

Thus, $868 is equivalent to $500 received 1 year from now plus another $500 received 2 years from now:

$$First\ year:\qquad \frac{\$500}{1.10} \doteq \$455$$

$$Second\ year:\qquad \$455 + \frac{\$500}{(1.10)^2} = \$455 + \frac{\$500}{1.21}$$
$$\doteq \$455 + \$413 \doteq \$868$$

The concept of equivalence is the cornerstone for time-value-of-money comparisons. To have a precise meaning, income and expenditures must be identified with time as well as with amount. A decision between alternatives having receipts and disbursements spread over a period of time is made by comparing the equivalent outcomes of the alternatives at a given date. Figure 2.1 shows the translation of $1000 at time zero (now) into equivalent alternative expressions of cash flow.

COMPOUND-INTEREST FACTORS

Cash flow is translated to a given point in time by determining either its present worth or its future worth. A present-worth calculation converts a single future sum or a series of future values to an equivalent amount at an earlier date. This date is not necessarily the present time. Future-worth calculations convert values occurring at any time to an equivalent amount at a later date.

Equivalent values could be determined by calculating the compound amount of each sum for each period. This tedious routine is avoided by using compound-interest tables for different present- and future-worth factors. There are two basic types of factors. The one we have already considered converts a single amount to a present or future value. The other type is for a series of

---

$1000 today is equivalent to $1791 received 10 years from now.

$1000 today is equivalent to $237.40 received at the end of each year for the next 5 years.

$1000 today is equivalent to $317.70 received at the end of years 6, 7, 8, 9, and 10.

$237.40 received at the end of each year for the next 5 years is equivalent to a lump sum of $1791 received 10 years from now.

$317.70 received at the end of years 6, 7, 8, 9, and 10 is equivalent to $1791 received 10 years from now.

$237.40 received at the end of each year for the next 5 years is equivalent to $317.70 received at the end of years 6, 7, 8, 9, and 10.

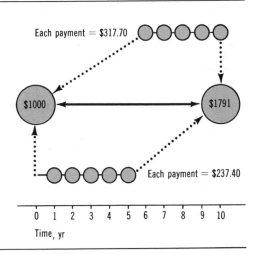

FIGURE 2.1
Equivalent outcomes with an interest rate of 6 percent compounded annually.

uniform values called an *annuity*. The tables in Appendix D are based on an annuity characterized by (1) *equal payments A,* (2) *equal periods between payments N,* and (3) *the first payment occurring at the end of the first period.* Annuity factors are used to convert a series of payments to a single future or present sum and to translate single sums into a series of payments occurring in the past or future.

Conversion Symbols

There are seven basic interest factors for discrete compounding. Names and notations for these factors are those suggested by the Engineering Economy Division of the American Society for Engineering Education.* Each factor is described by a name (for instance, one is the "compound-amount factor," used previously to find the future worth of a single payment) and two notational forms: (1) a mnemonic symbol (to assist memory by association) as in (*CA-i%-N*) for the compound-amount factor, and (2) a functional symbol to suggest the use of the interest factor) as in (*F/P, i%, N*), again for the compound-amount factor which is used to find *F* given *P*. Since the functional notation is most descriptive of the operation to be performed, it will be utilized for all the interest factors. Time-value conversions and associated factors are summarized in Table 2.2.

The symbols for the first six time-value conversions are abbreviations for the equivalent values sought (future worth *F*, present worth *P*, or uniform series amounts *A*) and the data given (*F, P,* or *A* with its associated interest rate *i* and number of compounding periods *N*). The arithmetic gradient conversion factor is used to convert a constantly increasing or decreasing series into a uniform series of amounts *A* which can then be an input to other

---

TABLE 2.2

Interest factors for discrete cash flow with end-of-period compoundings

Factor	To Find	Given	Symbol
Compound amount	Future worth, *F*	Present amount, *P*	(*F/P, i%, N*)
Present worth	Present worth, *P*	Future amount, *F*	(*P/F, i%, N*)
Sinking fund	Annuity amounts, *A*	Future amount, *F*	(*A/F, i%, N*)
Series compound amount	Future worth, *F*	Annuity amounts, *A*	(*F/A, i%, N*)
Capital recovery	Annuity amounts, *A*	Present amount, *P*	(*A/P, i%, N*)
Series present worth	Present worth, *P*	Annuity amounts, *A*	(*P/A, i%, N*)
Arithmetic gradient conversion	Annuity amounts, *A*	Uniform increase in amount, *G*	(*A/G, i%, N*)

---

* See *The Engineering Economist,* vol. 14, no. 2, Winter 1969.

interest factors. In the equation

$$F = \$1000(F/P, 10, 2)$$

$1000 is the known present amount, the interest rate is 0.10 per period (i is 10 percent) and F is the equivalent future worth after two periods ($N = 2$). The whole symbol stands for the numerical expression $(1 + 0.10)^2$, and the numerical value is found in Appendix D. To find the value for $(F/P, 10, 2)$, look for 2 in the N column of the 10 percent table and then read across to the compound-amount factor column to find 1.2100.

The conversion descriptions and symbols connote that certain factors are reciprocals of one another:

$$(F/P, i, N) = \frac{1}{(P/F, i, N)}$$
$$(A/F, i, N) = \frac{1}{(F/A, i, N)}$$
$$(A/P, i, N) = \frac{1}{(P/A, i, N)}$$

Other relationships are not so apparent from the abbreviations but are useful in understanding conversion calculations. The following equalities are verified during the development of the conversion symbols:

$$(F/P, i, N) \times (P/A, i, N) = (F/A, i, N)$$
$$(F/A, i, N) \times (A/P, i, N) = (F/P, i, N)$$
$$(A/F, i, N) + i = (A/P, i, N)$$

As an example of how notations vary, in Japan, $(F/P, i\%, N)$ is called the *final worth* factor and its notation is $[P \rightarrow F]_n^{i\%}$.

Development of Interest Formulas

A better understanding of the conversion process is achieved by studying the development of the interest-factor formulas. The symbols employed in the following discussion of the seven interest factors are the same as those described previously. It should be remembered that these factors are for discrete compounding and their numerical values are tabulated in Appendix D; corresponding factors for continuous compounding are presented in the next chapter.

Additional sample applications of the interest factors are provided in the review exercises at the end of this chapter.

The effect of compound interest on an investment was demonstrated in previous examples. The future worth of a present amount when interest is accumulated at a specific rate i for a given number of periods N, where $F1$ is the future

1 Compound-Amount Factor (Single Payment)

Use: To find F, given P
Symbols: (F/P, i%, N)
(CA-i%-N)
Formula: $F = P(1 + i)^N$
$= P(F/P, i, N)$

worth at the end of the first period and F_N is the future worth at the end of N years, is

$$F1 = P + Pi = P(1 + i)$$
$$F2 = P[(1 + i) + (1 + i)i]$$
$$= P(1 + i)(1 + i) = P(1 + i)^2$$
$$F3 = P[(1 + i)^2 + (1 + i)^2 i]$$
$$= P(1 + i)^2(1 + i) = P(1 + i)^3$$
$$F_N = P(1 + i)^N$$

The ratio of future worth to present amount is then expressed as

$$\frac{F}{P} = (F/P, i, N) = (1 + i)^N$$

2 Present-Worth Factor (Single Payment)

P is the present worth of a sum N periods in the future. Rearranging the single-amount future-value formula $F = P(1 + i)^N$ to express P in terms of F gives

Use: To find P, given F
Symbols: (P/F, i%, N)
(PW-i%-N)
Formula: $P = F\dfrac{1}{(1 + i)^N}$
$= F(P/F, i, N)$

$$P = F\frac{1}{(1 + i)^N}$$

Then the ratio of present worth to future value is

$$\frac{P}{F} = (P/F, i, N) = \frac{1}{(1 + i)^N}$$

That the present-worth factor is simply the reciprocal of the compound-amount factor is confirmed by applying it to the data given in Table 2.1, where the future worth of $1000 at 10 percent compounded annually was shown to be $1210. Equivalently, $P = F(P/F, i, N)$ is the expression for the present worth when the future worth is known. The numerical value of the present-worth factor $(P/F, 10, 2)$ is found in the 10 percent table of Appendix D at $N = 2$: 0.82645. Then,

$$P = \$1210(0.82645) = \$1000$$

3 Sinking-Fund Factor

Use: To find A, given F
Symbols: (A/F, i%, N)
(SF-i%-N)
Formula: $A = F\dfrac{i}{(1 + i)^N - 1}$
$= F(A/F, i, N)$

A fund established to accumulate a given future amount through the collection of a uniform series of payments is called a *sinking fund*. Each payment has a constant value A and is made at the end of an interest period.

The growth pattern of a sinking fund is illustrated in Table 2.3. Each end-of-year payment A is equal to $1000, and payments continue for 5 years. Interest is 8 percent compounded annually. It is assumed that each payment begins to draw interest as soon as it is deposited in the sinking-fund account. Thus, the first payment draws interest for 4 years and the last payment receives no interest.

TABLE 2.3

Compound amount of a uniform series of payments

Time of Payment (end of year)	Amount of Payment, A	Future Worth at the End of Each Year
1	$1000	$1000(1.08)^4 = $1360
2	1000	1000(1.08)^3 = 1260
3	1000	1000(1.08)^2 = 1166
4	1000	1000(1.08)^1 = 1080
5	1000	1000(1.08)^0 = 1000
		Annuity value F at the end of year 5 = $5866

A more general expression for the future worth of an annuity develops from the use of symbols to represent the values in Table 2.3. The first payment, earning interest for $N - 1$ periods, where N is 5 years in the example, increases to a future worth of

$$F = A(1 + i)^{N-1}$$

Each of the payments is treated in the same manner and collected to obtain the total amount F:

$$F = A(1 + i)^{N-1} + A(1 + i)^{N-2} + A(1 + i)^{N-3} + A(1 + i)^{N-4} + A(1 + i)^{N-N}$$

Factoring out A and letting the exponent $N - N = 0$, we have

$$F = A[(1 + i)^{N-1} + (1 + i)^{N-2} + (1 + i)^{N-3} + (1 + i)^{N-4} + 1]$$

Multiplying this equation by $1 + i$ results in

$$F(1 + i) = A[(1 + i)^N + (1 + i)^{N-1} + (1 + i)^{N-2} + (1 + i)^{N-3} + (1 + i)]$$

Subtracting the original equation from the last equation gives

$$F(1 + i) - F = -A + A(1 + i)^N$$
$$Fi = A[(1 + i)^N - 1]$$

Solving for A,

$$A = F\frac{i}{(1 + i)^N - 1}$$

we see the sinking-fund factor expressed as

$$(A/F, i, N) = \frac{i}{(1 + i)^N - 1}$$

Then, applying the sinking-fund factor to the data in Table 2.3, we have

$$A = \$5866(A/F, 8, 5)$$
$$= \$5866(0.17046) = \$1000$$

4   Series Compound-Amount Factor (Uniform Series)

Use: To find F, given A

Symbols: (F/A, i%, N)
              (SCA-i%-N)

Formula:

$$F = A\frac{(1 + i)^N - 1}{i}$$
$$= A(F/A, i, N)$$

From the development of the sinking-fund-factor formula,

$$Fi = A(1 + i)^N - 1$$

which is expressed in terms of F as

$$F = A\frac{(1 + i)^N - 1}{i}$$

Then the time value for the future worth of an annuity is

$$(F/A, i, N) = \frac{(1 + i)^N - 1}{i}$$

and the future worth of the annuity composed of five annual payments of $1000 each invested at 8 percent compounded annually, as portrayed in Table 2.3, is

$$F = \$1000(F/A, 8, 5)$$
$$= \$1000(5.8665) = \$5866$$

5   Capital-Recovery Factor

Use: To find A, given P

Symbol: (A/P, i%, N)
            (CR-i%-N)

Formula:

$$A = P\frac{i(1 + i)^N}{(1 + i)^N - 1}$$
$$= P(A/P, i, N)$$

The capital-recovery factor is used to determine the amount of each future annuity payment required to dissipate a given present value when the interest rate and number of payments are known. For instance, the amount of each annual payment made for 5 years in order to repay a debt of $3993 bearing 8 percent annual interest can be determined through the use of the capital-recovery factor. Table 2.4 shows that it would take five $1000 payments to repay the $3993 debt.

Using symbols to represent the conversions shown in Table 2.4, we find that the present worth of an annuity is

$$P = A[(1 + i)^{-1} + (1 + i)^{-2} + (1 + i)^{-3} + (1 + i)^{-4} + (1 + i)^{-N}]$$

and multiplying both sides of the equation by $(1 + i)^{-1}$ results in

TABLE 2.4		

Present worth of a uniform series of payments

Time of Payment (end of year)	Amount of Payment, A	Present Worth of Payments at End of Year
1	$1000	$1000(1.08)^{-1} = $ 926
2	1000	1000(1.08)^{-2} = 857
3	1000	1000(1.08)^{-3} = 794
4	1000	1000(1.08)^{-4} = 735
5	1000	1000(1.08)^{-5} = 681
	Present worth P of the 5-year annuity =	$3993

$$P(1 + i)^{-1} = A[(1 + i)^{-2} + (1 + i)^{-3} + (1 + i)^{-4}$$
$$+ (1 + i)^{-N} + (1 + i)^{-N-1}]$$

Subtracting the first equation from the second equation gives

$$P[(1 + i)^{-1} - 1] = A[(1 + i)^{-N-1} - (1 + i)^{-1}]$$

Converting $(1 + i)^{-1} - 1$ to $-i/(1 + i)$, multiplying both sides by $-(1 + i)$, and rearranging yields

$$P\frac{i(1 + i)}{1 + i} = A[(1 + i)(1 + i)^{-1} - (1 + i)(1 + i)^{-N-1}]$$
$$Pi = A[1 - (1 + i)^{-N}]$$
$$= A\frac{(1 + i)^N - 1}{(1 + i)^N}$$

or
$$A = P\frac{i(1 + i)^N}{(1 + i)^N - 1}$$

from which comes the expression for the capital-recovery factor,

$$(A/P, i, N) = \frac{i(1 + i)^N}{(1 + i)^N - 1}$$

As applied to the data in Table 2.4 where $P = \$3993$,

$$A = \$3993(A/P, 8, 5)$$
$$= \$3993(0.25046) = \$1000$$

The relation among time-value annuity factors is apparent from the way the capital-recovery factor can be converted to the sinking-fund factor by substituting $P = F(1 + i)^{-N}$ in the capital-recovery formula, as

$$A = P(A/P, i, N) = \frac{F}{(1 + i)^N}(A/P, i, N) = \frac{F}{(1 + i)^N}\frac{i(1 + i)^N}{(1 + i)^N - 1}$$

$$A = F\frac{i}{(1 + i)^N - 1} = F(A/F, i, N)$$

or

$$(A/P, i, N) = (A/F, i, N) + i$$

as indicated by

$$\frac{i(1 + i)^N}{(i + i)^N - 1} = \frac{i}{(1 + i)^N - 1} + i = \frac{i + i(1 + i)^N - i}{(1 + i)^N - 1} = \frac{i(1 + i)^N}{(1 + i)^N - 1}$$

6 Series Present-Worth Factor (Uniform Series)

> **Use:** To find P, given A
>
> **Symbols:** (P/A, i%, N)
>   (SPW-i%-N)
>
> **Formula:**
>
> $$P = A\frac{(1 + i)^N - 1}{i(1 + i)^N}$$
> $$= A(P/A, i, N)$$

The present value of a series of uniform end-of-period payments can be calculated in the cumbersome fashion shown in Table 2.4. The present worth is more readily determined by use of the series present-worth factor.

Expressing the known relationship

$$A = P\frac{i(1 + i)^N}{(1 + i)^N - 1}$$

in terms of P yields

$$P = A\frac{(1 + i)^N - 1}{i(1 + i)^N} = A(P/A, i, N)$$

which is the time value expression of the present worth of an annuity.

The reciprocal relationship between the capital-recovery factor and the series present-worth factor is demonstrated by the data from Table 2.4:

$$P = \$1000(P/A, 8, 5)$$
$$= \$1000(3.9926) = \$3993$$

which indicates the equivalence of having $3993 in hand and a firm contract to receive five year-end payments of $1000 each when the interest rate is 8 percent.

7 Arithmetic-Gradient Conversion Factor (to Uniform Series)

> **Use:** To find A, given G
>
> **Symbols:** (A/G, i%, N
>   (GUS-i%-N)
>
> **Formula:**
>
> $$A = G\left[\frac{1}{i} - \frac{N}{(1 + i)^N - 1}\right]$$
> $$= G(A/G, i, N)$$

Enough situations occur in which series of payments increase at equal increments to warrant a special conversion factor. A series of payments that increases at a rate of $200 per year is illustrated in Figure 2.2. The $200 periodic change is the gradient G, and the payment at the end of the first period is the base annuity value A'. The pattern of an arithmetic gradient is then

$$A', A' + G, A' + 2G, \ldots, A' + (N - 1)G$$

where N is the duration of the series ($N = 5$ in Figure 2.2).

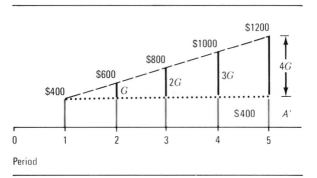

FIGURE 2.2
Uniform gradient series for five periods.

A uniformly increasing series can be evaluated by calculating F or P for each individual payment and summing the collection. Calculation time is reduced by converting the series into an equivalent annuity of equal payments A. The formula for this translation is developed by separating the series shown in Figure 2.2 into two parts: a base annuity designated A' and an arithmetic-gradient series increasing by G each period. The future worth of the G values in Figure 2.2 is calculated as

$$F = \$200(F/P, i, 3) + \$400(F/P, i, 2) + \$600(F/P, i, 1) + \$800$$

or

$$F = G(1 + i)^3 + 2G(1 + i)^2 + 3G(1 + i)^1 + 4G$$

Multiplying the latter expression by $(1 + i)^1$ gives

$$F(1 + i) = G(1 + i)^4 + 2G(1 + i)^3 + 3G(1 + i)^2 + 4G(1 + i)^1$$

Subtracting the last equation from the one above it yields

$$F - F(1 + i) = -G(1 + i)^4 - G(1 + i)^3 - G(1 + i)^2 - G(1 + i)^1 + 4G$$

Letting $4G = (N - 1)G$ and multiplying both sides of the equation by -1 gives

$$F(1 + i) - F = G[(1 + i)^4 + (1 + i)^3 + (1 + i)^2 + (1 + i) + 1] - NG$$

We have, in brackets, the series compound-amount factor $(F/A, i, 5)$ so that

$$Fi = G(F/A, i, N) - NG$$

To convert F to an annuity, both sides of the equation are multiplied by the sinking-fund factor $(A/F, i, N)$, which is the reciprocal of $(F/A, i, N)$, to get

$$Fi(A/F, i, N) = G - NG(A/F, i, N)$$

and since $A = F(A/F, i, N)$,

$$A = \frac{G}{i} - \frac{NG}{i}(A/F, i, N)$$

or

$$A = G\left[\frac{1}{i} - \frac{N}{i}(A/F, i, N)\right]$$

in which the bracketed expression is called the *artihmetic gradient conversion factor* with the symbol $(A/G, i, N)$.

For the cash flow diagramed in Figure 2.2, the equivalent uniform annuity calculated at an effective interest rate per period of 10 percent is

$$
\begin{aligned}
A &= A' + G(A/G, i, N) \\
&= \$400 + \$200(A/G, 10, 5) \\
&= \$400 + \$200(1.8100) = \$762
\end{aligned}
$$

which means that five end-of-period payments of $762 are equivalent to five payments starting at $400 and increasing by $200 each period.

The gradient factor may also be applied to a pattern of payments that decrease by a constant increment each period. The formula would then be

$$A = A' - G(A/G, i, N)$$

As an example, assume that an endowment was originally set up to provide a $10,000 first payment with payments decreasing by $1000 each year during the 10-year endowment life. What constant annual payment for 10 years would be equivalent to the original endowment plan if $i = 8$ percent? We have

$$
\begin{aligned}
A &= \$10,000 - \$1000(A/G, 8, 10) \\
&= \$10,000 - \$1000(3.8712) = \$6128.80
\end{aligned}
$$

Development of Geometric Series Formulas

A geometric series is a nonuniform progression that grows or declines at a *constant percentage rate* per period. The most familiar example is the effect of inflation or deflation on a cash flow stream. For instance, the rate of growth during a period of inflation could be 10 percent per period. An item priced at $1000 during the first year would then increase by 10 percent per year, as shown in Figure 2.3(a). A decline of 10 percent per year, starting from the same $1000 figure, would take the pattern shown in Figure 2.3(b).

The present worth of such series could be calculated by bringing each payment back to its present value by multiplying individual amounts by their appropriate $(P/F, i, N)$. For many applications it is more convenient to apply geometric gradient formulas. These formulas are based on the cash flow in period 1 (A'). Thus the cash flow N periods from time zero is $A'(1 + g)^N$, where g is the rate of growth or decline.

Where g is the rate of growth or decline and A' is the base value (first payment) in the geometric series.

Three different formulas are necessary to calculate the present worth P of a geometric series. Which one to use depends on the relation between the percentage rate g and the interest rate i.

For the case where g is larger than i, the present worth using $i*$ is calculated as

Case 1: When g is greater than i, define a special interest rate i* as

$$i* = \frac{1 + g}{1 + i} - 1$$

$$P = \frac{A'}{1 + i}(F/A, i*, N)$$

Consider the cash flow in Figure 2.3(a), where $g = 0.10$ and $A' = \$1000$. For $i = 0.08$, case 1 $(g > i)$ applies and

$$i* = \frac{1 + 0.10}{1 + 0.08} - 1 = \frac{1.10}{1.08} - 1 = 1.0185 - 1 = 1.85\%$$

which is the special interest rate that is used in the series compound-amount factor with $N = 5$. Since there are no tables in Appendix D for 1.85 percent, we must interpolate between tables to obtain a value for $(F/A, 1.85, 5)$.

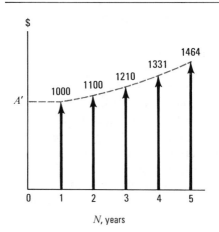

(a) Growth $(g = 10\%)$

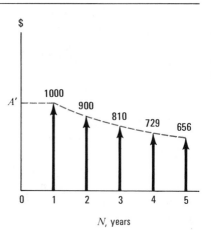

(b) Decline $(g = -10\%)$

FIGURE 2.3
Compound growth and decline patterns. Changes in cash flows occur at a constant *rate* in a geometric series, in contrast to changes of equal *increments* in an arithmetic series.

$$P = \frac{\$1000}{1 + 0.08}(F/A, 1.85, 5) = \$926(5.1882) = \$4804$$

Case 2: When g equals i, the growth and discount rates coincide to produce a present-value equation of

$$P = \frac{NA'}{1 + i}$$

If the interest rate for the 10 percent growth pattern in Figure 2.3(a) is also 10 percent ($i = g$), then

$$P = \frac{5(\$1000)}{1 + 0.10} = \frac{\$5000}{1.10} = \$4545$$

which demonstrates that an initial investment of \$4545 compounded at 10 percent attains an A' value of \$5000 in 1 year, an obvious growth of 10 percent: $g = i$.

Applying i^* in a situation where i is greater than g,

Case 3: When g is less than i, the special interest rate is

$$i^* = \frac{1 + i}{1 + g} - 1$$

$$P = \frac{A'}{1 + g}(P/A, i^*, N)$$

Again using the cash flow pattern in Figure 2.3(a), but this time assuming $i = 12$ percent ($g < i$), we have

$$i^* = \frac{1 + 0.12}{1 + 0.10} - 1 = \frac{1.12}{1.10} - 1 = 1.0182 - 1 = 1.82\%$$

and

$$P = \frac{\$1000}{1 + 0.10}(P/A, 1.82, 5) = \$909(4.7408) = \$4309$$

Case 3 applies to all declining cash flow streams because g is always negative during deflation. As applied to the cash flow pattern in Figure 2.3(b), where $g = -10$ percent, $A' = \$1000$, $N = 5$, and i is assumed to be 8 percent,

$$i^* = \frac{1 + 0.08}{1 + -0.10} - 1 = \frac{1.08}{0.90} - 1 = 1.20 - 1 = 20\%$$

$$P = \frac{\$1000}{1 - 0.10}(P/A, 20, 5) = \$1111(2.991) = \$3323$$

Use of a gradient series for studies that include inflation is demonstrated in Chapter 13.

In summary, cases 1 and 3 require the calculation of a special interest rate i^* that depends on the growth rate g and regular interest rate i. This special rate is used like any other interest rate in obtaining the values for applicable interest factors, F/A or P/A. After applying the appropriate formula for the case where g is greater than, equal to, or less than i, we obtain the present value of the geometric series. The value thus calculated can be converted to an equivalent cash flow by applying any other discrete compound-interest formula using the regular interest rate i.

Review Exercises and Discussions

The money earned from making a loan is evident in the contract (sometimes in fine print). A loan of $10,000 for 1 year at an interest rate of 10 percent earns the lender $10,000 × 0.10 = $1000. A borrower usually does not know in advance exactly how much will be gained from a loan to buy productive goods. It is often impossible to segregate precisely the receipts due a certain production operation when that operation is a small part of a much larger production system. In such cases an evaluation study may be made on the amount production costs of the system are decreased by improvements to the given operation, assuming the operation must be performed to maintain the total process. Then the earnings are in the form of "cost savings" in the system which are compared with the investment cost of acquiring and using assets to improve the operation.

Assume a machine is purchased for $10,000 with the loan mentioned above. The machine will be completely worn out by the end of the year and its operating costs will be $100 per month more than the costs of the present operation. How large a cost reduction must be provided by the machine for its purchase to earn a 15 percent return for the borrower?

The costs involved include repaying the loan plus interest charges for the loan, extra operating costs, and investment earnings resulting from the purchase of the machine.

Loan repayment (purchase price of the machine)	$10,000
Interest paid on the loan for 1 year = $10,000 × 0.10	1000
Additional operating cost incurred = $100/month × 12	1200
15% earnings on the $10,000 borrowed = $10,000 × 0.15	1500
Necessary cost reduction to support the investment	$13,700

If the cost reduction turned out to be only $13,700 − $1500 = $12,200, it would cover only expenses and nothing would be gained from the machine's purchase. However, if the company could afford $10,000 of its own money for the machine rather than borrowing it, a cost reduction of $12,200 would yield a 10 percent return on the investment, which is the rate the company could earn by simply lending its money at the "going" interest rate of 10 percent. A cost reduction of $13,700 would produce a return of

$$\frac{\$13,700 - \$10,000 - \$1200}{\$10,000} = \frac{\$2500}{\$10,000} = 100\% = 25\%$$

which is attractive to the company because it provides an added 15 percent beyond the cost of capital to support the project.

A loan of $200 is made for a period of 13 months, from January 1 to January 31 the following year, at a simple interest rate of 8 percent. What future amount is due at the end of the loan period?

Using ordinary simple interest, we find that the total amount to be repaid after 13 months is

$$F = P + PiN$$
$$= \$200 + (\$200)(0.08)(1 + 1/12)$$
$$= \$200 + (\$200)(0.0867)$$
$$= \$200 + \$17.34 = \$217.34$$

If exact simple interest is used, the future value (assuming the year in question is not a leap year) is

$$F = P + PiN$$
$$= \$200 + (\$200)(0.08)\left(1 + \frac{31}{365}\right)$$
$$= \$200 + (\$200)(0.0868)$$
$$= \$200 + \$17.36 = \$217.36$$

Exercise 3   A credit plan charges interest at the rate of 36 percent compounded monthly. What is the effective interest rate?

Solution 3   The nominal 36 percent rate constitutes monthly charges of 3 percent. From this statement we know that $r = 0.36$ and $m = 12$, so the effective interest rate can be calculated as

$$i = \left(1 + \frac{r}{m}\right)^m - 1 = \left(1 + \frac{0.36}{12}\right)^{12} - 1$$
$$= (1.03)^{12} - 1 = 1.4257 - 1 = 42.57\%$$

The same result can be obtained by recognizing that

$$(1.03)^{12} - 1 = (F/P, 3, 12) - 1$$

Then the tables in Appendix D can be used to find the value of the compound-amount factor at $i = 3$ percent and $N = 12$:

$$(F/P, 3, 12) = 1.4257$$

so $$1.4257 - 1 = 42.57\%$$

Exercise 4   How much would a person have had to invest 1 year ago to have $2500 available today, when the investment earned interest at the nominal rate of 12 percent compounded monthly?

Solution 4   It is first necessary to convert the nominal rate to its corresponding periodic rate: 12 percent compounded monthly means that an investment earns 1 percent per month. Next, it must be recognized that today's worth is a future worth in terms of when the investment P was made, 1 year previously. It is known that $F = \$2500$, $i = 1$ percent, and $N = 12$ (12 months have passed since the original investment); therefore,

$$P = F(P/F, 1, 12) = \$2500(0.88746) = \$2219$$

Exercise 5   What annual year-end payment must be made each year to have $10,000 available 4 years from now? The compound annual interest rate is 10 percent.

Solution 5   The 4-year annuity is a sinking fund which has a value at maturity of $F = \$10,000$. The necessary annual deposits equal

$$A = F(A/F, 10, 4) = \$10,000(0.21547) = \$2155$$

If you deposit $10,000 today, what equal amounts can you withdraw at the end of each year for the next 4 years when the interest rate is 10 percent?

The withdrawals form an ordinary annuity where $N = 4$ and $i = 10$ percent. Given $P = $10,000$, the capital invested is recovered by payments of

$$A = \$10,000(A/P, 10, 4) = \$10,000(0.31547) = \$3158$$

An ambitious saver plans to deposit $2000 in a money-market account starting 1 year from now and wants to increase annual deposits by $1000 each year for the next 6 years. Assuming that deposits earn 9 percent annually, determine what equal-payment annuity would accumulate the same amount over the 7-year period.

The first step in using the arithmetic-gradient conversion factor is to identify the base annuity A' and the gradient G. A' is the first payment of $2000 and G is the amount by which the payments increase each year, $1000. Then

$$A = A' + G(A/G, i, N) = \$2000 + \$1000(A/G, 9, 7)$$
$$= \$2000 + \$1000(2.6572) = \$2000 + \$2657 = \$4657$$

Thus seven equal payments of $4657 are equivalent to seven payments increasing by $1000 increments from $2000 for the first one to $8000 for the last one.

Receipts from an investment will decline by $150 each year for 5 years from a level of $1000 at the end of the first year. For an interest rate of 7 percent, calculate a constant annual series amount that is equivalent to the gradient over the 6-year period.

The base amount ($A' = \$1000$) is decreased by a uniform amount each year ($G = -\$150$). Given $i = 7$ percent and $N = 6$,

$$A = \$1000 - \$150(A/G, 7, 6) = \$1000 - \$150(2.3030) = \$654.55$$

The ambitious saver in Exercise 7 has changed plans. Instead of increasing each deposit by $1000, each deposit will be raised by 20 percent over the previous one. What equal payment annuity will accumulate the same amount over the 7-year period when deposits earn 9 percent annually?

The annual 20 percent change is a gradient where $g = 0.20$. Since g is greater than $i = 0.09$,

$$i^* = \frac{1 + g}{1 + i} - 1 = \frac{1 + 0.20}{1 + 0.09} - 1 = 1.10 - 1 = 0.10$$

Given $A' = \$2000$ and $N = 7$, the present worth of the series is

$$P = \frac{A'}{1 + i}(F/A, i^*, N) = \frac{\$2000}{1.09}(9.4870) = \$17,407.34$$

from which the equivalent uniform annuity is

$$A = \$17,407(A/P, 9, 7)$$
$$= \$17,407(0.19869) = \$3458.66$$

PROBLEMS

Problems 2.1 through 2.11 are adapted from Mathematics of Finance by L. L. Smail (McGraw-Hill, New York). This college text was published in 1925, and it shows that basic interest problems have not changed much over the years.

2.1   What sum must be loaned at 6 percent simple interest to earn $47 in 2 years?

2.2   How long will it take $800 to yield $72 in simple interest at 4 percent?

2.3   At what rate will $65.07 yield $8.75 in simple interest in 3 years, 6 months?

2.4   How long will it take any sum to double itself at a 5 percent simple interest rate?

2.5   Find the ordinary and exact simple interest on $3300 at 6 percent for 56 days.

2.6   If the interest on a certain sum for 3 months is $63.87 at 5 percent simple interest, what would it be at 6 percent?

2.7   Find the compound amount of $100 for 3 years at 5 percent compounded annually.

2.8   What is the compound amount of $750 for 5 years at 6 percent compounded quarterly?

2.9   Accumulate a principal of $600 for 5 years, 9 months at 6 percent compounded monthly. How much interest is earned?

2.10   Find the difference between the amount of $100 at simple interest and at compound interest for 5 years at 5 percent.

2.11   Find the compound amount of $5000 at 6 percent for 4, 6, 8, and 10 years, and compare the results. Does doubling the time double the amount?

2.12   Determine the effective interest rate for a nominal annual rate of 12 percent which is compounded:

(a) Semiannually.
(b) Quarterly.
(c) Monthly.
(d) Daily.

2.13   A personal loan of $1000 is made for a period of 18 months at an interest rate of $1\frac{1}{2}$ percent per month on the unpaid balance. If the entire amount owed is repaid in a lump sum at the end of that time, determine:

(a) The effective annual interest rate.
(b) The total amount of interest paid.

2.14   How is it possible to determine the numerical value of a capital-recovery factor (A/P) if the only table available is:

(a) (P/F)?
(b) (A/F)?
(c) (F/A)?

2.15   Determine the value of $(F/A, 4, 8)$ using only the table for (P/A).

2.16   What amount must be deposited in an account paying an effective annual rate of 10 percent in order to accumulate $2000 in

    (a) 5 years?
    (b) 10 years?
    (c) 20 years?

2.17 What nominal annual interest rate compounded monthly yields an effective annual rate of 19.56 percent?

2.18 A loan of $5000 is scheduled to be repaid in equal monthly installments over $2\frac{1}{2}$ years. The nominal interest rate is 6 percent. How large is each payment?

2.19 How much will a piece of property have to increase in value over the next 5 years if it is to earn 10 percent per year on the purchase price?

2.20 What simple interest rate must be applied to earn the same interest over 5 years as earned by an investment at 5 percent compounded semiannually?

2.21 A loan of $5000 is to be repaid in equal monthly payments over the next 2 years. Determine the payment amount if interest is charged at a nominal annual rate of 15 percent.

2.22 Compare the effective interest rates for 16 percent compounded quarterly and 15 percent compounded monthly.

2.23 How much less would it cost to pay off a $3000 loan in 1 year with 12 equal payments when interest is 12 percent compounded monthly, as opposed to making a single payment when the effective interest rate is 12 percent?

2.24 How much money could you borrow if you agreed to pay back $1000 at the end of each year for 5 years? The lender expects to earn 10 percent per year.

2.25 The net income from a newly purchased piece of construction equipment is expected to be $12,000 the first year and to decrease by $1500 each year as maintenance costs increase. The equipment will be used for 4 years. What annual annuity would produce an equivalent income when the interest rate is 8 percent?

2.26 A loan of $10,000 is to be repaid in a lump sum at the end of 4 years. Determine the amount of the payment if interest is charged at a nominal annual rate of 18 percent compounded monthly.

2.27 If $10,000 is deposited today in an account paying a nominal 12 percent per annum compounded quarterly, how many years will be required for the account balance to total $25,000?

2.29 Service records for a specific piece of production equipment indicate that a replacement machine will have first-year maintenance costs of approximately $1000 and these costs will increase by $200 per year for each additional year of service. Assuming the equipment is to be in service for 10 years and using an interest rate of 15 percent, determine the maximum amount which should be paid for a lifetime maintenance contract at the time the equipment is purchased.

2.30 Traffic flow over a new bridge is expected to be one million vehicles the first year it is in use. If the average traffic rate is expected to increase by 5 percent per year, determine:

    (a) The expected number of vehicles using the bridge in the twentieth year of service.
    (b) The total number of vehicles using the bridge during the 20-year period.

2.31 Assuming that a toll of $1 per vehicle is charged for use of the bridge in Problem 2.30, determine the present worth of all projected toll collections using an interest rate of 8 percent.

2.32 Maintenance costs on a selected piece of production equipment are expected to be $1000 the first year of operation and will probably increase at a rate of 20 percent per year throughout a 10-year service life. Using an interest rate of 15 percent, compute the present worth of the expected costs.

2.33 Compare the answers obtained from Problem 2.29 to those obtained from 2.32. What accounts for the significant difference in present worth values?

2.34 Fred Furd borrowed $1000 from the Friendly Finance Company at a nominal annual rate of 18 percent compounded monthly. Determine how long Fred will be required to pay off his debt if he makes the following monthly payments?
(a) $175.54
(b) $ 91.69
(c) $ 49.93
(d) $ 20.00
(e) $ 15.00

2.35 Tuition costs are expected to inflate at the rate of 8 percent per year. The first year's tuition is due 1 year from now and will be $2000. A fund is to be set up today to cover tuition costs for 4 years in an account that will earn interest at rate i. How large must the fund be if
(a) $i = 5$ percent.
(b) $i = 8$ percent.
(c) $i = 10$ percent.

2.36 Today's price for materials used in a production process is expected to hold constant for this year at $100,000. What is the present worth of 5 years' supply for the same amount of material used each year when the interest rate is 8 percent, if the price changes at a constant annual rate of
(a) $g = -5$ percent.
(b) $g = 0$ percent.
(c) $g = 5$ percent.
(d) $g = 8$ percent.
(e) $g = 15$ percent.

EXTENSION

2A Compounding Confusion

A decision to set aside a sum of money is simply the first step in starting a savings account. The money could be put in such places as a money market account, certificate of deposit, and an Individual Retirement Account (IRA). These savings instruments are distinguished, respectively, by interest rates that vary from period to period, commitments to leave the money invested for a specified time or paying a penalty for an early withdrawal, and deferred taxes on the interest earned in an IRA.

Within any savings instrument the method of compounding is likely to differ between institutions. The method employed can make a considerable difference in the future

TABLE 2.5

F given P and N when the compounding method varies for a nominal interest rate of 10 percent

Compounding Method	Value of An Initial $1000 Investment After			
	1 Year	5 Years	10 Years	20 years
Yearly	$1100.00	$1610.51	$2593.74	$6727.50
Quarterly	1103.81	1638.62	2685.06	7209.57
Monthly	1104.71	1645.31	2707.04	7328.07
Weekly	1105.06	1647.93	2715.67	7374.88
360 daily	1105.16	1648.61	2717.90	7387.00
365 daily	1105.16	1648.61	2717.91	7387.03
Continuous	1105.17	1648.72	2718.28	7389.06
365/360 daily	1106.69	1660.08	2755.88	7594.86

value of the account. The figures in Table 2.5 reveal how much a one-time investment of $1000 at 10 percent interest amounts to after various lengths of time under various compounding patterns.

As apparent in the first column, the effective interest rate climbs with an increasing number of compounding periods per year. The only feature of the table that deviates from the discussion in the chapter about effective interest rates is the division of a year into 360 days rather than the logical 365 days. This is comparable to ordinary simple interest being based on 360 days, except that the amount in the account is compounded 365 times a year at the rate of $0.10/360 = 0.0002777$. This plan (365/360 daily) yields a slightly higher return than does the "365 daily" method, in effect providing a "bonus" of five extra interest-paying days,

a feature which some institutions use as a lure to attract more savings accounts.

The interest-rate jungle gets even thicker when special phrases are used to characterize compounding plans. For instance, "interest paid daily" is not necessarily the same as "daily interest paid." The former conventionally means compounding daily; the latter means only that interest is credited to the account every day, but the amount credited could be based on monthly, quarterly, or even annual compounding. Such devices have led to calls for a truth-in-savings law similar to the truth-in-lending law enacted in 1968 (see Extension 6B).

QUESTIONS

2A.1 What do you feel would be the most readily understood way to quote the yield on savings?

2A.2 Complete the following table that compares the annual yield earned by an investment at the nominal interest rate of 6 percent under different compounding patterns.

Compounding Period (m)	Compound Amount Factor at r = 6 Percent	Effective Interest Rate (i)
Annually		0.06
Semiannually		
Quarterly	$(0.015)^4$	
Monthly		
Daily		0.0618305
Continuously		

CHAPTER 3

DISCOUNTED CASH FLOW
CALCULATIONS

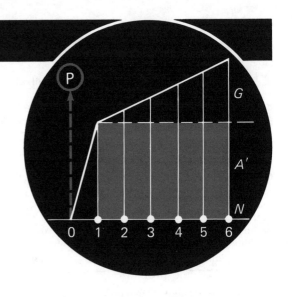

OVERVIEW

Concepts concerning the time value of money from the last chapter become working tools in this chapter. Compound-interest factors for both discrete and continuous interest are applied to a variety of cash flows. The purpose of the calculations is to develop skills in converting cash flow patterns into equivalent sums that are more useful in comparing investments.

Receipts and disbursements associated with an economic situation can be portrayed on a cash flow diagram. When it is so diagramed, the timing of the cash flow is more apparent and the chance of making careless errors is reduced. With practice, different cash flow patterns may be recognized and can suggest the most direct approach for analysis.

Calculations made with either continuous or discrete interest factors rely on the same reasoning and rules; only the effect of interest on the cash flow varies. Both discount a cash flow to remove the time effect and thereby reveal its equivalent value at a specific point in time.

CASH FLOW DIAGRAMS

During the construction of a cash flow diagram the structure of a problem often becomes distinct. It is usually advantageous to first define the time frame over which cash flows occur. This establishes the horizontal scale, which is divided into time periods, frequently but not always years. Receipts and disbursements are then located on the time scale in adherence to problem specifications. Individual outlays or receipts are designated by vertical lines; relative magnitudes can be suggested by the heights of lines, but exact scaling wastes time. Whether a cash flow is positive or negative (positive above the axis, and negative below) depends on whose viewpoint is portrayed.

Figures 3.1(a) and (b) represent the same transaction: a loan paid off in three installments. From the borrower's viewpoint in Figure 3.1(a), the receipt of the loan is a positive inflow of cash, whereas subsequent installment payments represent negative outflows. Flows are reversed when viewed from the lender's perspective in Figure 3.1(b).

Although cash flow diagrams are simply graphical representations of income and outlay, they should exhibit as much information as possible. It is useful to show the interest rate, and it may be helpful to designate the unknown that must be solved for in a problem. Figure 3.1 is redrawn in Figure 3.2 to represent specific problems. In Figure 3.2(a), the three equal payments are represented by a convenient convention that indicates a 3-year annuity in which the payment size A is unknown. A is circled to indicate the needed solution. The given interest rate ($i = 10$ percent) is entered in a conspicuous place, and the amount of the loan is placed on the arrow at time zero.

In Figure 3.2(b), numbered years are replaced with dates, and the size of A is given. The problem is to find the interest rate that makes the annuity equivalent to the loan value; i is circled. Sometimes it may clarify the situation to put in dashed lines for arrows that represent cash flows of unknown magnitude. This or a similar tactic is especially useful when a problem comprises several separate cash flow segments, each of which must be replaced by an equivalent value; these, in turn, are converted to a single sum representing all the segments.

The style of a cash flow diagram is important only insofar as it contributes to clarity.

The obvious requirements for cash flow diagraming are completeness, accuracy, and legibility. The measure of a successful diagram is that someone

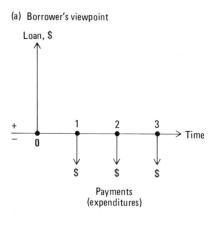

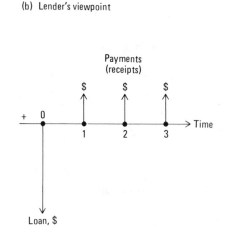

FIGURE 3.1

Perspectives for cash flow diagrams.

FIGURE 3.2
Different versions of cash flow diagrams.

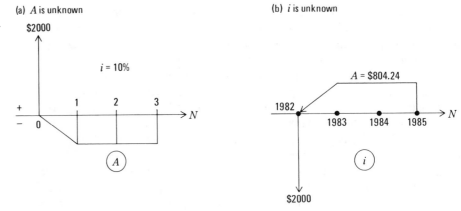

else can understand the problem fully from it. If it passes this test, you are unlikely to confuse yourself.

CALCULATION OF TIME-VALUE EQUIVALENTS

The purpose of time-value calculations is to translate receipts and disbursements of various amounts occurring at various times to a cash flow pattern that assists an economic evaluation. The translation is essentially mechanical in the same fashion a vector is routinely decomposed into component forces; rules of geometry direct vector operations, and time-value relations direct cash flow translations. Although errors in translation can arise from carelessness, mistakes due to incorrect problem formulations are the ones to guard against.

Cash flow diagrams assist problem formulation. They serve the same purpose as the free-body diagrams used to portray the effect of forces acting on a body. The intent is to isolate the factors pertinent to a problem and to display them clearly to observe what data are available and what calculations are needed.

The notable statements in a discounted cash flow problem are the interest elements: P, F, A, N, and i. Generally, three of the elements are known for each cash flow, and the problem entails solving for a fourth element. Several money-time translations may be required in one solution.

A variety of typical cash flow patterns are treated in the following pages. Examples are presented to put cash flow problems in a realistic setting, show different perspectives for the same type of problem, demonstrate the use of cash flow diagrams, and provide familiarity with the use of interest factors and tables. Although the examples all employ discrete compounding, the diagrams and solution formulations are equally applicable to continuous compounding. Specific applications of continuous interest factors are examined in the last section of this chapter.

Single-Payment Cash Flow

Translation of a future amount to its present worth, or the reverse from present to future, has been demonstrated previously for both discrete and continuous compounding. Sometimes both present and future amounts are known, so the problem is to find the value of i or N that makes them equivalent.

Example 3.1
Unknown Interest Rate

At what annual interest rate will $1000 invested today be worth $2000 in 9 years?

Solution 3.1

Given P = $1000, F = $2000, N = 9, years, find i.

$$\frac{F}{P} = (F/P, i, 9)$$

$$\frac{\$2000}{\$1000} = 2 = (F/P, i, 9)$$

$$i = 8\%$$

The interest rate i is determined by locating the interest rate at which the single-payment compound-amount factor is equal to 2.0 and N = 9 (a reciprocal relation using a present-worth factor could serve as well). The numerical value of i is found by leafing through the pages of interest rates and noting the appropriate factor values for the given number of periods.

It is usually necessary to interpolate from table values when N or i is unknown. The error introduced by linear interpolation is relatively insignificant for most practical applications. If the investment period for Example 3.1 had been 10 years instead of 9, the interest-rate calculation would have been

$$(F/P, i, 10) = \frac{F}{P} = \frac{\$2000}{\$1000} = 2.0$$

At i = 7 percent, $(F/P, 7, 10)$ = 1.9671; at i = 8 percent, $(F/P, 8, 10)$ = 2.1589. Then, by interpolation,

$$i = 0.07 + 0.01 \frac{2.0000 - 1.9671}{2.1589 - 1.9671}$$

$$= 0.07 + 0.01 \frac{0.0329}{0.1918}$$

$$= 0.07 + 0.0017$$

$$= 0.0717$$

or

$$i \doteq 7.2\%$$

Most of the inconvenience of interpolation is avoided with calculators that can be programmed, or are preprogrammed, for the interest-factor formulas. Then it is a simple operation to punch in various values of the unknown element until matching values are obtained. Nonetheless, the mechanics of interpolation are presented because the origin of an element's value is more easily observed with numerical examples, and there may be occasions when the use of interest tables is unavoidable.

Example 3.2
Unknown Number of Interest Periods

A loan of $1000 is made today under an agreement that $1400 will be received in payment sometime in the future. When should the $1400 be received if the loan is to earn interest at a rate of 8 percent compounded quarterly?

Solution 3.2

Given $P = \$1000$, $F = \$1400$, $i = r/m = 8\%/4 = 2\%$, find N (in quarters).

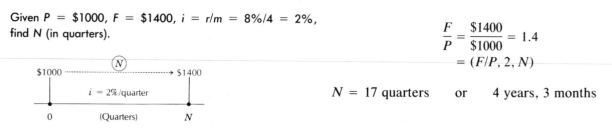

$$\frac{F}{P} = \frac{\$1400}{\$1000} = 1.4$$
$$= (F/P, 2, N)$$

$$N = 17 \text{ quarters} \quad \text{or} \quad 4 \text{ years, 3 months}$$

Example 3.2 is slightly deceptive because it involves a nominal interest rate. The problem is clarified by noting in the cash flow diagram that the time scale is in quarters of a year and that i is given as the rate per quarter. The problem could also have been solved by converting the nominal rate to its equivalent effective interest rate,

$$i = \left(1 + \frac{0.08}{4}\right)^4 - 1 = 1.082 - 1 = 0.082$$

and solving the following equation by interpolation between the 8 percent and 9 percent interest tables and N values between 4 and 5:

$$1.400 = (F/P, 8.2, N)$$
$$1.3604 + 0.2(1.4115 - 1.3604) = 1.3706 = (F/P, 8.2, 4)$$
$$1.4693 + 0.2(1.5386 - 1.4693) = 1.4831 = (F/P, 8.2, 5)$$
$$N = 4 + 1\frac{1.4000 - 1.3706}{1.4831 - 1.3706}$$
$$= 4.265 \text{ years}$$

The result differs from 4 years, 3 months only by round-off errors in the interest factors.

Multiple-Payment Cash Flows

Practical problems customarily involve both single payments and annuities. For instance, determining the equivalent present worth (cost) of owning a car for 3 years involves a series of payments to purchase the car and provide gas, repair, and maintenance costs at irregular intervals, and a single payment (receipt) when the car is sold. All these receipts and disbursements would be translated to "now" and summed to find the present worth—a single sum equivalent to ownership costs for 3 years.

Example 3.3
More Compounding Periods than Payments

"Now" is June 30, 1982. Three payments of $500 each are to be received every 2 years starting 2 years from now and deposited in a bank where they will earn interest at 7 percent per year. How large will the bank account be on June 30, 1990?

Solution 3.3

$$F = \$500(F/P, 7, 6) + \$500(F/P, 7, 4) + \$500(F/P, 7, 2)$$
$$= \$500(1.5007 + 1.3107 + 1.1449)$$
$$= \$500(3.9563)$$
$$= \$1978$$

The equal payments in Example 3.3 do not constitute an ordinary annuity because there are fewer payments than there are compounding periods. Therefore, each payment must be translated individually to the 1990 date and added to the worth of the other payments at that date to obtain the equivalent future worth of all three payments.

Example 3.4
Annuity with an Unknown *i*

A new machine that can be purchased for $8065 will annually reduce production costs by $2020. The machine will operate for 5 years, at which time it will have no resale value. What rate of return will be earned on the investment? (Alternative statement of the problem: At what interest rate will a cash flow of $2020 per year for 5 years equal a present value of $8065?)

Solution 3.4

$$\frac{P}{A} = \frac{\$8065}{\$2020} = 3.993 = (P/A, i, 5)$$
$$i = 8\%$$

A series of payments made at the beginning instead of the end of each period is sometimes referred to as an *annuity due*. Rather than create a special factor for this annuity pattern, we divide the series into two parts. If the first payment is translated separately, the remaining payments fit the pattern for an ordinary annuity beginning at the time of the first payment. The present worth of the series is the sum of the first payment plus the product of one payment times the series present-worth factor, where N is the number of payments minus 1.

Example 3.5
Annuity Due

What is the present worth of a series of 15 year-end payments of $1000 each when the first payment is due today and the interest rate is 5 percent?

Solution 3.5

$$P = A + A(P/A, 5, 14)$$
$$= \$1000 + \$1000(9.8985)$$
$$= \$1000 + \$9899$$
$$= \$10,899$$

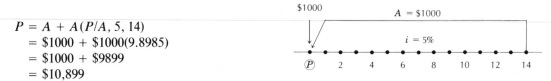

Another pattern for a series of payments, in which the first payment does not begin until some date later than the end of the first period, is called a *deferred annuity*. Like an annuity due, a deferred annuity is evaluated by dividing the time period into two parts. One portion is N, the number of payment periods, plus 1. This portion forms an ordinary annuity of N periods. The second portion is the number of periods left after subtracting $N + 1$ periods. A solution results from determining the present worth of the ordinary annuity and then discounting this value through the deferred period.

Example 3.6
Deferred Annuity

With interest at 6 percent, what is the worth on June 30, 1986 of a series of year-end payments of $317.70 made on June 30 from 1992 through 1996?

Solution 3.6

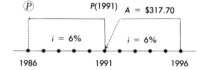

Starting from the known values of A, i, and N,

$$P(1991) = A(P/A, 6, 5)$$

P(1991) becomes the F value in calculating P(1986):

$$P(1986) = P(1991)(P/F, 6, 5)$$

Collecting terms,

$$P(1986) = A(P/A, 6, 5)(P/F, 6, 5)$$
$$= \$317.70(4.2123)(0.74726)$$
$$= \$1000$$

The results of Example 3.6 may be recognized as one of the equivalent outcomes presented without proof in Fig. 2.1. Another of the outcomes from the same figure is shown in Figure 3.3.

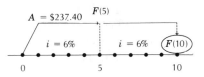

FIGURE 3.3
Cash flow diagram displaying the data required to calculate the future worth of a 5-year annuity. The data shown are based on cash flows given in Figure 2.1.

The value of the annuity after five payments of $237.40 is labeled $F(5)$. By letting the end of the fifth year be "now," $F(5)$ is treated as a present value, and the compound-amount factor is used to find the future value at the end of year 10, or $F(10)$:

$$F(5) = A(F/A, 6, 5) \quad \text{and} \quad F(10) = F(5)(F/P, 6, 5)$$
$$F(10) = A(F/A, 6, 5)(F/P, 6, 5)$$
$$= \$237.40(5.6370)(1.3382)$$
$$= \$1791$$

Example 3.7
Present Worth of an Arithmetic Gradient

A contract has been signed to lease a building at $20,000 per year with annual increases of $1500 for 8 years. Payments are to be made at the end of each year, starting 1 year from now. The prevailing interest rate is 7 percent. What lump sum paid today would be equivalent to the 8-year lease-payment plan?

Solution 3.7

The base annuity A' and the gradient G per period are shown on the cash flow diagram.

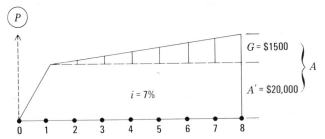

$$A = A' + G(A/G, 7, 8) = \$20,000 + \$1500(3.1463)$$
$$= \$20,000 + \$4719.45$$
$$= \$24,719.45$$

Then the annuity is translated to its present worth as

$$P = A(P/A, 7, 8) = \$24,719.45(5.9712)$$
$$= \$147,604.78$$

The first step in solving for the present worth of the lease-payment plan is to convert the increasing annual payments to a uniform series:

which is the amount in today's dollars equivalent to the lease contract that provides yearly payments of $20,000 with annual increases of $1500 for 8 years.

More extensive economic situations often include both income and outlay. Such situations are evaluated by calculating the net outcome at a certain point in time. A cash flow diagram incorporates receipts and disbursements by displaying income above the time line and outlays below the line. Other payment categories can be handled similarly.

Example 3.8
Income and Outlay

A boy is now 11 years old. On his fifth birthday he received a gift of $4000 from his grandparents which was invested in 10-year bonds bearing interest at 4 percent compounded semiannually. His parents plan to have $3000 available for the boy's nineteenth, twentieth, twenty-first, and twenty-second birthdays to help finance a college education. To assist the financing the grandparents' gift will be reinvested when the bonds mature. How much should the parents allocate to invest each year on the boy's twelfth through eighteenth birthdays to complete the education plan? All future investments will earn 6 percent annually.

Solution 3.8

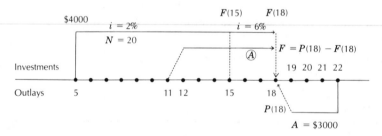

The cash-flow diagram for the problem is based on an evaluation date of the boy's eighteenth birthday. It would be equally valid to use any of his birthdays for the comparison date. The same value for the parents' payments, A, would result for any chosen date, but there is usually one time which reduces the number of required calculations.

As shown in the diagram, there are three payment plans involved: a series of outlays, a lump-sum investment, and a series of annual investments. After computing the difference between the amount available from the lump-sum investment and the amount required for the education expenses at a common date, the annual payment necessary to accumulate this difference can be determined.

The present worth of the education annuity on the boy's eighteenth birthday is

$$P(18) = A(P/A, 6, 4) = \$3000(3.4650) = \$10,395$$

The future worth of the grandparents' gift is a function of two investment rates: 2 percent per period for 20 periods (a nominal interest rate of 4 percent compounded semiannually for 10 years), followed by 6 percent per period for three periods:

$$F(18) = P(F/P, 2, 20)(F/P, 6, 3)$$
$$= \$4000(1.4859)(1.1910) = \$7079$$

The difference F between the amount required and the amount available at year 18 is

$$F = P(18) - F(18) = \$10,395 - \$7079 = \$3316$$

which is the required future value for the series of 18 − 11 = 7 payments of amount A beginning on the boy's twelfth birthday:

$$A = F(A/F, 6, 7) = \$3316(0.11914) = \$395$$

CALCULATIONS WITH CONTINUOUS INTEREST

In the discussion of nominal interest rates in Chapter 2, it was apparent that for a specific nominal rate, the effective interest rate gets larger as the compounding interval is shortened. The effective interest rate i for a nominal interest rate of 18 percent ($r = 0.18$) compounded semiannually ($m = 2$) was shown to be

$$i = \left(1 + \frac{r}{m}\right)^m - 1 = \left(1 + \frac{0.18}{2}\right)^2 - 1 = 0.188 \quad \text{or} \quad 18.8\%$$

and that $i = e^r - 1$ when the interest periods are infinitesimally small. For the same 18 percent nominal interest compounded continuously,

$$i = e^{0.18} - 1 = 1.1972 - 1 = 0.1972 \quad \text{or} \quad 19.72\%$$

Using continuous compounding in place of discrete compounding leads to an increase in the force of interest on the time value of money. A single payment earning continuous interest has a future value of

$$F = Pe^{rN}$$

where the continuous-interest compound factor e^{rN} corresponds to the $(1 + i)^N$ used in the F/P factor for discrete compounding.

Example 3.9
Comparison of Continuous and Discrete Compounding

A sum of money doubles in size after 9 years when invested at 8 percent interest compounded annually:

$$F = P(1 + 0.08)^9$$
$$2P \doteq P(1.999)$$

At what rate of continuous compounding will an amount double in only half the time taken at the 8 percent effective rate?

Solution 3.9

Given $F = 2P$ at $N = 9/2 = 4.5$ years. Substituting these values into the future-worth equation using continuous compounding gives

$$F = 2P = P(e^{rN}) \quad \text{or} \quad 2 = e^{4.5r}$$

The right-hand equation can be solved directly for r to get $r = 15.4$ percent.

Continuous-Compounding, Discrete Cash Flow

There are two versions of continuous interest. The *discrete cash flow* form applies continuous compounding to a payment whenever it is received, but the payment is assumed to be received in one lump sum. The other form assumes that cash flow is continuous. The future-worth formula $F = Pe^{rN}$ and its reciprocal $P = Fe^{-rN}$ are discrete cash flows compounded continuously.

Continuous-compounding factors with *discrete payments* can be similarly developed from the end-of-period annuity formulas. First recall from the discussion of effective interest that e^{rN} corresponds to $(1 + i)^N$ and that the effective interest rate is $i = e^r - 1$. Substitution of these expressions into the sinking-fund formula yields the fact that the discrete cash flow, end-of-period compounding, sinking-fund formula is equal to the discrete cash flow, continuous-compounding, *sinking-fund* formula when $i = e^r - 1$:

$$A = F\left[\frac{i}{(1 + i)^N - 1}\right] = \left[F\frac{e^r - 1}{e^{rN} - 1}\right]$$

and the reciprocal of the bracketed expression is the *series compound-amount factor* for continuous compounding.

By similar reasoning, the *capital-recovery* formula with continuous compounding and discrete payments is

$$A = P\left[\frac{e^{rN}(e^r - 1)}{E^{rN} - 1}\right]$$

and the reciprocal of the bracketed expression is the continuous-compounding *series present-worth factor*.

In every formula the A, F, and P values resulting from computations using either end-of-period or continuous compounding are identical for discrete payments when the continuous interest rate is equivalent to the effective interest rate. This relation is apparent when numbers are substituted into the sinking-fund formulas displayed above. Letting $i = 22.1$ percent, which corresponds to a nominal continuous interest rate of 20 percent ($0.221 = e^{0.2} - 1$), and applying the equivalent interest rates in the two sinking-fund formulas with $N = 2$ give

End-of-period compounding		*Continuous compounding*
$A = F\dfrac{0.221}{(1.221)^2 - 1}$	$=$	$F\dfrac{e^{0.2} - 1}{e^{(0.2)(2)} - 1}$
$= F\dfrac{0.221}{1.491 - 1}$	$=$	$F\dfrac{1.221 - 1}{1.492 - 1}$
0.45	$=$	0.45

which shows that the factors are equal when $i = e^r - 1$ and payments are discrete.

Example 3.10
Continuous Compounding of a Discrete-Payment Annuity

Each year a single payment of $1766 is deposited in an account that earns 6 percent compounded continuously. What is the amount in the account immediately after the fifth payment?

Solution 3.10

Given $A = \$1766$, $r = 0.06$, and $N = 5$, F is calculated as

$$F = A\frac{e^{rN} - 1}{e^r - 1} = \$1766\frac{e^{(.06)(5)} - 1}{e^{(.06)} - 1}$$

$$= \$1766\frac{1.3499 - 1}{1.0618 - 1} = \$1766(5.6618)$$

$$= \$9999$$

This future worth is not notably higher than is that earned by annual compounding of the same annuity

$$F = \$1766(F/A, 6, 5) = \$1766(5.6370) = \$9955$$

Continuous-Compounding, Continuous Cash Flow

The other version of continuous compounding occurs when the *total payment for a period is received in continuous, small, equal payments during that period*. Letting $\overline{A}$ designate this total amount of each payment in the series, which continues for N periods at interest rate r per period, we find the future worth directly by integration as

$$F = \overline{A}\int_0^N e^{rt}\, dt = \overline{A}\left[\frac{e^{rN} - 1}{r}\right]$$

The expression in brackets is called the *continuous-compounding series compound-amount factor* for continuous uniform payments.

Functional notations and formulas for the six basic continuous-compounding factors for continuous-flow payments are given in Table 3.1. The first two in the table are for a single continuous payment for one period, and the last four are annuity payments. Bars over the symbols for future, present, and annuity payments (respectively, $\overline{F}$, $\overline{P}$, and $\overline{A}$) represent the total amount accumulated from continuous small payments throughout each compounding duration.

The assumption of a continuous flow of disbursements and incomes throughout a year is rare as compared with the end-of-year payment pattern. However, the continuous-flow assumption is more revealing and applicable than is continuous compounding of discrete payments because the main reason for using continuous interest in an economic evaluation is to determine the effects of continuous cash flows, not just the continuous interest on discrete payments. Consequently, tables are provided in Appendix E for continuous-

TABLE 3.1

Symbols and formulas for continuous compounding of a continuous flow. $\bar{F}$, $\bar{P}$, and $\bar{A}$ designate total amounts accumulated in small equal payments during one compounding period N for $\bar{A}$, and over N periods for $\bar{F}$ and $\bar{P}$. The continuous interest rate is r per period.

Application	Formula	Application	Formula
$(F/\bar{P}, r\%, N)$	$F = \bar{P}\dfrac{e^{rN}(e^r - 1)}{re^r}$	$(F/\bar{A}, r\%, N)$	$F = \bar{A}\dfrac{e^{rN} - 1}{r}$
$(P/\bar{F}, r\%, N)$	$P = \bar{F}\dfrac{e^r - 1}{re^{rN}}$	$(\bar{A}/P, r\%, N)$	$\bar{A} = P\dfrac{re^{rN}}{e^{rN} - 1}$
$(\bar{A}/F, r\%, N)$	$\bar{A} = F\dfrac{r}{e^{rN} - 1}$	$(P/\bar{A}, r\%, N)$	$P = \bar{A}\dfrac{e^{rN} - 1}{re^{rN}}$

compounding, continuous-flow interest factors. These factors follow the functional notations given in Table 3.1 and are applied to continuous cash flows according to the same logic and procedures that govern discrete compound-interest calculations.

Example 3.11
Continuous Compounding of a Continuous-Payment Annuity

A savings plan offered by a company allows employees to set aside part of their daily wages and have them earn 6 percent compounded continuously. What annual amount withdrawn from pay will accumulate $10,000 in 5 years? (Alternative statement of the problem: What continous-flow annual annuity will yield a future value of $10,000 in 5 years when compounded continuously at 6 percent?)

Solution 3.11

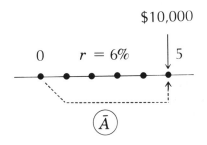

The effect of continuous compounding of continuous flow can be judged by comparing Example 3.10 with 3.11.

$$\bar{A} = F(\bar{A}/F, 6, 5) = F\frac{r}{e^{rN} - 1}$$

$$= F\frac{0.06}{e^{0.3} - 1} = \$10,000\frac{0.06}{1.35 - 1}$$

$$= \$10,000(0.1714)$$

$$= \$1714$$

Role of Continuous Compounding in Engineering Economic Studies

Some savings institutions advertise continuous compounding as an inducement to savers. The intent is to attract investors by paying higher effective interest than that of competitors while still adhering to the nominal interest rate set by regulation. This is an example of discrete continuous compounding.

Actual flows of funds in industry, minute-by-minute incoming receipts and outgoing disbursements, are essentially continuous. Since continuous cash flow closely approximates the pattern of business transactions, it would seem reasonable to compound that flow pattern in evaluating business proposals. It seldom is. Engineering economic studies primarily rely on discrete compounding because they typically focus on aggregate payments that are assumed to take place on specified dates. For instance, the gain or loss from owning an irregularly used machine is conventionally determined once a year from accounting records, and that summary figure is viewed as a single annual payment in evaluating the machine. Estimates of cash flow for the future use of a new machine would be even less precise than would be assessments of performance for existing machines; applying continuous compounding would add negligibly to the precision of the evaluation, even though cash flows are closer to being continuous than annual.

Discrete compounding is featured in this text because it is conceptually easier to grasp, provides adequate precision, and is the most widely applied method. Despite its lack of acceptance, continuous compounding is sometimes more appropriate, particularly when data show that the cash flow is indeed continuous. It is for these occasions that continuous-compounding, continuous cash flow calculations have been discussed. Furthermore, it is conceivable that continuous compounding may eventually be a prominent analytical tool for engineering economists.

Differences between continuous (r) and effective (i) interest are much more pronounced at higher rates, as is apparent in the following equivalent annual values:

i, %		r, %
5	=	4.9
10	=	9.5
20	=	18.2
30	=	26.2
40	=	33.7

Review Exercises and Discussions

Exercise 1

A manufacturing firm in a foreign country has agreed to pay $25,000 in royalties at the end of each year for the next 5 years for the use of a patented product design. If the payments are left with the foreign company, interest on the retained funds will be paid at an annual rate of 15 percent.

What total amount will be available in 5 years under these conditions?

How large would the uniform annual payments have to be if the patent owners insisted that a minimum of $175,000 must be accumulated in the account by the end of 5 years if they are to leave it with the company?

Solution 1

The annual payments form an annuity. Knowing that $A = $25,000 per period, $i = 15$ percent per period, and there are five periods, we calculate the future worth F

$$F = A(F/A, \ 15, \ 5) = \$25,000(6.7423) = \$168,558$$

If the patent owners insisted on an accumulated value of $175,000, the five end-of-year royalty payments would have to be

$$A = F(A/F, 15, 5) = \$175,000(0.14832) = \$25,956$$

Exercise 2   The management of a wearing-apparel firm is considering a proposal from a consulting group to introduce a new method of training inexperienced sewing-machine operators. The consultants claim that their program will produce savings of $7000 per year over the planned 5-year life of the project. Immediate costs to implement the program are $12,000. Annual training expenses will be $4000. The company uses 6 percent annual interest for cost comparisons. Do the anticipated savings warrant the expense of hiring the training consultants?

Solution 2   Assuming that the costs and savings occur at the end of the year, we find that

$$A = \text{annual savings} - \text{annual costs} = \$7000 - \$4000 = \$3000$$

For the proposal to be acceptable, the net return must be greater than the $12,000 initial cost. By translating the 5-year annuity to the present time, we compare the initial cost and the present worth of savings P directly:

$$P = A(P/A, 6, 5) = \$3000(4.2123) = \$12,637$$

The indicated total net savings exceed the initial cost by

$$\$12,637 - \$12,000 = \$637$$

which gives very little leeway for any error in cost or savings estimates.

The same conclusion results from a different approach with the same data. From the capital-recovery formula, the annual return on the gross savings A required for 5 years to meet a current obligation of $12,000 is

$$A = P(A/P, 6, 5) = \$12,000(0.23740) = \$2849$$

Comparing the required return with the expected annual gross savings shows annual net savings of

$$\$3000 - \$2849 = \$151$$

Exercise 3   A very successful engineer plans to endow a chair of engineering economics at her alma mater. The endowment will last for 7 years at $20,000 per year, commencing 3 years from now. In addition, she wants to award annually a graduate engineering economist scholarship for the last 4 years of the endowment period. The 1-year scholarships will start at $5000 and increase by $500 each year. Funds for the endowment are to be provided by three equal payments starting 1 year from now and the transfer to the university of a $90,000 bond the engineer owns that will mature in 6 years. How large will each of the payments be for the next 3 years to fund the engineering economics chair and scholarships, if all funds earn 7 percent interest?

Solution 3   If the bond does not pay dividends before its maturity and all university disbursements occur at the first of each funding year, the cash-flow diagram takes the pattern shown in

Figure 3.4, where the engineer's investments are considered positive cash flows and the university's disbursements for the chair and scholarships are negative cash flows.

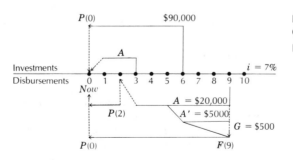

FIGURE 3.4
Cash flow diagram of multiple payments.

The future worth of the scholarships is calculated by determining the equivalent uniform annuity of the gradient and then its F value:

$$F(9) = [A' + G(A/G, 7, 4)](F/A, 7, 4)$$
$$= [\$5000 + \$500(1.4153)](4.4398) = \$25,340.82$$

The equivalent cost at the end of the second year for the seven annual disbursements to the holder of the engineering economics chair is

$$P(2) = A(P/A, 7, 7) = \$20,000(5.3892) = \$107,784$$

The current present worth of the two annuities less the present worth of the bond is

$$P(0) = F(9)(P/F, 7, 9) + P(2)(P/F, 7, 2) - \$90,000(P/F, 7, 6)$$
$$= \$25,340.82(0.54394) + \$107,784(0.87344) - \$90,000(0.66635)$$
$$= \$13,774 + \$94,143 - \$59,972 = \$47,945$$

The annuity of three annual payments to get the endowment started is then calculated as

$$A = P(0)((A/P, 7, 3) = \$47,948(0.38105) = \$18,271$$

Exercise 4

The inventor of an automatic coin-operated newstand called "Automag" believes that the economic evaluation of the invention should be based on continuous compounding of continous cash flow because income from the Automag will be essentially continuous and disbursements for services and purchases of materials (newspapers, paperbacks, news magazines, etc.) will occur regularly and frequently. The expected life of an Automag is 5 years. Annual income should average $132,000 at a good location, and total expenses for servicing materials are expected to average $105,000 per year. What initial price could be paid for an Automag (delivered and ready to operate) to allow a buyer to earn 15 percent on the investment if the income and disbursement estimates are accurate?

Solution 4

Net annual receipts from the Automag are expected to average

$$\$132,000 - \$105,000 = \$27,000$$

Since the cash flow is almost continuous, it is appropriate to apply continuous interest discounting. After recognition that the acceptable price of an Automag is the present-worth equivalent of a continuous annuity resulting from net receipts over 5 years, the value of the $P/\overline{A}$ factor is obtained from Appendix E for $N = 5$ and $r = 13.976$ percent, which corresponds to an effective interest rate of 15 percent.

$$\overline{P} = \overline{A}(P/\overline{A},\ 13.976,\ 5) = \$27,000(3.598) = \$97,146$$

PROBLEMS

3.1   The amount of $1200 per year is to be paid into an account over each of the next 5 years. Using a nominal interest rate of 12 percent per year, determine the total amount that the account will contain at the end of the fifth year under the following conditions:
   (a) Desposits made at the first of each year with simple interest.
   (b) Deposits made at the end of each year, interest compounded annually.
   (c) Deposits made at the end of each month, interest compounded annually.
   (d) Deposits made at the end of each month, interest compounded monthly.
   (e) Deposits made at the end of each year, interest compounded monthly.
   (f) Deposits made at the end of each year, interest compounded continuously.
   (g) Deposits made continuously, interest compounded annually.
   (h) Deposits made continuously, interest compounded continuously.

3.2   A building site for a new gasoline station was purchased 10 years ago for $50,000. The site has recently been sold for $120,000. Disregarding any taxes, determine what rate of interest was obtained on the initial investment.

3.3   The rights to a patent have been sold under an agreement in which annual year-end payments of $10,000 are to be made for the next 15 years. What is the current worth of the annuity at an interest rate of 10 percent.

3.4   At the beginning of 10 successive years, a premium of $120 is paid on a certain insurance policy. What is the worth of the sum of these payments at the end of the tenth year if computed at 4 percent compound interest?

3.5   A deferred annuity is to pay $500 per year for 10 years with the first payment coming 6 years from today. Determine the present worth of the annuity using an interest rate of 12 percent.

3.6   What nominal interest, compounded quarterly, is required to provide a 6 percent effective interest rate? A 12 percent effective interest rate?

3.7   How long will it take for $1 to double in value (disregarding any change in the buying power of the dollar) if:
   (a) The interest rate is 10 percent compounded annually?
   (b) The interest rate is 10 percent compounded semiannually?
   (c) The interest rate is 10 percent ordinary simple interest?

3.8   What is the future worth of each of the following investments?
   (a) $6300 in 6 years at 15 percent compounded annually
   (b) $2000 in 4 years at $4\tfrac{1}{4}$ percent compounded annually
   (c) $200 in 27 years at 6 percent compounded annually
   (d) $4300 in 6 years at 7 percent compounded semiannually
   (e) $500 in 17 years at 9 percent compounded quarterly

3.9   What annual interest rate increases an investment of $1400 to $2000 in 9 years?

3.10   How many years will it take for the balance left in a savings account to increase from $1000 to $1500 if interest is received at a nominal rate of 6 percent compounded semiannually throughout the period?

Solve Problems 3.11 to 3.18 with an annual interest rate of 5 percent.

3.11   What amount must be invested today to secure a perpetual income of $6000 per year?

3.12   A present expenditure of $50,000 is justified by what annual saving for the next 12 years?

3.13   What is the annual payment that will provide a sum of $20,000 in 20 years?

3.14   How much money can be loaned today on an agreement that $700 will be paid 6 years from now?

3.15   What payment can be made today to prevent a series of year-end expenses of $2800 lasting 15 years?

3.16   If the down payment on a piece of land is $7000 and the annual payments for 8 years are $2000 per year, what is the value of the land now?

3.17   What single payment 10 years from now is equivalent to a payment of $5500 in 3 years?

3.18   If $12,000 must be accumulated by equal annual payments in 15 years but the first end-of-year payment cannot be made until 4 years from now, how much will each payment be?

3.19   What semiannual cost beginning today is equivalent to spending lump sums of $2000 in 2 years, $4000 in 4 years, and $8000 in 8 years if the nominal interest rate is 8 percent compounded semiannually?

3.20   If investments of $2000 now, $2500 in 2 years, and $1000 in 4 years are all made at 4 percent effective interest, what will be the total in 10 years?

3.21   An inventor has been offered $12,000 per year for the next 5 years and $6000 annually for the following 7 years for the exclusive rights to an invention. At what price could the inventor afford to sell the rights to earn 10 percent, disregarding taxes?

3.22   A shady individual engaged in making small loans offers to lend $200 on a contract under which the borrower must pay $6.80 at the end of each week for 35 weeks in order to pay off the debt. What are:
    (a) The nominal interest rate per annum?
    (b) The effective interest rate?

3.23   A student borrowed $4000 3 years ago to help pay for her education, agreeing to repay the loan in 100 payments at an interest rate of 12 percent compounded monthly. She has just received a bonus to play professional tennis and desires to pay the principal in a lump sum. How much does she owe?

3.24   What annual expenditure for 10 years is equivalent to spending $1000 at the end of the first year, $2000 at the end of the fourth year, and $3000 at the end of the eighth year, if interest is at 8 percent per year?

Solve Problems 3.25 to 3.29 using a continuous effective interest rate of 8 percent.

3.25 How long is required for a single payment to double in value? Triple in value?

3.26 What discrete amount must be deposited at the end of each year in order to accumulate $10,000 in 10 years?

3.27 What total amount must be deposited each year in order to accumulate $10,000 in 10 years if the cash flow is continuous?

3.28 What amount deposited continuously throughout only the first year will be worth $10,000 in 10 years.

3.29 Determine the total amount accumulated in an account paying a continuous effective rate of 10 percent per year at the end of 5 years if payments total $1000 per year and are made:

(a) Discretely at year's end.
(b) Continuously through the year.

3.30 If the population of a certain suburb is 29,000 in 1989 and the average annual rate of increase is estimated at 7 percent, what will its population be in 1999 if the growth rate remains constant?

3.31 Derive the equation for calculating the future worth of a series of discrete payments when interest is compounded continuously. Start from the expression

$$F = A + Ae^r + Ae^{2r} + \cdots + Ae^{(n-2)r} + Ae^{(n-1)r}$$

and follow the procedure used in developing the sinking-fund factor for end-of-period compounding.

3.32 Net receipts from a continuously producing oil well add up to $120,000 over 1 year. What is the present worth of the well if it maintains steady output until it runs dry in 8 years if $r = 10$ percent?

3.33 Construct a table showing the amount of interest paid, principal repaid, and principal remaining to be repaid at the end of each year on a loan of $10,000 which is being repaid in five equal annual installments with interest at 12 percent.

3.34 A person owes debts of $3380 due in 4 years, with annual interest included at 9 percent, and $1200 plus interest due in 2 years on which the interest charge is 10 percent compounded annually. She now wishes to discharge these debts with two equal payments payable 1 and 2 years from now. What is the value of the installments if money is now worth 6 percent?

3.35 A recently developed orchard will come into full bearing in 6 years. Starting in the seventh year and continuing through a productive life of 20 years, the orchard is expected to produce an average net yield of $50,000. What is the equitable present cash value of the investment if money is worth 8 percent per annum?

3.36 A beachcomber bought a dune buggy, paying $1000 in cash and agreeing to pay

$500 every 6 months for 3 years. As she was driving her buggy down the beach one day, she spotted a bottle that held five moldy but cashable $500 bills. She had not yet made her first semiannual installment for the dune buggy. When should she make a payment of $2500 to discharge her obligation for the purchase in a lump-sum settlement if the interest rate is 12 percent compounded semiannually?

3.37 A family is planning to buy a vacation cabin for $30,000. The family intends to keep the cabin for 6 years and expects the annual unkeep and taxes to amount to $900 per year. Without any major repairs, the cabin should have a resale price of $26,000. What is the equivalent annual cost of owning the cabin if the family's acceptable interest rate is 6 percent?

3.38 The family depicted in Problem 3.37 could also do some renovations on the cabin during its vacation periods which would increase the resale value to $35,000. How much could the family afford to invest in materials each year if it hoped to receive $1200 per year for its labor?

3.39 Maintenance records of a certain type of machine indicate that the first-year maintenance cost of $80 increases by $30 per year over the 10-year replacement period of the machine. Answer the following if the maintenance cost is considered to occur at the end of the year and the firm's interest rate is 12 percent:

  (a) What equal annual payments could the firm make to a service organization to carry out the maintenance for 20 machines?
  (b) How much additional could be paid for a new type of machine with the same service life that required no maintenance during its life?

3.40 A new piece of materials handling equipment costs $10,000 and is expected to save $2500 the first year of operation. Maintenance and operating cost increases are expected to reduce the net savings by $200 per year for each additional year of operation until the equipment is worn out at the end of 10 years. Determine the net present worth of the equipment at an interest rate of 15 percent.

3.41 A postal clerk wants to accumulate $20,000 in order to take a year's vacation 10 years from now. She now has $6000 in savings certificates which earn interest at 6 percent compounded semiannually. She plans to invest an equal amount each year in an account that earns 6 percent annual interest. How large should this amount be to give her $20,000 in 10 years when it is combined with the future value of the savings certificates?

3.42 The postal clerk in Problem 3.41 figures that in addition to the annuity calculated to give her $20,000 in 10 years, she will also be able to save part of her annual raises in pay. If the portion saved increases by $300 each year starting at the end of the second year from now (no additional increment is possible at the end of this year), how much extra money will she have when she is ready to take her year's vacation?

3.43 Sam Smartt is purchasing a new car with a list price of $12,000. Sam is paying $2000 down and will finance the automobile through the dealer at a nominal annual rate of 18 percent, compounded monthly for 5 years.

  (a) Determine the amount of Sam's monthly payment.
  (b) How much total interest will be paid?
  (c) Determine the amount of principal that Sam will have paid off after 2 years.

3.44 Office equipment can be purchased for $6000 cash or a down payment of $1000 followed by 24 end-of-month payments of $220 each. At what effective interest rate are these terms equivalent?

3.45 If you invest $1000 at 10 percent compounded annually for the same length of time it takes an investment to triple in value at 12 percent simple interest, how much will you have?

3.46 Assume that you sold property today for $2421 (net value) and that you had purchased the property 4 years ago with $2000 withdrawn from your savings account. During the 4-year period your savings would have earned 6 percent compounded quarterly. For a comparison of the investments, calculate the nominal interest rate received from your property purchase.

3.47 To repay a loan for which interest charges are 12 percent compounded monthly, 40 monthly payments of $200 apiece are needed. What percentage of the amount borrowed has been repaid by the time the thirtieth payment is made?

3.48 After taking an engineering economics course, an optimist decides that she wants to be rich and that the way to become so is to make use of the earning power of money. If she is to be precisely a millionaire on her sixtieth birthday and she expects to start her millionaire-making plan on her twenty-second birthday, how much will she have to deposit on each birthday through her fifty-ninth (including her twenty-second) if her investment can earn 15 percent annually?

3.49 Joshua Jones was born on January 1, 1899. On that day his grandfather deposited $1000 in his name in an account that paid 7 percent ordinary simple interest. Joshua graduated from high school on June 30, 1916. How much was in his account, assuming no withdrawals were made?

3.50 Joshua had a monumental coming-of-age party, and when he awakened he found that he had volunteered for navy service on his eighteenth birthday. While in the Navy for 28 months he had $10 per month deducted from his pay and deposited in a bank that paid interest at the rate of 6 percent compounded monthly. How much did he have in the bank when he was discharged from the Navy?

3.51 Joshua used his savings, mustering-out pay, and other earnings to buy a business for $1500. It tripled in value by the time he sold it in 3 years. What nominal interest rate, compounded quarterly, did his investment earn?

3.52 On his thirtieth birthday, Joshua designed a 10-year savings plan to produce $25,000 on his fortieth birthday. The first payment of $1500 was made on his thirty-first birthday. He enlarged each of the remaining nine payments by an equal, increasing increment each year. The savings earned 8 percent compounded annually. How large was the increase each year?

3.53 In 1942, Joshua bought a "war bond." It cost $18.75 and could be cashed in for $25 after 10 years. What interest rate was earned?

3.54 By his fifty-fifth birthday, Joshua had accumulated $75,000. He invested it in a 10-year bond that earned 9 percent compounded annually. He planned to retire at age 65 and to have enough saved so that he could draw out $25,000 from his savings each year for the next 15 years, starting on his sixty-sixth birthday. What equal payments into a savings account that paid annual interest at 7 percent did he have to make, starting on his fifty-fifth birthday with the last payment made on his sixty-fifth birthday, to fund his retirement? The amount accumulated, including the maturity value of the bond (first cost plus interest), will also earn 7 percent during the retirement period.

EXTENSION

3A Puzzling Patterns of Interest-Earning Periods

Confusion about the many ways financial institutions calculate interest, discussed in Extension 2A, is further confounded by variations in the time periods over which interest is compounded. Not only do patterns differ among savings plans, still other patterns apply to loan-repayment plans. A few of the more common plans are described below. The examples confirm the value of investigating before investing or borrowing.

Compounding Patterns for Savings Accounts

To illustrate differences in four basic plans for calculating interest on savings, we base comparisons on an annual interest rate of 6 percent, in a year which has 365 days and is referenced to the deposit and withdrawal schedule given.

Transactions in a Savings Account during One Quarter

Transaction Date	Withdrawal	Deposit	Balance*
January 1			$3000
January 25	$2000		1000
February 1		$5000	6000
March 31	1000		5000

* Balances do not include earned interest

Low balance. Interest is based on the lowest balance in the account any time during the interest period. For the given schedule, the lowest balance is $1000, which occurred on January 25. At 6 percent compounded quarterly, interest earned during the first quarter of the year is

$$I = \$1000(0.06/4) = \$15$$

FIFO (first in, first out). Withdrawals are deducted from the balance at the start of the quarter. If the initial balance is depleted, deductions are then made from subsequent deposits. In the sample schedule, the two withdrawals totaling $3000 are deducted from the $3000 opening balance, in effect wiping it out. Thus interest is earned only on the $5000 deposit which was made on February 1. This amount earned interest for 2 months, or 1/6 of the year, to produce interest for the quarter of

$$I = \$5000(0.06/6) = \$50$$

LIFO (last in, first out). Withdrawals are deducted from the most recent deposit. The January 25 withdrawal is deducted from the opening balance, leaving $3000 − $2000 = $1000 to earn interest for all 3 months of the quarter. The same pattern applies to the $1000 deduction from the $5000 deposit, which leaves a balance of $4000 for 2 months. Combining the two balances gives

$$I = \$1000(0.06/4) + \$4000(0.06/6) = \$55$$

Day of deposit to day of withdrawal. Interest is paid on the actual number of days the money is in the account, with the exception that no interest is paid on the day of withdrawal for the amount withdrawn. This pattern is called *daily interest,* but it is not the same as *daily compounding,* where daily interest is added to the balance each day. As applied to the nominal 6 percent interest rate, the daily interest is $0.06/365 = 0.0001643$. Interest for the quarter is calculated by multiplying each balance level by the number of days it is in effect and then multiplying by the daily interest factor. For the given schedule,

$$I = [\$3000(24) + \$1000(7) + \$6000(58) + \$5000(1)] \times 0.0001643$$
$$= \$432,000 \times 0.0001643 = \$70.98$$

Compounding Plans for Loan Repayments

A procedure that is somewhat the reverse of the interest-payment procedure is used to determine how much is owed on loans in which the balance changes between loan-charge calculations. The formulas used to determine the amount owed on a credit account make a major difference in the financing charge. There are two basic methods of computing the interest owed on "revolving" accounts such as bank credit cards or charge accounts issued by stores. Actual practices and interest rates charged vary among lending institutions and the states within which they operate.

Adjusted-balance method. An example of how it works is as follows: Suppose that you charge a purchase for $200 on March 1 and the billing date for your account is March 2. You are given 25 days from the billing date to pay your full bill without incurring any interest charges. If the bill is not paid in full, you are charged interest on the unpaid portion figured from the billing date. Assume that you pay $100 on March 24, leaving a balance due of $100. On your April bill you will be charged interest on the adjusted balance, which amounts to ($200 − $100)(0.015) = $1.50 if the monthly interest is the typical 1½ percent.

Average-daily balance method. The system works as follows: If the full bill is not paid in the 25-day grace period, then the purchase of $200 on March 1 and the $100 payment on March 24 with the billing date of March 2 are interpreted to mean that you owed $200 from March 1 to March 24 and $100 from the latter date until the next billing on April 2. That is 22 days at $200 and 9 days at $100, which result in

$$\text{Average daily balance} = \frac{22(\$200) - 9(\$100)}{31} = \$170.97$$

Again assuming an interest rate of 1½ percent per month, we find that

$$\text{Interest owed} = \$170.97 \times 0.015 = \$2.56$$

which is $2.56 − $1.50 = $1.06 greater than the interest charge using the adjusted-balance method, a significant difference.

Thus, with the adjusted-balance method, the interest charges are determined solely by the amount that you pay. With the average-daily-balance system, the charges are also determined by when in the billing period you pay; the earlier you pay, the lower your interest. If you pay your entire bill within the first 25 days of the billing period, there is no interest charge in either system.

Considering the variety of compounding methods and the bewildering ways of describing them, it is wise to remember that whenever a payoff plan for either an investment or loan seems too good to be true, it probably is.

QUESTIONS

3A.1 For the schedule in Figure 3.5, which represents deposits and withdrawals in a savings account that pays 8 percent compounded quarterly, how much interest is earned during the year when calculated according to the following methods:
(*a*) Low balance.
(*b*) FIFO.
(*c*) LIFO.

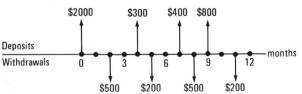

FIGURE 3.5

3A.2 Let interest be 18 percent compounded monthly, and the billing date be the second day of each month. Compare the total interest charges for April, May, and June, using both the adjusted-balance and average-daily-balance methods, for the following purchases on credit (−) and repayments (+):

March 1	− $200	April 29	− 300
March 24	+ 100	May 1	+ 200
March 29	− 300	May 7	− 150
April 1	+ 200	May 15	+ 100
April 14	− 150	May 21	− 50
April 21	+ 200	June 1	+ 350

CHAPTER 4

PRESENT-WORTH COMPARISONS

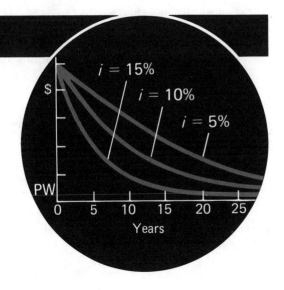

OVERVIEW

A student once answered the question, "What is the present worth?" as, "The value the person receiving the present attaches to it." Though hardly apropos for an engineering economics class, the answer deserves at least some credit because it translates anticipation of future pleasures to the present's present worth. The answer even implies interest, albeit a personal value rather than a rate of return. This chapter answers "What is the present worth?" by presenting the reasons for and methods of calculating the equivalent present value of cash flow.

Present-worth comparisons are made only between coterminated proposals, to ensure equivalent outcomes. Cotermination means that the lives of the involved assets end at the same time. When assets have unequal lives, the time horizon for an analysis can be set by a common multiple of asset lives or by a study period which ends with the disposal of all assets.

Net present worth, the difference between the present worths of benefits and cost, is the most widely used present-worth (PW) model. A capitalized cost model is used when an asset is assumed to have infinite life. Other PW models include deferred investments and valuation of property, stocks, and bonds.

Many economists prefer a present-worth analysis because it reveals the sum in today's dollars that is equivalent to a future cash flow stream, and PW models are less subject to misinterpretation. It is therefore important to become familiar with the assumptions upon which PW calculations are based and the types of comparisons that can be made.

CONDITIONS FOR PRESENT-WORTH COMPARISONS

Future-worth (FW) analysis comparable to the PW method is examined in Extension 4B.

The previous chapter on discounted cash flow calculations introduced present-worth comparisons. A present worth resulted from each calculation of P. In this chapter, the present worth of a cash flow stream is distinguished by the designation PW—the present worth of receipts and disbursements associated with a particular course of action. Each alternative is thus represented by a PW, a measure of its economic merit.

The ingredients of a present-worth comparison are the amount (in dollars) and timing (N) of cash flow, and the interest rate i at which the flow is discounted. Listed below are assumptions used in this and the next two chapters in introducing the basic comparison methods; also indicated are topics from subsequent chapters which make comparisons less restrictive.

1 *Cash flows are known.* The accuracy of cash flow estimates are always suspect because future developments cannot be anticipated completely. Transactions that occur now, at time zero, should be accurate, but future flows become less distinct as the time horizon is extended. Ways to evaluate riskier cash flows are discussed in Chapters 16 and 17.

2 *Cash flows are in constant-value dollars.* The buying power of money is assumed to remain unchanged during the study period. Ways to include the effect of inflation are presented in Chapter 13.

3 *The interest rate is known.* Different interest rates have a significant effect on the magnitude of the calculated present worth, as illustrated in Figure 4.1. The rate of return i required by an organization is a function of its cost of capital, attitude toward risk, and investment policy. Alternative courses of action for the same proposal are normally compared using the same interest rate, but different proposals may be evaluated at different required rates of return. The sensitivity of interest rates is explored in Chapter 14.

4 *Comparisons are made with before-tax cash flows.* Inclusion of income taxes greatly expands the calculation effort for a comparison and correspondingly increases reality. Since the workings of comparison methods are featured here and in the next few chapters, inclusion of income taxes is delayed until Chapter 11.

5 *Comparisons do not include intangible considerations.* "Intangibles" are difficult-to-quantify factors that pertain to a certain situation. For instance, the "impression" created by a design is an intangible factor in evaluating that design and an important one for marketing, but it would not be included in a present-worth comparison unless its economic consequence could be reasonably estimated. (If a dollar value can be assigned, a factor is no

longer intangible.) Ways to include intangible factors in the decision-making process are considered in Chapter 17.

6 *Comparisons do not include consideration of the availability of funds to implement alternatives.* It is explicitly assumed that funds will be found to finance a course of action if its benefits are large enough. Although financing is not a direct input for computations, the output computed can be appraised with respect to available funding. For instance, an old, inefficient machine could be kept in operation because there appears to be insufficient capital available in the organization to afford a replacement, but an engineering economic analysis might point out that the savings from replacing the bungling machine would be so great that the organization could not afford not to find funds for a replacement. Such capital-budgeting considerations are examined in Chapter 9.

Example 4.1
Present Worth by the 72-Rule

The present worth of a future amount drops off rapidly as the time between "now" and "then" increases, particularly at higher interest rates. The pattern of this decrease in present value is shown in Figure 4.1. Examination of the curves supports an interesting rule of thumb: the *72-rule* indicates the number of years N^* over which a future value loses half its present worth at the annually compounded interest rate i:

$$72\text{-rule}: N^* = \frac{72}{i}$$

Another way to view the 72-rule is that $(1 + i)^N$ doubles about every N^* years. Thus, as is apparent in Figure 4.1, the present worth is halved in about 14 years at $i = 5$ percent, 7 years at $i = 10$ percent, and 4 years at $i = 20$ percent.

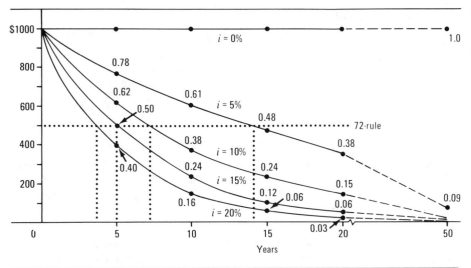

FIGURE 4.1
Present worth of $1000 as a function of time and interest rates when interest is compounded annually.

BASIC PRESENT-WORTH COMPARISON PATTERNS

The present worth of a cash flow over time is its value today, usually represented as time zero in a cash flow diagram. Two general patterns are apparent in present-worth calculations.

Present-Worth Equivalence

A do-nothing option is always an alternative, even if it results only from procrastination.

One pattern determines the present-worth equivalence of a series of future transactions. The purpose is to secure one figure that represents all the transactions. This figure can then be compared with a corresponding figure that represents transactions from a competing option or it can be compared with the option of doing nothing. More often there is a "go, no-go" situation where each alternative is selectively weighed to decide whether it is worth exercising. For instance, a series of expenses that will occur in the future can be discounted to obtain its PW, and then a decision can be made about whether an investment of the PW amount should be made *now* to avoid the expenses. Similar reasoning guides the equivalence comparison in Example 4.2.

Example 4.2
Equivalent PW of an Option

An investor can make three end-of-year payments of $15,000, which generates receipts of $10,000 at the end of year 4 that will increase annually by $2500 for the following 4 years. If the investor can earn a rate of return of 10 percent on alternative 8-year investments, is this alternative attractive?

Solution 4.2

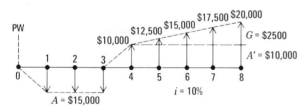

$$PW = -\$15,000(P/A, 10, 3) + [\$10,000$$
$$+ \$2500(A/G, 10, 5)](P/A, 10, 5)(P/F, 10, 3)$$
$$= -\$15,000(2.4868) + [\$10,000$$
$$+ \$2500(1.8100)](3.7907)(0.75132)$$
$$= -\$37,302 + \$41,368 = \$4066$$

From the cash flow diagram, it is apparent that the receipts and disbursements each constitute an annuity, one positive and the other negative. Both are discounted to time zero at 10 percent:

The interpretation of the $4066 present worth is that the transactions provide a return of 10 percent on the investment *plus* a sum of $4066. This investment is therefore preferable to one that would return exactly 10 percent over 8 years.

It is common practice to drop the negative sign when cash flow is predominately composed of costs. The calculated PW is then called the *present worth of costs*. This is purely a convention of convenience and is unlikely to cause confusion as long as the alternatives in a comparison are treated in the same way (see Example 4.6).

Net Present Worth

The second general pattern for PW calculations has an initial outlay at time zero followed by a series of receipts and disbursements. This is the most frequently encountered pattern, which leads to the fundamental relation

$$\text{Net present worth} = \text{PW(benefits)} - \text{PW(cost)}$$

The criterion for choosing between alternatives is to select the one that maximizes net present worth, or, simply, the one that yields the larger positive PW. A negative PW means that the alternative does not satisfy the rate-of-return requirement. Applications of net present worth to benefits and costs of public projects are developed in Chapter 10. A typical industrial application is given in Example 4.3.

Example 4.3
Net Present-Worth Comparison

Two devices are available to perform a necessary function for 3 years. The initial cost (negative) for each device at time zero and subsequent annual savings (positive) are shown in the adjacent table. The required interest rate is 8 percent.

Year	0	1	2	3
Device A	− $9,000	$4500	$4500	$4500
Device B	− $14,500	$6000	$6000	$8000

Solution 4.3

PW(device A) = − $9000 + $4500(P/A, 8, 3)
= − $9000 + $4500(2.5770) = $2597

PW(device B) = − $14,500 + $6000(P/A, 8, 2)
+ $8000(P/F, 8, 3)
= − $14,500 + $6000(1.7832)
+ $8000(0.79383)
= $2594

Both alternatives meet the minimum acceptable rate of return, because both are positive, and their net present worths are almost equal. In this case other considerations must be involved in the choice, such as the availability of the extra $5500 needed to purchase device B.

In Examples 4.2 and 4.3 the alternatives were naturally coterminated. That is, in Example 4.2 the investment periods were identical and in Example 4.3 the lives of the two devices were equal. When the lives of assets being compared are unequal, special procedures must be applied to coterminate the analysis periods.

COMPARISON OF ASSETS THAT HAVE UNEQUAL LIVES

The utilization of present-worth comparisons for *coterminated projects* implies that the lives involved have a common endpoint. The necessity for cotermination is readily apparent when a familiar decision is considered such as the choice between paying $30 for a 3-year subscription to a magazine and paying $40 for a 5-year subscription to the same publication. A simple comparison of $30 to $40 for a subscription is inaccurate, because the extra $10 buys two more years of issues. *Alternatives must be compared on the basis of equivalent outcomes.*

Several variations have been proposed to accommodate present-worth comparisons of unequal-life assets.* Two prominent methods are described below.

1 *Common-multiple method.* Alternatives are coterminated by selecting an analysis period that spans a common multiple of the lives of involved assets. For instance, if assets had lives of 2, 3, 4, and 6 years, the least common multiple is 12 years, which means that the asset with a life of 2 years would be replaced 6 times during the analysis period. The assets with the 3-, 4-, and 6-year lives would be replaced 4, 3, and 2 times, respectively.

   Legitimate use of a least common multiple of lives depends on the validity of the assumption that assets will be repeatedly replaced by successors having identical cost characteristics. This assumption is more often reasonable when the least common multiple is small. Then there is less likelihood that a technologically better asset will become available during the analysis period.

2 *Study-period method.* A more justifiable analysis is based on a specified duration that corresponds to the length of a project or the period of time the assets are expected to be in service. An appropriate study period reflects the replacement circumstances. Some of the options are to set the study period as the length of

- *The shortest life of all competing alternatives*—a protection against technological obsolescence

- *The known duration of required services*—a project philosophy in which each new undertaking is considered to start with new assets, continue to use like replacements, and dispose of used assets when the project is completed

- *The time before a better replacement becomes available*—an attempt to minimize cost by purposely upgrading assets as improvements are developed

A study-period comparison presumes that all assets will be disposed of at the end of the time period. Therefore it is usually necessary to estimate the income that can be realized from the sale of an asset which can still provide

* D. J. Kulonda, "Replacement Analysis with Unequal Lives," *Engineering Economist*, Spring, 1978.

useful service. This *salvage value* can be established somewhat arbitrarily by prorating the investment in the asset over its normal service life and then calculating the size of the unrecovered amount at the time of its disposal. A better estimate is obtained when it is possible to accurately appraise an asset's market value at the end of the study period. The most precise study would result from also having explicit knowledge of operating costs as a function of the asset's age and future capital costs for each successor. These conditions are explored more completely in Chapter 8.

Example 4.4
PW Comparisons of Alternatives with Unequal Economic Lives

Assets A1 and A2 have the capability of satisfactorily performing the required function. A2 has an initial cost of $3200 and an expected salvage value of $400 at the end of its 4-year economic life. Asset A1 costs $900 less initially, with an economic life 1 year shorter than that of A2, but it has no salvage value and its annual operating costs exceed those of A2 by $250. When the required rate of return is 15 percent, which alternative is preferred when compared by

a   The repeated-projects method
b   A 2-year study period (assuming the assets are needed for only 2 years)

Solution 4.4a

The repeated-projects method is based on the assumption that assets will be replaced by identical models possessing the same costs. Equivalent service results from comparing costs over a period divisible evenly by the economic lives of the alternatives; in this case the least common multiple is 12 years.

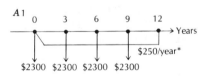

* Additional annual cost more than *A*2.

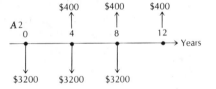

$$PW(A1) = -\$2300 - \$2300(P/F, 15, 3)$$
$$- \$2300(P/F, 15,6) - \$2300(P/F, 15, 9)$$
$$- \$250(P/A, 15, 12)$$
$$= -\$2300 - \$2300(0.65752)$$
$$- \$2300(0.43233) - \$2300(0.28426)$$
$$- \$250(5.4206) = -\$6816$$

$$PW(A2) = -\$3200 - \$2800(P/F, 15, 4)$$
$$- \$2800(P/F, 15, 8) + \$400(P/F, 15, 12)$$
$$= -\$3200 - \$2800(0.57175) - \$2800(0.3269)$$
$$+ \$400(0.18691) = -\$5642$$

The present-worth advantage of A2 over A1 for 12 years of service is $6816 − $5642 = $1174.

Solution 4.4b

A service-period comparison is utilized when a limited period of ownership of assets is set by specific operational requirements. A 2-year study period for A1 and A2 indicates that the service required from either asset will be needed only 2 years and will be disposed of at that time. If possible, estimates of the worth of assets at the end of the study period should be secured. These salvage values may be quite large when the service period is only a small fraction of the economic life. When it is difficult to secure reliable estimates of worth after use during the ownership period, minimum resale levels can be calculated to make the alternatives equivalent. Then only a judgment is needed about whether the market value will be above or below the minimum level.

For instance, assuming $S = 0$ for alternatives A1 and

A2 after 2 years of service, we have

$$PW(A1) = -\$2300 - \$250(P/A, 15, 2)$$
$$= -\$2300 - \$250(1.6257)$$
$$= -\$2707$$
$$PW(A2) = -\$3200$$

which show that A1 has the lower present worth of costs for the 2-year service period. The salvage value for A2 that would make PW(A2) is

$$\$2707 = \$3200 - S(P/F, 15, 2)$$
$$S = \frac{\$3200 - \$2707}{(P/F, 15, 2)} = \frac{\$493}{0.75614} = \$652$$

which means that A2 is preferred to A1 when the resale value of A2 at the end of 2 years is more than \$652 greater than the resale value of A1 at the same time. The advantage of calculating this *aspiration level* is to avoid making an estimate of S; only a judgment is required about whether S will exceed a certain amount, the aspiration value, which is \$652 in this case.

COMPARISON OF ASSETS ASSUMED TO HAVE INFINITE LIFE

The sum of first cost plus the present worth of disbursements assumed to last forever is called a *capitalized cost*. This type of evaluation is essentially limited to long-lived assets. It is not used as extensively as it once was, but it is still favored by some for studies of dams, railway rights of way, tunnels, and similar structures which provide extended service.

The calculation of capitalized cost is conducted in the same way as that of a present-worth comparison, where N equals infinity. This makes the analysis very sensitive to the selected rate of return. As with all present-worth calculations, the final figure is usually an impressive amount. As such, it could appear discouragingly high unless properly interpreted. Expressed as a formula,

$$\text{Capitalized worth} = P + \frac{A}{i}$$

where A is the uniform difference between annual receipts and disbursements. When there is no revenue, the formula becomes

$$\text{Capitalized cost} = P + \frac{\text{disbursements}}{i}$$

Example 4.5
An Asset that Lasts Forever

A \$500,000 gift was bequeathed to a city for the construction and continued upkeep of a music shell. Annual maintenance for a shell is estimated at \$15,000. In addition, \$25,000 will be needed every 10 years for painting and major repairs. How much will be left for the initial construction costs after funds are allocated for perpetual upkeep? Deposited funds can earn 6 percent annual interest, and the returns are not subject to taxes.

Solution 4.5

The total capitalized cost is known to be $500,000. From the capitalized-cost formula,

$$\text{First cost} = \text{capitalized cost} - \frac{\text{annual disbursements}}{i}$$

$$\frac{\text{First}}{\text{cost}} = \$500,000 - \frac{\$15,000 + \$25,000(A/F, 6, 10)}{0.06}$$

$$= \$500,000 - \frac{\$15,000 + \$25,000(0.07587)}{0.06}$$

$$= \$500,000 - \$281,613 = \$218,387$$

The equivalent annual disbursements are $15,000 for maintenance plus the annual payments necessary to accumulate $25,000 every 10 years. The funds will earn interest at 6 percent, so we have

which means that the interest earned on the amount left after allowing $218,387 for construction will cover all the anticipated upkeep indefinitely, provided that the interest rate continues at 6 percent or more.

COMPARISON OF DEFERRED INVESTMENTS

An occupational hazard for engineers is the habitual appeal to "get it done today, or preferably yesterday." The accustomed response to such an appeal is a workable solution that admittedly may not be the most economical long-run course of action. But keeping an operation going with a less-than-optimum solution is often less costly than is the wait caused by a search for something better. For instance, suppose that flooding at a construction site was the provocation for an engineer to remedy the situation quickly. An emergency purchase and installation of a pump plus sandbags for revetments cost $4000. The pumping facility was used off and on during the 2-year construction project for a total cost discounted to the time of purchase (assuming no salvage value and $i = 12$ percent) of

First cost of pump and revetment	$4000
Pumping cost and maintenance ($460/year):	
$460(P/A, 12, 2) = \$460(1.690)$	770
PW of emergency pumping operation	$4770

A post-project review shows that a smaller pump could have been purchased to perform the same work adequately at

First cost of smaller pump and revetment	$3100
Pumping cost and maintenance ($640/year):	
$640(P/A, 12, 2) = \$640(1.690)$	1082
PW of a lower-cost solution to flooding	$4182

However, if it had taken only a week longer to select and get delivery of the lower-cost pump, it is likely that the cost of 7 days of flooding would have exceeded the $4770 - $4182 = $588 potential saving.

Piecemeal additions to existing capacity are seldom an efficient expansion program. A planning horizon too short to accommodate growth is an example of myopic suboptimization.

A more typical way to analyze deferred investments is to determine the timing of capital expenditures to meet anticipated activity increases. Designs to accommodate growth usually involve the question of whether to acquire a full-sized facility now and absorb the temporary cost of unused assets or to acquire a smaller facility with a later addition and accept the extra cost of duplicated effort and dislocation inconvenience. For a given capacity, one large facility inherently has a lower-per-unit cost because it is designed specifically for that level of operation, but it increases the chance of technical obsolescence and idleness, owing to changing future conditions. The economic analysis of a deferred addition is usually conducted by a present-worth comparison of the options.

Example 4.6
Immediate and Deferred Investments for Identical Capacity

A small novelty manufacturing company must acquire storage space in order to reduce production costs by stabilizing employment. Ninety percent of the products produced are sold during the Christmas holiday season. A resource utilization study has shown that producing at a constant rate during the year and storing output will reduce the overall manufacturing costs.

The products produced by the novelty company have been well-received, and sales have increased each year. Increased capacity will be needed in the future; two alternatives have been identified. A large warehouse with sufficient space to meet all needs for 10 years can be leased for that period at $23,000 per year. Since there is some doubt about how much business will increase in the future and the company is reluctant to go into debt deep enough to build a warehouse as large as the one available for leasing, the other feasible alternative is to build a small warehouse now for $110,000 and make an addition to it in 3 years for $50,000. Annual cost for taxes, insurance, maintenance, and repairs are expected to be $1000 for the first 3 years and $2000 for the next 7 years. The added-to warehouse should have a resale value of $50,000 in 10 years. Based on a study period equal to the lease contract and a 12 percent cost for capital, which alternative is preferable?

Solution 4.6

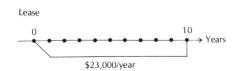

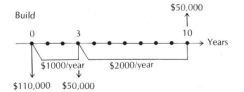

Where costs are considered to be positive,

PW(lease)   = $23,000(*P/A*, 12, 10)
                       = $23,000(5.6502) = $129,955
PW(build and add) = $110,000 + $50,000(*P/F*, 12, 3)
                       − $50,000(*P/F*, 12, 10)
                       + $1000(*P/A*, 12, 3)
                       + $2000(*P/A*, 12, 7)(*P/F*, 12, 3)

= $110,000 + $50,000(0.71178)
       − $50,000(0.32197)
       + $1000(2.4018)
       + $2000(4.5637)(0.71178)

= $138,389

The analysis indicates that the present worth of storage costs for the next 10 years will be $138,389 − $129,955 = $8434 less from leasing than building the required warehousing facility.

The same problem is examined from a different perspective in Chapter 17, where it is no longer assumed that future demands are known with certainty. If reliable probabilities for future capacity needs can be generated, they should be included in the analysis. The difficulty lies with finding a reliable way to peek into the future.

VALUATION

Value is a measure of the worth of something in terms of money or goods. A barter system uses a personally directed trade of goods to establish equivalent values. In an auction the measure of worth is established by competitive monetary bids. To some extent, selling price in an openly competitive market sets a value on goods as a result of the amount customers are willing to pay. Value is more difficult to determine before a transaction occurs. Expert appraisers are familiar with prices from previous transactions and interpolate to set values on goods that have not been exposed to the market. Works of art and land properties are subject to sometimes controversial evaluations.

Appraisals regularly encountered in financial practice are known by specific names. A *going value* is how much the assets of an organization are worth as an operating unit. It is opposed to a *liquidating value,* which is the amount that could be realized if the assets were sold separately from the organization that is using them. The going value is normally greater than the liquidating value, in recognition of the "organizational" value of a unit still in operating condition; accountants term this difference *goodwill.* The worth of an asset for accounting purposes is its *book value,* which may be quite different from its *market value,* the price at which it can be sold. Book value reflects historical cost, whereas market value is dependent upon earnings.

When the value of property depends upon its earning capacity, valuation results from discounting future probable earnings to their present worth. When risk is ignored, the "value" of such property is simply its present worth at the interest rate deemed appropriate by the appraiser. Examples of properties whose value tends to be a function of future cash flows are bonds, stocks, and rental assets.

Bond Valuation

A bond is sold by an organization to raise money. Bonds represent a debt to the bondholders rather than a share of ownership in the organization. Most bonds bear interest semiannually and are redeemable for a specified maturity value at a given date. There are many variations designed to make bonds more attractive to purchasers, such as an option to convert them to common stock under specified conditions, or to make the bond debt more manageable for the issuing organization, as in *callable* bonds that may be paid off prior to maturity according to a printed repurchase schedule. Some public organizations are allowed to issue bonds for which the interest payments are not taxable income to the bondholders.

The value of a bond of a given denomination depends on the size and timing of the periodic dividends and the duration before maturity. The bond valuation is thus the present worth* of the cash flow stream of dividends plus the discounted value of the redemption payment. The key to the valuation is the rate of return expected by the bond purchaser. Lower rates are reasonable when there is very little risk of default. For instance, a United States Treasury security would have less risk of nonrepayment than would one issued by the Fly-by-Nite Corporation; consequently, a lower discount rate would be appropriate.

Additional considerations concerning bond yields are discussed in Extension 4A.

Each bond has a stated redemption amount, called the *face value*, at which it can be redeemed at maturity. Bonds sell for less than their face value when buyers are not satisfied with the rate of interest promised, called the *bond rate*. Interest is paid in the form of regular premiums; the flow of premiums constitutes an annuity where

$$A = \text{premium} = (\text{face value})(\text{bond rate})$$

and the number of payments per year is a function of the bond rate structure (e.g., four payments per year for a quarterly bond rate). A present-worth purchase price for a bond that satisfies a buyer seeking a return greater than the bond rate is calculated from the bond's maturity value F and the annuity composed of premium payments A, both discounted at the buyer's desired rate of return.

Example 4.7
To Buy a Bond

A 10-year corporate bond has a face value of $5000 and a bond rate of 8 percent payable quarterly. A prospective buyer desires to earn a nominal rate of 12 percent on investments. What purchase price would the buyer be willing to pay?

Solution 4.7

The bond can be redeemed for $5000 ten years from now. During that period, four premium payments will be paid each year in the amount of $A = \$5000(0.8/4) = \$5000(0.02) = \$100$. The buyer desires a return of $12/4 = 3$ percent per period. To meet the buyer's requirement, the purchase price must be the present worth of income from the bond discounted at 3 percent for 40 periods:

$$\begin{aligned} \text{PW} &= \$5000(P/F, 3, 40) + \$100(P/A, 3, 40) \\ &= \$5000(0.30657) + \$100(23.144) \\ &= \$3847 \end{aligned}$$

Bonds are traded regularly through financial markets. Depending on the prevailing interest rate at a given time, a bond may sell for a price that is less than, more than, or equal to its face value. When the owner of a bond

---

* The procedure for calculating a value is sometimes called *capitalization of income.*

purchased previously seeks to sell it before maturity, the original purchase price and premiums already received have no bearing on its market value; only future cash flow has consequence.

Example 4.8
Evaluation of a Bond Purchase

A utility company sold an issue of 4 percent bonds 6 years ago. Each bond has a face value (value at maturity) of $1000, is due in 14 years, and pays interest twice a year (2 percent per period). Because interest rates on savings have climbed in recent years, the bond can now be sold on the bond market for only $760. If buyers expect their money to earn 8 percent compounded semiannually and they must pay a brokerage charge of $20 to purchase each bond, is the current selling price reasonable?

Solution 4.8

Semiannual interest payments amount to 2 percent of the face value of the bond, or $0.02 \times \$1000 = \$20$, and $1000 will be redeemed in 14 years, or 28 half-year periods. Therefore, the present value of the cash flow when the desired nominal interest rate is 4 percent per 6-month period is

Since the price of the bond is $760 + $20 = $780, prospective bond purchasers should look elsewhere to obtain their desired rate of return on investments.

$$PW = \$20(P/A, 4, 28) + \$1000(P/F, 4, 28)$$
$$= \$20(16.662) + \$1000(0.33348)$$
$$= \$666.72$$

Assuming that an original purchaser of the utility bond in Example 4.8 paid the face value ($1000) 6 years ago and accepted $760 for it now, we find that only the original investment was recovered and no interest was earned during the 6 years it was held:

$$\$1000 = \$20(P/A, i, 12) + \$760(P/F, i, 12)$$
$$= \$20(P/A, 0, 12) + \$760(P/F, 0, 12)$$
$$= \$20(12) + \$760(1)$$
$$= \$1000$$

The current market rate of interest strongly affects bond prices. Higher market rates tend to lower bond prices by decreasing the present worth of the future stream of payments promised by the bond. Figure 4.2 displays the valuation at different market rates of interest for a 20-year bond as described in Example 4.8 and a bond which pays the same dividends but matures in 3 years.

As is apparent in Figure 4.2, the longer the maturity of a security, the greater its price change in response to a change in the market rate of interest. This pattern explains why short-term bonds with the same risk of default as long-term bonds usually have lower dividend rates. Investments in short-term securities expose the investor to less chance of severe fluctuations in the market rate of interest than do comparable long-term investments.

FIGURE 4.2
Value of $1000 long-term (20-year) and short-term (3-year) bonds at different market interest rates. Both bonds promise semiannual dividend payments of $20.

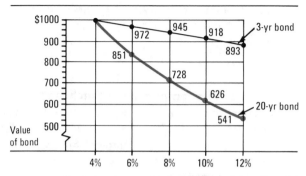

Nominal market rate of interest (compounded semiannually)

Stock Valuation

Stock in a company represents a share of ownership, as opposed to a bond, which is essentially a promissory note. There are many types of stocks and bonds, varying with respect to their degree of security and associated special privileges. In general, all bonds have claims on a company's assets before stock in case of a business failure. Stock is still a popular investment because it has the potential of increasing in value and may pay higher dividends than bonds when the company is very successful. Some stockholders have voting privileges which allow major investors to have some say in company policy.

Stock market concepts and vocabulary are examined in Extension 9A.

Preferred stocks usually entitle owners to regular, fixed dividend payments similar to bond interest. Characteristically, preferred stock has no voting rights and initially has a higher yield than bonds. The *par value* of a preferred stock is the amount due the stockholder in the event of liquidation, and the annual yield is often expressed as a percentage of the par value. Since preferred stock has no maturity date, it may be treated as a perpetuity whose value is

$$PW = \frac{\text{annual dividend on the preferred stock}}{\text{annual rate of return expected by the investor}}$$

Common stocks, the most common form of equity shares in a company, are more difficult to value than are preferred stocks or bonds because dividends and prices of common stocks are not constant; investors hope that they will increase over time. It is therefore necessary to forecast future earnings, dividends, and stock prices. If reliable forecasts could be made (and that is a highly questionable assumption), stock valuation would result from discounting the forecast cash flow.

To illustrate, suppose that a share of Sumplex, Inc., has a current market price of $40. It is expected to pay a $3 dividend by the end of 1 year. Since the company went public in 1967, the value of its stock has been rising at an average rate of 4 percent. Using a 1-year study period, we find that

$$\text{Present worth} = \frac{\text{dividend}}{(F/P, i, 1)} + \frac{\text{market price at end of year}}{(F/P, i, 1)}$$

$$\text{Present price} = \frac{\text{dividend}}{1 + i} + \frac{\text{present price} \times 1.04}{1 + i}$$

Rearranging and substituting numerical data gives

$$1 + i = \frac{\$3 + \$40(1.04)}{\$40} = \frac{\$44.60}{\$40} = 1.115$$

to reveal the discount rate

$$i = 1.115 - 1 = 0.115 \qquad \text{or} \qquad 11.5 \text{ percent}$$

If an investor is satisfied with an 11.5 percent rate of return after considering the risk involved, shares in Sumplex, Inc., could be purchased. The same expected rate of return could be calculated as

$$i = \text{dividend rate} + \text{growth rate} = \frac{\$3}{\$40} + 0.04 = 0.075 + 0.04 = 0.115$$

The partition of expected returns into dividends and growth is important when taxes are considered. Capital gains, the increase in market value between the purchase and sale of a stock, are taxed at a lower rate than is ordinary income. Thus, a very prosperous investor would be attracted to stocks that promise greater growth over more generous dividends, given equal risk and rate of return.

Review Exercises And Discussions

An entrepreneur intending to start a new business knows that the first few years are the most difficult. To lessen the chance of failure, a loan plan for start-up capital is proposed in which interest paid during the first 2 years will be at 3 percent, at 6 percent for the next 2 years, and at 12 percent for the final 2 years of the 6-year loan. How large a loan can be substantiated for proposed repayments at the end of years 2, 4, and 6 of, respectively, $20,000, $30,000, and $50,000?

Exercise 1

It is not unusual for interest rates to change during a study. Time-value equivalence is calculated by translating individual cash flows through each time period at the interest rate applicable during that period. The procedure is illustrated by considering each loan repayment independently and calculating its present worth, as

Solution 1

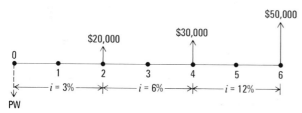

$$\begin{aligned}
\text{PW(\$20,000 payment)} &= \$20,000(P/F, 3, 2) = \$20,000(0.94260) \\
&= \$18,852
\end{aligned}$$

$$\begin{aligned}
\text{PW(\$30,000 payment)} &= \$30,000(P/F, 6, 2)(P/F, 3, 2) \\
&= \$30,000(0.89000)(0.94260) = \$25,167
\end{aligned}$$

$$\begin{aligned}
\text{PW(\$50,000 payment)} &= \$50,000(P/F, 12, 2)(P/F, 6, 2)(P/F, 3, 2) \\
&= \$50,000(0.79719)(0.89000)(0.94260) \\
&= \$33,439
\end{aligned}$$

The loan that can be repaid by the given repayment amounts at the given interest rates is then

$$\text{PW(total loan)} = \$18,852 + \$25,167 + \$33,439 = \$77,458$$

A graph of the loan's balance over the 6-year period is shown in Figure 4.3. From an initial value of $77,458 at time zero, the amount owed increases to $82,175 in 2 years with interest at 3 percent. Then the $20,000 payment is recorded, dropping the amount owed to $62,175. The steeper rates of increase during the later stages of the loan result from greater earnings generated by higher interest rates.

Exercise 2 A new rock pit will be operated for a construction project that will last 5 years. Rock can be loaded from an elevated box loader served by a conveyor from the pit, or by mobile shovel loaders. The box loader and conveyor have an initial cost of $264,000 and will have no salvage value at the end of the project. Two shovel loaders, each priced at $42,000, can provide the same capacity, but their operating costs together will be $36,000 per year more than the box loader. Normal service life for a shovel loader is 3 years with zero salvage value, but a 2-year-old machine can likely be sold for $10,000. Which alternative is preferred when the interest rate is 13 percent?

FIGURE 4.3
Loan balance during the 6-year repayment period.

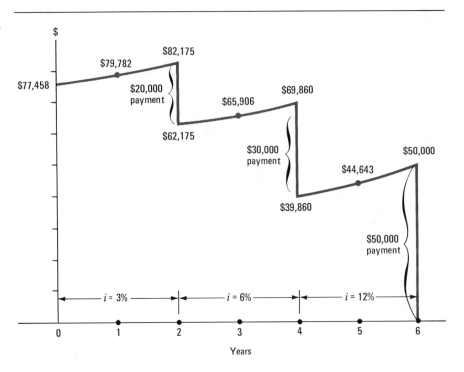

The study period is the duration of the project, 5 years. The present worth of the box-loader alternative is its initial cost: PW(box loader) = −$264,000.

Since the first two shovel loaders are replaced at the end of 3 years and the replacements will not be in service for their entire useful life, the estimated salvage value must be included as a cash inflow at the end of year 5. The additional operating costs incurred for the shovel-loader alternative form a 5-year annuity of annual $36,000 payments:

Solution 2

$$\begin{aligned}
\text{PW(shovel loaders)} &= 2(-\$42,000) + 2(-\$42,000)(P/F, 13, 3) - \$36,000(P/A, 13, 5) \\
&\qquad + \$20,000(P/F, 13, 5) \\
&= -\$84,000 - \$84,000(0.69305) - \$36,000(3.5172) \\
&\qquad + \$20,000(0.54276) \\
&= -\$84,000 - \$58,216 - \$126,620 + \$10,856 \\
&= -\$257,980
\end{aligned}$$

The shovel loaders appear to be more economical if the machines can be sold for the indicated salvage value after 2 years of use.

A bank has offered to loan the novelty company described in Example 4.6 the sum of $145,000 at 8 percent interest compounded annually to build a large warehouse immediately. The warehouse would be the same size as the small one plus the addition and would have annual expenses of $1500 per year. If its resale value at the end of 10 years is $50,000, would it be a better alternative than the other two in Example 4.6?

Exercise 3

The interest rate offered by the bank should not be used in the comparison, because the required rate of return for an organization includes more than simply the cost of borrowing. At $i = 12$ percent, the present worth of 10 years of storage from the construction of a large warehouse comparable in size to the other alternatives is

Solution 3

$$\begin{aligned}
\text{PW} &= \$145,000 - \$50,000(P/F, 12, 10) + \$1500(P/A, 12, 10) \\
&= \$145,000 - \$50,000(0.32197) + \$1500(5.6502) \\
&= \$137,375
\end{aligned}$$

This cost is lower than is that of the build-small-and-add alternative but is more than the leasing cost. Note that the cost of construction in two phases of the same-sized building exhibits the typical relation that the absolute cost (at $i = 0$ percent) of acquisition by parts is greater than is acquisition all at one time. However, a deferral may make the time value of acquisition by parts less expensive.

An investor has been investigating the stock performance of two companies: Withit and Righton. The Withit Corporation has consistently paid dividends that increase 10 cents per year while the selling price of the stock has averaged a 2 percent annual rise. Righton is a new glamour company that has paid no dividends because all earnings are retained for expansion, but its market price is expected to increase by $10 per year. Moreover, in about 5 years it is expected to start paying dividends equal to 2 percent of its price per share (a price-to-earnings ratio of 50:1). Current data about the two companies are summarized below.

Exercise 4

	Withit Corporation	Righton Company
Dividend	$2.25 (10¢/year increase)	0 (2% of market price after 5 years)
Market price	$28 (2% annual increase)	$65 ($10/year increase)
Capitalization rate	9%	12% (risk adjusted)

Since it is generally believed that Righton stock is less stable than is Withit's, the extra risk of investing in Righton is recognized by requiring a higher rate of return for the valuation, 12 percent versus 9 percent.

Disregarding tax effects and brokerage commissions to buy or sell, determine which stock has the greater valuation for an anticipated 10-year ownership.

Solution 4   A 10-year study period is used to calculate the present worth of each stock alternative, assuming that dividends are paid at the end of the year. The valuation of Withit stock at $i = 9$ percent is

$$PW(\text{Withit}) = [\$2.25 + \$0.10(A/G, 9, 10)](P/A, 9, 10) + \$28(F/P, 2, 10)(P/F, 9, 10)$$
$$= [\$2.25 + \$0.10(3.7976)](6.4176) + \$28(1.2189)(0.42241)$$
$$= \$16.87 + \$14.42 = \$31.29$$

which makes the current market price of \$28 appear attractive.

The valuation of Righton Company's stock based on a 12 percent desired rate of return and the assumption that dividends are paid annually after a 5-year wait is diagramed and calculated as

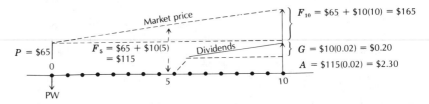

$$PW(\text{Righton}) = [\$2.30 + \$0.20(A/G, 12, 5)](P/A, 12, 5)(P/F, 12, 5)$$
$$+ \$165(P/F, 12, 10)$$
$$= [\$2.30 + \$0.20(1.7745)](3.6047)(0.56743) + \$165(0.32197)$$
$$= \$5.43 + \$53.13 = \$58.56$$

Since the \$60 market price for Righton stock exceeds the calculated valuation, it appears that Withit stock is a better investment opportunity, assuming that risk-return ratings and forecast cash flows are reasonably accurate.

PROBLEMS

4.1   The lease on a warehouse amounts to \$5000 per month for 5 years. If payments are to be made on the first of each month, what is the present worth of the agreement at a nominal annual interest rate of 18 percent, compounded monthly.

4.2   Determine the present worth of the lease in Problem 4.1 if the payments are to increase by \$250 per month for each year of occupancy.

4.3   Determine the present worth of an investment which returns a net profit of \$6000 the first year, \$5500 the second year, \$5000 the third year, with a continuing decline of \$500 per year until the return equals zero.

4.4   A company borrowed \$100,000 to finance a new product. The loan was for 20 years at a nominal interest rate of 6 percent compounded semiannually. It was to be repaid in 40 equal payments. After half the payments were made, the company decided to pay the remaining balance in one final payment at the end of the tenth year. How much was owed?

4.5 A proposed improvement in an assembly line will have an initial purchase and installation cost of $67,000. The annual maintenance cost will be $3500; periodic overhauls once every 3 years, including the last year of use, will cost $6000 each. The improvement will have a useful life of 12 years, at which time it will have no salvage value. What is the present worth of the lifetime costs of the improvement at $i = 8$ percent?

4.6 A small truck has a first cost of $11,000 and is to be kept in service for 6 years, at which time the salvage value is expected to be $1000. Maintenance and operating costs are estimated at $800 the first year and will increase at a rate of $200 per year. Determine the present worth of this vehicle using an interest rate of 15 percent.

4.7 Determine the present worth of the following series of cash flows using a discount rate of 10 percent.

Year	Flow, $	Year	Flow, $
0	−12,000	5	1600
1	2,000	6	1400
2	2,000	7	1200
3	2,000	8	1000
4	1,800	9	3000

4.8 The Amjay Company is currently renting a parking lot for employee and visitor use at an annual cost of $9000, payable at the first of each year. The company has an opportunity to buy the lot for $50,000. Maintenance and taxes on the property are expected to cost $2500 annually. Given the property will be needed for 10 more years, determine what sales price must be obtained at the end of that period in order for Amjay to break even when using an interest rate of 12 percent.

4.9 A small dam and irrigation system are expected to cost $300,000. Annual maintenance and operating costs are expected to be $40,000 the first year and will increase at a rate of 10 percent per year. Determine the equivalent present worth of building and operating the system with interest equaling 10 percent over a 30-year life.

4.10 A new automobile may be purchased with either a gasoline engine or a recently developed turbodiesel. The diesel has a first cost which is $1100 more than the gasoline option, but expected average fuel mileages are 26 miles per gallon and 19 miles per gallon, respectively, for the two engines. The cost of operation is also lower, with diesel fuel currently priced at $1.09 per gallon compared with $1.20 per gallon for unleaded gasoline. Assuming that the engine selected will have no effect on resale value at the end of a 5-year service life, and that fuel prices are predicted to increase at an annual rate of 6 percent, determine which option should be selected based on driving an average of 15,000 miles per year. Use an interest rate of 12 percent for the evaluation.

4.11 Machine A has a first cost of $9000, no salvage value at the end of its 6-year useful life, and annual operating costs of $5000. Machine B costs $16,000 new and has a resale value of $4000 at the end of its 9-year economic life. Operating costs for machine

B are $4000 per year. Compare the two alternatives on the basis of their present worths, using the repeated-projects assumption at 10 percent annual interest.

4.12 A commercial rental property is for sale at $100,000. A prospective buyer estimates the property would be held for 12 years, at the end of which it could be sold for $90,000. During the ownership period, annual receipts from rentals would be $15,000 and average disbursements for all purposes in connection with ownership would be $6000. If a rate of return of 9 percent is expected, what is the maximum bid the prospective purchaser should make to buy the property?

4.13 A manufacturer requires an additional 10,000 square feet (929 square meters) of warehouse space. A reinforced-concrete building added to the existing main structure will cost $440,000, whereas the same amount of space can be constructed with a galvanized building for $310,000. The life of the concrete building is estimated at 50 years with a yearly maintenance cost of $19,200. The life of the galvanized building is estimated to be 30 years, and the annual maintenance cost is $48,000. Average annual property taxes are 1.2 percent for the concrete building or 0.5 percent for the metal building. Assume that each type of warehouse would be replaced by the same model and that the salvage value equals the unrecovered balance of the purchase price at any time.

Compare the present worths of the two warehouse additions, using 12 percent interest, for a 50-year study period.

4.14 Two alternative structures are being considered for a specific service. Compare the present worth of the cost of 24 years of service, using an interest rate of 5 percent, when neither structure has a realizable salvage value. The pertinent data are as follows:

	Structure I	Structure II
First cost	$4500	$10,000
Estimated life	12 years	24 years
Annual disbursements	$1000	$720

4.15 Perpetual care for a small shrine in a cemetery is estimated to be available for $500 per year. The long-term interest rate is expected to average about 5 percent. If the capitalized cost is estimated at $15,000, what amount is anticipated for the first cost of the shrine?

4.16 A proposed mill in an isolated area can be furnished with power and water by a gravity-feed system. A stream high above the mill will be tapped to provide flow for water needs and power requirements by connecting it to the mill with a ditch-and-tunnel system or with a wood-and-concrete flume that winds its way down from the plateau. Either alternative will meet current and future needs, and both will utilize the same power-generating equipment.

The ditch-and-tunnel system will cost $500,000 with annual maintenance of $2000. The flume has an initial cost of $200,000 and yearly maintenance cost of $12,000. In addition, the wood portion of the flume will have to be replaced every 15 years at a cost of $100,000.

Compare the alternatives on the basis of capitalized costs with an interest rate of 6 percent.

4.17 What is the maximum amount that you could afford to bid for a bond with a face value of $5000 and a coupon rate of 8 percent payable semiannually if your minimum attractive rate of return were 10 percent. The bond matures in 6 years.

4.18   The Swampwater Flood Control District is selling 20-year tax-free bonds with a face value of $2 million and a stated yield of 6 percent with semiannual interest payments. Printing and legal costs associated with issuing the bonds have totaled $50,000, and costs associated with paying the dividends are expected to be $1000 per period. What is the minimum amount the bonds must sell for such that the utility district's cost of capital does not exceed a nominal 8 percent per annum?

4.19   A $5 million school-bond issue bearing interest at 7 percent payable on a semiannual schedule and maturing in 25 years sold at a price which provided a nominal 8 percent rate of return to investors. The brokerage fee for handling the sale was 0.4 percent of the total bond issue. What amount was realized from the sale to actually use for school construction?

4.20   On January 2, 1975, Consolidated Lintpickers sold an issue of 25-year bonds with a stated annual interest rate of 10 percent payable semiannually. On January 3, 1987, immediately after an interest payment has been made, Sam Smartt has an opportunity to purchase a series of these bonds with a face value of $5000. What is the maximum Sam should pay for the bonds if he wishes to obtain a nominal 12 percent rate of return on his investment?

4.21   Interest on a 7 percent, $1000 bond due in 25 years is payable semiannually by clipped coupons, with the first payment due 6 months from now. What must be the price of the bond to provide a rate of return to the buyer of 10 percent compounded semiannually?

4.22   Bonds of the Overightors Corporation are perpetuities bearing 7 percent annual interest. Their par value is $1000.

  (a) If bonds of this type currently are expected to yield 6 percent, what is the market price?
  (b) If interest rates rise to the level at which comparable bonds return a yield of 8 percent, what would the market price be for Overightor bonds?
  (c) How would the prices change in Problems 4.22a and 4.22b if the bonds had a definite maturity date in 20 years?

4.23   The XX Company is currently earning $2 million per year after taxes. Shareholders own 1 million shares. If investors require a rate of return of 15 percent on stocks in the same risk class as XX stock and the previous dividend was $1, what price will the stock sell for when future dividends are expected to grow at a constant rate of 5 percent per year? What is the price-to-earnings ratio? Since XX is not considered to be in a legitimate growth industry, any increases in price per share will result only from improved dividends.

4.24   An engineer working on a construction project has reason to believe that the stock of one of the companies supplying materials to the project is undervalued at the current selling price of $60 per share. The last dividend was $2.50. She believes that both the value of the stock and dividends will double in 10 years. Assuming that the increase occurs in constant increments each year and the engineer requires a 10 percent return on such investments, how much could she afford to pay per share for the stock?

4.25   A bond with a face value of $5000 pays quarterly interest of $1\frac{1}{2}$ percent each period. Twenty-six interest payments remain before the bond matures. How much would you be willing to pay for this bond today if the next interest payment is due now and you want to earn 8 percent compounded quarterly on your money?

4.26   The present worth of a bond is 70 percent of the face value when the total return to buyers is 12 percent compounded annually. What is the annual bond rate on this 10-year bond?

4.27   A company is considering the purchase of a new piece of testing equipment which is expected to produce $8000 additional before-tax profit during the first year of operation; this amount will probably decrease by $500 per year for each additional year of ownership. The equipment costs $20,000 and will have an estimated salvage value of $3000 after 8 years of use. For a before-tax interest rate of 25 percent, determine the net present worth of this investment.

4.28   A grant has been given by a corporation to support a summer music festival for 4 years. The first payment of $200,000 will be made a year from now, and payments will increase by 7 percent of the first payment each year to account for inflation. What is the present worth of this commitment today if the corporation uses a 12 percent annual interest rate for discounting its transactions?

4.29   To attract industry, a city has made an offer to a corporation. The city will install all roads and services for the plant site at no immediate cost to the corporation, but the corporation will be expected to bear the cost of operation and maintenance on a cost-sharing schedule whereby $65,000 is paid the first year and each subsequent payment will be $5000 less. The first payment will be made 2 years from now, and the corporation's obligation will end with the last $5000 payment. What is the present worth today of this agreement if the annual interest rate is 9 percent?

4.30   A marina has two alternative plans for constructing a small-boat landing on a lake behind their sales building; one is a wooden dock and the other is a metal and concrete wharf. Data for the two plans are as shown.

	Wood	Metal and Concrete
First cost	$25,000	$50,000
Period before replacement	20 years	30 years
Salvage value	$5000	0
Annual maintenance	$5500	$2800

Using a minimum attractive rate of return of 10 percent, compare the present worths of the two plans. Assume that both will provide adequate service and that replacement costs will be the same as the original cost.

4.31   A refining company entered into a contract for raw materials with an agreement to pay $600,000 now and $150,000 per year beginning at the end of the fifth year. The contract was made for 10 years. At the end of the third year, because of unexpected profits, the company requested that it be allowed to make a lump-sum payment in advance for the rest of the contract. Both parties agreed that 7 percent compounded annually was a fair interest rate. What was the amount of the lump sum?

4.32   A machine can be repaired today for $2000. If repairs are not made, the operating expenses will increase by $200 each year for the next 5 years. Assume that the expenses will occur at the end of each year and that the machine will have no value under either alternative at the end of the 5-year period. The minimum acceptable rate of return is 12 percent. Compare the present worths of the two alternatives.

4.33 The following alternatives are available to accomplish an objective of 12 years' duration:

	Plan A	Plan B	Plan C
Life cycle	6 years	3 years	4 years
First cost	$2000	$8000	$10,000
Annual cost	$3200	$700	$500

Compare the present worth of the alternatives using an interest rate of 7 percent.

4.34 The lining of a chemical tank must be replaced every 3 years at a cost of $1800. A new type of lining is available that is more resistant to corrosion. The new lining costs $3100. If the minimum rate of return required is 12 percent and taxes and insurance are 4 percent of the first cost annually, how long must the improved lining last to be more economical than the present lining?

4.35 A single underground transmission circuit is needed immediately, and load studies indicate the need for a second circuit in 6 years. If provision is made for a second conduit when the conduit for the first circuit is installed, there will be no future need for reopening, trenching, backfilling, and repaving. The cost of installing a single circuit with minimum preparation for the eventual second circuit is $850,000. The installation of the second circuit will be considered to cost $800,000 at the end of year 5 in order to be in operation by the end of year 6. If the second circuit is installed immediately, the total cost will be $1.4 million.

Levelized annual operating and maintenance costs of the circuits are 8 percent of the first cost and begin 1 year after the first costs are incurred. The average life of a circuit is 32 years. The required rate of return on such investments is 10 percent before taxes.

   (a) Compare the deferred investment with the immediate investment using a 32-year study period.
   (b) Compare the two conduit plans on a study period taken to eternity. Then compare this solution with the one in Problem 4.35a. Which is the most reasonable? Why?

4.36 The Deadeye Distillery is opening a new production facility at Sweetwater Springs. Two types of stills are being considered and the following cost estimates have been developed:

Alternative	Sourmasher	Cornsqueezer
First cost	$10,000	$15,000
Useful life	4 years	6 years
Salvage value	$0	$2,000
Maintenance costs		
First year	$800	$1,000
Increase each additional year	$200	$400

Assuming that all other operational costs are equal, compare the present worths of the alternatives using a before-tax interest rate of 25 percent.

4.37 On September 1, 1985, Susie Smartt contacted the registrar at Fairlygood State College and found that annual tuition was currently $2000. She expects this fee to increase at a rate of about 6 percent per year in the future. What single deposit must she place in a tax-free account paying 10 percent per annum if she hopes to accumulate an amount sufficient to pay for 4 years of tuition starting on September 1, 2003?

4.38 Everwhite, a small community with big ideas, envisions development of a nearby winter recreational area. The idea seems reasonable, it is located near a major population center and normally has snow of exceptional quality. The state government has agreed to guarantee bonds, which pay 6 percent interest and mature in 20 years, to be sold to fund the project. The main decision remaining is whether to build a modest facility at first and enlarge it later or to acquire all the land and do all the construction at one time.

Costs for a small lodge with an open ice rink and two ski runs with lifts on 400 acres of land will be $2 million. Construction will take 2 years. Once the resort opens, net revenues from this size operation are expected to average $300,000 per year, exclusive of bond interest and debt retirement.

A year-round resort with provisions for golf, tennis, and two more ski runs will boost the first cost to $5 million, require 3 years to build, and occupy 1500 acres. Net revenue (exclusive of debt payments) for a resort of this size should average $400,000 per year for the first 4 years of operation until the resort becomes well known, and should then increase by $100,000 per year for the next 13 years, at which time net revenue will level off.

The enlargement of the resort can be deferred 5 years, but the extra land required (1100 acres) should be purchased immediately for $500,000 to avoid the increased valuation expected when the resort opens. Construction can be done while the original facilities continue to operate. Money for the $3-million addition should be available at the end of the fifth year. When the additions are ready for use by the end of the seventh year, net revenue should at once jump annually by $150,000 over the revenue for the smaller version. This gradient increase should continue for 5 years and then drop to an annual increase of $100,000 over the previous year until the end of the twentieth year.

Depending on the development plan followed, bonds will be sold to accumulate $2, $2.5, or $5 million immediately, and $3 million at the end of year 5. Analyze the alternatives using a desired rate of return of 8 percent. Discuss your preferred solution with respect to the other options available.

EXTENSIONS

4A Things to Think About when You Are Buying Bonds

A decision to buy a bond rather than invest in a different savings instrument, such as stocks or a money market fund, is only the first decision. The next one is to decide which one to buy.

There are treasury, corporate, and municipal bonds. In general, corporate bonds offer a higher stated interest rate than do "treasuries," because they suffer greater risk of nonpayment than do those from the federal treasury. "Municipals" pay a much lower "coupon" rate because their interest is usually not taxable. Risk ratings are made for both corporate and municipal bonds by agencies that specialize in evaluating the credit worthiness of issuing organizations. Thus the rate of interest must be balanced against the possibility of default.

The yield on a bond includes both the interest paid and the difference between the purchase price and maturity value. Yield typically increases for longer-term bonds. For example, in 1984, the yield on a 5-year treasury bond was 12.5 percent, whereas the yields on 1-year and 20-year bonds were 10.6 and 12.9 percent, respectively. This relation is displayed by the yield curves in Figure 4.4. Note that the two yield curves differ significantly in shape as well as magnitude. A top-grade triple-A-rated 5-year municipal bond has a yield of 7.4 percent, which increases to 9.5 percent for a 20-year bond. Thus yields must be balanced against maturity dates.

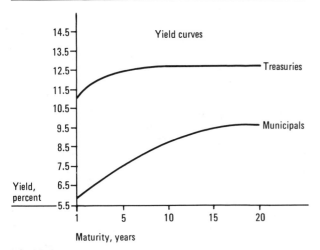

FIGURE 4.4
Relation of yields to maturities for treasury and municipal bonds.

Consideration of tax status, risk, coupon rate, and maturity date leads to an acceptable purchase price. This price (present worth) should then be balanced against expected changes in prevailing interest rates. When market interest rates rise, the price an investor is willing to pay goes down for a bond with a given coupon rate. The purchase price of a bond with a 2-year maturity, for instance, will drop about 10 percent after 1 year if the market interest rate has increased by 10 percent during the year.

Consider a bond issued 5 years ago. It originally sold at its $1000 face value with an annual coupon rate of 9 percent and will mature in 10 more years. If comparable

bonds are currently selling for 11 percent, an investor would be willing now to pay only

$$PW = \$1000(P/F, 11, 10) + \$1000(0.09)(P/A, 11, 10)$$
$$= \$1000(0.35219) + \$90(5.8892)$$
$$= \$882.22$$

When interest expectations are higher than is a bond's coupon rate, buyers are willing to pay more for shorter maturity dates. For the 9 percent bond above and an unchanging 11 percent expectation, the PW increases to $926.07 when the maturity decreases from 10 to 5 years. Conversely, when an investor believes that the market interest rate will be less than the coupon rate during the maturity period, longer-term bonds are awarded higher prices. The sample $1000 bond with a 9 percent coupon rate has a present worth of $1081.99 and $1140.47, respectively, 5 and 10 years before maturity, when an investor is satisfied with a 7 percent yield. From the foregoing calculations, it is evident that a two percentage point change in market interest rates for the 5-year bond creates the following potential gains and losses:

$$\text{Potential upside }\% \atop \text{(interest drops 2\%)} = \frac{\$1081.99 - \$1000}{\$1000}100\% = 8.20\%$$

$$\text{Potential downside }\% \atop \text{(interest rises 2\%)} = \frac{\$926.07 - \$1000}{\$1000}100\% = -7.39\%$$

To accurately compare the upside and downside potential of bonds with other types of investments, income taxes and capital gains taxes must be figured in. Chapter 11 tells how to do it.

QUESTIONS
4A.1 Does a bond with a longer maturity exhibit larger or smaller upside and downside potential? Why? Compare the potential gain and loss for a plus and minus two percentage point change in the bond described previously for 5-, 10-, and 20-year maturities.
4A.2 Complete the following table for $1000 municipal bonds that pay annual interest when the current date is 1/1/87 and taxes are not considered.

Coupon rate (%)	Maturity	Current* yield (%)	Offering yield (%)	Offering price ($)
5.7	1/1/88	5.77	7	_____
6.1	7/1/88	6.44	8	_____
10.5	1/1/92	___	___	$1058.34
	1/1/97	9.12	10	_____

* Current yield = coupon rate/offering price

4B   Future-Worth Comparisons

It is evident from the equivalence concept that a present worth can be translated to a future worth at any given time at a given interest rate. The mechanics of the translation are elementary, but the change in perspective by the analyst can be significant. For certain situations the future worth of cash flow may be more meaningful to a decision maker than the present worth. The amount of debt or surplus accumulated at a future date could be a milestone in a financial plan or a goal that directs current activities.

Attention to future cash flows may yield more accurate estimates of receipts and disbursements because thinking tends to be in "then-current" dollars. Because the purchasing power of a unit of currency is eroded by inflation, it may be advantageous to focus on how many currency units must be available to meet needs at a later date—the quantity of then-current dollars. Consequently, future-worth calculations are frequently utilized in escalation analyses, particularly for evaluating the effects of inflation.

The future-worth perspective can be illustrated by reference to the cash flow in Example 4.2. The 8-year cash flow in the example is composed of two annuities: annual $15,000 costs for the first 3 years followed by five equal-step receipts ranging from $10,000 to $20,000.

The future value of disbursements, treated as debts for which interest at 10 percent must be paid, builds in 8 years to a total indebtedness of

$$
\begin{aligned}
\text{FW(disbursements)} &= -\$15,000(F/A, 10, 3)(F/P, 10, 5) \\
&= -\$15,000(3.3099)(1.6105) \\
&= -\$79,959
\end{aligned}
$$

A complementary future worth at the end of year 4 would show the accumulated indebtedness before any income is generated as

$$
\text{FW(4)} = -\$15,000(F/A, 10, 3)(F/P, 10, 1) = -\$54,613
$$

This amount is reduced by the receipt of $10,000 at the end of year 4, but it reveals the maximum liability level.

The positive arithmetic-gradient cash inflow reaches a maximum at the end of year 8, amounting to

$$
\begin{aligned}
\text{FW(receipts)} &= [\$10,000 + \$2500(A/G, 10, 5)](F/A, 10, 5) \\
&= [\$10,000 + \$2500(1.8100)](6.1050) \\
&= \$88,675
\end{aligned}
$$

The balance on hand at the end of the study period is then

$$
\text{Net FW} = \$88,675 - \$79,959 = \$8716
$$

This same figure (differing only by rounding error from the interest tables) could be obtained from the previously calculated PW = $4066 as

$$
\text{Net FW} = \$4066(F/P, 10, 8) = \$4066(2.1435) = \$8715
$$

The relative usefulness of PW and FW comparisons depends on which figure is more valuable for decision making. If knowledge of the dollars available at some impending date is critical, then a future-worth calculation is called for. Knowing when to answer such a call is the mark of a competent engineering economist.

QUESTION

4B.1 For the cash flow in Example 4.2, calculate the time at which the amount owed equals the amount received, including interest. This is the breakeven point for the cash flow.

4B.2 What is the future worth of the loan agreement made by the entrepreneur in Review Exercise 1?

CHAPTER 5

EQUIVALENT ANNUAL-WORTH COMPARISONS

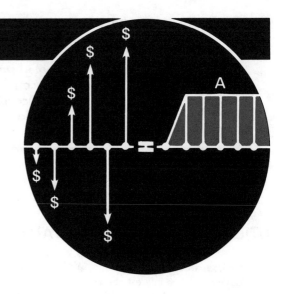

OVERVIEW

Comparisons of investment proposals by the equivalent annual-worth method complement the more widely applied present-worth and rate-of-return methods. Sometimes it is most meaningful to structure a solution in terms of annual payments, but more often the annual-worth model is used because it is a more convenient way to arrive at a present worth or rate of return. All three methods of analysis indicate the same preference among investment alternatives.

An annual charge to account for the repayment of invested capital plus interest earnings on the unrecovered balance of the investment is obtained as

$$\text{Equivalent annual cost} = \text{EAC} = (P - S)(A/P, i, N) + Si$$

where P = present value, or purchase price
   S = salvage value, or resale price
   N = study period, or economic life of the asset
   i = interest rate expected on investments

Alternatives are compared on the basis of net equivalent annual worth (EAW) when both income and disbursements are known, or on their equivalent annual cost (EAC) when they serve the same function but income resulting from that function is not known.

When comparisons are made among assets that have different lives, a repeated-projects assumption is usually made, but the study-period method can be used when enough information about replacements is known. Occasionally an assumption of perpetual life is appropriate.

A sinking fund may be established to protect investors by enforcing an orderly retirement of debt from current income. Periodic payments are made to an independent institution, where they earn interest at the institution's normal rate.

UTILIZATION OF THE EQUIVALENT ANNUAL-WORTH COMPARISON METHOD

Methods to calculate uniform payments when interest rates vary are given in Extension 5B.

With an annual-worth method all the receipts and disbursements occurring over a period of time are converted to an equivalent uniform yearly amount. It is a popular method because of the widespread inclination to view a year's gains and losses as a yardstick of progress. Cost-accounting procedures, depreciation expenses, tax calculations, and other summary reports are annual in nature. These yearly cost tabulations generally make the annual-worth method easier to apply and comprehend than do the other comparison methods.

Equivalent annual-worth comparisons produce results compatible with present-worth and rate-of-return comparisons. For a set of common assumptions, a preference for an alternative exhibited by one method will be mirrored by the other two. Annual-worth calculations are frequently a part of the computations required to develop present-worth and rate-of-return values, and parallel computations by different methods are useful for complementary comparisons that improve the clarity of an analysis.

The six conditions listed in Chapter 4 for basic present-worth comparisons also apply to basic annual-worth comparisons: cash flows and interest rates are known, cash flows are before taxes and in constant-value dollars, and comparisons include neither intangible considerations nor limits due to availability of financing. These restrictions are relieved in later chapters.

Structure of a Capital-Recovery Annuity

Annuity contracts for lifetime-income planning are discussed in Extension 5A.

The cornerstone of annual-worth calculations is the capital-recovery factor, which converts a lump sum into an equivalent annuity. This annuity usually represents an investment in an asset that is expected to generate a positive future cash flow, and the duration of the annuity is therefore the life of the asset. Since the cost of an asset is a cash outlay, the resulting annuity is a uniform series of negative payments. This negative cash flow is offset by the positive revenue produced by the asset in establishing the net equivalent annual worth of the investment.

The *capital-recovery factor* (A/P, i, N) accounts for both the repayment of invested capital P plus the interest earned on the unrecovered portion of the investment. Although the payments A are uniform in size, the proportion of capital recovered and interest earned changes each period. The structure of an annuity, in which varying amounts from the equal payments are allocated to capital recovery and interest, is best revealed by examining a sample application.

Assume that an asset is purchased for $40,000. It has an expected life of 4 years and no salvage value at the end of its life. The purchaser intends to recover the $40,000 investment over 4 years *plus* the interest the $40,000 would have earned if invested elsewhere. If an acceptable interest rate is 10 percent, the series of equal payments that would return the capital plus interest is computed as

$$\text{Equivalent annual payment} = A = P(A/P, 10, 4)$$
$$= \$40,000(0.31547) = \$12,619$$

Every year the proportion of a payment allotted to capital recovery and interest charges changes because interest is earned only on the amount of capital not yet recovered, and that amount changes each year. During year 1, before a payment is received, the $40,000 investment earns $40,000 \times 0.10 =$ $4000 interest. The first payment of $12,619 then reduces the unrecovered capital by $12,619 - \$4000 = \$8619$. During the second year, 10 percent interest is paid on $40,000 - \$8619 = \$31,381$, amounting to $3138. Therefore, of the second payment, $3138 is interest and $12,619 - \$3138 = \$9481$ is allocated to capital recovery. The complete sequence of changing proportions is given in Table 5.1

Capital-Recovery Calculations

As indicated in Table 5.1, the sum of the four annuity payments is $50,476, of which $10,476 is interest. This tabular format can be used to trace the capital

TABLE 5.1

Pattern of capital recovery and interest charges when the capital-recovery factor at $i = 10$ percent is applied to the purchase of a $40,000 asset with a life of 4 years and no salvage value

End of Year	Capital Not Recovered by End of Year	Interest Due on Unrecovered Capital	Amount of Capital Recovered	Annual Capital-Recovery Charge
0	$40,000			
1	31,381	$ 4000	$ 8619	$12,619
2	21,900	3138	9481	12,619
3	11,471	2190	10,429	12,619
4	0	1147	11,471	12,619
Totals		$10,475*	$40,000*	$50,476*

* The sum of $10,475 + $40,000 differs from $50,476 by rounding errors.

recovery; or, when only the amount of unrecovered capital at a certain time is sought, it can be determined directly from the present worth of the remaining payments:

$$\text{Unrecovered capital(year 3)} = A(P/A, 10, 1)$$
$$= \$12,619(0.9091) = \$11,472$$
$$\text{Unrecovered capital(year 2)} = A(P/A, 10, 2)$$
$$= \$12,619(1.7355) = \$21,900$$

Then,

$$\text{Recovered capital(year 3)} = \$40,000 - \$11,472 = \$28,528$$
$$\text{Interest due(year 4)} = \$12,619 - \$11,472 = \$1147$$
$$\text{Recovered capital(year 2)} = \$40,000 - \$21,900 = \$18,100$$
$$\text{Interest due(years 3 and 4)} = 2(\$12,619) - \$21,900$$
$$= \$3338$$

The equivalent annual payment equation can be modified to include a salvage value in two ways. If S = salvage value, then

$$\text{Equivalent annual cost} = (P - S)(A/P, i, N) + Si$$

or

$$\text{EAC} = P(A/P, i, N) - S(A/F, i, N)$$

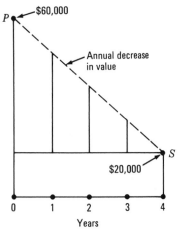

Both expressions yield the same EAC, but the first formula is used in this text because it is generally easier to manipulate. Consider an asset that costs $60,000 and has a $20,000 salvage value. The net amount that must be recovered from annuity payments is $P - S = \$60,000 - \$20,000 = \$40,000$; the remaining portion of the purchase price is returned by receipt of the salvage value, $20,000. However, the purchaser is deprived of the use of the $20,000 during the life of the asset, so interest is owed on this amount because it represents unrecovered capital. The Si term in the formula accounts for this interest payment.

If the life of the $60,000 asset with a $20,000 salvage value is 4 years and the interest rate is 10 percent, then

$$\text{EAC} = (\$60,000 - \$20,000)(A/P, 10, 4) + \$20,000(0.10)$$
$$= \$40,000(0.31547) + \$2000$$
$$= \$12,619 + \$2000 = \$14,619$$

The first term above ($12,619) is the same capital-recovery expression depicted in Table 5.1 and accounts for annual decline in the value of the asset. The second term ($2000) is the annual charge for the capital locked in the salvage value, S.

The amount of capital unrecovered at the end of year 3 is the sum of the salvage value plus the annual capital-recovery charge $((EAC - Si)$ discounted for 1 year:

$$
\begin{aligned}
\text{Unrecovered capital(year 3)} &= A(P/A, 10, 1) + S \\
&= (\$14{,}619 - \$2000)(0.9091) + \$20{,}000 \\
&= \$31{,}472
\end{aligned}
$$

from which is calculated

$$
\begin{aligned}
\text{Recovered capital(year 3)} &= \$60{,}000 - \$31{,}472 \\
&= \$28{,}528
\end{aligned}
$$

The pattern of capital recovery parallels that shown in Table 5.1, differing only by the effect of the salvage value.

The alternative formula for equivalent annual cost yields the same solution for the given data:

$$
\begin{aligned}
EAC &= P(A/P, 10, 4) - S(A/F, 10, 4) \\
&= \$60{,}000(0.31547) - \$20{,}000(0.21547) \\
&= \$18{,}928 - \$4309 = \$14{,}619
\end{aligned}
$$

Interest due on unrecovered capital	
at $N = 1$:	$ 6000
$N = 2$:	5138
$N = 3$:	4190
$N = 4$:	3147
	$18,475

and

$$
\begin{aligned}
\text{Unrecovered capital(year 3)} &= EAC(P/A, 10,1) + S(P/F, 10,1) \\
&= \$14{,}619(0.9091) + \$20{,}000(0.9091) = \$31{,}472
\end{aligned}
$$

which confirms that the two expressions for equivalent annual cost are indeed equivalent and can be applied interchangeably; in both situations the unrecovered capital at any time is the present worth of the remaining installments in the annuity, including the salvage value.

SITUATIONS FOR EQUIVALENT ANNUAL-WORTH COMPARISONS

The term *annual worth* suggests a positive value, but the calculations can just as well produce a negative value. A negative annual worth indicates that the equivalent value of negative cash flow for disbursements is greater than is the corresponding positive flow of receipts. Negative worths usually mean that an alternative is unacceptable. Exceptions occur when projects must be undertaken to satisfy certain requirements such as safety citations or building codes. Then the objective is to identify the alternative with the least equivalent cost (negative cash flow).

It is often very difficult, and not worth the required study time, to discover the income derived from one component in a complex system. For instance, the income produced by a copying machine is troublesome to derive exactly since its output is utilized by many people, often from different departments working on many projects. In this type of situation, alternatives to satisfy the copying needs are evaluated on the basis of their relative costs, because each alternative capable of meeting the requirements of the system will produce the same income to the system. When it is apparent that *only* costs are involved in an evaluation, it is convenient to ignore the negative sign convention and let comparison figures represent the absolute value of costs.

Equivalent annual cost (EAC) designates comparisons involving only costs.

Several situations for applying equivalent annual-worth calculations are described in the examples that follow.

Consolidation of Cash Flows

"What's it worth?" is the decisive query in the appraisal of a proposal. It is difficult to ascertain what to expect from a proposal until the myriad receipts and disbursements associated with its conduct are collectively analyzed. Improvement programs are prime examples. Organizations regularly engage in programs to improve productivity, reduce accidents, raise quality, and the like. Each is a worthwhile goal, expected to have positive rewards, but each has costs too. Consolidating the various costs into a pattern that can be compared with potential rewards may take the form of a net annuity.

Example 5.1
Equivalent Net Worth of Cash Flows

A consulting firm proposes to provide "self-inspection" training for clerks who work with insurance claims. The program lasts 1 year, costs $2000 per month, and professes to improve quality while reducing clerical time. A potential user of the program estimates that savings in the first month should amount to $800 and increase by $400 per month for the rest of the year. However, operation confusion and work interference are expected to boost clerical costs by $1200 the first month, but this amount should decline in equal increments at the rate of $100 per month. If the required return on money is 12 percent compounded monthly and there is a stipulation that the program must pay for itself within 1 year, should the consultants be hired?

Solution 5.1

$$i = \frac{r}{m} = \frac{0.12}{12} = 0.01/\text{period}$$

$$N = 12 \text{ periods}$$

Cash flow

$$
\begin{aligned}
\text{Equivalent monthly} \atop \text{worth of savings} &= \$800 + \$400(A/G, 1, 12) \\
&= \$800 + \$400(5.3682) \\
&= \$2947
\end{aligned}
$$

Equivalent monthly
worth of costs
$$= -\$2000 - [\$1200 - \$100(A/G, 1, 12)]$$
$$= -\$3200 + \$100(5.3682)$$
$$= -\$2663$$

Equivalent cash flow

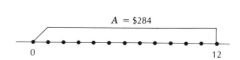

$A = \$284$

Equivalent net
monthly cash flow
$= \$2947 - \$2663 = \$284$

The program looks very promising because the equivalent monthly worth is positive during the first year and savings generated by the training should continue into the future.

Gradient factors are utilized in Example 5.1 to convert uniformly varying cash flows to their equivalent constant worths. Recovery of capital is not an issue, since no property ownership is involved. The comparison is made directly on the basis of expected income versus outgo.

Recovery of Invested Capital

"Will it pay off?" is the question investors want answered. An adequate payoff recovers the invested capital plus the desired rate of return. Since returns are spread over the life of the investment, it is convenient to convert capital-recovery costs to the same annual pattern. The result of consequence from combining uniform cost and revenue flows is a *positive, zero,* or *negative* series of payments that, respectively, categorize the investment as gratifying, adequate, or insufficient.

Example 5.2
Net Annual Worth of a Single Project

The purchase of a truck with an operator's platform on a telescoping hydraulic boom will reduce labor costs for sign installations by $10,000 per year. The price of the boom truck is $57,000 and its operating costs will exceed those of the present equipment by $100 per month. The resale value is expected to be $6000 in 12 years. Should the boom truck be purchased when the prevailing interest rate is 12 percent?

Solution 5.2

Net annual worth

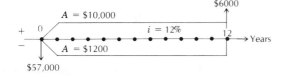

$A = \$10,000$

$i = 12\%$

$A = \$1200$

$\$57,000$

6000

Years

$$\text{EAW} = -\$57,000(A/P, 12, 12) + \$6000(A/F, 12, 12)$$
$$- \$100(12) + \$10,000$$
$$= -\$57,000(0.16144) + \$6000(0.04144)$$
$$- \$1200 + \$10,000$$
$$= -\$9202 + \$249 + \$8800 = -\$153$$

Equivalent annual-worth calculations indicate that the purchase and use of the boom truck will cause a loss equivalent to $153 per year for 12 years, compared with other investments that could earn a 12 percent return.

The solution to Example 5.2 was developed from the accompanying cash flow diagram. The capital-recovery factor leads to the same solution when capital-recovery costs are registered negatively and the net annual savings are positive:

$$EAW = \text{annual savings} - \text{capital-recovery costs}$$

$$= \$10,000 - \$1200 - [(P - S)(A/P, 12, 12) + Si]$$

$$= \$8800 - [(\$57,000 - \$6000)(0.16144) + \$6000(0.12)]$$

$$= \$8800 - (\$8233 + \$720) = \$8800 - \$8953 = -\$153$$

Net Cash Flow Comparison

"Which one is better?" may have conflicting answers. If the criterion is strictly economics, the alternative with the highest net worth is preferred, if worth is measured by revenues; but when worth is evaluated from costs, a low EAC is preferred.

Example 5.3
Comparison of Net Annual Worths

A supplier of laboratory equipment estimates that profit from sales should increase by $23,000 per year if a mobile demonstration unit is built. A large unit with sleeping accommodations for the driver will cost $71,000, while a smaller unit without sleeping quarters will be $55,000. Salvage values for the large and small units after 5 years of use will be, respectively, $8000 and $3500. Lodging costs saved by the larger unit should amount to $3000 annually, but its yearly transportation costs will exceed those of the smaller unit by $1300. With money at 15 percent, should a mobile demonstration unit be built, and if so, which size is preferable?

Solution 5.3

EAW of large mobile demonstration unit:

Annual increase in profit	$23,000
Savings in lodging costs over smaller unit per year	3,000
Extra transportation costs over smaller unit, per year	−1,300
Capital-recovery cost:	
($71,000 − $8000)(A/P, 15,5) + $8000i	
$63,000(0.29832) + $8000(0.15)	−19,994
Net AW =	$4,706

EAW of small mobile demonstration unit:

Annual increase in profit	$23,000
Capital-recovery cost:	
($55,000 − $3500)(A/P, 15, 5) + $3500i	
$51,500(0.29832) + $3500(0.15)	−15,888
Net AW =	$7,112

The net annual worths indicate that both alternatives will produce positive cash flows while concurrently repaying investment costs. The small mobile demonstration unit is preferred because its net cash flow is larger.

Since the profit increase expected from building either of the mobile labs is the same, the comparison could have been conducted by considering only the costs first:

	Small Unit	Large Unit
Capital-recovery cost:	$15,888	$19,994
Net extra cost of smaller unit: $3000 − $1300	1,700	
Total annual cost	$17,588	$19,994

and then evaluating the best resulting alternative in terms of the expected income.

CONSIDERATION OF ASSET LIFE

Translating cash flows to equivalent annuities is a mechanical process that becomes almost automatic with practice. Understanding the meaning of an economic comparison and being able to explain its significance to others are the critical skills. The discussion of economic asset life, as introduced in Chapter 4, is continued in this section to stress the importance of selecting an appropriate study period in equivalent annual-worth-comparisons.

Definitions of Asset Life

In time-value mechanics, N is simply the number of compounding periods appropriate for the analysis of cash flows. N takes on a special meaning when it represents the life of an asset that loses value as a function of use or time. The more frequently applied terms to describe the life of an asset are listed and defined as follows:

Ownership life or *service life* is the period of time an asset is kept in service by an owner or owners. Implied is a period of useful service from the time of purchase until disposal. Actually, under the vague expectation that it might somehow again prove useful, equipment is often retained beyond the point where it is capable of satisfying its intended function. A machine can have a *physical life* longer than its service life; the machine is still physically sound, but there is no useful function for it to perform.

Accounting life is a life expectancy based primarily on bookkeeping and tax considerations. It may or may not correspond to the period of usefulness and economic desirability.

Economic life is the time period that minimizes the asset's total equivalent annual cost or maximizes its equivalent net annual income. This period terminates when the asset is displaced by a more profitable replacement or the asset's service is no longer required. Economic life is also referred to as the *optimal replacement interval* and is the condition appropriate for most engineering economic studies.

Mark Twain's advice was "buy land, they're not making it any-more."

Land is not subject to a specified life or to capital recovery because it historically appreciates in value rather than depreciates with age. The cost of land ownership is the interest not received on funds invested in the property.

Comparisons of Assets with Equal and Unequal Lives

Examples 5.4 and 5.5 below are typical applications of EAC comparisons. The "lease or buy" question posed in Example 5.4 is raised with increasing regularity that corresponds to the rapid growth of leasing companies. It is now possible to lease almost any type of production equipment that is not custom-designed for narrowly specialized service. Important tax considerations involved in the lease-buy choice are discussed in Chapter 11.

Example 5.4
Alternatives with Equal Annual Costs

A machine needed for 3 years can be purchased for $77,662 and sold at the end of the period for about $25,000. A comparable machine can be leased for $30,000 per year. If a firm expects a return of 20 percent on investments, should it buy or lease the machine when end-of-year payments are expected?

Solution 5.4

Equivalent annual cost to buy
$$= (\$77{,}662 - \$25{,}000)(A/P, 20, 3) + \$25{,}000(0.20)$$
$$= \$52{,}662(0.47473) + \$5000$$
$$= \$30{,}000$$
Annual cost to lease $= \$30{,}000$

If the salvage value is considered reasonably accurate, the machine should be purchased, because the funds invested in it will earn 20 percent (assuming 20 percent is an attractive rate of return for the firm).

The two alternatives in Example 5.4 are compared on the basis of their costs because the income resulting from their contribution is not available and it is believed that both are capable of producing that contribution. The question is not whether to get a machine. It is known that the machine is necessary, so it is a choice of buying or leasing it. In selecting between the alternatives with equal equivalent annual worth, it should be recognized that they are equal only after ownership of the machine has earned 20 percent on the capital invested in it. Therefore, if the owners are satisfied with a 20 percent return on their money and no other considerations are involved, it is prudent to purchase.

Example 5.5
Comparison of Assets with Unequal Lives

Two models of machines can be purchased to perform the same function. Type I has a low initial cost of $3300, high operating costs of $900 per year, and a short life of 4 years. The more expensive type II costs $9100, has annual operating expenses of $400, and can be kept in service economically for 8 years. The scrap value from either machine at the end of its life will barely cover its removal cost. Which is preferred when the minimum attractive rate of return is 8 percent?

Solution 5.5

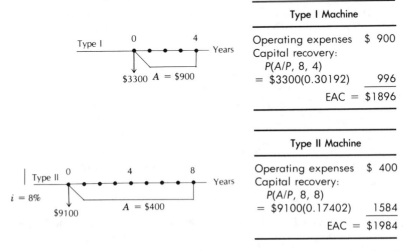

Type I Machine	
Operating expenses	$ 900
Capital recovery:	
$P(A/P, 8, 4)$	
$= \$3300(0.30192)$	996
EAC =	$1896

Type II Machine	
Operating expenses	$ 400
Capital recovery:	
$P(A/P, 8, 8)$	
$= \$9100(0.17402)$	1584
EAC =	$1984

The type I machine has a lower annual cost for service during the next 4 years and is therefore preferred.

The machines described in Example 5.5 exhibit the common feature that more expensive models, designed to serve the same function as less expensive versions, are expected to operate more economically and/or last longer. (If a costlier machine also produces better-quality products, the benefits from improved quality must be included in the analysis to make the outcomes comparable.) The difficulty in comparing alternatives with unequal lives is to account for the service provided during the period in which one outlasts the other.

The implied assumption in Solution 5.5 is that two machines of type I will be purchased consecutively to provide the same length of service as one type II machine. The equivalent annual cost for 8 years of service from two type I machines is, of course, the same as is calculated in the solution above:

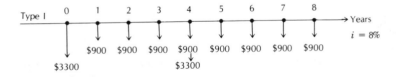

$$EAC = \$3300(A/P, 8, 8) + \$3300(P/F, 8, 4)(A/P, 8, 8) + \$900$$
$$= \$3300(0.17402) + \$3300(0.73503)(0.17402) + \$900$$
$$= \$574 + \$422 + \$900 = \$1896 \text{ (as calculated previously)}$$

The "repeated projects" assumption for evaluating assets with different lives is reasonable and widely used. Unless reliable forecasts can be made about future operating conditions and the probability of technical advances, the assumption that today's conditions will exist in the future is plausible. When future conditions can be estimated with confidence, these valuations are the data for equivalent annual-worth calculations. For instance, a confident prediction that current developmental work on the type I machine will produce refinements within 4 years to reduce operating costs by one-third while increasing the purchase price by one-half leads to a revised economic analysis:

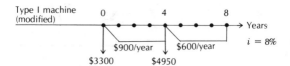

The annual worth is calculated by (1) translating the cash flow in the last 4 years to a present value at year 4 and discounting this value to year 0; (2) computing the present value of the first 4 years' cash flow; (3) adding the two present worths; and (4) converting the total to an 8-year annuity:

$$EAC(\text{type I modified}) = \{[\$600(P/A, 8, 4) + \$4950](P/F, 8, 4)$$
$$+ \$900(P/A, 8, 4) + \$3300\}(A/P, 8, 8)$$
$$= \{[\$600(3.3121) + \$4950](0.73503) + \$900(3.3121)$$
$$+ \$3300\}(0.17402)$$
$$= [\$6937(0.73503) + \$2981 + \$3300](0.17402)$$
$$= (\$5099 + \$6281)(0.17402) = \$1980$$

The modified type I machine now has nearly the same equivalent annual cost as does the type II model. If a decision maker has confidence in the forecasted data, the choice between types can go either way, but it will more likely swing to type I because the lower purchase price and shorter life mean that less capital is committed for a shorter period. This conservative philosophy provides some protection from unexpected developments. It was assumed in the comparison that the need for the machine would exist for 8 years. Many things can happen in 8 years to thwart the most carefully conceived plans—new designs, changing markets, successful competition, etc. The opportunity to reevaluate tactics in 4 years, as is possible with the type I machine, is a subtle but valued attribute.

Perpetual Life

Occasionally an asset is treated as if it will last forever. The assumption of infinite life in terms of capital recovery is slightly more reasonable than is the physical interpretation. Nothing made by humans, even Egyptian pyramids or the great wall of China, lasts forever, but the difference between infinity and 100 years in the numerical value of the capital-recovery factor is quite small:

$$(A/P, 10, \infty) = \frac{0.10(1.1)^\infty}{(1.10)^\infty - 1} = 0.10 \frac{\infty}{\infty} = 0.10$$
$$(A/P, 10, 100) = 0.10001$$

Therefore, in an economic comparison involving an asset with an infinite life, *such as land,* the interest rate replaces the capital-recovery factor. The human-made assets most closely approaching perpetual life are dams, tunnels, canals, aqueducts, and monuments. The nature of very long-lived assets relegates them mostly to public projects, and there the trend has been to set a study period of 50 years or so in recognition of changing public needs and technological advances that generate new ways to fulfill the needs.

The similarity between the perpetual-life assumption for calculating equivalent annual worth and the capitalized cost method associated with present-worth models should be apparent. Both assume infinite life and therefore have very limited application.

USE OF A SINKING FUND

The sinking-fund factor was discussed previously as an alternative means of calculating capital-recovery costs:

$$(P - S)(A/P, i, N) + Si = P(A/P, i, N) - S(A/F, i, N)$$

As is apparent in the equation, the sinking-fund factor is applied to compute the annuity required to accumulate a certain future amount. Organizations are sometimes obligated by legislated or contractual agreements to establish a fund, separate from their internal operations, to accumulate a specified amount by a specified time. This accumulation is called a *sinking fund.*

A provision for a sinking fund requires that an organization set aside a portion of the income derived from sales or taxes each year in order to retire a bond issue (or, in some cases, an issue of preferred stock). Failure to meet the sinking-fund payments forces the bond issue to be thrown into default, causing serious credit and credibility problems. The payments are a direct cash drain on the organization. That is the purpose of the sinking fund—to protect investors by enforcing an orderly retirement of debt from current income.

A debt can be retired by regular payments from the sinking fund or by letting money accumulate in the sinking fund until the debt is due and then paying it in full (plus interest). Periodic payments are associated with *callable bonds.* A call provision in a bond gives the issuing organization the right to

Amortization plus interest, a capital recovery method based on the sinking fund concept, is explained in Extension 5C.

pay off a bond before its maturity date. Funds reserved from income are allocated each year to retire a portion of the bond issue. Sinking-fund payments thus utilized earn at the rate at which interest payments are avoided on the bonds retired.

When a sinking fund is established to accumulate sufficient money to meet the bond cost at maturity, annual payments are normally required to be invested in a savings institution. The interest rate for these "savings" typically is at a rate lower than the organization earns on its own capital, but the sinking fund is less an earning device than a way to ensure that funds will not be diverted to other ventures.

For example, a firm borrows $1 million at 9 percent simple interest because it believes that the amount borrowed can be utilized within the firm to earn double the borrowing rate, 18 percent. If the $1 million is acquired by issuing 20-year bonds with the stipulation that a sinking fund be set up, the firm has to set aside a payment each year and put it in an external account. This account probably earns less than the 18 percent internally earned on investment funds, but it represents a very secure investment. Assuming that the account pays 6 percent annual compound interest, we find that the annual payments into the sinking fund would be

$$A = \$1,000,000(A/F, 6, 20)$$
$$= \$1,000,000(0.02719) = \$27,190$$

to make total annual debt repayment on the principal plus interest of $27,190 + $1,000,000(0.09) = $117,190.

Review Exercises and Discussions

Exercise 1   An asset was purchasd 5 years ago at a price of $52,000. It was expected to have an economic life of 8 years, at which time its salvage value would be $4000. If the function that the asset was serving is no longer needed, what price must it be sold for now to recover the invested capital when $i = 12$ percent?

Solution 1   The expected annual cost of the asset over its 8-year life was

$$EAC = (P - S)(A/P, 12, 8) + S(0.12)$$
$$= (\$52,000 - \$4000)(0.20130) + \$4000(0.12)$$
$$= \$48,000(0.20130) + \$480 = \$10,142.40$$

The unrecovered capital at the end of the fifth year is the present worth of the last 3 years of the EAC annuity plus the discounted salvage value:

$$\text{Unrecovered capital(year 5)} = \$10,142.40(P/A, 12, 3) + \$4000(P/F, 12, 3)$$
$$= \$10,142.40(2.4018) + \$4000(0.71178)$$
$$= \$24,360 + \$2847 = \$27,207$$

This sum is the required price.

An alternative solution method provides the same result:

$$\text{Selling price(year 5)} = \$48,000(A/P, 12, 8)(P/A, 12, 3) + \$4000$$
$$= \$23,207 + \$4000 = \$27,207$$

Exercise 2

Experience with a conventional mechanical log debarker suggests that its service life is 4 years. A newly designed hydraulic debarker costs one-third more than a mechanical debarker but makes much less noise. Both debarkers have about the same operating costs and no salvage value. What will the optimal replacement interval of the new hydraulic debarker have to be to make its cost comparable with that of the mechanical debarker at $i = 12$ percent?

Solution 2

$$\text{AW(hydraulic debarker)} = \text{AW(mechanical debarker)}$$
$$P(\text{hd})(A/P, 12, N) = P(\text{md})(A/P, 12, 4) \quad \text{and} \quad P(\text{hd}) = 1.33P(\text{md})$$
$$(A/P, 12, N) = \frac{P(\text{md})}{1.33P(\text{md})}(0.32924) = 0.24693$$
$$N = 5 + 1\frac{0.27741 - 0.24693}{0.27741 - 0.24323} = 5 + 0.89 = 5.89 \text{ years}$$

The less tangible advantage of noise reduction associated with the hydraulic debarker should also be considered to make the outcomes of the alternatives comparable.

Exercise 3

A short concrete canal can be constructed as part of a flood-control project; the placement of a large galvanized culvert will serve the same function. The cost of the canal, which will last indefinitely, is $75,000, and maintenance costs will average $400 per year. A culvert, which will have to be replaced every 30 years, will cost $40,000 and have annual maintenance costs of $700. Salvage values are negligible for both alternatives, and the government interest rate is 6 percent. Which alternative has the lowest equivalent annual cost?

Solution 3

Annual-cost comparison of a canal with perpetual life and a culvert with an economic life of 30 years:

Canal

Annual maintenance	$ 400
Interest on investment: $75,000(0.06)	4500
Equivalent annual cost	$4900

Culvert

Annual maintenance	$ 700
Capital recovery: $40,000(A/P, 6, 30) = $40,000(0.07265)	2906
Equivalent annual cost	$3606

The culvert has the advantage of a lower equivalent annual cost.

Exercise 4

A grain elevator was built 15 years ago at a cost of $230,000. It was supposed to have a salvage value after 30 years of 10 percent of its first cost. Capital recovery is by a sinking fund held by a local bank that has paid interest during the period at 4 percent compounded annually. The owners now want to add a second grain elevator which will cost $400,000. How much additional capital will they need if they apply the capital-recovery reserves from the first elevator toward construction of the second?

Solution 4 Annual payments made into the sinking fund for capital recovery of the first grain elevator were

$$A = (P - 0.1P)(A/F, 4, 30) = (\$230,000 - \$23,000)(0.01783)$$
$$= \$3691$$

The amount accumulated in the sinking fund after 15 payments is

$$F(15) = \$3691(F/A, 4, 15) = \$3691(20.023) = \$73,901$$

which can be deducted from the amount needed to finance a second grain elevator, $400,000. Then

$$\text{Additional capital needed} = \$400,000 - \$73,901 = \$326,099$$

PROBLEMS

5.1 Megabitt Electronics is considering the purchase of a new programmable circuit tester in order to improve their product quality. The equipment has a first cost of $55,000 and the salvage value is predicted to be $6000 after a service life of 5 years. Maintenance and operating costs are expected to be $8000 the first year of operation and to increase by $1500 per year for each additional year of use. Using a before-tax interest rate of 20 percent, determine what annual savings must be obtained through the use of this equipment to make it economically justifiable.

5.2 A pilot plant has been constructed to convert garbage to oil, natural gas, and charcoal. The plant cost $4 million to build and will have a 20-year life with no salvage value. The cost of processing 1 ton (907 kilograms) of garbage is $14. From each ton are derived products which can be sold for $21 ($0.023 per kilogram). However, 20 percent of the end products are consumed as energy during the processing. The cost of delivery, sorting, and shredding a ton of garbage is $2.50 ($0.0028 per kilogram), after deducting the value of salvageable materials recovered. How many tons (kilograms) of garbage must be processed each year to recover the cost of the plant when the interest rate is 9 percent?

5.3 A standby electric power generator was purchased 6 years ago for $8,000. At that time it was expected that the equipment would be used for 15 years and would have a salvage value of 10 percent of the first cost. The generator is no longer needed and is to be sold for $2500. Using an interest rate of 15 percent, determine the difference beween the anticipated and actual equivalent annual capital costs.

5.4 A company can purchase a piece of equipment for $20,000 and sell it for $4,000 at the end of an 8-year service life or lease the unit for the same period by making first-of-the-year payments of $3000. Compare the equivalent annual costs of the alternatives using an interest rate of 15 percent.

5.6 Riprock Sand and Gravel Company is considering the feasibility of purchasing a piece of land for a small quarrying operation. The following cost estimates have been developed for evaluating the venture:

Cost of land	$2,000,000
Site clearing and road preparation	200,000
Annual operating costs:	
First year	400,000
Increase for each additional year of operation	50,000
Site cleanup prior to resale	100,000

The quarry will probably have a useful life of 10 years, and the reclaimed site should have a resale value of $1 million. Using a before tax interest rate of 25 percent, determine the equivalent annual cost of this operation.

5.7   Five years ago, a car owner bought an automobile for $5600. A trade-in of $600 was allowed on the purchase of the new car. The old one had been driven 70,000 miles (112,651 kilometers). If the owner's other investments earn 6 percent annually, indicate the cost per mile (kilometer) for capital recovery plus interest during the period of ownership.

5.8   Today is January 24, 1987. Suppose that 10 years ago you had started putting $100 per month in the bank. You made payments continually for 6 years and then stopped, but the accumulated deposit was left in the bank. On January 24, 1991, you plan to open a mousetrap factory. The money in the savings account will be used to advertise your new mousetrap. If you use $1000 per month in advertising, how many months can you continue before the fund is exhausted? The money is invested at 6 percent compounded monthly.

5.9   An investment in a cherry orchard is being considered. The asking price for the orchard is $120,000, half of which is accounted for by the land value. The cherry trees and harvesting equipment are past their prime and are expected to have no realizable value after 10 years. Income from the orchard, after deducting operating expenses and taxes, should be about $20,000 the first year, declining by $1000 each year. To obtain this net income, a working fund of $15,000 is required to maintain operations (the working fund is essentially invested capital because it is not available for other investments). Since the orchard is near a growing city, the land is expected to increase in value by 50 percent within 10 years. What is the annual worth of the orchard investment over 10 years when money is expected to earn 9 percent annually?

5.10   Laser beams are to be used on a major construction project to ensure the exact alignment of components. Two types of laser alignment systems, with the costs shown below, are suitable for the project.

	IC System	UC System
First cost	$5000	$3200
Salvage value	1000	0
Annual operating cost	600	950
Additional taxes and insurance	180	0

(a) If both systems have a life of 4 years and the minimum rate of return is 15 percent, which offers the lowest equivalent annual cost?

(b) How much longer would the economic life of the IC system have to be in order to make the equivalent annual costs of the two systems equal?

5.11   In Problem 4.13, costs were given for a decision between a concrete and a galvanized warehouse building. Now another option is available. The manufacturer can lease a structure which is located about half a mile (0.805 kilometer) from the main plant and has approximately the same floor space as do the warehouses being considered for construction. It can be leased for $50,000 per year. In addition to the lease expense, it will likely cost $2000 per month in extra material-handling expense to transport materials to and from the leased building.

How much can be saved by leasing when a 12 percent interest rate is applicable?

5.12   Compare the annual costs of the two alternatives described in Problem 4.32

5.13   Two methods of supplying water and sewage treatment for a housing development outside the districts where water and sewage-disposal services are provided by the city are described by the accompanying cash flows for a 40-year study period:

Years	Method 1	Method 2
0	− $350,000	− $735,000
1–10	− 11,000	− 8000
10	− 25,000	+ 100,000
11–20	− 13,000	− 13,000
20	− 150,000	+ 100,000
21–30	− 15,000	− 15,000
30	− 25,000	− 75,000
31–40	− 18,000	− 15,000

Both methods have the same absolute cash flow amounts over the study period ($1,120,000), and both provide comparable quality of service. At an interest rate of 8 percent, determine which method has the lower equivalent annual cost.

5.14   A family that enjoys outdoor activities can purchase a small run-down cabin in a desirable mountain location for $6000. To make it livable they will have to repair the roof and construct a new outhouse at a cost of $1800 for materials and transportation. If the cabin is to be usable for skiing in the winter, they will have to add a fireplace or stove and insulation that will cost at least $2000. Other annual expenses for electricity, taxes, maintenance, etc., will average $550. They expect to own the cabin 5 years before a job transfer will force them to sell it. Assuming that the repairs are done at once, heating arrangements are added by the end of the second year, and the cabin can be sold for $7000, determine how much the family could afford to spend each year for other outings in place of going to their cabin. Their savings earn 7 percent. What intangible factors might affect their decision to buy the cabin.

5.15   A bond issue for $75,000 has been passed by voters to buy six minibuses for a senior citizens' transportation service. It is anticipated that the revenue from the bus service will yield a rate of return of 7 percent on the investment. A provision in the bond issue was that a sinking fund be established through a local bank to accumulate enough money to recover the $75,000 in 6 years. The bond issue is to pay 8 percent simple interest (due in a lump sum at maturity), and the local bank pays annual interest of 6 percent.

(a) What annual payment is required for the sinking fund at the bank?
(b) What annual return is required to recover the capital invested plus profit.

5.16   A company borrows $100,000 under an agreement to set up a sinking fund to pay back the amount owed in 5 years. The interest earned on the sinking fund is 7 percent. Annual interest payments of $100,000 \times 0.09 = \$9000$ are to be paid by the company to the lenders. The company expects internally invested funds to earn twice the interest rate paid on the external sinking fund. What is the annual cost to the company to repay the loan?

5.17 A sheltered workshop requires a lift truck to handle pallets for a new contract. A lift truck can be purchased for $17,000. Annual insurance costs are 3 percent of the purchase price, payable at the first of each year. An equivalent truck can be rented for $667 per month, payable at the end of each month. Operating costs are the same for both alternatives. What minimum number of months must a purchased truck be used on the contract to make purchasing more attractive than leasing. Interest is 12 percent compounded monthly. (Assume that the purchased truck has no salvage value at any time.)

5.18 A clever investor purchased a piece of land 5 years ago for $175,000. Its appraised value is now twice the original cost. Year-end taxes have been 4 percent of the purchase price until now. A street assessment of $17,500 was paid 2 years ago. A land developer is willing to buy the land at the appraised price plus the investor's ownership costs on a 4-year contract at 9 percent compounded annually. How large should the developer's four annual payments be if the investor uses an annual interest rate of 15 percent in figuring the time value of ownership costs?

5.19 An asset is expected to depreciate in market value at a constant rate from its purchase price of $20,000 to a zero salvage value during its 8 years of physical life. Operating costs are expected to be $8000 the first year and to increase at a 10 percent compound rate as the asset gets older. What is the annual cost of ownership if the asset is replaced every 3 years? The required rate of return is 12 percent.

5.20 A sinking fund was set up to pay off a debt of $100,000 in 10 years. The payments earn 7 percent compound annual interest. How much of the debt has been accounted for after the seventh payment has been made?

5.21 The present wooden bridge over Deep Deep Bay is in danger of collapse. The highway department is currently considering two alternatives to alleviate the situation and provide for expected increases in future traffic. One plan is a conventional steel bridge and the other is a tunnel under the bay. The department is familiar with bridge construction and maintenance but has no experience with maintenance costs for tunnels. The following data have been developed for the bridge:

First cost	$17,000,000
Painting every 6 years	1,000,000
Deck resurfacing every 12 years	3,000,000
Structural overhaul at the end of 30 years	4,000,000
Annual maintenance	300,000

The tunnel is expected to cost $24,000,000 and will require repaving every 12 years at a cost of $2,000,000. If both designs are expected to last 60 years with negligible salvage value, determine the maximum equivalent annual amount for maintenance that could be permitted for the tunnel while holding the total EAC to that of the bridge. Let $i = 6$ percent.

5.22 An earth compactor costs $18,000 and has an economic life of 9 years. However, the purchaser needs it only for one project that will be completed in 3 years. At the end of the project, it can be sold for half its purchase price. What is the annual cost to the owner if the required rate of return is 20 percent.?

5.23 The athletic department of East Ragbag State University is proposing that a new general-purpose stadium be constructed on campus. A design utilizing a combination earthwork bowl with a steel upper deck and pressbox is being considered. The following cost estimates have been developed:

First cost of complete construction	$32,000,000
Paint steel structure every 6 years	2,000,000
Replace wooden seats every 10 years	4,000,000
Repave parking facilities and ramps every 12 years	3,000,000
Annual maintenance	1,500,000

Assuming a 60-year life and negligible salvage value, determine the equivalent annual cost of the project using a tax-free interest rate of 7 percent.

5.24 Two types of power converters are under consideration for a specific application. An economic comparison is to be made using an interest rate of 20 percent, and the following cost estimates have been obtained:

Alternative	Alpha	Beta
Purchase price	$10,000	$20,000
Estimated service life	5 years	9 years
Salvage value	$0	$5,000
Annual operating cost	$2,500	$1,200

(a) Determine the annual equivalent costs of the alternative systems.
(b) Determine a salvage value for the beta system such that it would have an equivalent annual cost equal to the alpha system.
(c) Using a study period of 4 years, determine the present worth of the system.
(d) If the systems are to be compared using the study period approach as in part (c), what are the implied salvage values of each at the end of 4 years?

5.25 Dirtydirt Mining and Excavation uses a large number of light pickup trucks for crew transport and general utility duties. The trucks have a first cost of $8000 and owing to the generally rugged use lose value at a rate of approximately 40 percent per year. Operation and maintenance costs for two-shift use amount to $3000 the first year and increase at about $1000 per year for each additional year of service. Current company policy is to keep the vehicles for 5 years before they are sold. The maintenance supervisor has suggested that they be sold one year earlier in order to reduce maintenance expense. Using an interest rate of 15 percent, determine the equivalent annual cost effect of implementing the supervisor's suggestion.

5.26 The Frozen Fin Fish Company is planning an expansion to a cold storage facility. Four alternative site-design proposals are being considered using an interest rate of 30 percent. Plans A and B require an expenditure of $200,000 for land and plans C and D require $300,000 for land. These real estate investments are assumed to be permanent. The buildings are expected to last for 30 years, the compressors and related equipment will last 10 years before requiring replacement, and energy costs will increase throughout the building's life. Neither the buildings nor the equipment are expected to have any salvage value. With this information and the data provided below, use an annual cost comparison to determine which proposal is preferred.

Proposal	A	B	C	D
Building and insulation	$600,000	$700,000	$400,000	$500,000
Compressors	70,000	50,000	70,000	50,000
Expected energy costs:				
First year	45,000	30,000	55,000	40,000
Increase for each				
additional year	3,000	2,000	3,000	2,000
Annual maintenance expense	20,000	10,000	50,000	40,000

EXTENSIONS

5A   Annuity Contract for a Guaranteed Income

Annuities are often associated with retirement plans. In this context, an *annuity* implies a payment received each year according to a purchased contract. In practice, annuity payments are more likely to be made monthly to the receiver, or *annuitant*.

An annuity contract is essentially the reverse of a life insurance policy. In the latter, an insurance company pays a stipulated sum to heirs based on the amount of payments made during the policyholder's lifetime. In the classical annuity situation, an individual pays a stipulated sum, or number of payments, to a company and then receives regular income payments starting at a designated date and continuing for life. Annuities of this form were used very early in recorded history; mortality tables for computing their values were compiled by Romans and have been found in Egypt and Babylon.

A vast variety of annuity contracts are available. A *straight life annuity* is the classic pattern: Income payments terminate with the death of the annuitant. Since many people are reluctant to purchase such an annuity because they fear that they will die prematurely, causing their investment in the annuity to be wasted, companies offer plans that guarantee a refund to heirs under specified conditions, such as an unusually early death. Annuities may have variable rates that are linked to inflation indexes, gold prices, or foreign currencies. The purpose of indexing is to conserve the buying power of annuity income. As with any investment, the annuity buyer must balance risks against returns.

QUESTIONS

5A.1   More frequent income payments to the annuitant increase the purchase price of the annuity. That is, an annuity that pays $100 a month will cost more than an annuity of $1200 paid once a year. List three reasons why this is so.

5A.2   Assume that your advice is sought by a 65-year-old person who has $200,000 in a savings account and is considering investing that amount in an annuity which will provide a fixed monthly income for life, starting immediately. List five factors that the prospective annuitant should consider in deciding whether to buy an annuity or to draw money out of the $200,000 account whenever it is needed.

5B   Equivalent Uniform Payments when Interest Rates Vary

Interest rates have fluctuated widely in recent years as a response to different degrees of inflation. Rates also vary according to the type and size of investment; larger investments are often awarded higher interest rates, and riskier investments demand higher returns. It is therefore occasionally appropriate to assign different interest rates to specific periods of cash flow.

Present and future worths of cash flows with changing

TABLE 5.2

End of Year	Deposit or Withdrawal	i during Year	Balance in the Account at End of Year
0	+$5000		
1		6%	$5000(F/P, 6, 1) = \$5300$
2	-$2000	7%	$5300(F/P, 7, 1) - \$2000 = \$3671$
3		8%	$3671(F/P, 8, 1) = \$3964.68$
4		9%	$3964.68(F/P, 9, 1) = \$4321.50$

interest rates are calculated by translating each transaction backward or forward in time according to the prevailing interest rate in each period of the translation. For example, assume that a deposit of $5000 was made 4 years ago and a withdrawal of $2000 was made 2 years ago. The prevailing interest rate during the first year was 6 percent and the rate was increased by 1 percent each year. The amount in the account at the end of each year is shown in Table 5.2.

By the same approach, the present worth at time zero of the cash flow is

$$PW = \$5000 - \$2000(P/F, 6, 1)(P/F, 7, 1) = \$3236.64$$

A uniform series of four payments equivalent to the given cash flow with changing interest rates is calculated as

$$P = A(P/F, 6, 1) + A(P/F, 6, 1)(P/F, 7, 1)$$
$$+ A(P/F, 6, 1)(P/F, 7, 1)(P/F, 8, 1)$$
$$+ A(P/F, 6, 1)(P/F, 7, 1)(P/F, 8, 1)(P/F, 9, 1)$$
$$\$3236.64 = A[(0.94340) + (0.9340)(0.93458)$$
$$+ (0.94340)(0.93458)(0.92593)$$
$$+ (0.94340)(0.93458)(0.92593)(0.91743)]$$
$$= A(3.3904)$$

$$A = \$3236.64/3.3904 = \$954.65$$

A corresponding calculation of a series of equal payments A from the future worth is carried out as

$$F = A + A(F/P, 9, 1) + A(F/P, 8, 1)(F/P, 9, 1)$$
$$+ A(F/F, 7, 1)(F/P, 8,1)(F/P, 9,1)$$

and

$$A = \frac{\$4321.50}{1 + 1.09 + 1.08(1.09) + 1.07(1.08)(1.09)}$$
$$= \frac{\$4321.50}{4.5268} = \$954.65$$

Thus, four year-end payments of $954.65 yield a future amount of $4321.50 when interest rates progress from 6 percent to 7 percent to 8 percent to 9 percent during the 4-year span, and the worth of this annuity is equivalent to a cash inflow of $5000 at time zero and an outflow of $-$2000$ at the end of year 2.

QUESTION

5B.1 What series of equal annual payments is equivalent to the cash flow shown below?

End of year	0	1	2	3	4	5
Interest rate		7%	7%	10%	11%	9%
Receipts	$10,000		$10,000		$10,000	
Disbursements		$3000		$5000		$10,000

5C   Amortization Plus Interest

Some government agencies and utility companies use a method of capital recovery referred to as *amortization plus interest*. *Amortization* means money put aside at intervals for gradual payment of a debt. The agencies utilizing this method normally employ the sinking-fund factor to calculate depreciation on their assets; hence, amortization is associated with a sinking fund in this context. The "interest" part of the method occurs as a charge on the first cost of the asset.

The rationale for applying the "sinking fund plus interest on first cost" method runs somewhat as follows: "I pay out a certain sum P for an asset, and this asset decreases in value to S by the end of its economic life N. To account for this devaluation, I will make payments into a fund that earns interest at rate i to accumulate $P - S$ dollars in N years. However, during this period I am deprived of the use of amount P because this amount is invested in the asset and unavailable to me. Therefore, I will charge interest at rate i on P for the entire period as compensation for not having the use of that money."

QUESTIONS
5C.1 For an asset described by $P = \$15,000$, $S = \$5000$, $i = 5$ percent, and $N = 20$ years, calculate its annual cost of ownership according to amortization plus interest:

$$AC = (P - S)(A/F, i, N) + Pi$$

5C.2 Apply the capital-recovery factor to obtain the annual cost for the asset in Question 5C.1. Use the formulas for the interest factors to prove that the two methods for recovering capital plus a return must produce the same annual cost.

5C.3 An advantage sometimes cited for the amortization-plus-interest method is that different interest rates can be used in the calculation. For instance, when the sinking fund for an asset earns interest at a rate of 5 percent, the asset's owner might want 10 percent compensation for the investment funds that are tied up in the asset. Consequently, $i = 5$ percent when used in the sinking fund, and $i = 10$ percent when applied to the first cost. With reference to Questions 5C.1 and 5C.2 and the discussion of the use of sinking funds in the chapter, comment on the use of two different interest rates in making an economic comparison.

CHAPTER 6

RATE-OF-RETURN COMPARISONS

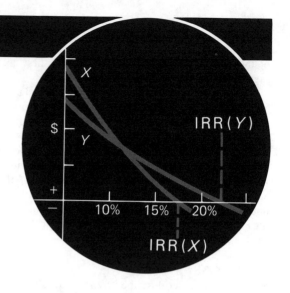

OVERVIEW

The rate of return is the last of the discounted cash flow comparison methods considered here. It is favored by some who claim that it provides the most easily understood comparisons and that it produces percentage ranking figures that are very useful.

A minimum acceptable rate of return (MARR) is the lowest level at which an alternative is still attractive. It varies among and within organizations. Although there is a wide variety of recommendations for determining this lowest level of acceptability, it is generally agreed that the level should be no lower and most likely considerably higher than the cost of capital. How much higher depends on the circumstances, objectives, and policies of the organization. The purpose of establishing a minimum acceptable rate of return is to ration capital to the most deserving proposals.

Calculation of an internal rate of return (IRR) may begin with either an EAW or PW formulation, the latter being more prevalent because by definition the internal rate of return yields a present worth equal to zero by assuming that the IRR is the reinvestment rate. There is no way to avoid trial-and-error computations in calculating IRR for complex formulations, but the structure of the cash flow offers clues about where to begin.

Consistent results are obtained from EAW, PW, and IRR comparisons. A rate of return can be calculated for proposals that involve only costs, as well as income-producing proposals. PW and IRR calculations may appear to produce contradictory results when an alternative with heavy cash flows in early years is compared with an alternative having major flows in later years. The contradiction is resolved by basing selection on the minimum acceptable rate of return.

A cumulative cash flow pattern that reverses signs (from negative to positive, or vice versa) more than once may have more than one rate of return. An applicable IRR is determined by applying an external rate of return to a portion of the cash flow that eliminates sign reversals. This external reinvestment rate is primarily a function of the MARR.

An external rate of return (ERR) is utilized by applying an explicit interest rate (*i* percent), often equal to the MARR, to calculate the future worth of receipts and then equating this FW to the future worth of expenditures compounded at the ERR. This ERR calculation is not subject to multiple rates of return, avoids most trial-and-error solutions, and recognizes realistic limits on reinvestment rates.

RATE-OF-RETURN METHODS

Rate of return is the most celebrated method of comparing investment alternatives. It provides a percentage figure that indicates the relative yield on different uses of capital. Since interest rates are well understood throughout the world of commerce, there should be little danger of misinterpreting rate-of-return figures. Another minor advantage is that it avoids the necessity of knowing a required or minimum rate of interest before calculations can be conducted; the calculations produce a percentage figure that can be compared directly with other investment proposals. These features are achieved at the expense of more tedious calculations, a minor but frustrating drawback.

Three rates of return appear frequently in engineering economic studies:

- *Minimum acceptable rate of return—MARR—*is the rate set by an organization to designate the lowest level of return that makes an investment acceptable.

- *Internal rate of return—IRR—*is the rate that yields a present worth of zero by assuming that all cash flows are reinvested at the IRR.

- *External rate of return—ERR—*is the rate that yields a future worth of zero by assuming that all positive cash flows are reinvested at the MARR.

All three of these rates of return are discussed in the following sections. The IRR and ERR comparison methods are examined without considering the effects of income taxes, which are introduced in Chapter 11, but the discussion includes practical difficulties in conducting rate-of-return comparisons and understanding the results.

MINIMUM ACCEPTABLE RATE OF RETURN

MARR is also known as the minimum *attractive* rate of return.

The *minimum acceptable* rate of return is a lower limit for investment acceptability set by organizations or individuals. It is a device designed to make the best possible use of a limited resource, money. Rates vary widely according to the type of organization; and they vary even within the organization. Historically, government agencies and regulated public utilities have utilized lower required rates of return than have competitive industrial enterprises. Within a given enterprise, the required rate may be different for various divisions or activities. These variations usually reflect the risk involved. For instance, the rate of return required for cost-reduction proposals may be lower than that required for research and development projects in which there is less certainty about prospective cash flows.

There is a wealth of literature on the subject but a poverty of agreement. It is generally accepted that the lower bound for a minimum required rate of return should be the *cost of capital*. The constitution of this cost is also subject to controversy. As discussed in Chapter 9, the cost of capital for competitive industries must reflect the expense of acquiring funds from various sources; and, as discussed in Chapter 10, the cost of safe government bonds is a basis for the lower bound of interest rates used to evaluate public investments.

How much above the cost of capital to set the minimum required rate of return depends on an organization's circumstances and aspirations. A small company strapped for cash and burdened by a low credit rating must have a very attractive proposal before it can consider investing. Larger established companies tend to view the rate as a realistic expectation of how much their capital can earn when invested. This MARR is a typical figure promised (and later substantiated) for a large number of high-quality investment proposals available to the firm; it is assumed that the proceeds earned from current projects can be reinvested at comparable rates in future proposals. The rate so derived is sometimes called the *opportunity cost of capital* because any proposal funded to earn a lower rate precludes the opportunity to earn the minimum attractive rate of return.

Extension 6A explores the economic theory behind the MARR.

The effect of establishing a minimum required rate of return is to ration capital. It is rationed to divisions of an organization and to the whole organization as a function of time. The purpose is to avoid unproductive investments in marginal activities, perhaps favored for political reasons, and to conserve capital during periods when fewer "attractive" proposals are submitted. Through such considerations the MARR becomes a management tool to stimulate or restrain certain developments. Establishing a minimum acceptable rate of return is therefore a complex and significant issue.

INTERNAL RATE OF RETURN

The IRR is the best-known and most widely used rate-of-return method and is the one applied predominately in this text. It is also known as the *true rate-*

of-return and the *discounted cash flow method*. The latter term is indicative of the way interest rates were interpreted in previous chapters; the internal rate of return, represented by *i*, is the rate of interest earned on the unrecovered balance of an investment. Stated another way, in the IRR method all receipts are assumed to be reinvested at the calculated IRR.

Table 5.1 illustrates the pattern of capital recovery at an IRR of 10 percent.

   The internal rate of return can be calculated by equating either the annual or present worths of cash flows to zero and solving for the interest rate (IRR) that allows the equality. Although both the EAW and PW approaches are legitimate, the rate of return is generally defined in terms of present worth:

> The rate of return over cost is that rate which when employed in computing the present worth of all costs and the present worth of all the returns will make these two equal. Or, as a mathematician would prefer to put it, the rate which when employed in computing the present worth of the whole series of differences between two income streams (some differences being positive and others negative) will make the total zero.*

> The interest rate at which the present worth of the cash flows on a project is zero.†

Because rate-of-return computations begin with a problem expressed in terms of present worth or annual worth, it is necessary to heed the guidelines for the EAW and PW methods. In particular, alternatives must be compared on the basis of equivalent outcomes. As in the previous discussions of discounted cash flow, we initially investigate the rate-of-return methods without considering the effects of income taxes.

Calculation of IRR

The rate of return for a single proposal is determined by setting the present worth (or EAW) of receipts equal to the present worth (or EAW) of disbursements. Then an interest rate is sought that makes the discounted flows conform to the equality:

Find *i* so that: PW(receipts) = PW(disbursements)

The same relation obviously occurs when the discounted flows are subtracted from each other to equal zero:

For either PW formulation, the calculation of *i* is usually a trial-and-error procedure.

Find *i* so that: PW(receipts) − PW(disbursements) = 0

---

* I. Fisher, *The Theory of Interest,* Kelley and Millman, New York, 1930.
† *The Engineering Economist,* Spring 1972.

When a single proposal is for a cost-reduction project, the receipts take the form of net savings from the method of operation used before the cost-reduction investment.

Example 6.1
Income-Producing Proposal

A parcel of land adjacent to a proposed freeway exit is deemed likely to increase in value. It can be purchased now for $80,000 and is expected to be worth $150,000 within 5 years. During that period it can be rented for pasture at $1500 per year. Annual taxes are presently $850 and will likely remain constant. What rate of return will be earned on the investment if the estimates are accurate?

Solution 6.1

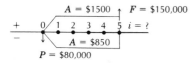

The conditions of the proposal are depicted in a cash flow diagram. The income (positive cash flow) and disbursements (negative cash flow) can be equated according to their equivalent present worths as

$$\$150,000(P/F, i, 5) + \$1500(P/A, i, 5)$$
$$= \$80,000 + \$850(P/A, i, 5)$$

or the positive and negative cash flows can be subtracted as

$$\$150,000(P/F, i, 5) - \$80,000 + \$1500(P/A, i, 5)$$
$$- \$850(P/A, i, 5) = 0$$

which reduces to

$$\$150,000(P/F, i, 5) - \$80,000 + \$650(P/A, i, 5) = 0$$

The value of i that conforms to the equation above is the rate of return on the $80,000 investment. Its value is determined by trial and error.

A quick preliminary check to see if the relation has a positive rate of return results from letting $i = 0$. At $i = 0$,

$$\$150,000 - \$80,000 + \$650(5) = \$70,000 + \$3250$$
$$= \$73,250$$

The positive value indicates that the investment will produce a positive rate of return because the total income is much greater than the outgo. The check also gives a very rough idea of how large the rate of return might be. For instance, the 72-rule suggests that a sum doubles in value every $72/i$ years. Since the $80,000 almost doubles in value in 5 years, i should be near $72/5 = 14.4$ percent.

Letting $i = 15$ percent as the first trial, we have

$$\$150,000(P/F, 15, 5) - \$80,000 + \$650(P/A, 15, 5) \overset{?}{=} 0$$
$$\$150,000(0.49718)$$
$$- \$80,000 + \$650(3.3521) = -\$3244.14$$

The negative value indicates that the trial i used was too large. Now it is known that i lies between 0 and 15 percent.

Letting $i = 14$ percent gives

$$\$150,000(P/F, 14, 5)$$
$$- \$80,000 + \$650(P/A, 14, 5) \overset{?}{=} 0$$
$$\$150,000(0.51937) - \$80,000 + \$650(3.4330) = \$136.95$$

which shows that 14 percent $< i <$ 15 percent. The approximate value of i is determined by linear interpolation from

i	PW
14%	$ 136.95
?	0
15%	-$3244.14

$$\text{Range of } i = 15\% - 14\% = 1\%$$

$$\text{Range of PW} = \$136.95 - (-\$3244.14)$$
$$= \$3381.09$$

The amount by which i is greater than 14 percent is equal to the proportion of the PW range to the point were PW = 0:

$$i = 14\% + 1\%\frac{\$136.95 - 0}{\$3381.09} = 14\% + 1\%(0.041)$$

$$\text{IRR} \doteq 14\%$$

A characteristic worth noting in the previous calculations is that whenever the present worth turns out to be positive, the next trial should employ a higher interest rate to approach the desired zero outcome. Conversely, lowering the interest rate in the present-worth formulation increases the resulting outcome.

A small degree of error is introduced by linear interpolation between interest-table values that are not linearly related. To keep the error as small as possible, interpolations should be conducted between adjacent interest tables. The error is naturally less between lower-interest-rate tables, separated by $\frac{1}{2}$ percent, than at increments of 10 percent for the largest interest rates. For the purposes of this book, interpolated rates of return computed to the nearest tenth of a percent are adequate. The slight error that may be thus introduced will very seldom influence the choice among alternatives; this error is probably much less significant than are actual deviations from the cash flows estimated in the comparisons.

Example 6.2
Cost-Reduction Proposal

Subassemblies for a model IV scope are purchased for $71 apiece. The annual demand is 350 units and is expected to continue for 3 years, at which time the model V scope now under development should be ready for manufacturing. With equipment purchased and installed for $21,000, the production costs to internally produce the subassemblies should be $18,500 for the first year and $12,250 each of the last 2 years. The equipment will have no salvage value. Should the company make or buy the subassemblies?

Solution 6.2

The savings expected in a cost-reduction proposal are treated as income. Assuming the transactions occur at the end of each year, we create the following cash flow diagram:

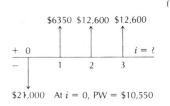

$21,000   At $i = 0$, PW = $10,550

Present annual cost = 350 × $71 = $24,850

Net savings (year 1) = $24,850 − $18,500 = $6350

Net savings (years 2, 3) = $24,850 − $12,250
$$= \$12,600$$

$$PW = -\$21,000 + \$6350(P/F, i, 1)$$
$$+ \$12,600(P/F, i, 2) + \$12,600(P/F, i, 3) = 0$$

Trying successively higher rates of return gives, at $i = 10$ percent,

$$PW = -\$21,000 + \$6350(0.90909) + \$12,600(0.82645)$$
$$+ \$12,600(0.75132)$$
$$= \$4652.62$$

at $i = 15$ percent,

$$PW = -\$21,000 + \$6350(0.86957) + \$12,600(0.75614) \\ + \$12,600(0.65752)$$

$$= \$2333.89$$

at $i = 20$ percent,

$$PW = -\$21,000 + \$6350(0.83333) + \$12,600(0.69445) \\ + \$12,600(0.57870)$$

$$= \$333.78$$

and at $i = 25$ percent,

$$PW = -\$21,000 + \$6350(0.80000) + \$12,600(0.64000) \\ + \$12,600(0.51200)$$

$$= -\$1404.80$$

By interpolation, the rate of return on the $21,000 investment is

$$IRR = 20\% + 5\% \frac{\$333.78 - 0}{\$333.78 - -\$1404.80} \\ = 20\% + 0.96\% = 21\%$$

The answer to the make-or-buy question depends on how large a return the firm expects on its invested capital. Conditions for accepting the proposal to manufacture the subassemblies internally are displayed by the graph in Figure 6.1, wherein the present worth of the proposal is shown as a function of the MARR. For any required rate of return lower than 21 percent, the firm should view the proposal favorably.

FIGURE 6.1
Present worth of cash flow for the proposal in Example 6.2 at different interest rates. At any MARR below 21 percent, the present worth of the proposal is positive and therefore acceptable. For a minimum acceptable rate of return greater than 21 percent, the proposal exhibits a negative present worth and is thereby rejected.

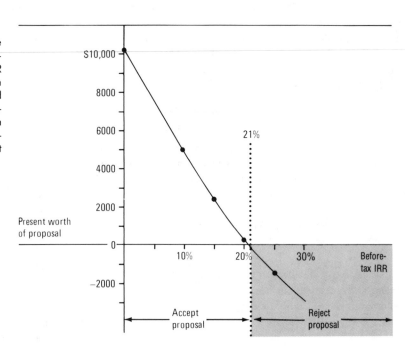

Clues for IRR Calculations

There is no way to avoid the trial-and-error search procedure for determining the IRR in problems with complex cash flows. But a little preplanning may narrow the area of search. An even neater tactic, if available, is to use a computer with a canned rate-of-return program.

The first maneuver to avoid unnecessary computations is to sum the cash flows. A negative total indicates that the proposal being considered cannot meet even an $i = 0$ percent requirement, thereby eliminating it from further consideration whenever a positive IRR is required. The size of a positive net sum, with respect to amount and length of investment, gives a rough suggestion of the rate of return. For instance, the net cash flow at $i = 0$ percent is $10,550 in Example 6.2, and the initial investment is $21,000 for 3 years. This 50 percent return on investment over 3 years suggests a substantial IRR.

A quick proximity fix on the IRR is possible for problems that have their major cash flows at the beginning and end of the study period, or those consisting largely of constant cash-flow streams. When the salvage value is close to 100 percent of the first cost, the net annuity divided by the first cost gives a close approximation to i; that is, $A/P \doteq i$. As demonstrated in Example 6.1, when the total net income at $i = 0$ sums to about twice the initial outlay, $i \doteq 72/N$.

The more variation in cash flows, the more difficult the guessing game gets. Sometimes irregular cash flows can be rounded off to approximate an ordinary annuity, or individual transactions within short time intervals can be lumped together to allow gross preliminary calculations that suggest the vicinity of the IRR. For instance, in Example 6.2 the irregular receipts could be approximated by an average A of, say, $10,000. Then, $A/P = \$10,000/\$21,000 = 0.4761$. With this figure as an entry to an interest table for a capital-recovery factor at $N = 3$, $(A/P, 20, 3) = 0.47473$ gives a good place to begin RR trial computations.

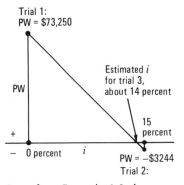

Data from Example 6.1 shows how a rough sketch provides a clue to the IRR.

CONSISTENCY OF THE IRR WITH OTHER ECONOMIC COMPARISON METHODS

In practically all situations, the acceptability of alternative courses of action will be identical whether evaluated according to their annual worth, present worth, or IRR. (Some exceptions are noted in the next section.) The important points to understand are the meaning of the measures of acceptability and the assumptions upon which they are based. Sample calculations applying all three comparison methods to the data in Table 6.1 reveal the consistency of results. They also illustrate how the rate of return is calculated to compare investments when only the disbursements are known and how to interpret the outcomes.

Suppose that a certain function is currently being performed at an annual labor expense of $20,000. One alternative, plan A, is to leave the operation unchanged. In effect, this is the do-nothing alternative which is almost always present in a decision situation. A second alternative, plan B, is to invest $30,000 in layout modifications which will allow the function to be performed at a reduced labor cost of $15,000. The expense of the renovations must be recovered in 10 years according to operating policy. Plan C is a proposal to install a labor-saving device which will cut the labor cost to $12,000. The

TABLE 6.1

Estimated cash flows for alternative operating plans

Year	Plan A	Plan B	Plan C
0		$30,000	$25,000
1	$20,000	15,000	12,000
2	20,000	15,000	12,000
3	20,000	15,000	12,000
4	20,000	15,000	12,000
5	20,000	15,000	37,000
6	20,000	15,000	12,000
7	20,000	15,000	12,000
8	20,000	15,000	12,000
9	20,000	15,000	12,000
10	20,000	15,000	12,000

device has a first cost of $25,000, will be worn out in 5 years, and has no salvage value. Table 6.1 is a year-by-year tabulation of the cash flow for the three plans. If the company's minimum acceptable rate of return is 8 percent, which plan offers the greatest economic benefit?

AW and PW Comparisons

Using the already familiar procedures for computing the equivalent annual worth (cost) of a cash flow stream, we find that the annual cost for the current operating method, plan A, is read directly from the table:

$$\text{EAC(plan } A) = \text{labor expense} = \$20,000$$

In plan B the initial investment is spread over the 10-year study period and added to the annual labor expense to get

$$\text{EAC(plan } B) = \$30,000(A/P, 8, 10) + \$15,000$$
$$= \$30,000(0.14903) + \$15,000 = \$19,471$$

Since the study period constitutes two cycles of the 5-year economic life of the labor-saving device in plan C, the annual cost will be the same over each 5-year period and is equal to

$$\text{EAC(plan } C) = \$25,000(A/P, 8, 5) + \$12,000$$
$$= \$25,000(0.25046) + \$12,000 = \$18,262$$

Thus, plan C, with the lowest annual cost, is preferred. Compared with the currently existing plan A, an investment of $25,000 in a labor-saving device will yield a return of 8 percent per year plus the equivalent receipt of $20,000 − $18,262 = $1738 each year from savings in labor expense for 5 years.

The equivalent present worth (cost) of the three plans is calculated by simply multiplying each EAC by the uniform-series present-worth factor, $(P/A, 8, 10) = 6.710$:

$$PW(\text{plan } A) = \$20,000(6.710) = \$134,200$$
$$PW(\text{plan } B) = \$19,471(6.710) = \$130,650$$
$$PW(\text{plan } C) = \$18,262(6.710) = \$122,538$$

Since a lower present cost is preferred, plan C again gets the nod, as advertised. This means that over a 10-year period when money is worth 8 percent, plan C is expected to cost $\$134,200 - \$122,538 = \$11,662$ less in *today's dollars* to accomplish the same operation now being done under plan A. This total saving is, of course, the present worth of the annual gain beyond the 8 percent return calculated in the EAC comparison:

$$\$1738(P/A, 8, 10) = \$1738(6.710) = \$11,662$$

IRR Comparison

The given data provide examples of IRR comparisons that do not have natural cash flows which are positive. The positive cash flow stream is developed from the savings generated by each additional increment of investment. A more detailed discussion of incremental analysis is presented in the next chapter, but it is sufficient here to understand that *each increment of capital expended must be justified of itself.*

The smallest investment over the do-nothing alternative in plan A is the $\$25,000$ outlay for the labor-saving device in plan C. The "earnings" from this investment are the annual reductions in labor expense: $\$20,000 - \$12,000 = \$8000$. The present worths of plans A and C are equated as first cost to initiate plan C minus present worth of annual savings using plan C equals 0 at IRR.

$$\$25,000 - \$8000(P/A, i, 5) \overset{?}{=} 0$$
$$(P/A, i, 5) = \frac{\$25,000}{\$8000} = 3.125$$

The repeated 5-year cost pattern for plan C leads to a 5-year study period for PW calculations.

which, by interpolation between the 15 percent and 20 percent interest tables, furnishes

$$\text{IRR} = 15\% + 5\%\frac{3.352 - 3.125}{3.352 - 2.991} = 15\% + 5\%\frac{0.227}{0.361} = 18.1\%$$

proving that plan C is an acceptable alternative when the required rate of return is 8 percent.

Although it is already known from the EAC and PW comparisons that plan C is preferred to plan B, it can be checked by determining the IRR for the $30,000 investment in plan B over the do-nothing alternative, plan A:

$$PW(\text{plan } B) = PW(\text{plan } A) \qquad \text{at IRR}$$
$$\$30,000 + \$15,000(P/A, i, 10) \stackrel{?}{=} \$20,000(P/A, i, 10)$$

$$(P/A, i, 10) = \frac{\$30,000}{\$20,000 - \$15,000} = 6.0$$

from which, by interpolation between 10 percent and 11 percent,

$$IRR = 10\% + 1\%\frac{6.1445 - 6.0000}{6.1445 - 5.8892} = 10.6\%$$

It is evident from comparing IRR(plan C) = 18.1 percent to IRR(plan B) = 10.6 percent that plan C deserves its preference.

IRR IRREGULARITIES

The consistency of AW and PW comparisons is above reproach, and both *generally* agree with IRR evaluations. However, there are two situations in which calculations obscure or contradict the preferences shown by the other two comparison methods. Both situations involve distinctive cash flow patterns that provide a clue to possible confusion.

Ranking Reversal

Let two projects have the cash flows indicated in Table 6.2. Both proposals require the same $1000 initial investment. The contrasting net annual returns are conspicuous; project X starts low and increases, whereas project Y has a high first-year flow followed by constant lower flows.

The two projects are first compared by their present worths when the minimum required rate of return is 10 percent:

TABLE 6.2					

Cash flows for two projects with 4-year lives and no salvage value

	End-of-Year Cash Flow				
Project	0	1	2	3	4
X	− $1000	$ 100	$350	$600	$850
Y	− 1000	1000	200	200	200

$$\text{PW}(X) = -\$1000 + [\$100 + \$250(A/G, 10, 4)](P/A, 10, 4)$$
$$= -\$1000 + [\$100 + \$250(1.3810)](3.1698)$$
$$= \$411.56$$
$$\text{PW}(Y) = -\$1000 + [\$1000 + \$200(P/A, 10, 3)](P/F, 10, 1)$$
$$= -\$1000 + [\$1000 + \$200(2.4868)](0.90909)$$
$$= \$361.27$$

This *ranks project X higher than project Y.*

When an IRR comparison is made, the rankings switch, as shown by the following calculation:

For project X,

$$\text{PW} = -\$1000 + [\$100 + \$250(A/G, i, 4)](P/A, i, 4) \gtreqless 0$$

At $i = 20$ percent,

$$\text{PW} = -\$1000 + [\$100 + \$250(1.2742)](2.5887) = \$83.51$$

and at $i = 25$ percent,

$$\text{PW} = -\$1000 + [\$100 + \$250(1.2249)](2.3616) = -\$40.66$$

to give

$$\text{IRR}(X) = 20\% + 5\% \frac{\$83.51 - 0}{\$83.51 - -\$40.66} = 23.4\%$$

For project Y,

$$\text{PW}(Y) = -\$1000 + [\$1000 + \$200(P/A, i, 3)](P/F, i, 1) \gtreqless 0$$

is solved by trial and error to obtain $\text{IRR}(Y) = 34.5$ percent, which *ranks project Y ahead of project X.*

The net present-worth profiles for the two projects are shown in Figure 6.2. Indicated from the previous calculations, the present worth of X exceeds that of Y at $i = 10$ percent, and the internal rate of return is higher for Y than X when PW $= 0$. The point of intersection of the two curves is the incremental rate of return as calculated in Table 6.3.

The intersection is the pivot point for selecting the superior alternative. Select X when the required rate of return is less than 13 percent, and select Y when the required rate is more than 13 percent but less than 35 percent. This selection rule abides by the PW comparison at $i = 10$ percent, which indicated a preference for X. A similar PW comparison at $i = 14$ percent indicates a preference for Y:

FIGURE 6.2
Relationship of net present worth
and different discount rates, showing how the rankings for two projects can switch.

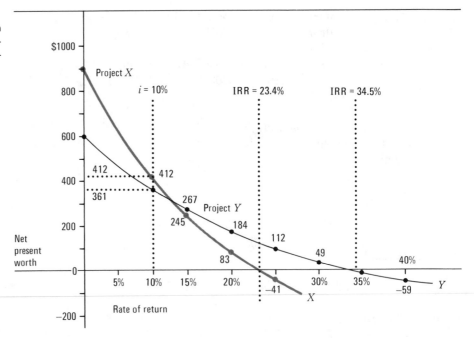

TABLE 6.3

Incremental rate of return that makes project X equivalent to project Y. The
−$900 difference in year 1 leads to additional income the following 3
years, the rate of return being approximately 13 percent per year.

	Project		
	X	Y	X − Y
Year 0	−$1000	−$1000	$0
Year 1	100	1000	−900
Year 2	350	200	150
Year 3	600	200	400
Year 4	850	200	650

PW = −$900 + $150(P/F, 13, 1) + $400(P/F, 13, 2) + $650(P/F, 13, 3) ≑ 0

$$PW(X) = -\$1000 + [\$100 + \$250(A/G, 14, 4)](P/A, 14, 4)$$
$$= \$265.20$$
$$PW(Y) = -\$1000 + [\$1000 + \$200(P/A, 14, 3)](P/F, 14, 1)$$
$$= \$284.49$$

As another example to support the selection procedure, assume that an
investment of $100 returns $200 at the end of 1 year; hence,

$$PW(\$100 \text{ investment}) = -\$100 + \$200(P/F, i, 1)$$
$$= 0 \quad at \; IRR = 100\%$$

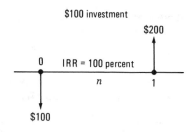

$100 investment

$200

0   IRR = 100 percent

n   1

$100

Let another investment of $1000 return $1500 a year later; hence,

$$PW(\$1000 \text{ investment}) = -\$1000 + \$1500(P/F, i, 1)$$
$$= 0 \quad at \; IRR = 50\%$$

Thus the $100 investment is more attractive according to the internal rate-of-return criterion. However, if the MARR is only 10 percent,

$$PW(\$100 \text{ investment}) = -\$100 + \$200(P/F, 10, 1)$$
$$= \$81.82$$

and

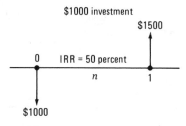

$1000 investment

$1500

0   IRR = 50 percent

n   1

$1000

$$PW(\$1000 \text{ investment}) = -\$1000 + \$1500(P/F, 10, 1)$$
$$= \$363.64$$

which indicates a preference for the $1000 investment. The reason that the $1000 investment is superior may be better understood by considering the IRR on the additional $900 increment of investment:

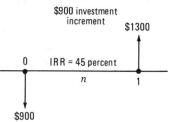

$900 investment increment

$1300

0   IRR = 45 percent

n   1

$900

$$PW(\$900 \text{ increment}) = [-\$1000 - (-\$100)]$$
$$+ (\$1500 - \$200)(P/F, i, 1) = 0 \quad at \; IRR$$
$$= -\$900 + \$1300(P/A, 45, 1) \doteq 0$$

Thus, the additional $900 investment has an internal rate of return of 45 percent, which well exceeds the minimum required rate of 10 percent, making it acceptable and confirming the preference shown by the previous present-worth comparison.

Multiple Rates of Return

When the *cumulative* cash flow of a project switches from negative to positive (or the reverse) *more than once,* the project may have more than one rate of return. In such situations, relatively rare in practice, no single percentage is immediately available to rank the alternative; two or more IRR figures are equally correct, as is demonstrated in Example 6.3.

Cumulative cash flow is the algebraic summation of receipts and disbursements from time 0 to N.

Example 6.3
Two Solutions for an IRR Evaluation

One of the alternatives for improving an operation is to do nothing to it for 2 years and then spend $10,000 on improvements. If this course of action is followed, the immediate gain is $3000 followed by 2 years of breakeven operations.

Thereafter, annual income should be $2000 per year for 4 years. What rate of return can be expected from following this course of delayed action?

Solution 6.3

The cash flow diagram suggests that there might be a sign reversal in the flow pattern. This is confirmed by tabulating

End of Year	Cumulative Cash Flow	End of Year	Cumulative Cash Flow
0	+$3000	4	−3000
1	+3000	5	−1000
2	−7000	6	+1000
3	−5000		

the cumulative cash flow, where it is clear that total transactions reverse from positive to negative at year 2 and again reverse signs at year 6. Since a double sign reversal is not always accompanied by dual rates of return, trials can be conducted at arbitrarily selected interest rates to determine the general PW profile. Based on

$$PW = \$3000 - \$10,000(P/F, i, 2) \\ + \$2000(P/A, i, 4)(P/F, i, 2) \gtreqless 0$$

At $i = 0$, PW is the value of cumulative cash flow over the total study period, shown in the table at year 6, or calculated

as $PW = \$3000 - \$10,000 + \$2000(4) = \$1000$

At $i = 10$ percent,

$$PW = \$3000 - \$10,000(0.82645) \\ + \$2000(3.1698)(0.82645) = -\$25$$

which indicates by the sign reversal from PW at $i = 0$ that one IRR is slightly less than 10 percent.

At $i = 51$ percent,

$$PW = \$3000 - \$10,000(0.43906) \\ + \$2000(1.5856)(0.43906) = \$2$$

where the second sign reversal confirms there is a second IRR.

The profile of present worths over the discounting range of 0 to 51 percent is shown in Figure 6.3. The equivocal answer for the proposal's IRR is that it is either 9.4 percent or 51 percent when returns can be assumed to be reinvested at either rate.

FIGURE 6.3
Net present worth of a proposal with multiple rates of return.

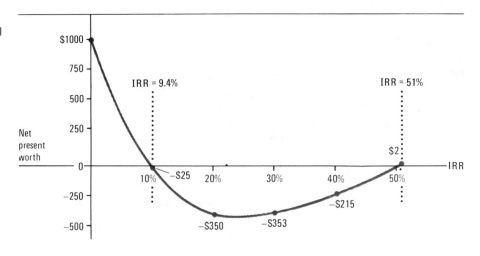

A more specific answer to the IRR question in Example 6.3 is developed by applying an explicit interest rate to a limited portion of the cash flow that will disturb the total cash flow pattern as little as possible while eliminating one of the sign reversals. The *explicit reinvestment* rate may be the minimum attractive rate of return employed by the organization or a rate suggested by the PW profile. The significance of the choice is apparent from comparisons in Table 6.4, wherein the $3000 receipt is assumed to be invested at an explicit interest rate for 2 years. Under this assumption, one sign reversal is avoided and the present worth of the modified cash flow is

An explicit reinvestment rate is a designated interest percentage appropriate for a specific application.

$$PW = \$3000(F/P, i\%, 2) - \$10,000 + \$2000(P/A, i, 4) \overset{?}{=} 0$$

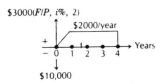

where $i\%$ is the explicit interest rate applied for 2 years.

From Table 6.4 it is apparent that the proposal is attractive only when the explicit reinvestment rate is below 9.4 percent and above 51 percent. When funds can be invested externally at, say, 15 percent, there is no incentive to

TABLE 6.4

Different IRR percentages resulting from explicit reinvestment rates used on a limited portion of the cash flow to convert a dual rate of return to a single IRR

Explicit Reinvestment Rate Applied to the $3000 Payment for 2 Years, %	IRR on Net Investment when an Explicit Reinvestment Rate Is Utilized, %	
0	5.5	
5	7.5	
9.4	9.4	
15	12.4	Range where the project's
20	18.9	IRR is lower than the
30	22.7	explicit interest rate
40	33.4	
51	51	
60	82	

invest in a proposal that returns only 12.4 percent, as does the described alternative when the first $3000 payment is invested at 15 percent. However, if only a 5 percent return can be confidently obtained on external investments, then the $3000 invested at that rate for 2 years would amount to

$$FW(\text{year } 2) = \$3000(F/P, 5, 2) = \$3000(1.1025) = \$3307$$

which is subtracted from the $10,000 disbursement at that date to produce the cash flow, shown below, which has one sign change.

Year	0	1	2	3	4	5	6
Cash flow	0	0	− $6693	$2000	$2000	$2000	$2000

Now the IRR for the modified cash-flow pattern is simply the interest rate that makes an initial investment of $6693 equivalent to a $2000 annual annuity for 4 years:

$$PW = -\$6693 + \$2000(P/A, i, 4) = 0 \qquad \text{at IRR}$$

$$(P/A, i, 4) = \frac{\$6693}{\$2000} = 3.3465$$

and
$$IRR = 7\% + 1\%\frac{3.3871 - 3.3465}{3.3871 + 3.3121} = 7.5\%$$

The use of an explicit interest rate may seem to be an artificial device to alleviate a mathematical difficulty, but it is both realistic and reasonable. Funds received from an ongoing project are indeed reinvested in new projects that have passed the minimum-acceptable-rate-of-return test. These funds then earn at least the required rate, but it would be unrealistic to expect them to earn an enormously higher rate, such as the 51 percent suggested by the sample problem. However, if 51 percent were the actual external reinvestment rate, then the given cash flow pattern would also meet that criterion.

The reasons given here invite the use of the ERR method described in the next section.

EXTERNAL RATE OF RETURN

The occasional occurrence of multiple rates of return is avoided by using the external rate-of-return method. The main appeal of the ERR method, however, is its pragmatic assumption that receipts are actually reinvested at a generally available interest rate. This rate is typically taken to be the MARR. Another advantage of the ERR method is that it can usually be computed directly rather than indirectly by trial and error.

An external rate of return is calculated by equating the future worth of receipts (positive cash flows) compounded at an explicit interest rate to the future worth of disbursements (negative cash flows) compounded at the ERR. If the explicit interest rate is represented by $i\%$, then

FW(receipts compounded at $i\%$) = FW(disbursements compounded at i)

and ERR is the value of i that conforms to the equality. When i % is the MARR and ERR exceeds i %, the investment is attractive because it promises a yield greater than the lower limit of acceptability.

Example 6.4
Example 6.3 Revisited with an ERR Evaluation

Evaluate the cash flow described in Example 6.3 when receipts are reinvested at the MARR of 15 percent.

Solution 6.4

Given i % = 15 percent, the future worth of receipts and disbursements are equated as

$$\$3000(F/P, i\%, 6) + \$2000(F/A, i\%, 4)$$
$$= \$10,000(F/P, i, 4)$$

from which

$$(F/P, i, 4) = \frac{\$3000(2.3130) + \$2000(4.9933)}{\$10,000}$$
$$= \frac{\$6939 + \$9986.60}{\$10,000}$$
$$= 1.69256$$

which, by interpolation, produces an external rate of return of 14.8 percent. Since the ERR is less than the MARR, the investment is unacceptable.

The calculated value of ERR leads to a rejection of the project for a MARR of 15 percent, as happened when the same explicit reinvestment rate was used in the IRR analysis; see Table 6.4. But the explicit interest rate in the IRR calculation was applied differently than it was in the ERR formulation, so results must be interpreted differently. Both methods are useful. The ERR method offers an alternative approach for economic comparisons that is consistent and recognizes realistic limits on the interest rate at which incoming funds can be reinvested.

Review Exercises and Discussions

Exercise 1

A $1000 utility bond with 14 years remaining before maturity can now be purchased for $760. It pays interest of $20 each 6-month period. What rate of return is earned by purchasing the bond at the current market price plus a brokerage charge of $20?

Solution 1

This is a variation of the bond problem in Example 4.8. In the original version the question was how much could be paid for the described bond in order to earn a rate of return of 8 percent compounded semiannually. The solution revealed that the PW of the bond at i = 4 percent per period was $666.76. So we know that the actual rate of return must be smaller than 8 percent compounded semiannually.

Trying i = 5 percent compounded semiannually, or $2\frac{1}{2}$ percent per period for 28 periods, gives

$$PW = \$1000(P/F, 2\tfrac{1}{2}, 28) + \$20(P/A, 2\tfrac{1}{2}, 28) - (\$760 + \$20) \overset{?}{=} 0$$
$$= \$1000(0.50089) + \$20(19.964) - \$780 = \$120.17$$

which indicates that i should be greater than $2\frac{1}{2}$ percent per period. At i = 6 percent compounded semiannually, again for 28 periods,

$$PW = \$1000(P/F, 3, 28) + \$20(P/A, 3, 28) - \$780 \overset{?}{=} 0$$
$$= \$1000(0.43709) + \$20(18.763) - \$780 = \$32.38$$

which is closer, but i is still too low. The present worth of the bond at $i = 4$ percent per period is \$666.76; the difference between this PW and the selling price is \$666.76 − \$780 = −\$113.24. By interpolation between the PWs at 3 percent and 4 percent,

$$i = 3\% + 1\% \frac{\$32.38 - 0}{\$32.38 - -\$113.24} = 3\% + 0.2\% = 3.2\%$$

which means that the bond purchased for \$780 will earn 6.4 percent compounded semiannually.

Exercise 2   An old hotel was recently damaged by a fire. Since it has a desirable location in the old part of the city that is currently being rejuvenated by an urban-renewal project, it will be rebuilt and renovated as either a showroom and office building or a modern apartment building. Estimated receipts and disbursements for the 30-year life of the refurbished structure are shown.

	Offices	Apartments
First cost of renovation	\$340,000	\$490,000
Increase in salvage value from renovation	120,000	190,000
Annual receipts	212,000	251,200
Annual disbursements	59,100	88,000
Present value of fire-damaged building	485,000	485,000
Expected salvage value of the fire-damaged building after 30 years	266,000	266,000

If the required rate of return is 12 percent, which renovation plan is preferable?

Solution 2   Investigating first the lowest-cost alternative, we check to see if an office building will be profitable at $i = 0$.

$$PW = \underbrace{\frac{-(\$485,000 + \$340,000)}{P = -\$825,000}} + \underbrace{\frac{(\$120,000 + \$226,000)(1)}{S = \$386,000}}$$

$$+ \underbrace{(\$212,000 - \$59,100)(30) = \$4,148,000}_{A = \$152,900}$$

Knowing a positive cash flow exists, we find a rough estimate of the IRR evident from the more significant flows of P and A:

$$(A/P, i, 30) \doteq \frac{A}{P} \doteq \frac{\$152,900}{\$825,000} \doteq 0.1853$$

which falls between the 15 percent and 20 percent interest tables. Then, by trial and error,

$$PW = -\$825,000 + \$386,000(P/F, i, 30) + \$152,900(P/A, i, 30) \overset{?}{=} 0$$

At IRR = 18 percent,

$$PW = -\$825,000 + \$386,000(0.00698) + \$512,900(5.5168)$$
$$= \$21,213$$

At IRR = 19 percent,

$$PW = -\$825,000 + \$386,000(0.00560) + \$152,900(5.2478)$$
$$= -\$20,450$$

From these, by interpolation, IRR = 18.5 percent.

The conversion of the fire-damaged hotel into a showroom and office building is thus an acceptable alternative; the 18.5 percent IRR is greater than the required 12 percent.

The alternative plan to convert to an apartment has incremental additional values of

First cost: $\$490,000 - \$340,000 = \$150,000$
Salvage value: $\$190,000 - \$120,000 = \$70,000$
Net annual returns: $\$251,200 - \$88,000 - \$152,900 = \$10,300$

The incremental rate of return is calculated as

$$PW = -\$150,000 + \$70,000(P/F, i, 30) \qquad + \$10,300(P/A, i, 30) \geq 0$$

At IRR = 6 percent,

$$PW = -\$150,000 + \$70,000(0.17412) + \$10,300(13.764)$$
$$= \$3958$$

At IRR = 7 percent,

$$PW = -\$150,000 + \$70,000(0.13137) + \$10,300(12.409)$$
$$= -\$12,991$$

By interpolation,

$$IRR = 6\% + 1\% \frac{\$3958 - 0}{\$3958 - (-\$12,991)} = 6.2\%$$

The IRR lower than the required 12 percent rate disqualifies the additional investment needed to proceed from the office plan to the apartment plan. It should be noted that the apartment plan still has a total IRR greater than the minimum required 12 percent:

$$PW = -\$975,000 + \$456,000(P/F, i, 30) + \$163,200(P/A, i, 30) \geq 0$$

At IRR = 16.7 percent,

$$PW = -\$975,000 + \$456,000(0.00972) + \$163,200(5.9298)$$
$$= -\$2834$$

At IRR = 16.6 percent,

$$PW = -\$975,000 + \$456,000(0.00998) + \$163,200(5.96399)$$
$$= \$3178$$

Even with a 17 percent rate of return, the apartment plan is not an acceptable alternative because the additional investment required for its implementation does not meet the minimum rate-of-return standard. The results are summarized in the table.

	Office Plan ⟶	Increment ⟶	Apartment Plan
First cost	$825,000	$150,000	$975,000
Salvage value	$386,000	$70,000	$456,000
Annual returns	$152,900	$10,300	$163,200
Rate of return	18.5% ⟶	6.2% ⟶	16.7%

Exercise 3   Expected cash flows for a strip-mining project are estimated as shown in the cash flow diagram.

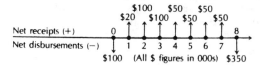

A start-up cost is incurred immediately. Then income exceeds outlays for the next 7 years. During the eighth year the major cost is for landscape improvement. Does the strip-mining project appear to be a profitable investment?

Solution 3   The − to + to − cash flow pattern suggests dual rates of return. The suspicion is verified by trial-and-error calculations based on

$$PW = \$100,000 + \$20,000(P/F, i, 1) + \$100,000(P/F, i, 2) + \$100,000(P/F, i, 3) \\ + \$50,000(P/A, i, 4)(P/F, i, 3) - \$350,000(P/F, i, 8) \gtreqless 0$$

to identify the IRRs.

End of Year	Cash Flow	PW AT 2.9%		PW AT 44.5%	
		Factor	Amount	Factor	Amount
0	− $100,000	1.00	− $100,000	1.00	− $100,000
1	20,000	0.97	19,000	0.69	14,000
2	100,000	0.94	94,000	0.47	47,000
3	100,000	0.92	92,000	0.33	33,000
4	50,000	0.89	45,000	0.22	11,000
5	50,000	0.86	43,000	0.15	7,500
6	50,000	0.83	42,000	0.10	5,000
7	50,000	0.81	40,000	0.07	3,500
8	− 350,000	0.78	− 275,000	0.06	− 21,000
Net PW			0		0

A more comprehensive picture of the proposed strip-mining venture is provided by the discounted income and cost curves shown in Figure 6.4. It is apparent that the project can be considered profitable only when the interest rate used in the present-worth calculations is below 3 percent or above 45 percent. This means that the strip-mining project will be attractive only when the organization is satisfied with a modest 3 percent

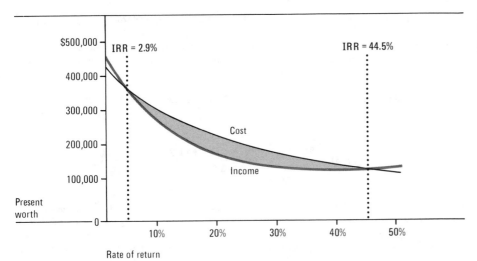

FIGURE 6.4
Present worths of income and cost
streams at different rates of return.
The curves intersect where net PW
= 0.

rate of return or is optimistic enough to accept the premise of a reinvestment rate of 45
percent or greater. Between the points where the curves intersect in Figure 6.4, the PW
is negative. Thus, a minimum attractive rate of return of 10 percent rejects the project
when it is used as an external reinvestment rate to remove the sign reversal in the cash
flow, as demonstrated below.

Let i = 10 percent to compound the values of the first two transactions ($-$100,000
and $20,000); the cash flow diagram is revised as

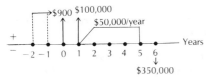

When

$$FW(\text{year } 0) = \$100,000 + \$20,000(F/P, 10, 1) - \$100,000(F/P, 10, 2)$$
$$= \$900$$

then, at year 0,

$$\$900 = \$100,000 + \$20,000(F/P, 10\%, 1) + -\$100,000(F/P, 10\%, 2)$$
$$= \$100,000 + \$22,000 - \$121,000$$

to reveal that the net PW is negative even when i = 0 percent:

$$PW = \$900 + \$100,000(P/F, 0, 1) + \$50,000(P/A, 0, 4)(P/F, 0, 1)$$
$$- \$350,000(P/F, 0, 6)$$
$$= \$900 + \$100,000 + \$200,000 - \$350,000 = -\$49,100$$

Exercise 4   Calculate the ERR for the cash flow in Example 6.2 when $i \% = $ MARR $= 20$ percent.

Solution 4   Given receipts (savings) of $6350, $12,600, and $12,600 at the end of years 1, 2, and 3, respectively, and an initial disbursement (investment) of $21,000 at time 0, the future worths of the positive and negative cash flows are equated as

$$\$6350(F/P, 20, 2) + \$12,600(F/P, 20, 1) + \$12,600 = \$21,000(F/P, i, 3)$$

from which

$$(F/P, i, 3) = \frac{\$6350(1.4400) + \$12,600(1.200) + \$12,600}{\$21,000}$$
$$= 1.7554$$

that produces an $i = $ ERR $= 20.1$ percent. This ERR makes the project barely acceptable when MARR $= 20$ percent. The same conclusion was reached using the IRR method. The closeness of the two rates of return results from the explicit reinvestment rate (MARR) being only slightly less than the 21 percent reinvestment rate implied by the calculated IRR of 21 percent.

PROBLEMS

6.1   Sometimes objects of art are respectable investments. In 1975 a marble bust of Benjamin Franklin, from which the engraving was made for $100 bills, was auctioned for $310,000. It was sculpted in France in 1778 by Jean-Antoine Houden. In 1939 the same bust sold for $30,000. What rate of return was earned by the collector who owned the statue from 1939 to 1975?

6.2   A $5000 bond matures in 10 years and pays 3 percent interest twice a year. If the bond sold for $5050, what is the actual investment rate?

6.3   A construction firm can lease a crane required on a project for 3 years for $180,000 payable now, with maintenance included. The alternative is to buy a crane for $240,000 and sell it at the end of 3 years for $100,000. Annual maintenance costs are expected to be $5000 the first 2 years and $10,000 the third year (payable at the end of each year). At what interest rate would the two alternatives be equivalent?

6.4   Additional parking space for a factory can be either rented for $6000 per year on a 10-year lease or purchased for $90,000. The rental fees are payable in advance at the beginning of each year. Taxes and maintenance fees will be paid by the lessee. The land should be worth at least $60,000 after 10 years. What rate of return will be earned from the purchase of the lot?

6.5   Proposal 1 has an initial cost of $1500 and a positive cash flow that returns $200 the first year and increases by $200 each of the following years until the end of the 5-year study period. Proposal 2 also has a 5-year life and an initial cost of $1500. Its positive cash flow is constant at $200 for the last 4 years. It also has another receipt in year 1. All receipts occur at the end of the year.
  (a) What is the rate of return on proposal 1?
  (b) If the two proposals are equally attractive at $i = 14$ percent annually, how large is the first payment in proposal 2?

6.6 In 1980 Sam Smart purchased a small apartment building for $110,000. Receipts from rent have averaged $19,200 a year; taxes, maintenance, and repair costs have totaled $5600 annually. Sam intends to hold the property until he retires in 1990. If at that time he sells the property for $110,000, what rate of return will he obtain on the investment?

6.7 An investor has an opportunity to purchase ten $1000 bonds paying interest at a nominal rate of 8 percent with semiannual payments. The bonds are currently available for $7200 and will mature in 15 years. What effective annual rate of return will this investment provide?

6.8 Stock in the Gonzo Corporation was purchased 10 years ago for $80 per share. For the first 6 years, the stock paid annual dividends of $11 per share and the market price climbed to $120. However, for the past 4 years annual dividends have been only $5 per share and the price of the stock has dropped to $100.

   (a) What rate of return would have been obtained by an investor who sold the stock at the end of the first 6 years?
   (b) What rate of return would be obtained if an investor purchased the stock 4 years ago and sold it today?
   (c) What rate of return would be obtained by an investor who purchased the stock 10 years ago and sold it today?

6.9 A company can purchase a new desk-top laser printer for $7500 or lease the machine for 3 years with annual payments of $3000. Determine at what interest rate leasing and purchasing costs would be equivalent.

   (a) If lease payments were due at the first of each year.
   (b) If lease payments were due at the end of each year.

6.10 Determine the annual effective interest rate at which the lease and purchase costs in Problem 6.9 would be equivalent if a payment of $250 were due at the first of each month over the 3-year period.

6.11 An investor has an opportunity to purchase a commercial rental property for $300,000. The current occupants have signed a 10-year lease at a constant annual rent of $48,000. Maintenance costs and taxes on the structure are currently $12,000 and are expected to increase at a rate of $1500 per year over the 10-year period. Assuming that the property can be sold for at least the purchase price when the current lease expires, determine the investor's minimum expected rate of return.

6.12 Bubblewater Brewery is financing an expansion through the sale of $1,000,000 of 9 percent 20-year bonds on which interest is to be paid semiannually. Printing costs and other expenditures involved with issuing the bonds have totaled about $30,000, and clerical and accounting expenses are expected to be approximately $5000 each dividend period. A bond brokerage firm has bid $950,000 for the bonds and Bubblewater has tentatively accepted the bid.

   (a) What effective annual interest rate represents the cost of this money to the brewery?
   (b) The brokerage firm hopes to sell the bonds to private investors at a discounted value of $970,000 within 1 month. If they are able to do this while incurring direct costs of $5000, what nominal annual rate of return do they obtain?
   (c) What rate of return will the private investors in Problem 6.12b obtain if they hold the bonds until maturity?

6.13   A student renting an unfurnished apartment has decided to purchase some furniture from Fred's Fine Furniture. The total purchase price of the three-room set is $495. However, after a down payment of $95 Fred will finance the balance through a 2-year series of end-of-the-month payments of $19.98. Determine the nominal and effective annual rates of interest paid by the student.

6.14   The student in Problem 6.13 has an opportunity to purchase a magazine entitled *Popular Furniture Repair*. Subscriptions, payable in advance, are as follows:

1 year	$10
2 years	$18
3 years	$26

Assuming that the student wishes to read the magazine for at least the next 3 years, determine which subscription plan should be selected if a 20 percent rate of return is desired on the investment?

6.15   Two alternative investment proposals are under consideration for a vacant lot owned by the Urban Development Corporation. Plan A would require an immediate investment of $120,000 and a first-year expenditure for property taxes, maintenance, and insurance of $4000, with this amount expected to increase at a rate of $1000 per year. Plan B would have a first cost of $170,000 and total first-year expenses of $9000, with an increase of $1000 per year. The economic life of each project is forecast to be 10 years; and at the end of this time, only the facilities from plan B with a value of $50,000 are expected to be salvaged. During the life of the project, the facility in plan A is expected to produce $34,000 annually, whereas plan B is expected to produce $42,000.

(a) Determine the rate of return of each plan.
(b) Determine the rate of return of the additional investment required in plan B compared with plan A.
(c) Which plan should urban development select if the company uses a minimum attractive rate of return of 12 percent?

6.16   Two types of productivity programs are being considered for funding. Both have an initial cost of $10,000 for training equipment and consulting contracts. Program A promises to produce constant net revenues of $4000 per year for 5 years. Net revenues from program B are expected to be $10,000 the first year and $2000 per year for the next 4 years. All revenues are considered end-of-year receipts.

(a) Which program is preferable at IRR = 10 percent?
(b) Which program is preferable at IRR = 20 percent?
(c) Draw a graph of PW versus IRR for the two proposals and state a decision rule for selecting between the two proposals.

6.17   You have loaned $1000 to a friend whom you consider a good credit risk at a nominal interest rate of 12 percent compounded monthly. The loan period is 1 year. You plan to take each monthly installment received from your friend and invest it on the day received in a savings account that pays you interest at the nominal rate of 6 percent compounded monthly. What nominal interest rate are you receiving on the total return from the loan plus the reinvestment of the monthly repayment installments? All installment transactions occur at the end of each period.

6.18   Two mutually exclusive projects are being considered: Project X requires $500 and results in a return amounting to a one-time-only profit of $1000 five years from now. Project Y also requires $500, but will return $170 per year for each of the next 5 years. Determine at what IRR the two projects are equivalent and explain the significance of the intersection point of the two PW profiles for the selection of the preferred alternative.

6.19   A 5-year subscription can be purchased today for $48. One-year subscriptions, payable in advance year by year, could be purchased to provide the same number of issues. The current price of a 1-year subscription is $10, but it will probably increase in price at the rate of $2 per year. What IRR is earned by the 5-year subscription?

6.20   The cash flow for a project is shown in the table below.

Year	0	1	2	3	4	5
Cash flow	$3000	$1000	− $5000	− $5000	$2000	$5000

The first two payments (in thousands of dollars) represent advance payments for distribution rights on a motion picture. The next 2 years show a negative cash flow from production costs, and the last 2 years are net receipts from the finished picture. If advance payments can be invested at the external interest rate of 7 percent, what rate of return can be expected from the project?

6.21   Owing to perennial complaints by students and faculty about the lack of parking spaces on campus, a parking garage on university-owned property is being considered. Since there are no university funds available for the project, it will have to pay for itself from parking fees over a 15-year period. A 10 percent minimum rate of return is deemed reasonable for consideration of how large the structure should be. Based on the income and cost data shown below, determine how many levels should be built.

Number of Levels	Cumulative Construction Costs	Annual Operating Cost	Income per Year
1	$ 600,000	$35,000	$100,000
2	2,200,000	60,000	350,000
3	3,600,000	80,000	570,000
4	4,800,000	95,000	810,000

6.22   A business property can be purchased today for $90,000; the expected resale value after 20 years is $60,000. If annual rental income is $11,800 and expenses are $4700, what before-tax rate of return would be earned by purchasing the property?

6.23   A cash flow pattern shows an income of $250 at the end of year 1 between expenditures of $100 now and $156 at the end of year 2.
   (a) Calculate the dual rates of return by trial and error.
   (b) Use the quadratic formula to solve for the two rates. As a reminder, the quadratic formula is $(-b \pm \sqrt{b^2 - 4ac})/2a$, and the coefficients in this case are $a = -100$, $b = 250$, and $c = -156$.

6.24   The owner of a truck-weighing and lumber-scaling station has agreed to lease the facility for 15 years at $10,000 per year under an agreement that the scales and other

TABLE 6.5

Present worth of a secondary oil-recovery project at different discount rates (cash flows are in thousands of dollars)

Year	Cash Flow	Present Worth of Cash Flow at						
		10%	20%	28%	30%	40%	49%	50%
1	$ 200	$ 182	$167	$156	$154	$143	$134	$133
2	100	83	69	61	59	51	45	44
3	50	38	29	24	23	18	15	15
4	−1800	−1229	−868	−671	−630	−469	−365	−356
5	600	373	241	175	162	112	82	79
6	500	282	167	114	104	66	46	44
7	400	205	112	71	64	38	24	23
8	300	140	70	41	37	20	12	12
9	200	85	39	21	19	10	5	5
10	100	39	16	8	7	3	2	2
Total	$ 650	$ 198	$ 42	0	−$ 2	−$ 8	0	$ 1

equipment will be overhauled and repaired by the owner at the end of the eighth year at a cost not to exceed $150,000.

  (a) What rate(s) of return will the owner receive for the station lease with the equipment-repair agreement?

  (b) After negotiations on the above lease, caused by concern that the equipment needed overhauling before 8 years, it was agreed that the owner would pay up to $150,000 for repairs at the end of year 4 instead of making the repairs at the end of year 8. What rate(s) of return will the owner receive under the revised agreement?

  (c) Check the revised agreement to see if it is acceptable at an IRR of 20 percent. Use the 20 percent as an external reinvestment rate to determine a single IRR for the lease and repair agreement in Problem 6.24b. How would you explain the expected IRR to the owner?

6.25 An interesting article about the reasons why Continental Oil Company switched in 1955 to the discounted-cash method for evaluating investments was written by John G. McLean, then vice-president for international and financial operations.* One of the applications described was a water-flood project that exhibited dual rates of return.

  The problem was to determine the profitability of acquiring a small oil-producing property in which the primary reserves were nearly exhausted. The owner of the property would receive a royalty of $12\frac{1}{2}$ percent on all oil produced from the property, and the company would agree to water-flood the reservoir at an expected cost of $2.5 million. The injection of water into a reservoir is a method of "secondary recovery" that often increases the total amount of oil recovered after the free-flowing oil supply has diminished.

  Estimated cash flows for the 10-year project are shown in the second column of Table 6.5. The primary reserve of oil yields returns the first 3 years. The water flood is then

---

* "How to Evaluate New Capital Investments," *Harvard Business Review*, November-December 1958.

expected to boost the company's income to $700,000 in the fourth year while it invests $2.5 million, for a net outlay of $1,800,000 that year. Thereafter, income decreases annually by $100,000. The present worths of the yearly cash flows are indicated for different discount rates. At $i = 28$ percent and $i = 49$ percent, the present worth of the cash flow is zero. Between these rates the present worth of the venture is negative.

Instead of settling for two rates of return, assume that the cash flows from the first 3 years can actually be reinvested at an annual interest rate of only 15 percent. What is the IRR on the resulting net investment when this portion of the total is reinvested at the external 15 percent rate? How would you explain your solution to a group of investors not very familiar with discounted cash flow analysis?

EXTENSIONS

6A Theory and Practices behind Interest Rates

A minimum attractive rate of return sets a price on money. Prices are rationing devices. The rate (or price) an organization sets for the use of its money reflects its internal economic conditions, such as the amount of capital accumulated from successful operations, and external factors such as the total supply of loanable capital in the whole economy and current or anticipated government fiscal policies.

A supply-and-demand relationship is responsible for a "market" rate of interest under traditional capital theory. This single theoretical rate is the percentage return yielded by any riskless bond or other security. Since capital is assumed to be subject to the law of diminishing returns, its demand curve D takes the general shape shown in Figure 6.5. At a certain time, the demand curve is intersected by an inelastic supply line S that represents the amount of capital then available. The intersection point is the temporary interest rate at which projects yielding that rate of productivity (measured as a percentage) are funded. This short-run

equilibrium point implies that all projects with greater net productivity have already been funded and those that will yield a lower rate must await the accumulation of additional capital. The graph also suggests that interest rates will decline as the pool of capital increases; this contention has not been supported by recent history.

The relationship of supply and demand in Figure 6.5 is probably more representative of an individual firm than of the economy as a whole. Shifting expectations, technological innovations, inflation, and other factors impinge on traditional capital theory to thwart the stability of conceived investment relations. Yet the theory does offer a simplified explanation of the purpose served by interest rates. Money is "hired" by businesses to buy capital goods. The productivity of these goods must at least pay the "wages" of the money used to acquire the goods, and attaining this wage level sets a lower limit on accepting investment proposals.

Although it is convenient to theorize a single interest rate for the whole economy, there exist, of course, many

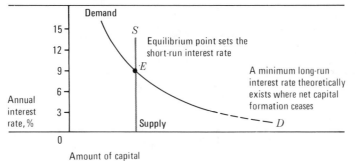

FIGURE 6.5
Traditional capital theory suggests that a current rate of interest is a function of the amount of capital available and the productivity of capital expected at that level of supply.

different interest rates. Reasons for the differences among loans include:

1 Higher interest charges are applied when there is greater risk that a loan will not be repaid.
2 Charges for long-term loans are usually greater than those for short-term loans because a lender forgoes the opportunity to take advantage of alternative uses of the money for a longer period and incurs greater risk from potential changes in the economic environment.
3 Administrative costs of lending are a higher percentage of the loan value of a smaller than a larger loan.
4 Local money markets vary, owing to regional differences; interest rates are often higher in small towns than in large cities because borrowers find it less convenient to shop for better rates.

Whatever theory one adopts to explain how interest rates depend on the supply of and demand for capital, no one can deny the dominant influence of government on those rates. In effect, an interest rate is an administered price. Fiscal policies of government affect the supply of capital through the power to issue money and the demand through massive borrowing. The reason government purposely influences interest rates is to direct the economy toward national priorities. For instance, other things being equal, a lowering of interest rates will stimulate business expenditures and create more employment. However, other things are seldom equal, so fiscal policies do not always achieve the desired objectives in themselves. Interacting forces in the national economy (and indirect international influences) are awesomely complex; money management is only one contributor.

An indication of the extent of interest rate fluctuations is given in Figure 6.6.

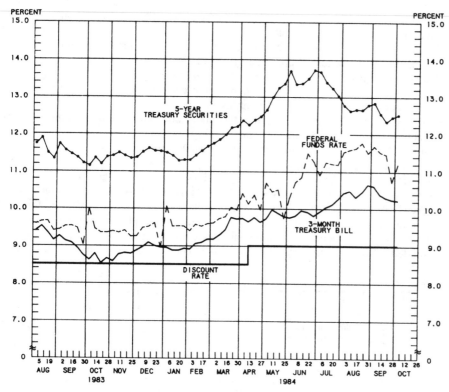

FIGURE 6.6
Pattern of variations in interest rates in the United States during a 15-month period ending October 5, 1984. (Source: *U.S. Financial Data*, prepared by Federal Reserve Bank of St. Louis, Missouri.)

QUESTION

6A.1 Determine the interest charges or rate of return earned in your locality for each of the following loans or investments. Discuss reasons for the differences.

- $10,000 money-rate certificate for 6 months

- $10,000 certificate of deposit for 30 months

- $1000 government bond that matures in 20 years

- $1000 Treasury note due in 90 days

- $4000 borrowed from an auto-maker's credit corporation to buy a new car, with repayment over 3 years

- $4000 borrowed from a local commercial bank to buy a new car, with repayment over 3 years

- $30,000 FHA mortgage loan for a new house, with repayment over 20 years

- $50 pawnshop loan for 1 month

- $300 borrowed from a small-loan company to be repaid in 6 months

6B   Truth in Lending

Consumer credit reforms of the 1960s culminated with truth-in-lending legislation passed in 1969. It took the pro-consumer element in Congress almost a decade to overcome the opposition of creditors who feared government intervention and were reluctant to advertise true interest rates.

Typical disclosure to satisfy the federal truth-in-lending act: This loan is repayable in 24 monthly installments of $28.71 each. Amount financed is $500.13. Finance charge: $188.91. Annual percentage rate: 32.87%. Total of payments $689.04 (per schedule).

Before truth in lending, creditors had no obligation to explain how interest charges were determined. It was usually difficult for consumers to compare financing costs because installment purchases had differing terms and the language of the interest charges was vague, at best.

When interest rates were quoted in the pre-truth-in-lending era, deceptive "add-on" quotes were often used. For instance, a $1000 purchase might have quoted interest charges of 12 percent for 2 years. But the 12 percent would be charged on the full $1000 amount—$120 each year and $240 for the 2 years—rather than on the unpaid balance. A contract would be signed to pay back $1240 for the $1000 loan, in 24 monthly payments of $51.67. The true interest is, as any engineering economist would know, almost 24 percent.

After truth in lending became effective, interest charges had to be quoted on an annual basis—the effective interest

rate. Several examples of potentially misleading loan arrangements are provided in the following questions.

QUESTIONS

6B.1 A gullible borrower went to the Fast-Buck Loan Company expecting to borrow money at the advertised rate of "one percent per month." He agreed to repay a loan of $500 in 12 equal end-of-month payments of $47.50 each. The amount of the monthly payments was calculated as follows:

Principal to be repaid (amount borrowed)	$500
Interest at 1%/month (12 months × 0.01 × $500)	60
Credit investigation (check on credit rating)	10
Total cost of loan	$570

Monthly payments = $\dfrac{\$570}{12 \text{ months}}$ = $47.50/month

What effective interest rate was charged?

6B.2 A prepaid interest plan has the interest deducted from the amount of the loan. For a 1-year loan of $1000 at a $1\frac{1}{2}$ percent interest charge per month, the interest of 12 months × 0.015 × $1000 = $180 is deducted from the amount loaned to give the borrower only $1000 − $180 = $820 to spend. If monthly payments are $1000/12 = $83.33, what is the actual nominal interest rate?

6B.3 "Balloon" payments occur when the last loan-repayment charge is exceptionally large compared with the constant payments up to the final one. This repayment plan may not be as troublesome with respect to unusually high interest charges as with respect to the difficulty of making that last balloon payment; inability to meet the final payment may lead to repossession of the item purchased or the necessity to refinance, often at a higher interest rate. For example, a loan of $1000 for a used car could lead to a repayment plan of $60 per month for 11 months with a final year-end payment of $460: that is, $460 + $60 × 11 = $1000 + 0.01 × $1000 × 12. What effective interest rate is charged? Discuss why such plans might be offered and why they may be accepted by the borrower.

6B.4 Sometimes it is possible to buy an item for less by paying the purchase price in one lump sum because the seller prizes immediate cash over a time contract for a series of future payments. Assume that you can purchase a $10,000 building lot on a 6-year contract with annual payments at 7 percent compounded annually on the unpaid balance. As you consider the purchase, you learn that someone else purchased an essentially identical lot from the same seller for $8500 in cash. In effect, what total interest rate are you paying if you elect to make the purchase on a 6-year contract?

SECTION TWO

ECONOMIC ANALYSES

Economics is reputed to be an academic, not a vocational, subject, whereas engineering is a practical subject that applies science to workaday problems. Combining economics with engineering blends theory with practice. It leads to the application of economic principles to the practical problems of efficiently allocating and utilizing scarce resources. It is the foundation for deciding which engineering proposals are financially worthy of support. Economic principles ignored in decision making are wasted, but unprincipled economic decisions are criminal.

An engineering economic analysis can be likened to food for the decision maker. It nourishes healthy decision making. A recipe for its preparation could be

- Gather facts and sift well to eliminate extraneous data.

- Blend facts with financial practices.

- Carefully stir according to established methodologies.

- Season with creativity to recognize unusual relations.

- Garnish with graphics and serve with confidence.

The menu in this section includes the structural analysis of alternatives, replacement analysis, financial analysis, and benefit-cost analysis. Each form of analysis is associated with a particular class of problems likely to confront an engineering economist.

* Campbell R. McConnell, *Economics*, 6th ed., p. 2, McGraw Hill, New York, 1975.

CHAPTER 7

STRUCTURAL ANALYSIS OF ALTERNATIVES

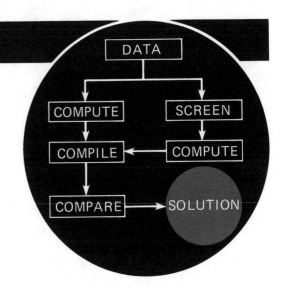

OVERVIEW

From previous chapters we know that alternatives can be compared according to their present worths, equivalent annual worths, and internal rates of return. The selection of a preferred proposal from a set of alternatives depends on the structure of the set.

Given proposals A and B and unlimited capital, the options are to select any, all, or none, when the alternatives are *independent*—i.e., when acceptance of one has no effect on the acceptance of any other alternative in the set.

When selection is narrowed to the do-nothing option, or A, or B, the alternatives are *mutually exclusive*. The alternative with the highest present worth based on the MARR is preferred. An identical choice results from applying an incremental IRR comparison; the preferred alternative is the largest total investment in which each additional increment of capital above the last acceptable investment level has an incremental IRR greater than the MARR.

When alternatives are related to each other in any way that influences the selection process, they are said to be *dependent*. For instance, proposal B would be dependent if it could be selected only *after* proposal A was accepted; then the choice would be limited to doing nothing, selecting A, or accepting A and B. The case of mutually exclusive reinvestment-dependent alternatives is discussed in Extension 7A.

DEVELOPMENT OF ALTERNATIVES

An *alternative* in engineering economics is an investment possibility. It is a single undertaking with a distinguishable cash flow. Alternatives vary from doing nothing—leaving the existing cash flow intact—to very elaborate and lengthy projects. It is necessary to understand both the composition of an alternative and its structural relation to other investment options.

Identifying Alternatives

Many people besides engineers develop and evaluate investment alternatives. However, few other professions are so intimately and regularly involved with selecting the best device or way to do a particular operation. And the operation usually has long-term consequences. That is why it is important for engineers, and others engaged in similar decision situations, to be competent economic analysts.

Analysis starts with the identification of alternatives. A need to do something originates from asking, "What must be done?" "What can be done?" "What should be done?" A general idea for an undertaking evolves into a family of alternatives through further questioning on the feasibility of alternative solutions and of how to do them and when. Answers to such probes may suggest several ways and means to accomplish the same mission, or other missions that deserve attention.

A balance was suggested in Example 1.2, wherein the cost of a less-than-ideal solution was weighed against the cost of making the decision.

Perhaps the most bothersome question in the search for alternatives is: "How many is enough?" Decision making can be paralyzed by a continuing search for a still better option. At the other extreme, acting on the first option that comes to mind may ensure a fast decision but seldom a wise one. The governing objective is to develop a set of alternatives large enough to include the best possible solution.

Defining Alternatives

From previous discussions of comparison methods we know that the basic data required for an economic analysis are the timing and amount of cash flows. These come from an understanding of the present situation and comprehension of future situations. Besides the current MARR and today's cost commitments, knowledge is needed about future objectives, resources, and operational constraints.

Proposals for investments typically exceed available funding. Some proposals can be sliced from the list of alternatives owing to their excessive size. Others may be suspect from their assumptions about future resources, especially when their attraction stems from unproven technology or "iffy" cost estimates. Part of this uncertainty can be handled by techniques presented in Chapters 16 and 17, but for now all the data will be considered reliable. Therefore, cash flows are treated as certain to occur as estimated, and financial relations among alternatives are assumed to be known.

Alternatives can be classified as *independent, mutually exclusive,* or *dependent.* The classification establishes how a proposal is evaluated in the selection process.

CLASSIFICATION OF ALTERNATIVES

On the first pages of this text the role of engineering economists was defined to include the identification of alternatives and analysis of their worth. Since then, alternative courses of action have been evaluated for many different situations. Alternatives that share certain features can be categorized together to facilitate analysis. The alternatives in Table 7.1 will be utilized to illustrate the classifications; for computational convenience all proposals have the same life and no salvage value.

An *independent alternative* is not affected by the selection of another alternative. Each proposal is evaluated on its merit and is approved if it meets the criteria of acceptability. Comparisons of independent investment proposals are designed to determine which proposals satisfy a minimum level of economic value. All those that surpass the minimum level may be implemented so long as sufficient capital is available. For example, if the proposals in Table 7.1 were independent and the criterion of acceptability was a return of 5 percent on any investment, proposals *N, W,* and *S* would be satisfactory investments because they possess positive net PW at the required interest rate.

Alternatives are *mutually exclusive* when the selection of one eliminates the opportunity to accept any of the others. Most of the comparisons made in previous chapters were among mutually exclusive alternatives. Operational problems normally fit into this category, because a single course of action is sought to solve a particular, often urgent, problem. When the best solution is determined, the problem is theoretically resolved by implementing the indicated course of action.

Multiple alternatives are often associated with levels of development for

TABLE 7.1

Cash flows for proposals having the same life and no salvage value

Proposal	Investment	Life	Net Annual Cash Flow	Net PW at $i = 5\%$
N	−$1000	5	+$300	+$299
E	−2000	5	+400	−268
W	−3000	5	+900	+896
S	−4000	5	+1000	+329

When the cash flows for *any combination* of alternatives in a set of proposals can be *freely combined*, the proposals are said to be *mutually independent*.

the same asset or activity. Each level is an alternative, and the purpose of the analysis is to select the most promising level. For instance, the proposals in Table 7.1 would be considered mutually exclusive if they represented different sizes of the same design (e.g., diameters of pipes or thicknesses of insulation) or incremental levels of resource application (e.g., number of crews assigned or number of units ordered).

Let the proposals in Table 7.1 represent investments in insulation to reduce heat loss (equivalently, to conserve energy) and the cash flows the savings from various thicknesses compared with the cost of heating without insulation. The net present worth of proposal W,

$$PW(W) = -\$3000 + \$900(P/A, 5, 5)$$
$$= -\$3000 + \$900(4.3294) = -\$3000 + \$3896 = \$896$$

is greater than that of any of the other proposals and indicates that the thickness of insulation it represents is the one to use.

It is often instructive to observe the incremental returns for mutually exclusive alternatives. Savings resulting from each additional $1000 of investment are indicated in Table 7.2.

It appears that proposal W is again far superior. However, the increment of investment from $N(I = \$1000)$ to $W(I = \$3000)$, from one acceptable level of investment to the next *acceptable* level, increases the savings from $300 to $900 per year to produce a ratio of

$$\frac{\Delta S}{\Delta I} = \frac{\$900 - \$300}{\$3000 - \$1000} = \frac{\$600}{\$2000} = \$300/\$1000$$

which is identical to the return earned on N. This means that money invested in alternative N earns at the same rate as an investment in W. The only reason W would be preferred to N is that a larger investment at a given rate of return earns a larger total amount, ignoring other influencing factors such as the future availability of energy. The actual percentage rate of return earned on incremental investments will be discussed in the next section.

Occasionally, individual investment opportunities are linked to other

---

TABLE 7.2

Incremental annual savings in relation to increments of investment

Levels of Investment Proposals	Increment of Investment, ΔI	Incremental Savings per Year, ΔS	$\Delta S/\Delta I$
No investment to N	$ 0–$1000	$ 0–$ 300	$300/$1000
Investment N to E	1000– 2000	300– 400	100/ 1000
Investment E to W	2000– 3000	400– 900	500/ 1000
Investment W to S	3000– 4000	900– 1000	100/ 1000

alternatives through legal, administrative, political, or physical requirements. Then the acceptance of one alternative depends on the simultaneous acceptance of one or more related alternatives. Cash flows are also affected by relational factors, as in a flood-control proposal where the benefits of a levee of a given height depend on the acceptance of proposed dams on the headwaters of the river. These are *dependent alternatives*.

Return again to Table 7.1 and assume that proposal W could be implemented only if E were also funded. This linkage simply consolidates two investments into one alternative with an initial cost of $2000 + $3000 = $5000, which is expected to earn $400 + $900 = $1300 annually. The evaluation then proceeds with E + W treated as an individual alternative. The end effect of relational considerations is a listing of grouped, internally dependent alternatives which are collectively unrelated. An impetus for investigating relational connections is to avoid the selection of a course of action that cannot be implemented without additional commitment of resources.

Mutually exclusive alternatives which are *reinvestment-dependent* are discussed in Extension 7A.

IRR ANALYSIS OF SEVERAL MUTUALLY EXCLUSIVE ALTERNATIVES

The aim of a cash flow analysis is to put all competing alternatives into a comparable investment perspective. It acts as a screen. The alternative with the most promising future from one evaluation is then pitted against winners of other evaluations where intangible and financial considerations become critical. No organization has sufficient capital to fund all conceivable worthwhile proposals within its province. Care must be taken in all evaluations to ensure correct and consistent selections. Two logical-sounding selection criteria that sometimes lead to inaccurate conclusions are

1 Selecting the alternative that offers the highest rate of return on total investment
2 Selecting the alternative with the largest investment that meets the minimum required rate of return

The first criterion may bypass alternatives that earn lower rates of return which are still higher than other options available. The second criterion could lead to a larger investment than is desirable, which prevents a portion of the funds from earning higher returns available through substitute investments. A decision procedure to avoid inaccurate selections is shown by the flowchart in Figure 7.1. The procedure implicitly assumes that any size investment is possible. Whether the selected proposal is actually funded is a financial decision that relates available capital to all types of investment proposals and their effect on the funding organization. The IRR comparison rates the alternatives as inputs to the final decision process.

See Example 7.4 for samples of both types of mistakes.

FIGURE 7.1
Flow-chart for a selection among several mutually exclusive investment alternatives when a minimum rate of return is required and capital is assumed to be unlimited.

Step 3 is graphically portrayed in Extension 7B.

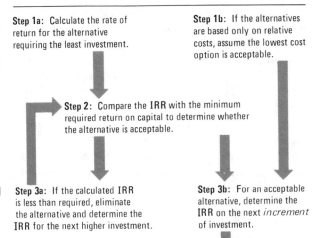

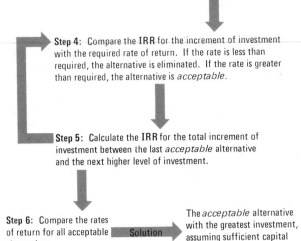

Example 7.1
Comparison of Several Mutually Exclusive Alternatives

Four designs for a product with their associated revenue and cost estimates have been presented to top management for a decision. A 10-year study period was used. A minimum rate of return of 10 percent before taxes, a rate expected from other investments with similar risk, is required. Based on the projected cash flows in Table 7.3, which of the four alternative designs appears the most attractive?

Solution 7.1

The alternatives are arranged in order of increasing investment requirements in Table 7.3. A supplementary table of incremental values is given on page 163. The annual net returns are calculated by subtracting the annual disbursements from the annual receipts for each alternative and then obtaining the differences between alternatives. For example,

TABLE 7.3

Ten-year cash flows

	Design			
	A	B	C	D
Initial investment	$170,000	$260,000	$300,000	$330,000
Annual receipts	114,000	120,000	130,000	147,000
Annual disbursements	70,000	71,000	64,000	79,000

Annual net return of increment $A{\rightarrow}B$
$$= (\$120,000 - \$71,000)$$
$$\quad - (\$114,000 - \$70,000)$$
$$= \$5000$$

	Increment		
	$A{\rightarrow}B$	$B{\rightarrow}C$	$C{\rightarrow}D$
Additional investment	$90,000	$40,000	$30,000
Annual net return	5,000	17,000	2,000

Using the steps in Figure 7.1, we calculate the IRR of design A from

$$-\$170,000 + (\$114,000 - \$70,000)(P/A, i, 10) \overset{?}{=} 0$$

and

$$(P/A, i, 10) = \frac{\$170,000}{\$44,000} = 3.8636$$

which, by interpolation, gives a rate of return of 22.5 percent. It is thereby an acceptable alternative because IRR = 22.5 percent > 10 percent required.

The increment of investment from design A to design B is $260,000 − $170,000 = $90,000, and the net annual returns are $5000. These annual returns will obviously not pay back the investment in 10 years: $5000 × 10 < $90,000. Therefore, design B is eliminated from contention because its incremental IRR is lower than the required 10 percent.

Design C *is evaluated by comparing it with the last acceptable alternative*, which in this case is design A. Then the

incremental comparison $A{\rightarrow}C$ progresses as

$$-\$300,000 - (-\$170,000) + (\$130,000 - \$64,000$$
$$- \$44,000)(P/A, i, 10) \overset{?}{=} 0$$

$$(P/A, i, 10) = \frac{\$130,000}{\$22,000} = 5.9090$$

to obtain

$$\text{IRR} = 10\% + 1\%\frac{6.1445 - 5.9090}{6.1445 - 5.8892} = 10.9\%$$

With its rate of return greater than 10 percent on the indicated extra increment of investment, design C becomes an acceptable alternative.

The next additional increment of investment to afford design D shows

$$(P/A, i, 10) = \frac{\$30,000}{\$2000} = 15$$

which is unacceptable by inspection. Therefore, assuming that sufficient capital is available, we prefer design C. Its rate of return on total capital invested is calculated from

$$(P/A, i, 10) = \frac{\$300,000}{\$130,000 - \$64,000} = 4.5454$$

which leads to an IRR of 18.9 percent.

The results of the evaluation of alternative designs in Example 7.1 are supplemented and summarized in Table 7.4.

TABLE 7.4					

IRR relationships for Example 7.1

Design	Total Investment	Annual Return	IRR on Total Investment, %		Incremental IRR
A	$170,000	$44,000	22.5	....	A→B: (−)
B	260,000	49,000	13.5	....	A→C: 10.9%
C	300,000	66,000	18.9	....	C→D: (−)
D	330,000	68,000	15.9		

One type of mistake would have been to select design A, which has the greatest rate of return on total investment, as the preferred alternative. This choice would prevent the additional investment of $300,000 − $170,000 = $130,000 in design C, which returns 10.9 percent. Since 10.9 percent is higher than the 10 percent expected from other investments with similar risks, a loss of about 1 percent on $130,000 would occur. However, it is prudent to keep in mind the high returns possible through an investment in design A in case there is insufficient capital to fund design C or if there is another independent opportunity for investing $130,000 at an IRR greater than 10.9 percent.

Another type of error would be to select the largest investment that still meets the 10 percent rate-of-return requirement. The unsatisfactory IRR for the extra investment in design D over design C was apparent in the incremental analysis. Therefore, putting $330,000 − $300,000 = $30,000 into design D forces this amount of capital to earn less than the 10 percent it could receive if invested elsewhere.

Example 7.2
Compatibility of PW and IRR Incremental Analysis

Given the same data for a selection between mutually exclusive alternatives, PW and EAW methods agree with the choice indicated by the rate-of-return method, *as long as the IRR is applied to increments instead of the whole investment*. Computing the present worths of the four projects described in Example 7.1 provides the values shown in Table 7.5.

Comparing Table 7.5 with Table 7.4 confirms that PW calculations based on *total* investment point to the same selection as IRR calculations for *incremental* investments. Both indicate a preference for design C. The supplementary calculation of the present worths of incremental investments also agrees with IRR results. Note that

$$PW(C) = PW(A) + PW(A{\to}B) + PW(A{\to}C)$$
$$\$105,537 = \$100,358 - \$59,278 + \$64,457$$

The incremental PW($A{\to}B$) is negative, confirming its unacceptableness. Combining PW($A{\to}B$) with PW($B{\to}C$) produces a net positive gain that corresponds to IRR($A{\to}C$) > MARR, indicating the acceptability of design C. Thus the $90,000 + $40,000 = $130,000 investment increment that affords design C over design A earns 10 percent plus a present sum of $105,537 − $100,358 = $5179, or, equivalently, it earns an internal rate of return greater than the required 10 percent— 10.9 percent. This increment is preferable to investing the $130,000 in another investment that will earn only the minimum acceptable rate of return—10 percent.

Comparing the IRR of a single investment with the MARR to determine its acceptability may appear to be in violation of the incremental IRR criterion. It is a legitimate comparison because the increment between a do-nothing alternative and

TABLE 7.5

Present worths of designs from Example 7.1 when the MARR is 10 percent

Design	Investment Total	Investment Increment	Annual Return Total	Annual Return Increment	PW at 10% Total	PW at 10% Increment
A	$170,000		$44,000		$100,358	
		$90,000		$5,000		−$59,278
B	$260,000		$49,000		$41,080	
		$40,000		$17,000		$64,457
C	$300,000		$66,000		$105,573	
		$30,000		$2,000		−$17,711
D	$330,000		$68,000		$87,826	

a do-something proposal is the same as the total investment. Therefore, the incremental IRR equals the IRR on the total investment, thereby satisfying the incremental analysis procedure.

ANALYSIS OF INDEPENDENT ALTERNATIVES

The demarcation between dependent and independent alternatives gets fuzzy at times. Consider the plans for different product designs in Example 7.1. When only one new product design is sought, the four product proposals are certainly mutually exclusive. Similarly, four plans for remodeling an office are surely mutually exclusive. The best product proposal and the best office plan appear to be completely independent of each other. Yet both contribute to the profit status of the organization.

The product proposal is for a new product that will bring in revenue in excess of costs in order to increase profit, and the office plan is designed to produce improved working conditions that will lower costs to increase profit. At least tangentially, both plans are related through their contribution to the firm's financial standing and by their reliance for initial funding on the firm's capital. In effect, they may be mutually exclusive.

A respected caveat is that a choice which minimizes cost may not be the one that maximizes profit.

It is seldom worthwhile to trace tenuous links that may relate various investment proposals, particularly during initial comparison screenings. Diligent detective work could probably uncover links between any and all alternatives, but their effect on early economic comparisons would be negligible. Most mutually exclusive options are clearly evident via coinciding functions; and conditional dependencies are conspicuous by physical relations, such as the condition for adding a second floor to a building is that of constructing the first floor. Such obvious relations are natural and necessary recognitions for most economic evaluations. The question of how to finance the alternatives may or may not enter the initial comparisons.

Independent proposals can be collected in various combinations and evaluated as grouped to determine how well each combination meets the

Proposals with mixed dependencies are examined in Extension 7B.

investment objectives. The groupings selected depend on the conditions set for the evaluation. For instance, one division of a firm might be allowed to fund two proposals while other divisions are allowed only one. Then each combination would include two proposals from the favored division with one from each of the other divisions.

Several programming techniques, such as linear programming, are adaptable to systematic selections among numerous combinations. A direct search method is used to illustrate the conversion of independent proposals to mutually exclusive combinations because it is conceptually simple and adequate for most routine situations. Consider the four independent proposals listed in Table 7.6, from which any combination with first costs totaling $5000 or less can be selected. After appropriate groupings that satisfy the specified capital limitation are identified, each combination is evaluated by one of the discounted cash flow comparison methods. Since only one combination can be selected, the decision is between mutually exclusive alernatives.

A check-off procedure for identifying appropriate combinations of proposals is shown in Table 7.7. The leftmost columns indicate by a 1 when a proposal is included in a combination, and by a 0 when it is not. The first combination is the do-nothing alternative; the four zeros in the first row indicate that no proposals are included. Altogether there are nine combinations that have total first costs of $5000 or less.

Tangible ratings for a managerial decision on the allocation of capital are clearly presented in the table. The highest IRR is earned by allocating only $1000 to fund proposal I. The next highest IRR results from an investment of $4000 in proposals I and III. If the full $5000 is to be allocated, 17.3 percent is the best IRR that can be expected. The final allocation is a management prerogative relying on economic and political objectives.

Much more complex comparisons can be conveniently accomplished by the iterative approach represented by Table 7.7. However, without constraints such as a limit on investment capital, the number of combinations grows rapidly as more proposals are included, and computations grow correspondingly

---

TABLE 7.6

Three-year cash flows for four independent proposals, all of which have passed a screening based on a minimum required rate of return of 10 percent

Proposal	First Cost	End-of-Year Cash Flow		
		Year 1	Year 2	Year 3
I	−$1000	$ 550	$ 550	$ 550
II	−2000	875	875	875
III	−3000	1400	1400	1400
IV	−4000	1665	1665	1665

TABLE 7.7

Binary numbers 0 and 1 represent the proposals included in a combination. An IRR is calculated for each combination to rank the alternatives. Note that the extra $1000 investment required for alternative 9 over alternative 7 earns less than 0 percent rate of return.

I	II	III	IV	Alternative	0	1	2	3	IRR, %
		Proposals				End-of-Year Cash Flow			
0	0	0	0	1	0	0	0	0	0
1	0	0	0	2	−$1000	$ 550	$ 550	$ 550	30
0	1	0	0	3	−2000	875	875	875	15
0	0	1	0	4	−3000	1400	1400	1400	18.8
1	1	0	0	5	−3000	1425	1425	1425	20
0	0	0	1	6	−4000	1665	1665	1665	12
1	0	1	0	7	−4000	1950	1950	1950	21.7
1	0	0	1	8	−5000	2215	2215	2215	15.7
0	1	1	0	9	−5000	2275	2275	2275	17.3

cumbersome. Computerized and programmed assistance then becomes a near necessity.

Review Exercises and Discussions

Data for three alternative investment plans are listed below.

Exercise 1

Alternative	Investment	Salvage Value	Life, Years	Annual Net Cash Flow
X	$ 6,000	$ 0	3	$2600
Y	12,000	3000	6	2500
Z	18,000	0	6	4000

When the minimum attractive rate of return is 10 percent, which alternative(s) should be selected under each of the following decision conditions?

a   Individual alternatives are mutually exclusive.
b   Individual alternatives are independent.

The present worths of the three alternatives displayed in the table could be calculated as

Solution 1

$$PW(X) = -\$6000 + \$2600(P/A, 10, 3) = -\$6000 + \$2600(2.4868)$$
$$= \$465$$

$$PW(Y) = -\$12,000 + \$2500(P/A, 10, 6) + \$3000(P/F, 10, 6)$$
$$= -\$12,000 + \$2500(4.3552) + \$3000(0.56448)$$
$$= \$582$$

$$PW(Z) = -\$18,000 + \$4000(P/A, 10, 6) = -\$18,000 + \$4000(4.3552)$$
$$= -\$579$$

a The deception that alternative Y is more attractive than alternative X occurs because the present worths are not based on equal time durations. When both investments are compared over a 6-year study period, implicitly assuming that a second investment in X can be made at the end of 3 years at the original cost, the present worth of repeated investments in X (X_1X_2) is

$$PW(X_1X_2) = -\$6000 - \$6000(P/F, 10, 3) + \$2600(P/A, 10, 6)$$
$$= -\$6000 - \$6000(0.75132) + \$2600(4.3552) = \$816$$

Therefore, when the individual alternatives are mutually exclusive, X is preferred.

b Under the assumption that funds are available for all alternatives that meet the minimum attractive rate of return, investments in X and Y are indicated by their positive present worths. Both satisfy the 10 percent return criterion.

Exercise 2 The five investment proposals shown below have been investigated carefully and are deemed to be equally safe. An investor will be satisfied if a minimum before-tax return of 9 percent is realized.

Proposal	Investment P	Life N, Years	Salvage Value S	Net Annual Cash Flow
Alpha	$30,000	5	$ 0	$ 7,500
Beta	60,000	5	10,000	13,755
Gamma	20,000	5	0	5,000
Delta	40,000	5	10,000	10,000
Epsilon	30,000	5	5,000	7,500

Using EAW calculations, determine:

a Which proposal is preferred if only one can be selected?
b How much should be invested if the proposals are independent and unlimited capital is available?
c Which proposals should be utilized if an investor wishes to allocate exactly $60,000 of investment capital, and any funds not put into one or more of the proposals will be used to purchase bonds that pay annual dividends of $7\frac{1}{2}$ percent on the amount invested?

Solution 2 The annual worth of each proposal is calculated as follows:

$$EAW(Alpha) = \$7500 - \$30,000(A/P, 9, 5) = \$7500 - \$30,000(0.25709)$$
$$= \$7500 - \$7713 = -\$213$$

$$EAW(Beta) = \$13,755 + (-\$60,000 + \$10,000)(A/P, 9, 5) + (-\$10,000)(0.09)$$
$$= \$13,755 - \$50,000(0.25709) - \$900$$
$$= \$13,755 - \$13,755 = 0$$

$$EAW(Gamma) = \$5000 - \$20,000(A/P, 9, 5) = \$5000 - \$20,000(0.25709)$$
$$= \$5000 - \$5142 = -\$142$$

$$\begin{aligned}
\text{EAW(Delta)} &= \$10,000 - (\$40,000 - \$10,000)(A/P, 9, 5) - \$10,000(0.09) \\
&= \$10,000 - \$30,000(0.25709) - \$900 \\
&= \$10,000 - \$8613 - \$1387
\end{aligned}$$

$$\begin{aligned}
\text{EAW(Epsilon)} &= \$7500 - (\$30,000 - \$5000)(A/P, 9, 5) - \$5000(0.09) \\
&= \$7500 - \$25,000(0.25709) - \$450 \\
&= \$7500 - \$6875 = \$625
\end{aligned}$$

a If only one proposal can be accepted, Delta is preferred because it has the largest net annual worth on the capital invested.

b With unlimited capital, Delta and Epsilon are obvious selections, owing to their positive net annual worths. Beta should also be accepted, even with its zero annual worth, because the $60,000 invested will still earn 9 percent; any positive AW indicates a return *beyond* the minimum attractive rate, 9 percent. Therefore, $60,000 + $40,000 + $30,000 = $130,000 should be invested from the unlimited treasury.

c An investment capital limitation of $60,000 allows three logical choices:

   (1) Invest the full amount in Beta where it will earn exactly 9 percent.

   (2) Invest in Delta ($40,000) plus Gamma ($20,000) to earn 9 percent plus $1387 − $142 = $1245.

   (3) Invest in Delta plus a $20,000 bond. This combination is obviously superior to the similar combination of Epsilon plus a $30,000 bond. The annual worth of Delta plus the bond is calculated from the cash flow diagram below, where annual returns from the bond are 0.075 × $20,000 = $1500.

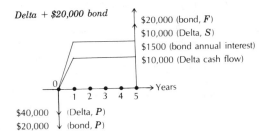

$$\begin{aligned}
\text{EAW} &= \$10,000 + \$1500 - (\$40,000 + \$20,000)(A/P, 9, 5) \\
&\qquad\qquad\qquad\qquad\qquad + (\$20,000 + \$10,000)(A/F, 9, 5) \\
&= \$11,500 - \$60,000(0.25709) + \$30,000(0.16709) \\
&= \$11,500 - \$15,425 + \$5013 = \$1088
\end{aligned}$$

Thus the EAW of the $60,000 investment in Delta plus the bond is better than the $60,000 Beta investment proposal, but less rewarding than investing in Delta and Gamma.

Exercise 3

The four independent proposals in Table 7.6 are still under consideration, and new conditions have been added. Proposal I is so attractive that it must be included in any accepted combination. Another proposal with the following cash flow has been added to the list:

Proposal	End-of-Year Cash Flow			
	0	1	2	3
IIA	− $2500	$1150	$1150	$1150

TABLE 7.8

Proposal	Alternatives											
	1	2	3	4	5	6	7	8	9	10	11	12
I	1	1	1	1	1	1	1	1	1	1	1	1
II	1	1	1	1	0	0	0	0	1	1	1	1
IIA	0	0	0	0	1	1	1	1	1	1	1	1
III	0	1	0	1	0	1	0	1	0	1	0	1
IV	0	0	1	1	0	0	1	1	0	0	1	1

In addition to proposal I, either proposal II or IIA must be in any acceptable combination. If the management decision is to fund the combination possessing the highest IRR regardless of the amount of capital required, which proposals will be funded?

Solution 3   All possible combinations that satisfy the investment criteria are delineated in Table 7.8.

From knowing the IRRs for individual proposals already listed in Table 7.7, and by calculating IRR(IIA) = 18 percent, we can eliminate all combinations which include proposal II because IRR(II) = 15 percent; compared with proposal IIA, the inclusion of proposal II has to lower any combination's IRR below 18 percent. This narrows the choice to alternatives 5, 6, 7, and 8 in Table 7.8.

Again noting from Table 7.7 that the IRR of proposal III is above 18 percent and that of proposal IV is below, we conclude that the best combination must be alternative 6. This combination has a first cost of $1000 + $2500 + $3000 = $6500 and annual returns of $550 + $1150 + $1400 = $3100, which lead to an IRR of 20.3 percent. Utilizing dominating relationships can bypass a lot of computations.

PROBLEMS

7.1 A spare-parts service department for a construction-equipment supplier must be established in Alaska to meet contractual commitments. The firm has eliminated the option of building a new field office because it expects to conduct business in Alaska for only the next 10 years, during the construction boom. At best, the field office will meet expenses and make a small profit contribution. Only three structures that are available for purchase meet the space needs and have the desired location. A major portion of the investment will likely be recovered by appreciation in land values. A minimum acceptable rate of return of 15 percent before taxes is expected on such investments. What rate of return will be earned by each increment of investment from the anticipated cash flows shown?

	Site 1	Site 2	Site 3
Purchase price	$140,000	$190,000	$220,000
Resale value	125,000	155,000	$175,000
Net revenue	24,000	31,000	41,000

7.2 Compare the rates of return for the following plans, and select the preferable alternative. The minimum acceptable rate of return is 6 percent.

	Plan 1	Plan 2	Plan 3
First cost	$40,000	$32,000	$70,000
Salvage value	$10,000	$6,000	$20,000
Economic life	7 years	7 years	7 years
Annual receipts	$18,000	$18,000	$24,500
Annual disbursements	$11,000	$14,500	$14,500

7.3 A temporary water line is required to supplement the water supply at a plant until city water becomes available in a new industrial area. Three alternative pipe sizes with associated pumping facilities will satisfy the water requirements:

	Pipe Size		
	14 in (35.6 cm)	16 in (40.6 cm)	18 in (45.7 cm)
First cost	$18,000	$25,000	$34,000
Annual pumping cost	6,400	4,400	2,800

The pipeline and pumping stations will be in the same locations for all three alternatives. The planning period is 5 years, and pipe can be recovered at the end of the period. It is expected to yield 40 percent of its first cost when recovered, and the cost of recovery will be $2000 regardless of pipe size. Compare the rates of return for the alternatives when a 9 percent return before taxes is desired.

7.4 Three alternative investments to be compared by the rate-of-return method are described below. The MARR is 9 percent.

Alternatives	X	Y	Z
First cost	$20,000	$16,000	$10,000
Annual expense	$5,000	$3,000	$4,000
Annual income	$11,500	$11,500	$7,000
Economic life	4 years	2 years	4 years

(*a*) Which alternative has the highest overall IRR, and what is it?
(*b*) Which incremental investment has the greatest rate of return, and what is it?

7.5 Four requests for expenditures are listed below.

Proposal	First Cost	Net Annual Savings
Hot Shot	$ 50,000	$20,000
Serendipity	60,000	26,100
Slam Bang	75,000	32,400
Big Bucks	100,000	41,700

All proposals have a life of 3 years, and the minimum acceptable rate of return is 12 percent.

(a) Which proposal(s) should be accepted if they are independent?

(b) Which proposal should be accepted if they are mutually exclusive?

7.6   An estimating service for contractors plans to provide an additional service for its customers. Besides the normal routine for making estimates, an additional follow-up service will be offered to review the records of each project and analyze actual costs versus estimated costs. Based on the expected workload for the next 3 years, the proposal could be conducted by:

1   One secretary and a clerk with total direct and indirect wages of $28,000 per year, using hand-operated equipment with an initial cost of $2800

2   One secretary with total annual wages of $15,000, using computerized equipment which is priced at $16,000 and has maintenance costs of $300 per month

3   Subcontracting part of the work at a monthly cost of $1400 and hiring a part-time secretary for $10,000 per year

Any investment must be written off in 3 years with no salvage value, and an 8 percent rate of return is required. Assuming that all three alternatives provide work of equal quality, determine which one should be selected. Use an EAC comparison and check the incremental IRR.

7.7   Based on the data in Exercise 2 and the investment conditions in part c (where the investment capital is limited to $60,000), explain the following characteristics of the solution given on page 168.

(a) The reasoning that eliminated the Epsilon proposal without recourse to calculations when it is compared with the Delta proposal as part of the $60,000 investment in a proposal plus a bond

(b) The reason why the annual worth of Delta plus a $20,000 bond paying $7\frac{1}{2}$ percent interest is less than the AW of the Delta proposal alone

7.8   A firm finds that it will be necessary to air-condition a rather large area for its computers and data-processing equipment. An engineering study revealed that the more money spent on insulating the walls and ceiling area, the less money is required for the air-conditioning unit. The engineer's estimates are as follows:

	1	3	3	4
First cost of insulation	$35,000	$45,000	$60,000	$80,000
First cost of air-conditioning equipment	52,000	45,000	38,000	32,000
Annual power cost	6,500	5,100	4,100	3,500

The study also estimated that the insulating material would have a life of 20 years with zero salvage value, and the air-conditioning equipment would have a life of 10 years with no salvage value. Taxes and insurance are expected to be 2 percent of the first cost per year. Which alternative should be selected if the firm requires a 15 percent rate of return before taxes? Use a PW comparison, and check the incremental IRR.

7.9   An asset can be purchased for $105,000 if payment is in cash. It can also be purchased by a payment of $45,000 now plus a payment of $75,000 in 3 years or it can be purchased with 10 yearly payments of $15,000 each with the first payment made now.

Calculate the incremental rates of return for the alternatives, and state which one is preferred if MARR = 9 percent.

7.10 The Advanced Product Engineering Group has developed a list of potential, mutually exclusive project proposals as possible new investments. All have a 10-year life and no salvage value. Using IRR calculations and the data shown below, determine which project, if any, should be accepted. Use MARR = 10%.

Project	First Cost	Annual Cash Flow
A	$100,000	$16,980
B	65,000	13,000
C	20,000	2,710
D	40,000	6,232
E	85,000	16,320
F	10,000	1,770

7.11 The four alternative investment proposals described below are being evaluated.

Proposal	Initial Investment	IRR for Proposal	Incremental IRR Compared with Proposal		
			A	B	C
A	$100,000	19%			
B	175,000	15%	9%		
C	200,000	18%	17%	23%	
D	250,000	16%	12%	17%	13%

(a) If the proposals are mutually exclusive, which one should be selected if the MARR is 15 percent?

(b) If the proposals are independent, which should be selected if the MARR is 15 percent?

(c) If the proposals are independent, what is the highest rate of return that can be gained when investment capital is limited to $275,000 and any funds not allocated to the given proposals can be invested at 15 percent?

7.12 The think-tank group at Imagineering Inc. has recently come up with six proposals for consideration. Each of the alternative projects has an estimated life of 10 years with no salvage value. Their potential impacts are given below.

	Project					
	A	B	C	D	E	F
Required capital	$8000	$4000	$1000	$3000	$1500	$9000
Annual cash flow	$1100	$800	$200	$715	$250	$1400

(a) Assuming Imagineering Inc. has unlimited capital and the projects are independent, determine which should be accepted. The MARR is 12 percent.

(b) Given that Imagineering has only $9000 of investment capital available, determine which of the independent alternatives should be selected to maximize the overall rate of return. Assuming that any capital not used for project funding can be invested elsewhere at 15 percent, indicate what is the best return obtainable on the $9000.

(c) Assuming that the projects are mutually exclusive and the MARR is still 12 percent, determine which, if any, of the proposals should be accepted.

7.13 Cash flows from five investment proposals are itemized below.

Proposal	Investment	Annual Net Cash Flows Each Year				
		1	2	3	4	5
A	−$15,000	$ 4,500	$ 4,500	$ 4,500	$ 4,500	$ 4,500
B	−25,000	12,000	10,000	8,000	6,000	4,000
C	−20,000	2,000	4,000	6,000	8,000	10,000
D	−30,000	0	0	15,000	15,000	15,000
E	−10,000	4,500	4,500	−2,500	4,500	7,500

All investment opportunities are evaluated using a 15 percent rate of return and are based on a 5-year study period.

(a) If the proposals are mutually exclusive, which one is the best investment based on its present worth?

(b) If investment capital is limited to $30,000, which proposals should be accepted?

7.14 A half dozen cost-reduction proposals have been forwarded by the industrial engineering department. A 20 percent rate of return is expected and all equipment investments are to be written off in 5 years with no salvage value.

Proposal	Equipment Cost	Net Annual Savings
Combine	$25,000	$ 9,500
Rearrange	60,000	21,500
Modify	20,000	7,000
Eliminate	20,000	9,000
Up quality	40,000	17,000
Make safer	30,000	10,000

(a) Which proposal should be selected if only one can be accepted?

(b) Which proposals should be selected if all proposals are independent and there is effectively unlimited capital available for cost-reduction projects?

(c) What is the highest IRR that can be earned from any combination of independent alternatives that has a total investment cost of at least $50,000 but less than $100,000?

(d) What combination of alternatives would produce the greatest total savings on an investment of not over $100,000?

(e) Would you recommend the combination identified in Problem 7.14c or 7.14d? Why?

7.15 Six mutually exclusive alternatives are being considered. They are listed in order of increasing first costs in the following table, where the rate of return for the overall investment in each and the incremental IRR for every increment are given. All the alternatives have the same lives and comparable intangible values.

Alter-native	IRR on Overall Investment	IRR on Increments of Investment Compared with Alternative:				
		I	II	III	IV	V
I	1%					
II	8%	21%				
III	11%	15%	12%			
IV	15%	22%	19%	17%		
V	13%	19%	16%	15%	9%	
VI	14%	21%	18%	16%	14%	21%

(a) If one of the alternatives *must* be implemented but there are insufficient funds available to afford any of the last three alternatives, which one should be selected?

(b) What would you recommend if none of the alternatives is mandatory (do nothing is an acceptable alternative); there are still insufficient funds for alternatives IV, V, and VI; and the minimum attractive rate of return is 12 percent? Why?

(c) If the selection of one of the alternatives is mandatory, at what range of MARR is alternative I the proper choice?

(d) At what minimum required rate of return is alternative IV the correct choice?

(e) If funds are almost unlimited and no minimum rate of return is required (although the cost of capital is 9 percent), which alternative would you select? Why?

(f) If the minimum attractive rate of return is 12 percent, what would be the opportunity cost, expressed as a rate-of-return percentage, when a shortage of funds causes alternative III to be selected?

7.16   Three independent proposals have passed a preliminary screening to confirm that all are acceptable at a minimum IRR of 15 percent. Each one has an economic life of 4 years. The cash flows are given in the table, with the salvage values included in the final year's income.

Proposal	First Cost	End-of-Year Cash Flows			
		1	2	3	4
A	$17,000	$10,000	$8,000	$ 6,000	$ 4,000
B	22,200	4,000	7,000	10,000	13,000
C	20,700	8,000	8,000	8,000	8,000

(a) Which combination of proposals should be selected if sufficient capital is available to fund any choice, so long as the rate of return is 20 percent or greater? Why?

(b) Which combination of proposals should be selected when the minimum must be 15 percent but an income of at least $14,000 per year is necessary?

7.17 Routine Investments Incorporated (RII) is currently investigating six alternative investment opportunities. Each of the proposals has an expected life of 20 years with negligible salvage value. The following data have been collected:

Alternative	Investment Required	Net Annual After-Tax Cash Flow
A	$ 40,000	$ 4,380
B	80,000	12,100
C	10,000	2,100
D	55,000	8,800
E	130,000	17,400
F	30,000	4,000

Use the appropriate rate-of-return analysis to answer the following:
    (a) Which proposals should be accepted if they are independent and RII must obtain a minimum after-tax return of at least 13 percent? How much total capital will be required and what will be the overall rate of return on the investment.
    (b) Which proposals should be accepted if they are independent and RII has only $110,000 to invest. What will be the overall rate of return on the investment?
    (c) Which proposal should be selected in that they are mutually exclusive and RII must obtain a rate of return of at least 10 percent?

7.18 The Smallwater Public Utility Board is considering an addition to the existing pumping system. Alternative designs are being studied at two sites and the following cost estimates have been developed:

Site	Design	Equipment Cost	Annual Operations and Maintenance Expense
Cloverdale	A	$2,000,000	$400,000
	B	2,500,000	320,000
Apple Hill	C	3,000,000	200,000
	D	4,000,000	150,000

The Cloverdale site will require an expenditure of $2,000,000 for land and right-of-way easements and the Apple Hill plans would require $3,000,000. The land will represent a permanent investment, whereas the equipment will require replacement at a 30-year interval.
    (a) Determine which of the mutually exclusive proposals should be selected through a rate-of-return comparison based on a tax-free MARR of 7 percent.
    (b) Determine the capitalized cost of each alternative using a tax-free interest rate of 7 percent.

7.19 The township of Swampy Bottom has been troubled by periodic flooding since its settlement in 1886. The city fathers have at last decided to rectify the situation with a small flood control and drainage system. The following cost estimates have been provided by a local consulting firm.

Plan	Capacity of System, ft³/s	Initial Cost	Annual Maintenance	Average Annual Damage Due to Flooding
A	Do nothing	$ 0	$ 0	$180,000
B	12,000	300,000	30,000	90,000
C	15,000	400,000	40,000	65,000
D	20,000	600,000	50,000	42,000
E	25,000	1,000,000	80,000	27,000
F	28,000	2,000,000	100,000	13,000
G	30,000	2,500,000	110,000	10,000

Determine which of the designs should be selected in order to minimize total cost if the project is expected to remain in service for 50 years and is to be financed through 6 percent tax-free bonds.

7.20 The Clinker Creek Coal Company is considering purchase of new pumps for use in their open pit mine. Occasionally, excess rainfall causes minor flooding which reduces production while pumping is being carried out. Several alternatives are to be considered on a before-tax annual cost basis using an i of 20 percent. The service life of the pumps is assumed to be 10 years and the portable piping and other equipment is expected to last 5 years with no salvage value for any of the items.

	Type of Pump			
	Waterhog	Pumpsalot	Mudmucker	Chugalug
First cost of pump	$20,000	$30,000	$50,000	$75,000
Piping and auxiliary equipment	10,000	10,000	20,000	20,000
Annual energy and maintenance costs	7,000	6,000	5,000	4,000
Annual expected cost of lost production while pumping	30,000	25,000	20,000	12,000

Determine which of the pumping systems should be selected.

EXTENSIONS

7A   Analysis of Mutually Exclusive, Reinvestment-Dependent Alternatives

An investment made today could allow special investment opportunities in the future that would not otherwise be available. Such an investment is *reinvestment-dependent*. If one or more alternatives in a set of mutually exclusive proposals is reinvestment-dependent, the special reinvestment opportunities must be included in the analysis.

The PW and incremental-IRR methods described in this chapter are sufficient when future investments are independent of current investments. In this case, *current* proposals are accepted or rejected on the basis of the cash flows they generate, and *future* investments are evaluated according to their cash flows. Reinvestment-dependency can also be ignored when future investments are expected to earn less than the MARR, because such future investments would be rejected owing to their negative PW.

Reinvestment-dependency typically results from a current investment that permits a higher rate of return to be earned on reinvested receipts or allows the investment of new capital at a higher rate than would otherwise be possible. For instance, the purchase of a computer might yield an IRR of 30 percent *and* provide the opportunity for future purchases of peripheral equipment that would return 40 percent on subsequent investments; the latter investments would not be possible without the former.

Analysis of reinvestment-dependent alternatives is conducted by (1) estimating the rate of return applicable to future cash flows from reinvestments and additional capital investments; (2) determining the amount, timing, and future

worth of the dependent cash flows; and (3) discounting the FW back to the present at the MARR. The study period during which funds can be reinvested must be the same for all alternatives *and* no less than the longest life of any proposal in the mutually exclusive set. The preferred alternative is the investment plan with the maximum net present value.

To illustrate a comparison involving reinvestment-dependency, we will assume that the three investments in Table 7.9 are mutually exclusive. Investment A provides no special reinvestment opportunities. Investment B allows its annual receipts to be reinvested each year at IRR = 20 percent. Proposal C recaptures its original investment plus interest in 1 year and permits another investment at year 2 equal to twice the original investment, to earn a 30 percent rate of return. Reinvestment opportunities are continued over a 4-year study period. The MARR is 10 percent.

Since receipts from proposal A are assumed to be invested at the MARR,

$$PW(A) = -\$1000 + \$400(P/A, 10, 4) = \$268$$

Of course, the same PW(A) results from FW(A) discounted back to time zero:

TABLE 7.9

Comparison of proposals B and C that permit future investments at a rate of return greater than the MARR = 10 percent with proposal A that allows no special reinvestment opportunities. Proposal C is superior for the 4-year study period.

End of Year	Alternative A	Alternative B	Alternative C
0	−$1000	−$1000	−$1000
1	400	600*	1100
2	400	600*	− 2000†
3	400	0	0
4	400	0	0
Reinvestment rate	MARR	20%	30%
PW at IRR = 10% (ignoring reinvestments)	$268	$41	0‡
FW(reinvested receipts)	0	$1901	$4844
PW at IRR = 10% (including reinvestments)	$268	$298	$665

* Receipts reinvested at 20 percent.
† New capital invested at 30 percent.
‡ Based only on original $1000 investment.

$PW(A) = [-\$1000(F/P, 10, 4)$
$\qquad + \$400(F/A, 10, 4)](P/F, 10, 4)$
$\qquad = \$392(P/F, 10, 4) = \$268$

The two \$600 receipts in proposal B can be reinvested at 20 percent as soon as they are received and therefore have a future worth in 4 years of

$FW(\text{reinvestments}) = \$600(F/P, 20, 3)$
$\qquad + \$600(F/P, 20, 2) = \$1901$

Discounting the future worth back to time zero at MARR $= 10$ percent then makes the cash flows of the alternatives comparable:

$\qquad PW(B) = -\$1000 + \$1901(F/P, 10, 4) = \$298$

In proposal C the \$1100 capital-recovery payment is assumed to earn only the MARR of 10 percent when reinvested, but the original investment allows a new \$2000 investment to earn 30 percent for 2 years. The terminal value of the two investments is

$FW(\text{reinvestments}) = \$1100(F/P, 10, 3)$
$\qquad + \$2000(F/P, 30, 2) = \$4844$

The present worth of proposal C is calculated by discounting the future worth at 10 percent for 4 years and subtracting the present worths of the \$1000 and \$2000 outlays:

$PW(C) = -\$1000 - \$2000(P/F, 10, 2)$
$\qquad + \$4844(P/F, 10, 4) = \$665$

Outcomes of PW comparisons with and without reinvestment are shown in Table 7.9. The change in preference between alternatives resulting from opportunities for reinvestment amply demonstrate the importance of recognizing the existence of reinvestment-dependent options and accommodating dependent cash flows in the analysis.

QUESTIONS

Two mutually exclusive proposals have the cash flows shown. All receipts from proposal J and K can be reinvested at 20 percent and 25 percent, respectively. The reinvestment rates are expected to continue throughout the 5-year comparison period. The minimum acceptable rate of return for all investments is 15 percent.

Year	Proposal J	Proposal K
0	-\$100,000	-\$100,000
1	60,000	60,000
2	50,000	15,000
3	40,000	60,000
4	30,000	0
5	0	0

7A.1 What is the PW of each proposal when reinvestment opportunities are ignored?

7A.2 What are the PWs when receipts are reinvested at the given rates?

7A.3 Proposal L also requires an initial investment of \$100,000 but permits no special reinvestment possibilities. What uniform annual cash flow would make independent proposal L as attractive as the preferred reinvestment-dependent alternative?

7A.4 Under what conditions would it be sufficient to base the choice among mutually exclusive reinvestment-dependent alternatives on FWs alone?

7A.5 A moneylender has \$2000 to loan. Three people want to borrow \$1000 apiece. All three are equally good credit risks, but the first one is willing to pay annual interest of 15 percent, the second 20 percent, and the third 25 percent. Since the the moneylender can invest his money safely in a bank at only 10 percent, all three loans are attractive.

Another option occurs when the moneylender finds that he can borrow an additional \$1000 at a 17 percent annual interest rate. He muses that he could borrow the \$1000, loan it to the person who is willing to pay 25 percent, and use his own money for the other two loans. Should he do it? Why?

7B   Illustrated Incrementalism

Many decision situations involve mutually independent proposals that have alternatives which become mutually exclu-sive when resources are limited. Since these situations are usually settled by marginal analysis—examining each additional activity increment—proposals can be displayed graphically to show the relation of alternatives and their relative acceptability.

Analyzing Mixed Dependencies with Increment Graphs

Three government departments have indicated a desire to hire engineering students on a part-time basis: Highway (H), Motor Vehicle (MV), and Internal Management (IM) departments. All have forwarded their requests to the executive department stating the number of students wanted and the savings expected from hiring them to work on specific problems that would not otherwise be addressed. The proposals are shown in Table 7.10, where "gross margin" is the expected value of savings prorated by month for each project the students might work on.

Employment proposals from the respective departments are independent alternatives for funding by the executive department, but the number of positions that can be filled in each department is mutually exclusive.

The incremental relations of positions within each department are shown in the three frames of Figure 7.2. Indicated in each frame by a dashed line is the monthly cost of employment, which amounts to $1000 per student. Consideration of this employment expense leads to the elimination of unqualified alternatives, such as having only one student in the Internal Management department; apparently the combination of two students working together will produce collective savings of $3800, which is $1900 apiece, but one student working alone can generate savings of only

$500. The skip over the unacceptable $500 alternative to the acceptable $3800 margin for two positions is indicated by a dotted line labeled with the incremental savings, $1900 per student when two are employed.

Marginal Maneuvers

If the executive department can budget only enough to fill one position, it should go to Department H, where the gross margin from hiring one student is highest, $2100. But if two students can be hired, they should both be assigned to Department IM, where their average gross margin is $1900. When three can be hired, one should go to H and the other two to IM for a total gross margin of $5900, which amounts to $1967 per person.

This example shows the importance of having a complete array of alternatives available for analysis. If each department had gone through a preselection process to submit only its best plan, as frequently happens in decentralized organizations, a less-than-optimal allocation results whenever three or more positions are filled, based on the given data. A preselection policy may thus cause suboptimization when resources are limited. This is a capital-budgeting problem, which is examined in more detail in Chapter 9.

QUESTIONS

7B.1 Identify the net margins (savings minus costs) expected from the most efficient allocation plan for each employment level from 4 to 9 students based on data in Figure 7.2.

7B.2 A dispatcher has four crews available to service three routes. As indicated in the table below, the gross margin on each route depends on the number of crews allocated to it. (Gross margin as used here implies weekly revenue before deducting crew costs.)

TABLE 7.10

Possible positions in three departments with associated savings (gross margin) expected from filling each position. Not listed but implied is the possibility that none of the positions will be filled.

Department	Number of student positions	Gross margin
H	1	$2100
	2	3700
	3	5000
	4	6100
MV	1	$1800
	2	2600
	3	5500
IM	1	$ 500
	2	3800

Number of crews	Weekly Gross Margin		
	Route A	Route B	Route C
1	$ 8,800	$ 800	$ 6,400
2	13,600	8,000	12,000
3	16,800	10,400	14,400

7B.2(a) Draw graphs for each of the routes and determine how many crews should be allocated to each when crew costs are $2800 per week?

7B.2(b) If additional crews can be hired and trained to service the routes, each at a cost of $3000 per week, to which routes should they be assigned?

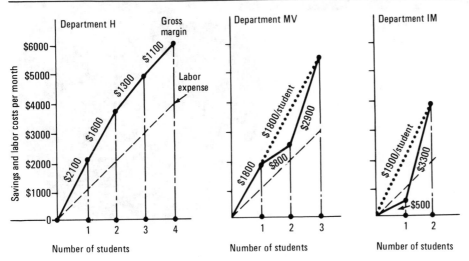

FIGURE 7.2

Graphs of employment alternatives. Solid lines connect the gross margins for departmental employment levels. For Department H these lines form a convex polygon which is a common pattern that occurs when successive alternatives have diminishing marginal returns. Departments MV and IM have listed positions that do not cover the cost of employment unless an additional employee is also hired; incremental gain from combining the two positions is shown by a dotted line. The dashed lines indicate employment expenses accumulated at the rate of $1000 per student per month; these cost levels establish a minimum acceptable gross margin, equivalent to a required rate of return for investments, since the margin must at least recover costs.

CHAPTER 8

REPLACEMENT ANALYSIS

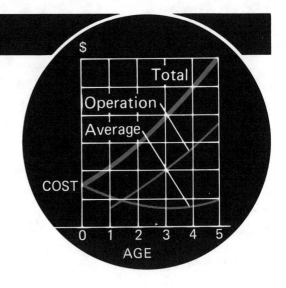

OVERVIEW

A replacement analysis is conducted to determine if and when an asset currently in service (the *defender*) should be replaced by a more economical alternative (a *challenger*). The suggested analysis procedure is to calculate the equivalent annual cost of each alternative. The current value of the defender is considered to be the capital cost for the alternative of keeping the existing equipment in service.

The analysis is based on a *study period* when there is a known limit on the time a defender's service is required. A *repeated-life analysis* may be used when the asset's services are needed indefinitely. Then a replacement should be made when the EAC of the challenger is lower than the defender's cost for the coming year *and* its EAC for its remaining service life.

Existing assets may be replaced because of deteriorating performance, obsolescence, or inadequate capacity. The current value of a defender is its market value, and the first cost of the challenger should include all expenses required to make it operational. Salvage value is an estimate of the future worth of an alternative at the time its services are no longer needed and should incorporate all disposal expenses. Cyclic replacement by identical assets should take place at the interval which minimizes equivalent annual cost.

REPLACEMENT STUDIES

Replacement refers to a broad concept embracing the selection of similar but new assets to replace existing assets, and evaluation of entirely different ways to perform an asset's function. For instance, old trucks could be replaced with new models that operate similarly but have advanced features that improve their performance. The trucks could also be replaced with a conveyor system, an overhead crane, a subcontract for hauling, or even manual labor, if any of these methods serves the needed function at a lower total cost.

Replacement decisions are critically important to a firm. A hasty decision to "get rid of that junk" because a machine is temporarily malfunctioning, or a decision to faddishly buy the latest model because "we take pride in being very modern," can be a serious drain on operating capital. A firm hard pressed for operating funds may go to the other extreme by adopting a policy that postpones replacements until there is no other way to continue production. A policy of postponement places a firm in the dangerous position of becoming noncompetitive. Reliance on inefficient equipment and processes that lead to higher long-run operating costs or low quality, while competitors enjoy declining costs and better quality gained from modern machinery, is a delaying action that eventually must be paid for, perhaps in bankruptcy. Engineers bear responsibility to recognize when an asset is no longer employed efficiently, what replacements should be considered, and when replacement is economically feasible.

A folksy replacement policy is "if it ain't broke, don't mess with it."

A replacement decision is a choice between the present asset, sometimes called the *defender,* and currently available replacement alternatives, sometimes called *challengers.* The defender may or may not be at the end of its economic life. An asset is *retired* when its owner disposes of it, but it may still serve other owners as second-hand equipment before it is scrapped. Unsatisfactory performance and inability to meet current capacity needs are the main causes of retirement. The challenger may or may not perform the function of the defender in the same way.

Replacement studies are usually made as equivalent annual-cost calculations to take advantage of data traditionally collected as annual charges: depreciation, maintenance costs, operating expenses, salaries, taxes, etc. Other considerations, such as price escalations and cost estimating, are developed in subsequent chapters.

Tax effects, which may be significant in replacement studies, are discussed in Chapter 11.

Current Salvage Value of the Defender

The salvage value of an asset being evaluated for replacement is the cost of keeping it in service. That is, the best estimate of the defender's worth is the capital cost of the no-change alternative, which is usually taken to be its market value at the time of the study.

An alternative procedure is that of considering the defender's salvage value as a receipt (positive cash flow) that offsets part of the purchase price (negative cash flow) of each challenger. Then the net differences between the cash flows of the defender and each challenger are compared. To illustrate the two approaches, consider the choice between a defender that has a current market value of $5000 and a challenger that can be purchased for $7500. Both

TABLE 8.1

Costs for replacement alternatives D and C

Year	Defender (D)	Challenger (C)	Difference ($D - C$)
0	$P = \$5000$	$P = \$7500$	$-\$2500$
1	1700	500	1200
2	2000	1100	900
3	2500	1300	1200

have a service life of 3 years with no salvage value expected at the end of that time. Their operating costs are shown in Table 8.1.

Operating costs typically increase as an asset gets older. Since the defender has already provided prior service, its annual costs are usually higher than the challenger's. Also, the first-year operating cost for a challenger may be very low when a warranty is issued by the seller.

If the market value of the defender is considered to be the capital cost of its continued service with the MARR at 12 percent, then

A trade-in value that exaggerates the real market value of a defender is described in Example 8.3.

$$EAC(D) = [\$5000 + \$1700(P/F, 12, 1) + \$2000(P/F, 12, 2)$$
$$+ \$2500(P/F, 12, 3)](A/P, 12, 3)$$
$$= (\$5000 + \$4892)(0.41635) = \$4119$$

The equivalent annual cost of the $7500 challenger that reduces the yearly operating costs is

$$EAC(C) = [\$7500 + \$500(P/F, 12, 1) + \$1100(P/F, 12, 2)$$
$$+ \$1300(P/F, 12, 3)](A/P, 12, 3)$$
$$= (\$7500 + \$2249)(0.41635) = \$4059$$

which indicates that it should replace the defender. A PW comparison would show the same preference, of course, by indicating a lower present worth of costs over the 3-year period for the challenger:

$$PW(D) = -\$9892 \quad \text{and} \quad PW(C) = -\$9749$$

The other way to make the comparison is to subtract the market value of the defender from the first cost of the challenger ($7500 − $5000 = $2500) to establish the net purchase price. Then the differences in annual operating costs are determined as the savings (positive cash flow) associated with the replacement. Using the differences ($D - C$) from Table 8.1 gives

$$EAW(D - C) = [-\$2500 + \$1200(P/F, 12, 1) + \$900(P/F, 12, 2)$$
$$+ \$1200(P/F, 12, 3)](A/P, 12, 3)$$
$$= (-\$2500 + \$2643)(0.41635) = \$60$$

Since identical cost estimates were used and the asset lives were the same, the equivalent annual worth of replacing the defender equals the difference between the previously calculated EACs:

$$EAC(D) - EAC(C) = \$4119 - \$4059 = \$60$$

It is important to remember that in the recommended procedure of calculating an EAC for *each* alternative, the defender's salvage value at replacement time cannot be both a capital cost for the defender *and* a reduction in the challenger's purchase price; it is the former, not the latter.

Example 8.1
Inflated Salvage Value of a Defender

A second challenger competes with the defender described in Table 8.1. This challenger (CX) has a purchase price of $9000, but $6000 is offered for the defender as a trade-in and the seller guarantees that operating costs will be no more than $800 per year. Should the offer be accepted when the required rate of return is 12 percent and no salvage value is expected at the end of challenger CX's 3-year life?

Solution 8.1

It is not unusual for a vendor to offer more than the market value for a used asset, to make the sale; the seller hopes to obtain a steady buyer and likely has accounted for the discount with an inflated selling price. Therefore, *the difference between a known market value and a seller's higher offer is treated as a reduction in the challenger's purchase price.* The discount of $6000 - $5000 = $1000 is applied to the $9000 first cost of challenger CX, reducing its effective purchase price to $8000. Then, with the yearly $800 operating cost,

$$EAC(CX) = (\$9000 - \$1000)(A/P, 12, 3) + \$800$$
$$= \$8000(0.41635) + \$800 = \$4131$$

This value is compared with EAC(D) based on the defender's real market value ($5000) and EAC(C) of the first challenger's actual purchase price; the new challenger is rejected because its EAC(CX) is higher, even with the purchase-price discount included.

Defender and Challengers with Different Lives

Ways to make comparisons of assets having different lives were introduced in Chapters 5 and 6. There are two basic methods: study period and repeated life. The former is applicable when an asset is required for only a specified period, and the latter is appropriate when the asset is needed indefinitely. In both cases the most economical time for the replacement to occur should be checked.

The duration of a *study period* is set by the period of known need for an asset's service. When this time period is less than or equal to the remaining service life of the defender, and the challengers have longer lives, the salvage values at the time of service termination must be estimated. These salvage values are then inserted for S in the equivalent annual capital-cost formula, $EAC = (P - S)(A/P, i, N) + Si$, where N is the number of years in the study

See Exercise 2.

period. It may be necessary to check several termination points to secure the lowest total EAC (sum of equivalent capital and operating costs), because the replacement can take place during any year of the study period, or not at all.

When a known service need extends beyond the defender's remaining service life, it is usually assumed that the defender will be replaced by the best challenger. This policy applies to both the original defender and a replacement of the original challenger, if the study period is longer than the challenger's life. In some studies it may be feasible to anticipate that a better challenger will become available later in the study period; then future replacement plans will include this best-possible challenger. The cash flow of each alternative during a study period thus includes the original capital cost, net purchase price of repeated (or best available) replacements, salvage value at the termination of the study period, and annual operating costs.

A *repeated-life* analysis is based on assumptions that an asset is replaced with an identical asset and that there is a continuing requirement for its services. Identical replacement is a reasonable assumption for a challenger, but unrealistic for the defender. A defender is typically an older piece of equipment exhibiting decreasing efficiency, and it is doubtful that an identical piece could be found, even if it were wanted. However, a repeated-life comparison is still appropriate for the continuous-service case because it indicates whether a challenger promises a lower EAC than the existing asset. If so, the next question is when to make the replacement.

Consider the defender described in Table 8.1 and a challenger with a first cost of $12,000, salvage value of $2000 at the end of its 5-year economic life, and annual operating costs of $700. Services provided by the equipment will be needed indefinitely. When MARR = 12 percent, repeated identical replacements of the challenger lead to

$$\text{EAC(challenger)} = (\$12{,}000 - \$2000)(A/P, 12, 5) + \$2000(0.12) + \$700$$
$$= \$2774 + \$240 + \$700$$
$$= \$3714$$

Since this equivalent annual cost is significantly lower than the previously calculated EAC(defender) = $4119, replacement is advisable. The rationale for this decision is that keeping the defender for 3 years costs the equivalent of $4119 per year, *after which it will be replaced by the minimum-cost replacement;* the best current challenger will be that minimum-cost replacement unless there is reason to believe that an even better challenger will become available in 3 years.

Given a challenger with a lower EAC than the defender and no expectation of a better challenger in the foreseeable future, the remaining question is when to make the replacement. The replacement normally takes place when the cost of one more year's service by the defender exceeds the equivalent annual cost of the challenger. An exception occurs in the rare instance when a defender's market value does not progressively decline with age or its operating costs do not progressively increase.

As an example of the normal capital and operating cost pattern, assume that the defender in Table 8.1 loses $2000 in value from one more year of service. Then its salvage value at the end of the year is $3000, making its EAC for next year

See also Exercise 1.

$$\begin{aligned} \text{EAC}(D_{one\ yr}) &= (\$5000 - \$3000)(A/P,\ 12,\ 1) + \$3000(0.12) + \$1700 \\ &= \$2000(1.12) + \$360 + \$1700 \\ &= \$4300 \end{aligned}$$

which exceeds EAC(challenger) = $3714. The replacement should be made immediately.

Additional examples of replacement analysis are provided in the following sections, where comparison procedures are related to the causes of replacement: deterioration, obsolescence, and inadequacy.

REPLACEMENT DUE TO DETERIORATION

Deterioration is manifested through excessive operating costs, increased maintenance cost, higher reject rates, or a combination of added equipment costs. As costs climb, it soon becomes apparent that a replacement study is warranted. Successive studies may be required to determine when the costs for operating another period without change become greater than the annual cost expected from the replacement. The assumption is that the challenger, when acquired, will be kept its full economic life, but it will not be acquired until its equivalent annual cost is lower than next year's cost for the defender. Before final acceptance, the challenger must also pass the test that its equivalent annual cost is smaller than the equivalent annual cost of the defender, based on the defender's remaining years of service.

Example 8.2
Annual Costs for Next Year and Remaining Years of Ownership of an Existing Asset Compared with the Equivalent Annual Cost of a Proposed Replacement

An existing machine is worth $2500 today and will lose $1000 in value by next year plus $500 per year thereafter. Its $8000 operating cost for this year is predicted to increase by $1000 annually, owing to deterioration. It will be retired in 4 years, when its salvage value will be zero.

A new, improved machine that satisfactorily performs the same function as the existing machine can be purchased for $16,000 and is expected to have relatively constant annual operating costs of $6000 to the end of its 7-year economic life, at which time the salvage value will be $1500. No major improvements are expected in designs for machines of this type within 7 years.

If the minimum attractive rate of return is 12 percent, should the existing machine be replaced? If so, when?

Solution 8.2

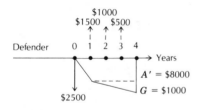

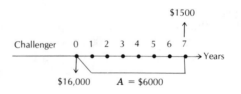

Equivalent annual cost of challenger

Operating cost	$6000
Capital recovery: $(P - S)(A/P, 12, 7) + Si$	
($16,000 − $1500)(0.21912) + $1500(0.12)	3357
	$9357

One-year cost of defender

Operating cost	$8000
Capital recovery: $(P - S)(A/P, 12, 1) + Si$	
($2500 − $1500)(1.12) + $1500(0.12)	1300
	$9300

Since the defender has a lower annual cost for next year, it should be retained. To anticipate the future replacement budget, conditions for service during year 2 should also be checked. At the beginning of year 2 the existing asset has a value of $1500, which declines to $1000 at the end of the year. Operating costs for the year are $8000 + $1000 = $9000.

Equivalent annual cost of defender during year 2

Operating cost	$9000
Capital recovery:	
($1500 − $1000)(1.12) + $1000(0.12)	680
	$9680

If the cost estimations are deemed reasonable, purchase of the challenger should be anticipated 1 year from now.

It would be a mistake to base the decision on comparison of the equivalent annual cost of the existing machine for its 4 remaining years before retirement, as follows:

Equivalent annual cost of defender based on a 4-year life

Operating cost: $A' + G(A/G, 12, 4)$	
$8000 + $1000(1.3588)	$ 9,359
Capital recovery: $(P - S)(A/P, 12, 4) + Si$	
= $2500(0.32924)	823
	$10,182

The challenger would then have been purchased while the defender's cost of one more year of ownership was lower than the replacement's equivalent annual cost. A comparable calculation for the conditions expected 1 year from now confirms the decision to make the replacement then:

Equivalent annual cost of defender based on the last 3 years before retirement

Operating cost: $A' + G(A/G, 12, 3)$	
$9000 + $1000(0.9246)	$ 9,925
Capital recovery: $P(A/P, 12, 3)$	
= $1500(0.41635)	625
	$10,550

REPLACEMENT DUE TO OBSOLESCENCE

Each new development or refinement of an older asset makes the previous way of accomplishing an objective less appealing. Improved methods and machines affect replacement studies in two opposing ways. Improvements usually result in higher first costs for a challenger and reduce the salvage value of a defender; together these two capital costs tend to favor retention of the existing asset. However, dramatic operating-cost reductions or impressive quality gains provided by technologically advanced challengers are the typical causes of obsolescence, and these annual savings tend to exceed the capital-cost advantage of a defender.

Progressive design enrichments can outdate an existing asset well before its economic life has expired. When this happens, the owned asset probably retains considerable resale value, unless the new design is revolutionary. The replacement study should be based on the current *market* value of the defender, *not* on the accounting records or original estimates of resale value, because these figures rely on data acquired before the existence of the improved challenger became known. The remaining years of service life for an obsolete asset usually conform to the original estimate of N.

Rapid developments in computer-aided machines create complex replacement questions, as discussed in Extension 8A.

Example 8.3
Selection of a Salvage Value for an Obsolete Asset

A low-volume office copying machine was purchased 2 years ago for $700. At the time of purchase it was believed that the machine would have an economic life of 5 years and a salvage value of $100. Operating costs over the first 2 years for material, labor, and maintenance have averaged $4200 annually and are expected to continue at the same level. Some type of copier will be needed for the next several years.

The same company that manufactured the presently used copying machine has a new model which costs $1000 but will perform the current workload with operating costs of $3500 per year. They are offering $500 for the old model as a trade-in on the new machine. The expected salvage value for the new model is $200 at the end of 10 years.

Another company has a different type of copier which is available only on a lease basis. The company claims that leasing their copier at $750 per year will reduce the operating expense for the present amount of work to $2750. Since they do not accept trade-ins, the machine now in use would have to be sold on the open market, where it is expected to bring only $300.

If the minimum acceptable rate of return is 10 percent before taxes, should the defending copier be replaced by one of the challengers?

Solution 8.3

The market value of the present copier seems to depend on the alternative to which it is being compared, but in either case the remaining period of expected service is 3 years and the salvage value is $100. Trade-in values are often placed unrealistically high to entice a buyer. What the bloated trade-in actually amounts to is a discount from the selling price. For the given data, it appears that the true market value is $300 rather than $500. The difference is a discount of $500 − $300 = $200, which makes the effective purchase price = $1000 − $200 = $800. In order to compare both challengers with the defender under equitable conditions, the apparent market value ($300) is used, and the price for purchasing a copier is discounted accordingly.

Equivalent annual cost of defender

Operating costs	$4200
Capital recovery: $(P - S)(A/P, 10, 3) + S(0.10)$	
$(\$300 - \$100)(0.40212) + \$100(0.1)$	90
	$4290

Equivalent annual cost of the challenger from the same vendor

Operating costs	$3500
Capital recovery: $(P - S)(A/P, 10, 10) + S(0.10)$	
$[(\$1000 - \$200) - \$200](0.16275) + \$200(0.1)$	117
	$3617

Equivalent annual cost of the leased challenger

Operating costs	$2750
Annual lease contract	750
	$3500

The annual-cost calculations indicate that both challengers are preferred to the defender. Leasing is more attractive than buying, if it can be assumed that lease charges will not increase. Another consideration is the possibility that future improvements will make the current $1000 challenger obsolete in the near future. A lease is less expensive, for next year at least, and provides flexibility to take advantage of future cost-reduction developments if they occur.

Using the trade-in value for the existing asset in Example 8.3 does not alter the solution appreciably:

Equivalent annual cost of defender
when P = trade-in value = $500

Operating costs	$4200
Capital recovery:	
($500 − $100)(0.40212) + $100(0.1)	171
	$4371

Equivalent annual cost of purchased challenger
when P = $1000

Operating costs	$3500
Capital recovery:	
($1000 − $200)(0.16275) + $200(0.1)	150
	$3650

The disparity between annual worths using the trade-in value ($4371 − $3650 = $721) and the market value for the defender ($4290 − $3617 = $673) is $721 − $673 = $48. It is a result of the difference in the capital-recovery periods of the defender and challenger. Equal reductions in the present values of the challenger and defender favor the defender in a comparison because it almost always has a shorter life. Thus the use of effective value is a more conservative approach.

Car sales are the classic illustration of selling tactics that use bloated trade-in offers or discounts for cash payment as two paths to the same sticker price.

Viewing a replacement decision from a stranger's outlook may clarify the situation and provide impartiality. A stranger viewing the choice of a copier described in Example 8.3 could place the decision in the following context: "I need a copying machine. From one manufacturer I can buy a new machine for $1000 which will have a $200 salvage value at the end of 10 years. The annual operating costs for this machine will be $3500. From the same manufacturer I can also get a used copier for $500 which will last 3 years. It will have a $100 salvage value and annual operating expenses of $4200. On the other hand, I can lease a machine for $750 a year, with expected operating costs of $2750, or buy a used machine for $300 which will last 3 years and have annual operating costs of $4200. I will select the alternative which provides the lowest equivalent annual cost."

REPLACEMENT DUE TO INADEQUACY

When current operating conditions change, an older asset occasionally lacks the capacity to meet new requirements. Sometimes a similar asset can be purchased to supplement the old asset, as in the case of placing a new generator alongside an old one to meet new power demands. New layouts, building additions, and design changes are examples of possible modifications to increase capacity.

Alternatives or supplements to an existing asset are usually compared with a challenging new asset which may perform an equivalent function in an entirely different manner. Replacing a wood stove with an oil furnace, for example, provides an entirely different way to heat a home. Even though a challenger is deemed a desirable replacement, the defender may still have value as usable equipment. In such cases it can be sold or retained for standby purposes. It is also possible that secondary uses can be found for assets replaced from their primary function; the wood stove replaced by an oil furnace might be used as an incinerator.

Example 8.4
Upgrading versus Demolition and Replacement

A small bridge leading to a proposed industrial park has a load limit of 10,000 pounds (4536 kilograms). A manufacturing firm will lease a building site in the park if the capacity of the bridge is raised to 60,000 pounds (27,216 kilograms). The developers of the land have two alternatives. They can reinforce the old bridge or they can tear it out and fill in the low area, leaving a culvert to carry away surface water.

The present bridge has no realizable salvage value. Reinforcement would cost $30,000 and should provide adequate access for 10 years without any major additional work. The salvage value from added materials would be $8000 in 10 years.

A culvert-and-fill approach to the park would cost $60,000 and should meet all requirements for the next 50 years. There would be no salvage value. In addition, it will cost $2000 to remove the old bridge. Maintenance costs are expected to be $2200 per year less than upkeep for a bridge.

Annual property taxes and insurance on the improvements will be 1 percent of the first cost. The required return on investments is 8 percent before taxes. If the developer feels that a new approach to the park is required, which alternative should be selected?

Solution 8.4

Since a replacement for the old bridge is definitely necessary, there is no distinct defender-challenger relationship. The lower-initial-cost alternative, reinforcement, would best fit the role of defender. It's equivalent annual cost is the sum of capital recovery, extra maintenance costs, and taxes plus insurance.

Equivalent annual cost of reinforcing the bridge

Additional maintenance costs	$2200
Capital recovery: $(P - S)(A/P, 8, 10) + S(0.08)$	
($30,000 − $8000)(0.14903) + $8000(0.08)	3919
Taxes and insurance: $30,000(0.01)	300
	$6419

Equivalent annual cost of the culvert and fill

Capital recovery: $P(A/P, 8, 50)$	
($60,000 + $2000)(0.08174)	$5068
Taxes and insurance: $60,000(0.01)	600
	$5668

From strictly a cost viewpoint, the culvert and fill has a clear advantage of $6419 − $5668 = $751 per year. Other management considerations could influence the final choice. For instance, the developers might be willing to forgo the annual $751 benefit in order to be allowed an opportunity to change plans in 10 years when the reinforced bridge will again need a replacement study. They might be short of capital at the present time or have other possible investments with a greater potential rate of return than that earned on the extra increment of investment ($62,000 − $30,000 = $32,000) required for the culvert and fill. (Note that the extra increment earns 8 percent *plus* $751 a year.)

ECONOMIC LIFE FOR CYCLIC REPLACEMENTS

Costs of automobile ownership are indicated in Extension 8B.

Many mechanical items used in service agencies and manufacturing are replaced by essentially the same machine when the original wears out. Informal rules may be used to establish cyclic replacement times. One government department replaces a car whenever it exceeds total mileage of 65,000 miles (104,605 kilometers) or 5 years of service, or when the cumulative maintenance cost equals the purchase price. Such rules recognize that automobiles, trucks, typewriters, and similar machines become less efficient and accumulate higher and higher repair bills as they age. Conversely, the longer they are kept in operation, the lower will be their average annual capital cost because the purchase price is spread over more years. The sum of these two types of cost is the total cost of providing the machines' services. As shown in Figure 8.1, total lifetime cost continues to increase with age, but average annual cost passes through a minimum.

Cyclic Replacement Analysis when MARR = 0 Percent

The objective of a cyclic replacement study is to determine the pattern of replacement that will minimize the average annual cost. Data for future costs must be estimated, but cost figures for most assets that are used enough to warrant a cyclic replacement study are available from internal or suppliers' records. Simplifying assumptions made to introduce the calculation procedure include a replacement price that remains constant and no interest charges for the use of capital. However, relaxing these assumptions does not affect the optimum cycle time very much unless there are dramatic changes in price or exceptionally high interest rates.

Calculations to determine the minimum-cost age follow an iterative routine in which the capital cost (decrease in salvage value to a certain age) is added to the accumulated operating cost (maintenance plus use expenses) for that same period. Each sum is divided by the relevant age to give an average cost for that replacement interval. This routine is illustrated in Table 8.2 for a city delivery service that owns a fleet of small trucks for store-to-home deliveries. The purchase price per truck is $15,000, and the anticipated schedule of future operating costs and salvage values is

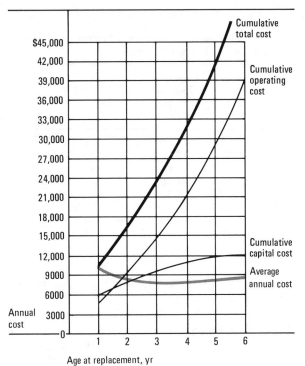

Annual cost

Age at replacement, yr

FIGURE 8.1
Cost pattern for increasing age of an asset.

	Year					
	1	2	3	4	5	6
Operating cost	$4500	$4800	$5700	$6900	$8400	$10,200
Resale value	9000	6900	5100	3900	3300	2700

These costs are inputs to the tabular solution method which reveals that trading the trucks in every 4 years for new ones is the minimum-cost replacement cycle. The cost data from Table 8.2 were graphed in Figure 8.1

Cyclic Replacement Analysis when MARR = 10 Percent

Essentially the same procedure for calculating the minimum-cost replacement age can be used when interest charges are included. This calculation determines the *economic life* of an asset subject to cyclic replacement—its *least-cost replacement interval*.

Data from Table 8.2 are again utilized to illustrate the discounted cash flow approach. Setting the MARR at 10 percent and assuming each year is the

TABLE 8.2

Average-annual-cost calculations for replacing an asset when the purchase price is constant and no interest is charged on capital

Age of Truck at Replacement, Years (1)	Cumulative Operating Costs (2)	Cumulative Capital Cost* (3)	Total Cost [(2) + (3)] (4)	Average Annual Cost [(4) ÷ (1)] (5)
1	$ 4,500	$ 6,000	$10,500	$10,500
2	9,300	8,100	17,400	8,700
3	15,000	9,900	24,900	8,300
4	21,900	11,100	33,000	8,250
5	30,300	11,700	42,000	8,400
6	40,500	12,300	52,800	8,800

* Cumulative capital cost equals $15,000 less resale value.

economic life N of the asset, we calculate the equivalent annual cost of owning the asset N years from

$$\text{EAC} = \text{capital recovery} + \text{equivalent annual operating cost}$$
$$= (P - S)(A/P, i, N) + Si + \text{FW(operating costs for } N \text{ years)}(A/F, i, N)$$

The annual cost for 1 year of ownership when $i = 10$ percent is

$$\text{EAC}_{N=1} = (\$15,000 - \$9000)(A/P, 10, 1) + \$9000(0.10) + \$4500(A/F, 10, 1)$$
$$= \$6000(1.10) + \$900 + \$4500(1.0000) = \$12,000$$

For successive years,

$$\text{EAC}_{N=2} = (\$15,000 - \$6900)(A/P, 10, 2) + \$6900(0.10)$$
$$+ [\$4500(F/P, 10, 1) + \$4800](A/F, 10, 2)$$
$$= \$8100(0.57619) + \$690 + [\$4500(1.10) + \$4800](0.47619)$$
$$= \$5358 + \$4644 = \$10,002$$

$$\text{EAC}_{N=3} = (\$15,000 - \$5100)(A/P, 10, 3) + \$5100(0.10)$$
$$+ [\$4500(F/P, 10, 2) + \$4800(F/P, 10, 1) + \$5700](A/F, 10, 3)$$
$$= \$9900(0.40212) + \$510$$
$$+ [\$4500(1.21) + \$4800(1.10) + \$5700](0.30212)$$
$$= \$4491 + \$4962 = \$9453$$

$$\text{EAC}_{N=4} = (\$15,000 - \$3900)(A/P, 10, 4) + \$3900(0.10)$$
$$+ [\$4500(F/P, 10, 3) + \$4800(F/P, 10, 2)$$
$$+ \$5700(F/P, 10, 1) + \$6900](A/F, 10, 4)$$
$$= 11,100(0.31547) + 390$$

$$+ [\$4500(1.331) + \$4800(1.21) + \$5700(1.10) + \$6900](0.21547)$$
$$= \$3891 + \$5379 = \$9270$$

$$
\begin{aligned}
EAC_{N=5} = \ & (\$15{,}000 - \$3300)(A/P, 10, 5) + \$3300(0.10) \\
& + [\$4500(A/P, 10, 4) + \$4800(F/P, 10, 3) \\
& \quad + \$5700(F/P, 10, 2) + \$6900(F/P, 10, 1) + \$8400](A/F, 10, 5) \\
= \ & \$11{,}700(0.26380) + \$330 + [\$4500(1.4641) + \$4800(1.331) \\
& \quad\quad\quad\quad + \$5700(1.21) + \$6900(1.10) + \$8400](0.16380) \\
= \ & \$3417 + \$5874 = \$9291
\end{aligned}
$$

which indicates that the economic life of the truck and the minimum-cost replacement period is 4 years. This is the same replacement interval indicated in Table 8.2, where the analysis was conducted at $i = 0$. Note that equivalent annual capital costs decline as the ownership period increases and at the same time, the equivalent annual operating costs increase owing to deteriorating performance. The pattern of changes in capital and operating costs should be checked periodically, when like-for-like replacements are in effect, to confirm that the economic life remains the same.

The effects of inflation on replacement prices and operating costs are examined in Chapter 13.

PRECAUTIONS FOR REPLACEMENT STUDIES

The value of a replacement study, like other economic evaluations, is directly proportional to the validity of the data. Some costs have more effect in a comparison than do others. Replacement decisions are very sensitive to recurring cash flows, especially operating costs. Unfortunately, operating expenses are more difficult to extrapolate into the future than are other periodic cash flows such as taxes and insurance.

The classic MAPI method of replacement analysis is presented in Extension 8B.

A conservative approach when estimates are highly uncertain is to give every advantage to the defender. This is accomplished by assuming that the present differential between the operating costs of the defender and challenger will remain constant during the study period and that there are no capital-recovery costs for the defender. If a challenger still looks good under these handicaps, it is truly a valid contender.

Salvage values are necessarily subject to question because they occur at the most distant point in a replacement study. The basic principle is to use the best *current* estimate of the future, regardless of previous estimates. Appraisals may change from one study to the next, owing to price fluctuations, availability, and needs. An often-neglected cost associated with salvage is the expense of getting an old asset ready for the new purchaser. These expenses could include dismantling, overhaul, painting, crating, cartage, and repairs to the area vacated by the disposed asset. When such costs exceed the disposal price, the salvage value is a loss and is treated as a minus quantity (S is negative) in capital-recovery calculations.

Another commonly ignored cost is the expense associated with putting a

Additional features that favor a defender in purchasing decisions are presented in Extension 12A as "incumbency theory."

new asset in operating order. Special wiring, piping, guard rails, foundations, and other facilities may be needed before new equipment can operate. Radically different or complex equipment often requires more "debugging" than is provided by the supplier. In a replacement study these once-only operational-type costs should be treated as capital costs.

The bridge-replacement example demonstrates a case in which there was no choice about making a replacement study. More commonly an asset performs its intended function without obvious financial loss. If an acceptable challenger goes unnoticed, the accumulated yearly losses from failing to recognize the need for a replacement can be substantial. One clue to the replaceability of an asset is its economic life. As an asset nears the end of its original life estimation, it becomes a more likely candidate for a replacement study. Other clues include an awareness of new developments which could lead to asset obsolescence and the deterioration of performance, as indicated by reject rates or frequent repairs. Since machines age less obviously than do humans and cannnot complain about their frailties, it is an engineering function to diagnose infirmities and prepare remedies.

Review Exercises and Discussions

Exercise 1 A grinder was purchased 3 years ago for $40,000. It has provided adequate service, but an improved version is now available for $35,000 which will reduce operating costs and cut inspection expenses. Costs and salvage values for the two machines are shown below. Should a replacement be made if the required rate of return is 15 percent and the services of a grinder will be needed for only 4 more years?

Year	Defender (D) Operating cost	Defender (D) Salvage value	Challenger (C) Operating cost	Challenger (C) Salvage value
0		$12,000		$35,000
1	$3400	7,000	$ 200	30,000
2	3900	4,000	1000	27,000
3	4600	2,500	1200	24,000
4	5600	1,000	1500	20,000
5			2000	17,000
6			2600	15,000

Solution 1 The first calculation determines the EAC when replacement is made immediately. This allows consideration of only the first 4 years of the challenger's 6-year economic life. Its salvage value after 4 years is $20,000. The defender's current salvage value, $12,000, is its market worth now; the original purchase price of $40,000 has no bearing on the replacement decision.

$$\text{EAC}(D) = [\$12,000 + \$3400(P/F, 15, 1) + \$3900(P/F, 15, 2)$$
$$+ \$4600(P/F, 15, 3) + \$5600(P/F, 15, 4) - \$1000(P/F, 15, 4)](A/P, 15, 4)$$
$$= \$23,560(0.35027)$$
$$= \$8252$$

$$EAC(C) = (\$35,000 - \$20,000)(A/P, 15, 4) + \$20,000(0.15)$$
$$+ [\$200(P/F, 15, 1) + \$1000(P/F, 15, 2)$$
$$+ \$1200(P/F, 15, 3) + \$1500(P/F, 15, 4)](A/P, 15, 4)$$
$$= (\$15,000 + \$2577)(A/P, 15, 4) + \$20,000(0.15)$$
$$= \$17,577(0.35027) + \$3000$$
$$= \$9157$$

A replacement should not be made now, nor should one be made anytime during the remaining period of needed service, as indicated below by the EACs for the other three periods of possible ownership.

Periods of Service	Defender	Challenger
EAC(if used in years 2 to 4)	$7399	$ 9,170
EAC(if used in years 3 and 4)	7061	9,543
EAC(if used only in year 4)	7475	10,450

The headquarters building owned by a rapidly growing company is not large enough for current needs. A search for enlarged quarters revealed only two alternatives that provide sufficient room, enough parking, and the desired appearance and location. One can be leased for $144,000 per year and the other can be purchased for $800,000, including a $150,000 cost for land.

Exercise 2

The study period for the comparison is 30 years, and the desired rate of return on investments before income taxes is 12 percent. It is believed that land values will not decrease over the ownership period, but the value of a structure will decline to 10 percent of the present worth in 30 years. Property taxes are 4 percent and rising. For comparison purposes, annual tax payments should be uniformly close to 5 percent of the purchase price.

The present headquarters building is already paid for and is now valued at $300,000. The land it is on is appraised at $60,000. An engineer suggests that consideration be given to remodeling the present structure. The engineer estimates that an expenditure of $300,000 will provide the necessary room and improve the appearance to make it comparable to the other alternatives. However, the remodeling will occupy part of the existing parking lot. An adjacent privately owned parking lot can be leased for 30 years under an agreement that the first year's rent of $9000 will increase by $500 each year. If upkeep costs are the same for all three alternatives, which one is preferable?

The cash flows are summarized in the following three diagrams:

Solution 2

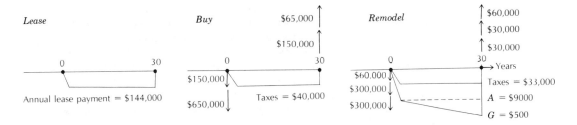

Equivalent annual cost of leasing

Rent for building and parking area $144,000

Equivalent annual cost of buying

Capital recovery: $(0.9P)(A/P, 12, 30) + (0.1P + land)(0.12)$
= $(0.9)(\$650,000)(0.12414) + [(0.1)(\$650,000) + \$150,000](0.12)$ $ 98,422

Taxes: $800,000(0.05)   40,000

     $138,422

Equivalent annual cost of remodeling:

Parking-lot rent: $A' + G(A/G, 12, 30)$ = $9000 + $500(7.2974) $ 12,649

Capital recovery: $(0.9)(\$300,000 + \$300,000)(A/P, 12, 30)$
            $+ [(0.1)(\$300,000 + \$300,000) + \$60,000](0.12)$
= $540,000(0.12414) + $120,000(0.12)   81,436

Taxes: $660,000(0.05)   33,000

     $127,085

Remodeling the presently occupied building is the preferred course of action. Note that the only capital recovery on land value is for interest on the amount invested in land and that the already-paid-for value of the present headquarters building is included in the capital-recovery charge.

PROBLEMS

8.1 At the end of half its expected economic life, a 4-year-old machine has a book value of $5800 from its original cost of $9200. Estimated operating costs for the next year will amount to $6000. An equipment dealer will allow $3600 if the machine is traded in now and $2800 if it is traded in a year later. The dealer proposes the purchase of a new machine to perform the same function; it will cost $14,000 installed. This machine will have an estimated operating cost of $4500 per year and a salvage value of $3000 at the end of 4 years. Is it profitable to replace the existing machine now if the minimum return on investments is 15 percent before taxes?

8.2 Machine A was installed 6 years ago at a total cost of $8400. At that time it was estimated to have a life of 12 years and a salvage value of $1200. Annual operating costs, excluding depreciation and interest charges, have held relatively constant at $2100. The successful marketing of a new product has doubled the demand for parts made by machine A. The new demand can be met by purchasing an identical machine which now costs $9600 installed. The economic life and operating costs for the two machines will be the same. The salvage value for the second A-type machine will be $1600.

Machine B, a different type, costs $17,000 installed but has twice the capacity of machine A. Its annual operating costs will be about $3100, which should be relatively constant throughout its 10-year economic life. Salvage is expected to be $4000. The present machine can be used as a trade-in on the new machine B. It is worth $3000.

Compare the two alternatives on the basis of equivalent annual cost when the interest rate is 10 percent.

8.3 Compare the alternatives of replacing two type-A machines with one type-B machine, as described in Problem 8.2 by the rate-of-return method. Does the resulting IRR have any advantage over the results obtained from an annual-worth comparison when it is necessary to explain the evaluation to someone not familiar with discounted cash flow comparisons? Explain.

8.4 A firm purchased a pump and motor for $1925 installed. It was later discovered that the pump had been improperly selected for the required head and discharge. As a result, the power bill for operating the pump was $900 for a year.

A new pump, suited to the requirements, is available for $2450 installed, with a guarantee that power costs will not exceed $500 annually. The original pump and motor can be sold for $375.

Assume a 10-year study period with zero salvage value for both pumps at the end of the period. The firm uses a minimum attractive rate of return of 10 percent before taxes. Based on present-worth calculations, should the pump be replaced?

8.5 Machine 1 will do a required operation adequately; it can be delivered immediately at a price of $11,250. Its operating costs are $9500 annually, and it will have no realizable salvage value at the end of its 5-year economic life. This machine is being compared with another presently available machine; machine 2 has a first cost of $30,000, annual operating costs of $6500, and a salvage value of $3000 in 10 years.

The much-improved machine 3 performs the same function as do the other two contenders; it has a first cost of $14,500 with an expected salvage value of $4250 after 5 years. Its major advantage is a lower operating cost of $5000 per year. The only drawback is that the machine is so much in demand that there is a 2-year wait for delivery. Since the firm cannot delay acquiring one of the machines, it must choose between the currently available alternatives 1 and 2. It uses a 15 percent rate of return before taxes for economic comparisons.

  (a) Compare the alternatives using the repeated-projects method of analysis.
  (b) Compare the alternatives using a 10-year study period.
  (c) Compare the alternatives using the best-possible-replacement approach.

8.6 Assume that your present car is a clunker that could be sold to a used-car dealer for $500. It will probably keep running for 2 more years, at which time it will have a scrap value of $100. Operating costs next year are expected to be $2400 and will likely be $2950 the following year. Its resale value will decrease by $250 during the coming year.

The lowest-priced new car that you can find costs $6500. If you buy it, you plan to keep it for 7 years, at which time it will have a market value of $500. The car dealer has offered you $750 for your clunker as a trade-in on the new one and claims that your operating costs should drop to $1200 the first year and should increase by about $250 in each of the following years. Your annual interest rate is 11 percent.

  (a) What is the equivalent annual cost of the challenger?
  (b) What are the equivalent annual costs for your old car?

8.7 Assume that a friend of yours is thinking about buying a new car. Her presently owned car can be sold for $1600 or can provide adequate service for 2 more years at an operating cost of $1500 next year and $1700 the year afterward. She has been offered $2000 for her car as a trade-in on a demonstrator that has a purchase price of $6000. The demonstrator is expected to have an economic life of 6 years, with operating costs of $800 the first year that increase by $150 per year thereafter. The value of her car in 2 years will be the same as the new car in 6 years: $600. An 8 percent interest charge is appropriate. Assuming that she will need a car indefinitely, decide what advice you can offer based on equivalent annual costs.

8.8 Using a 12 percent minimum acceptable rate of return, calculate the present worth of the costs for the two alternatives described below, based on a best-possible-replacement policy, when the need for the equipment is expected to continue indefinitely.

The present equipment could be sold now for $5000, but in 3 years it will have no salvage value owing to radically improved equipment expected to be available then; it will cost $1000 to remove the old equipment in 3 years. This alternative (A) has annual expenses of $3900.

A greatly improved version of the equipment which is to be introduced in 3 years will have annual operating costs of $1900, a first cost of $10,000, and a salvage value of $2000 at the end of its 3-year life.

A presently available replacement has a price of $13,500 and a salvage value of $500 at the end of its 6-year economic life. This alternative (B) has operating costs of $3000 per year.

8.9 A small retail business is housed in a structure built 10 years ago and heated by a central gas furnace. Current heating costs average $500 per month and the furnace is in need of an overhaul, including replacement of the fan motor and filter system. Cost of this work, which is expected to be $1600, will improve the furnace efficiency by approximately 10 percent. The furnace could also be replaced by either of two new units. The first is a modified version of the current system, which would cost $4200 and is expected to reduce energy costs by 25 percent. The second is a pulse-jet system costing $7000 and providing a 35 percent saving. All three systems would be expected to last 10 years with no maintenance costs other than routine cleaning and filter replacement. At the end of that time, the current system would have no salvage value and the new system would probably be worth about 20 percent of the purchase price. The current system has a salvage value of $500 which could be applied as a trade-in. Using a MARR of 20 percent, determine which alternative should be selected.

8.10 A question has arisen with respect to whether it is more economical to replace a forklift truck or to rebore the engine and completely rebuild the present one. The original cost of the present truck 10 years ago was $9000. To rebore the engine and completely rebuild the truck will extend its life another 5 years and will cost $7200. A new forklift truck of the same capacity will have a first cost of $12,300 and will have an estimated life of 10 years. Fuel and lubricants for the rebuilt truck will be about $4000 per year. Similar costs for the new truck will be about 15 percent less. Repairs for the new truck are expected to be about $300 per year less than for the rebuilt truck. Neither unit will have any realizable value when retired. State your assumptions and determine the equivalent present worths of the two alternatives using an interest rate of 8 percent.

8.11 A necessary function can be performed by any of the assets described by the accompanying data:

Alternative	P	N	S	Annual Cost
I	$6000	6	$2000	$ 800
II	3000	3	1000	1000
III	2000	3	0	1200

When investment funds are expected to earn 12 percent annually, what is the equivalent annual cost of the alternatives according to each of the following comparison procedures?

(a) Repeated-life method.

(b) Best possible replacement as determined in a 9-year evaluation period.

8.12 A new delivery truck has a sticker price of $6895 but the dealer will sell it for $6300 cash. A 3-year-old van can be traded for $2700 on the new truck purchased at the sticker price or sold to a used-car dealer for $2500. The resale value of the van is expected to decrease annually by 40 percent of the previous year's value each year. Operating costs for the van will be $1020 next year and will increase each following year by $400.

A new truck is expected to have operating costs of $700 per year for the next 2 years; these costs will then increase by $300 each year. After 6 years a truck can likely be sold for 10 percent of its sticker price. It is the policy to retire trucks and vans after 6 years of service and to earn 8 percent annually on invested funds.

Assuming the price of a new truck will not increase in the next few years, determine when the van should be replaced.

8.13 A specialized sewage treatment pump has an initial cost of $10,000 and first-year operating and maintenance expenses of $4000. These annual costs are expected to increase at an arithmetic rate of $800 per year, but the salvage value of the pump is dependent on material recovery only and will remain constant at $500 regardless of how long it remains in service. Assuming a tax-free interest rate of 8 percent, determine how long the pump should be retained in service before replacement.

8.14 Two years ago an automated testing machine was installed to monitor product quality. The equipment cost $25,000 and is capable of testing 400 units per day at a total operation and maintenance cost of $0.26 per unit. Product sales have now increased to the point that it is necessary to increase production to 700 units per day and the company must choose between replacing the current equipment or buying an additional machine of the same type. A larger machine with a capacity of 800 units per shift would cost $45,000 and have operating and maintenance costs of $0.22 per unit. A new 400 unit per shift machine would cost $27,000 and have inspection costs identical to the existing equipment. It is expected that the testing equipment will be required for 5 more years, at which time the salvage value will be approximately equal to 10 percent of the first cost. An equipment supplier has offered a $10,000 trade-in for the existing equipment on the high-capacity unit. Assuming the company continues on a single-shift 250-day-per-year operation, determine which alternative should be selected using a MARR of 20 percent.

8.15 A mechanical testing machine with an initial cost of $8000 closely follows the cost pattern shown below.

Year	Operating Cost	Salvage Value
1	$3500	$6000
2	3800	5000
3	4200	4300
4	4600	3900
5	5100	3500
6	5800	3200
7	6700	3000

Using 0 percent interest and assuming that replacement equipment will follow the same cost pattern, determine the economic life.

8.16 A new type of testing machine to perform the same function as that depicted in Question 8.15 is now available at a purchase price of $14,000. The new machines cost more, but they have twice the capacity of the older models. A company plans to replace four of the older types with the new design. The four candidates for replacement are now 2 years old. They will be replaced when the expected cost of keeping them in service one more year is greater than the average annual cost for new machines of comparable capacity. The expected cost pattern for the new type of testing machine is:

Year	Operating Cost	Salvage Value
1	$ 7,200	$10,800
2	7,600	8,600
3	8,400	7,500
4	9,500	5,500
5	11,000	4,000

When should the replacement occur if no interest is charged?

8.17 The Uptown Urban Transit District is considering the purchase of new midsized buses for use in expanding service to suburban areas. The new units have a first cost of $105,000 and are expected to have a service life of 6 years. Maintenance and operating costs are estimated at $40,000 for the first year of use and will probably increase at an arithmetic rate of $2000 per year thereafter. The projected intermediate salvage values are as follows:

	End of Year					
	1	2	3	4	5	6
Salvage value	$80,000	$50,000	$40,000	$30,000	$10,000	$5000

(a) Determine when replacement should occur if interest is charged at a rate of 15 percent annually.
(b) Assume that the transit district has purchased several of the buses described above and used them for 3 years. A new model with comparable service characteristics is made available on a lease basis for a cost of $60,000 per year including all maintenance and operating expense. Should the existing buses be sold and replaced by a lease agreement? If so, should the change be made now or at the end of 1, 2, or 3 years?

8.18 Three years ago, Dynamic Mechanics purchased a desk-top testing machine for $7000 and since that time, the resale or salvage value has been decreasing at a linear rate of $1000 per year. Maintenance and operating costs were $1400 the first year and have been increasing at a rate of $400 per year for each additional year of service. No exact projections are available, but it appears that both these trends should continue for at least the next 3 years. At the end of that time, the equipment could be replaced by another unit with similar cost patterns. However, a new machine with a service life of 10 years is now available for a purchase price of $10,000 and no salvage value. A factory maintenance plan is also available at a fixed cost of $1700 per year on a contract basis.

Using a before-tax MARR of 25 percent, determine what replacement policy Dynamic should follow with regard to this equipment. Assume that a need for this type of machine will exist for at least 10 more years.

8.19   Tip Top Testers Inc. (TTT Inc.) is considering the purchase of a new testing machine. The following cost estimates have been obtained:

Purchase price of equipment   = $90,000
Installation cost for equipment = $10,000

End of year	1	2	3	4	5	6
Market value	$70,000	$50,000	$40,000	$30,000	$10,000	$5000

The expected service life is 6 years. Operating expenses will be $10,000 the first year, increasing by $2000 per year each following year.

(a) When MARR = 10 percent, determine the EAC of owning and operating the equipment over the expected service life.
(b) Assume TTT Inc. has purchased the tester described above and used it for 3 years. A new model is now available at a cost of $90,000 which includes installation charges. This unit is expected to have negligible salvage value after an 8-year life, and operating expenses are expected to be a constant $15,000 per year. Should the new equipment be purchased now if the old machine can be sold at its market value, the need for a tester will continue indefinitely, and the MARR is still 10 percent?

8.20   A production machine has an initial cost of $3750 and has no salvage value at any time. The costs of inferior performance and maintenance are $300 the first year and increase by $300 each successive year the machine is used. Develop a total-cost formula that does not include interest charges for the special case in which operating expenses increase each year by the amount of the first year's cost. Then use the formula to solve for the minimum-cost ownership period.

8.21   A factory uses 11 machines of identical design. Each one costs $5000. In operation, maintenance costs are $1000 the first year and increase by 20 percent each year (for example, the third-year maintenance cost is $1000 × 1.2 × 1.2). Since the machines are of special design, they have no salvage value at any time. The MARR is 10 percent. What is the economic life for cyclic replacement of the machines?

8.22   The Shortcircuit Co. is considering the purchase of a newly developed Sparksalot chip tester. The following data have been provided by the manufacturing engineering group for a cost evaluation: The purchase price of the equipment is $45,000 + $5000 for installation. The installation cost for each replacement is $5000. The maximum life is 6 years, with the following intermediate salvage values: year 1, $30,000; year 2, $20,000; year 3, $15,000; year 4, $11,000; year 5, $8000; and year 6, $6000. Operating expenses will be $20,000 for year 1, with an annual increase of $4000 in future years.

Based on a required rate of return of 20 percent and an assumption of cyclic replacement, determine the economic life of the tester.

EXTENSIONS

8A Adventures in Advanced-Technology Replacements

By definition, an adventure is "an enterprise involving financial risk." Investments in high-technology apparatus fit the definition. Because uncertainties are associated with all new developments, estimates of receipts and disbursements are always suspect. Performance of a conceptually advanced machine does not always live up to its billing. There may even be doubt about whether to make a replacement now or wait until a still-more-advanced challenger becomes available.

Advancements in technology are most evident in manufacturing and process industries. *Flexible manufacturing systems* (FMS) is the latest round in factory automation that began in the 1950s with numerically controlled machine tools. Then came *computer-aided design* (CAD) and *computer-aided manufacturing* (CAM), which replaced the drafting board with the CRT screen and the numerical control tape with the computer. The newest systems integrate all these elements. But the cost is high—around $25 million for computer controls, a half-dozen machine centers, and accompanying transfer robots—and refinements to further increase efficiency are introduced continually.

Engineering economists must evaluate the merit of replacing individual machines with updated versions or with models that perform the same function in an entirely different way, perhaps requiring whole new support systems. An added challenge is to coordinate replacements into a schedule that maintains production capabilities while building up to an integrated manufacturing system.

Robots are the most controversial replacements in the system. They often replace people. Two developments are behind this alarming exchange: (1) computers shrank in size

and cost to allow construction of robots, and (2) wage inflation raised the cost of labor to a level that made the use of robots feasible. A $100,000 robot may be an economically acceptable replacement for a human worker who earns only $5 an hour because the robot can work three shifts a day without breaks or vacations and produce consistently high quality. However, a robot is far less flexible than is a person, being unable to solve problems, innovate, and self-adjust.

A robot replacement study may result from a *problem-pull* condition or a *technology-push* situation. The former attempts to correct a known difficulty and the latter is an experiment with promising technology. Solving production problems with piecemeal robotics is unlikely to yield maximum savings because each introduction is an independent occasion. Comparably, installing a robot only because it is fashionable is hardly justifiable. An analysis should incorporate conventional cash flow categories plus reasonable estimates of difficult-to-quantify factors such as quality benefits and losses owing to inflexibility. Problem solutions can then pull in appropriate technology to push financially responsible modernization.

QUESTION
8A.1 List 10 factors that should be considered in an economic analysis of a proposed robot installation. For which ones would data be relatively easy to obtain, of uncertain accuracy, or difficult to quantify? For example, the first cost of a robot is readily available, its salvage value is uncertain, and the monetary worth of experience gained from employing a robot is difficult to estimate. Now name 10 more factors.

8B Replacement Planning by Individuals and Organizations

A machine does not have to rattle or wheeze or look like a battle-scarred veteran to be a candidate for replacement. Almost-new equipment can be economic liabilities, whereas elderly machines may be as cost-effective as are items out

of the latest catalog. Detecting the difference is part of replacement planning for both individuals and organizations.

The first step is that of listing the candidates. For most individuals the list will be short: larger household utensils, automobiles, roofs, pipes, etc. Companies have more possibilities. In both cases it is economically impractical to attempt to closely track the performance of all equipment. Only items that represent a substantial investment or are critical to operations deserve monitoring. What is considered "substantial" and "critical" is, of course, a matter of judgment.

Many companies do not have accurate inventory data on everything they own, even when the records are computerized. Once a machine is written off the books for accounting purposes (an asset still in service beyond its tax-deductible life), it tends to disappear. Machinery that is moved around often may also vanish from inventory records. Recorded data typically include year of acquisition, year of manufacture, maintenance and repair history, and hours of productive use per month. Such information is the basis for replacement decisions.

MAPI Method of Replacement Analysis

One of the most elaborate replacement analysis procedures is the MAPI (Machinery and Allied Products Institute) method. It has been refined over several years to provide a complete package of formulas, worksheets, and charts applicable to most economic evaluations. MAPI is basically a rate-of-return test that compares a defender with a challenger and with an improved challenger. Obsolescence is cleverly handled by factors for "accumulated inferiority" of a present machine with respect to improvements expected in future machines. Three variations of the basic MAPI formula represent different patterns of inferiority accumulation.

The standard-projection MAPI formula based on sum-of-digits tax depreciation is shown below.*

$$C = \frac{n(Q^n - w^n)(Q-1)^2 - (1-b)P[(Q^n-1) - n(Q-1)]}{nQ^n(Q-1) - (Q^n-1)} - (Q-1)$$

with $\qquad P = w^n\left[1 - w + py + \frac{(1-P)z}{1-b}\right]$

and $\qquad\qquad Q = 1 + i - bpy$

where C = next year's capital consumption as a ratio of P
$\quad\quad\;\; n$ = useful life
$\quad\quad\;\; w$ = ratio of salvage value in year $n + 1$ to salvage value in year n

$\quad\; w^n$ = salvage value expressed as a decimal fraction of original cost
$\quad\quad p$ = fixed ratio of debt to equity funding = 25%
$\quad\quad y$ = fixed interest rate on debt = 3%
$\quad\quad b$ = fixed income tax rate = 50 %
$\quad\quad z$ = fixed rate of return on equity after taxes = 10%
$\quad\quad i$ = fixed interest rate for capitalizing future sums = 8.25%

Prepared charts allow the formula to be solved graphically. When the assumptions built into the MAPI model fit a firm's operations and objectives, it offers a convenient and tested method for analyzing replacement alternatives.

Long-Term Costs of Owning and Operating an Automobile

When to replace a car is one of the most arduous replacement questions faced by many people. Few follow a systematic replacement policy, relying instead on emotions, current state of prosperity, and chance events. As a result, people seldom realize how expensive it is to operate a car and are shocked at the replacement cost.

Spending on passenger cars in 1983 amounted to 12.7 percent of United States personal income. Over the previous 10 years, the average price for a new car rose from $3404 to $9179, up 170 percent. During the same period, typical family income rose 156 percent and inflation climbed 126 percent.

A 1984 report by the Federal Highway Administration projected costs of buying and running selected United States made cars for 120,000 miles over 12 years—the mileage and length of time new autos were expected to last. Cost data in seven categories for five vehicle sizes are shown in Table 8.3. Depreciation cost does not include any charge for the use of capital; for example, an average large sedan which sold for $11,554 in 1984 was considered to have per-mile depreciation of $11,554/120,000 miles = $0.096/mile. Even these optimistic cost estimates, which may be understated by 40 percent or more when capital costs and a shorter ownership period are included, show the economic impact of replacement decisions.

QUESTIONS

8B.1 Without considering the time value of money, but based on a purchase price of $10,320 for an intermediate-sized car which will be driven 10,000 miles per year for 5

---

* From G. Terborgh, *Business Investment Policy*, Machinery and Allied Products Institute, Washington, D.C., 1958.

TABLE 8.3							

Cost per mile of owning and operating a 1984 domestic automobile, as calculated by the U.S. Federal Highway Administration. No interest charges are included in the calculations.

Type of Automobile	Depre-ciation	Mainte-nance	Gas, Oil	Parking Tolls	Insur-ance	Taxes	Total
Van	10.7¢	6.9¢	9.1¢	0.9¢	8.9¢	2.7¢	39.2¢
Large car	9.6	6.0	7.0	0.9	4.9	2.2	30.6
Intermediate	8.6	5.2	5.7	0.9	5.6	1.8	27.8
Compact	7.3	4.6	4.6	0.9	4.3	1.6	23.3
Subcompact	5.9	5.1	4.4	0.9	5.0	1.4	22.7

years and then sold for $1500, determine the driving-plus-ownership cost per mile.

8B.2 If the owner's money can earn 12 percent annually, what is the per-mile cost for the car described in 8B.1? Would this cost change if the owner borrowed the money at 12 percent to buy the car?

CHAPTER 9

FINANCIAL ANALYSIS

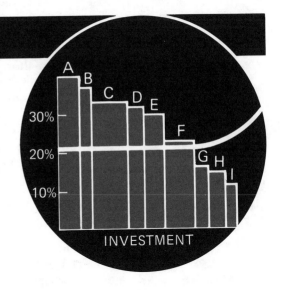

OVERVIEW

Economic comparison methodologies developed in previous chapters implied financial considerations, but our concern was directed mainly toward the equivalency of cash flow streams from competing alternatives. We skirted the question of financing the proposals. In this chapter the sources and cost of capital are examined, a comparison method is introduced that does not rely on discounting, and capital budgeting is discussed as a procedure for selecting proposals when investment capital is limited.

Investment capital is acquired *internally* from depreciation-derived funds and retained earnings, and *externally* by borrowing money and selling equity. The *weighted cost of capital* is a composite interest rate which represents the cost of all acquired funding for an organization. It can be calculated from the cost of debt and equity capital weighted according to the proportion of funds from each source. The resulting figure sets a lower bound for the minimum acceptable rate of return by which proposals are evaluated.

The *payback method* (first cost divided by net annual savings) indicates how much time will elapse before the amount invested is recovered from the net earnings it generates. In its simplest form, the recovery period is based on actual dollar flows; in more sophisticated applications the flow is discounted. The payback method is popular because it is simple to apply and stresses the turnover of capital, but it may show a preference for an alternative that is inferior by a discounted cash flow comparison.

Proposals to be funded are selected by a *capital-budgeting* process. The *capital-inventory* approach matches the rates of return from different investments against the cost of acquiring capital; all proposals with returns greater than capital costs are acceptable. In *capital rationing* a limit is set on the amount of investment capital, and the best proposals are funded until the supply of capital is exhausted. Objective capital budgeting is difficult, owing to the need to consider numerous combinations of proposals affecting all parts of the organization, dependencies among requests for expenditures, various degrees of riskiness, intangible aspects, mandatory or "urgency" ratings for certain investments, and politic factors.

SOURCES OF FUNDS

Effective use of funds in the public sector is discussed in Chapter 10.

Financial management has traditionally been concerned with obtaining funds and learning how to use them. Although sources of funding are obviously important considerations in any private enterprise, since securing capital is a prerequisite to implementing an acceptable proposal, the primary focus of this text remains on the effective use of funds. Engineers and managers of operating units have responsibilities for proposing and evaluating investments that support the productivity of operations, be they highway construction, steel production, office services, mail delivery, or any other productive function. Proposals are usually generated and evaluated on the assumption that a given level of funding is available to carry out the best ones.

The size of a pool of investment funds available at a particular time is a function of many financial decisions made previously. As shown in Figure 9.1, the pool is fed by gross revenue resulting from sales, such other income as returns from investments and capital obtained by borrowing or selling equity in the organization; *debt* and *equity* sales are *external* sources of funds. A portion of the inflow is allocated to *depreciation,* a charge made for accounting purposes that spreads the cost of owning assets over several years. One effect of depreciation charges is that of increasing the net throughput of funds by decreasing the percentage of income paid in taxes. The flow attributable to depreciation plus the net income remaining after all outflows—*retained earnings*—is the sum available from *internal sources* to satisfy investment proposals. The best way to cultivate funding sources is to have profitable operations that yield large net earnings, or activities that strongly promise future profits, thereby easing borrowing costs and boosting stock prices.

Depreciation accounting is explained in Chapter 11.

The management of an industrial enterprise works for the owners. In smaller firms the managers may be the owners. In the role of either an owner or someone representing owners, management has the responsibility of conducting operations in ways that financially benefit the equity holders. Owners receive gains from successful corporate operations in two forms: dividend payments and increases in the value of their stock. Each of the four main sources of funds shown in Figure 9.1 has its own effect on returns to equity holders.

Depreciation charges are the least controversial source of funds for new

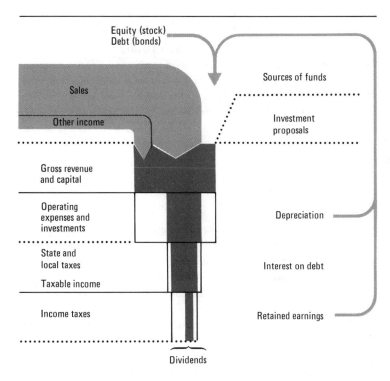

FIGURE 9.1
Flow of funds into, through, and out of an industrial organization. New funds are derived from borrowing or selling equity. Sources of internal funds are depreciation charges and retained earnings.

investments. How a firm allocates the depreciation reserve shown on accounting records is not specified by legal requirements. However, there is an implied obligation that depreciation charges be applied to purchases that maintain the production capabilities of the firm. Investments in replacements for worn assets and new processes to update production functions are customary practices.

Retained earnings come from after-tax income that could be distributed to shareholders as dividends. Expenditures of retained earnings are judged by the effect they have on the market price of stock. Stock prices typically reflect current earnings per share and the growth potential of a firm. If the stock market is confident that a firm is building up to a more profitable position, the stock price can rise even when the earnings are modest and no dividends are declared over a prolonged period.

The stock market is examined in Extension 9A.

Equity financing results from selling part ownership in a firm. When additional shares of stock are issued and sold and profits do not increase, there is a dilution of earnings that decreases the capitalization rate.

Debt financing does not dilute ownership, and it offers certain tax advantages, but the incurred indebtedness can be an onerous drain on finances during periods of low economic activity. Example 9.1 illustrates some of the considerations involved in an equity-versus-debt financing decision.

Example 9.1
A Decision to Borrow Funds or to Sell Equity

Two brothers own and operate a vegetable cannery valued at $300,000. They have an opportunity to buy a firm which now supplies most of their warehousing and distribution services. Only an insignificant amount of the to-be-acquired firm's assets is depreciable because the warehouses and equipment are leased. An annual saving of $28,000, exclusive of financing costs, should result from the acquisition. There are insufficient equity funds to meet the $100,000 purchase price.

The needed funds can be secured from a bond issue or a sale of 25 percent control of the firm. The 20-year bonds would carry 8 percent annual interest. Which source of funds would provide the greater return to the brothers on their equity when their effective income tax rate is 40 percent?

Solution 9.1

The bond expense would amount to an annual interest charge of $100,000(0.08) = 8000 and a commitment to have $100,000 available in 20 years to retire the debt. If the brothers' tax rate is assumed to remain unchanged, the annual after-tax return from the acquisition would be

$$
\begin{aligned}
\text{Net annual return} &= \text{before-tax returns} \\
&\quad - \text{interest payments} - \text{taxes} \\
&= \$28,000 - \$100,000(0.08) \\
&\quad - (\$28,000 - \$8000)(0.4) \\
&= \$28,000 - \$8000 - \$8000 \\
&= \$12,000
\end{aligned}
$$

which provides an *additional*

$$
\text{Return on equity} = \frac{\$12,000}{\$300,000} = 0.04 \quad \text{or} \quad 4\%
$$

when the acquisition is made from funds obtained by borrowing.

If they sell one-quarter interest in their organization for $100,000 and invest that amount in the firm, the brothers' equity is still ($100,000 + $300,000)(0.75) = $300,000$. The annual $28,000 saving now comes from ownership funds for which no interest is owned and no tax deductions are allowed:

$$
\text{Net annual return} = \$28,000 - \$28,000(0.4) = \$16,800
$$

The brothers' share is then $16,800(0.75) = $12,600$ which provides an additional

$$
\text{Return on equity} = \frac{\$12,600}{\$300,000} = 0.042 \quad \text{or} \quad 4.2\%
$$

Both return-on-equity figures require interpretation. The return provided by borrowing $100,000 does not include provisions for paying off the debt. A sinking fund that draws interest at the same rate as the loan would require annual payments from the revenue earned by the investment of

$$
\begin{aligned}
\text{Annual redemption reserve} &= \$100,000(A/F, 8, 20) \\
&= \$2185
\end{aligned}
$$

which reduces the net return to $12,000 - $2185 = 9815. This figure is less than the net return for the brothers when they sell one-quarter ownership, but it represents *only* the return from the investment in the distribution service. The brothers would also have to share the profit from their regular cannery operations with the new owners.

If it is assumed that the brothers had an after-tax return on investment of 15 percent before acquiring the distribution service, the total return on equity when $100,000 is borrowed would be

$$
\begin{aligned}
\text{Return on equity} \atop \text{by debt financing} &= \frac{\$300,000(0.15) + \$9815}{\$300,000} \\
&= 0.183 \quad \text{or} \quad 18.3\%
\end{aligned}
$$

The preacquisition profit due the brothers is reduced by one-fourth through the sale of equity, because the new owners share the total returns. The resulting proportion of the brothers' share of total returns after the acquisition would be

$$
\begin{aligned}
&\text{Return on equity after selling 25\% ownership} \\
&= \frac{\$45,000 - \$45,000(0.25) + \$12,600}{\$300,000} \\
&= 0.155 \quad \text{or} \quad 15.5\%
\end{aligned}
$$

which is slightly better than the preacquisition return on equity but less than that afforded by debt financing. In addition, the brothers would have to share their management prerogatives with the new owners. The effect on managerial decision making is not a major consideration in selling a small block of shares in a large corporation, but the sale of a controlling interest or even a significant minority interest can create an uneasy leadership situation. Therefore, the brothers would probably be wise to employ debt financing.

COST-OF-CAPITAL CONCEPTS

The cost of capital is derived from the composition of the capital pool. The term *pool* is appropriately suggestive of capital from many sources, pooled for funding purposes. The proportion of capital from different sources obtained at different costs to a firm is represented by a *weighted cost of capital*. This weighted cost sets the lower bound for the minimum acceptable rate of return (MARR), as discussed in Chapter 6.

The actual rate of return expected from new investments is normally greater than the cost of capital. How much greater depends on the risk involved. Riskier proposals are subject to higher discount rates to compensate for the chance that they will not meet net-return expectations. But even a minimum-risk investment, such as the purchase of government bonds, must yield a return greater than the interest rate a firm is charged on its debts; otherwise, it would be sensible to direct currently available funds toward debt retirement in anticipation of more rewarding future investment opportunities.

The appealing notion that all proposals can be funded if they exceed the cost of borrowing is a fallacy, because all borrowers have a maximum level of indebtedness based on their reputation and resources.

The weighted cost of capital k_w may be calculated from the formula

$$k_w = p_1 k_1 + p_2 k_2 + \cdots + p_n k_n$$

where there are n types of financing sources in proportions p of total capital, each source with its own cost k. For instance, if a firm is financed from bonds, preferred stock, and common stock with costs of 7, 9, and 12 percent and in proportions 20, 30, and 50 percent, respectively, then

$$
\begin{aligned}
\text{Weighted cost of capital} = k_w &= 7\%(0.2) + 9\%(0.3) + 12\%(0.5) \\
&= 1.4\% + 2.7\% + 6\% \\
&= 10.1\%
\end{aligned}
$$

The financial structure of a firm is comprised of debt and equity, usually in the form of bonds, promissory notes, preferred stock, common stock, and retained earnings. Each carries an obligation for monetary returns. From the discussion of valuation methods for preferred and common stocks in Chapter 4, we know that

Debt issues are easily accommodated because the interest rates are stated in the contractual agreements.

$$k_{\text{preferred stock}} = \frac{\text{dividend}}{\text{price}}$$

and

$$k_{\text{common stock}} = \frac{\text{dividend next year}}{\text{price}} + \text{growth rate}$$

Although these formulas are simple compared with sophisticated formulations that account for special considerations, they provide a reasonable approximation of return rates.

The cost of capital is troublesome to estimate, despite the apparent precision of its formula. There are differences in opinions about both costs and proportions and, naturally, a number of variations to the general formula have been proposed. Another cost of equity indicator is

When k_w is based on after-tax costs, long-term debt is reduced by the effective tax rate. For example, at a 40 percent tax rate, 8 percent interest on bonds would be reduced to
$$k_{\text{bond}} = 8\%(1 - 0.4)$$
$$= 4.8\%$$

$$\text{Capitalization rate} = \frac{\text{earnings per share}}{\text{market price per share}}$$

where

$$\text{Earnings per share} = \frac{\text{annual dividends} + \text{retained earnings}}{\text{number of shares of stock}}$$

and the market price is the amount bid for a share of stock. The capitalization rate is the "value" of earnings to an individual stockholder before taxes on personal income are imposed. For instance, if corporate after-tax earnings are $5 per share and the stock is selling for $50 per share, then

$$\text{Capitalization rate} = \frac{\$5}{\$50} = 0.10 \quad \text{or} \quad 10\%$$

which suggests that cash kept for internal operations rather than distribution to stockholders should have a reinvestment rate of return of at least 10 percent.

The proportions of debt and equity financing can be determined from the existing proportions stated in the firm's accounting records, expected future proportions, or current market values for the firm's securities. *Current market value* is widely recommended because it avoids the uncertainties of forecasting and reflects investors' current assessments of the financial state of the firm.

Example 9.2
Capital Cost as a Weighted Average of the Cost of Debt, Equity, and Retained Earnings

One way to measure the cost of capital is to assume that the return on any project funded at the weighted average rate will not change the market price of the firm's stock. That is, if the yield from a project equals the composite cost of funds from other sources, the money generated will neither raise nor lower the stock price. Since changes in the price of stock act as a rough index of a firm's economic health and managers do not

care to see the index dip, the cost of capital effectively sets a bottom limit for the rate of return required from future investments.

With k representing the discount rate that aligns future cash flows with the present value of debt and equity funding, the cost of capital from each source can be calculated according to the valuation procedures for stocks and bonds. If 20 percent of a firm's capital comes from $1000 bonds, now selling for $926, which mature in 10 years and pay annual dividends of $50, then

$$\$926 = \$50(P/A, k, 10) + \$1000(P/F, k, 10)$$

from which k is computed by trial and error to be 6 percent.

Then, if 50 percent of the firm's capital is represented by common stock which now sells for $65 per share and pays no dividends but is expected to continue to increase in price by $7 per year for the next 5 years,

$$\$65 = [\$65 + \$7(5)](P/F, k, 5)$$

from which k is found to be 9 percent.

With the remaining 30 percent of total capital derived from retained earnings which, at a minimum, should be valued the same as common stock because stockholders are being denied dividends from retained earnings, the weighted cost of capital can be determined as shown in Table 9.1.

TABLE 9.1

Calculation of the cost of capital as a weighted average of the funding from debt, equity, and retained earnings

Source of Capital	Percentage of Total (1)	Cost, % (2)	Weighted Cost, % [(1) × (2)]
Debt (bonds)	20	6	1.2
Equity (stock)	50	9	4.5
Retained earnings	30	9	2.7
Cost of capital			8.4

PAYBACK COMPARISON METHOD

The *payback* method (sometimes called the *payout* method) avoids the need to calculate the cost of capital and still recognizes financial concern for limited resources. It guards against unexpected price (cost) increases and utilizes a cutoff criterion by requiring proposals to return their original investment from the savings they generate in a specified period of time, usually of short duration. It is typically applied to relatively small investment proposals that originate from operating departments. A department manager often has the authority to accept proposals up to a given ceiling without subjecting them to outside review.

The formula for obtaining a rough measure of the time an investment takes to pay for itself is simple to use and understand:

$$\text{Payback period} = \frac{\text{required investment}}{\text{annual receipts} - \text{annual disbursements}}$$
$$= \frac{\text{first cost}}{\text{net annual savings}}$$

Data utilized in applying the formula are usually direct, *not discounted*, cash flow amounts, and no salvage values are included. The resulting figure tells how long before the amount invested is recovered in actual dollars.

Claims such as "this investment will pay for itself in 18 months" are commonly heard in industry; they usually indicate anxiety about the elapsed time before a proposed investment begins to show a profit. The payback period is an extremely popular investment criterion in the United States and throughout the world (see Extension 9B). Polls consistently reveal that the payback method is used more than any other comparison method by United States industry to rate investments, particularly proposals from operating units for relatively small capital expenditures to improve operations.

In actual practice the simple payback formula is sometimes modified to recognize capital recovery through depreciation charges and to include some discounted values. The simple payback formula is still widely used without elaborations, although it yields ratings that may lead to incorrect conclusions unless carefully interpreted. Its deficiencies arise from failing to give recognition to savings that occur *after* the payback period has passed and ignoring the time value of money.

For instance, as an extreme illustration of payback-period deception, an investment of $1000 in an asset with a life of 1 year and an associated net return of $1000 would yield

$$\text{Payback period} = \frac{\$1000}{\$1000/\text{year}} = 1 \text{ year}$$

Another investment of $1000 promises to return $250 per year during its economic life of 5 years and yields

$$\text{Payback period} = \frac{\$1000}{\$250/\text{year}} = 4 \text{ years}$$

Favoring the alternative with the shortest payback period would rate the first alternative best, but this alternative actually earns nothing: $1000 − $1000 = 0. Meanwhile, the spurned second alternative would have provided an annual return of 8 percent: $P/A = \$1000/\$250 = 4 = (P/A, 8, 5)$.

If the results of payback calculations are questionable, why are they used? There are at least two apparent reasons. One, they are simple calculations. Since both depreciation and interest effects are usually ignored, calculations are quick and easy and the results are intuitively logical. The other reason stems from a preoccupation with the flexibility of capital. If the money spent on an improvement is recovered rapidly, the funds can be allocated again to other desired projects. This concept tends to engender a false sense of security, with reasoning such as, "If the project can quickly pay for itself, it must be good" or, "Only the best projects can meet our short-payback-period requirement."

When investment proposals are ranked by the payback method, shorter is better.

The payback period for uniform savings is given by N in $(P/A, i, N) = $ first cost/annual savings when the cost of capital is considered.

An implied requirement is that the proposal's payback period must be shorter than its economic life.

Only when independent proposals have uniform returns and equal lives do the payback and IRR comparison methods indicate the same preferences.

Even though the payback-period criterion is not always appropriate, it does address the problem of working-capital management by attempting to protect a firm's liquidity position. During times of constricted income, when a firm may have trouble meeting operating expenses and have very limited capacity for funding new investments, an extremely short (as low as 6 months) payback period ensures that only quick-profit projects will be endorsed. In such exceptional situations, cash-availability considerations may be equal to or more important than total earnings. As an auxiliary criterion, requiring a short payback period guards against the chance of losses due to new technological developments.

> Since the payback criterion has serious weaknesses, it should never be applied alone, only as an auxiliary aid in decision making.

Example 9.3
Payback-Period Comparisons for Alternatives with Different Lives

The supervisor of a small machine shop has received three suggestions for reducing production costs. Suggestion A is to buy new jigs and fixtures; B is to rebuild an existing machine to improve its performance; and C is to purchase a new machine to replace some manual labor. Estimates have been made for the three alternative investments, as shown in the accompanying table.

The supervisor selects alternative B, explaining that because of limited capital for investments, shorter payback periods are preferable. With alternatives A and B having the same payback period, B is favored because the annual savings are greater than those for A. What are the fallacies in this reasoning?

	Alternative		
	A	B	C
First cost	$1800	$2350	$4200
Economic life, years	3	4	8
Net annual saving	$645	$840	$1100
Payback period, years	2.8	2.8	3.8

Solution 9.3

Based on only the payback-period criterion, consistent attention to the condition of limited capital would dictate a preference for alternative A over B because A requires less investment capital. Based on the rate of return on each investment, C is the most attractive, as attested to by

$$IRR(A): \quad (P/A, i, 3) = 2.8 \quad \text{or} \quad i = 3.5\%$$
$$IRR(B): \quad (P/A, i, 4) = 2.8 \quad \text{or} \quad i = 15\%$$
$$IRR(C): \quad (P/A, i, 8) = 3.8 \quad \text{or} \quad i = 20\%$$

The flaws in the reasoning stem from a strict reliance on the payback criterion. Too short a period can obviously block acceptance of some high-return alternatives. If successive investments in shorter-lived alternatives produce a smaller rate of return than does investment in one longer-lived alternative, flexibility is purchased by the losses in total earnings. In the example, successive utilization of alternatives A and B limits the maximum investment level to $2350 during 7 years, but it also limits the rate of return to less than half that achievable by accepting alternative C.

CAPITAL BUDGETING

It could be said that the central theme of this book is *capital budgeting—the process of analyzing and determining optimum capital expenditures*. Successful capital budgeting is vital for the long-run prosperity of any industrial organization. Funding decisions affect all departments within a firm, and mistakes in forecasting or fulfilling asset requirements for any type of operation can damage the others. Both immediate and long-term needs must be considered. And these needs must be evaluated in conjunction with questionable future developments in unique markets and in the economy as a whole. The combined effects of internal pressures and external uncertainties make the capital-budgeting decision at once the most important and difficult kind of decision to make in financial management.

The basic role of an engineering economist in capital budgeting is to uncover, examine, and prepare analyses of profitable ways to invest capital. The prepared analyses are acted upon by screening groups that vary in size and composition for different organizations; successive screenings by departmental groups may be forwarded to a board of top-level managers, all proposals may be put through a committee-type screening mechanism before a final allocation by select executives, or some other management device may be designed to guarantee a *careful, objective scrutiny* of requested expenditures. Major outlays for strategic objectives usually await approval of the board of directors or top management.

Capital Inventory

It is almost certain that in any ambitious firm there will be more requests for expenditures than there is capital to satisfy them. An abundance of proposals is a healthy sign that operating personnel, engineers, and managers are actively pursuing improvements. A firm might increase the amount of capital available in a budget period by external financing or by retaining a larger-than-normal share of earnings, but considerations of debt obligations and stockholders' interests limit the amount procurable. Unless unusually attractive investment opportunities become apparent, the amount of capital available is essentially set by the long-term financial policies of the firm. Asset expansion is necessarily related to future sales, because a decision to invest in a fixed asset is made in anticipation of future returns from sales, and future sales are the basis of capital formulation.

Besides capital limitations, investment proposals are also affected by interrelationships, such as securing funds for the second phase of a project for which the start-up capital is being budgeted now.

The *capital-inventory* approach for selecting which proposals to fund matches the cost of capital to the returns expected from investments. As portrayed in Figure 9.2, the shaded blocks form a ladder of investment proposals promising various rates of return; IRR_1 is a smooth curve representing the blocks. Curves IRR_2 and IRR_3 represent other configurations of investment opportunities that might become available during a budgeting period. The cost of each additional dollar acquired for purposes of making capital expenditures is given by the bold MCC (*marginal cost of capital*) curve. This curve is relatively flat up to an amount which exhausts the normal sources of capital; beyond this point, it rises sharply as more expensive sources are tapped.

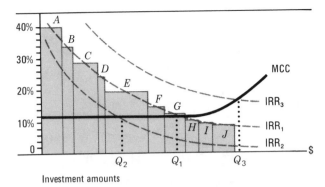

FIGURE 9.2
Capital-inventory graph showing the marginal cost of capital (MCC) and three investment-proposal schedules (IRR$_1$, IRR$_2$, IRR$_3$). The smooth-curve schedules represent discrete investment projects, as indicated by the lettered ladder of proposals represented by IRR$_1$.

If all proposals that promise a rate of return greater than the marginal cost of capital are funded in schedule IRR$_1$, capital allocations would be made to proposals *A* through *G*. These investments total Q_1. The same criterion would indicate investments in the amount Q_2 for the IRR$_2$ schedule; Q_2 is smaller than Q_1 because the investment opportunities of schedule IRR$_2$ are not so lucrative as are those of IRR$_1$. A wealth of proposals with high rates of return might induce the firm to secure additional capital to allow the Q_3 level of investment for the IRR$_3$ schedule. Any proposals that have a rate of return smaller than the marginal cost of capital are rejected by the investment-inventory criterion.

Capital Rationing

The practice of placing a fixed limit, or limited range, on total investments rations capital to the best proposals that can be funded within the budget limits. The floor for capital rationing is the cost of capital; that is, if the available investment capital exceeds the requested expenditures, no request would be granted that yields a rate of return smaller than the cost of capital. Such requests should not even be made if economic evaluations are conducted properly, because the minimum rate of return for an acceptable alternative is always equal at least to the cost of capital.

Figure 9.3 shows a *cutoff rate* that conforms to a ceiling on capital. The point on the *cutoff index* is set by the intersection of a line from the amount of capital available for investments Q_1 and the IRR$_1$ curve, which represents proposals laddered according to the cutoff index. For the condition shown, the cutoff point limits acceptable proposals to a number less than what would be admissible using a capital inventory. Rationing capital to amount Q_1 rather than allowing all investments Q_2 which exceed the marginal cost of capital (MCC) increases the *average* rate of return on accepted investments. However, the firm may be missing an opportunity to increase its *total* profit by not funding all proposals that earn more than the cost of acquiring the capital for funding.

FIGURE 9.3
Cutoff rate established by the point where the investment-opportunity schedule (IRR₁) crosses a projection of the amount of capital available Q_1. Capital rationing limits acceptable proposals to a higher cutoff index than is allowed by Q_2.

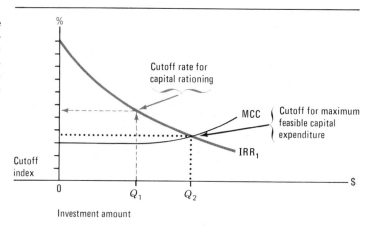

The cutoff rate for a budget ceiling becomes the *minimum acceptable rate of return* for the budgeting period when the cutoff index is expressed as a percentage. The laddering of proposals could also be accomplished by rating proposals according to

$$\text{Present-worth index} = \frac{\text{PW(receipts} - \text{disbursements)}}{\text{PW(investments)}}$$

$$= \frac{\text{PW(net savings)}}{\text{first cost}}$$

PW rankings do not recognize that shorter-lived assets make capital available earlier for more attractive reinvestment than do longer-lived assets.

when the entire investment occurs at time zero. Ranking proposals and accepting only the best ones until all the available capital is allocated works reasonably well, but care should be exercised to apply the concept adroitly and without bias.

Example 9.4
Capital Rationing Using a Cutoff Rate of Return and a Present-Worth Cutoff Index

Six independent alternatives are listed in Table 9.2. None of the proposals is expected to have any realizable salvage value. If the marginal cost of capital is 12 percent and the amount of investment capital is limited to $100,000, which requests for expenditure should be granted?

Solution 9.4

The rate of return for proposal A is calculated as

$$PW(A) = -\$30,000 + \$12,600(P/A, i, 9) \doteq 0$$

At IRR = 40 percent,

$$PW(A) = -\$30,000 + \$12,600(2.3790) = -\$25$$

or IRR $\doteq$ 40 percent. Rates of return for the other proposals are shown below.

Proposal	A	B	C	D	E	F
IRR	40%	15%	28%	22%	11%*	20%

* Unacceptable because the rate of return is smaller than the cost of capital.

TABLE 9.2

Proposal	Request for Expenditure	Investment	Net Savings	Useful Life, Years
A	Remodel loading dock	$30,000	$12,600	9
B	Modernize office	35,000	12,250	4
C	Purchase shredder	10,000	4,450	4
D	Install conveyor	25,000	10,025	4
E	Purchase press	30,000	9,700	4
F	Construct storage shed	45,000	11,150	9

Since present-worth calculations are based on a given interest rate, the minimum cost of capital, 12 percent, is assumed reasonable for budgeting and is used to determine the present-worth index of proposal A as

$$\text{Present-worth index}(A) = \frac{\text{PW(net savings)}}{\text{investment}}$$
$$= \frac{\$12,600(P/A, 12, 9)}{\$30,000}$$
$$= \frac{\$12,600(5.3282)}{\$30,000}$$
$$= 2.24$$

The remaining present-worth indexes are given below.

Proposal	A	B	C	D	E	F
PW index	2.24	1.06	1.35	1.22	0.98*	1.32

* Unacceptable because the index is smaller than 1.0.

The proposals, ranked according to their index values and cumulative investment totals, are tabulated to determine the cutoff level, as shown in the accompanying tables.

The rankings produced by the rate-of-return and present-worth calculations are different. The reasons for the difference lie with the assumptions embodied in the two methods. The rate-of-return method measures the *rate* of excess income accumulation over outgo. A single *amount* by which income exceeds outgo is measured by the present-worth method based on an explicit discounting rate. In addition, the present-worth

Rate-of-Return Rankings

Ranked Proposals	Rate of Return, %	Investment	Cumulative Investment
A	40	$30,000	$ 30,000
C	28	10,000	40,000
D	22	25,000	65,000
F	20	45,000	110,000
B	15	35,000	145,000
E	11	30,000	Not acceptable

Cutoff level (between D and F)

Present-Worth Rankings

Ranked Proposals	PW Index	Investment	Cumulative Investment
A	2.24	$30,000	$ 30,000
C	1.35	10,000	40,000
F	1.32	45.000	85,000
D	1.22	25,000	110,000
B	1.06	35,000	145,000
E	0.98	30,000	Not acceptable

Cutoff level (between F and D)

index does not fully account for differences in the lives of the assets; it is assumed that the returned excess from a shorter-lived asset draws 0 percent interest for the period from the end of its life to the end of the life of a longer-lived asset.

For the proposals in Example 9.4, and often in other capital-budgeting situations, the two methods lead to the same selections. It might appear that the cutoff rate is 22 percent because the inclusion of proposal *F* would push the total investment over the $100,000 limit. The combination of proposals A, C, and D uses only $65,000 of the available capital. Assuming that the remaining $35,000 is invested in proposal B, which still has a return greater than the marginal cost of capital, we find that the overall rate of return is

$$\text{IRR}(A + C + D + B)$$
$$= \frac{\$30,000(0.4) + \$10,000(0.28) + \$25,000(0.22) + \$35,000(0.15)}{\$100,000}$$
$$= 0.256 \quad \text{or} \quad 25.6\%$$

A different and better combination is that of funding proposals A, F, and D. This combination gives

$$\text{IRR}(A + F + D) = \frac{\$30,000(0.4) + \$45,000(0.2) + \$25,000(0.22)}{\$100,000}$$
$$= 0.265 \quad \text{or} \quad 26.5\%$$

The most profitable combination is also identified by the net present worth:

$$\text{PW}(A + C + D + B) = -\$100,000 + \$12,600(P/A, 12, 9)$$
$$+ (\$4450 + \$10,025 + \$12,250)(P/A, 12, 4)$$
$$= \$48,302$$
$$\text{PW}(A + F + D) = -\$100,000 + (\$12,600 + \$11,150)(P/A, 12, 9)$$
$$+ \$10,025(P/A, 12, 4)$$
$$= \$56,993$$

Qualifications for Capital Budgeting

The discussion of capital inventory and rationing was simplified by disregarding several considerations that frustrate actual applications:

1 Determining a cutoff rate, or an index cutoff point, does not automatically identify the best combination of proposals to take advantage of the limited capital, as evidenced in Example 9.4. Several programming models have been computerized to assist the rationing exercise, but each model has its own assumptions that limit general applicability.

2 *Independent* projects are not always *completely* independent. Sometimes a proposal from one department is made on the assumption that a proposal from another department will be funded. A proposal with a very high rate of return might be made for an addition to an activity that is only marginally profitable; the highly profitable addition could be voided if the marginal activity is modified or eliminated. An *interdependency* could exist as an informal policy of the capital budgeters to try to spread investments along geographical or organizational lines.

Larger capital budgeting problems of this nature may be conveniently formulated by linear programming techniques (LP). Computer programs for LP applications are readily available.

Even tenuous linkages between proposals deserve attention, but excessive investigation can lead to paralysis by analysis.

Mandatory projects are effectively exempt from capital rationing when failure to fund them can shut down operations. Government regulations, such as required pollution controls, are the cause of many *mandatory investments*. These investments may have a negative rate of return, but they still top the list of requests for expenditure. Other projects that are not compulsory but are rated urgent may be ranked by management edict above proposals with higher rates of return.

3   All the proposals in Example 9.4 were treated as if they had identical *risk*. Rarely do different courses of action have the same probability of success. Riskier ventures are typically expected to yield a higher rate of return. Adding a percentage of the required rate of return to compensate for risk is a reasonable practice, but it is a conservative tactic that dilutes the objectivity of investment laddering.

One large corporation divides its investment-opportunity schedule into distinct categories: New projects must have a rate of return greater than 25 percent to be even considered for funding; improvements or modifications to existing projects have a floor at 17 percent; and investments that increase labor utilization are accepted if they produce a 10 percent return. These minimum required percentages reflect the risk involved and management views about the most effective use of capital. An amount of capital is allocated for each category and is rationed to proposals within the category according to the rankings. Projects considered mandatory, such as safety improvements, are funded from the top of the capital pool, irrespective of their return on investment.

4   Subtle forms of persuasion may intimidate or prejudice the workings of the decision-making process. *Politic* considerations surround most decisions, and capital-allocation evaluations are not exempt. Perhaps it is important to reward the originator of an idea by accepting his or her most recent proposal, after several previous ones were rejected, even if it has a modest IRR; the encouragement may spur more activities of the type desired. Successful proposal sponsors gain stature, and the types of proposals currently accepted tend to guide the submission of future proposals. Susceptibility to the "squeaky wheel" principle that directs oil to the noisiest wheel may bias judgment, but the aggressiveness exhibited in seeking attention (and change) may be in the best interest of the firm. A manager's relation with peers and subordinates can be affected by success in pushing proposals through, but should personal relations be a factor in determining the merits of proposals? Though the questions are many and the answers are not easy, competent capital budgeting is at stake.

Politic means shrewd or sagacious management, as opposed to *political*, which indicates a sensitivity to politics.

Review Exercises and Discussions

The owners of land with geothermal potential are seeking capital to develop the property. They feel that $10 million is necessary. That amount should yield annual revenue of $5 million for a prolonged period of time. Potential financing is from the sale of common

Exercise 1

stock on which buyers will expect a return of about 12.5 percent, preferred stock on which the annual dividend rate would have to be 10 percent, or a loan on which interest would be 8 percent. The effective tax rate will be 50 percent.

Evaluate the financing options.

Solution 1 The relative impact of the financing plans on income to the owners of the land is shown in Table 9.3. Since the capitalization rate for common stock is expected to be 12.5 percent, stock dividends would consume $10,000,000 \times 0.125 = \$1,250,000$ annually, which requires before-tax revenue of $\$1,250,000/0.50 = \$2,500,000$. Given the expectation of a $5 million annual revenue, the common stockholders must own 50 percent of the company to accommodate their 12.5 percent return. The original owners of the land would then receive $1,250,000 per year for the use of their property.

Returns to the owners would be higher if they issue preferred stock or borrow. Preferred stock is considered less risky and it may command a lower dividend than does common stock. The $850,000 advantage of a loan over common stock is derived from a tax saving, owing to interest on debts being deductible while dividends are not ($800,000 \times 0.5 = \$400,000$), and a difference between dividend and interest rates, $\$10,000,000(0.125 - 0.08) = \$450,000$. However, the comparison does not include provisions for repaying the loan.

A more complete analysis would investigate the sensitivity of the financing alternatives to differences in revenue.

Exercise 2 Go-Go was incorporated 3 years ago. The following data have been collected to determine its current cost of capital:

Outstanding Debt, Preferred Stock, and Common Stock (in millions of dollars)		
	Market Value	Interest Rate, %
Bonds (9½% due in year 2000)	16	9.1
Other debt	24	9.6
Preferred stock ($36 per share)	17	7.4
Common stock ($40 per share)	159	

Common Stock Record for 3 Years				
	Year 1	Year 2	Year 3	Average Growth Rate, %
Dividend per share	$1.55	$1.60	$1.67	3.8
Earnings per share	$2.49	$2.70	$2.98	9.5

What weighted cost of capital best represents Go-Go's financial status?

Solution 2 The total debt of $16 + \$24 = \$40$ (million) has a weighted average interest rate of 9.3 percent. This rate can be adjusted to an after-tax rate through the use of Go-Go's tax rate of 48 percent (interest payments are deductible from profits, so the actual outlay for interest is effectively 52 percent of the amount paid when the corporate tax rate is 48 percent):

$$\text{After tax interest rate on debt} = 9.3\%(1 - 0.48) = 4.8\%$$

Assuming a common stock dividend for next year of $1.76, we get

TABLE 9.3

Comparison of financing through $10 million obtained by sales of common stock, sales of preferred stock, and a loan at an interest rate of 8 percent. Preferred stock pays dividends of 10 percent, and new common stock accounts for 50 percent of the company's shares.

	$10 Million from Common Stock	$10 Million from Preferred Stock	$10 Million from a Loan
Before-tax cash flow	$5,000,000	$5,000,000	$5,000,000
Annual interest expense	0	0	− 800,000
Taxable income	$5,000,000	$5,000,000	$4,200,000
Taxes at 50%	− 2,500,000	− 2,500,000	− 2,100,000
Earnings after taxes	$2,500,000	$2,500,000	$2,100,000
Preferred dividend	0	− 1,000,000	0
Earnings on common equity	$2,500,000	$1,500,000	$2,100,000
Income to new stockholders	− 1,250,000	0	0
Income to original owners	$1,250,000	$1,500,000	$2,100,000

$$k_{\text{common stock}} = \frac{\$1.76}{\$40} + 0.038 = 0.082 \quad \text{or} \quad 8.2\%$$

This rate appears to be somewhat low for stockholders' expectations with respect to the yield on Go-Go bonds. Since the dividend rate is low because the company is reinvesting its earnings to support future growth, the growth rate for earnings per share (9.5 percent) reasonably seems to be a more representative figure for investor's expected return. Accepting this reasoning gives

$$k'_{\text{common stock}} = \frac{\$1.76}{\$40} + 0.095 = 0.139 \quad \text{or} \quad 13.9\%$$

A weighted cost of capital of 11.9 percent results from the calculations that are shown in Table 9.4.

TABLE 9.4

	Amount	Proportion	Rate, %	Weighted Rate, %
Debt	$ 40	0.194	4.8	0.94
Preferred stock	7	0.034	7.4	0.25
Common stock	159	0.772	13.9	10.73
Totals	206	1.000		11.92

Another possible cutoff index is the net difference between the present worths of income (or savings) and expenditures. Apply this index to obtain a ranking for the six proposals in Example 9.4.

Exercise 3

a Which proposals should be funded when capital is limited to $100,000?

b Which proposals should be funded when the capital budget is only $75,000? Compare the combinations indicated by the net-present-worth method and the rate-of-return method for this budget level.

Solution 3 The net present worth of proposal A at the minimum discount rate of 12 percent is

$$PW(A) = -\$30,000 + \$12,600(P/A, 12, 9) = \$37,135$$

The remaining PWs are calculated similarly to yield the amounts by which the proposals are ranked.

Ranked Proposals	Net PW	Net Savings	Investment	Cumulative Investment
A	$37,135	$12,600	$30,000	$ 30,000
F	14,409	11,150	45,000	75,000
D	5,449	10,025	25,000	100,000
C	3,516	4,450	10,000	110,000
B	2,207	12,250	35,000	145,000
E	− 538	9,700	30,000	Not acceptable

a The ranked proposals suggest the same combinations of investments as identified in Example 9.4: Fund proposals A, F, and D.

b When available capital is limited to $75,000, the rankings by net present worth suggest allocations to proposals A and F, which yield an average rate of return based on calculations in Example 9.4:

$$IRR(A + F) = \frac{\$30,000(0.4) + \$45,000(0.2)}{\$75,000} = 0.28 \quad or \quad 28\%$$

But a more profitable combination includes proposals A, C, and D, which consume $65,000 of the $75,000 available. Assuming the remaining $10,000 can always be invested to earn at least the cost of capital, we find that

$$IRR(A + C + D + \$10,000 \text{ invested at } 12\%)$$
$$= \frac{\$30,000(0.4) + \$10,000(0.28) + \$25,000(0.22) + \$10,000(0.12)}{\$75,000}$$
$$= 0.287 \quad or \quad 28.7\%$$

Rating by the net present worth emphasizes the discounted amount of total benefits. Unless corrected for unequal lives among proposals, the higher ratings naturally go to the proposals which yield returns for a longer period of time, even when these returns are at a lower rate than is obtainable with shorter-lived projects. Some analysts feel that the total return from long-lived assets is a legitimate comparison criterion, especially for public projects (discussed in the next chapter).

PROBLEMS

9.1 Given the following data, calculate the weighted cost of capital.

Source	Market value
Bonds: Value now = $95 per $100 maturity value in 7 years; bond rate = 8% compounded quarterly	$9.5 million
Preferred stock: Value now = $100 per $100 par value; 7% annual interest	$2.0 million
Common stock: $21 per share, having increased from $13 five years ago; next year's dividend = $1.80; 5-year growth rate = 5.8%	$20.0 million
Retained earnings: Expected to earn at the same rate as stock price growth	$1.5 million

9.2 A troubled firm is assessing its cost of capital. Its proportion of debt to equity is 1:2. The debt is in the form of long-term securities for which the interest rates average 5.9 percent. Until the past year the firm's common stock had increased in value at an 11 percent compound rate and dividends grew at a 13 percent rate. A recent report by an investment advisory service, critical of the company's management, contributed to a drop in its stock value from $65 per share to $45. The report predicted that the company would grow at a 6 percent rate for the next several years. The dividend payable next year ($5) conforms to the historical growth rate. What weighted cost of capital seems most reasonable. Why?

9.3 A company's current capitalization structure is 35 percent equity and 65 percent debt. The estimated weighted average of all debt financing over an extended future period is 4.2 percent. The corresponding cost of all outstanding equity financing is 10.77 percent.
  (a) What is the cost of capital?
  (b) Using the firm's cost of capital as its required rate of return, determine how much it could afford to pay for an investment that promised to return $10,000 for three consecutive years starting 2 years from now.
  (c) Should the firm make the investment (in Problem 9.3b) if the risk is considered quite small? Why?

9.4 Four basic forms of debt and equity financing are described in Table 9.5.
  (a) Discuss the relative merits of investing in each security from the viewpoint of a private investor.
  (b) Discuss the relative merits of each source of funds from the viewpoint of a small company that recently went public (owners sold stock to the public), is in desperate need of cash, has not been able to sell all the initial stock offering, and has mortgages on 10 percent of its physical assets.

9.5 A company expects a new machine to save $15,000 per year for 8 years and to have a salvage value of $2000 at the end of its economic life. Eighty percent of the firm's capital is represented by common stock which sells for $30 per share and pays annual dividends of $2.70 but has not increased in selling price over the last 4 years. The other 20 percent of its capital comes from long-term debt on which the annual interest rate averages 12 percent. How much can the company pay for the new machine if the investment is to earn twice the cost of capital?

9.6 Machine A saves $2000 each year of its 3-year economic life and has a simple

TABLE 9.5

Name of Security	Type	Market Value	Life	Obligation to Pay Return	Return	Vote
First-mortgage bond	Mortgage on physical assets	Relatively stable	30–35 years	First	Lowest (fixed)	None
Debenture bond	Unsecured obligation of the company	Relatively stable	10–50 years	Second	Second lowest (fixed)	None
Preferred stock	Part owner of company	Variable	Usually perpetual	Third	Second highest (usually fixed)	Usually no
Common stock	Part owner of company	Variable	Perpetual	Last	Highest (variable)	Yes

payback period of 2 years when calculated without using discounted cash flow. Machine B has an economic life of 4 years, earns the same rate of return, and has the same annual savings as machine A. What is its simple nondiscounted payback period?

9.7   Use the payback criterion to select a preferred investment proposal from the six cash flow patterns given in Problem 7.10.

9.8   Eight investment proposals that have identical economic lives are given below.

Proposal Number	Investment	IRR, %	Proposal Number	Investment	IRR, %
1A	10	12	3A	15	2
1B	15	14	3B	20	9
2A	5	11	4A	10	8
2B	15	10	4B	20	15

(a) If all proposals are independent and the capital rationing cutoff rate is 10 percent, which ones should be accepted?

(b) If the two levels of each numbered proposal are mutually exclusive and the cutoff rate is still 10 percent, which ones should be accepted?

(c) If the marginal cost of capital is 9 percent and investment funds are limited to 60, what is the best average rate of return that can be earned if all proposals are independent?

9.9   Six independent investment proposals have the same payback period based on the after-tax cash flows shown below.

	Year					
Proposal	0	1	2	3	4	5
1	− $1000	$200	$300	$500	$300	$200
2	− 900	300	300	300	300	300
3	− 800	0	400	400	800	0
4	− 700	0	0	700	400	400
5	− 600	300	200	100	300	500
6	− 500	300	100	100	200	0

(a) Which alternative would be preferred when the cash flows are discounted to their present worths at a discount rate of 8 percent?

(b) Comment on the cash flow patterns that are most likely to cause the payback criterion to disagree with a preference indicated by the discounted cash flow criterion.

(c) What cash flow pattern is most likely to make the payback and discounted cash flow comparisons agree?

(d) Which alternatives would be selected by a capital-inventory approach when the cost of capital is 10 percent?

(e) If the available capital is limited to $2000 and any unused funds can be invested at 10 percent, what combination of investments should be selected? What average rate of return results from this combination?

9.10 Ten requests for expenditure have been received. All are for investments to reduce operating costs, and all have about the same risk. The assets involved have a useful life of 4 years and no salvage value. Investment amounts and the uniform cash flows are shown in Table 9.6.

TABLE 9.6

After-tax cash flows expected to result from capital expenditures for the listed proposals. All alternatives have the same useful life and risk.

Proposal	Investment	After-Tax Cash Flow
1A	$10,000	$ 3,087
1B	20,000	7,434
1C	30,000	10,722
1D	40,000	14,012
2A	10,000	4,159
2B	20,000	8,020
2C	30,000	11,151
3A	10,000	3,863
3B	20,000	6,864
4A	30,000	9,877

(a) If the alternatives are independent and the cost of capital is 13 percent, how much should be invested?

(b) If the alternatives are independent and capital is limited to $100,000, which proposals should be accepted? What is the cutoff rate? What is the average rate of return?

(c) Answer the questions in Problem 9.10b when the only change is that proposal 4A is mandatory.

(d) Use the present-worth index to answer Problem 9.10a.

(e) Let the numbers in the proposal identifications represent four different departments in the organization. Determine the answers to the questions in Problem 9.10b if it is company policy to accept at least one proposal from every department.

(f) Assume that the policy in Problem 9.10e is changed to the capital-inventory approach in which departments 1 and 3 have a 15 percent limit on cost of capital and the other two departments have a 20 percent minimum. What is the average rate of return resulting from this policy?

(g) Let the numbers in the proposal identifications represent four categories in which the lettered alternatives (A, B, C, and D) are mutually exclusive. Which proposal should be selected from which categories if capital is limited to $70,000? What is the average rate of return? (*Hint:* Calculate incremental rates of return.)

(h) What combination of proposals would yield the highest average rate of return for the condition in Problem 9.10g if the capital is limited to $50,000? Any unallocated capital can earn 13 percent. Does a lower limit placed on capital always raise the average rate of return of accepted proposals?

(i) Again assuming that lettered alternatives within numbered categories are mutually exclusive, determine what average rate of return results when only $40,000 is available in the capital budget.

9.11 Determine present worth indexes for the six proposals presented in Problem 7.12. Rank the proposals according to their index values using a MARR of 12 percent and determine which requests for expenditure should be granted if available funds total only $15,000.

9.12 Nine proposals have been received from the Advanced Products Group of the Chemco Company. These proposals fall into three major groups of three alternatives each and are based on a project life of 5 years with no salvage value.

Alternative	Investment Required	Uniform Annual After-Tax Revenue
1A	$10,000	$ 2,650
1B	18,000	5,000
1C	30,000	8,950
2A	$20,000	$ 5,965
2B	25,000	6,965
2C	35,000	10,194
3A	$10,000	$ 3,350
3B	30,000	8,950
3C	45,000	11,270

(a) Assuming that all of the projects are independent, decide which should be selected if the MARR is 13 percent. What overall rate of return will result from this selection?

(b) If the projects are independent and only $50,000 is available for investment, which projects should be selected in order to maximize the overall rate of return? Assume that any capital not used may be invested at 13 percent outside the company.

(c) Given that the projects within each group are mutually exclusive, but the groups are independent, set up a table showing the order in which capital should be invested and the cutoff rate of return for each incremental investment.

(d) Assuming that one, and only one, project must be selected from each group, state which projects should be selected in order to maximize the overall rate of return. What amount of capital will be required and what rate of return will be obtained?

9.13 A manufacturing manager has received several proposals for future expenditures to improve material handling in four production areas. The proposals within each area are mutually exclusive, but the acceptance of a proposal in one area has no effect on the economic attractiveness of a proposal in any other area.

Area	Proposal	Initial Cost	Net Annual After-Tax Revenue
A	A1	$ 50,000	$ 8,200
	A2	60,000	15,000
	A3	75,000	17,000
B	B1	100,000	20,000
	B2	120,000	25,000
	B3	130,000	29,000
	B4	140,000	30,000
C	C1	80,000	15,000
	C2	120,000	24,000
D	D1	60,000	14,000
	D2	80,000	15,000
	D3	90,000	21,000

Using an estimated project life of 10 years with no salvage value and a corporate after-tax MARR of 12 percent, answer the following:

(a) Which proposals should be selected if unlimited funds are available? What overall rate of return would be obtained on the investment.

(b) If only $160,000 of internal funding is available, which proposals should be selected? What effect will this have on the overall rate of return?

(c) If additional funds can be borrowed at an overall after-tax cost of 15 percent, should more than $160,000 be invested in these proposals?

(d) Corporate management has decided that only two capital expenditure projects are to be funded. Which two should be selected, and what overall rate of return will be obtained?

9.14 Several requests for expenditure are being considered by the management of the Crumbco Cookie Company.

Proposal	Initial Investment	Uniform Annual After-Tax Revenue	Expected Life
A	$20,000	$4800	10
B	40,000	6700	15
C	10,000	2800	10
D	30,000	6300	15
E	40,000	7400	10

(a) Assuming that the alternatives have no salvage value, rank the economic attractiveness of the proposals using the present worth index with a MARR of 13 percent.

(b) Assuming that the alternatives have no salvage value, rank the economic attractiveness of the proposals using the rate of return as the criterion.

(c) Based on the results from Problem 9.14a and 9.14b, which projects would you fund if a total of $70,000 were available?

(d) What effect would using a 10-year study period have on your answer to Problem 9.14c?

EXTENSIONS

9A Stock-Market Vocabulary and Workings

A corporation is a legal entity. Like a person, it can acquire resources, own assets, incur debts, and sue or be sued. Yet it is separate from the individuals who own it. The owners (stockholders) have limited liability for the corporation's actions, risking only the amount they have paid for its stock. These stock purchases are made in anticipation of monetary gains from the organization's performance. The amount paid for a share of stock reflects the expected level of gain and the confidence of that expectation.

Organized stock exchanges facilitate the transfer of securities among buyers and sellers. Individuals can spread their risk by buying shares in a variety of corporations and can change their holdings easily and quickly. Ten of the more common terms involved in these transactions are listed below.

Broker. An agent who executes orders to buy and sell shares of stock and other securities for a commission.

Common stock. Also called equities, because holders have a share in assets and control of the company, but few fully utilize their voting rights to exercise control.

Preferred stock. A part of the capital of a business, usually having fixed dividends and no voting rights.

Dividends. Payments made by a company to stockholders—can be cash or stock—that are usually paid quarterly.

Listed stock. Stocks that are traded on a stock exchange, such as the New York, American, Pacific, or other regional exchanges.

Over-the-counter. Stocks that are not listed on exchanges but are traded by direct negotiation between brokers representing buyers and sellers.

Margin. Stocks that investors can purchase on margin by not putting up the full purchase price; they borrow the balance from their broker.

Selling short. A technique used by investors for selling stock that they do not own. They borrow stock from the broker, sell it, and hope the price will go down so they can buy shares at a lower price, return the shares to the broker, and pocket the difference.

Price-earnings ratio. The relation of the price of a stock to its annual per share earnings. To calculate it, divide the latest price of the company's stock by the company's latest earnings per share. Earnings per share are calculated by taking the company's latest annual profits and dividing them by the total number of company shares outstanding.

Yield. The return on a security figured by dividing the annual dividend rate by the price of the stock. A stock selling at $50 with an annual dividend of $5 per share has a yield of 10 percent.

Deciphering Stock Listings

Most newspapers carry weekly summaries of stock movements at the different exchanges. An example of a stock listing and explanation of the abbreviations used are given below for the fictitious *Engrecon* Company.

Additional notations may be used to show special conditions, such as an "x" in front of the dividend to indicate that the stock is selling *ex-dividend*—that is, the buyer will not receive the latest declared dividend.

Interpreting the Dow Jones Industrial Average

When someone asks how the stock market is doing, a common answer is that the Dow is up or down so many points. The reference is to the Dow Jones industrial average, an index based on 30 highly respected stocks. It is the most

Yearly						Vol				Net
High	Low		Div	Yld	P/E	100s	High	Low	Last	Chng
20½	16⅛	Engrecon	1.52	8	11	1723	19¼	18½	19	+ ½

Yearly **High Low**	The highest and lowest price paid for the stock in the latest 52-week period. In this case, Engrecon's high was 20½ and its low was 16⅛. Stock prices are quoted in fractions as low as ⅛, or 12.5 cents.
Div	Dividend paid on an annual basis: $1.52 per year.
Yld	Yield, defined previously.
P/E	Price-earnings ratio, defined previously.
Vol **100s**	The volume, or number, of shares traded during the week. This is printed in hundreds, so the actual total is 172,300 shares of Engrecon stock.
High Low	Two columns that show high and low prices for the week.
Last	The price of the stock at the final trade of the last trading day.
Net **Chng**	The net change from the close of the previous week. Engrecon stock closed the week selling at $19 per share, a gain of 50 cents from the close of the previous week.

quoted and scrutinized stock market proxy despite its narrow base and the presence of more representative indices, such as the Standard and Poor's (S&P) 500, which is based on 500 stocks.

The Dow Jones average is only the sum of current prices for the 30 corporations divided by an adjustment factor to account for dividends and newly issued shares; the divisor was 1.132 in late 1984. The Dow is not a true indicator of the *average* stock's performance because its 30 blue-chip industrials are certainly not representative of the several thousand issues currently being traded, and even these 30 issues are price-weighted to give proportionately more weight to high-priced stocks than to cheaper ones.

Even though the Dow Jones index has shortcomings, it is nonetheless a widely watched signal of stock market movement.

QUESTION

9A.1 The cost of capital for an organization is partly determined by the sources of funds available for investment. The proportion of corporate earnings paid out in dividends versus the amount retained is a decision that affects both managers and stockholders. Discuss factors that influence the decision and the way the outcome affects the cost of capital.

9B   A World of Investment Criteria

Industrial organizations in every nation face similar problems in screening out the most advantageous investments to improve production. Discounted cash-flow analysis is the

most celebrated screening device in this country and is widely used elsewhere. The payback period also has international acceptance, apparently more so than the theoretically sounder discounting methods. In developing countries where capital is severely limited, the emphasis on capital turnover in the payback-period criterion is more appreciated. In centralized socialist societies, the payback criterion in various versions is broadly applied because it conforms nicely to ideological beliefs concerning the social harm of interest charges.

Paul Samuelson observed in his ninth edition of *Economics:*

Recent studies show that the social engineers of the Soviet Union are anxious not to be denounced as capitalistic apologists; yet they need some form of interest rate (or "discount factor," or "payoff period") for making efficient investment calculations. As a result, about a dozen different accounting methods are in vogue there for introducing a thinly disguised interest-rate concept into Soviet planning procedures.

This observation is supported by Kljusev in his writings:

The methods for estimating the efficiency of the investment which are characteristic of the circumstances of capitalist production cannot automatically be used in a socialist economy because the stimulus and interests of the investment are different.*

Kljusev also notes that the efficiency criterion (known also as the "lucrativity index") is

$$E = \frac{\text{difference in annual cost between two alternatives}}{\text{difference in investment between two alternatives}}$$

or its reciprocal (the "period of return") is the most widely used economic evaluation method in the USSR and other socialist countries. But it still is considered inadequate because it is viable only for the relative economic efficiencies of alternative ways to produce the same article; it does not account for the apportionment of new revenue between continued investment to broaden production and amounts made available for personal and social expenditures.

Several extensions of basic economic comparison models have been forwarded by writers from different countries. The suggestions generally reflect particular communal concerns for strategic or tactical effects of capital expenditures, as indicated by the following samples.

1 Special attention should be given to investments involving convertible currency: Encourage projects that conserve or build foreign-exchange credits.
2 A weighting method should be used to evaluate the territorial impact of investments: Direct industrial development toward underdeveloped areas.
3 The time between appropriation and realization should be considered in comparing investments: Proposals that promise quick results should be rewarded.
4 Consider the total returns expected from a completely developed program rather than piecemeal returns in deciding which alternatives are most beneficial: Variations due to budgetary constraints and general economic conditions may mask the total effect of a program.
5 Develop ways to quantitatively include the social effects as well as the economic effects: Supplement economic models with a quantitative mechanism to express societal worth.
6 Stress considerations such as maintainability of equipment, flexibility of use, and variability of inputs when evaluating tactical investments: Detailed anticipation of future supply and demand conditions should influence present choices among long-lived assets, possibly causing a sacrifice of immediate profits in expectation of higher later benefits.

QUESTIONS
9B.1 Discuss the suggested extensions in terms of their applicability to engineering economic studies. What current trends in this country (e.g., consumerism, recycling, environmental protection) are likely to affect future investment decisions, and how might their influence be shown quantitatively in economic analyses?
9B.2 Comment on the quote below by Raymond Mayer. Do you agree with the statements? Why? Does this perspective account for the popularity of the payback-period criterion?

Theoretical methods of capital investment analysis, although sound and well established, have proven to be unacceptable to many industrial managers. The typical business man agrees that the relevant factors in the evaluation of an investment proposal are: the dollar investment, the cost of money, the alternative's life and salvage value, and its operating costs and revenues. But he also knows that accurate estimates of their values are not usually possible. Given this risk of error,

* N. Kljusev, *About Criteria and Methods of Valuation of Economic Effectiveness of Investment,* Ekonomiski Institut of Skopje, Yugoslavia.

it follows that he cannot appreciate the need for tedious calculations involving obscure compound interest factors. Even more important, he feels that the use of these interest factors gives the calculated results an aura of precision which simply is not there due to the approximate nature of the estimates.*

* R. Mayer, ''Capital Investment Analysis Another Way,'' *Industrial Engineering*, July 1970.

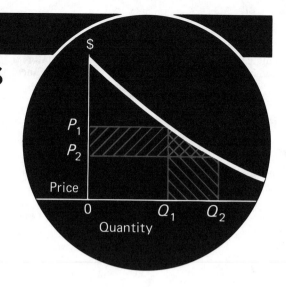

CHAPTER 10

ANALYSIS OF PUBLIC PROJECTS

OVERVIEW

Benefit-cost analysis is a well-rooted method of evaluating public projects. The basic measure of acceptability is a benefit-to-cost ratio greater than 1.0 or, equivalently, a positive net difference between benefit and cost. Incremental B/C ratios should also exceed unity for added increments of investment. B/C > 1 is a standard of minimum approval but is not an effective ranking criterion. Discounting procedures, computational practices, and capital-rationing considerations are essentially the same in the private and public sectors.

The appropriate interest rate for the evaluation of public projects is a much-discussed issue. Social projects are particularly sensitive to the rate of discount applied, because most major expenditures occur early, whereas the benefits extend many periods into the future; thus, a high discount rate tends to reduce the relative proportion of benefit to cost. Low discount rates are recommended by those who feel it is government's social responsibility to undertake projects that do not necessarily provide immediate returns comparable with those earned in the private sector. Counterarguments claim that resources are wasted when transferred from private uses, where returns are high, to low-yield public investments. One measure of the *social discount rate* results from computing the total cost of government borrowing.

Social projects often provide *public goods*—facilities or services available to all if available to one. The actual amount paid for a public good may not be a true measure of its worth. A *consumers' surplus* exists when a good is worth more to the user than the price paid for it. *Spillover benefit* or *cost* occurs when an activity affects third parties not directly involved in a project or program. These special properties of social involvement, which complicate the evaluation of benefits and costs, are discussed in the Extensions.

THE PUBLIC SECTOR

The public and industrial sectors of the United States economy interact as mutually supportive, but occasionally contradictory, forces. Government functions to facilitate the operation of a free-enterprise system may thwart or make difficult certain actions deemed advantageous by certain industrial subsystems. Despite claims of "government is best which governs least," the private sector simply does not have the means to provide all the necessary social goods and services. Government activities regulate and support the legal framework of the marketplace, establish and collect taxes for the redistribution of income, and allocate resources to programs and projects believed beneficial to society. Ideally, these activities will mirror the disposition of the citizenry with reasonable accuracy, while not completely satisfying the loudly vocal minorities calling either for no governing or for governmental solutions to all the perceived ills of society. The diversity of public-sector activities designed to serve so many interests complicates the evaluation of their effectiveness.

The concept of a *public good,* one which is available to all if it is available to anyone, is explored in Extension 10A.

Comparison methods discussed in previous chapters are suitable for the evaluation of government activities. Some considerations, such as taxes and payoff periods, are less significant in the public sector; and other considerations, such as the secondary effects of benefits, are more prominent. In this chapter, attention is directed toward the evaluation of specific projects rather than the general economic functions of government. Thus, the role of government to provide national defense, education, and dams is not questioned; subjects of evaluation are explicit investment proposals to implement these government functions.

Government projects nominally fall within the categories of protection, infrastructure development, cultural enhancement, and resource management.

Evaluations take the form of *benefit-cost* (sometimes called *cost-benefit*) analyses. Operationally, the technique is similar to the present-worth index. Conceptually, an analysis considers the worthiness of shifting resources from the private sector to the public sector and the extent to which a public project should be pursued when its benefits exceed its costs. Many factors besides a benefit-cost analysis influence the final decision about funding public projects, and weaknesses of benefit-cost analyses as traditionally conducted have cast doubts on the resulting ratings; but the analytical procedures still invite credibility and objectivity in appraisals of public expenditures.

BENEFIT-COST ANALYSIS

The U.S. Department of Defense, in DOD *Instruction 7041.3* (1972), defines benefit-cost analysis as

An analytical approach to solving problems of choice. It requires the definition of objectives, identification of alternative ways of achieving each objective, the identification, for each objective, of that alternative which yields the required level of benefits at the lowest cost. This same analytical process is often referred to as cost-effectiveness analysis when the benefits or outputs of the alternatives cannot be quantified in terms of dollars.

The need for formal evaluation of public expenditures can be traced back to the 1844 writings of J. Dupuit in France.* The Rivers and Harbor Act of 1902 in this country stipulated that a board of engineers must report on the merits of river and harbor projects of the Army Corps of Engineers. The reports were to include the amount of commerce benefited with respect to the estimated cost. A later act required a statement of local benefits to facilitate sharing of project costs with local interests which would benefit from the project. Government participation in public projects was extended by the Flood Control Act of 1936, which justified improvements to waterways for flood control *"if the benefits to whomsoever they may accrue are in excess of the estimated costs."* In the 1940s this principle was expanded to justify other projects or programs for social welfare, and it is now an accepted measure of desirability of projects at the federal, state, and local levels of government.

Benefit-to-Cost Criteria

In comparing benefit (B) to cost (C), several different perspectives are reasonable. Consider the simplified data in Table 10.1 that describe the alternatives for a small flood-control project. The current average annual damage from flooding is $200,000. Three feasible options are available to reduce the damages; each larger investment of public funds provides greater protection.

The three projects and the do-nothing alternative are mutually exclusive. Data for selecting the most attractive alternative are shown below, where the figures are in thousands of dollars.

Alter-native	Annual Benefit, B	Annual Cost, C	Total B/C	Total B − C	Incremental ΔB	Incremental ΔC	Incremental ΔB/ΔC	Incremental ΔB − ΔC
A	$ 0	$ 0	0	$ 0				
B	70	40	1.75	30	$70	$40	1.75	$30
C	160	120	1.33	40	90	80	1.125	10
D	190	160	1.19	30	30	40	0.75	−10

The following criteria indicate different plausible preferences among the alternatives:

* J. Dupuit, "On the Measurement of Utility of Public Works," *International Economic Papers,* vol. 2.

TABLE 10.1

Annual costs and benefits from different levels of investment in a small flood-control project

Alternative	Equivalent Annual Cost of Project	Average Annual Flood Damage	Annual Benefit
A: No flood control	$ 0	$200,000	$ 0
B: Construct levees	40,000	130,000	70,000
C: Small reservoir	120,000	40,000	160,000
D: Large reservoir	160,000	10,000	190,000

1 Minimum investment: Choose *alternative A*. If funds are severely limited, this may be the only possible choice.
2 Maximum benefit: Choose *alternative D*. Flooding would occur only during extremely wet seasons.
3 Aspiration level: Depends on the threshold set for cost or benefit. If, for instance, the aspiration level is to reduce flood damage by 75 percent, *alternative C* should be chosen because it meets the aspiration with a lower cost than does *alternative D*. Similarly, an annual-cost threshold of $100,000 indicates a preference for *alternative B*.
4 Maximum advantage of benefits over cost (B − C): Choose *alternative C*.
5 Highest benefit-to-cost ratio (B/C): Choose *alternative B*.
6 Largest investment that has a benefit-to-cost ratio greater than 1.0: Choose *alternative D*.
7 Maximum incremental advantage of benefit over cost ($\Delta B - \Delta C$): Choose *alternative B*.
8 Maximum incremental benefit-to-cost ratio ($\Delta B/\Delta C$): Choose *alternative B*.
9 Largest investment that has an incremental B/C ratio greater than 1.0: Choose *alternative C*.

A liberal interpretation of the wording of the 1936 Flood Control Act suggests that the highest investment in which benefits exceed costs is the most desirable. A more realistic interpretation is that the act sets a lower limit on acceptability, with respect to both total project amount and the incremental amounts in multilevel alternatives. By the latter interpretation, alternative *D* is rejected because of its negative $\Delta B - \Delta C$ component and its $\Delta B/\Delta C$ ratio less than 1.0. Then alternatives *B* and *C* are both acceptable, so other considerations should enter into the final choice. These considerations include the availability of investment funds, capital-rationing criteria, and the special features of social merit and economic objectives that affect most public projects.

Benefit-to-Cost Comparisons

The mechanics of benefit-cost comparisons are straightforward and simple—deceptively so. The basic comparison formulas are

$$B/C = \frac{\text{present worth of benefits}}{\text{present worth of costs}}$$

$$= \frac{\text{equivalent annual benefits}}{\text{equivalent annual costs}}$$

and

$$\text{Present value of net benefit } (B - C) = \text{PW(benefits)} - \text{PW(costs)}$$

The rationale and mechanics for constant, or real, dollar evaluations are examined in Chapter 13.

where, according to *Circular No. A-94* (1972) from the Office of Management and Budget, Executive Office of the President, the present worth of benefits is the discounted *constant* dollar value of goods and services expected to result from a project or program for each of the years it is in effect. Estimates may reflect changes in the *relative* prices, when there is a reasonable basis for estimating such changes, but should not include any forecast change in the general price level during the planning period.

The present worth of costs is the discounted annual value in *constant* dollars of resources, goods, and services required to establish and carry out a project or program. All economic costs, including acquisition, possession, and operating costs, must be included whether or not actually paid by the government. Such costs, *not* generally involving a direct payment by the government, include imputed market values of public property and state and local property taxes forgone.

It is apparent from the B/C and B − C formulas that benefit-to-cost comparisons are simply reformulated PW comparisons. Consequently, the precautions applicable to PW calculations also pertain to benefit-to-cost calculations. For instance, a preference for $X2$ is indicated by a comparison of the two present-worth levels given below for project X according to the incremental differences for benefits and costs:

Alternative	B	C	B/C	B − C	ΔB	ΔC	ΔB/ΔC	ΔB − ΔC
X1	4	2	2	2				
X2	7	4	1.75	3	3	2	1.5	1

Although $X1$ has a higher B/C than does $X2$, the incremental ratio for the extra benefit and cost exceeds 1.0, making $X2$ an acceptable alternative. If it can be assumed that sufficient capital is available to fund either alternative, $X2$ is selected because it provides greater benefits. However, when the total costs for projects exceed the resources allocated to the proposing agency, capital-budgeting procedures are required and intangible effects may influence the selection.

Example 10.1
Incremental Benefit-to-Cost Evaluation

A number of small earthen dams are contemplated for the headwaters of a drainage system. Four tributaries originate in a national forest and flow together to form a river which passes through private lands. Each year there is some flooding, and every few years a major inundation occurs. Construction of one or more dams will ease the threat of high water. Dams on all the tributaries would largely eliminate the chance of a major flood.

In addition to the damage to private lands, floods also ruin fire and logging roads in the forest. Other benefits from the dams include the value of the impounded water for fire protection and recreational use. The following benefit and cost estimates have been developed for the only topologically feasible combinations of dams:

A 40-year life and no salvage value is assumed for earthwork dams. An interest rate of 4 percent is deemed appropriate for the investment. This rate reflects the low risk involved and is in line with the historical interest rate for bonds issued by the federal agency to finance public projects.

Based on B/C ratios, which of the four alternatives should be selected?

Dam Sites	Construction Costs	Annual Maintenance and Operation	Annual Flood Benefits	Annual Fire Benefits	Annual Recreation Benefits
1	$1,200,000	$20,000	$200,000	$20,000	$30,000
1 and 2	1,500,000	35,000	190,000	40,000	30,000
1, 2, and 3	2,700,000	50,000	280,000	60,000	60,000
1, 2, 3, and 4	3,500,000	60,000	300,000	70,000	70,000

Solution 10.1

A B/C ratio based on equivalent annual values for each alternative is calculated from

$$\text{B/C ratio} = \frac{\text{annual flood and fire savings} + \text{recreation benefits}}{\text{equivalent annual construction costs} + \text{maintenance cost}}$$

where

Equivalent annual construction cost
$$= (\text{initial construction cost})(A/P, 4, 40)$$

The incremental B/C ratio is determined from the additional benefits returned by an increment of cost above the last acceptable alternative (B/C ratio > 1.0).

Dam Sites	Annual Benefits	Annual Cost	Increments of Benefit	Increments of Cost	Total B/C Ratio	Incremental B/C Ratio
1	$250,000	$ 80,630			3.10	
1 and 2	260,000	110,770	$ 10,000	$30,140	2.34	0.32
1, 2, and 3	400,000	186,400	140,000	75,630	2.14	1.42
1, 2, 3, and 4	440,000	236,820	40,000	50,420	1.85	0.79

For an accurate evaluation, the requirement that annual benefits must equal annual costs should be applied to each separable increment of project costs. Site 1 is compared with the alternative "no action" and yields a total B/C ratio = incremental B/C ratio = 3.10, which qualifies it as an acceptable alternative. The next increment that meets the B/C-ratio standard is the location of dams at sites 1, 2, and 3 compared with a dam at site 1:

$$B/C(3) = \frac{\$10,000 + \$140,000}{\$30,140 + \$75,630} = 1.42$$

Adding another dam site fails to produce an acceptable B/C ratio:

$$B/C(4) = \frac{\$40,000}{\$50,420} = 0.79$$

Therefore, preference is indicated for the third alternative, dams at sites 1, 2, and 3.

Without an incremental analysis, the last alternative (four dams) might have been selected, because it does possess a B/C ratio greater than 1.0 and offers the greatest total benefits. Another mistake would be to eliminate all other alternatives because the dam at site 1 has a larger benefit-to-cost ratio than do the rest of the options. The reasoning errors behind such conclusions are the same as those examined for incremental rates of return (Chapter 7). The conclusion to accept the three-dam alternative based on the given data would also result from a rate-of-return or present-worth evaluation.

It is interesting to note the *sensitivity* of the selection to changes in the data. Using a required interest rate of 7 percent instead of 4 percent would change the choice to the first alternative (one dam) because all the added benefit and cost increments would produce B/C ratios smaller than unity. Including only the flood-control benefits would also make site 1 the only alternative with an acceptable B/C ratio. The number of spillover benefits to include in an analysis and the monetary rating given to less tangible benefits can significantly influence decisions.

Irregularities in B/C Comparisons

No investment criterion seems to be able to escape all theoretical objections. The benefit-cost criterion has been subjected to considerable criticism based on its oversimplification of complex inputs, susceptibility to misinterpretation, and potential for misuse. Some of the more prominent objections are discussed below.

1 Insensitivity to Priorities

A B/C ratio includes the cost and benefit of a particular project but no indication of how valuable the benefit is, as compared with other projects, or the relative amount of resources involved.* As an extreme example, two projects could have the same B/C ratio and have costs of $1 million and $1 billion. The billion-dollar project might have long-range consequences for improvement of the environment in a sparsely populated region whereas the million-dollar project would serve a critical inner-city need. Eventually both projects might serve the same number of people, but the smaller project would have far greater immediate benefits for the taxpayers who are funding it. A third $100,000 project, with the same B/C as the other two, could be to preserve a site that will be destroyed if not subsidized immediately. If there are not

* Discussed in A. Maass, "Benefit-Cost Analysis: Its Relevance to Public Decisions," *Quarterly Journal of Economics*, vol. LXXX, May 1966.

enough resources available to undertake all three projects, the B/C ratios are of little help in setting priorities.

A related objection is that the B/C standard assumes that benefits are equally valuable to the rich and poor. A project to push a new freeway through a densely populated low-cost-housing neighborhood would thus use money partially drawn from the poor to force them to relocate to make room for an expressway that would mainly benefit wealthier suburban commuters.

The argument against the criticisms above is that the B/C criterion was not designed to rank projects; it sets only a minimum level of acceptability and does not pretend to identify the source of investment funds. Its use as a guide to decision makers is poignantly described in a compendium of papers submitted to the Subcommittee on Priorities and Economy in Government of the United States Congress:

Conceptually, public funds are expended to equalize opportunities among the citizenry.

If benefit-cost analysis is to be implemented and used to its fullest potential, renewed efforts must be made by policy makers in both the executive and legislative branches of government. The economics profession has made significant advances in the level of sophistication of their analyses which should aid this task, but one thing is clear— benefit-cost analysis does not make decisions.

Analysis can provide an important and helpful tool for making decisions, but it is no more than a tool. Problems involving social policy and value judgments must be considered and weighed in conjunction with the results of benefit-cost analysis and the final decision made by the human policymaker.†

Spillover costs and benefits which deserve consideration in B/C comparisons are discussed in Extension 10B.

The numerator of the B/C ratio is usually taken to mean the net benefit—the sum of all benefits minus all disbenefits. If the disbenefits were treated as costs in the denominator, the ratio would obviously change. For instance, if the present worth of benefit is 10, disbenefit is 3, and cost is 5, then

2 Variations Due to Formula Inputs

$$B/C = \frac{10 - 3}{5} = 1.40$$

But if the disbenefit is counted as a cost, then

Tolls and other user fees are typically treated as disbenefits.

$$B/C = \frac{10}{5 + 3} = 1.25$$

This variation presents no difficulty when it is remembered that the purpose of the B/C > 1 criterion is simply to distinguish between acceptable and unacceptable alternatives. There is no way in which shifting of disbenefits from the numerator to the denominator can cause a B/C ratio to change from greater than 1.0 to less than 1.0. Equivalently, the new present value (benefit minus disbenefit minus cost) must be the same regardless of the order of

---

† *Benefit-Cost Analysis of Federal Programs*, Government Printing Office, Washington, D.C., 1973.

subtractions. Nor will an incremental analysis indicate any difference in the absolute acceptability of alternatives owing to the location of disbenefits in the B/C ratio.

Two alternatives are evaluated below, first by subtracting disbenefits from benefits and then by adding them to costs. In both cases, even though the value of the ratios differ, the extra increment of investment is found to be acceptable according to the $\Delta B/\Delta C > 1$ standard.

Alternative	Present Worth of Benefit, Disbenefit, and Cost				Disbenefits Subtracted from Benefits				Disbenefits Added to Costs			
	B	D	C	B − D − C	B/C	ΔB	ΔC	ΔB/ΔC	B/C	ΔB	ΔC	ΔB/ΔC
X	10	3	5	10−3−5=2	7/5=1.4	9−7=2	6−5=1	2/1=2	10/8=1.25	13−10=3	10−8=2	3/2=1.5
Y	13	4	6	13−4−6=3	9/6=1.5				13/10=1.3			

The relative sizes of B/C ratios provide an indication of investment productivity that may be misleading. Consider a project $P1$, in which an investment of 10 and annual costs of 95 produce yearly benefits of 100 for 20 years: At $i = 10$ percent,

$$B/C(P1) = \frac{100(P/A, 10, 20)}{10 + 95(P/A, 10, 20)} = \frac{100(8.5135)}{10 + 95(8.5135)}$$

$$= \frac{851}{10 + 809} = 1.04$$

Compare this project with $P2$, which also has an initial investment of 10, no annual costs, and a yearly benefit of 1.25 for 20 years: At $i = 10$ percent,

$$B/C(P2) = \frac{1.25(8.5135)}{10} = \frac{10.6}{10} = 1.06$$

Both projects are barely acceptable, and $P2$ has a slightly higher B/C ratio, but $P1$ will "pay back" the initial investment in $10/(100 - 95) = 2$ years, while $P2$ requires $10/1.25 = 8$ years before the sum of annual benefits equals the investment. Stated another way,

$$\text{IRR}(P1): \quad (100 - 95)(P/A, i, 20) - 10 = 0 \quad \text{at } i = 50\%$$
$$\text{IRR}(P2): \quad 1.25(P/A, i, 20) - 10 = 0 \quad \text{at } i = 10.9\%$$

The rate of return of $P1$ far exceeds that of $P2$, although their B/C ratios were nearly the same.

This apparent conflict is caused by the inability of the B/C formula to distinguish between cash flow patterns; it tends to underrate the quality of a project with high annual costs. This same characteristic was discussed in Chapter 9, where ranking reversals were noted for alternatives evaluated by PW and IRR comparisons when the discount rate was raised to a certain level. In the example above, a discount rate of 11 percent would make $P2$ an unacceptable alternative:

$$\text{B/C}(P2) = \frac{1.25(P/A, 11, 20)}{10} = \frac{1.25(7.9633)}{10} = \frac{9.95}{10} = 0.995$$

whereas $P1$ remains acceptable at the 11 percent rate:

$$\text{B/C}(P1) = \frac{100(P/A, 11, 20)}{10 + 95(P/A, 11, 20)} = \frac{796}{10 + 757} = 1.038$$

THE DISCOUNT-RATE QUESTION

The appropriate interest rate to use in evaluating public investments is a voluminously debated issue. Although numerous suggestions have been forwarded, none has been awarded complete acceptance. About the only near consensus is that discount rates established for early benefit-cost analyses in the depressed 1930s (around 2½ to 3¼ percent) are not valid today. Studies made of the Corps of Engineers' projects reveal that 80 percent of the projects authorized in 1962 would have been rejected at a discount rate of 8 percent, and California's big Feather River project would be uneconomical at a 6 percent rate.* The effect of the discounting rate on benefit-cost ratios for projects with different lives is shown in Figure 10.1. The five curves represent B/C ratios at different discount rates for an initial investment of $1 that returns a total of $2 in equal installments over N years. Therefore, at a 0 percent discount rate and $N = 10$,

$$\text{B/C} = \frac{(2/10)(P/A, 0, 10)}{1} = \frac{0.2(10)}{1} = 2$$

but at $i = 10$ percent for 10 years,

$$\text{B/C} = \frac{(2/10)(P/A, 10, 10)}{1} = \frac{0.2(6.1445)}{1} = \frac{1.23}{1} = 1.23$$

* See J. A. Stockfish, "The Interest Rate Applicable to Government Projects," in H. H. Hinricks and G. M. Taylor, eds., *Program Budgeting and Benefit-Cost Analysis,* Goodyear, Pacific Palisades, Calif., 1969.

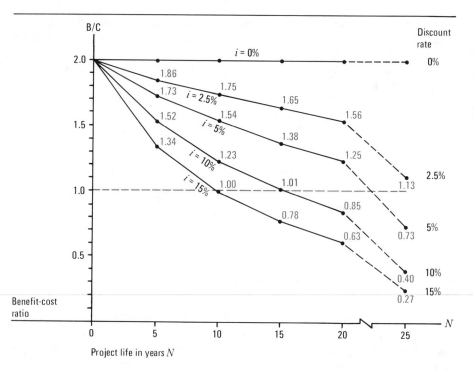

FIGURE 10.1
Graph of benefit-cost ratios at five discount rates for different project lives of a proposal to invest $1 in order to receive $2 in equal installments prorated over the project's life.

Range of Discount Rates

The government borrowing rate is an obvious measure of financing costs for public projects. It is akin to the *cost-of-capital* concept by which a minimum attractive rate of return is determined for investments in the private sector. The cost of capital to the government can also be taken as a minimum discount rate for public projects. In recent years the interest rate for government borrowing has varied between 7 percent and 15 percent, depending on the level (federal, state, or local), tax-exemption privileges offered, current economic conditions, and length of the loan. Since government bonds are generally considered to be risk free, any public project should promise a minimum return that is at least equal to the bond rate. Otherwise, the resources that could be used to fund projects should instead be applied to debt repayment.

The rates of return expected in the private sector are higher than are the interest rates on government bonds. This is a reasonable condition, because a firm would invest in risk-free government bonds if it did not expect larger returns from industrial investments. In the opinion of W. A. Baumol,* "The

---

* W. A. Baumol, "On the Appropriate Discount Rate for the Evaluation of Public Projects," in ibid.

correct discount rate [for public projects] is the percentage rate of return that the resources would otherwise provide in the public sector." His view recognizes the opportunity cost to general welfare that occurs when resources are used to produce benefits smaller than could have been obtained if the resources were applied elsewhere; specifically, a public project that barely exceeds B/C = 1 when evaluated at a low discount rate would, if carried out, produce fewer benefits than could have been obtained by leaving the amount invested in the private sector.

Sometimes overlooked are the benefits derived from private investments that develop a nation's infrastructure, conserve resources, and promote culture.

An argument against charging higher rates for public projects is that certain socially desirable programs would never meet higher evaluation standards but are nonetheless worthwhile. This line of reasoning emphasizes the responsibility of the present population to ensure adequate living conditions for future generations. Since most people are more concerned with satisfying their current needs and wants than with saving for the future, proponents of low discount rates say only government action can enforce investments for future benefits and this transfer of resources necessarily yields a lower return than do resources used for current consumption.

Counterarguments question the legitimacy of allowing a few policy makers to decide what is best for general welfare, especially when their allocations appear to contradict public preferences shown in a free market. It is generally agreed that no generation has the right to completely consume resources that cannot be replaced by its successors (for example, erosion of soil, extinction of a wildlife species, destruction of natural phenomena).The dispute is with subsidizing future generations by resource expenditures acceptable only if evaluations are based on artificially low discount rates; such special assistance is not warranted, critics say, because the future is also served by public and private investments that earn returns comparable to current resource commitments. From a historical perspective that indicates each generation has been wealthier than its predecessor, it appears imprudent to justify "wasteful" resource allocations on the basis of protecting future beneficiaries.

Biased recommendations may result when local governments treat federal funds as "free" money, often causing scandals by whimsical cost justifications and fanciful benefit claims.

Social Discount Rate

The rate of interest used in evaluating public projects is often referred to as the *social discount rate*. It reflects the cost of capital obtained from the private sector and the opportunity cost of resources applied to public projects rather than private investments. One approach to the calculation of a social discount rate based on the cost of government borrowing is given on page 246;* it is a summation of contributing cost components that provides a composite percentage figure.

* Adapted from E. B. Staats, "Survey of Use by Federal Agencies of the Discounting Technique in Evaluating Future Programs," in ibid.

Since interest rates for borrowing vary markedly over time (see Figure 6.6), the numbers in the sample calculations are not always applicable, but the cost components remain appropriate.

Sample Calculation of a Social Discount Rate
Based on the Cost of Government Borrowing

Cost Components	Rate, %
1 Approximate interest rate for long-term low-risk borrowing	6.0
2 Corporate taxes forgone by the government, under the assumption that the average corporate return is 13% before taxes, the fraction of money borrowed by the government that would have gone into corporate investment is 65%, and the marginal corporate tax rate is 40%: $(0.13)(0.65)(0.4) =$	3.4
3 Personal taxes forgone by the government, under the assumption that the 35% not put into corporate investment earns 10% for individual taxpayers and the marginal tax rate is 30%: $(0.10)(0.35)(0.3) =$	1.0
4 Taxes forgone from dividends and bond interest generated by the 65% of government-borrowed dollars that would have gone into corporate investment:	

a Under the assumption that individuals are taxed at 30%, dividend payments are 40% of corporate profits, and 60% of corporate earnings are after-tax profits:

$$(0.3)(0.4)(0.6)(0.13)(0.65) = 0.6\%$$

b Under the assumption that interest on corporate bonds is 7% and the individual marginal tax rate is still 30%:

$$(0.07)(0.30)(0.65) = 1.4\%$$

If there is an equal weight for taxes forgone from bond interest and dividend payments: $(0.6\% + 1.4\%)/2 =$	1.0
5 Income taxes collected on interest payments from government borrowing are a negative cost of capital, if the borrowing rate is 6% [from (1) above] and the composite tax rate for corporate and individual bondholders is 35%: $(0.06)(0.35) =$	−2.1
Social discount rate	9.3%

An alternative aggregate method for determining the discount rate based on the cost of government borrowing assumes a composite personal and corporate tax rate of 50 percent and a taxable return of 12 percent on any money not borrowed by the government:

Interest cost of borrowing	6.0%
Taxes forgone = $(0.50)(0.12)$	6.0%
Less taxes on bond interest: $(0.06)(0.35)$	−2.1%
Social discount rate	9.9%

The calculations above do not include any consideration of *risk*. It is well known in the financial market that government bonds are perceived to have low risk, which is rewarded by the lower interest rate needed to attract capital. The inference for public projects is that the government is in a better position than a private enterpreneur to cope with risk; governments undertake so many projects that their operations are essentially immune to risk, on the average. However, critics point out that there have been some obvious and major failures in government-funded projects and proposals. Therefore, as is the case with investments in the private sector, the social discount rate should include a risk factor. A surcharge of ½ to 1 percent has been suggested, but this across-the-board addition fails to account for differences in the nature of risk for various classes of projects.

BENEFIT-COST APPLICATIONS

Many of the charges of misuse of benefit-cost analyses are the result of specific projects that have gone awry. The well-publicized cases of dams that failed to stop flooding as promised and project overruns that boosted costs well above anticipated benefits could be considered as evidence of the need for more exacting benefit-cost analysis, rather than of deficiencies in the B/C criterion.

In addition to the theoretical critiques of the applicable discount rate and the fairness of benefit measurements, criticism has been leveled at the way the analyses are conducted. It is claimed that unrealistically high values have been assigned to intangibles to compensate for low monetary benefits in projects that would be unacceptable when evaluated only on quantifiable data. Indirect or intangible costs seldom seem to get the same recognition as do nonmonetary benefits. Costs have been underestimated, say critics, because the local impact of major federal projects was not anticipated; labor and material prices went up because of increased local demand caused by the project to the detriment of both project expenses and consumers in the community.

Methods for quantifying essentially intangible benefits are explored in Extension 10C.

Projects are sometimes undertaken without the support of the residents in the area. Planners are thus forced into the position of telling residents what is good for them, which requires a selling campaign that may create costly delays or added community-relations costs. Part of the problem could be alleviated by exposing a project to a public vote. However, difficulties of informing the voters about the issues involved and the remoteness of many projects from their base of funding (the taxpayers) limit the workability of decisions by vote. A choice between a football stadium and an art museum in a city would probably arouse enough interest by those affected to ensure a careful appraisal and representative turnout of voters. Yet an equal expenditure to develop a wildlife refuge might inspire less interest, even if all the affected voters could be given a chance to vote. On all questions of national objectives, there will likely be a few opponents, and many of these will be in the geographical locality where a project is to be carried out. Local protests can naturally be expected from people whose lives are disrupted by a project designed to serve the general welfare. It is the duty of policy makers to give consideration and adequate compensation to the local interests while supporting regional and national social interests.

Two benefit-cost analyses are described in the following pages. The applications are digested to point out certain highlights; the actual studies and techniques employed were much more exhaustive. Example 10.2 illustrates a feasibility study that disqualified public expenditures in solving a local problem. A postaudit to determine the actual benefit and cost of a program is featured in Example 10.3.

Example 10.2
Feasibility Investigation of a Water-Resource Project*

"The Corps of Engineers was directed by the Congress of the United Sates to make a study of Marys River Basin. The purpose of the study was to determine what could be done to reduce flood damages and to conserve, use, or develop the basin's water resource."

Six alternatives were investigated: (1) floodproofing individual structures, (2) building levees, (3) improving river channels, (4) instituting a system of land-owner-constructed dams, (5) implementing a system of small tributary reservoirs, and (6) erecting multipurpose reservoirs. The costs of floodproofing, levees, and channel improvements were found to be much greater than the benefits, mainly because of the lack of secondary benefits. Landowner-constructed dams built with government financial support would have some localized benefits but would require an unreasonable amount of land for the storage obtained, and problems of balancing outflows made the alternative impractical.

The last two alternatives also proved to be economically infeasible, as shown in Figure 10.2. Flood-control benefits were based on historic flood damages updated to 1974 prices, estimates of flood-stage reductions by operation of the projects investigated, and projections of economic growth in the area. Recreation benefits were based on user-day projections. The cost of the cheapest alternative water supply was used in the analysis as a measure of the water-supply benefits that could reasonably be expected. Data provided by the U.S. Fish and Wildlife Service showed that fish and wildlife benefits resulting from any of the project alternatives would be negligible. Costs included construction and, as appropriate, land; relocation of roads, railroads, and utilities; fish-passage facilities; wildlife-migration features; recreation developments; design costs; and construction supervision costs.

Example 10.3
Follow-up Study to Evaluate Benefit versus Cost for a Human-Resource Program†

The Roswell Employment Training Center was established in 1967 under contract with the Bureau of Indian Affairs. The Center was designed as a residential employment training program for Indian families, solo parents, and single adults. Emphasis is placed on training and counseling intended to aid trainees in adjusting to typical work situations and to living off the reservation.

Using earnings differentials to measure benefits from human-resource investment programs is a standard method of evaluation. The major resource costs are staff services, where salaries are taken as the measure of their value in an alternative

use; other direct costs such as supplies; and the opportunity costs of training—the earnings forgone as a result of undertaking training.

In June 1970, a follow-up study was conducted by mail of all trainees who had entered Roswell Center between March 1968 and February 1970. The calculation of a benefit-cost ratio using the data from this survey is shown below. Of the group of 170 returning adequate survey forms, 70 were currently employed at an average hourly wage of $2.25 after training, whereas 82 members of the same group were employed before training was given. Because the average wage rate

---

* Adapted from *Summary Report on Marys River Basin Water Resource Study*, United States Corps of Engineers, 1975.
† Adapted from B. F. Davis, "Using Benefit-Cost Analysis to Assess a Human Resource Investment Program," in ibid.

	Noon Dam	Wren Dam	Tumtum Dam	Tributary Dam System
Costs, total				
Construction	$51,352,000	$52,944,000	$37,024,000	$19,090,000
Investment†	57,386,000	59,165,000	41,374,000	21,333,000
Costs, annual				
Interest and amortization‡	3,380,000	3,485,000	2,437,000	1,257,000
Operation, maintenance, and replacement	210,000	180,000	170,000	135,000
Total average annual cost	$ 3,590,000	$ 3,665,000	$ 2,607,000	$ 1,392,000
Benefits, average annual				
Flood control	$ 945,400	$483,100	$258,500	$126,300
Recreation	573,000	345,000	401,000	163,000
Irrigation water supply	85,000	85,000	85,000	85,000
Municipal and industrial water supply	90,600	63,400	8,800	0
Fish and wildlife enhancement	0	0	0	0
Total average annual benefits	$1,694,000	$976,500	$753,300	$374,300
Benefit-cost ratio	0.47:1	0.27:1	0.29:1	0.27:1

†Construction cost plus interest during construction.
‡The annual cost equivalent to the total cost spread over a 100-year period computed at 5 7/8 percent interest rate.

FIGURE 10.2
Benefit-cost analyses for alternative reservoir projects.

after training was significantly higher than was the average wage before training, there was still a positive earnings differential associated with training, equal to $573 per year per trainee:

Additional annual earnings/trainee

$$= \frac{(\$4680 \times 70) - (\$2808 \times 82)}{170} = \$573$$

Benefit of program. PW of extra annual earnings over 41 years (the number of remaining years in the labor force for trainees whose average age was 24) at a 6% discount rate
$8674

Cost of program. Contracted cost for an average length of training of 8 months at $658 per month per trainee = $5264. Forgone earnings (number of trainees working before training times their monthly earnings times the months of training, divided by the total number of trainees)

$$= [(82)(\$2808/12)(8)]/170 = \$902$$

Then, PW of total cost = $5264 + $902 = $6166

$$B/C = \frac{\$8674}{\$6166} = 1.41$$

The benefit-cost analysis described above is the simplest possible for a human-resource investment program. Earnings before and after training are compared and differentials are projected, at a constant amount, over the years former trainees can be expected to remain in the labor force. The present value of these amounts are then compared with direct program costs plus the opportunity costs of trainees to estimate benefit-cost ratios. Every step of the analysis can be challenged on theoretical and empirical grounds. Pretraining earnings may be understated; perhaps earnings differentials should be projected at growing rather than constant amounts in future years; the lack of complete follow-up data may bias the estimated earnings differentials; a control group should actually be established; maybe 6 percent is not the "right" discount to use; an estimate of the value of the physical facilities in an alternative should actually be included as a cost—all this and probably more could be charged. To focus on these niceties of analysis would be to miss an important aspect of program analysis. Analysis should not only ask the question, "Is this a good program?" or "Has the benefit-cost ratio been appropriately and accurately estimated and is it greater than unity?" but should provide program managers with the information they need to change particular aspects of their program in an effort to improve performance.*

Cost effectiveness is an analysis technique associated with complex government systems, procurement policies, and development projects. It attempts to systematize evaluation procedures.

Review Exercises and Discussions

Exercise 1   A proposal has been made to modify certain navigational aids that will decrease the cost of operation by $10,000 per year for the next 30 years. A loss in benefit during the same period amounts to $1000 per year. Conduct a benefit-cost evaluation with a discount rate of 10 percent, assuming that the modifications will be completed with little extra cost as part of the routine maintenance.

Solution 1

$$B/C = \frac{-\$1000(P/A, 10, 30)}{-\$10,000(P/A, 10, 30)} = 0.1$$

The criterion of B/C > 1.0 for acceptability is not applicable when both the benefit and cost of a project are negative. The ratio of savings to costs is obviously 10:1 in favor of the proposal. Thus, when both the numerator and denominator of a B/C ratio are negative, a ratio of less than 1.0 indicates acceptability. Moreover, a project with a positive B and a negative C is automatically accepted.

Exercise 2†   The present worth of benefits and costs for two mutually exclusive "base" proposals are shown below, along with data for three supplementary projects which can be combined with either base to yield additional benefits. The supplementary projects are not mutually exclusive. Which combination of projects is preferred when resources are limited to $400,000?

| Basic-Project Proposals | | | | Supplementary Projects | | |
Project	B	C	B/C	Project	B	C
P1	$300,000	$150,000	2.0	S1	$ 75,000	$ 50,000
P2	450,000	250,000	1.8	S2	140,000	100,000
				S3	300,000	150,000

* Ibid.
† Adapted from G. A. Fleischer, *Benefit-Cost Analysis: An Introductory Exposition*, Tech. Report 72-1, Department of Industrial and Systems Engineering, University of Southern California.

Since P1 has a higher benefit-cost ratio than does P2, it is a logical base for combinations. Only combinations that approach but do not exceed the $400,000 limit are considered, because the objective is to maximize benefits with the resources available.

Solution 2

Combination	Benefit	Cost	B − C	B/C
P1 + S1 + S2	$515,000	$300,000	$215,000	1.72
P1 + S3	600,000	300,000	300,000	2.0
P1 + S1 + S3	675,000	350,000	325,000	1.93
P1 + S2 + S3	740,000	400,000	340,000	1.85

Of these combinations, the one that spends all the available resources, P1 + S2 + S3, provides the greatest benefits ($740,000), yields the highest net benefits ($340,000), has a B/C ratio (1.85) greater than 1.0, and gives an incremental benefit-cost ratio exceeding unity,

$$\Delta B/\Delta C = \frac{\$740,000 - \$675,000}{\$400,000 - \$350,000} = \frac{\$65,000}{\$50,000} = 1.3$$

is the most satisfying.

However, the best possible combination is one that utilizes the other base, P2. The combination of P2 + S3 has a total benefit of $450,000 + $300,000 = $750,000 and a total cost of $250,000 + $150,000 = $400,000 to yield

$$(B - C)_{P2+S3} = \$750,000 - \$400,000 = \$350,000$$

and

$$(B/C)_{P2+S3} = \frac{\$750,000}{\$400,000} = 1.875$$

This combination might be overlooked by the preselection error of eliminating P2 from consideration because its benefit-cost ratio is less than the B/C for P1. It is also interesting to note that in this case the B/C ratio provides an accurate ranking of top contenders because both have the same total cost.

Benefit-cost ratios for two mutually exclusive alternatives have been calculated with disbenefits assigned to both numerator and denominator as shown:

Exercise 3

Alter-native	Present Worth			B/C = (B − D)/C	B/C = B/(C + D)
	B	D	C		
A1	150	0	100	(150 − 0)/100 = 1.5	150/(100 + 0) = 1.5
A2	200	20	120	(200 − 20)/120 = 1.5	200/(120 + 20) = 1.43

What reply would you make to a claim that A1 is economically superior because its B/C ratio is equal to or better than that obtained from A2, regardless of the way disbenefits are handled?

The net benefit of A2 is 200 − 20 − 120 = 60 versus 150 − 100 = 50 for A1. If any preference is shown by the benefit-cost criterion, it is thus bestowed on A2. An incremental analysis conducted either way,

Solution 3

	Disbenefit in Numerator				Disbenefit in Denominator		
	B	C	ΔB/ΔC		B	C	ΔB/ΔC
A1	150	100	30/20 = 1.5	A1	150	100	50/40 = 1.25
A2	180	120		A2	200	140	

confirms that A2 is an acceptable project because its incremental B/C ratio is consistently greater than 1.0.

Exercise 4   Consider a freeway project that proposes a new four-lane highway leading from the downtown core of a city to a major arterial bypass in the suburbs. What factors besides the construction requirements could be considered in the total project cost?

Solution 4   Benefit-cost analyses for some major highway-construction projects consume hundreds of pages. A few of the considerations include:

1 Air pollution from traffic flow
2 Expenditures for traffic control
3 Accident frequency and severity, past and forecast
4 Parking expectations in affected areas
5 Time and cost consequences of congestion, before and after
6 Shift of business competition
7 Accessibility of outlying jobs to central-city poor
8 Changes in property values due to accessibility
9 Relocation of dwellings in the highway path; associated disruption of neighborhoods
10 Visual and auditory impact
11 Temporary effects of construction: wages, price of materials, living costs, employment, etc.

PROBLEMS

10.1 Seven mutually exclusive proposals are to be compared using the benefit-cost criterion. Each of the projects has an expected life of 50 years with negligible salvage value and is to be evaluated with a tax-free interest rate of 10 percent.

	Proposal						
	A	B	C	D	E	F	G
Initial cost	$100	250	700	80	500	50	150
Annual maintenance cost	6	8	30	2	20	2	6
Estimated annual benefits	16	40	110	9	70	11	35

(a) Rank-order the proposals based on increasing first cost and calculate the respective B/C ratios treating annual maintenance as a cost.
(b) Calculate the respective B/C ratios treating annual maintenance as a disbenefit.
(c) Determine which alternative would be selected if funds are severely limited.

(d) Indicate which alternative would be selected if the sole criterion were largest available benefit.

(e) Decide which alternative would be selected if the sole criterion were largest net benefit.

(f) Develop an incremental B/C analysis and select the appropriate alternative.

(g) Determine which alternative would be selected if an incremental rate of return analysis with MARR equal to 10 percent was employed.

10.2 Several alternative projects involving water supply systems are under consideration by the Leakydam Public Utility Board. The following data have been summarized for your consideration.

	Project				
	A	B	C	D	E
First cost	$2000	$100	$700	$1200	$500
Annual operating cost	70	4	60	100	30
Annual recreation benefits	150	0	55	170	25
Annual increase in agricultural production	200	30	50	160	90

All costs are given in thousands of dollars and negligible salvage values are assumed at the end of a 50-year life. Since annual operating costs are to be paid from area property tax revenues, they are treated as a disbenefit. Using a tax-free interest rate of 7 percent, determine:

(a) Which projects should be selected if the alternatives are independent?

(b) Which project should be selected if the alternatives are mutually exclusive?

(c) Which of the mutually exclusive projects should be selected if a rate of return analysis is used?

10.3 Transportation service is currently provided across Backwater Bay by the Bilgewater Boating and Ferry Corporation. The State Department of Transportation is negotiating a new 30-year service lease with Bilgewater and it appears that the bid will be $4 million per year. Several public hearings have also been held to investigate the feasibility of replacing the ferry service with a bridge. User costs associated with delays encountered with the ferry service are estimated as averaging $5 per vehicle, and a low-level bridge located farther up the bay would reduce the total user cost to $3. A high-level bridge located near the present ferry crossing would further reduce this cost to $1.50 per vehicle. However, the cost of a high-level bridge would be $35 million as compared with $20 million for the low-level alternative. Annual maintenance expense on the high- and low-level bridges would be $2.5 million and $1.8 million, respectively. It is assumed that either project would have a residual value of 40 percent of first cost at the end of 30 years.

(a) Develop a benefit-cost analysis of these proposals using an 8 percent interest rate and the assumption that an average of one million vehicles per year would use either of the bridges or the ferry.

(b) Develop a benefit-cost analysis of these proposals using an 8 percent interest rate and the assumption that average annual traffic will increase at a rate of 3 percent per year.

10.4   A major intersection has been the scene of numerous property damage and some personal injury accidents over the past 3 years. With a single traffic light and no sheltered turning lanes, vehicles are delayed an average of 0.4 minute each trip through the area. A redesigned intersection with advanced left turns and lane-control devices is expected to increase the delay to 0.7 minute but should decrease the number of accidents from 80 to 30 per year.

The average cost per accident is expected to remain constant at $4500, and maintenance costs associated with the new design are forecast to increase by $12,000 per year. The cost of the improvement, which is expected to total $300,000, will probably remain effective for the next 10 years. Estimated average user cost per vehicle delayed in route is $11 per hour, and average annual traffic count totals nearly two million vehicles. Determine the benefit-cost ratio for this proposal using a tax-free interest rate of 8 percent.

10.5   A 2.2-mile (3.54-kilometer) stretch of highway is known locally as "Fog Hollow." In an attempt to reduce accidents, the shoulders along Fog Hollow could be widened at a cost of $103,000 per mile ($64,000 per kilometer) and large electric warning signs could be installed at both ends of the dangerous stretch. The signs would be controlled by a computer hooked to sensing devices to display recommended speeds for various fog conditions. The fog-warning signs could be installed for $165,000 and would have annual maintenance costs of $7000 during their 10-year life.

If the average accident cost in Fog Hollow amounts to $2385, how many accidents per year would have to be avoided by the project to make it acceptable for a 20-year study period? The discount rate used for such projects is 6 percent.

10.6   Two projects from each of four departments have been submitted for evaluation. The projects from each department are mutually exclusive.

Department	Project	Benefit	Cost
A	A1	$100,000	$ 70,000
	A2	112,000	80,000
B	B1	70,000	55,000
	B2	76,000	60,000
C	C1	160,000	100,000
	C2	184,000	120,000
D	D1	110,000	75,000
	D2	122,000	85,000

(a) Which projects should be funded according to the B/C criterion if one project must be selected from each department?

(b) Which projects should be funded if only $300,000 is available?

10.7   A state-sponsored "forest protective association" is evaluating alternative routes for a new road into a formerly inaccessible region. Different routes for the road provide different benefits, as indicated in the following table:

Route	Construc-tion Cost	Annual Saving in Fire Damage	Recrea-tional Benefits	Timber Access	Annual Mainte-nance Cost
A	$185,000	$5000	$3000	$ 500	$1500
B	220,000	5000	6500	500	2500
C	310,000	7000	6000	2800	3000

The roads are assumed to have an economic life of 50 years; the interest rate normally required is 3 percent per year.
  (a) According to a B/C comparison, which route should be selected?
  (b) Would the choice be changed if the interest rate were doubled?
  (c) Would the choice be changed if annual maintenance costs were not included?

10.8   The Dry Springs Flood Control District has applied for a federal grant to eliminate a persistent flooding problem. Two alternative projects are being considered, and the following data have been collected.

	Plan 1	Plan 2
First cost	$23,000,000	$50,000,000
Average annual reduction in flood damage	1,300,000	2,800,000
Annual irrigation benefits	350,000	600,000
Annual recreation benefits	100,000	300,000
Annual operating and maintenance costs	210,000	400,000

A social discount rate of 6 percent is recommended. The life of both plans is estimated at 50 years and no salvage value is expected.
  (a) Calculate the benefit-cost ratio using disbenefits in the numerator and compare it with B/C when disbenefits are added to costs in the denominator.
  (b) Determine the incremental benefit-cost ratio of plan 1 with respect to that of plan 2 using both methods from Problem 10.8a.

10.9   Three proposals have been submitted for developing a recreational facility along a scenic highway. The proposals are:

1 Build a picnic ground with 25 picnic sites, each including a table, parking space, and charcoal grill.
2 Build a picnic and camping facility with 20 picnic sites and 10 camp sites.
3 Build a picnic, camping, and cabin facility with 15 picnic sites, 10 camp sites, and 5 cabins.

The expected demand per site, first cost, and annual maintenance cost for each type of site are given on page 256. In addition, the estimated value to the public per visit (above charges) is given for each type of facility.

Type of Unit	Expected Visits/Year/Unit	First Cost/Unit	Annual Mainte-nance Cost/Unit	Perceived Value/Visit
Picnic site	800	$1000	$200	$0.50
Camping site	200	2000	250	2.50
Cabin site	200	5500	400	7.00

Based on a 10-year study with no residual value anticipated after 10 years and a discount rate of 5 percent, determine which proposal, if any, should be accepted.

10.10 The State Highway Department has collected data for a proposal to construct an overpass and extra lanes at a busy, dangerous highway intersection. Estimated land-acquisition, demolition, and construction costs are:

Land and demolition	$410,000
Additional lanes [0.6 mile at $1 million/mile (0.966 kilometer at $621,400/kilometer)]	$600,000
Overpass construction	$2,740,000

and annual maintenance is expected to average $105,000 per year during the 30-year life of the project.

Improved safety is the primary benefit expected from the project. A road count revealed that 12,000 vehicles per day pass through the intersection. An average of 2.4 fatalities per year have occurred at the present usage rate. Usage is expected to rise to 15,000 vehicles per day by the time improvements are completed, and accidents would be expected to rise commensurately. For every fatal accident there are 19 nonfatal accidents and 160 property-damage accidents. Settlements for fatal, nonfatal, and property accidents are calculated, respectively, at $65,000, $5000, and $1000. The improved intersection is expected to reduce all accidents by 90 percent.

Expenses associated with patrolling, maintaining the traffic signals, and directing traffic at the intersection are currently $31,000 per year; these would be eliminated by the project. Twenty-one percent of the vehicles using the intersection now must stop, with an average wait of 0.8 minute. The average cost of a stop for a vehicle's operation is estimated at 6 cents, and waiting time is assumed to be worth $2.40 per hour. When the overpass is in operation, all stops will be avoided, but 40 percent of the vehicles will have an added travel distance of 0.2 mile (0.322 kilometers). A vehicle operating cost of 18 cents per mile (11.2 cents per kilometer) is assigned as an average value.

The overpass will reduce pollution. This reduction is valued at $50,000 per year. Annual taxes forgone from property condemned for the expansion amount to $8000. The discount rate for the study is 8 percent.

(a) What is the benefit-cost ratio based only on the primary benefit of improved safety and only the building costs?

(b) What is the B/C based on secondary benefits and the building cost?

(c) What is the total B/C? What other factors might be considered in the project?

10.11 A dam on Roily River is being evaluated. A 50-year life and a 9 percent discount rate are to be used. Estimates have been obtained for the following benefits and costs:

Flood losses prevented in the Roily River area	$ 900,000/year
Flood losses reduced in downstream rivers	800,000/year

Increases in property values along the Roily	
River (present worth of values)	1,000,000
Income from electric power produced	3,200,000/year
Construction of dam and access roads	25,000,000
Cost of powerhouse and transmission facilities	10,000,000
Interest costs during construction	2,000,000
Operating and maintenance costs	100,000/year

Assume that construction costs occur at time zero but the benefits do not begin until the start of the fifth year and interest charges during production are prorated evenly over the 4-year period.

(a) What is the benefit-cost ratio?

(b) Assume that opponents of the dam are questioning the study on the basis that no costs were included for the destruction of a popular recreation area that would be inundated by the dam's reservoir. The land is all federally owned, but a very popular park that features mineral-water springs will be inundated. What value would have to be placed on this park and associated recreational areas to make the project unacceptable? Could a reasonable case be developed to abandon the project? What other benefits and disbenefits might be considered?

10.12 Two alternative routes for a new expressway are being evaluated. One follows the valley along a river and the other takes a shortcut through a range of hills.

The river route has a length of 30 miles (48.28 kilometers) and a first cost of $14,250,000. Its annual cost of maintenance will be $5000 per mile ($3106 per kilometer), and a major overhaul will be required every 10 years at a cost of $1,500,000.

The hilly route will be 7 miles (11.27 kilometers) shorter than the river route, but it will cost $19 million. Annual maintenance costs will be $8000 per mile ($4972 per kilometer), and the major overhaul and surfacing every 10 years will cost $1,150,000.

Traffic on either expressway is predicted to average 6000 vehicles per day, one-fourth of which will be commercial traffic. Anticipated average speed on either route is 50 miles per hour (80.5 kilometers per hour). Time is valued at $14 per hour for commercial traffic and $5 per hour for other vehicles. The average operating costs for commercial and noncommercial traffic are, respectively, 70 cents and 20 cents per mile (43.6 cents and 12.4 cents per kilometer) for the river route and 80 cents and 22 cents (24.8 cents and 13.6 cents per kilometer) on the steeper hilly route.

Compare the alternative routes according to an incremental benefit-cost ratio based on a 30-year life and a discount rate of 7 percent.

10.13 When the owner of an old mansion in a residential neighborhood died, she left a will deeding 4 acres (16,187 square meters) to the city if the city agreed to use it for a park and maintain the ornate gardens of the site for 40 years. If the city chose not to accept the offer, the land would be sold to a developer at a price of $260,000. The city uses its municipal-bond rate of 7 percent for evaluating recreational projects.

Modifying the garden to make it into a suitable park would require an investment of $110,000. Annual maintenance, service, and policing would cost $21,000. Forty adjacent homes would increase in value by $2000 each if the land were used as a park. Property taxes in the area are 2.4 percent of the assessed value.

The park would be used mostly for passive recreation such as walking, picnicking, and resting. No team sports would be allowed, and playground equipment for children would be minimal. It is estimated that 30,000 visitors per year would spend an average of 1 hour in the park.

(a) What value per hour would visitors to the park have to place on their visits to have the benefit equal cost?

(b) Is the value per visit calculated in Problem 10.10a a reasonable figure? Compare it with the cost of roughly equivalent experiences which people pay for.

(c) What action would you recommend that the city take?

10.14 A proposal to develop a new type of mine-shaft borer is being evaluated by a government agency. If approved, the project will be carried out as a joint venture with a private company which will share the research-and-development investment with the government. The new borer is an adaptation of a tunnel-boring machine made by the cooperating company to perform as a coal-mining machine. Benefits and costs are described below.

Direct benefits. (Advantage over conventional shaft-sinking techniques). The conventional cost for sinking a 1000-foot (304.8-meter) shaft, which takes an average of 18 months, is $1.7 million per shaft. The cost for the new borer to sink a 1000-foot (304.8-meter) shaft is:

$$\text{First cost prorated over a life of 15 shafts} = \$137,000/\text{shaft}$$
$$\text{Operation (cutter and maintenance costs)} = \$152,000/\text{shaft}$$
$$\text{Labor (crew} \times \text{wages} \times \text{6 months to sink shaft)} = \$420,000/\text{shaft}$$

Time benefits. Because a borer will complete a shaft in 6 months instead of the conventional 18 months, production profits are available 1 year sooner. If a typical shaft produces 250,000 tons (226,795,000 kilograms) of coal, average profit is $2.80 per ton and financing costs are 10 percent for the coal-mining companies, a company will benefit both from obtaining profits sooner and enjoying reduced financing charges because the investment in a mine is recovered 1 year sooner when a new borer is used.

Spillover benefits. Savings expected from other mining operations besides coal mining which can utilize the advanced technology available from the borer are $3.9 million per year.

Derived benefits. Additional benefits from improved safety are likely but are not included in the analysis.

Costs. The government share of the 3-year development costs are, respectively, $100,000, $500,000, and $500,000 for years 1, 2, and 3. An equal amount will come from the private sector for deployment and demonstration activities.

There are 27 new borers expected to be built and sold. There will be 5 in operation in year 4, 14 in year 5, 24 in year 6, and 27 thereafter. Each borer is expected to sink two shafts per year. Using a 10-year study period and a discount rate of 10 percent, determine the B/C ratio for the borer project. State all assumptions required in your analysis.

10.15 As one part of a model program to demonstrate ways to rehabilitate a decaying portion of a city, a proposal has been made to provide swimming facilities for residents of the model area. This can be accomplished by constructing swimming pools in small parks or by busing individuals to a nearby lake. Pertinent data are listed below.

1   Population in the 5-square-mile (12.95-square-kilometer) area that might take part in the swimming program is 35,000.
2   Average daily usage for a pool could vary from 1 percent to 5 percent of the population living within 1½ miles (2.41 kilometers) of a pool.
3   Minimum standard for the surface area of water per swimmer is 20 square feet (1.86 square meters).
4   A standard swimming pool has 7000 square feet (650.3 square meters) of water surface and requires 2 acres (8094 square meters) of land (part of the land serves also as a small park). Construction costs including equipment would be $400,000. Operating costs, including lifeguard wages, would be $70,000 during the swimming season. The life of a pool is 15 years.
5   The swimming season lasts 100 days each year.
6   Land prices in the area are $160,000 per acre ($39.54 per square meter). Taxes forgone on land acquired by the government would average $3600 per acre per year ($0.89 per square meter per year). Demolition and relocation expenses would amount to $140,000 per acre ($34.60 per square meter).
7   School buses could be used during the swimming season at a cost of $210 per bus per day including drivers' wages.
8   Four round trips could be made each day per bus during the swimming season to carry swimmers from the model area to a nearby lake. A bus can carry 70 people. The same number of people are expected to participate in the swimming program if either the pools are constructed or busing is provided.
9   Extra lifeguards and operating expenses at the city park on the lake to handle an influx of additional swimmers would be $400 per day. Sufficient space is available at the lake to accommodate any foreseeable usage.
10   A "recreation leader" would be hired for each 100 people expected to take part in the busing arrangement. A recreation leader would cost the program $50 per day, including fringe benefits. Four administrators would receive $8000 each during the season if busing is used.
11   A $1 admission fee is charged for pools in other parts of the city, and the same charge is reasonable for bus fare for a round trip to the lake.

Analyze the swimming program based on a 10 percent discount rate.
(a)   What is the equivalent annual cost per swimmer per day under each plan? Assume that no fees are paid for swimming. State your assumptions.
(b)   Assuming that a $1 fee per day is charged each swimmer, calculate the amount of additional benefits for the swimming pools that would have to be counted to allow them to pass the B/C > 1 criterion. Discuss the benefits in terms of consumers' surplus, spillover effects, and derived or intangible benefits. (See Extensions 10A and B)
(c)   Conduct a sensitivity analysis. Is there any combination of readily quantifiable factors than can provide a B/C > 1 for the swimming pools? State your assumptions.
(d)   Based on your calculations above, what decision would you make if you were the government policy maker who had to decide whether or not a summer swimming program should be part of the model-city rehabilitation demonstration? Write a short summary supporting your conclusions.

EXTENSIONS

10A Consumers' Surplus

According to the theorems of economics, a person is indifferent to the choice of having an additional unit of any commodity or paying the equilibrium price of that commodity. Many benefits and costs applicable to public projects can be determined from the going market price of commodities. However, there is a class of *public goods* that are available to all if they are available to a single individual. Public parks, drinking water, and clean air are examples of public goods. The value that individuals attach to such goods is elusive because the benefits may be taken for granted and individuals feel that the goods will be available whether or not they contribute an amount equal to the value received. Also, in many cases there is no marketplace in which to measure the value of a public good, even if it could be assumed that everyone enjoys a public good in the same amount and would willingly pay for it.

The concept of *consumers' surplus* has been utilized to explain the difference between what consumers actually pay for a public good and what they might be willing to pay. As shown in Figure 10.3, the demand curve indicates the value (expressed in dollars) placed by consumers on various quantities of a public good: The willingness to pay is a function of quantity. The shaded area is the *consumers' surplus,* representing the surplus value a consumer receives over the amount paid for at price P. This surplus is difficult to measure, because demand curves are difficult to develop; but some measure of the value of goods above price P is desirable in determining the worth of certain kinds of benefits derived from public projects.

Assume that electricity is to be supplied to a region where it was never before available. A hydroelectric project produces power priced at $\$P$ per kilowatthour to residents of the region. Benefits from the project might be based on the market value of the power sold, $P \times Q$, when the residents buy Q kilowatthours. True benefits are then understated because the market value is based on the price of the *last* unit of power purchased. In general, a consumer would willingly pay an amount significantly greater than the market price for the first kilowatthour furnished, a little less for the second kilowatthour, and so on. This bonus to the consumer represents an extra lump sum that the consumer would theoretically be willing to pay for the opportunity to buy electric power at price P. The correct measure of the benefit provided by the project should include the surplus in addition to the actual dollar value of power sold.

A consumers'-surplus analysis should take into account several conceptual questions concerning appropriateness to a given project. Exact quantitative measures may be difficult to justify. However, one frequently encountered case is easily handled. When a project provides a quantity of public good at a lower price than was previously possible, the increase in usage is the difference between the previous and present quantities consumed. In Figure 10.4, usage has increased from Q_1 to Q_2 as a function of a price decrease from P_1 to P_2 realized from a new public project. The new dollar amount of consumption is P_2Q_2, and the increase from the previous level is $P_2Q_2 - P_1Q_1$. But the benefit from the new price should also include the incremental change in the consumers' surplus. If the demand curve can be assumed to be linear over the range of Q_1 to Q_2, then the change in consumers' surplus is calculated as

$$\Delta(\text{consumers' surplus}) = \frac{(P_1 - P_2)(Q_1 + Q_2)}{2}$$

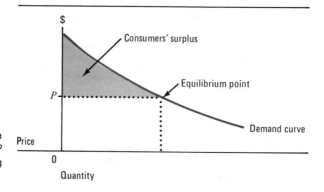

FIGURE 10.3
Consumers' surplus as the extra amount over the market price P that consumers would be willing to pay for quantity Q.

which is added to the dollar value of increased consumption to provide a measure of total benefit.

As a sample application of the consumers'-surplus concept, let the conditions in Figure 10.4 represent a road-construction project. The original unimproved road provided benefits to users that are shown in Figure 10.4 by the enclosed area xyE_1Q_1. Paving, grading, and straightening can improve the value of the road until its benefits are represented in the figure by the area xyE_2Q_2. Because the new road completely replaced the existing one, the total benefit claimed is the entire area xyE_2Q_2. Is this the correct benefit for the road-improvement project?

The claimed benefit overstates the value added by the road-improvement project. The original benefit xyE_1Q_1 produced by the old road exists regardless of improvements. The new benefit from the project that can be counted is represented by the area $Q_1E_1E_2Q_2$, which can be calculated as

$$\begin{aligned}\text{Added}\atop\text{benefit} &= P_2Q_2 - P_1Q_1 + \frac{(P_1 - P_2)(Q_1 + Q_2)}{2}\\ &= P_2Q_2 - P_1Q_1 \\ &\quad + \tfrac{1}{2}(P_1Q_1 + P_1Q_2 - P_2Q_1 - P_2Q_2)\\ &= \tfrac{1}{2}(P_2Q_2 - P_2Q_1 + P_1Q_2 - P_1Q_1)\\ &= \tfrac{1}{2}(P_2 + P_1)(Q_2 - Q_1)\end{aligned}$$

QUESTIONS

10A.1 A public good is represented by a linear demand schedule ranging from equilibrium at $1 for the first unit available to zero when 1 million units are available. The current level of availability and consumption is 300,000 units.

(a) What is the dollar volume of current consumption?

(b) What is the dollar volume of the current consumers' surplus?

(c) If a public project can decrease the cost to 40 cents per unit, what additional total benefit does it produce?

(d) What would the consumers' surplus be if the project were carried out?

(e) What is the dollar volume of consumption after the project is completed?

(f) Calculate by formula the increase in consumers' surplus resulting from the project. Check the figure obtained with respect to the answers to the previous problems.

(g) Would it be possible for a demand schedule to be such that total consumption would decrease from a previous level after implementation of a project that allowed the quantity consumed to increase as a function of a decreased price? Illustrate your answer. Comment on the change in consumers' surplus.

10A.2 A public good has a linear demand function such that the first unit available is valued at $100,000, but when 50,000 units are available, the per unit value drops to $2.

(a) What is the consumers' surplus when the demand is 30,000 units?

(b) What would be the change in the consumers' surplus if a project increased demand from 30,000 to 50,000 units?

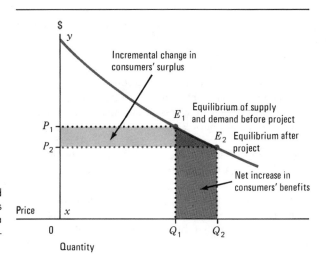

FIGURE 10.4
Change in cosumers' surplus caused by a new project that decreases the price from P_1 to P_2 with a resulting increase in quantity consumed from Q_1 to Q_2.

10B   Spillover Benefits and Costs

Spillovers (or externalities) exist when there is a divergence between private and social benefit or between private and social cost. An example of spillover benefits is the increase in land values brought about by the construction of a new highway or park. Spillovers are easily overlooked and are sometimes tricky to quantify, owing to filtering effects, but engineering economists should be conscious of spillover considerations for public projects because they constitute a significant source of benefits and costs. Cognizance is necessary to avoid B/C ratios bloated by exaggerated spillover benefits. Similarly, spillover costs deserve close inspection since their magnitude may be exaggerated by damage reports from injured parties. It is reasonable to view any spillover claim with some suspicion, if only to trace its logic.

Enumeration of spillovers may suggest the most effective direction for public expenditures. For example, in the case of production plants that are dumping pollutants, a tax on the effluent which is a measure of the external costs of pollution can make private costs identical with social costs. Or a subsidy could be granted to the offending plants to assist them in reducing the undesirable discharges. Both courses of action are probably better alternatives than is public ownership of the production facilities. In other cases the spillovers may be so extensive that public ownership is preferable, as in a multipurpose river-basin project. It would be more practical for the government to plan, undertake, and operate the entire river-basin development than it would be to create a system of market incentives capable of directing the combined resources of private investors toward the desired social objectives.

QUESTION
10B.1   There are public schools and private schools. From consideration of spillover benefits, why is the public investment in education logical?

10C   Quantification of Project Costs and Benefits

Costs for all resources required to achieve the stated objectives of a project should be included in the analysis. Each alternative in multilevel proposals should be self-contained in terms of cost to avoid double counting or overlapping expenses. A few cost considerations of particular significance to benefit-cost analyses are:

1   Imputed costs of existing assets employed on a project should be included when there are alternative uses for the assets. For instance, land or facilities desired for use in another project besides the one being analyzed would be treated in both projects as costs based on fair market values. Neglecting such costs would be similar to building houses on inherited land and then basing the selling price of the homes on only the construction costs because the land was free.

2   Preliminary costs of investigation and technical services required to get a project started are part of the project's budget. Once the project is under way, interest charges during the construction period before earnings begin are applicable expenditures. Management costs, whether incurred by the sponsoring agency or contributed by the beneficiaries, are part of the total project cost.

3   Spillover costs constitute all significant adverse effects caused by the construction and operation of a project and are expressed in terms of market prices whether or not actual outlays for compensation are made. For instance, an irrigation project could reduce the quality of water downstream from the area to be serviced; the additional treatment facilities necessary to restore the quality of the water or provisions for water from another source would be costs to the irrigation project.

Public projects are typically conceived to provide a certain benefit or family of benefits. Many of the difficulties of quantifying these benefits have already been discussed. The nature of some social benefits often precludes their measurement by direct market pricing, because there is no market in the private sector which offers the benefits. Recreational benefits are a good example of derived values. Factors affecting recreational values include the number and distribution of the population in the project area; socioeconomic characteristics including disposable income, occupation, education, age, and mobility of the relevant population; and the population's leisure time and recreational habits as indicated by trends in hunting and fishing license sales, sales of recreational equipment, and total demand. Such factors suggest the potential participation rate but do not put dollar amounts on the value of the participation.

To facilitate valuation, two types of recreation days are recognized, a *recreation day* being defined as a unit of use consisting of a visit of one individual to a recreation development during a reasonable portion of a 24-hour period.

Type of Recreation Day	Value of 1 Day's Use
General Type of activity attracting the majority of outdoor recreationists—requires the development of convenient access and maintenance of adequate facilities (for example, swimming, picnicking, studying nature, tent and trailer camping, canoeing)	$4.50–$13.50
Specialized Type of outdoor activity for which opportunities are limited and intensity of use is low and which often involve a large personal expense by the user (for example, cold-water fishing, big-game hunting, wilderness pack trips)	$18.00–$54.00

Indications of the values that users put on such experiences can be obtained from charges for similar activities in the private sector, time and cost of travel required to partake in an activity, cost of alternative uses of the facilities, and questionnaires about how much a user would be willing to pay for various experiences. A suggested composite value for a visitor day to any area is simply the gross national product per day divided by the total population.

Some attributes of a public project are *intangible*. In the 1940s, aesthetic values were considered of immeasurable worth, so any cost that made them available to the public was acceptable. This line of reasoning meant that every project that promised aesthetic values automatically had a B/C ratio of at least 1.0. An equivalent conclusion results from placing an infinite value on irreplaceable assets such as wilderness areas.

A more systematic approach to the valuation of intangible characteristics defines categories of attributes and employs a subjective rating system to rank the quality of different projects. Formal environmental quality-evaluation procedures have been developed to compare intangible benefits and disbenefits for alternative water- and land-use plans. The procedure starts with the identification of the environmental resources involved in a proposed project. Resources include open spaces, stream systems, beaches, wilderness areas, historical and cultural features, and ecological systems. These resources are then evaluated according to the following criteria:

1 *Quantity*   Specific environmental features are enumerated and measured in units such as acres of wilderness; miles of white water; area of lakes; and number of waterfalls, animals, scenic attractions, and significant sites.
2 *Quality*   The desirability of an environmental feature is rated by assigning a number from 1 to 10 after comparing the features with known or projected conditions at other locations.
3 *Human influence*   Subjective 1-to-10 ratings are made to indicate the degree to which people would use the resource identified; the degree to which it is protected for continued use; and the degree to which it contributes to education, scientific knowledge, and human enjoyment.
4 *Uniqueness*   A 1-to-10 rating measures the frequency of occurrence of a specific resource in the project relative to its occurrence elsewhere.
5 *Significance*   Irreversible damage to a resource is rated on a 1-to-10 scale that indicates the magnitude of adverse effects on the environment. The rating reflects the scarcity of supply of a resource and the capability of its returning to its original or natural state after the proposed project is implemented.

The ratings for intangible benefits are typically reported as an appendix to the monetary comparisons. A narrative may accompany the presentation of intangible considerations to explain how the ratings were developed, to discuss opposing views, and to provide supplementary information.

QUESTIONS

10C.1 In recent years many cities have constructed large municipal stadiums. The sales pitch to get voter approval for funding often appealed to civic pride and implied that the stadium would be a self-supporting enterprise as well as an attraction for visitors whose purchases would increase revenues for many merchants. After completion, some stadium projects failed to meet advertised expectations and caused difficulties that the voters had not anticipated.

Disregarding misjudgments of construction costs, which

far exceeded original estimates in many cases, and inaccuracies which overestimated the income that would be received from promotions booked into a stadium, discuss what disbenefits and spillover costs could have been logically anticipated and accounted for in a thorough benefit-cost analysis. List 10 considerations and discuss how values could be obtained for the factors involved.

10C.2 Assume that a modest-sized city park with which you are familiar is the end product of a public project. That is, a project produced the present park based on an accepted benefit-cost analysis. Further assume that the land became available for the park because a school on the site burned down and since there was sufficient classroom space available in nearby schools, the city decided not to rebuild the facility. Instead, it decided that the land would be used for other city purposes or sold to add money to the city treasury. One of the proposals for the land was to build the park that now exists.

(a) Based on the information given above, list the benefit and cost categories that should have been included in the evaluation of the park project.

(b) Insofar as possible, quantify the benefit and cost categories in terms of current values to complete the benefit-cost analysis. Use a 30-year life and a discount rate of 7 percent. Discuss the intangible benefits involved.

SECTION THREE

MANAGEMENT OF MONEY

"The name of the game is business, and the score is kept with money" is a catchy old saying that contains a lot of truth. Perhaps business is too serious to be called a game and money is too restrictive a measure, but there are plenty of players and they are surely interested in finances. Even nonprofit organizations and governments are subject to economic yardsticks. Very few individuals or organizations are immune to the economic evaluation criteria.

Professor Han of the University of Novi Sad in Yugoslavia observed an intriguing semantic quality associated with the quest for money. In English we say that we "make money" in the notion of manufacturing it, or "earn money" as just compensation for labor. Oriental languages roughly translate as "seeking" money. In German one "serves" for money, and in Russian one "works" for it. Hungarians "search" for money. The French and Italian versions are inspired by the notion of "winning" money. An appropriate but apparently unused phrase would be to "manage" money. Engineering economists manage money by directing and controlling its movement.

Only the very brash or very naïve would enter a chess competition without knowing the rules and practices of the game. Similar inadequacies may fail to bar the unprepared and the unsuspecting from economic competition. Many are lulled into complacency by an honest but often erroneous belief that they possess an innate money sense that will carry them through. In the last section we dealt with some of the mechanics of economic competition. In this section we will investigate the special rules that govern the business game (depreciation and taxes) and the way it is played by experienced competitors (industrial practices, consideration of inflation, and sensitivity analysis).

265

CHAPTER 11

DEPRECIATION AND INCOME TAX CONSIDERATIONS

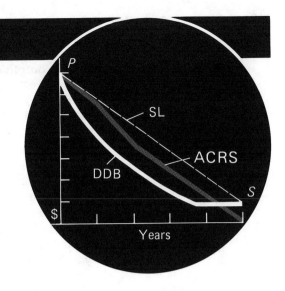

OVERVIEW

If, indeed, "the name of the game is business and the score is kept with money," then the rules are set by government, the scorekeepers are accountants, and the referee is the tax collector. An engineering economist acts as a playmaker.

Provisions for recovering capital invested in income-producing assets are made by charging depreciation against current income. These depreciation charges and the regular operating expenses are deductible from gross income in determining taxable income. The type of depreciation method used for tax purposes affects the timing and amount of tax payments, which in turn affect the after-tax worth of proposals.

Depreciation charges are not actual cash flows and the true decrease in market value of an asset may not correspond to the allowable deductions. Tables and formulas approved by the Internal Revenue Service establish the pattern of depreciation. Major changes in depreciation accounting occurred with the passage of the Economic Recovery Act of 1981, which introduced the *Accelerated Cost Recovery System (ACRS)*. This system specifies percentages for the recovery of investments in various property classes that determine the amount that can be deducted each year for tax purposes. Pre-1981 depreciation methods, such as *straight-line*, *sum-of-digits*, and *declining balance*, which are based on the useful life of individual assets rather than classes of property, can still be used in certain situations. Since many situation-specific conditions are regularly involved in investment recovery and the rules governing depreciation are voluminous, appropriate IRS publications should be consulted for all depreciation decisions.

Taxes are a major factor in any profit-seeking venture. *Property, sales, excise,* and/or *income taxes* affect net returns for both individuals and corporations. Types and amounts of taxes vary as governments pursue their *fiscal policies*. In the past few years, *surcharges* and *investment income tax credits* have been enacted, rescinded, and reinstated. Other features change too, such as provisions for *capital gains and losses*, options to *carry amounts forward* and *backward*, and depreciation allowances.

An *effective income tax rate* that represents total corporate tax liability can be developed for after-tax economic evaluations. Special tax provisions and charges for depreciation and/or interest are applied to the before-tax cash flow to determine *taxable income*. The taxable income is then multiplied by the effective tax rate and the resulting product is subtracted from the before-tax cash flow to obtain the *after-tax cash flow* for the year. A tabular format, convenient for these calculations, provides information equivalent to a cash flow diagram. Once the after-tax data are tabulated, economic comparisons are conducted as described in previous chapters.

An after-tax analysis defines the *actual* cash flow expected from a proposal. It may reveal the tax advantages of one proposal over another—advantages which are not part of a before-tax comparison. For instance, one proposal might qualify for an investment credit, which would increase its attractiveness over a comparable alternative that does not qualify for the tax deduction. However, comparisons that include tax effects are more complicated because tax laws are very complex. Precise tax considerations require expert assistance, but attention to basic tax provisions provides an adequate evaluation for most situations in the province of engineering economics.

Benjamin Franklin once observed that "in this world nothing is certain but death and taxes." He might have added that attention to both increases your chances of survival and prosperity.

DEPRECIATION

Depreciation means a decrease in worth. Most assets are worth less as they get older. Newly purchased production assets have advantages of possessing the latest technical improvements and operating with less chance of breakdown or need for repairs. Except for possible antique value, production equipment gradually becomes less valuable through wear. This lessening in value is recognized in accounting practices as an expense of operating. Instead of charging the full purchase price of a new asset as a one-time expense, the outlay is spread over the life of the asset in the accounting records. This concept of amortization may seem to disagree with the actual cash flow for a particular transaction, but for all transactions taken collectively it provides a realistic representation of capital consumption.

Very seldom is a depreciation fund actually established to accumulate money earmarked for the replacement of a specific asset. Instead, "depreciation reserves" are used to fund the most attractive proposals put forward to improve operations, some of which are to replace wornout and obsolete assets. Recovered capital is thus reinvested in a general way to maintain a company's physical plant and to pursue new ventures.

Example 11.1
Where Did the Green Go?

Sandy worked for several years as a general handyman in construction. He inherited $50,000 and decided to start a landscaping service. Combining his inheritance with a $40,000 loan, he bought a used tractor, truck, and hauling rig. Business was good. A gross income of $75,000 per year covered annual operating and loan-repayment expenses of $31,000. The rest he spent personally. After 5 years of good living Sandy found that the loan was repaid but the equipment was worn out. Because he had made no provisions to compensate for the loss in value, he was left with almost worthless assets and no money to renew them. What did he do wrong?

Solution 11.1

Sandy got off to a good start by investing his inheritance in productive assets, but he failed to conserve his capital base, apparently not realizing that depreciation was occurring. He should have known that his investment was, in effect, a prepayment of operating costs that had to be recovered from the revenues they helped create. By not charging for the use of his inherited capital, Sandy forfeited income tax deductions, probably underpriced his services, and tricked himself into believing he had more money to spend than his landscaping service actually produced.

Causes of Declining Value

Assets depreciate in value for several reasons. Their decreasing worth may be attributed to any of the following conditions:

1 *Physical depreciation.* The everyday wear and tear of operation gradually lessens the physical ability of an asset to perform its intended function. A good maintenance program retards the rate of decline but seldom maintains the precision expected from a new machine. In addition to normal wear, accidental physical damage can also impair ability.

"Wear and tear" is an obvious cost of production.

2 *Functional depreciation.* Demands made on an asset may increase beyond its capacity to produce. A central heating plant unable to meet the increased heat demands of a new building addition no longer serves its intended function. At the other extreme, the demand for services may cease to exist, as with a machine which produces a product no longer in demand.

3 *Technological depreciation.* Newly developed means of accomplishing a function may make the present means uneconomical. Steam locomotives lost value rapidly as railroads turned to diesel power. Current product styling, new materials, improved safety, and better quality at lower cost from new developments make old designs obsolete.

Since changes in needs and obsolescence cannot be accurately anticipated, their direct effect is more on replacement decisions than on tax deductions.

4 *Depletion.* Consumption of an exhaustible natural resource to produce products or services is termed *depletion.* Removal of oil, timber, rock, or minerals from a site decreases the value of the holding. This decrease is compensated for by a proportionate reduction in earnings derived from the resource. Theoretically, the depletion charge per unit of the resources removed is

$$\frac{\text{Present value of resource}}{\text{Remaining units of resource}} = \text{depletion rate (\$/unit)}$$

Samples of depletion percentages and calculations of cost depletion are shown in Review Exercise 1.

In practice, the depletion rate is largely set by the percentage of a year's income allowed for a depletion allowance by the Internal Revenue Service. Allowances for depletion vary with the type of resource and with the most recent legislation. Highest allowances are theoretically allowed for the resources which require the greatest expenditures for discovery and development.

5 *Monetary depreciation.* A change in price levels is a subtle but troublesome cause of decreases in the value of owned assets. Customary accounting practices relate depreciation to the original price of an asset, not to its replacement. If prices rise during the life of an asset, as in the case of high inflation rates during the early 1980s, a comparable replacement becomes more expensive. This means that the capital recovered will be insufficient to provide an adequate substitute for the wornout asset. It also suggests that the selling price of the product being produced by the asset does not accurately reflect the cost of production. Because the depreciation is actually happening to the invested capital representing the asset instead of to the asset itself, monetary depreciation is very difficult to accommodate. It cannot be charged as an operating expense for tax purposes.

The "nonresponsive" nature of depreciation accounts and its effect on a proposal's rate of return during periods of high inflation are discussed in Chapter 13.

Vocabulary of Depreciation Accounting

Depreciable property is property that is allowed a depreciation deduction or can be amortized. It may be *tangible* or *intangible.* Tangible property is any property that can be seen or touched. Intangible property, such as a copyright or franchise, is not tangible. Depreciable property may also be *real* or *personal.* Personal property is property that is not real estate, such as machinery or equipment. Real property is land and generally anything that is erected on, growing on, or attached to land. However, *land itself is never depreciable.*

The given definitions conform to those in IRS publication 534: Depreciation.

   Property is depreciable if it meets the following requirements:

1 It must be used in business or held for the production of income.
2 It must have a determinable life and that life must be longer than 1 year.
3 It must be something that wears out, decays, gets used up, becomes obsolete, or loses value from natural causes.

In general, if property does not meet all three of these conditions, it is not depreciable. Most engineering economic problems are concerned with tangible personal property.

Recovery property, as defined for the Accelerated Cost Recovery System (ACRS), is depreciable tangible property used in trade or business or held for the production of income, *after 1980.* It does not include certain public utility property or property that is depreciated on some basis other than years, such as wear on a machine based on the number of units produced by that machine. The *unadjusted basis* for recovery is essentially the first cost of the property plus additional investments associated with preparation and installation. Unadjusted basis minus the capital recovered from depreciation charges is known as the *adjusted basis*, or *book value.*

Unadjusted basis is also called cost basis.

Depreciation methods used for property placed in service *before 1981* were based on *useful life*—length of time an asset might reasonably be expected to be useful in the production of income. Such methods as straight-line and sum-of-digits depreciation are designed to recover the unadjusted basis minus the *salvage value*—estimated market value at the end of the asset's useful life. A *recovery period,* as used in *post-1980* depreciation accounting, is the increment of years over which the unadjusted basis is recovered according to an asset's assigned *property class*—3-, 5-, 10-, or 15-year property.

ACCELERATED COST RECOVERY SYSTEM (ACRS)

The *Economic Recovery Act of 1981* had a pervasive impact on taxpayers, personal and corporate, big and small. Cuts in personal tax rates partially offset "bracket creep" caused by years of inflation. Estate tax provisions eased the tax burden on inheritances. Modernization of facilities and equipment was encouraged by larger investment tax credits and more lenient rules for research expenses. But the most sweeping revisions were in depreciation rules. The 1981 Act's intent was "to encourage economic growth through reduction of the tax rates for individual taxpayers, acceleration of capital cost recovery of investment in plant, equipment and real property, and incentives for savings, and other purposes."

The Internal Revenue Code of 1954, as amended, is the fundamental law governing federal taxes in the United States.

Accelerated Cost Recovery System (ACRS) was the new method prescribed by the Economic Recovery Act of 1981 for the depreciation of all property placed in service after 1980.

ACRS Recovery Property

According to ACRS, recovery property is classified as either Section 1245 or Section 1250 class property. *Section 1245* property generally includes all personal property, elevators and escalators, and special-purpose structures and storage facilities. *Section 1250* property generally includes all real property

The ACRS acronym is informally voiced as "acres."

that is subject to an allowance for depreciation and is not or never has been depreciable personal property.

A recovery period is obtained by assigning an asset to a 3-, 5-, 10-, or 15-year property class:

3-year property. Section 1245 class property with a present class life of 4 years or less or used in connection with research and experimentation. Examples include machinery and equipment used in research and development, automobiles and light trucks, and certain special tools.

5-year property. Section 1245 class property that is not 3-, 10-, or 15-year public utility property. Examples include machinery and equipment *not* used in research and development, office furniture, and fixtures. This class of property appears frequently in engineering economic studies.

The unadjusted basis of all 3- or 5-year property placed in service during a tax year may be grouped together to determine the deduction.

10-year property. Section 1250 class property with a present class life of 12.5 years or less. It includes certain public utility property. Other examples include mobile homes and railroad cars.

15-year real property. Section 1250 class property that has a present class life of more than 12.5 years. It includes all real property, such as buildings, other than that designated as 5- or 10-year property. This class is separated into low-income housing and all other real property, with each division having its own schedule of ACRS percentages. (See Table 11.2.)

ACRS Deductions

The allowable ACRS deductions for tax purposes are simply calculated by multiplying the unadjusted basis of property by the appropriate percentages. Table 11.1 contains the recovery percentages for four classes of property.

Letting $DC(n)$ be the depreciation charge in year n and $p(n)$ be the permissible deduction percentage for year n, we find that

$$DC(n) = p(n) \times P$$

where P is the unadjusted basis. Then the amount of the investment unrecovered at the end of year n is

$$BV(n) = P - \sum_{1}^{n} DC(n) = P\left(1 - \sum_{1}^{n} p(n)\right)$$

where $BV(n)$ is the book value, or unrecovered balance, at the end of year n.

Unlike the 3-, 5-, or 10-year classes of property, and the 15-year public utility property, the percentages for 15-year real property depend on when the property is placed in service during the tax year. As shown in Table 11.2, each month is represented by a column. The depreciation deduction for each

TABLE 11.1

ACRS recovery percentages for property placed in service *after 1985*

Year (n)	Applicable Recovery Percentages p(n) For			
	3-year Property	5-year Property	10-year Property	15-year Public Utility Property
1	33	20	10	7
2	45	32	18	12
3	22	24	16	12
4	0	16	14	11
5		8	12	10
6		0	10	9
7			8	8
8			6	7
9			4	6
10			2	5
11			0	4
12				3
13				3
14				2
15				1
16				0

TABLE 11.2

ACRS recovery precentages for 15-year real property, other than low-income housing

Year	Month in Which 15-Year Real Property Is Placed in Service											
	1	2	3	4	5	6	7	8	9	10	11	12
1	12%	11%	10%	9%	8%	7%	6%	5%	4%	3%	2%	1%
2	10	10	11	11	11	11	11	11	11	11	11	12
3	9	9	9	9	10	10	10	10	10	10	10	10
4	8	8	8	8	8	8	9	9	9	9	9	9
5	7	7	7	7	7	7	8	8	8	8	8	8
6	6	6	6	6	7	7	7	7	7	7	7	7
7	6	6	6	6	6	6	6	6	6	6	6	6
8	6	6	6	6	6	6	5	6	6	6	6	6
9	6	6	6	6	5	6	5	5	5	6	6	6
10	5	6	5	6	5	5	5	5	5	5	6	5
11	5	5	5	5	5	5	5	5	5	5	5	5
12	5	5	5	5	5	5	5	5	5	5	5	5
13	5	5	5	5	5	5	5	5	5	5	5	5
14	5	5	5	5	5	5	5	5	5	5	5	5
15	5	5	5	5	5	5	5	5	5	5	5	5
16	—	—	1	1	2	2	3	3	4	4	4	5

year of the recovery period is specified by the percentages in the column determined by the month in which the asset's use begins. Comparable percentages for low-income housing are given in IRS publication 534.

Many rulings surround the basic elements of ACRS presented in this brief coverage. The purpose here is to simply introduce concepts and communicate the major features, not to provide explanations detailed enough to construct depreciation accounts. Details are given in publications obtainable from IRS Forms Distribution Centers. The following three examples illustrate the general approach for calculating ACRS deductions under different investment conditions.

Example 11.2
ACRS 3-Year Property Depreciation and Book Values

Three vehicles were placed in service in 1986: a car in January for $10,000, a light truck in May for $20,000, and a used automobile in December for $7000. What is the depreciation schedule for these vehicles? What would happen to the schedule if the vehicles were sold sometime in 1988?

Solution 11.2

All three purchases are 3-year recovery property and all were placed in service during the same year, 1986. Therefore, the total unadjusted basis for the year is $10,000 + $20,000 + $7000 = $37,000. The ACRS deductions over the recovery period are shown below.

If the three vehicles were sold any time in 1988, no deduction would be allowed for that year, but disposal after 1988 would have no effect on the deduction schedule. Note that the book value (undepreciated portion of the original purchase price) is $8140 for the year in which the sale is anticipated.

Year	ACRS Deduction	Cumulative Depreciation	Book Value at End of Year n	Year (n)
1986	$37,000 \times 0.33 = $12,210	$12,210	$24,790	1
1987	37,000 \times 0.45 = 16,650	28,860	8,140	2
1988	37,000 \times 0.22 = 8,140	37,000	0	3

An asset that is sold for an amount greater than its adjusted basis (book value) produces income that must be considered for tax purposes.

The percentages applied to the $37,000 unadjusted basis in Example 11.2 are from the 3-year property column in Table 11.1. The month in which the 3-year property is put into service is irrelevant to the deduction percentage. Similarly, any prospective salvage value *beyond* the recovery period has no effect on the unadjusted basis or the percentages. Book values, representing the unrecovered investment, show that 67 percent of the investment remains unrecovered by the end of the first year—BV(1) = $24,790. The vehicles are fully depreciated by the end of 3 years, which produces a book value of zero for 1989, assuming that the vehicles are still in service at that date.

Example 11.3
ACRS 5-Year Property Purchased with a Trade-In

An old typewriter that has an adjusted basis of $125 is traded in on a new model that has a fair market value of $925. The new typewriter was purchased in December 1986 by a cash payment of $800, but the machine was not delivered until January 1987. What is the ACRS deduction for 1987?

Solution 11.3

Since the typewriter was not put into service until 1987, the first deduction takes place in that year. Typewriters are classified as 5-year property, so the first year percentage is 20 percent, as indicated in Table 11.1. The unadjusted basis of the new machine is the sum of the cash outlay ($800) plus the value of the trade-in as specified by its adjusted basis ($125): $925. Therefore, the ACRS deduction for 1987 is $925 × 0.20 = $185.

Example 11.4
ACRS 15-Year Real Property Depreciation and Resale

An engineering consulting firm uses the calendar year as its tax year. The firm purchased an adjacent building in March 1986 for $230,000, not including land, and converted it into office space at a cost of $50,000. Renovations were completed by September and personnel from the firm moved in during that month. What are the permissible deductions for the first 3 years of occupancy? What deduction would be allowed in 1989 if the building were resold for $300,000 on May 1, 1989?

Solution 11.4

An office building is classified as 15-year real property. Deductions in this class are indexed to the month the property is placed in service. The consulting firm's building was placed in service during the ninth month, at which time its unadjusted basis was $230,000 + $50,000 = $280,000. The percentage for the first year is then 4 percent, allowing a 1986 deduction of $280,000 × 0.04 = $11,200. The same column, month 9 in Table 11.2, indicates percentages of 11 and 10 percent for years 2 and 3, respectively, that produce deductions of $30,800 in 1987 and $28,000 in 1988.

If the building were sold on May 1, 1989, the allowable deduction for 1989 would be 4/12 (4 months of the year have passed) of the percentage designated for the fourth year of service: $280,000 × 0.09 × 4/12 = $8400. The price obtained from the sale has no effect on the deduction, but it must be included in income tax calculations.

ALTERNATIVE ACRS METHOD

In lieu of using the percentages given in Tables 11.1 and 11.2, deductions may be calculated using a straight-line adaptation of the ACRS method. When this alternative method is utilized, the unadjusted basis of the recovery property is divided by the number of years in a recovery period selected from the options given in Table 11.3.

Straight-line percentages are calculated from the fraction $1/N$, where N is the elected recovery period. A percentage is applied to the unadjusted basis as

$$\text{Full-year deduction} = P \times p(n)$$
$$= \text{unadjusted basis} \times 1/N$$

Straight-line percentages for $p(n)$:

$N = 3$:	0.33333
$N = 5$:	0.20000
$N = 10$:	0.10000
$N = 12$:	0.08333
$N = 15$:	0.06667
$N = 25$:	0.04000
$N = 35$:	0.02857
$N = 45$:	0.02222

TABLE 11.3

Permissible recovery periods for the alternative ACRS method that uses a straight-line approach to calculate annual deductions

ACRS Recovery Property Classification	Optional Recovery Periods (N) for Deductions Based on the Alternative Straight-Line Method
3-year	3, 5, or 12 years
5-year	5, 12, or 25 years
10-year	10, 25, or 35 years
15-year real	15, 35, or 45 years
15-year public utility	15, 35, or 45 years

where $p(n)$ is constant over the recovery period. No salvage value is allowed in the calculation. Once a value of N is elected, the same value must be applied to all property in a 3-, 5-, or 10-year class placed in service during the same year; this percentage cannot be changed during the recovery period. However, different percentages may be elected for a property class in subsequent years.

The fraction of months in the first and last year that a 15-year real property is in service determines its deduction for those years.

Only a half year of depreciation can be claimed the first year that the property is placed in service. A full year's deduction can be taken for the rest of the recovery period, and then another half-year's depreciation is applied in the year following the recovery period. If the property is sold during the recovery period, no deduction is allowed in the year of the sale.

Example 11.5
ACRS Straight-Line Deductions with the Half-Year Convention

A new machine is placed in service at a research lab in December 1986. The purchase price is $10,000. The company elects to use a 5-year recovery period even though the machine is classified as 3-year recovery property. What is the ACRS deduction schedule for the machine if it is kept in use for 6 years?

Solution 11.5

The 5-year straight-line percentage is $100\%/5 = 20\%$. Regardless of the month an asset is placed in service (other than 15-year property), the half-year convention applies:

$$P \times 1/N \times 1/2$$

Thereafter, the full deduction is registered for the remaining years of the elected recovery period—in this case, 4 years. Then the final half-year deduction is taken in the sixth year, as shown below.

Year	ACRS Straight-Line Deduction	Cumulative Depreciation
1986	$10,000 \times 0.20 \times \frac{1}{2} = \$1000$	$ 1,000
1987 through 1991	$10,000 \times 0.20 = \$2000$	9,000
1992	$10,000 \times 0.20 \times \frac{1}{2} = \$1000$	$10,000

OTHER DEPRECIATION METHODS

Before ACRS was enacted, other methods were used to figure depreciation. Assets placed in service prior to 1981 and property that does not qualify for ACRS must still use these methods, but they cannot be applied to property that now qualifies for ACRS. Whereas many different methods of figuring depreciation are acceptable, it is the taxpayer's responsibility to show that a method is "reasonable," if it is questioned.

To be acceptable to IRS, a depreciation method must be used *consistently* and depreciate property *no faster than* the double-rate-declining-balance method.

Three of the most prominent depreciation methods are examined in the following pages: *straight-line, sum-of-digits,* and *declining balance.* They require input information about an asset's basis, useful life, and salvage value expected at the end of its useful life. Different annual depreciation charges result from the application of each method, and all three differ from ACRS. These methods are illustrated by reference to the problem data given in the table for Examples 11.6 to 11.8

The symbols used in the development of the formulas are

P = purchase price (unadjusted basis) of asset
S = salvage value or future value at end of asset's useful life
N = useful life of asset
n = number of years of depreciation or use from time of purchase
DC = annual charge for depreciation
BV = book value shown on accounting records (adjusted basis)

The basis for depreciation is the same as that used for figuring the gain on a sale. The original basis is usually the purchase price.

Straight-Line Method

Straight-line depreciation is the simplest method to apply and the most widely used of the depreciation methods. The annual depreciation is constant. The book value is the difference between the purchase price and the product of the number of years of use times the annual depreciation charge:

$$DC = \frac{P - S}{N}$$

$$BV(\text{end of year } n) = P - \frac{n}{N}(P - S)$$

Data and problem statement for Examples 11.6 through 11.8

Small computers were purchased by a public utility at a cost of $7000 each. Past records indicate that they should have a useful life of 5 years, after which they can be sold for an average price of $1000. The company currently has a cost of capital of 7 percent. Determine:

a The depreciation charge during year 1
b The depreciation charge during year 2
c The depreciation reserve accumulated by the end of year 3
d The book value of the computers at the end of year 3

Example 11.6
Straight-Line Depreciation Applied to the Basic Data

Solution 11.6

a and b Since the annual depreciation cost is constant, the charges for both the first and second year are

$$DC = \frac{P - S}{N} = \frac{\$7000 - \$1000}{5} = \$1200/\text{year}$$

c The depreciation reserve at the end of the third year is the sum of the annual depreciation charges for the first 3 years and is equal to $3 \times \$1200 = \$3600$.

d
$$BV(3) = \$7000 - \frac{3}{5}(\$7000 - \$1000)$$
$$= \$7000 - 0.6(\$6000) = \$3400$$

or, with the book value considered as the difference between the purchase price and the amount accumulated in the depreciation reserve,

$$BV(3) = \$7000 - 3(\$1200) = \$3400$$

Sum-of-Digits Method

This method can be applied only to property that meets requirements for double-declining-balance depreciation.

The sum-of-digits method provides a larger depreciation charge during the early years of ownership than it does during the later years. The name is taken from the calculation procedure. The annual charge is the ratio of the digit representing the remaining years of life $(N - n + 1)$ to the sum of the digits for the entire life $(1 + 2 + 3 + \cdots + N)$ multiplied by the initial price minus the salvage value $(P - S)$. Thus, the annual charge decreases each year from a maximum the first year:

$$DC = \frac{N - n + 1}{1 + 2 + 3 + \cdots + N}(P - S)$$

$$= \frac{2(N - n + 1)}{N(N + 1)}(P - S)$$

$$BV(n) = \frac{2[1 + 2 + \cdots + (N - n)]}{N(N + 1)}(P - S) + S$$

Example 11.7
Sum-of-Digits Depreciation Applied to the Basic Data

Solution 11.7

a The sum of digits for the 5-year useful life is

$$1 + 2 + 3 + 4 + 5 = 15$$

or

$$\frac{N(N + 1)}{2} = \frac{5(5 + 1)}{2} = 15$$

which is the denominator of the formula for

$$DC(1) = \frac{N - n + 1}{15}(P - S)$$

$$= \frac{5 - 1 + 1}{15}(\$7000 - \$1000)$$

$$= \frac{5}{15}\$6000 = \$2000$$

b After the first year, only 4 years remain in the useful life. Therefore, with $N - n + 1 = 5 - 2 + 1 = 4$,

$$DC(2) = \frac{4}{15} \$6000 = \$1600$$

c The ratio for calculating the depreciation reserve has a numerator equal to the sum of digits representing the years during which the reserve was built up:

$$\frac{\text{Depreciation reserve}}{\text{at end of year 3}} = \frac{5 + 4 + 3}{15} \$6000 = \$4800$$

d $BV(3) = P -$ depreciation reserve at end of year 3
$= \$7000 - \$4800 = \$2200$

or, by formula,

$$BV(3) = \frac{2[1 + 2 + \cdots + (N - n)]}{N(N + 1)}(P - S) + S$$

$$= \frac{2[1 + (5 - 3)]}{5(5 + 1)} \$6000 + \$1000$$

$$= \frac{6}{30} \$6000 + \$1000 = \$1200 + \$1000 = \$2200$$

Declining-Balance Method

The declining-balance method is another means of amortizing an asset at an accelerated rate early in its life, with corresponding lower annual charges near the end of service. An important point with this method is that salvage value must be greater than zero. A depreciation rate is calculated from the expression

$$\text{Depreciation rate} = 1 - \left(\frac{S}{P}\right)^{1/N}$$

which requires a positive value for S in order to be feasible. This constant rate is applied to the book value for each depreciation period. Since the undepreciated balance decreases each year, the depreciation charge also decreases, and

$$BV(n) = P(1 - \text{depreciation rate})^n$$

$$= P\left\{1 - \left[1 - \left(\frac{S}{P}\right)^{1/N}\right]\right\}^n$$

$$= P\left(\frac{S}{P}\right)^{n/N}$$

$$DC(n) = BV(n - 1)\left(1 - \sqrt[N]{\frac{S}{P}}\right)$$

A much more widely used version of the declining-balance method, allowed by the income-tax code since 1954, is based on a depreciation rate which does not depend on the S/P ratio. Under certain circumstances a rate is allowed that is twice as great as would be proper under the straight-line method. Under other circumstances, the rate is limited to 1.5 or 1.25 times that of the straight-line method.

To qualify for depreciation by the double-declining-balance method, property must be
● Tangible and personal
● $N \geqslant 3$ years
● Placed in service before 1981
● Acquired new or built after 1953

When the maximum rate is used,

$$\text{Depreciation rate}_{max} = \frac{200\%}{N}$$

It is called the *double-declining-balance method* of depreciation. It has the same characteristics as does the declining-balance method.

Example 11.8
Double-Declining-Balance Depreciation Applied to the Basic Data in Table 11.4

Solution 11.8

a Given that

$$\text{Depreciation rate}_{max} = 200\%/5$$
$$= 40\% \text{ or } 0.4$$
$$DC(1) = P(0.4) = \$7000(0.4) = \$2800$$

$$DC(1) + DC(2) + BV(2)(0.4)$$
$$= \$2800 + \$1680 + \$2520(0.4)$$
$$= \$4480 + \$1008 = \$5488$$

b $DC(2) = BV(1)(0.4)$
$$= (\$7000 - \$2800)(0.4)$$
$$= \$4200(0.4) = \$1680$$

d $BV(3) = P -$ depreciation reserve
$$= \$7000 - \$5488$$
$$= \$1512$$
or $\quad BV(3) = P(1\text{-rate})^3$
$$= \$7000(0.6)^3$$
$$= \$1512$$

c The depreciation reserve at the end of year 3 is the sum

A difficulty may arise with the use of double-declining-balance depreciation because the salvage value is not included in the calculation of depreciation charges. Continuing Example 11.8 to determine the book value at the end of year 5, we find that

$$BV(5) = P(1 - \text{depreciation rate})^N = \$7000(0.6)^5 = \$544$$

which is well below the anticipated salvage value of $1000.

The IRS requires an estimate of a "reasonable" salvage value. This may be a *net* value: resale value minus what it costs to remove or dispose of the asset.

Since the IRS does not permit depreciation charges that drop the book value below the salvage value, it is necessary to halt depreciation when $BV = S$. Referring again to Example 11.8, we find that

$$DC(4) = \$1512(0.4) = \$605$$

causing $BV(4) = \$1512 - \$605 = \$907$, which is less than $S = \$1000$. Therefore, the depreciation charge in year 4 is limited to $\$1512 - \$1000 = \$512$, and no depreciation charge is made in year 5. This pattern of double-declining-balance depreciation is shown in Figure 11.1

It is not uncommon for the book value calculated by double-declining-

FIGURE 11.1
Book values of a $7000 asset re-
sulting from the application of five
methods of depreciation.

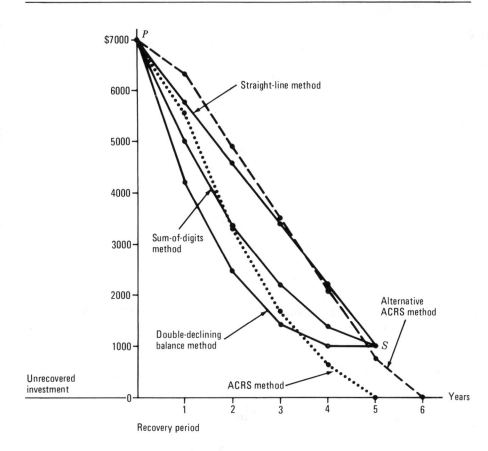

balance depreciation to exceed the asset's value at the end of its life. This situation *always* occurs when $S = 0$. Then it is usually advantageous to switch to straight-line depreciation. Since the IRS allows a switch in any year, the preferred time is the one which provides a present-worth tax advantage by deferring taxes to later years (see Table 11.4). The time to switch from double-declining-balance to straight-line depreciation is *when the straight-line depreciation charge on the undepreciated portion of the asset's value exceeds the double-declining-balance allowance.* The undepreciated portion is the difference between the asset's book value in a given year and its salvage value: straight-line DC $= [BV(n) - S]/(N - n)$. The procedure is demonstrated in Example 11.9.

For property with a useful life of 3 years or longer, a salvage value less than 10 percent of basis may be ignored. This is called the "ten percent rule."

Example 11.9
Switch from Double-Declining-Balance to Straight-Line Depreciation

An asset has a first cost of $7000, a 5-year useful life, and no salvage value. Determine an accelerated depreciation schedule in which $BV(N) = 0$.

TABLE 11.4

Depreciation pattern for a switch from the double-declining-balance (DDB) method to the straight-line (SL) method. The switch takes place at the end of year 3 when the SL depreciation charge exceeds the DDB charge.

End of Year	DDB Depreciation Charges	Book Value with DDB	SL Depreciation on the Undepreciated Balance	Book Value, DDB→SL
0		7000		$7000
1	$2800	4200	$1500	4200
2	1680	2520	1050	2500
3	1008	1510	840	1510
4	605	907	755	755
5	363	544	755	0

Solution 11.9

Applying the double-declining-balance (DDB) method, as was done for the same P and N values in Example 11.8, we know that

$$BV(5) = \$7000(0.6)^5 = \$544$$

which is higher than the zero salvage value. Therefore, a switch to the straight-line (SL) method is advisable to get the ending book value down to zero. The pattern of depreciation charges and book values resulting from the DDB method and the composite method of starting with DDB and switching to SL (DDB→SL) is shown in Table 11.4.

At the end of year 2 the book value resulting from DDB depreciation is $2520, which equals the undepreciated balance because $S = 0$. Then the SL charges for the last 3 years would be

$$DC_{SL} = \frac{\$2520 - 0}{3} = \$840$$

Since this annual charge is less than the DDB charge for year 3 ($1008), accelerated depreciation is continued another year. Then BV(3) = $1510 and the SL depreciation charge for each of the last 2 years is $1510/2 = $755. This is larger than the DDB depreciation charge for year 4 ($605) and signals the time to switch.

EVALUATION OF DEPRECIATION METHODS

Much of the choice among depreciation methods disappeared with enactment of the Economic Recovery Act of 1981. Although most property is now subject to ACRS, it is still informative to explore the intent and mechanics of depreciation as represented by different methods. An engineering economist should understand the reasons for depreciation, the equivalence of different methods, and the way that the choice of a method influences the timing—and time value—of income tax payments.

Each depreciation method has unique features which appeal to different management philosophies. A method by which the bulk of the money invested is recovered early in the life of an asset is a popular conservative view. An early write-off guards against sudden changes which could make the equipment less valuable and shifts some taxes toward later years. Methods in which the annual charge is constant simplify the accounting procedure. In general, the desirable features of a depreciation method are that it (1) recovers the capital invested in an asset, (2) is easy to apply, and (3) is acceptable to the Internal Revenue Service.

Depreciation based on usage rather than time is appealing because it tends to maintain book values closer to market value.

Broad patterns of capital recovery for the five methods being considered are shown in Figure 11.1. The curves are based on the basic data given for Examples 11.6 to 11.8 and indicate the book value any time during the life of the asset. The steeper the curve, the more rapid the depreciation. Both the sum-of-digits and declining-balance methods recover a large share of the initial investment early in the depreciable life. ACRS yields a zero book value in 5 years, whereas the alternative ACRS method stretches the recovery period to 6 years. The path of the unrecovered investment obviously varies widely with the different methods.

Equivalence of Depreciation Methods when Interest Charges Are Included

The time-value equivalence of different methods is an interesting aspect of depreciation accounting. None accounts for the return that would be received if the funds were not invested in the depreciable asset. *If the return on invested capital is included, the different patterns share a common, equivalent annual cost.* The importance of this condition is that it allows one basic formula to represent any depreciation method when alternatives are being compared.

Equivalence is demonstrated by example in Table 11.5. Two depreciation methods are compared according to the present worth of annual capital recovery plus a return on the unrecovered balance of the investment. The return may be thought of as the "profit" that would result if the undepreciated portion of the funds for an asset were invested at a given rate of return.

ACRS applied to the same data yields an identical PW after correcting for its recovery of the entire purchase price (P = $7000) instead of $P - S$ = $6000.

The calculations in Table 11.5 are based on the basic problem data used previously. They demonstrate that the total present worth of payments plus a theoretical interest charge on the unrecovered balance total to the same amount for the two depreciation methods. This total, $6287, is obtained by applying the capital-recovery factor to determine the annual cost of ownership as

$$(P - S)(A/P, 7, 5) + S(0.07) = (\$7000 - \$1000)(0.24389) + \$1000(0.07)$$
$$= \$1533.34$$

from which the total present worth of all payments is

$$\text{Present worth of five capital-recovery payments} = \$1533(P/A, 7, 5)$$
$$= \$1533(4.1001) = \$6287$$

TABLE 11.5

Capital recovery plus return at 7 percent interest for straight-line and sum-of-digits depreciation

Year, N	BV at Beginning of Year N (1)	Return on Unrecovered Capital at 7% Interest [0.7 × (1)] (2)	Depreciation Charge in Year N (3)	Capital Recovery plus Return [(2) + (3)] (4)	(P/F, 7, N) (5)	PW of Payments [(4) × (5)] (6)
			Straight-Line Depreciation Method			
1	$7000	$490	$1200	$1690	0.93458	$1580
2	5800	406	1200	1606	0.87344	1403
3	4600	322	1200	1522	0.81630	1242
4	3400	238	1200	1438	0.76290	1097
5	2200	154	1200	1354	0.71399	965
				Total PW of payments plus interest on unrecovered balance		$6287
			Sum-of-Digits Depreciation Method			
1	$7000	$490	$2000	$2490	0.93458	$2327
2	5000	350	1600	1950	0.87344	1703
3	3400	238	1200	1438	0.81630	1174
4	2200	154	800	954	0.76290	728
5	1400	98	400	498	0.71399	355
				Total PW of payments plus interest on unrecovered balance		$6287

The significance of the agreement in PW of the totals in Table 11.5 and that determined by the standard discounted cash-flow model is that it proves *discounted cash-flow comparisons provide equivalent results regardless of the depreciation method utilized in bookkeeping.* The only exception is the after-tax evaluation discussed in the next section.

After-Tax Effect of Depreciation Methods

According to IRS Publication 534, any reasonable method that is consistently applied may be used to compute depreciation. The total tax paid under any of the methods is the same over the life of the asset for a given set of tax rates, income, and expenses. But the timing of the tax payments is different. The more rapid capital recovery by accelerated depreciation methods allows a higher rate of return on invested capital because the present worth of higher early after-tax returns is greater than are the same total returns when a larger proportion comes later in the life. The time value of after-tax returns is illustrated in Table 11.6.

In general, tax benefits accrue from faster and shorter depreciation schedules, making straight-line a less attractive method.

The table is based on an annual income of $500 after deducting operating expenses, but before deducting depreciation and income taxes, resulting from ownership of an asset that cost $1500 when new and has a useful life of 5 years with no salvage value. The effective tax rate is 40 percent. As is apparent in column 6, the total tax paid is the same whether depreciation is calculated

TABLE 11.6

Comparison of taxes that result from depreciating an asset ($P = \$1500$, $N = 5$, $S = 0$) with straight-line, sum-of-digits, and ACRS methods when income is $500 per year. The depreciation methods lead to the same total tax payments over the life of the asset, but the present worth of the taxes differ in favor of the accelerated recovery methods.

End of Year N (1)	Income before Taxes and Depreciation (2)	Annual Depreciation Charge (3)	Income less Depreciation [(2) − (3)] (4)	Income-Tax Rate (5)	Income Tax (4) × (5) (6)	Present-Worth Factor (P/F, 10, N) (7)	PW of Income Tax [(6) × (7)] (8)
			Straight-Line Depreciation Method				
1	$ 500	$ 300	$ 200	0.4	$ 80	0.90909	$ 73
2	500	300	200	0.4	80	0.82645	66
3	500	300	200	0.4	80	0.75132	60
4	500	300	200	0.4	80	0.68302	55
5	500	300	200	0.4	80	0.62092	50
Total	$2500	$1500	$1000		$400		$304
			Sum-of-Digits Depreciation Method				
1	$ 500	$ 500	$ 0	0.4	$ 0	0.90909	$ 0
2	500	400	100	0.4	40	0.82645	33
3	500	300	200	0.4	80	0.75132	60
4	500	200	300	0.4	120	0.68302	82
5	500	100	400	0.4	160	0.62092	99
Total	$2500	$1500	$1000		$400		$274
			ACRS Depreciation Method				
1	$ 500	$ 300	$ 200	0.4	$ 80	0.90909	$ 73
2	500	480	20	0.4	8	0.82645	7
3	500	360	140	0.4	56	0.75132	42
4	500	240	260	0.4	104	0.68302	71
5	500	120	380	0.4	152	0.62092	94
Total	$2500	$1500	$1000		$400		$287

by ACRS, straight-line, or sum-of-digits method. However, as shown in column 8, the total present worth of the series of tax payments is less for the accelerated capital-recovery pattern of the sum-of-digits method. The difference when money is worth 10 percent favors the sum-of-digits method over straight-line depreciation by $304 − $207 = $30 and ACRS by only $13 during the 5-year recovery period.

TAX CONCEPTS

Everyone has an opinion about taxes. Some of the more famous ones include

The art of taxation consists in so plucking the goose as to obtain the largest possible amount of feathers with the smallest possible amount of hissing (*attributed to Jean Baptiste Colbert, 1665*).

To tax and to please, no more than to love and be wise, is not given to men (*Edmund Burke in a speech, "On American Taxation," in 1774*).

When I catch myself resenting not being immortal, I pull myself up short by asking whether I should really like the prospect of having to make out an annual income tax return for an infinite number of years ahead (*Arnold J. Toynbee, in Saturday Review, 1969*).

The Internal Revenue Code covered 3250 pages in 1984, almost double the size of a decade earlier.

The U.S. Constitution did not permit the federal government to tax incomes earned by individuals. This was changed by the Sixteenth Amendment in 1913. Since then Congress has enacted laws that have made tax considerations a major influence in investment analysis and the operation of a business. The Internal Revenue Service (IRS) collects taxes and issues regulations that interpret and implement legislation passed by Congress.

Corporate income taxes are featured in this chapter. They are a significant factor in the cash flow of any investment proposal. All the analyses in previous chapters were made on a before-tax basis. In most cases before- and after-tax analyses indicate the same order of preference among competing alternatives because the alternatives usually have similar characteristics. However, when some proposals are subject to special tax treatment and others are not, the order of preference can switch abruptly in an after-tax comparison. Also, the after-tax analysis reveals the actual cash flow that results from a proposal.

Noncorporate taxes, with major emphasis on individual taxpayers, are discussed in Extension 11A.

Tax laws are extremely intricate and subject to frequent changes. Consequently, the intent in this chapter is to present basic tax concepts, not the details on how to calculate the specific amount of taxes due. Furthermore, the discussion is limited to federal income taxes because of the diversity of tax laws among different states and municipalities.

Types of Taxes

Federal, state, and sometimes city or county taxes are imposed on income, property, and/or transactions. The transfer of wealth through the taxing mechanism is a major concern of governments, and the payment of those taxes is a major concern of income producers, both corporate and individual. The principal types and their relevance to engineering economic studies are described below.

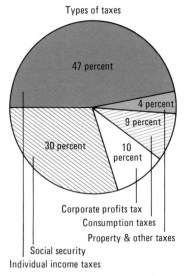

Types of taxes

47 percent

4 percent

9 percent

30 percent

10 percent

Corporate profits tax
Consumption taxes
Property & other taxes
Social security
Individual income taxes

Approximate percentages of total Federal income obtained from different sources in 1983.

1 *Property taxes* are charged by local governments on land, buildings, machinery and equipment, inventory, etc. The amount of the tax is a function of the appraised value of the assets and the tax rate. Property taxes are usually not a significant factor in an engineering economics study because of their small magnitude compared with income taxes and their similar effect on competing proposals.

2 *Excise taxes,* imposed on the production of certain products such as tobacco and alcohol, rarely affect economic comparisons. Other taxes that are not normally relevant but may become so in specific situations are *sales tax* on retail products, *user's tax, value-added tax, unemployment tax,* and *social security contributions.*

3 *Income taxes* are levied on personal and corporate income at increasingly higher rates for higher incomes. They are based on net income after deductions allowed for permissible "expenses." The tax effects of different types of expenses on the cash flow of proposals have significant influence on their acceptability. The rest of this chapter is devoted to the examination of income tax effects.

Changing Taxes

The federal government controls the monetary and fiscal policy of the nation to influence the level of economic activity. *Monetary policy* influences the availability and cost of credit, and *fiscal policy* deals with government receipts and expenditures. Taxation is the key instrument in fiscal policy. The principal methods for altering government receipts are (1) changing the tax rate, (2) changing the depreciation requirements, and (3) allowing tax credits.

Tax rates imposed on incomes may be raised to dampen the level of economic activity when rapid expansion threatens inflationary consequences. In theory, the reduction in disposable incomes reduces the purchasing power of individuals and thereby decreases demand for goods and services. An associated reduction in after-tax profits by corporations reduces the funds available for new investments and discourages expansion. The reverse, a tax-rate cut, theoretically encourages purchasing and expansion when the fiscal policy attempts to stimulate a depressed economy characterized by high unemployment.

There are changes in tax laws almost every year. Some alter the basic structure, as in the case of sweeping changes in depreciation requirements made in 1954, 1962, 1971, and 1981, which allowed more rapid write-offs of depreciable assets through accelerated schedules and shorter useful lives. More often the tax rates are changed in response to current fiscal needs. In 1968 Congress raised taxes via a *surcharge* (a percentage added to the amount of the taxes computed by an existing method): 10 percent in 1969 and 5 percent in 1970. Then tax rates were reduced by varying amounts, and rebates were given in 1974. The whole corporate tax schedule was lowered in 1978, and new investment credits were added in 1979. Many notable revisions were legislated in 1981.

Several of the tax-reducing measures legislated in 1981, such as very generous tax credits and "safe-harbor" leasing rules that allowed ailing corporations to sell their unused tax breaks, were curtailed or eliminated by the *Tax Equity and Fiscal Responsibility Act of 1982.* More changes were made by the *Deficit Reduction Act of 1984,* which reformed fringe-benefit and

Total taxes collected in the United States in 1983: $984.6 billion, up 149.4 percent in 10 years.

Possible future tax reforms and their potential effects are explored in Extension 11B.

real estate taxation, among other amendments. In the future there will assuredly be many more fine-tuning adjustments and major revisions to the tax code, which means that preparers of economic justifications and investment proposals based on after-tax analyses must be alert to changes in deductions, exemptions, credits, deferrals, preferences, shelters, and the gamut of rulings that affect returns on investment.

Engineering economic studies would ideally be based on the tax rates in effect during the lives of the assets being evaluated, but this is an unrealistic expectation, so current or "typical" rates are stated. The rates utilized in this chapter are representative but not necessarily currently correct.

CORPORATE INCOME TAXES

To be classified as a corporation, a professional service organization must be *organized* and *operated* as a corporation.

A corporation, for federal income tax purposes, includes associations, joint stock companies, insurance companies, and trusts and partnerships that actually operate as associations or corporations. Organizations of doctors, lawyers, engineers, and other professional people are generally recognized as corporations. Such organizations have the following characteristics:

- Associates organized to carry on business

- Gains from the business that are divided

- Continuity of life and centralized management

- Limited liability and free transferability of interests

A major shift in the overall United States tax burden from corporations to individuals is evident below.

Percent of Federal Income from	Years	
	1950	1983
Corporations	26.5	6.2
Individuals	39.9	48.1

Organizations possessing a majority of these characteristics must file corporate tax returns. Income taxes are due from corporations and businesses whenever revenue exceeds allowable tax deductions. Revenue includes sales to customers of goods and services, dividends received on stocks, interest from loans and securities, rents, royalties, and other gains from ownership of capital or property. Deductions embrace a wide range of expenses incurred in the production of revenue: wages, salaries, rents, repairs, interest, taxes, materials, employee benefits, advertising, etc. Also deductible, sometimes under special provisions, are losses from fire and theft, contributions, depreciation and depletion, bond interest, research and development expenditures, outlays to satisfy legislated objectives such as pollution control, etc. The difference between the revenue and deductions is taxable income. In general,

Taxable income = gross income − expenses − interest on debt − depreciation − other allowable deductions

Federal Income Tax

Ordinary domestic corporations are subject to the tax rates shown in Table 11.7. The rates are *progressive,* since the lower rates apply to the smaller

TABLE 11.7

Marginal tax rates (tax on each additional taxable dollar) for corporations

If Taxable Income Is		The Corporate Tax Rate Is	
Over	But not Over	This Amount	Of the Income Over
$ 0	$ 25,000	15%	$ 0
25,000	50,000	$ 3,750 + 18%	25,000
50,000	75,000	8,250 + 30%	50,000
75,000	100,000	15,750 + 40%	75,000
100,000		25,750 + 46%	100,000

increments of taxable income, becoming progressively higher with larger incomes. The top rate is 46 percent, the same maximum level in effect since 1978. For many years prior to 1978, a rate of 22 percent was applied to the first $25,000 of taxable income and 48 percent to corporate income above $25,000.

The average federal tax rate actually paid by corporations in 1983 was less than 20 percent. This figure includes all corporations, large and small, winners and losers.

Example 11.10
Progressively Taxing Income Progress

A corporation enjoyed an increase of $22,500 in taxable income over the amount earned the year before, which was $68,000. At what rate was the added income taxed?

Solution 11.10

Taxes for the previous year, based on the rates in Table 11.7, were $8250 + 0.3($68,000 − $50,000) = $13,650, and taxes in the following year were $15,750 + 0.4($68,000 + $22,500 − $75,000) = $21,950. The rate paid on the extra $22,500 was ($21,950 − $13,650)/$22,500 = 0.3689, which is the weighted average of the proportion at 30 percent ($7000) and at 40 percent ($15,500). The average rate on the total taxable income of $90,500 is 24.425 percent.

Effective Income Tax

Most states and some cities also impose a corporate income tax. The federal and state taxes can be combined into an *effective income-tax rate* that represents the total corporate tax liability. This one rate for all tax obligations is convenient to use in economic studies. It is based on the conditions expected during the study period. Recognizing that state income taxes are always deductible from federal taxes but that the opposite is rarely true, we have

State and local taxes increased at an annual average rate of 9.7 percent between 1972 and 1984, which is faster than the federal rate.

Effective tax rate = state rate + (1 − state rate)(average federal rate)

For example, if a firm expects a taxable income of $100,000 during a study period, and state taxes are 6 percent of this amount, then

$$\text{State tax due} = \$100,000 \times 0.06 = \$6000$$

$$\text{Taxable income for federal tax} = \$100,000 - \$6000 = \$94,000$$

$$\text{Federal tax due} = \$22,950$$

$$\text{Total tax due} = \$22,950 + \$6000 = \$28,950$$

$$\text{Effective tax rate} = \frac{\$28,950}{\$100,000} = 0.2895 \quad \text{or} \quad 28.95\%$$

Federal taxes on $1 million are $439,750, which is a 43.975 percent average rate.

Using the effective tax formula with corporate rates from Table 11.7 on taxable income of $1,000,000 gives an approximate

$$\text{Effective tax rate} = 0.06 + (1.00 - 0.06)(0.44)$$

$$= 0.4736 \quad \text{or} \quad 47.36\%$$

An effective income tax rate is appropriate for making economic comparisons but should *not* be utilized for computing annual tax returns without a very careful examination. There are many special provisions in the income tax regulations that should be evaluated in preparing tax returns. Some of the considerations are described below. These are only selected features and are not intended to be definitive. A collection of all the tax regulations that affect corporations and interpretations of their significance would occupy a book several times the size of this one.

Example 11.11
Effective Tax Rate when Federal Taxes Are Deducted from State Taxes

What is the effective income tax rate when the state rate is 5 percent, the federal rate is 46 percent, and the federal income tax is deductible in computing the state income tax?

Solution 11.11

Let the federal income tax be X (in dollars). Then the income subject to state tax is $T - X$, where T represents taxable income. The state income tax, at a 5 percent tax rate, is

$$0.05(T - X) = 0.05T - 0.05X$$

The income subject to federal tax is

$$T - 0.05T + 0.05X = 0.95T + 0.05X$$

The federal income tax is

$$X = 0.46(0.95T + 0.05X)$$

$$X - 0.023X = 0.437T$$

$$X = 0.4473T$$

$$\text{or} \quad 44.73\% \text{ of taxable income}$$

Then

$$\text{State tax} = 0.05(T - 0.4473T)$$

$$= 0.0276T$$

$$\text{or} \quad 2.76\% \text{ of taxable income}$$

Therefore,

$$\text{Effective income tax rate} = 0.0276 + 0.4473$$

$$= 0.4749 \quad \text{or} \quad 47.49\%$$

If the federal tax had not been deductible from state taxes, the formula on page 289 would be appropriate:

$$\text{Effective income-tax rate} = 0.05 + (1.0 - 0.05)(0.46)$$

$$= 0.05 + 0.437$$

$$= 0.487 \quad \text{or} \quad 48.7\%$$

Investment Credit

Tax credits in varying amounts have been given for several years to encourage business investment. The credit has been used as an incentive to increase investment when the national economy was weak, making it an on again–off again proposition. Recently, larger credits have been allowed with the intent to encourage investments that raise productivity and provide more energy.

An investment credit is deducted directly from the tax liability, not from income.

The exact deduction allowed as an investment credit depends on the current tax percentages, useful life of the asset, whether it is new or used, tax liability of the purchaser, and the purpose of the investment. In recognition of the turbulent history of investment credits and the tendency of lawmakers to continually tinker with economic stimulants, the wise course is to check deductions with a tax expert before claiming them in an after-tax analysis. Key features of investment credit in effect in 1985 are indicated below.

1   To be eligible for investment credits an investment must be in *recovery* property depreciable under ACRS; *tangible nonrecovery* property that has a useful life of at least 3 years; *tangible personal* property (except air conditioning or heating units); other *tangible* property (except buildings and their structural components) used in manufacturing, production, extraction, etc., and placed in service during the year in which credit is claimed for business and income-producing activities conducted primarily in the United States.

Claiming the investment credit does not affect the basis on which the designated property is depreciated.

2   The regular credit is 10 percent of the eligible investment but is limited in a single year to $25,000 plus 85 percent of the tax liability that is more than $25,000. A credit that cannot be used in the year it arises can be carried back 3 years and carried over 15 years.

See page 296 for an explanation of carrybacks and carryovers.

3   The amount of an investment in qualifying property that is eligible for the credit depends on the class of ACRS recovery property or the useful life of nonrecovery property. The proportions are shown in Table 11.8.
4   Used property can qualify for a regular credit (except energy property) up to a total investment of $150,000, subject to constraints on source, place of use, and other property considerations.
5   The business energy investment credit is 10, 11, or 12 percent of the qualifying investment, depending on the type of energy generation (biomass, hydroelectric, solar, wind, geothermal, etc.) and is subject to numerous restrictions.

Example 11.12
Available and Usable Investment Credits

A firm invested $1.2 million in property during the tax year to increase its profit-producing potential. Its tax liability for that year was $50,000. Investments were made in 10-year ACRS recovery property ($400,000), buildings ($600,000), and nonrecovery property with a useful life of 6 years ($300,000). What investment tax credit can be claimed for the year?

TABLE 11.8

Amount of the investment in qualifying property—ACRS recovery and nonrecovery—that is eligible for investment tax credit

Qualifying Property for Investment Credit	Proportion Eligible
ACRS recovery property class	
3-year	60%
5- and 10-year	100%
15-year public utility	100%
15-year real	None
Nonrecovery property life	
0, 1, and 2 years	None
3 and 4 years	1/3
5 and 6 years	2/3
7 or more years	Full

Solution 11.12

Assuming that there are no carryovers from previous years, we calculate tentative investment credit as

$$\frac{\text{Investment}}{\text{credit}} = \$400,000(0.10) + \$300,000(2/3)(0.10)$$
$$= \$60,000$$

where the $600,000 investment in buildings is not included because real estate is ineligible for the credit and only two-thirds of the investment in nonrecovery property with a 6-year life is eligible, as indicated in Table 11.8.

The maximum credit that can be claimed in the current year is determined to be

$$\frac{\text{Immediately}}{\text{usable credit}} = \$25,000 + (\$50,000 - \$25,000)(0.85)$$
$$= \$46,250$$

Since the applicable investment credit for the tax year exceeds the allowable limit for that year by $60,000 − $46,250 = $13,750, the excess can be carried backward or forward to be applied against past or future tax liabilities.

Example 11.13
Limited Investment Credits

Assume that the ACRS recovery property in Example 11.12 was in the 3-year class rather than the 10-year class and that the nonrecovery property was acquired used instead of new.

How much does the investment credit change from these reclassifications?

Solution 11.13

Investment credits are limited to 60 percent of the unadjusted basis for 3-year recovery property, as shown in Table 11.8, and to the first $150,000 invested in used property. Therefore, the $400,000 and $300,000 investments provide

$$\frac{\text{Investment}}{\text{credit}} = \$400,000(0.6)(0.10) + \$150,000(2/3)(0.10)$$
$$= \$34,000$$

which can be used to reduce the current year's tax liability.

TABLE 11.9

Percentages of the investment credit for which a *recapture tax* must be paid when ACRS recovery property is disposed of prior to its class life

Recovery Property Disposed of or Ceasing to Be Qualifying Property within	Recovery Percentage for	
	3-Year Property	5-, 10-, or 15-Year Property
First full year	100	100
Second full year	66	80
Third full year	33	60
Fourth full year	0	40
Fifth full year	0	20
After fifth full year	0	0

Recapture of Investment Credit

If recovery property is disposed of, or ceases to be qualifying property, before the end of its depreciation period for tax purposes, it may be necessary to *recapture* a portion of the investment credit. The credit recaptured is figured by multiplying a recapture percentage by the original investment credit. This computation determines how much the tax liability is increased in the year of property disposition. Table 11.9 supplies the percentages used in applying the "recapture rule" to investment credits for ACRS recovery property.

"Recapture of depreciation" is discussed in a later section on after-tax replacement study.

See IRS Form 4255 for recapture percentages for nonrecovery property.

Example 11.14
When Part of an Earlier Credit Becomes This Year's Tax Liability

In June 1986 two machines were purchased for $6000 and $5000. Both are 5-year recovery property and were put into service immediately after purchase. The 1986 tax return for the owner of the machines showed an investment credit of $1100 against an income tax liability of $10,000. In December 1987 the $6000 machine was sold. Apply the recapture rule.

Solution 11.14

Since the machine was sold within the second full year after it was placed in service, 80 percent of its $600 investment credit must be recaptured: $600(0.80) = $480. This decreases the original credit on the $6000 investment to $120, causing the 1987 tax liability to increase by $480.

Section 179 Expense Deduction

Taxpayers were permitted by Section 179 of the 1981 Tax Act to treat the cost of certain qualifying property as an expense rather than as a capital expenditure. That is, a portion of the investment can be deducted in the year of purchase

Allowable expense deductions rose from $5000 in 1982 to $7500 in 1984 to $10,000 in 1986.

instead of depreciating the total amount over the applicable recovery period. The maximum expense deduction is $10,000 per year. It can be claimed for tangible recovery property that qualifies for the investment tax credit. The amount allowed as an expense deduction must be subtracted from the unadjusted basis before depreciation deductions are calculated.

The election to expense a portion of the investment in Section 179 property must be exercised in the tax year in which the property is placed in service. That is, use it or lose it. The election, once made, is irrevocable. Since the annual limit is $10,000, Section 179 expense deductions are valuable mostly to small businesses.

Deductions are assumed to occur at the end of the property's first year in after-tax analyses.

Example 11.15
Relation of Expense Deduction, Depreciation, and Investment Credit

This year a timber operator purchased a saw for $1500 and a tractor for $24,000. These were the only investment properties placed in service during the year. The operator elected to expense the entire saw and to use the rest of the expense deduction for the tractor. What are the operator's tax deductions and investment credit for the tax year?

Solution 11.15

The operator is allowed to expense $10,000, an amount prorated at $1500 for the saw and $8500 for the tractor. The total is deducted from this year's income. Since the saw is fully "expensed," no depreciation or investment credit can be claimed for it. If there is a 5-year ACRS recovery period for the tractor, the first year depreciation percentage of 20 percent, as shown in Table 11.1, is applied to the adjusted basis of $24,000 − $8500 = $15,500. The same adjusted basis is also used in calculating the investment credit. Collectively, the investments provide the following tax adjustments for the timber operator:

Deductions from current income (assuming that income exceeds $13,100)

Expense deduction:

	$1500 for the saw	
	$8500 for the tractor	$10,000

First year depreciation for the tractor:

	$15,500(0.20)	$ 3,100

| Total deduction from curent income | | $13,100 |

Deduction from current taxes (assuming that tax liability exceeds $1550)

Investment credit for the tractor:

	$15,500(0.10)	$ 1,550

Capital Gains and Losses

Most tax decisions on capital gains hinge on the eligibility of capital assets for the preferential tax rate; legal counsel is advised for questionable cases.

The calculation of capital gains and losses is very complex because so many special provisions are involved. The tax rate assigned to capital returns is usually less than that for ordinary income. This preferential rate is allowed in recognition of the risk involved in holding capital assets over a prolonged period of time. Because taxpayers continually discover ingenious ways to advantageously utilize capital-gain taxation, familiarly called "loopholes," the provisions are continually altered by the IRS to constrict the loopholes. As a result, the rules are intricate, and our discussion is limited to an overview.

Everything owned is a capital asset *except:*

- Property held primarily for sale to customers, including accounts or notes receivable acquired in the course of normal business

- Depreciable and real property used in a trade or business

- A copyright, a literary, musical, or artistic composition, a letter or memorandum, or similar property, created by personal effort or acquired from a person who created the property

- Free publications and certain short-term government obligations

Qualified capital assets are classified as *short-term* if they are held for a year or less and as *long-term* if held longer than a year; exceptions are made for inventions and certain property acquired from an ancestor. A *gain* is the excess amount realized from the sale of property over its adjusted basis (original cost plus or minus certain additions and deductions). A *loss* occurs when the adjusted basis exceeds the selling price. In general, short-term capital gains or losses are treated as ordinary income and long-term gains are taxed at the preferential rate.

Net short- and long-term corporate gains and losses are calculated separately for tax purposes and then combined to determine the tax treatment. For instance, assume that a net short-term gain of $1000 and a net long-term capital loss of $3000 is recorded during the year; the consolidated amount is a loss of $1000 − $3000 = −$2000, which can be carried back or over to offset capital gains in other years. If the net values were reversed—a $1000 short-term loss and a $3000 long-term gain—the consolidated gain would be subject to a 28 percent capital gains tax, if the alternative tax is computed. When short-term gains exceed a long-term loss, the consolidated amount is taxed as ordinary income. And when there are both short- and long-term gains in a year, short-term gains are taxed as ordinary income whereas long-term gains may be taxed at the 28 percent rate. The decision about whether to use the 28 percent capital gains tax rate depends on the income of the corporation; when less tax is owed from treating capital gains as ordinary income, there is obviously no advantage in applying the 28 percent rate.

> Capital losses can be carried back 3 years and forward 5 years.

> Tax on the excess of long-term capital gain over short-term loss at a 28 percent rate is known as the *alternative tax on capital gains.*

Example 11.16
Calculation of an Alternative Tax on Capital Gains

A corporation has a taxable income of $70,000 that includes the following capital gains and losses:

Long-term capital gain $13,000
  Less: Long-term capital loss . . . . . −3,600
    Net long-term capital gain . . . .$ 9,400

Short-term capital gain . . . . $ 2,000
  Less: Short-term capital loss   −5,400
    Less: Net short-term capital loss . . . . −3,400

Excess of net long-term capital gain over short-term capital loss . . . .$ 6,000

Determine the taxes owed by using tax computations and by applying the alternative tax computation.

Solution 11.16

The federal tax owed on $70,000 of taxable income by the regular method, based on the tax rates given in Table 11.7, is calculated as

$$\text{Ordinary income tax} = \$8250 + 0.3(\$70,000 - \$50,000)$$
$$= \$14,250$$

By the alternative tax computation where the excess long-term capital gain is separated from the rest of the income as

$$\text{Taxable capital gain} = \$6000$$
$$\text{Remaining taxable income} = \$70,000 - \$6000$$
$$= \$64,000$$

the alternative tax computation proceeds as

$$\text{Ordinary income tax} = \$8250 + 0.3(\$64,000 - \$50,000)$$
$$= \$12,450$$

$$\text{Capital gains tax} = 0.28(\$6000) = \$1680$$

$$\text{Total tax} = \$12,450 + \$1680$$
$$= \$14,130$$

which yields a tax savings of $120 from applying the alternative tax instead of the regular tax computation.

Carrybacks and Carryovers

Tax offsets allowed in a current year that exceed the year's income can be applied to previous and future years' incomes. This occurs when a major loss is sustained and/or allowable deductions are unusually large. In general, ordinary income losses can be carried back to the 3 preceding years and carried over to the 15 succeeding years. The usual sequence is to apply a loss first against income earned 3 years previously. Any amount of the loss not used to offset income in the third preceding year is carried over to the second preceding year and then to the last year. If the loss is not entirely used to offset previous income, it is applied to future years' incomes in order of their occurrence. Any balance left over after offsetting incomes in 15 future years is forgone as a tax deduction.

Negative income taxes in economic analyses are assumed to be carrybacks or carryovers to positive taxable income. However, the actual application of carrybacks and carryovers can be quite complex when different sources of income are involved and the losses or deductions are subject to special provisions. In such cases, as with all unusual tax situations, a tax expert should be consulted.

AFTER-TAX ECONOMIC COMPARISONS

All the economic analyses in previous chapters were based on before-tax cash flows. In many situations, before-tax analyses provide adequate solutions. When the alternatives being compared are to satisfy a required function and are affected identically by taxes, the before-tax comparison yields the proper preference. Evaluations of public projects rarely include tax effects and are conducted as before-tax analyses.

Tax effects occasionally cause the preference to switch among alternatives between before- and after-tax evaluations. The principal causes are differences in depreciation schedules, deductions for interest payments, and special tax regulations such as applicable investment credits. The net return after taxes, the amount actually available, is often the main concern. This is usable capital, not subject to downward revision when taxes are withdrawn.

A simple adjustment of the rate of return calculated without regard for taxes gives a reasonable approximation to the after-tax rate of return:

$$IRR_{after-tax} \doteq IRR_{before-tax} (1 - \text{effective income-tax rate})$$

Thus, the after-tax rate of return resulting from a before-tax IRR of 15 percent and an effective income-tax rate of 40 percent would be

$$IRR_{after-tax} \doteq (1 - 0.4) \doteq 9\%$$

This figure is exact only when the assets involved are nondepreciable, financing is by equity capital alone, and no special tax provisions are applicable.

After-Tax Cash Flow

A tabular approach is convenient for modifying the before-tax cash flow to show the effects of taxes. It is normally sufficient to assume that tax payments occur at the end of each period, in the same way that the other cash flows are assumed to occur collectively at the year's end. The number of entries in the table depends on the number of tax considerations involved. The most common are depreciation and interest deductions. Table headings based on these tax effects are shown below.

End of Year (1)	Before-Tax Cash Flow (2)	Depreciation Charges (3)	Interest on Loan (4)	Taxable Income [(2) − (3) − (4)] (5)	Taxes [(5) × tax rate] (6)	Loan Cash Flow (7)	After-Tax Cash Flow [(2) − (6) + (7)] (8)

After-Tax Comparison of Proposals

An after-tax evaluation can be made using any of the comparison methods: EAW, PW, or IRR. All the precautions discussed for before-tax comparisons are applicable to after-tax analyses. Once the tax effects on cash flows have been determined, the computational procedures and interpretation of results are the same.

Example 11.17
After-Tax Evaluation of a Depreciable Asset

The budget includes $45,000 for the purchase of a new testing machine requested by the maintenance department. All the major investments in the budget are being checked to see if they meet the new required rate of return that has been raised to 12 percent *after taxes.*

The testing machine is 5-year ACRS recovery property.

Over 5 years, it is estimated to save $23,000 per year in maintenance costs with annual operating costs being $7300. It will be depreciated by the ACRS method. The firm has an effective income tax rate of 42 percent. Does the proposal to buy the testing machine satisfy the firm's new minimum acceptable rate of return?

Solution 11.17

A quick check by the after-tax IRR approximation can be made to see if the proposal is promising. Based on the cash flow diagram, where savings are considered to be income,

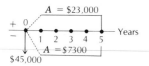

$$PW = -\$45,000 + (\$23,000 - \$7300)(P/A, i, 5) \gtreqless 0$$

At $i = 22$ percent,

$$PW = -\$45,000 + \$15,700(2.8701) = \$61$$

Using 22 percent as the before-tax rate of return, we find that

$$IRR_{\text{after-tax}} \doteq 22\%(1 - 0.42) \doteq 12.8\%$$

which is slightly above the required minimum.

After-tax computations are based on the values in Table 11.10. Depreciation charges result from applying the ACRS yearly recovery percentages.

Utilizing the after-tax cash flow data in column 6, we find that the present-worth formula for calculating the rate of return is

$$PW = -\$45,000 + \$12,886(P/F, i, 1)$$
$$+ \$15,154(P/F, i, 2) + \$13,642(P/F, i, 3)$$
$$+ \$12,130(P/F, i, 4) + \$10,618(P/F, i, 5)$$

At $i = 13$ percent,

$$PW = -\$45,000 + \$12,886(0.88496)$$
$$+ \$15,154(0.78315) + \$13,642(0.69305)$$
$$+ \$12,130(0.61332) + \$10,618(0.54276)$$
$$= \$929$$

At $i = 14$ percent,

$$PW = -\$45,000 + \$12,886(0.87719)$$
$$+ \$15,154(0.76947) + \$13,642(0.67497)$$
$$+ \$12,130(0.59208) + \$10,618(0.51937)$$
$$= -\$131$$

So

$$IRR_{\text{after-tax}} = 13\% + 1\%\frac{\$929}{\$131 + \$929} \doteq 13.9\%$$

The proposal apparently meets the after-tax-rate-of-return criterion. Note that the after-tax IRR is about 63 percent of the before-tax IRR and exceeds the after-tax approximation based on the before-tax IRR.

A couple of features of Table 11.10 merit review. The before-tax cash flow in column 2 corresponds to the cash flow diagram displayed in the solution. A comparable after-tax cash flow diagram, represented in column 6, is the one on which the calculations are based. The depreciation charges in column 3 *do not* represent a cash flow; they are shown only to accommodate the calculation of income taxes.

TABLE 11.10

Tabulated after-tax cash flow based on ACRS depreciation and an effective income-tax rate of 42 percent

End of Year, N (1)	Before-Tax Cash Flow (2)	ACRS Deductions (3)	Taxable Income (4)	Taxes (at 42%) (5)	After-Tax Cash Flow (6)
0	−$45,000				−$45,000
1	15,700	$ 9,000	$ 6,700	$2814	12,886
2	15,700	14,400	1,300	546	15,154
3	15,700	10,800	4,900	2058	13,642
4	15,700	7,200	8,500	3570	12,130
5	15,700	3,600	12,100	5082	10,618

Example 11.18

After-Tax Present Worth Evaluation when an Asset Is Eligible for an Investment Credit and an Expense Deduction

Further investigation of the testing-machine proposal described in Example 11.17 reveals that its purchase allows an investment credit and it is eligible for an expense deduction of $10,000, since it is the only investment made during the year. What is the revised after-tax return?

Solution 11.18

The expense deduction is taken first to establish an adjusted basis of $45,000 − $10,000 = $35,000 for depreciation calculations. ACRS recovery percentages are then applied to the adjusted basis to determine the ACRS deductions. The investment credit is also based on the adjusted basis, being calculated as $35,000(0.10) = $3500; this amount is subtracted from the taxes owed in the first year.

Since the combination of the expense deduction plus the first-year depreciation deduction exceeds the net income for that year ($15,700 − $10,000 − $7000 = −$1300), the taxable income is negative, which leads to a "negative tax" of −$1300(0.42) = −$546. When this amount is added to the investment credit, after-tax cash flow is $15,700 − (−$546) + $3500 = $19,746. The rest of the figures in Table 11.11 correspond closely to those in Table 11.10.

The after-tax cash flow is evaluated by discounting at the required after-tax rate of return, 12 percent.

$$PW = -\$45,000 + \$19,476(P/F, 12, 1)$$
$$+ \$13,810(P/F, 12, 2) + \$12,634(P/F, 12, 3)$$
$$+ \$11,458(P/F, 12, 4) + \$10,282(P/F, 12, 5)$$
$$= -\$45,000 + \$19,476(0.89286)$$
$$+ \$13,810(0.79719) + \$12,634(0.71178)$$
$$+ \$11,458(0.63552) + \$10,282(0.56743)$$
$$= \$5507$$

The investment is thus justified for an after-tax MARR of 12 percent.

As a comparison, the after-tax cash flow from Example 11.17 has a present worth of $2030 at $i = 12$ percent. That the present worth is much larger when investment incentives are applied to the same property should be no surprise. The intent of expense deductions and investment credit is to encourage certain types of investments by making them financially more attractive.

Corporations frequently pay interest for the use of borrowed funds and occasionally receive interest on investments. Interest earned on invested funds

TABLE 11.11

After-tax cash flow when an expense deduction and an investment credit are applicable

End of Year, N	Before-Tax Cash Flow	Expense Deduction	ACRS Deductions	Taxable Income	Taxes (at 42%)	Investment Credit	After-Tax Cash Flow
0	− $45,000						− $45,000
1	15,700	− $10,000	$ 7,000	− $ 1,300	− $ 546	$3500	19,746
2	15,700		11,200	4,500	1890		13,810
3	15,700		8,400	7,300	3066		12,634
4	15,700		5,600	10,100	4242		11,458
5	15,700		2,800	12,900	5418		10,282

Funding sources and the effects of different loan repayment plans are discussed in Chapter 12.

Theoretical aspects of interest were discussed in Extension 6A and practical aspects of loans are discussed in Extension 13B.

is usually treated as earned income. Interest paid on corporate debt is often a significant expense, affecting taxable income and, consequently, taxes. Both the timing of the receipt of a loan and its repayment schedule affect after-tax cash flows.

Interest paid for money borrowed to carry on a trade or business is deductible from revenue in determining taxable income. It is treated as an expense, being deducted in the year it is paid. The deduction is essentially a federal subsidy for borrowing and thereby reduces the cost of a loan.

Borrowing rates have varied wildly in recent years, ranging from less than 10 percent to well over 20 percent. These fluctuations have led lenders to offer variable rate loans, especially for mortgages, since neither borrowers nor lenders want to be caught with long-term loans at unfavorable interest rates. Variable rates are typically adjusted at specified times to conform with current market rates.

Example 11.19
After-Tax Evaluation when Investment Capital Is Borrowed

An alternative to the testing machine favored in Example 11.17 can be purchased for $67,000. It will have a useful life of 5 years, $20,000 salvage value, and net savings of $22,000 per year. Compare it with the machine in Example 11.17 using 5-year ACRS depreciation. The additional $67,000 − $45,000 = $22,000 required for the purchase must be borrowed at 9 percent annual interest with the principal due at the end of the fifth year.

Solution 11.19

Table 11.12 contains the figures employed in calculating the after-tax cash flow for the more expensive testing machine. The seventh column indicates the amount of the loan and the annual interest payments, which are tax deductible. The extra row at the bottom indicates the salvage value received and the loan repayment at the end of year 5. Note that taxes must be paid on the salvage value which is considered to be ordinary income received in year 5. The last column in the table is the sum of the before-tax and loan cash flows minus taxes. For the annual loan interest of $22,000 × 0.09 = $1980, the after-tax cash flow for year 1 is $22,000 − $1980 − $2780 = $17,240. In year 5 the after-tax cash flow is a combination of the positive cash value resulting from net income and the negative flow from taxes owed on the salvage value:

$$\$13,863 - \$10,400 = \$3463$$

TABLE 11.12

After-tax cash flow involving both borrowed and equity funds

End of Year, N (1)	Before-Tax Cash Flow (2)	ACRS Deductions (3)	Interest on Loan (4)	Taxable Income (5)	Taxes (at 42%) (6)	Loan Cash Flow (7)	After-Tax Cash Flow (8)
0	− $67,000					$22,000	− $45,000
1	22,000	$13,400	$1980	$ 6,620	$2780	− 1,980	17,240
2	22,000	21,440	1980	− 1,420	− 596	− 1,980	20,616
3	22,000	16,080	1980	3,940	1655	− 1,980	18,365
4	22,000	10,720	1980	9,300	3906	− 1,980	16,114
5	22,000	5,360	1980	14,660	6157	− 1,980	13,863
5	20,000			20,000	8400	− 22,000	− 10,400

The present worth of the after-tax cash flow resulting from the loan obtained and repaid to purchase a more expensive testing machine is

$$PW = -\$45,000 + \$17,240(P/F, 12, 1)$$
$$+ \$20,616(P/F, 12, 2) + \$18,365(P/F, 12, 3)$$
$$+ \$16,114(P/F, 12, 4) + \$3,463(P/F, 12, 5)$$
$$= -\$45,000 + \$15,393 + \$16,435$$
$$+ \$13,072 + \$10,241 + \$1965$$
$$= \$12,106$$

Recalling that the less expensive testing machine evaluated in Example 11.17 had a present worth of $2030, we conclude that the more expensive version secured with borrowed funds is obviously preferred. The preference is owed mainly to the increase in annual net savings generated by the $67,000 machine over the $45,000 version: $22,000 − $15,700 = $6300.

After-Tax Replacement Study

Before- and after-tax replacement evaluations are conducted in the same way once the tax effects have been imposed on the cash flow patterns. There are, however, a couple of points that occasionally cause confusion. One is the use of the book value as the worth of the presently owned asset in a replacement study.

The *book value* is the asset's purchase price less its accumulated depreciation charges. This value is the result of accounting practices and is unlikely to be the same as the asset's value in an open market. The difference between the book value and the present realizable value is the *sunk cost*. As depicted in Figure 11.2, the actual amount obtainable for a defender in a replacement study is less than its value shown by the accounting records. If the asset is replaced, the sunk cost is a loss for the year, charged against profit. However, part of the loss is absorbed by a reduction in taxes for that year. For instance, let the book value of the asset in Figure 11.2 be $960 and the market value be $300, to produce a sunk cost of $960 − $300 = $660. Taxable income for the

FIGURE 11.2
Sunk cost as a function of book value and actual price for an asset sold 2 years before the end of its 5-year cost-recovery period.

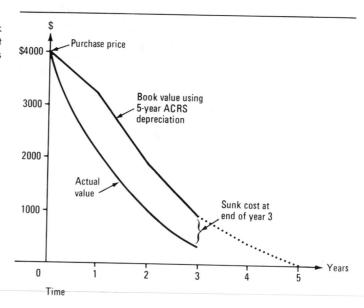

On rare occasions the actual value will exceed the book value.

year would be reduced by $660, which "saves" $660 × 0.40 = $264 in taxes when the effective income-tax rate is 40 percent. Thus, the effective loss is only $660 − $264 = $396.

The mental static generated by sunk costs is hard to comprehend unless there is personal involvement. A large sunk cost can be detrimental to the current year's profit picture and may cause censure of the current managers, even if they had nothing to do with the original purchase and the depreciation schedule that caused it. There may also be a feeling that "we've got so much invested in it that we can't afford to sell it for what it would bring." The weakness of this argument is that the owners should be looking ahead to determine the course of action that will minimize *future* costs, not the course that will make past expenditures look good.

Two *incorrect* ways to conduct replacement studies are to pretend the book value is always the defender's market value and to add the sunk cost to the challenger's purchase price. Both retard replacement.

The correct way to perform an after-tax replacement study is to determine the defender's current market value and use that figure as its worth at time zero. Then operating costs are estimated for the remaining useful life and depreciation charges based on the original cost basis are continued until the recovery period has lapsed. Taxes are levied as applicable and the resulting cash flow is compared with the challenger's after-tax cash flow by one of the standard discounting methods: EAC, PW, or IRR

Example 11.20
After-Tax Comparison of a Defender and a Challenger when Revenue Is Unknown

A machine has a book value of $10,000 and 3 years remaining on its 5-year ACRS recovery period. Its operating costs are $19,000 per year. The function performed by the machine will be needed for five more years. It is proposed that a $5000 overhaul be performed which will reduce operating costs to $16,000 per year for the remaining 5 years of use, at which time it will have no salvage value.

A new machine of advanced design can perform the same function as the existing machine for $12,000 per year. This machine is priced at $36,000 and will have a salvage value of $6000 when disposed of at the end of the 5-year use period. Because of the technological advances, the old machine can be sold now for only $8000.

The company employs ACRS depreciation and has a 50 percent effective tax rate. It requires an after-tax rate of return of 8 percent. Which machine will have the lower equivalent annual cost?

Solution 11.20

The defender's capital cost for the replacement study is its current market value, $8000, and this amount is considered to be an outlay at time zero for the comparison displayed in Table 11.13. However, during the next 3 years it will still have yearly deductions of 24, 16, and 8 percent according to its original depreciation schedule. Because the asset will have been depreciated to a zero book value by the end of the third year, no depreciation charge can be made for the additional 2 years it is to stay in use. Operating costs, including the $5000 overhaul in year 1, constitute the before-tax cash flows for years 1 through 5. Because no income is attributed to the machine's operation, all before-tax cash flows are negative, except the salvage value of the challenger.

Annual expenses and depreciation are deductible from income and are, effectively, "tax savings." For the defender in year 1, the sum of operating and overhaul costs ($16,000 + $5000 = $21,000) provides a tax deduction at the 50 percent rate of $21,000 × 0.5 = $10,500. When depreciation is included with the expenses, $21,000 + $5000 = $26,000, the tax savings amount to $26,000 × 0.5 = $13,000, which reduces the after-tax cash flow to −$21,000 − (−$13,000) = −$8000. The remaining charges are recorded similarly with the ACRS deductions based on the defender's unadjusted basis, calculated as $10,000/(0.20 + 0.32) = $19,231, occurring only in the first 3 years of the study period.

The challenger's 5-year recovery period has deductible charges composed of operating costs and ACRS deductions.

Its $6000 value in year 5 exceeds the zero book value in that year and is treated as ordinary income. This results in a positive after-tax cash flow of $3000 in year 5 when the tax rate is 50 percent.

The negative after-tax cash flows represent disbursements that are treated as positive costs in the following equivalent annual cost comparison:

$$\text{EAC}_{\text{after-tax}}(\text{defender})$$
$$= [\$8000 + \$8000(P/F, 8, 1) + \$6334(P/F, 8, 2)$$
$$+ \$7116(P/F, 8, 3) + \$8000(P/F, 8, 4)$$
$$+ \$8000(P/F, 8, 5)](A/P, 8, 5)$$
$$= [\$8000 + \$8000(0.92593) + \$6334(0.85734)$$
$$+ \$7116(0.79383) + \$8000(0.73503)$$
$$+ \$8000(0.68059)](0.25046)$$
$$= \$9470$$

$$\text{EAC}_{\text{after-tax}}(\text{challenger})$$
$$= [\$36,000 + \$2400(0.92593) + \$240(0.85734)$$
$$+ \$1680(0.79383) + \$3120(0.73503)$$
$$+ (\$4560 - \$3000)(0.68509)](0.25046)$$
$$= \$10,799$$

Since the defender exhibits a lower equivalent annual cost, it is preferred over the challenger.

The after-tax replacement study follows the procedures described in Chapter 8. The market value is considered to be the "purchase price" of the asset already possessed, and therefore it represents the best current estimate of ownership cost. A book-value loss, such as the sunk cost in Example 11.20 of $10,000 − $8,000 = $2000, has no bearing on the current situation, although previously instituted depreciation charges remain in effect. Note also from the example that overhaul and repair expenses are treated as current expenditures

TABLE 11.13

Comparison of after-tax costs for a defender and challenger at an effective tax rate of 50 percent when 5-year ACRS depreciation is used

End of Year (1)	Before-Tax Cash Flow (2)	ACRS Depreciation (3)	Deductible Charges (2) − (3) = (4)	Tax Savings at 50% (4)(0.5) = (5)	After-Tax Cash Flow (2) − (5) = (6)
			Defender		
0	− $ 8,000				− $8000
1	− 21,000	$5000	− $26,000	− $13,000	− 8000
2	− 16,000	3333	− 19,333	− 9,666	− 6334
3	− 16,000	1667	− 17,667	− 8,834	− 7166
4	− 16,000		− 16,000	− 8,000	− 8000
5	− 16,000		− 16,000	− 8,000	− 8000
			Challenger		
0	− $36,000				− $36,000
1	− 12,000	$ 7,200	− $19,200	− $9,600	− 2,400
2	− 12,000	11,520	− 23,520	− 11,760	− 240
3	− 12,000	8,640	− 20,640	− 10,320	− 1,680
4	− 12,000	5,760	− 17,760	− 8,880	− 3,120
5	− 12,000	2,880	− 14,880	− 7,440	− 4,560
5	6,000				3,000

deductible from current income; they are *not* capital costs of ownership. Overhaul and repair expenses can affect capital cost only by increasing the salvage value as a result of more conscientious care.

The reverse of a sunk cost occurs when the amount received for an asset at the end of its useful life is greater than the estimated salvage value. As already observed, the sale of an asset *below its book value* is treated as an *ordinary loss*. When the selling price is *greater than the book value but less than the original purchase price*, the difference is taxed as *ordinary income*. That portion of the selling price that is *above the original capital cost* is considered to be a *capital gain* and is taxed accordingly. The loss or gain is declared in the year in which the transaction takes place.

Because ACRS depreciation recognizes no salvage value, it is often necessary to *recapture depreciation* when the actual salvage value is greater than zero, as was the case in Example 11.20.

Review Exercises and Discussions

Exercise 1 A tract of timber was purchased for $500,000. Cruise data placed a value of $420,000 on 2 million board feet of standing timber. Land value was appraised at $80,000. What is the depletion allowance for the first year if 400 mbf (thousand board feet) of timber are removed from the parcel?

Solution 1 The reduction in value of the investment can be calculated by *cost depletion* according to the following general formula:

TABLE 11.14

Percentage of gross annual income deductible for depletion of mineral deposits

Deposits	Percent
Sulfur and uranium; and, if from deposits in the United States, asbestos, lead, zinc, nickel, mica, and certain other ores and minerals	22
Natural gas—regulated and sold under a fixed contract—with several limitations	22
Oil and gas—independent producers and royalty owners—with several limitations	15
If from deposits in the United States, gold, silver, copper and iron ore, and oil shale	15
Coal and sodium chloride	10
Clay and shale used in making sewer pipe or bricks or used as sintered or burned lightweight aggregates	$7\frac{1}{2}$

$$\text{Depletion allowance} = \frac{\text{cost of property} \times \text{units sold during year}}{\text{total number of recoverable units}}$$

Deducting the value of the land from the total investment ($500,000 − $80,000 = $420,000) sets the property cost for the 2000 mbf of recoverable timber. When 400 mbf are removed in one year,

$$\text{Depletion allowance (year 1)} = \frac{\$420,000 \times 400 \text{ mbf}}{2000 \text{ mbf}} = \$84,000$$

An alternative depletion method is to apply an allowable depletion percentage to the property. The percentage is based on the type of exhaustible resource and is applicable to gross income derived from the property. Percentage rates are subject to change, but in general the amount deducted cannot exceed 50 percent of the taxable income derived from the property. Rates applicable to certain deposits in 1985 are shown in Table 11.14.

Exercise 2

Compare the pattern of book values as a function of asset age resulting from the application of ACRS, straight-line, sum-of-digits, and double-declining-balance depreciation methods to an asset that has a probable useful life of 6 years, initial cost of $22,000, and an estimated salvage value of $1000. Assume that the asset qualifies as Section 1245 class property with a 5-year recovery period.

Solution 2

The ACRS method uses the percentages from Table 11.1 to fully depreciate the asset in 5 years, irrespective of the anticipated useful life and salvage value. The three other methods may be applied to property that does not qualify for ACRS.

The annual charge by straight-line depreciation is ($22,000 − $1000)/6 = $3500, the amount by which the book value decreases each year.

First year depreciation by the sum-of-years method is ($22,000 − $1000)(6)/21 = $6000. The annual charge decreases by $21,000/21 = $1000 each year.

FIGURE 11.3

Pattern of book values from the application of straight-line (——), sum-of-digits (– –), double-declining-balance (– – – –), and ACRS (═══) depreciation.

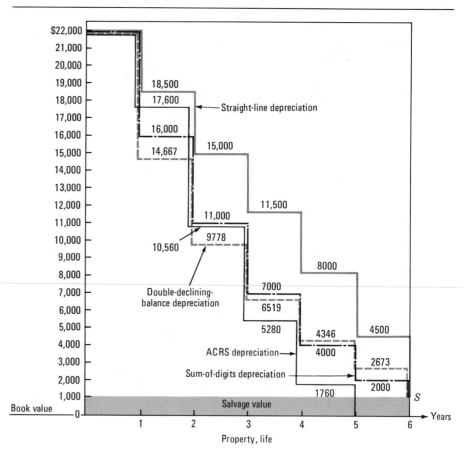

By the double-declining-balance method the depreciation charge is $2/6 = 1/3$ of the book value each year. In year 1 it is $\$22{,}000/3 = \$7333$. By year 6 the book value would be $BV(6) = \$22{,}000(1 - 1/3)^6 = \$1931$, which is well above the expected salvage value. Therefore, a switch to straight-line depreciation is advisable. It should take place at the end of year 4, when the book value from DDB depreciation for the first 4 years produces a book value of $\$4346$. Then the annual depreciation charge by the straight-line method over the last 2 years is $(\$4346 - \$1000)/2 = \$1673$.

The ladder pattern of book values resulting from the application of each depreciation method is shown in Figure 11.3. The adjusted basis for each year is labeled on the ladder. The ACRS method understates the salvage value, if the $\$1000$ estimate indeed turns out to be the actual market value in year 6.

Exercise 3   Log-loading equipment and logging trucks will be purchased to remove timber from the tract described in Review Exercise 1. Used assets can be purchased for $\$240{,}000$. The logging should be completed in 5 years, at which time the assets are expected to have a salvage value of $\$80{,}000$. A depreciation method is sought that is based on the actual usage of the equipment and trucks. The method selected is *units of production* depreciation, a recognized but not widely used form of depreciation. Its main limitation, shared by all special depreciation methods, is that total allowances in any year, during the first two-thirds of useful life, must not exceed those that would have resulted from the application of DDB depreciation.

If the logging schedule calls for the consecutive removal of 400, 600, 200, 600, and 200 mbf of timber during the next 5 years, what is the depreciation schedule?

Solution 3

The annual depreciation charge for the units of production method is based on the proportion of production in a given year to the total production expected from the asset:

$$\text{Depreciation charge} = (P - S)\frac{\text{units produced during year}}{\text{total units produced during useful life}}$$

For the given data, where the assets will be used to log 2000 mbf,

End of Year	Depreciation Charge	Book Value
0		$240,000
1	($240,000 − $80,000)(400/2000) = $32,000	208,000
2	$160,000(600/2000) = $48,000	160,000
3	$160,000(200/2000) = $16,000	144,000
4	$160,000(600/2000) = $48,000	96,000
5	$160,000(200/2000) = $16,000	80,000

This schedule abides by the DDB limitation.

Exercise 4

A transaction to buy real estate was closed on May 15. It was purchased with $55,000 in cash and a loan for $120,000. The new owner finally placed it in service on the first of August, electing to use the ACRS straight-line option over a 35-year recovery period. The calendar year is the owner's tax year. If the property is sold 10 years and 2 months later, what is the book value at the time of sale?

Solution 4

The alternative ACRS method for 15-year real property allows straight-line depreciation over recovery periods of 15, 35, or 45 years, as indicated in Table 11.3. Like the 15-year real property percentages shown in Table 11.2, the alternative percentages for straight-line deductions are based on the number of months the property is in service during the first and last years of ownership. Unlike the election of straight-line depreciation for 3-, 5-, and 10-year property, the *half-year convention* does not apply.

The unadjusted basis of the real property is $55,000 + $120,000 = $175,000. Since it was placed in service on August 1, depreciation is charged for 5 months during the first year. Based on annual straight-line depreciation of $175,000/35 years = $5000 per year, the charge for the first year is $5000(5/12) = $2083.33. When the real estate is sold on October 1, 10 years later, it would have been in service 9 months during its last year, making its book value at the time of disposal:

$$BV = \$175,000 - \$5000(10) - \$5000(9/12) = \$119,167$$

A firm sustained a loss of $175,000 in 1986. It had taxable income and paid taxes in each of the 3 preceding years as indicated below:

Exercise 5

Year	1983	1984	1985
Taxable income	$50,000	$75,000	$25,000
Taxes paid	8,250	15,750	3,750

How large a tax refund can the firm expect?

Solution 5   The firm would submit amended tax returns to the IRS deducting as much of the $175,000 loss as possible each year. The amended returns would indicate the following cash flows, where the loss is deducted first from the 1983 taxable income and then, successively, from incomes in 1984 and 1985:

	1983	1984	1985	Totals
Original taxable income	$50,000	$75,000	$25,000	$150,000
Less part of 1986 loss	(50,000)	(75,000)	(25,000)	(150,000)
Amended taxable income	0	0	0	0
Original taxes paid	$8,250	$15,750	$3,750	$27,750
Tax refund	$8,250	$15,750	$3,750	$27,750

The firm is eligible for an immediate tax refund of $27,750 and has an unused portion of the loss of $175,000 − $150,000 = $25,000 that can be used to offset otherwise taxable income in 1987, or in successive years if income in 1987 is less than $25,000.

Exercise 6   An experimental wave tank will be needed for only 3 years. It qualifies as Section 1245 property with a 3-year recovery period, has an initial cost of $460,000, and should have a salvage value of $100,000. The tank is expected to generate annual income of $250,000. A loan of $200,000 at 9 percent interest will help fund the purchase. It must be repaid in three equal installments including interest. What is the present worth of the after-tax cash flow for the tank when the minimum after-tax rate of return for ventures of this type is 15 percent. ACRS depreciation is employed, and the firm's effective income-tax rate is 40 percent.

Solution 6   Loan repayments can take several forms. The equal annual payment plan is a more common type of business loan than the plan in Example 11.19. Annual payments, which include interest charges at 9 percent on the unpaid balance, are

$$\text{Loan repayments} = \$200,000(A/P, 9, 3) = \$200,000(0.39506)$$
$$= \$79,012$$

Annual interest on the unrecovered portion of the loan is

$$\text{Year-1 interest} = \$200,000(0.09) = \$18,000$$
$$\text{Year-2 interest} = [\$200,000 - (\$79,012 - \$18,000)](0.09)$$
$$= \$12,509$$
$$\text{Year-3 interest} = [\$200,000 - \$61,012 - (\$79,012 - \$12,509)](0.09)$$
$$= \$6524$$

After-tax cash flow is calculated in Table 11.15.

TABLE 11.15

After-tax cash flow when a loan is repaid with uniform payments.
After-tax cash flow = before-tax cash flow − taxes + loan cash flow

End of Year	Before-Tax Cash Flow	Depreciation Charges	Interest on Loan	Taxable Income	Taxes at 40%	Loan Cash Flow	After-Tax Cash Flow
0	− $460,000					$200,000	− $260,000
1	250,000	$151,800	$18,000	$ 80,200	$32,080	− 79,012	138,908
2	250,000	207,000	12,509	30,491	12,196	− 79,012	158,792
3	250,000	101,200	6,524	142,276	56,910	− 79,012	114,078
3	100,000				40,000		60,000

$$PW = -\$260,000 + \$138,908(P/F, 15,1) + \$158,792(P/F, 15, 2)$$
$$+ (\$114,078 + \$60,000)(P/F, 15, 3)$$
$$= -\$260,000 + \$138,908(0.86957) + \$158,792(0.75614) + \$174,078(0.65752)$$
$$= \$95,319$$

The venture easily meets the required rate of return.

Exercise 7

A machine was purchased 5 years ago for $50,000. It was classified as Section 1250 property with a 10-year recovery period. The firm that owns the machine no longer has a need for the function served by the machine, but a buyer is now willing to pay $60,000 for it because machines of this type are no longer available. If the applicable tax rate for the firm is 46 percent, what tax is owed on the abandonment transaction?

Solution 7

The current book value for the machine is $50,000 − $50,000(0.70) = $15,000. The difference between the book value and the original purchase price is taxed as ordinary income at the 46 percent rate. That portion of the selling price above the original cost is a capital gain and subject to a 28 percent tax rate. Total tax for the transaction is

$$0.46(\$50,000 - \$15,000) + 0.28(\$60,000 - \$50,000) = \$18,900$$

The net gain on the $45,000 by which the disposal value exceeds the book value is $45,000 − $18,900 = $26,100.

PROBLEMS

11.1   A truck purchased on January 1, 1980 cost $30,000 and had an estimated service life of 8 years, with an estimated salvage value of $6000. Determine the depreciation charge for 1986 and the book value at the end of that year using:
  (a) Straight-line depreciation
  (b) Sum-of-digits depreciation
  (c) Double-declining-balance depreciation

11.2 Assuming that the true resale value of the truck in Problem 11.1 decreases by 20 percent per year for each year it is kept in service, determine which depreciation method most closely approximates the true resale value in years 1983 and 1986.

11.3 A materials testing machine was purchased for $20,000 and was to be used for 5 years with an expected residual salvage value of $5000. Graph the annual depreciation charges and year-end book values obtained by using:

    (a) Straight-line depreciation
    (b) Sum-of-digits depreciation
    (c) Double-declining-balance depreciation
    (d) ACRS with 5-year property recovery

11.4 An asset cost $400 when purchased prior to 1981. A scrap value of $50 was expected at the end of a 7-year service life. Determine the book value at the end of 4 year's use and the depreciation charge for the fifth year by:

    (a) Straight-line depreciation
    (b) Sum-of-digits depreciation
    (c) Double-rate-declining-balance depreciation

11.5 A production machine had a first cost of $40,000 in 1978 and was expected to have a useful life of 15 years with negligible salvage value. Determine the book value at the end of 1988 using:

    (a) Straight-line depreciation
    (b) Sum-of-digit depreciation
    (c) Double-rate-declining-balance depreciation

11.6 A central air conditioning unit was installed in 1980 at an initial cost of $65,000 and was expected to have a salvage value of $5000 after a service life of 12 years.

    (a) What amount had accumulated in a straight-line depreciation reserve at the end of 1985?
    (b) Using double-rate-declining-balance depreciation, determine the book value at the end of 1989.
    (c) If the equipment was sold in 1986 for $10,000, what sunk cost would result if sum-of-digits depreciation was being used?

11.7 Prior to 1981, Adventure Airlines purchased several baggage-handling conveyors for $200,000 each. They were expected to last for 15 years and have negligible salvage value. Determine the depreciation charge on one of the conveyors for the first year of life and the book value at the end of 4 years using:

    (a) Sum-of-digits depreciation
    (b) Straight-line depreciation
    (c) Straight-line depreciation assuming a 10 percent investment tax credit was taken on the purchase
    (d) Double-rate-declining-balance depreciation

11.8 Assume that the equipment described in Problem 11.7 was purchased in 1986 and was classed as ACRS 5-year property. Determine the depreciation charge for the first year of life and the book value at the end of 4 years using:

    (a) The appropriate ACRS schedule
    (b) ACRS alternative straight-line method

11.9 An asset with a life of 7 years was purchased 4 years ago at a cost of $10,000. It now has a book value of $5200 based on straight-line depreciation. What is its expected salvage value?

11.10 An asset with an original cost of $10,000 and no salvage value has a depreciation charge of $2381 during its second year of service when depreciated by the sum-of-digits method. What is its expected useful life?

11.11 The objective of most tax-paying companies is to maximize the present worth of total depreciation charges. If an asset has a first cost of $20,000 and an expected salvage value of $1000 at the end of its 5-year useful life, would it be advisable to switch from the double-declining-balance to straight-line-depreciation method anytime during the 5-year ownership period? If so, when?

11.12 The equipment described in Problem 11.3 is expected to produce an annual increase in taxable income of $8000. Assuming that the company is taxed at an effective rate of 40 percent and no expense deduction is available, use all four methods of depreciation to compare the net increase in after-tax income resulting from the second year of operation.

11.13 An aircraft assembly fixture has a purchase price of $90,000 and is classed as a 5-year property. Use of the fixture is expected to result in an annual before-tax savings of $30,000 for a period of 6 years, at the end of which it will be obsolete and virtually worthless. Applying the appropriate ACRS schedule and assuming that no tax credits or expense deductions are applied, determine:

    (a) The before-tax present worth of the investment using an interest rate of 40 percent

    (b) The after-tax present worth of the investment using an effective tax rate of 40 percent and an interest rate of 20 percent

11.14 Determine the first-year after-tax cash flow resulting from installation of the equipment in Problem 11.13 if a $10,000 expense deduction is applied.

11.15 The Amdex Company purchased three utility vans for $18,000 each in March 1986. One of the vans was sold in November 1988, but the others will be retained until March 1992. Using the appropriate ACRS schedule, determine the deductions allowed for each year from 1986 to 1992, inclusive. Disregard any investment tax credits.

11.16 Assuming that the vans purchased in Problem 11.15 qualify for an investment tax credit, state what would be the net effect of the credit in 1986.

11.17 Assuming that the vans purchased in Problem 11.15 qualified for an investment tax credit, determine the tax liability effect of selling one of them in 1988.

11.18 Megabit Electronics purchased a new logic test unit for their microprocessor research laboratory in 1986. The equipment cost $60,000 and is expected to have a useful life of 8 years. Assuming that Megabit applies the maximum possible expense deduction, use the appropriate ACRS schedule and a tax rate of 40 percent to determine the net tax liability effect of this equipment for each year of the service life.

11.19 Three corporations have the following data for 1 year of operation:

	Macro, Inc.	Meso, Ltd.	Micro Co.
Sales	$20,000,000	$2,000,000	$200,000
Expenses	15,250,000	1,000,000	140,000
Other revenue	750,000	250,000	50,000
Depreciation	3,000,000	750,000	60,000

(a) If all three corporations operate in areas where no state taxes are charged, what is their federal income tax?

(b) If state taxes average 5 percent and the federal income tax is not an allowable deduction in state tax return, what is the effective income tax rate of each corporation?

11.20 A corporation has an effective tax rate of 54 percent. Gross revenue for the past year was $8 million. Operating expenses and depreciation accounted for $6 million. Interest on outstanding debts was $1.2 million. What amount is left for dividends and surplus after taxes?

The following basic data pertain to a proposal that is analyzed under the conditions presented in Problems 11.21 through 11.27.

A proposed investment in a depreciable income-producing asset is expected to produce an annual net savings of $20,000. The asset has a first cost of $60,000 and an estimated life of 8 years with no salvage value. The asset qualified as a 5-year ACRS property and the company has an effective income tax rate of 40 percent. Unless otherwise stated, it is assumed that a cash flow occurs at the end of a year, equity funds are used for the purchase, and any deductions beyond the $20,000 net savings can be applied against other income generated within the organization.

11.21 What is the before-tax rate of return for the asset described in the basic data? Determine the *approximate* after-tax IRR from the before-tax figure.

11.22 Compute the after-tax rate of return when straight-line depreciation is charged on the 8-year useful life of the asset described in the basic data.

11.23 Compute the after-tax rate of return when straight-line depreciation is charged on the basis of a 5-year life even though the equipment is still used for 8 years.

11.24 Compute the after-tax rate of return when ACRS-based depreciation for a 5-year property is applied.

11.25 Compute the after-tax rate of return when the alternative ACRS straight-line method is used to calculate annual deductions for depreciation.

11.26 Compute the after-tax rate of return when ACRS-based depreciation is used and the property is eligible for an investment tax credit.

11.27 Compute the after-tax rate of return if the equipment is purchased with $20,000 of equity funds and $40,000 of borrowed capital. The borrowed funds are to be repaid in four equal end-of-year payments with interest calculated at 10 percent per annum. ACRS depreciation is to be used with no investment tax credits available.

11.28 A sales representative made an agreement to purchase a building lot by making six annual payments of $900 each. She was then transferred to a different territory before she could build a home. When she returned 2 years later, she found that she could buy an equivalent lot for $3000 because land values had decreased during her absence. She now feels that she will lose the $1800 she has already paid if she drops her contract to buy the equivalent lot. Assuming that she will suffer no penalty for reneging on the original contract, discuss the "sunk-cost" concept as it applied to the situation. What would you advise her to do? Does an assumption that she can invest her savings at 7 percent have any bearing on the situation?

11.29 Rimrock Construction Co. is purchasing a new piece of equipment for $34,000. The unit is expected to produce annual revenue of $21,000 for each of the next 4 years and will be sold at the end of that time for an expected salvage value of $4000. Maintenance expenses on the equipment are expected to be $2000 for the first year and to increase by $500 per year for each successive year of operation.

Rimrock is purchasing the equipment by paying $14,000 down and financing the balance at an effective interest rate of 10 percent to be repaid in three equal end-of-year payments. The company will be using ACRS depreciation with a 3-year life for computing income taxes. It has an effective tax rate of 40 percent and requires an after-tax MARR of 10 percent. What is the present worth of the proposed purchase?

11.30 Rarefind, Inc. has been in business for 5 years and is planning a new production area to manufacture a line of artificial antiques for sale to tourists. A primary piece of equipment for this line has a first cost of $9000 and is expected to have a life of 5 years with no salvage value. The use of this machine is expected to produce a direct net before-tax cash flow of $5000 for each year of operation. Rarefind can either purchase this machine outright with equity funds or buy it on a time-payment plan consisting of $4000 down and three equal end-of-year payments calculated at an interest rate of 10 percent. Assuming an effective income tax rate of 40 percent and ACRS depreciation, determine the net after-tax present worth of the two purchase plans using a 15 percent MARR.

11.31 An outmoded paper machine was purchased for $300,000 twelve years ago. Its annual operating costs are $48,000. It is expected to last 8 more years, at which time it will have zero salvage value. A newer model of the paper machine will cost $470,000, have a life of 15 years, and have a negative salvage value of $10,000. The new machine is expected to have average operating costs of $30,000 per year and to be eligible for a 10 percent investment credit.

The old machine is being depreciated via straight-line depreciation, but the new machine will utilize ACRS depreciation. The required after-tax rate of return is 10 percent and the effective tax rate is 40 percent.
  (a) Assuming that the old machine has a market value equal to its book value, what is its after-tax EAC?
  (b) What is the after-tax EAC of the challenger?

11.32 An investment proposal is expected to have the following characteristics:

	Year 1	Year 2	Year 3
Gross income provided	$17,000	$22,000	$19,000
Investments needed	16,000	14,000	0
Operating expenses	6,000	8,000	11,000
Depreciation charges	10,000	10,000	10,000

Investments occur during the year and allow the indicated earnings, but both the income and investments are considered to have been made at the end of each year for analysis purposes. In any year in which deductions exceed income, the tax benefits can be considered as additional income because the organization can apply the deductions against income earned from other operations.
  (a) Determine the rate of return before taxes.

(b) If the effective income-tax rate is 42 percent, what is the approximate after-tax rate of return?

(c) What is the actual after-tax rate of return?

11.33 An asset purchased for $10,000 that did not qualify for an investment tax credit will have a salvage value of $2000 at the end of its 6-year life. Net earnings attributable to the asset are $3200 per year. Assuming that it was purchased with equity funds and the corporation has an effective tax rate of 40 percent, determine the after-tax IRR resulting from the use of a 5-year ACRS depreciation schedule.

11.34 A new piece of chemical processing equipment was purchased in 1971 at a cost of $450,000. Its useful life was 15 years and negligible salvage value was assumed for straight-line depreciation purposes. Annual before-tax net operating income from the equipment was $75,000 and it was shut down in 1986. Surprisingly, the equipment was sold for a $60,000 value. Using an effective tax rate of 40 percent and a MARR of 15 percent, determine the equivalent after-tax annual worth of the equipment over its service life.

11.35 A consulting engineer in acoustical design has purchased and equipped a mobile sound lab at a cost of $50,000. She borrowed $30,000 on a 3-year contract at 10 percent interest compounded annually, with the loan to be repaid in three equal end-of-the-year installments. Her average annual gross income over the next 5 years is expected to be $100,000 and expenses are expected to be $30,000 annually. The sound lab can be depreciated as a 5-year ACRS property and its salvage value at the end of that time is expected to be $10,000. Determine the present worth of the engineer's net after-tax income for the 5 years using a MARR of 15 percent and an effective tax rate of 40 percent.

EXTENSIONS

11A Noncorporate Taxes: What They Are and Why

Taxes have been levied on individuals throughout recorded history, and probably before. Modest incomes were taxed at about 40 percent at time zero (about 2000 years ago), and the name for a tax collector (publican) in that era was synonymous with "robber." Excessive taxation in England led to the Magna Carta in 1215, and a few hundred years later inequitable taxes contributed to the independence movement in the American colonies. There was no income tax in the United States prior to 1914. In that year 370,000 individuals paid taxes at an average amount of $94 each. By 1983 the nation's tax bill had risen to $984.6 billion, 64 percent collected by the federal government and the rest by state and local collectors. As shown in Figure 11.4, total taxes on a per capita basis amounted to $4195 in 1983.

Federal Taxes on Individuals

Table 11.16 shows how federal income tax rates climb with rising income levels for a single taxpayer. Taxable income is the taxpayer's adjusted gross income minus standard exemptions and itemized deductions. For most wage earners, tax calculations are relatively staightforward, but they quickly become more complicated when income is derived from several sources. Then provisions akin to those discussed for corporate taxes become applicable. What individual taxpayers tend to forget is that their tax, like a corporation's tax, is simply a cost of operation—another disbursement similar to bills for clothing and shelter.

Most of the regulations governing corporate taxation apply to individuals in somewhat different forms. For instance, individuals may register short- and long-term capital gains and losses which receive essentially the same tax treatment as do corporations, except the rates differ, as apparent in the following 1984 capital-gains regulations for individuals:

• *Short-term gains* fully taxable as income.

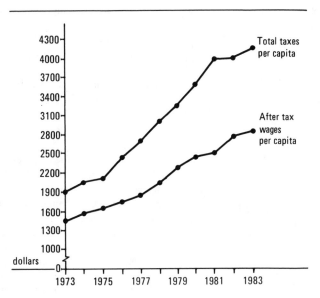

FIGURE 11.4
Ten-year comparison of per person take-home pay versus taxes paid. During the period consumer prices inflated 124.2 percent, wages grew by 114.9 percent, taxes increased 123.5 percent, and net wages after taxes went up only 96.3 percent. Total taxes include all government's net receipts: sales, income, social security, corporate taxes, etc.

TABLE 11.16

1984 federal tax-rate schedule for single taxpayers

SCHEDULE X—Single Taxpayers

If Taxable Income Is		The Tax Is	
Over	But Not Over	This Percent	Of the Amount Over
$ 0	$ 2,300	0	
2,300	3,400	. . . . 11%	$ 2,300
3,400	4,400	$ 121 + 12%	3,400
4,400	6,500	241 + 14%	4,400
6,500	8,500	535 + 15%	6,500
8,500	10,800	835 + 16%	8,500
10,800	12,900	1,203 + 18%	10,800
12,900	15,000	1,581 + 20%	12,900
15,000	18,200	2,001 + 23%	15,000
18,200	23,500	2,737 + 26%	18,200
23,500	28,800	4,115 + 30%	23,500
28,800	34,100	5,705 + 34%	28,800
34,100	41,500	7,507 + 38%	34,100
41,500	55,300	10,319 + 42%	41,500
55,300	81,800	16,115 + 48%	55,300
81,800	. . . .	28,835 + 50%	81,800

- *Short-term losses* fully deductible up to 50 percent of the loss, or a maximum of $3000.

- Forty percent of *long-term gains* included in taxable income; an individual in the 50 percent bracket would pay 50% × 0.4 = 20% on the incremental long-term gain.

- *Long-term losses* 50 percent deductible up to $3000 per year with the balance carried forward to next year; two dollars of long-term loss are necessary to offset one dollar of income.

Other Taxes on Individuals

Cities and states received $351.1 billion in taxes in 1983, an average of $1216 per person, and a rise of 153.4 percent over 1973. Their largest source of income was the sales tax. This tax is termed "proportional" because the same percentages apply to all purchases. A sales tax has a high rating for simplicity and efficiency, but has mixed ratings for fairness. It is considered more equitable when subsistence expenditures are untaxed.

The property tax is another controversial type of taxation. It has become an emotional issue owing to sharp increases caused by escalating housing prices and mushrooming costs of local government. There is less concern about the fairness of property tax than about the inequity of its administration. Although there is merit in arguments that ownership of property does not necessarily correspond to an ability to pay taxes, more distress is caused by properties of similar value being assessed quite differently for tax purposes.

There are numerous other forms of taxation, ranging from ad valorem taxes to import tariffs. All are subject to criticism and reform movements. Since a "tax reform" for most people means getting their own tax burden lightened, a universally appealing tax system is as unlikely as a universal solvent; both are subject to leaks.

QUESTIONS

11A.1 Determine the federal tax owed by an individual with a taxable income of $40,000, according to Table 11.16. Compare the individual's tax bill with a corporation's tax for the same taxable income, using Table 11.7.

11A.2 Individuals can control to some extent the amount of their capital gain (or loss) for a year by selecting which assets to sell, assuming that they own assets that have recorded short- *and* long-term gains *and* losses. Based on the given capital-gains regulations, discuss the tax advantages or disadvantages that would result from each of the following combinations:

- (*a*) A $1000 short-term loss combined with a $1000 long-term gain
- (*b*) A $1000 long-term loss combined with a $1000 short-term gain
- (*c*) A $1000 long-term loss combined with a $1000 long-term gain

11B   Potential Tax Reforms

Nearly everyone who pays taxes has an idea about how to improve the tax system, and it usually involves shifting the tax burden to someone else. Another popular improvement is to simplify the tax-calculation process. Less discussed but more controversial is the way current tax laws discourage capital formation by *double taxation:* taxing money before it is put into savings and then taxing the income generated by the savings and, equivalently, taxing corporate profits when they are earned and again taxing them when they are paid out as dividends.

Ideal Taxation

Many changes have been made in the United States tax laws since 1914, in an attempt to perfect the tax system. There may be no perfect system of taxation, but economists generally agree on the desirable properties that one should possess. A tax should be:

- *Fair and equal.* People in equal situations should pay equal amounts (horizontal equity), and people in unequal situations should pay according to their capacity (vertical equity).

- *Efficient.* A low ratio of administrative cost to revenue produced is most desirable.

- *Neutral.* The tax should have minimal effect on the total economic system.

- *Certain and predictable.* A person should know what is owed and when.

- *Simple*. Tax laws should be clearly understood and easily complied with.

The vexing problem is to design a system that at once possesses all the desired properties (a no-tax system is eliminated by definition).

The most controversial issue is fairness and equity. An obviously fair tax charges people for services in proportion to the benefits received. This is fine for golf fees and university tuition, but it is difficult to set fixed charges for national defense, the judicial system, etc. Another measure of fairness is the taxpayers' ability to pay. The federal income tax uses "progressive" rates (higher percentages at higher income levels) to make the tax burden equitable. Whereas most people agree in principle about the the fairness of progressive income taxes, there is far less agreement on how steeply the rates should rise.

Tax Reforms

Disturbing though they may be, tax codes are not easy to reform. Minor modifications are made with regularity, but substantial restructuring is rare, because any major change has very vocal opponents. For example, groups that benefit from each of the deductions shown below would noisily oppose any new limitations.

1984 Deductions from Taxable Income	Lost Tax Revenue
Exclusion of pension contributions and earnings	$56 billion
Investment tax credit	30 billion
Accelerated depreciation	27 billion
Mortgage interest deductions	25 billion
Deductions for state and local taxes paid	22 billion
Exclusion of employer-paid health benefits	20 billion
Capital-gain exclusion	19 billion

Legislation to close any of the loopholes above would be difficult to enact. Nonetheless, a few reforms have enough backers to remain in contention as reasonable taxation alternatives. Several enduring suggestions for revising the income tax code are explored in the following paragraphs.

Tax Indexing. The 1981 Congress approved tax indexing to eliminate *bracket creep*—people being pushed into higher tax brackets simply because inflation bloats personal income. The purpose of annual adjustments to the brackets is to prevent ever-higher taxes on the same level of buying power.

In accordance with the 1981 Tax Act, the 1985 indexation was based on a 4.08 percent average rise in consumer prices for the 12 months preceding October 1984. The effect can be seen by comparing the tax brackets in Table 11.17 with those in Table 11.16.

Flat-Rate Tax. The concept of a tax structure with an expanded tax base, fewer deductions and credits, and lower rates has many advocates. They claim a flattened tax would be less expensive to administer, easier for taxpayers to compute, and less subject to tax evasion. However, it would be fought by special-interest groups protecting their "sanctified" tax breaks.

Cash Flow Expenditure Tax. A tax based on *consumed income* is an offshoot of the flat tax that closes loopholes and lowers rates yet allows individuals to subtract income from savings and investment. These deductions avoid most double taxation but are considered regressive.

TABLE 11.17

Tax brackets for a single taxpayer adjusted to account for a 4.08 percent inflation rate. The standard deduction for which no tax is owed increases from $2300 to $2390. When taxable income is unchanged at $25,000 from the year before, the tax bill drops by $120 as a result of indexing: Tax without indexation = $4565 against tax with indexation = $4445.

	Taxable Income after Indexation	
Tax Rate, %	Over	But Not More Than
0	$ 0	$ 2,390
11	2,390	3,540
12	3,540	4,580
14	4,580	6,760
15	6,760	8,850
16	8,850	11,240
18	11,240	13,430
20	13,430	15,610
23	15,610	18,980
26	18,980	24,460
30	24,460	29,970
34	29,970	35,490
38	35,490	43,190
42	43,190	57,550
48	57,550	85,130
50	85,130	

Value-Added Tax (VAT). Proven to be effective in Europe, VAT is simply a levy on each stage of production and distribution. Each seller in the chain from raw-material purchases to finished-product sales collects the VAT on every transaction, paying the government a percentage. Besides generating huge revenues, VAT is relatively cheap to administer, difficult to cheat on, and good for investors. But it is regressive and creates a one-shot surge of inflation when started.

Other Tax Options. With all the barriers confronting major restructuring, most future tax revisions will likely be minor modifications. *Surtaxes* have been applied frequently to boost tax revenues for a short time period, so claimed when imposed but subject to stretching. *Loophole-squeezing* is a favored tinkering tactic that fails to address fundamental difficulties, such as collecting taxes on the underground economy. A *federal retail sales tax* would raise vast amounts without stifling capital formation, but it is regressive and intrudes on the form of taxation traditionally left to cities and states. Higher *excise taxes* on such items as tobacco and liquor, known as *sin taxes,* could be used to fund financially troubled programs such as Medicare. There is then the naked option of simply abolishing the tax reductions installed by the Tax Act of 1981; what government giveth can be revoked, and often is.

QUESTIONS

11B.1 An argument against steep progressive tax scales is that excessive rates discourage people from trying to earn more, thereby making them less productive then they might be. Steep rates are also characterized as disincentives for saving and investing. On the other hand, it is said that the degree of acclivity depends on the bracket from which it is observed. Discuss the progressive tax design as a "cost of operation."

11B.2 Compare corporate and individual income-tax rates with respect to the properties listed as desirable in a tax.

11B.3 Assuming a tax increase is necessary to raise federal revenue by 10 percent, and that any change in the tax code could be enacted, discuss what revision you would suggest. Why?

11B.4 The changes wrought by the Economic Recovery Tax Act of 1981 were the most significant in two decades. They were supposed to cause a surge of saving and investment, update production facilities, and stimulate business. By 1985 it was apparent that capital spending had indeed increased and business did recover from the 1981–1982 recession, but the savings rate had declined. From an engineering economist's perspective about the acceptance of investment proposals, what effect did the Tax Act of 1981 have? What types of investments were favored?

CHAPTER 12

INDUSTRIAL PRACTICES

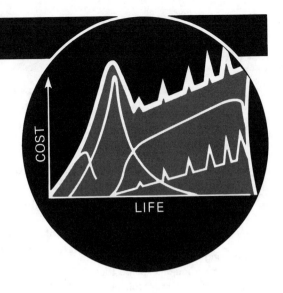

OVERVIEW

Economic decision models bereft of numbers are like empty cups: they must be filled before they are useful. In the previous chapter we saw how taxes and depreciation are factored into cash flows. In this chapter we will explore the sources of cash flow data, procedures companies use to organize and evaluate data, and financial considerations for funding approved expenditure proposals.

Most cash flow data can be secured from familiar sources: competitive bids, price quotes, accounting records, standard times and costs, cost indexes, etc. Two estimates to watch closely are first cost and overhead cost.

Many industrial organizations utilize customized economic evaluation procedures, often complemented by worksheet forms. A *request for expenditure* (RFE) includes supporting analysis documents and a letter of justification which explains the proposed investment. *Postauditing* is a systemized follow-up on approved RFEs to appraise the accuracy of previous estimates in order to improve future estimations.

Tax deductions allowed for interest payments enhance the attractiveness of debt financing. A higher *leverage factor* (ratio of debt to total assets) generally produces higher after-tax returns on equity when business conditions are favorable, but it lowers returns when conditions are unfavorable. An aggressive tactic of relying on short-term debt keeps financing charges lower by incurring greater risk of running out of working capital during periods of constricted cash flow.

The "moment of truth" in engineering economics is the final decision to fund or reject a proposed investment. Industrial practices attempt to make the judgment convenient, consistent, and credible.

DATA DIFFICULTIES

Economic analyses are built from data as a house is built of bricks, but an accumulation of data is no more an analysis than a pile of bricks is a house. There are piles of data everywhere. They fill the shelves of libraries and accumulate limitlessly from the unending parade of statistics and information issued by all kinds of agencies, public and private. Yet, when a particular fact is crucial to an analysis, it may be elusive or impossible to procure. Data difficulties range from an overabundance, when the problem is to digest information into a usable form, to a scarcity that may force guesses to substitute for facts.

Securing reasonably accurate data can be a major expense for an innovative proposal. In a planning phase, ±30 percent may be adequate accuracy. At the request stage, ±5 percent is reasonable.

Information is both the raw material and the finished product of an engineering economic evaluation. The evaluation process is similar to familiar production processes. Bits of data, the raw materials, are received from many sources and inspected for accuracy before being accepted. They are fed into the engineering economics "machinery," where they are translated and transformed into the desired product: an economic comparison. As in any product process, it is necessary to know the raw materials needed, the best sources, and the ways to determine whether they are adequate.

ESTIMATING TECHNIQUES

A new project may be identical in design and mission to one just completed, but the cash flows for the two will probably differ considerably. This variation from one study to the next is characteristic of economic analyses. Unlike physical laws which consistently follow an orderly cause-and-effect relation, economic laws depend on the behavior of people, and they are erratic. However, past behavior is still a respectable clue to future behavior.

Cost sources and assumptions used in estimating should be quoted in the proposal. Include sample computations for unusual cost categories and projections.

Engineers are well prepared to judge the future performance of materials and machines. Past performance can be extrapolated, and modified if necessary, to predict performance on the next project. The monetary values for the future conditions are less reliably predicted. Strikes, shortages, competition, inflation, and other factors that affect unit costs cannot always be anticipated. Therefore, estimating starts with the calculation of design amounts based on known physical relations and concludes with the assignment of comparatively less known monetary measures for the design conditions.

Sources of Cost Data

Figure 12.1 shows the traditional cost and price structure for a manufactured product. Most of the categories are also appropriate for service functions. The principal variation occurs in the relative sizes of the cost categories; manufacturing usually has a higher material-to-labor ratio than does service. Public services would be represented by the model by eliminating the top two tiers.

Unit cost components are discussed in Chapter 15.

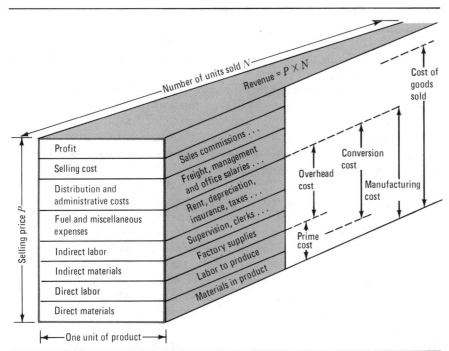

FIGURE 12.1
Composition of costs traditonally used in accounting for the price of a manufactured product.

The quantities and types of materials required can be determined quite accurately from past records and engineering design documents (e.g., the bill of materials and parts list that accompany blueprints.) Current prices are also readily assessable. Difficulties creep in as the price quotes are pushed farther into the future. Experts from the purchasing department in a large firm can assist with predictions through their experience with commodity futures and knowledge of the firm's policies toward *speculating* (buying in large quantities to take advanced of price fluctuations, or buying contracts for future deliveries at prices set now, called *hedging*). A vital part of any estimating process is to be aware of the organization's economic policies and practices.

When purchasing experts are not available, many types of *cost indexes* can furnish information. A cost index provides a comparison of cost or price changes from year to year for a fixed quantity of goods or services at particular locations. For building construction alone, there are a dozen major United States indexes compiled for different parts of the country and different types of construction. The Bureau of Labor Statistics is the primary source of national data.

A cost index is a dimensionless number that indicates how cost or price for a class of items changes with time.

Labor costs are a function of skill level, labor supply, and time required. The saying that "time is money" is especially true in estimating the cost of work. Standards for the amount of output per labor hour have been developed for many classes of work. Time standards are based on a "normal" pace for an activity, with allowances included for personal relief and unusually severe working conditions. Standard times combined with expected wage rates provide a reasonable estimate of labor costs for repetitive jobs. Labor costs for specialized work can be predicted from bid estimates or quotes by professionals and agencies offering the service. A project involving a new type of work will likely have higher labor costs at first, while experience is being gained.

A checklist of labor cost categories is shown in Table 12.2.

It should be remembered that the cost of labor includes more than direct wages. Fringe benefits may amount to more than half the base wage in accounting for accident, health, and unemployment insurance; vacation and retirement pay; and special agreements such as guaranteed annual wages or sabbatical leaves. Extra costs due to relocation allowances, accidents, and sickness should also be considered.

Maintenance costs are the ordinary costs required for the upkeep of property and the restoration required when assets are damaged but not replaced. Items under maintenance include the costs of inspecting and locating trouble areas, replacement of minor parts, power, labor, materials, and minor changes in or rearrangements of existing facilities for more efficient use. Maintenance costs tend to increase with the age of an asset because more upkeep is required later in life and the trend of wages and material prices is upward.

Property taxes and insurance are usually expressed as a percentage of first cost in economic comparisons. Although the value of property decreases with age, taxes and insurance seldom show a corresponding decrease. Therefore, a constant annual charge is a realistic appraisal of future expense.

The *first cost* of acquiring a major asset may rise well above expectation, especially if the asset originates from a new design or untried process. Cost "overruns" are legendary in government projects, and they haunt the private sector too. After a design is set, direct acquisition costs are readily determined from manufacturers' quotes or competitive bidding. When errors occur, they are typically on the low side, as a result of using incomplete listings of desired features and neglecting less obvious costs such as the following:

Most new projects require working capital: a revolving fund which is drawn on to pay operating costs. It is a sum of money that neither earns interest nor is consumed during a project. Unless a proposal requires a major increase, working capital is not an included cost.

Materials (freight, sales tax, storage costs, damage)

Installation (extra costs for special arrangements and unconventional designs)

Interest, taxes, salaries, and insurance during a design or construction phase

Change orders during construction (installation costs for additions to or deletions from the original plans)

Investigation, exploratory, and legal fees

Promotional costs

Engineering and associated fees

Debugging and start-up costs

Most construction projects include a "contingency cost" category to account for undefined costs that will assuredly crop up. Although such expenses are ticklish to tally, they can drain capital as thoroughly as can design changes for physical assets.

Current levels of costs used in estimating investments may not be applicable to future conditions. Generally this is not a serious problem, but two situations bear watching. One is when an old alternative is resurrected for reconsideration. Old cost estimates may be outdated by new methods or different price levels. The other situation is when current levels reflect abnormal or temporary prices. If prices are adjusted for first costs, similar adjustments should be investigated for other costs.

First costs can immediately eliminate some alternatives. Insufficient capital is a genuine reason to turn down an investment proposal even though it has a handsome rate of return. An investment that would be wise for a firm with adequate capital could be a futile or even disastrous course of action for a firm with limited finances and big ambitions.

Overhead cost is by definition that portion of the cost which cannot be clearly associated with particular operations or products and must be prorated among all the cost units on some arbitrary basis. The reason for this catchall category is the prohibitive expense of assigning and charging to each product a specific proportion of such costs as wages of supervisors, factory heat and light, janitorial services, secretarial help, and incidental supplies.

Fixed cost and *indirect cost*, as explained in Chapter 15, are generally taken to be synonymous with overhead cost.

Several methods are used to allocate the composite overhead expense to a product or operation. The versions differ among companies because of differences in the nature of production; it is not uncommon to find different overhead rates used within one firm. Overhead costs may be applied on the basis of

1 Direct labor, direct materials, or prime costs
2 Machine hours or direct labor time
3 Fixed and variable cost classifications (for instance, overhead such as indirect material and power costs would be treated as varying with output, whereas property taxes, depreciation, and indirect labor would be considered as fixed costs)

Once a base is selected, annual overhead as determined from accounting records is divided by the annual cost or usage time of the base category. The resulting ratio is multiplied by the base cost for the alternative being evaluated. For instance, if overhead is charged on the basis of direct labor,

$$\text{Direct labor ratio} = \frac{\text{total annual overhead cost}}{\text{total annual direct labor}}$$

and the overhead charge for one unit of product X is

$$\frac{\text{Overhead charge}}{\text{Unit of product } X} = \text{direct labor ratio} \times \frac{\text{direct labor cost}}{\text{unit of product } X}$$

which is included as a cost in determining the selling price of product X. By the same reasoning, a machine used in the production of product X could be assigned an overhead burden based on the operator's wages:

$$\frac{\text{Annual overhead charges}}{\text{Machine } A \text{ producing } X} = \text{direct labor rate} \times \frac{\text{annual wages of operator}}{\text{machine } A \text{ producing } X}$$

An important condition to recognize in economic comparisons is that *overhead costs are associated with a certain level of output*. This can be a critical factor in a comparison such as the purchase of either a $200,000 numerically controlled milling machine or a standard $20,000 general-purpose milling machine. If the labor rate of operators of both machines is $9 per hour and the burden rate is based on a direct labor ratio of 300 percent, the machine-hour costs of both machines would be $27 per hour. This figure has to be an incorrect machine-hour costing because the investment in the numerically controlled machine is 10 times that in the standard model. Fortunately, overhead costs often have an identical effect on several alternatives. That is, the same value for overhead costs would apply for different alternatives being compared, making their inclusion redundant in the comparison.

Sources of Income Data

In an industrial setting, revenue is the money received from customers for the services or products sold to them. There are many patterns of revenue flow. A retail store has an essentially continuous influx of revenue during working hours. Plumbers are often paid after each service call. Utility services are paid for by the month. Farmers usually get their money after a crop is harvested. A homebuilder has to wait until a house is sold before receiving revenue. The timing of the revenues associated with an alternative may have a significant bearing on its acceptability. One of the main reasons that new businesses often fail is the time lag between incurred first costs and the establishment of an expected level of revenue.

Revenue often takes the form of a value-increasing benefit to people served by an investment, particularly in public sector projects. As discussed in Extension 10C, a single benefit must not be counted twice as both a cost reduction and a value increase.

Revenue is somewhat harder to estimate than costs for many industrial projects. If a new investment is to serve the same purpose as an existing asset, historical data provide a reliable estimate of future revenue. When the investment is destined to satisfy a new function, revenue estimates are less certain. What looks like a sure bet on the drawing board may end as a miserable flop in the market, where it is exposed to the buying whims of the public.

Occasionally, a precise measure of revenue contribution is impossible. Sums spent on customer *goodwill* and improved employer-employee relations are at best extremely difficult to measure in terms of revenue increments. This situation often results in setting aside a certain sum for "public relations" or investment in intangible returns. Then the sum is divided among projects rated according to their perceived worth in satisfying certain intangible values.

Funds for government or public activities result from various types of taxes, charges for specific services, and borrowing. Everyone is familiar with income taxes and charges for services such as mail delivery. Both are revenue sources secured from the public for benefits expected from governing agencies. Because of their mandatory nature and long history, they can be estimated quite accurately. And as a supplement, when the perceived demand for public expenditures exceeds public revenue, governments at all levels tend to engage in deficit financing. The money so borrowed is a pledge that future members of society will pay for projects undertaken by the present society. (The economics of public projects was examined in Chapter 10.) Although the base for public revenues is broad and solid, the portion justified for a specific public project is as difficult to determine as it is with industrial projects.

Privately owned public utility companies occupy a position midway between government and industry. This position results from the great amount of invested capital required to provide public services such as electricity, water, telephone communications, and railroad transportation. Only by developing a high usage factor can an acceptable return on invested capital be obtained from low service rates. In order to ensure both reasonable returns and reasonable rates, an exclusive geographical franchise is allotted to a utility company. Coupled with the grant of a monopolistic position are regulations controlling rates and standards of service. A regulatory body seeks to set a quality of service which satisfies the customers while permitting rates that allow the utility to earn an acceptable rate of return. Because of this control, revenue can be quite accurately forecast.

Example 12.1
Generation of a Cogeneration Proposal

Cogeneration—production of electricity and process steam in one system—accounted for about 7 percent of the United States power supply in 1984. For instance, in a cascade system, leftover heat from a steel rolling mill furnace (3500 to 2000°F) would be input to a gas turbine (2000 to 1000°F) from which the output would flow to a steam turbine (1000 to 400°F) where the exhaust would supply process steam. A reverse cycle could begin at a lower temperature, where generation of electricity is the initial stage and the waste heat is used for industrial or commercial heating.

An engineer has received an assignment to investigate the feasibility of installing a turbine system for cogeneration. Where should the study start?

Solution 12.1

Assuming that the engineer already understands the cogeneration process, we find that a logical start is to list the information needed to conduct the study. Sixteen necessary items for the list are:

1   Total investment cost
2   Purchased power cost
3   Fuel cost
4   Turbine's rated output
5   Operating time per year
6   Load factor
7   Turbine cycle efficiency
8   Boiler efficiency
9   Production tons per year
10 Annual fixed costs
11 Maintenance and chemical costs
12 Power unit price (turbine plus generator)

13   Years of operation expected
14   Depreciation schedule
15   Company effective tax rate
16   After-tax MARR

An exploratory trial compares rough estimates of total investment cost to net savings in annual electricity costs. If the preliminary figures look promising, more detailed and reliable estimates are obtained from resident or outside experts. Alternatives are developed, such as installing a condensing turbine or a noncondensing back-pressure turbine. Then best-estimate data are collected for different configurations and operating situations. When data trustworthiness is doubtful, a sensitivity analysis may be required, as explained in Chapter 14.

DECISION-MAKING PRACTICES FOR INDUSTRIAL INVESTMENTS

Once the needed data have been determined, a comparison of investment alternatives follows the general procedures previously introduced. Specific procedures vary among industrial organizations. Some are content to compare proposals simply by the payback criterion. Others have specially developed procedures to account for their unique situations. Many utilize preprinted forms to standardize the comparisons.

The economic decision process gets more structured as the size of an organization increases. Larger firms specify channels of successive screenings for proposals and provide printed instructions and worksheets for analyses. Examples and descriptions of typical industrial practices are presented in the following sections.

Request for Expenditure

Proposals for investments normally originate with operating personnel or staff analysts. These personnel observe deficiencies in the present system and conceive ways to overcome them. Improvements may be directed toward *cost reduction* or *profit expansion,* and may or may not require commitments of capital. For instance, a training program for cost reduction is "expensed," or classified as an operating cost, when no durable assets are required to conduct it. Similarly, a change in packaging styles designed to increase sales could be an expense if only cosmetic alterations are involved, or a capital cost when new equipment is needed to produce redesigned containers. A list of proposed objectives is given in Table 12.1.

After preliminary data have been collected to confirm that an idea for cost reduction or profit expansion is worth pursuing, a proposal for expenditure is typically turned over to a staff engineer or analyst. If the initial assessment of the proposal is encouraging, firmer cost and demand figures are generated. These figures are based on estimates from engineering design, operations, purchasing, marketing, and other departments that would be involved in the proposed project in some way. Checklists of easily overlooked costs are often provided as reminders. An example is shown in Table 12.2.

TABLE 12.1

Objectives of investment proposals for industrial operations

Categories of Proposed Expenditures

1 Cost reduction	7 Process modernization
2 Profit improvement	8 Replacement of equipment or materials
3 Additional capacity, new plant	9 Energy conservation
4 Redesign or new product	10 OSHA conformance*
5 Product improvement	11 EPA conformance*
6 Plant modernization	12 Labor productivity improvement

* OSHA is the Occupational Safety and Health Act, and EPA is the Environmental Protection Agency.

A form to show the advantage of a proposed replacement or cost-reduction plan is given in Figure 12.2. In addition to the comparison of operating costs required on the form, the present and proposed facilities or equipment are described by model number, anticipated usage, net value or installed costs, and estimates of useful life and salvage value.

TABLE 12.2

Checklist of cost categories to be accounted for in the preparation of a request for expenditure

Cost and Operating Specifications for Expenditure Requests

1 *Land.* Acquisition cost, legal and escrow fees, title search, etc.
2 *Buildings.* Architectural design costs, contractor construction billings, transfer costs, waste treatment, power, fire control, heating and ventilating, elevators, loading docks, etc.
3 *Ground.* Grading and drainage, sidewalks, parking lots, art, fencing, utility tunnels, lighting, sprinklers, landscaping and landscape-architect fees, etc.
4 *Machinery and equipment.* Invoice prices, taxes, freight, installation costs, utility connections, consultant fees, building modifications, material-handling systems, spare parts, down time during replacement, maintenance costs, output rates over time, insurance, utility usage rates and other operating costs, etc.
5 *Dies, molds, fixtures.* Invoice price, freight, internal and external labor costs, modifications, associated materials, etc.
6 *Materials and supplies.* Type, price and discounts, delivery time and reliability of deliveries, build-up-capacity costs, storage, equipment modifications, handling requirements, protection, associated equipment required, inspection costs, etc.
7 *Labor.* Number, wages, skills, training costs, tools needed, affirmative action, overtime, and typical employment costs as indicated below.

Standard Wage	Incentives	Hourly Additives	Vacation, Shift Premiums, Allowances	Fringe Benefits	Employment Cost
$7.184	$1.058	$2.504	$2.624	$3.60	$16.97

OPERATING COST COMPARISON: REPLACEMENT OR COST REDUCTION

	PRESENT	PROPOSED	(PRESENT – PROPOSED) DIFFERENCE	EFFECT ON VOLUME IN UNITS
Direct Labor _____ @ _____ Hrs. Rate	_____	_____	_____	
Indirect Labor _____ @ _____ Hrs. Rate	_____	_____	_____	INCREASE DECREASE
Other Labor _____ @ _____ Hrs. Rate	_____	_____	_____	_____ _____
Subtotal Labor _____	_____	_____	_____	_____ _____
Benefits @ _____ % of Labor	_____	_____	_____	OTHER
	_____	_____	_____	
Maintenance Costs _____	_____	_____	_____	
Tooling Costs _____	_____	_____	_____	
Materials & Supplies _____	_____	_____	_____	
Down Time Cost _____	_____	_____	_____	Attach Supporting Documents
Utilities (power, air, etc.) _____	_____	_____	_____	_____
Floor Space (in feet) _____	_____	_____	_____	
Subcontracting Costs _____	_____	_____	_____	COST BENEFITS IN SUBSEQUENT YEARS.
Inventory (incr/decr) _____	_____	_____	_____	YR 2 $ _____
Safety _____	_____	_____	_____	YR 3 _____
Start-up Costs _____	_____	_____	_____	YR 4 _____
Training Costs _____	_____	_____	_____	YR 5 _____
Software Costs _____	_____	_____	_____	YR 6 _____
Other _____	_____	_____	_____	YR 7 _____
_____	_____	_____	_____	
_____	_____	_____	_____	
_____	_____	_____	_____	
TOTALS	$ _____	$ _____	_____	Prepared by Date
Cost Benefits of Project—First Year (Present Minus Proposed)			$ _____	

FIGURE 12.2
Cost-analysis worksheet for a replacement or cost-reduction proposal.

Investment Analysis

When no capital outlays are required, a request for expenditure (RFE) is based on only operating costs of the type shown in Figure 12.2, plus estimates of expanded profits, if appropriate. Some firms use only the payback criterion for relatively small investments, say up to $750; if the cost data meet the before-tax payback criterion, the RFE is transmitted to the screening group that decides which requests are funded.

Larger investments are usually subjected to a discounted cash flow analysis. Again the type of analysis and what is included vary among organizations. For instance, some analyses use continuously compounded interest, and firms that process large quantities of raw materials may require a "working-capital cost" to be charged against any investment that increases the amount of inventory held.

Two methods of analyzing investments in use by large corporations in different industries are shown in Figures 12.3 and 12.4. They are designed to ensure the uniformity of RFE preparations and to provide sufficient information for deciding which requests are most deserving. In essence, the forms are programmed routines for conducting the after-tax computations described in Chapter 11; only minor variations are added to customize the procedure.

Figure 12.3 is a cash flow worksheet that utilizes the data collected from the "operating advantage" form shown in Figure 12.2. Irregular expenses (column 4) and depreciation or depletion charges (column 5) are subtracted from the annual savings in column 3 to determine the taxable profit (column 6). Columns 7, 8, and 9 are the tax effects and lead to the after-tax cash flow in column 10. When the amount invested is entered as a negative value in the "zero point" row, the cumulative total in column 11 reveals the payback period—the number of years before the cumulative cash flow switches from negative to positive. This payback is the flow in actual rather than time-value dollars.

A convenient worksheet for calculating the rate of return, called the *profitability index,* is shown in Figure 12.4. The after-tax cash flow as developed in the last column of Figure 12.3 is the input for the "actual amount" column in the "operating benefits" section of Figure 12.4. Receipts and disbursements that occur irregularly before the nominal implementation date, the zero point, are listed in the negative-year rows; the regular cash flow is listed below the zero-point row. Similarly, regular investments are listed in the "capital costs" section. Then values in the "actual amount" column (PW at 0 percent interest rate) are multiplied in trial 2 by the present-amount factors for 10 percent. As long as the A/B ratio at the bottom of the trial column is less than 1.0, another trial is needed. The present-amount factors for each trial are multiplied by the "actual amount" figures to obtain trial values. The given interest factors are based on midperiod continuous compounding and are consequently different from the factors given in Appendix D.

The profitability index is determined from the sums of the discounted

CASH FLOW FROM OPERATIONS

1	2	3	4	5	6	7	8	9	10	11
						Taxes Paid			Cash Flow	
Calendar Year	Year No.	Operating Advantage Form xxx	Expense Portion of Project	Tax Depreciation & Depletion	Taxable Profit Col 3 Less 4 & 5	Normal Tax	Tax Credits	Total Col 7 Less 8	Annual Col 3 Less 9	Cumulative
	-3									
	-2									
	-1									
Zero Point										
	1									
	2									
	3									
	4									
	5									
	6									
	7									
	8									
	9									
	10									
	11									
	12									

Date: _____ , 19 _____

Prepared by: _____

Division: _____

Plant: _____

RFE No. _____

Payback: _____

FIGURE 12.3
Worksheet for calculating the after-tax cash flow and payback period before discounting.

CALCULATION OF PROFITABILITY INDEX (INTEREST RATE)

Interpolation Chart

Profitability Index (%) — vertical axis: 0, 10, 20, 30, 40, 50
Ratio (A/B) — horizontal axis: 0, 0.5, 1.0, 1.5, 2.0

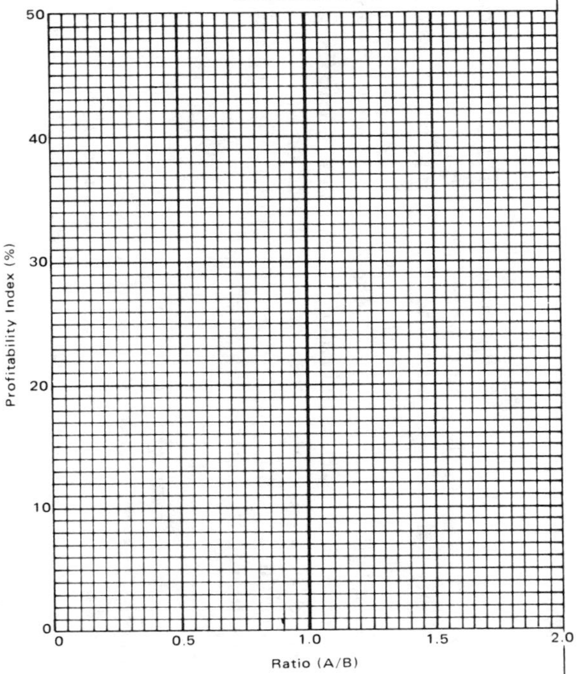

PROFITABILITY INDEX = _____

RFE No. _____ Date: _____

Plant: _____ Div.: _____

Capital Costs

Cash Outflow

Year (Actual)	Relative to Zero	Trial No. 1 — 0% Interest Rate — Actual Amount	Trial No. 2 — 10% Interest Rate — Factor	Present Worth	Trial No. 3 — 25% Interest Rate — Factor	Present Worth	Trial No. 4 — 40% Interest Rate — Factor	Present Worth
	-3		1.285		1.873		2.736	
	-2		1.162		1.459		1.834	
	-1		1.052		1.136		1.230	
	Zero Point		1.000		1.000		1.000	
	1		.952		.885		.824	
	2		.861		.689		.553	
	3		.779		.537		.370	
	4		.705		.418		.248	
	5		.638		.326		.166	
Totals A								

Operating Benefits

Cash Inflow

Relative to Zero	Trial No. 1 — 0% Interest Rate — Actual Amount	Trial No. 2 — 10% Interest Rate — Factor	Present Worth	Trial No. 3 — 25% Interest Rate — Factor	Present Worth	Trial No. 4 — 40% Interest Rate — Factor	Present Worth
-3		1.285		1.873		2.736	
-2		1.162		1.459		1.834	
-1		1.052		1.136		1.230	
Zero Point		1.000		1.000		1.000	
1		.952		.885		.824	
2		.861		.689		.553	
3		.779		.537		.370	
4		.705		.418		.248	
5		.638		.326		.166	
6		.577		.254		.112	
7		.522		.197		.075	
8		.473		.154		.050	
9		.428		.119		.034	
10		.387		.093		.023	
11		.350		.073		.015	
12		.317		.057		.010	
Totals B							
Ratio A/B							

FIGURE 12.4

Worksheet for determining the profitability index of an investment proposal. The present worth of capital costs and operating benefits are summed to obtain A/B ratios at different interest rates. The A/B ratios are then graphed to indicate the profitability-index percentage.

capital cost A and the operating benefits B. A/B ratios are calculated for each trial interest rate up to a rate which produces a ratio greater than 1.0. The profitability index lies between the interest rate which produced the last A/B ratio less than 1.0 and the next larger one. The actual index percentage is determined graphically by plotting the A/B ratios on the "interpolation chart." The profitability index is indicated by the point where the line connecting the plotted A/B ratios crosses the vertical 1.0 line on the interpolation chart. The return on investment is the percentage obtained by reading across from that intersection to the vertical scale on the left.

Example 12.2
Systematic Incremental Investment Evaluation

An electronics manufacturer performs economic evaluations on preprinted forms. A study starts with a form on which operational costs of two alternatives are compared. Included on the sheet are costs for labor (hourly labor rates times direct and indirect labor hours), benefits (35 percent added to labor cost to account for social security, holidays, vacations, etc.), property tax and insurance ($2\frac{1}{2}$ percent per year on installed cost), maintenance, rejects (based on labor and material content at the rejection point), tooling, training, utilities, materials, supplies, licenses, consulting, programming, system support, inspection, waste disposal, and other costs. Assume that the yearly differences in operational costs between two alternatives are

	Year				
	1	2	3	4	5
Cost savings	$30,000	$33,000	$29,000	$24,000	$15,000

After-tax cash flow (ATC) is calculated by considering depreciation, corporate taxes, and tax credits in conjunction with the cost savings. This is the same procedure as that described in Chapter 11 and results in the following ATC:

	Year				
	1	2	3	4	5
ATC	$36,000	$30,000	$26,000	$20,000	$10,000

Incremental investment is the difference in capital cost plus expenses between the two alternatives. To be on the conservative side, the manufacturer uses a zero investment cost when one of the alternatives is the defender in a replacement study. Expenses may include design costs by consultants, freight, floor space required, work-in-process inventory, operator's work-space, required aisles, start-up training, software development, sales or use tax, and spare parts. The incremental difference between the sum of these expenses and the purchase price for the alternatives currently being evaluated is $66,000.

Use the company's preprinted form to calculate the IRR and net present value when the MARR is 25 percent.

Solution 12.2

The completed capital-investment evaluation sheet is shown in Figure 12.5. Trial 0 in the first column is a listing of the 5 years of after-tax incremental cash flow. The bottom row is the ratio of incremental capital investment ($X = $66,000$) to ATC. The f figures are values of the present-worth, discrete-compounding interest factors found in Appendix D. Trials 1 through 3 discount ATC at 10, 25, and 40 percent. The calculated X/ATC ratios are plotted on the interpolation chart.

A smooth line connecting the plotted points indicates the rate of return at the point where it intersects the bold 1.0 line; read left from the intersection to the vertically scaled discount rate. Net present value based on a 25 percent hurdle rate is calculated by subtracting X from ATC$_2$; it can be either positive or negative. For the given data, the present worth is computed as $72,760 - $66,000 = 6760.

PROJECT NAME: **PREPARED BY:**

   EXT: **DATE:**

STEP NUMBER 1 – DISCOUNTING THE AFTER TAX CASH FLOWS (ATC)

YEAR	TRIAL NO. 0 AFTER TAX CASH FLOW (ATC)	TRIAL NO. 1 10% INTEREST RATE Factor f_1	Present Worth $(ATC)(f_1)$	TRIAL NO. 2 25% INTEREST RATE Factor f_2	Present Worth $(ATC)(f_2)$	TRIAL NO. 3 40% INTEREST RATE Factor f_3	Present Worth $(ATC)(f_3)$	TRIAL NO. 4 60% INTEREST RATE Factor f_4	Present Worth $(ATC)(f_4)$
1ST	36,000	.91	32,760	.80	28,800	.71	25,560	.63	
2ND	30,000	.83	24,900	.64	19,200	.51	15,300	.39	
3RD	26,000	.75	19,500	.51	13,260	.36	9,360	.24	
4TH	20,000	.68	13,600	.41	8,200	.26	5,200	.15	
5TH	10,000	.62	6,200	.33	3,300	.19	1,900	.10	
6TH		.56		.26		.13		.06	
7TH		.51		.21		.10		.04	
8TH		.47		.17		.07		.02	
9TH		.42		.13		.05		.01	
10TH		.39		.11		.04		.01	
TOTAL ATC	122,000	TOTAL ATC_1	96,960	TOTAL ATC_2	72,760	TOTAL ATC_3	57,320	TOTAL ATC_4	
RATIO X/ATC	0.54	RATIO X/ATC_1	0.68	RATIO X/ATC_2	0.91	RATIO X/ATC_3	1.15	RATIO X/ATC_4	

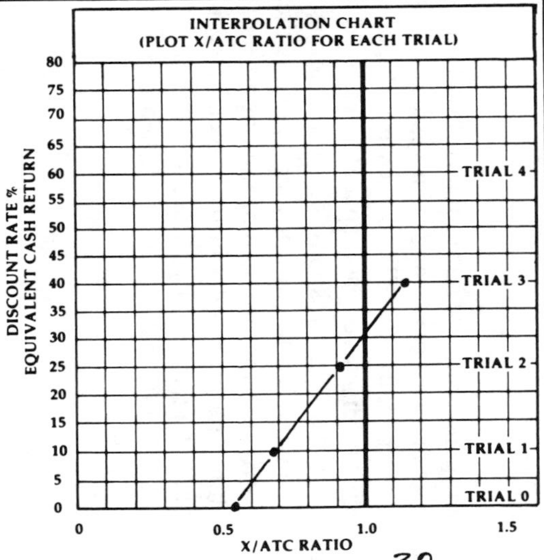

INTERPOLATION CHART
(PLOT X/ATC RATIO FOR EACH TRIAL)

DISCOUNT RATE % EQUIVALENT CASH RETURN

X/ATC RATIO

INTERNAL RATE OF RETURN _32_ %

Net Present Value = ATC discounted at 25% and subtracted from the incremental investment.

$ATC_2 = $ _72,760_   x = $ _66,000_

NET PRESENT VALUE $ _6,760_

FIGURE 12.5
Determining the internal rate of return and net present value of an investment increment by using a preprinted worksheet and an interpolation chart.

Letter of Justification

Most industrial organizations require submission of an explanatory letter along with the economic analysis of a proposed investment. A typical summary sheet for a request for expenditure (RFE) is shown in Figure 12.6. The key figures from the analysis are displayed, and space is provided for approvals from the staff engineers and managers responsible for budgeting.

The accompanying written justification explains the reasons for the RFE. It describes the origin of the problem or the opportunity for which the RFE is submitted and how the proposal will be implemented if approved. Intangible factors not included in the analysis are discussed. Supplementary data on related effects of the proposal are also supplied if pertinent to the decision.

Postaudits

Following the progress of a design from the drawing board through construction to on-the-job performance is a natural progression to engineers. They take pride in their designs, welcome the opportunity to make modifications to improve performance, and realize they will be able to make future designs better by knowing what was previously successful and what failed. Follow-ups on capital investments tend to be neither as extensive nor as rigorous. Engineering economists would be well served by the same monitoring instinct displayed in other engineering pursuits.

The purpose of auditing investments is not to punish those who approved the proposals any more than quality-control inspections are intended to penalize a production process. The aim is to improve future analyses. How else can analysts learn if their estimated cash flows are realistic? Many factors influence estimated receipts and disbursements; some are easily overlooked unless brought to someone's attention by postaudits.

Cost-reduction proposals tend to be more accurately estimated than profit-expansion proposals because costs are more controllable than are sales volume and market prices. Yet the savings generated by implementing a lower-cost process or by replacing a higher-cost asset cannot be checked, because the performance of the replaced assets cannot be determined for the latest operating conditions. A good indication is provided, however, by the comparison of estimated operating costs with the actual costs of the replacement.

A major advantage of a systematized postaudit program is the availability of the most recent cost data for categories not itemized in cost-accounting records. Maintenance costs for various types of machines, operating efficiency during training periods, labor costs for specific activities such as installation or set-up activities, and other data utilized in preparing requests for expenditure would be available from postaudits. The more complete computer-based management information systems that are spreading throughout industry promise to make such information conveniently accessible.

LETTER OF JUSTIFICATION					RFE No.		

Description of Request:

Budget Spending 19 _____ $ _____   Budget Total 19 _____ $ _____

	Total Investment	19 _____	19 _____	19 _____	19 _____	19 _____
Investment: Capital						
Expense						
Total Authorized Working Capital						
Future Obligations						
Total Justified						
Benefit: Sales						
Operating Profit after Tax						
Cash Flow from Operations						
% Return on Capital Employed after Tax						

Type of RFE	PI	Payout

Most Critical Assumptions

Prepared by _____   Sponsored by _____

Date	Dept.	Staff Review Signature	Approvals Signature	Date
_____	_____	_____	_____	_____
_____	_____	_____	_____	_____
_____	_____	_____	_____	_____
_____	_____	_____	_____	_____

FIGURE 12.6
Summary page for a letter of justification to support a request for expenditure.

FINANCIAL CONSIDERATIONS FOR INDUSTRIAL INVESTMENTS

Investment proposals may be financed with equity capital (retained earnings and owner's funds) or from debt (borrowed funds). Factors affecting the choice between equity funding and borrowing were examined in Example 9.1. In this section the examination is continued with attention to the optimal proportion of debt-versus-equity financing and the effects of long-term-versus-short-term debt.

Debt-Versus-Equity Financing

The *financial structure* of a firm describes the way that its assets are financed: via long-term debt, short-term credit, preferred stock, and/or common equity. Debt and credit represent obligations to repay borrowed amounts and to send regular interest payments outside the firm. Both loans and preferred stock represent claims on the income of a firm before common stock. Common equity includes common stock, capital surplus, and retained earnings.

Financial leverage is the ratio of total debts to total assets. A firm having assets of $50 million and debts of $20 million has a *leverage factor* of 0.4. Higher leverage means that more money is owed outside the firm; this allows a higher rate of return on the amount owners have invested but also increases the firm's *financial risk*—uncertainty about future returns to a firm's owners as a result of the financing of assets by debt or preferred stock.

The effects of different leverage factors on the percentage return on equity when earnings vary as a function of economic conditions are portrayed in Table 12.3. Three states of financial leverage are shown for total assets of $1 million with equity proportions of $1 million, $600,000, and $300,000. When all the assets are composed of equity holdings, the after-tax rate of return on investment is equal to half the before-tax rate of return, with the effective income-tax rate at 50 percent. Under unfavorable business conditions that cause low sales, the percentage return on equity drops as the leverage factor increases. The trend is reversed when favorable business conditions allow high sales. At average sales, when the before-tax rate of return equals the interest rate for borrowed funds, all three leverage positions result in the same after-tax return on equity investments. In general, *whenever favorable business conditions allow a before-tax rate of return greater than the interest rate on debt, the higher the leverage factor, the higher the after-tax percentage return on equity.*

A highly leveraged firm produces very attractive return-on-investment figures during periods of prosperity, yet it can quickly become a target for management criticism or a victim of bankruptcy during a recession. Lenders are often a determinant of the degree of leverage a firm can attain. The credit standing of a borrower is reduced by excessive borrowing, and lenders attempt to protect their loans by limiting the leverage to certain norms. The norms vary among industries with respect to the stability of incomes, and within an

TABLE 12.3

Equity holders' returns under different leverage factors and economic conditions when assets total $1 million (shown in units of $1000)

Financial Leverage (in $1000 Units)	Leverage Factor = 0: Equity = 1000 and Debt = 0			Leverage Factor = 0.4: Equity = 600 and Debt = 400			Leverage Factor = 0.7: Equity = 300 and Debt = 700		
Economic Conditions	Low Sales	Average Sales	High Sales	Low Sales	Average Sales	High Sales	Low Sales	Average Sales	High Sales
Rate of return before taxes and interest	2%	8%	14%	2%	8%	14%	2%	8%	14%
Net earnings before interest and taxes	$20	$80	$140	$20	$80	$140	$20	$80	$140
Interest on debt at $i = 8\%$	0	0	0	$32	$32	$32	$56	$56	$56
Taxable income	$20	$80	$140	−$12	$48	$108	−$36	$24	$84
Taxes at effective tax rate of 50%*	$10	$40	$70	−$6	$24	$54	−$18	$12	$42
Available for equity returns	$10	$40	$70	−$6	$24	$54	−$18	$12	$42
Percentage return on equity	1%	4%	7%	−1%	4%	9%	−6%	4%	14%

* Assumes losses are carried back and result in tax credits.

industry according to the confidence lenders have in a firm's management; service industries and public utilities typically have very high debt-to-asset ratios (0.6+), and manufacturing has relatively low leverage (0.3 to 0.4). A thriving business can expand faster with the infusion of debt funding, and a sick business may recover more rapidly with the aid of borrowed funds, but a firm weathers brief spells of adversity better when it is less leveraged and not obligated to pay off large debts.

Long-Term Debt Versus Short-Term Debt

After an organization has established its strategic leverage policy, there is still a question of long-term versus short-term debt. This question is most critical to working-capital management—the administration of a firm's short-term assets, such as cash and accounts receivable, to meet day-by-day financial obligations. The question also arises in investment decisions. The answer depends on the cost of loans and the nature of operations.

Short-term (under 1 year) loans conventionally carry a lower interest rate than do long-term loans. The actual percentages vary from year to year,

Wide swings in recent years in the *prime interest rate*—the rate banks charge their best customers—have occasionally raised short-term interest rates above long-term rates. See Figure 13.2.

depending mostly on the national monetary policy. Typical values might be 9 percent for a 90-day note, 12 percent for a 3-year loan, and 14 percent for a 20-year bond. In taking advantage of the lower interest rates allowed for short-term debt, a firm incurs the risk of being unable to repay the loan as it matures. Temporary revenue deficiencies could prevent a firm from refunding its debts on time and cause it to negotiate renewals from its creditors. The creditors may perceive the difficulties as a decrease in credit worthiness and raise the interest rates or even refuse new loans. Long-term (over 5 years) loans reduce the risk of failure to refund by decreasing the number of times debts mature.

A conservative tactic of relying completely on long-term debt is compared with a more aggressive short-term debt tactic in Table 12.4. Both approaches are based on an equity holding of $500,000, a leverage factor of 0.5, and earnings before taxes and interest of 14 percent. The question of relying completely on long-term or short-term credit is narrowed to a choice between taking a 2 percent loss on the percentage of equity return and taking the associated risk of loan-due dates falling in periods of low earnings.

The allowable degree of leverage and long- or short-term credit policy may also influence an economic comparison of proposals. Particularly in a small firm, where the use of credit is decided more by operating requirements than by a strategic financial policy, the evaluation of individual proposals is based on the current level of equity or debt funding available and the interest rate required for loans with different maturities. The following example

TABLE 12.4

Effect of debt maturity on the percentage return on equity when assets are $1 million and earnings before tax and interest are 14 percent; the conservative tactic, holding only long-term loans, yields a lower return on investment than does the riskier, aggressive tactic of using all short-term credit

	Conservative Tactic	Aggressive Tactic
Equity	$ 500,000	$ 500,000
Debt	500,000	500,000
Total debt and equity	$1,000,000	$1,000,000
Short-term debt, 6%	0	500,000
Long-term debt, 10%	500,000	0
Earnings before interest and taxes	140,000	140,000
Less interest on debt	50,000	30,000
Taxable income	$ 90,000	$ 110,000
Less taxes at 50% rate	45,000	55,000
Available for return on equity	$ 45,000	$ 55,000
Percentage return on equity	9%	11%

demonstrates the difference in the cost of an asset acquired with debt financing by loans of contrasting durations.

Example 12.3
Comparison of Costs for Purchasing an Asset by Different Financing Plans

A construction company must obtain a piece of heavy equipment to work on a long-term project. The initial cost is $100,000, and it will have no salvage value at the end of its 10-year useful life. The company uses ACRS depreciation and a minimum after-tax rate of return of 10 percent. Bank loans of 3 years at an annual interest rate of 8 percent and of 8 years at 9 percent are being considered for financing the purchase. Both loans require 10 percent down payment. The construction company has an effective tax rate of 40 percent.

Which financing plan would be the most economical?

Solution 12.3

The discounted cash flows for the purchase of the construction equipment by both funding plans are shown in Table 12.5. The longer loan offers a lower total discounted cash flow than does the shorter loan, even though the 8-year loan has a higher interest rate. A casual observation that interest charges on the longer loan exceed those on the shorter loan by $40,088 − $14,769 = $25,319 might leave the impression that the 3-year financing plan is superior. The tax savings generated by the additional stretched-out interest and loan charges make the longer financing plan preferable.

TABLE 12.5

Calculation of the present worth of alternative financing plans. An asset with $P = \$100,000$, $S = 0$, and $N = 10$ is depreciated (1) by the ACRS method for 10-year property. Annual payments (4) repay a $90,000 loan plus interest charges (2). Tax savings (3) result from multiplying the deductible charges for interest plus depreciation $[(1) + (2)]$ by the effective income tax rate, 40 percent. The cash flow at time zero is the $10,000 down payment, and the cash flow for successive years is the sum of the loan payments and tax savings $[(3) + (4)]$. The present worth of this flow (5) is determined by multiplying each year's cash flow by the appropriate present-worth factor, $(P/F, 10, N)$.

End of Year, N	Annual Depreciation Charge (1)	Three-Year Loan at 8%				Eight-Year Loan at 9%			
		Interest Charge (2)	Tax Saving* (3)	Annual Loan Payment (4)	PW of Cash Flow (5)	Interest Charge (2')	Tax Saving* (3')	Annual Loan Payment (4')	PW of Cash Flow (5')
0					−$10,000				−$10,000
1	$ 10,000	$ 7,200	$ 6,880	−$ 34,923	−28,221	$ 8,100	$ 7,240	−$ 16,261	−8,201
2	18,000	4,982	9,193	−34,923	−21,265	7,366	10,146	−16,261	−5,054
3	16,000	2,587	7,435	−34,923	−20,652	6,565	9,026	−16,261	−5,436
4	14,000		5,600		3,825	5,692	7,877	−16,261	−5,726
5	12,000		4,800		2,980	4,742	6,697	−16,261	−5,938
6	10,000		4,000		2,258	3,705	5,482	−16,261	−6,084
7	8,000		3,200		1,642	2,575	4,230	−16,261	−6,174
8	6,000		2,400		1,120	1,343	2,937	−16,261	−6,216
9	4,000		1,600		679		1,600		679
10	2,000		800		308		800		308
Totals	$100,000	$14,769	$45,908	−$104,769	−$67,326	$40,088	$56,035	−$130,088	−$57,842

* Tax savings are considered to be equivalent to net revenues and are consequently represented as positive cash flows.

Review Exercises and Discussions

Exercise 1 Graph the leverage factors in Table 12.3 to display the relation between returns on total assets and owners' assets. Discuss the significance of the intersection point.

Solution 1 Table 12.3 provides a range of estimates based on scenarios for debt-to-asset ratios of 0.0, 0.4, and 0.7. Lines representing the effect of these ratios on the after-tax rate of return owners can expect from a range of rates earned before taxes on total assets are shown in Figure 12.7. The intersection of the three ratio lines is at the point where return on total assets equals the interest cost of debt, 8 percent. At this point, the return on net worth (common equity) at a tax rate of 50 percent is 4 percent regardless of the degree of leverage. When operations afford a return on assets greater than 8 percent, the debt-financed portion of the assets pays its interest cost and contributes a surplus to the owners of the firm.

FIGURE 12.7
Relation between rates of return on total assets and owners' assets under different leverage factors.

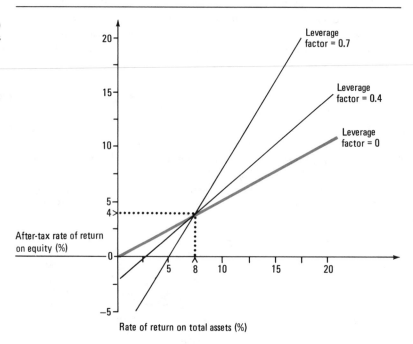

Exercise 2 An equipment sales representative claims that the installation of a new machine will allow equal quality and quantity of output with one fewer worker. Using the firm's required rate of return (15 percent) and direct labor ratio (0.6), the representative substantiates the claim with the following figures:

Annual savings for the elimination of one worker

Direct labor: 1 worker × $10,000/worker-year	$10,000
Overhead: $10,000 × 0.60	6,000
Total saving	$16,000

Annual cost of new machine (P = $20,000, S = 0, N = 10)

Capital recovery: ($20,000 − 0)(A/P, 15, 10) = 20,000(0.19925)	$ 3,985
Operation	9,000
Maintenance	600
Taxes and insurance	400
Total cost	$13,985
Net saving	$ 2,015

Does the sales representative's evaluation appear valid?

Is the sales representative's evaluation valid? Maybe. If the estimated life of the machine and its operating costs are accurate, there is still a question of whether or not overhead costs will be reduced by $6000. A one-worker reduction in the work force would probably have little effect on total overhead costs. Since the main purpose of an overhead ratio is to allocate indirect costs to products, it is not an exact measure of indirect wages. Therefore, the quoted savings in overhead should be investigated.   **Solution 2**

PROBLEMS

12.1 The owners of a prosperous partnership have profits of $40,000 that they wish to invest outside their firm. One partner wants to invest in a venture which will yield a rate of return of 9 percent. The other partner wants to put $20,000 in bonds and the remainder in a venture for which the returns are difficult to estimate. The returns from the bonds (4 percent per year) are nontaxable, but other returns will be taxed at the partnership's tax rate of 35 percent. Assuming that the risks involved are approximately the same, determine what rate of return the other half of the bond venture must earn in order to make the returns from the two alternatives equivalent.

12.2 A cooperative which pays a 50 percent tax rate on gross profit made two investments which have a total future value of $300,000 in 10 years. One of the investments (taxable) has a present worth of $50,000 and will yield before-tax returns of 16 percent compounded semiannually. The other investment is in nontaxable municipal bonds. One hundred bonds, each with a face value of $1000 and semiannual interest periods for 10 years, were purchased at a price to yield the same after-tax return as the other investment. What interest was stated on the bonds if it is assumed that dividends can be reinvested at the rate of return for taxable investments?

12.3 The owner of a drive-in restaurant believes that there is a good potential for another drive-in at a nearby town. She has $40,000 of owned capital with which to expand the business. Using this amount she can open a new operation which should yield a gross income before taxes of $8000. She feels that a conservative study period of 10 years is reasonable, and she will use straight-line depreciation for tax purposes. The salvage value is expected to be $10,000.

After checking with a bank, she finds that she can borrow $20,000 at 6 percent interest per year on the unpaid balance. The loan extends for 10 years and would be repaid in 10 equal annual installments.

If her effective income-tax rate is 40 percent for the period, which means of financing is the most attractive?

12.4 A corporation plans to buy out a supplier and needs $2 million in new capital to do so. The proportion of equity to debt financing will be 40:60 for the acquisition. The current cost of equity capital is 14 percent after taxes, and it is 12 percent before taxes for debt financing. The effective income tax rate is 55 percent.

(a) What is the minimum amount of before-tax earnings necessary per year from the purchase to justify raising the required capital?

(b) What minimum after-tax earnings should be expected?

12.5 Would you rather have net earnings before taxes of $70,000 when you have no debt or $45,000 before interest payments and taxes when you are leveraged at 0.8? Assume that the interest on debt is 10 percent and in both cases the total assets are $500,000 and the effective tax rate is 50 percent. What is the return on investment under both conditions?

12.6 A corporation has an effective tax rate of 45 percent. Its annual earnings total $2 million. For a leverage factor of 0, the after-tax percentage return on equity is 6.6 percent. If the leverage factor had been 0.6 and the interest rate for debt 9 percent, what would the after-tax percentage return on equity be?

12.7 A firm needs $300,000 to install secondary waste-recovery equipment. Since this is a mandatory investment to meet federal pollution-control regulations, the firm may use internal funds and forgo other investment opportunities or borrow the entire amount from outside sources. Internal funds are expected to earn a minimum after-tax rate of 12 percent, while the cost of borrowing money on a 15-year loan is 9 percent. The effective tax rate of the firm is 50 percent.

(a) Compare the present worths of full debt and equity financing when the MARR is 12 percent. Assume that the debt is repaid in 15 equal installments of $37,000 each, of which $17,000 is the constant annual interest charge and depreciation is by the optional 15-year straight-line method.

(b) What amount would annual income need to increase through use of the secondary waste-recovery equipment in order to obtain a 15 percent rate of return when the purchase is funded 50 percent from equity and 50 percent from borrowed funds? Assume that the given loan installment payments are reduced by half in the 50:50 financing plan.

12.8 A consulting engineer has a 10-year contract to monitor the water quality of discharges from several mills located on a small bay. Since the contract also calls for a hydrographic study of the area, a boat will be needed. A properly equipped craft can be purchased for $30,000. At the end of the 10-year study it can be sold for 10 percent of its purchase price. Moorage and maintenance costs will average $1000 per year. It will be depreciated as a 10-year ACRS property.

A loan for the entire purchase amount requires 10 annual payments of $3900. Interest charges during each year of the loan are shown below.

					Year				
1	2	3	4	5	6	7	8	9	10
					Interest				
$1500	$1300	$1260	$1140	$1020	$840	$690	$540	$390	$240

It would also be possible to lease an adequate boat from a marina on the bay. A 10-year lease that allows use of a boat whenever needed, if a 2-day warning is given, will cost $4500 per year and includes maintenance and moorage fees.

The engineer's effective income rate is 40 percent. Compare the after-tax present worths of leasing and buying using an interest rate of 10 percent.

12.9 A request for expenditure for a paint-drying system is being prepared. The investment in new assets includes $36,000 for the oven, $23,000 for ventilation and controls, and $14,000 for the filter assembly. The system is to be depreciated as an ACRS 10-year property with no salvage value, but it will be kept in operation for 12 years.

The major reasons for a forced drying system are (1) to comply with new federal health standards, (2) to lower costs by reducing worker hours, (3) to dry trucks and trailers at the same rate as that of the painting and shot blast operations, and (4) to reduce stenciling costs and to be able to better utilize outside storage facilities. The last reason is the least substantial, but painting will be a bottleneck for any other expansion plans unless the fast-dry system is installed.

Direct savings from the installed system will result from an approximate 9800 hours reduction in operator time cost at $11 per hour and $35,000 annually from reduced material expense. The company has an effective combined federal-state tax rate of 52 percent. A tax credit of 10 percent is allowed to a maximum of $10,000 for investments of the type anticipated.

    (a) Use the worksheet form in Figure 12.3 to calculate the after-tax cash flows.
    (b) What is the payback period indicated on the worksheet?
    (c) What is the after-tax present worth of the proposal using a MARR of 15 percent.

12.10 An asset has a purchase price of $50,000, a 10 percent salvage value, and a useful life of 5 years. It is to be depreciated by the sum-of-digits method. Operating costs will be the same under any of the financing plans to be considered. A discount rate of 10 percent is to be used in the comparisons. The effective income tax rate is 52 percent.

    (a) What is the present worth of ownership if the asset is purchased with equity funds?
    (b) What would the annual lease cost have to be to make leasing as attractive as cash purchase?
    (c) A 7 percent 5-year loan can be obtained if a down payment of $25,000 is made. What is the present worth of costs under this plan?

12.11 A cost-reduction proposal to equip a locomotive with remote radio control is being developed. Because the radio-control equipment, which costs $28,000, can be considered an operating expense, no depreciable assets are involved.

Two operators are now employed to shuttle scrap-weighing cars between the scrap yard and the steel-making furnace. By equipping the locomotive with remote radio-control equipment, the need for one operator each shift is eliminated. The locomotive is utilized 4000 hours per year. Operators are paid $9 per hour including fringe benefits. Installation costs for the controls are expected to be $2000. There have been significant improvements in remote-control systems since the company had its unpleasant experience with a radio-controlled locomotive 10 years ago. Justify the request for expenditure.

12.12 Measurement of space consumed by aisles in a company-owned warehouse reveals that there are 200 feet (61 meters) of 12-foot-wide (3.66-meter-wide) aisles and 750 feet

(228.6 meters) of 10-foot (3.05-meter) aisles. The 10-foot (3.05-meter) aisles are used because a 6000-pound (2722-kilogram) capacity fork truck must operate in them to handle some heavy equipment. All other loads run under 4000 pounds (1814 kilograms). If the heavy items are concentrated in one storage area, the 6000-pound (2722-kilogram) truck will not need to operate in all aisles. The layout could be arranged as follows:

	Width	Length
Central aisle for 6000-lb (2722-kg) truck access	12 ft (3.66 m)	30 ft (9.14 m)
Heavy storage area	10 ft (3.05 m)	75 ft (22.9 m)
Other central aisles	10 ft (3.05 m)	175 ft (53.3 m)
All other aisles cut to 7 ft (2.13 m)	7 ft (2.13 m)	670 ft (204.2 m)

The 7-foot (2.13-meter) aisles are possible only if three new 4000-pound (1814-kilogram) narrow-aisle fork trucks are purchased at a cost of $20,000 apiece. They will have a useful life of 4 years and no salvage value. The replaced 6000-pound (2722-kilogram) fork trucks will be kept for standby and scavenger purposes. Operating costs will increase by $7000 per year with the new trucks.

Rented storage space can be reduced by the amount of square footage saved in the new arrangement. Savings will be at the rate of $20 per square foot per year ($215.28 per square meter per year).

(a) Utilize the format shown in Figure 12.3 to calculate the after-tax cash flow. What are the payback period and rate of return?

(b) Of what value is the payback-period calculation as used in Problem 12.12a?

(c) Prepare a summary page for a letter of justification to support the aisle arrangement and fork-truck purchase.

(d) Assume that the savings in warehouse rental space are responsive to a 6 percent inflation rate. Use the format of Figure 12.4 to calculate the after-tax rate of return. Compare the IRR obtained with the answer to Problem 12.12a and comment on the comparison.

(e) To analyze the problem more completely, let the worth of storage space vary between −10 percent and +20 percent of the given values. Operating costs could double under unfavorable conditions. The fork trucks might last as long as 6 years or only for 3 years.

Calculate the before-tax rate of return under favorable, given, and unfavorable conditions. Make a recommendation based on the company's minimum acceptable before-tax rate of return of 20 percent.

12.13   The decision between buying and leasing a loader depends on the contract terms set by the seller. Identical loaders will be acquired by either method, and operating costs will be the same over the 3-year period that the asset will be used. The following information is available:

- The MARR is 10 percent.

- ACRS depreciations will be charged for 3 years.

- The salvage value is expected to be 40 percent of the purchase price.

- The effective corporate tax rate is 45 percent.

- Cash will be paid if the loader is purchased.

- A 3-year lease for the loader will cost $90,000, payable as $40,000, $30,000, and $20,000 at the beginning of years 1, 2, and 3.

What is the maximum price that can be paid for the loader to make purchasing more attractive than leasing?

12.14 A small construction company does not have sufficient capital to purchase equipment with cash. A needed piece of earth-moving equipment can be either leased or purchased on a plan that requires a 15 percent down payment on a 5-year loan for which interest charges are 8 percent of the total value each year; interest charges are paid each year, and the amount borrowed is repaid at the end of the fourth year.

The machine costs $40,000 and has a 25 percent salvage value at the end of 5 years. It is to be depreciated by ACRS depreciation. The company currently has an effective tax rate of 40 percent and expects a return on investment of 12 percent after taxes.

  (a) Conduct an after-tax analysis to determine how much could be paid annually for a constant-value lease that would produce the same present value as the loan-purchase plan.

  (b) Calculate the annual lease payment that would make the lease and loan plans equivalent if the loan were repaid in four equal installments (the interest rate is still 8 percent). Compare the lease amount to that calculated in Problem 12.14a.

EXTENSIONS

12A   Life Cycle Costing

A study that gives special attention to both direct and indirect cash flows over the *complete* life of a project is called *life cycle analysis*. The intent of life cycle analysis is to direct attention to factors that might be overlooked—especially inputs that occur during the inception stage, to get a project underway, and activities associated with the termination phase. The aspect of life cycle analysis introduced in this extension is life cycle *costing*. Product life cycle analysis is examined in Extension 14A.

Life cycle costing (LCC) has been a strength of engineering economics for over 50 years, but it was singled out for special recognition in the 1960s when the concept was adopted by United States government agencies as a means of enhancing the cost effectiveness of equipment procurement. The use of LCC has since spread to the commercial sector, for product development studies and project evaluations.

LCC is expected to reduce total cost by selecting the correct designs and components to minimize the *total* cost of service, not only the first cost. For instance, additional expenditures for the preliminary design might lower operating costs and thereby reduce total costs.

Expenditures during the life of most projects roughly follow the pattern shown in Figure 12.8. The first stage, design, accounts for research, engineering design, administration, and financing costs. The development stage takes the basic plan and converts it to hardware or services through charges for fabrication, installation, delivery, training, trial runs, and material purchases. After the process is established, operating costs required to keep it going include personnel, consumable supplies, overhead, maintenance, and services.

Design to cost is a variation of LCC that sets a limit on the total lifetime cost and forces designers to work backward from the disposal phase to the operating expenses through production costs and finally to initial design charges. For example, a government request for proposals for training services might specify a limit on the funds available. Submitted proposals that start from this upper limit would be expected to designate costs for course development, set-up

FIGURE 12.8
Cost-time distributions represent typical stages of life cycle costs for a product or project. Each stage has its own pattern. Both stage and aggregate effects are considered in life cycle analysis.

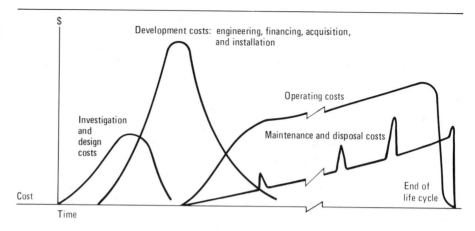

arrangements for the instruction, educational services and supplies per student, follow-up training, and evaluation procedures for determining the value of training provided. Proposals so structured provide costs by phases, which makes monitoring easier.

Projects, products, and systems of a similar nature tend to share common life cycle patterns. Once identified, the distributions of cash flow as a function of time that are exhibited by one undertaking may be used to estimate costs for a similar undertaking in the future. The cost-time distributions may also be helpful during the planning phase when trade-offs are evaluated, such as the balance between higher construction costs that allow lower operating and maintenance costs over the life of an investment. Studies of military hardware systems show that approximately two-thirds of life cycle costs are firmly established during the design stage. The LCC profile that results from aggregating individual cost-time distributions is subjected to standard discounted cash flow analysis to determine its economic attractiveness.

QUESTIONS

12A.1 Life cycle analysis is intuitive for most buyers when they are considering a major purchase. Very few customers look only at the sticker price of an automobile or the asking price for a house. They also examine the upkeep costs and resale value.

Construct a series of cost-time distributions that represent the life cycle costs of automobile ownership. Base it on a medium priced car ($10,000) that is driven 15,000 miles annually for 6 years. Assume that the car is purchased on a 4-year contract with a 20 percent down payment.

12A.2 Applications of LCC are described in "Life Cycle Costing and Associated Models," by W. J. Kolarik, in the *Proceedings* of the 1980 AIIE Conference:

The Columbian Rope Company of Auburn, New York, claims that by using a LCC design approach on their $7 million plant, they will save $5.5 million in terms of a 20 year life cycle for the plant. Estimated savings in buildings usually result from increasing the initial investment to curb energy consumption in the structure. However, many designers, architects, and engineers still find it difficult to convince developers and owners to consider LCC, rather than initial cost alone. Sometimes, as in the case of the 38 story Federal Reserve Bank of Boston, a LCC approach may actually reduce initial cost as well as operating and maintenance costs by forcing designers to become extremely cost conscious in all aspects of their design.

Discuss how a closer focus on all the costs associated with a product or service could lead to a more valuable economic analysis. Consider LCC with respect to resource shortages and environmental concerns that might affect "twilight" costs late in a life cycle.

12B Incumbency Theory

Conventional replacement studies focus on obvious cash flows: Comparison of an increment of investment to the savings it provides. There are other less apparent costs that influence the decision. Some of them are subjective. For example, prolonged irritation or poor performance during a critical period by a defender could cause so much frustration that every challenger would be preferred, regardless of cost. Conversely, a comfortable experience with a defender makes an owner reluctant to accept any challenger. More costs are objective. Consideration of the quantifiable factors that favor an incumbent asset over an unfamiliar challenger is the essence of *incumbency theory*.

The concept of an inherent automatic advantage attributable to an incumbent is intuitively logical. Unless there is clear displeasure, users are reluctant to change brands or do things in a new way. Challengers recognize this inertia and attempt to overcome it by price-cutting and advertising to create dissatisfaction.

Incentives to switch to a challenger are required because a satisfactorily performing incumbent has proven itself; its risk of poor performance is low. Only when a challenger can demonstrate that it will provide significantly better performance, which is difficult to do without a trial, can it gain favor. Longer periods of acceptable performance reinforce confidence in the incumbent, making it evermore difficult to dislodge, since a rational decision maker strives to minimize risk.

It has been estimated that a price differential of 2 to 6 percent is needed to overcome the practical advantages of staying with an adequate incumbent. Although many of the vantages are difficult to measure, most can be quantified. Incumbents usually enjoy benefits of

- *Standardization.* Lower costs for issuing orders for parts, repairs, and replacements because purchasing procedures and the suppliers' response are known; fewer documents to catalog and spares to store, less chance of damage owing to confusion of operators; knowing what to do when an unusual incident occurs because someone in the system has prior knowledge about the "standard" asset.

- *Familiarity.* Produces savings owing to reduced operating time gained from experience, such as immediate recognition of what causes a failure, the ability to make temporary repairs, and avoidance of bad habits exposed over time; no learning time needed.

- *Productivity.* Group gains attributable to less disorder and disruption avoided by retaining known practices; higher morale gained from confidence in the incumbent's performance; reliability of the system maintained by not introducing new products which might have unanticipated deficiencies that could cause damage.

Incumbency theory thus engages factors that are inconveniently reduced to dollar values and probabilities but still deserve attention. Some companies apply an informal rule that challengers in specified categories must have a purchase price lower than the incumbent by a certain percentage, say 3 to 5 percent, before they are even considered as replacements. In general, incumbency theory is pertinent more for supplies and low-cost products than for major investments.

QUESTIONS

12B.1 A consulting firm furnishes personal computers for all its engineers. The same computer model has been used for 3 years with excellent performance. Now a sales representative from a competing computer manufacturer offers to replace all the incumbents at a price 10 percent lower than that by which they can be replaced from the original manufacturer. The two makes of computers are not compatible. What conditions does incumbency theory suggest for evaluating the proposal?

12B.2 Describe three situations in which an incumbent is widely believed to hold a distinct advantage over challengers, such as the advantage held by elected office holders over challengers trying to unseat them. Comment on the reasons why the advantages exist in each case by considering risk, cost, familiarity, productivity, and standardization.

CHAPTER 13

EFFECTS OF INFLATION

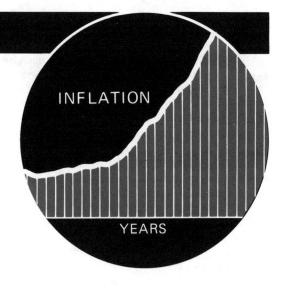

OVERVIEW

Inflation causes prices to rise and decreases the purchasing power of a unit of money with the passage of time. Deflation has the opposite effect. Inflation rates are measured by the Wholesale Price Index, Implicit Price Index, and Consumer Price Index, the last being the most quoted. The impact of inflation has become quite severe in the early 1980s in the United States, suggesting the desirability of including inflation effects in analyses. *Real*-dollar (constant purchasing power) or *actual*-dollar (future exchange) cash flows can be used in an analysis. The latter is recommended for its practicality, understandability, and versatility. The MARR used in a study must be consistent with the method of handling inflation. The same solution results from calculations using either real or actual dollars.

The most convenient way to account for inflation is to estimate real-dollar cash flows and apply a uniform escalation rate to them to convert them to actual dollars. However, since not all prices escalate uniformly or at the same rates, explicit rate estimates for certain cash flow components may be necessary. Actual-dollar cash flow for all components may be converted back to real dollars by discounting at the general inflation rate or at a customized composite rate.

An after-tax evaluation provides a more accurate assessment of the effects of inflation because it accounts for cash flow components that are not responsive to inflation—loan repayments, leases, and depreciation charges. Leasing versus buying is a frequently encountered example of an inflation-sensitive decision. In an inflation-prone economy it becomes even more imperative to conduct inflation-sensitive evaluations in order to detect proposals that promise savings in high-inflation operations.

CONCEPTS OF INFLATION

Inflation is a general increase in the price level. Equivalently, inflation results in a decline over time in the purchasing power of a unit of money. An individual perceives inflation as higher prices for food, cars, and other purchased commodities and services. In the case of food, inflation is a creeping increase in the cost of a necessity. It is worrisome but tends to seem inevitable. For larger purchases made at longer intervals, escalations are more startling and possibly more dismaying. In both cases inflation has eroded the purchasing power of savings and earnings, if interest rates and salary raises have not kept pace with general price trends. The same effects are felt by business and government.

Deflation is the opposite of inflation. The last time prices declined in the United States was in 1954. During the next 30 years prices rose by 289 percent. A dollar worth 100 cents in purchasing power in 1954 shrank to a worth of only 26 cents by 1984. And the decline continues. The causes and consequences of high inflation rates are examined in Extension 13A.

The awkward term disinflation *characterizes a condition where prices rise at a progressively lower rate.*

MEASURING INFLATION

Inflation is difficult to measure because the prices of different goods and services do not increase or decrease by the same amount, nor do they change at the same time. During a recent 15-year period, the average cost of a day's stay in a hospital increased by 796 percent and the price of a pound of hamburger rose by 305 percent. Meanwhile, the price of a 19-inch black-and-white TV set decreased by 17 percent. The calculation of a general inflation rate is further complicated by geographical differences in prices and changeable buying habits of consumers. Government statisticians attempt to overcome these difficulties by collecting data that profile the types and amounts of expenditures made by a middle-income family. Prices for these goods are obtained monthly and averaged according to demographic distributions. Then the prices are weighted according to the expenditure proportions of the typical family. The result is the *Consumer Price Index* (CPI).

The federal Department of Commerce (Bureau of Economic Analysis) and Department of Labor (Bureau of Labor Statistics) compile several indexes to measure inflation. The Consumer Price Index reveals the effect of retail price changes on a selected standard of living. The *Wholesale Price Index* (WPI) measures inflation at the wholesale level for both consumer and industrial goods, but not services. The *Implicit Pride Index* (IPI) is designed to show the effect of general price-level changes on the Gross National Product (GNP), the total market value of all goods and services produced by a nation's economy. The CPI is the most-used measure of prices in the United States and indicates about the same rates as the IPI. A 30-year history of the CPI is shown in Figure 13.1.

Seven categories are measured in the CPI "fixed market basket of goods;" these are:
1 Foods and beverages
2 Housing
3 Apparel and upkeep
4 Transportation
5 Medical care
6 Entertainment
7 Other—education, personal care, etc.

Many benefit programs are tied to the Consumer Price Index. For example, a higher CPI triggers increases in social security and COLA (cost-of-living adjustment) payments.

FIGURE 13.1
A 30-year climb in the Consumer Price Index and a 12-year tabulation of annual inflation rates.

Numerical values of the
Consumer Price Index are listed
in Table 13.7.

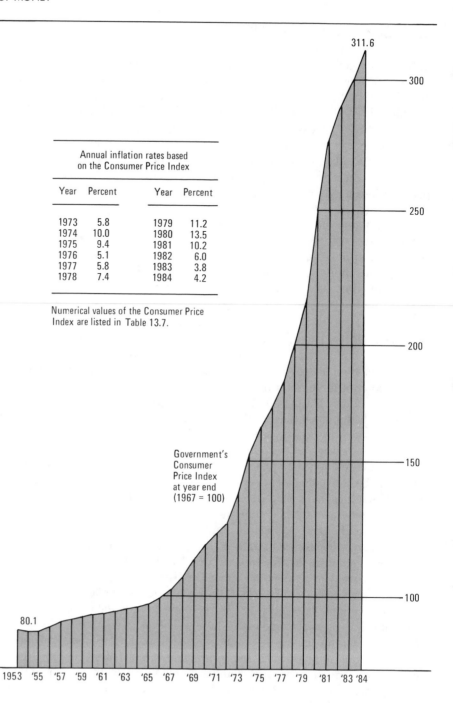

Annual inflation rates based
on the Consumer Price Index

Year	Percent	Year	Percent
1973	5.8	1979	11.2
1974	10.0	1980	13.5
1975	9.4	1981	10.2
1976	5.1	1982	6.0
1977	5.8	1983	3.8
1978	7.4	1984	4.2

Numerical values of the Consumer Price
Index are listed in Table 13.7.

311.6

300

250

200

Government's
Consumer
Price Index
at year end
(1967 = 100)

150

100

80.1

1953 '55 '57 '59 '61 '63 '65 '67 '69 '71 '73 '75 '77 '79 '81 '83 '84

Although the indexes measure price changes that have already occurred, they are useful in projecting future price trends. The historical data suggest the general movement of costs. If an index moves from 200 to 216 in 1 year, the rate of increase is $(216 - 200)/200 = 0.08$, or 8 percent. Since a trend over several periods is usually a better indicator for inflation expectation, an annual compound rate of growth is calculated. For an index that has risen from 176 to 216 over the past 3 years, the price trend or inflation rate f is

$$176(1 + f)^3 = 216$$

$$(1 + f)^3 = 216/176 = 1.2273$$

$$1 + f = \sqrt[3]{1.2273} = 1.071$$

$$f = 0.071 \quad \text{or} \quad 7.1\%$$

IMPACT OF INFLATION ON ECONOMIC EVALUATIONS

Years ago when inflation was a modest 2 to 4 percent per year, it was generally ignored in economic evaluations of proposals. It was argued that all proposals were affected similarly by price changes and there was too little difference between current and future costs to influence the order of preference. These arguments lose substance when inflation reaches double-digit levels and some goods and services escalate much more rapidly than others.

Once analysts recognize that inflation has an impact on most investment opportunities and therefore deserves consideration in their appraisals, they must decide on the most appropriate method with which to include it. There are two basic methods, with a number of refinements available for each.

1 Eliminate inflation effects by converting all cash flows to money units that have constant purchasing power, called *constant*, or *real*, dollars. This approach is most suitable for before-tax analysis, when all cash flow components inflate at uniform rates.
2 Estimate cash flows in the amount of money units actually exchanged at the time of each transaction. These money units are called *future, then-current*, or *actual* dollars. The actual-dollar approach is generally easier to understand and apply and is more versatile than the real-dollar method.

In an analysis by either method it is critical that the assumptions made in determining the cost of capital, and hence the minimum acceptable rate of return, correspond to inflation rates used in the study. As we observed in Chapter 9, the weighted average of interest rates representing all sources of funds acquired by an organization sets the lower bound for the organization's

MARR. To this is added a premium to account for the risk of investing in a proposal. Anticipation of future inflation affects expectations of returns on invested capital and the degree of risk that is acceptable. In effect, this anticipation is accounted for by adding an increment to the cost of capital to offset the effects of inflation. This offset is reflected in the MARR.

The impact of inflation on interest rates is discussed in Extension 13B.

Until the middle 1960s, 7 percent was considered to be a reasonable cost of money for a corporation. What little inflation occurred prior to that period was largely compensated for by productivity increases. As inflation increased, investors sought higher returns as compensation for buying power lost to price escalation. Since a firm's cost of money is intimately linked to the rate of return that investors deem satisfactory for their investment, the cost of capital rises in accordance with experienced and expected inflation.

The inflation adjustment to the cost of capital, which manifests itself in the minimum acceptable rate of return, must be consistent with the estimates of future cash flows used in economic evaluations. If the MARR includes an increment for 5 percent inflation, but estimates of future costs are based on a higher inflation rate, a study would be biased toward accepting proposals. Conversely, if cash flow is stated in real dollars and the MARR contains an inflation adjustment, a proposal is penalized by excessive discounting. The bias results because inflating cash flows and discounting them are counteracting procedures.

Example 13.1
Inconsistency between Inflation Estimates for the Cost of Capital and for Cash Flow

A proposal with an initial cost of $2000 is expected to produce net returns of $850 per year for 3 years in *real* dollars. The minimum acceptable rate of return has been raised from 10 percent to 15 percent in response to an adjustment in the cost of capital based on anticipation of 5 percent inflation during the next few years. Should the proposal be accepted?

Solution 13.1

If the cash flow were estimated in real dollars of constant purchasing power and the MARR were adjusted for inflation, the proposal would be rejected because its present worth would be negative:

$$PW = -\$2000 + \$850(P/A, 15, 3)$$
$$= -\$2000 + \$850(2.2832) = -\$59$$

This evaluation unfairly penalizes the proposal because the MARR is based on the assumption that the cash flow will state the *actual* amount received each year. Real dollars can be converted to actual dollars by inflating them to an amount that is equivalent in purchasing power to their value today.

For 5 percent annual inflation, the future cash flow equivalent to constant purchasing power of $850 a year is

End of Year	Real Dollars (No Inflation)	5% Inflation		Actual Dollars (Inflated Cash Flow)
0	−$2000			
1	850	× 1.05	=	$893
2	850	× $(1.05)^2$	=	937
3	850	× $(1.05)^3$	=	984

The inflated cash flow indicates that it would take $984 three years from now to acquire goods that could be purchased today for $850. When the inflated receipts are discounted at the inflation-adjusted MARR, the proposal has an acceptable present worth:

$$PW = -\$2000 + \$893(P/F, 15, 1) + \$937(P/F, 15, 2)$$
$$+ \$984(P/F, 15, 3)$$
$$= -\$2000 + \$893(0.86957) + \$937(0.75614)$$
$$+ \$984(0.65752)$$
$$= \$132$$

The proposal merits approval.

BEFORE-TAX CONSTANT-VALUE COMPARISONS

Actual-dollar flow indexed to a base year is frequently used to compare economic performance in different years. For instance, 1982 could be the base year for measuring the productivity of a firm. If the Consumer Price Index were used to convert actual-dollar flow in future years to real-dollar amounts in the base year, the output and input figures in 1983 would be deflated to their 1982 worth by using a CPI "deflator":

$$CPI \text{ deflator}(1983 \rightarrow 1982) = 1 + \frac{CPI(1983) - CPI(1982)}{CPI(1982)}$$
$$= 1 + \frac{299.0 - 288.1}{288.1} = 1.038$$

Then an output of, say, $402,744 in 1983 actual dollars would be deflated to $402,744/1.038 = $388,000 to compare it on an equivalent basis with an output of, say, $377,000 in 1982. Output has thus increased by $388,000 - $377,000 = $11,000 for the year *in constant terms*.

It is generally easier to estimate future costs in constant dollars because the estimator is familiar with today's values. It is a simple matter to convert estimates in real-dollar flow to actual-dollar flow when inflation is assumed to be a constant rate. Consider the two proposals shown below for which the estimates have been made in real dollars.

For several years up to the 1970s, inflation rates were quite low and stable. Since then few analysts put much faith in long-range inflation-rate forecasts. The only safe prediction is that they will not be negative.

| | Cost | Cash Flow, Real Dollars | | | | PW at |
	Year 0	Year 1	Year 2	Year 3	Year 4	12%
Proposal A	−$10,000	$4000	$4000	$4000	$4000	
Proposal B	−14,000	5500	5500	5500	5500	
Net difference	−$4,000	$1500	$1500	$1500	$1500	$556

The net difference might also have been estimated in actual dollars to show what receipts would have to be in years ahead to equal today's purchasing power. If inflation during the next 4 years is expected to be 6 percent per year, a consistent estimator would place the net difference as

	Cost	Cash Flow, Actual Dollars			
	Year 0	Year 1	Year 2	Year 3	Year 4
Net difference be- tween proposal A and proposal B	−$4000	$1590	$1685	$1787	$1894

The procedure for evaluating proposals stated in actual dollars is to convert them to real dollars. Doing so to the actual dollars above, by applying the $(P/F, 6\%, N)$ factor, naturally converts them to the previously given real dollars. By either approach the present worth of the additional investment in proposal B is $556.

The same present worth is obtained by determining a combined discount rate that represents both the minimum required rate of return and the inflation rate. Let i = rate of return and f = inflation rate; then the combined rate is $(1 + i)(1 + f) - 1$. For the rates from the previous examples, where $i = 12$ percent and $f = 6$ percent,

$$\text{Combined interest-inflation rate} = (1.12)(1.06) - 1 = 1.1872 - 1$$
$$i_f = 0.1872 \quad \text{or} \quad 18.72\%$$

Applying this rate to the given net difference expressed in actual dollars produces the same present worth as previously calculated and confirmed in Table 13.1.

TABLE 13.1

Combined interest-inflation rate applied to an actual-dollar cash flow to obtain the before-tax present worth

Year, N	Cash Flow in Actual Dollars	$(P/F, 18.72, N)$	Present Worth
0	−$4000	1.00000	−$4000
1	1590	0.84237	1339
2	1685	0.70964	1195
3	1787	0.59786	1068
4	1894	0.50372	954
			$ 556

Example 13.2
Equivalence of Real-Dollar and Actual-Dollar Cash Flow in a Before-Tax Analysis

A productive asset can be purchased for $100,000. It will have no salvage value at the end of its 5-year useful life. Operating costs will be $12,000 per year while it provides a revenue of $40,000 annually. Estimates are based on current economic conditions without consideration of price or cost escalations. Evaluate the proposed purchase according to the real-dollar data and actual-dollar cash flow when the inflation rate is 8 percent. The MARR is 15 percent without an adjustment for inflation, and taxes are not included in the analysis.

Solution 13.2

The real-dollar cash flow is composed of an immediate $100,000 outlay followed by net receipts of $40,000 − $12,000 = $28,000 at the end of each of the next 6 years. This flow, discounted at 15 percent, yields

$$PW = -\$100,000 + \$28,000(P/A, 15, 5)$$
$$= -\$100,000 + \$28,000(3.3521) = -\$6.141$$

Real dollars are converted to actual dollars by applying the inflation factor $(1 + f)^N$ to each of the annual receipts, where N equals the year in which the receipt occurs. The resulting actual-dollar cash flow is discounted by the combined interest-inflation rate i_f to obtain the present worth of the "then-current" dollars; $i_f = (1.15)(1.08) - 1 = 0.242$. The calculations are shown in Table 13.2.

The present worths from Table 13.2 and the PW formula agree, of course, because

$$(P/F, 24.2, N) = (P/F, 8, N)(P/F, 15, N)$$

The combined interest-inflation rate thus represents an inflation-adjusted MARR applied to an inflation-adjusted cash flow.

TABLE 13.2

Present worth of actual-dollar cash flow when discounted by the combined interest-inflation factor to determine its real-dollar equivalence

End of Year, N	Real-Dollar Cash Flow	Inflation Factor, $f = 8\%$ $(F/P, 8, N)$	Actual-Dollar Cash Flow	Combined Interest-Inflation Factor $(P/F, i_f, N)$	Present Worth of Cash Flow
0	−$100,000		−$100,000		−$100,000
1	28,000	1.0800	30,240	0.80516	24,348
2	28,000	1.1664	32,659	0.64827	21,172
3	28,000	1.2597	35,272	0.52196	18,410
4	28,000	1.3604	38,091	0.42025	16,008
5	28,000	1.4693	41,141	0.33838	13,921
					−$6,141

AFTER-TAX ACTUAL CASH FLOW COMPARISONS

Two weaknesses limit the usefulness of the preceding approach: Tax effects are ignored and no provision is made for differences in escalation rates among price and cost components.

Variable interest rate loans are occasionally used, as in adjustable-rate mortgages (See Extension 13B).

Tax effects are significant because deductions allowed for depreciation and loan interest are *not responsive* to inflation. That is, depreciation is based strictly on the purchase price of an asset, not its inflation-elevated replacement price, and interest payments on loans are set by contract in actual dollars that are not subject to correction for inflation. Because borrowed money is received in real dollars and is repaid in actual dollars, borrowers benefit from inflation unless lenders require an inflation-adjusted interest rate.

Price instability is a condition in which prices for goods and services do not change proportionately over time. Earlier in this chapter differences in escalation rates were observed for hospital services, hamburger, and TV sets. When price instability is significant for factors in an economic study, it can affect the preference among alternatives. For instance, two alternatives with identical real-dollar cash flows would be equally promising, but if the revenue for one of them resulted from energy savings that inflated 20 percent annually, while revenue for the other came from labor savings that escalated 10 percent per year, the energy-saving proposal would be preferred.

Nonresponsive Charges in After-Tax Analysis

Several financial commitments are made in real dollars but are not responsive to inflation, primarily loans, leases, and depreciation. Loan and leases that specify actual-dollar cash flows benefit borrowers and lessees when inflation rises faster than anticipated in the agreements. The reverse is true during disinflation, when interest rates and costs drop below those agreed to in contracts. The significance of nonresponsive cash flows is demonstrated in Example 13.3.

Example 13.3
Difference in After-Tax Present Worth Caused by Nonresponsive Cash Flows during Inflationary Periods

The proposal described in Example 13.2 is to be subjected to an after-tax analysis. Both earnings and expenses are responsive to the general inflation rate of 8 percent. Depreciation is by the ACRS method, and the tax rate is 40 percent. Calculate the after-tax present worth when the unadjusted MARR is still 15 percent.

Solution 13.3

Actual-dollar before-tax cash flow (BTCF) based on 8 percent inflation is shown in column 2 of Table 13.3. Depreciation charges in column 3 are subtracted from column 2 to set the taxable income for the year in column 4. Taxes are calculated in column 5 and deducted from the BTCF to reveal the actual-dollar after-tax cash flow (ATCF) in column 6. It is then necessary to convert the actual dollars back to real dollars, as shown in columns 7 and 8. Finally, the real-dollar ATCF is discounted at the inflation-free 15 percent rate to yield the present worth, column 10.

TABLE 13.3

PW calculation procedure for an after-tax analysis which includes
components that are not responsive to inflation

End of Year, N (1)	BTCF, Actual Dollars (2)	Depreciation (3)	Taxable Income $(2)-(3)=(4)$	Taxes at 40% (5)	ATCF, Actual Dollars $(2)-(5)=(6)$
0	−$100,000				
1	30,240	$20,000	$ 10,240	$4,096	26,144
2	32,659	32,000	659	264	32,395
3	35,272	24,000	11,272	4,509	30,763
4	38,091	16,000	22,091	8,836	29,255
5	41,141	8,000	33,141	13,256	27,885

End of Year, N (1)	Inflation Factor, (P/F, 8, N) (7)	ATCF, Real Dollars $(6)\times(7)=(8)$	Discount Factor, (P/F, 15, N) (9)	Present Worth $(8)\times(9)=(10)$
0		−$100,000		−$100,000
1	0.92593	24,208	0.86975	21,050
2	0.85734	27,774	0.75614	21,001
3	0.79383	24,421	0.65752	16,057
4	0/73503	21,503	0.57175	12,294
5	0/68059	18,978	0.49718	9,435
				−$20,158

The same PW solution would have resulted from discounting column 6 by the combined interest-inflation factor, 24.2 percent, as was done in Table 13.2. However, columns 7 to 9 descriptively display how the actual-dollar ATCF is converted to real dollars to allow discounting at the unadjusted MARR, 15 percent. It should be apparent that inflation has not increased the real value of the net revenue; column 2 shows the number of actual dollars required in each year to equal the buying power of $28,000 per year in real dollars. Therefore, actual-dollar cash flow must be discounted by an inflation-adjusted MARR.

The present worth of the proposal is lower when inflation is accounted for because the depreciation charges do not escalate in tandem with revenue. The result is higher annual taxes. Consequently, a more realistic evaluation is obtained from actual-dollar after-tax data when inflation is significant.

Price Instability in an After-Tax Analysis

There are often one or more components in a cash flow stream that have escalation rates prominently different from the general inflation rate. The flow patterns for such components may be stated as specific estimates of actual-dollar transactions or in terms of specific inflation rates that vary from the general rate. Specific enumeration of annual flows permits year-by-year variations in the rate of escalation, whereas assigning a stable inflation rate commits the cash flow to a continuous growth pattern. Both options are accommodated by the procedure demonstrated in Example 13.4.

Example 13.4
Evaluation of a Proposal in which Cash Flow Components Escalate at Different Rates

The conditions for the proposal described in Example 13.3 remain the same, except for the inflation rate of revenue. It is now believed that revenue will escalate at a 20 percent rate while expenses will continue to increase at the general inflation rate of 8 percent. Determine the present worth under the new assumption.

Solution 13.4

Recall from Example 13.2 that the net receipts of $28,000 per year resulted from annual revenue of $40,000 and annual expense of $12,000; then actual-dollar revenue in Table 13.4 (column 2) is obtained by applying an inflation factor of $(F/P, 20, N)$ for each N year, and actual expense (column 3) in year N is $12,000(F/P, 8, N)$. Taxable income is the difference between revenue and the sum of expense plus depreciation (column 6). After-tax actual dollars are discounted at the combined interest-inflation rate of 24.2 percent (col-umn 9) to determine the present worth of the revised proposal (column 10).

It is no surprise that the proposal's PW increases notably, compared with Example 13.3, from the disproportionately rapid rise in revenue. The rationale for using the general inflation rate to discount actual dollars to their real-dollar equivalency is that the overall inflation rate is the weighted average of all price escalations and is therefore a reasonable standard for uniformly deflating all cash flow components.

The general inflation rate may not be the most appropriate discount factor to use in certain studies. A unique inflation-factor index, a composite inflation rate for a given firm or area, could be calculated for any organization from the weighted average of price changes that the organization expects for its operations. Because escalations vary with geography and product mix, a customized inflation factor would more accurately represent future cash flow expectations. The composite rate is calculated from forecasted inflation rates for critical cash flow components weighted according to the proportion of utilization of each component. An example of the computation and use of a custom-built composite inflation index is given in Review Exercise 3.

AFTER-TAX MODIFIED CASH FLOW COMPARISON

It is also possible to utilize real dollars for an after-tax evaluation under the conditions stipulated for the proposal in Example 13.4. To do so, it is necessary to modify all components that inflate or deflate at a rate different from the general inflation rate. In Example 13.4, revenue escalates faster than the general inflation rate, and depreciation charges are constant (which means they deflate with respect to the general inflation rate). Therefore, these revenue and depreciation components must be modified to make their value correspond to the real-dollar value of expense, which is the only component expected to escalate at the general inflation rate.

TABLE 13.4

Calculation of the present worth of a proposal when revenue inflates at 20 percent and expenses at the general inflation rate of 8 percent. After-tax actual dollars are deflated at the combined interest-inflation rate, $i_f = 24.2$ percent.

End of Year, N (1)	BTCF in Actual Dollars			
	Revenue at $f=20\%$, $40,000 (F/P, 20, N) (2)	Expenses at $f=8\%$, $12,000 (F/P, 8, N) (3)	Net Cash Flow (2)−(3)=(4)	Depreciation (5)
0		$100,000	− $100,000	
1	$48,000	12,960	35,040	$20,000
2	57,600	13,997	43,603	32,000
3	69,120	15,116	54,004	24,000
4	82,944	16,325	66,619	16,000
5	99,532	17,632	81,900	8,000

End of Year, N (1)	Taxable Income (4)−(5)=(6)	Taxes at 40% (7)	ATCF, Actual Dollars (4)−(7)=(8)	Discount Factor, $i_f = 24.2\%$ (P/F, 24.2, N) (9)	Present Worth (8)×(9)=(10)
0			− $100,000		− $100,000
1	$15,040	$ 6,016	29,024	0.80516	23,369
2	11,603	4,641	38,962	0.64827	25,257
3	30,004	12,002	42,002	0.52196	21,923
4	50,619	20,248	46,371	0.42025	19,487
5	73,900	29,560	52,340	0.33838	17,711
					$7,747

Real-dollar equivalence can be calculated with the geometric-series factor introduced in Chapter 2. An explicit growth rate g higher than the general inflation rate f conforms to case 1:

$$i^* = \frac{1 + g}{1 + f} - 1$$

When g is less than f, case 3 applies:

$$i^* = \frac{1 + f}{1 + g} - 1$$

The special interest rate i^* is then used in the expression $(1 + i^*)^N$ to inflate

TABLE 13.5

Modified real-dollar analysis of the proposal described in Example 13.4. The special interest rate for revenue is $(1.20/1.08) - 1 = 0.1111$ (where $g = 20$ percent and $f = 8$ percent) and i^* for depreciation charges is $(1.08/1.00) - 1 = 0.08$ (where $g = 0$ and $f = 8$ percent). Revenue is inflated at 11.11 percent annually (column 2), and depreciation charges are deflated at 8 percent per year (column 5). After-tax cash flow (column 8) is discounted at MARR $= 15$ percent to obtain the present worth (column 9).

	BTCF in Modified Real Dollars							
End of Year, N (1)	Revenue, $i^*=0.1111$, $\$40{,}000(1 + i^*)^N$ (2)	Expenses (3)	Net Cash Flow (4)	Modified Depreciation, $i^* = 0.08$, Depr.$(1 + i^*)^{-N}$ (5)	Taxable Income $(4) - (5) = (6)$	Taxes at 40% (7)	AFTC in Modified Real Dollars $(4) - (7) = (8)$	Present Worth, MARR $= 15\%$: $(8)(P/F, 15, N)$ (9)
0			−$100,000				−$100,000	−$100,000
1	$44,444	$12,000	32,444	$18,519	$13,920	$ 5,570	26,874	23,369
2	49,382	12,000	37,382	27,435	9,947	3,979	33,403	25,258
3	54,869	12,000	42,869	19,052	23,817	9,527	33,342	21,923
4	60,965	12,000	48,965	11,760	37,205	14,882	34,083	19,487
5	67,738	12,000	55,738	5,445	50,293	20,117	35,621	17,710
								$ 7,747

or deflate applicable cash flow components. The resulting after-tax cash flow is discounted at a MARR that has *not* been adjusted for inflation.

Table 13.5 shows a real-dollar analysis that corresponds to the actual-dollar analysis given in Table 13.4. Both methods produce the same present worth for the identical data. Either approach can be utilized, but the actual-dollar procedure is more descriptive, is easier to comprehend, facilitates irregular cash flows, and is generally more convenient to apply.

LEASE OR BUY?—AN INFLATION-SENSITIVE DECISION

Over 3000 years ago, Phoenician shipowners engaged in leasing. Today virtually any asset can be leased. The leasing of capital equipment has grown enormously in the past decade. There are many reasons why leasing may be more attractive than purchasing. These include:

- Use of an asset's services if a firm lacks cash or borrowing capacity to purchase the asset

- Less effect on future borrowing capacity than debt financing

- Imposition of fewer financial restrictions than accompany a loan for the purchase of a comparable asset

- Possible reductions in the risk of obsolescence and escalation in ownership costs due to inflation

- Tax advantages under certain conditions

The most frequently cited reasons for leasing are income tax advantages and that of freeing working capital by enabling the acquisition of needed equipment without going into debt.

From a tax angle, a lease is treated as an annual expense directly deductible from income; a purchase is an investment subject to a multiyear depreciation schedule. Many tax factors, such as investment credits and carryovers, make investing more or less attractive in comparison with leasing. Tax implications of inflation also affect the lease-buy question. Like a fixed-rate loan and depreciation charges, a lease contract that specifies fixed payments is insensitive to price escalations.

A *make-or-buy* decision, as depicted previously in Figure 1.6, is similar to the lease-or-buy question. Making something is an expense; buying it is an investment.

Types of Leases

A lease is a contract between the owner of an asset, called a *lessor,* and a *lessee,* who makes periodic payments for the right to use the asset. When there is also a lender involved who furnishes capital to the lessor to purchase the asset, the lease is said to be *leveraged.*

An *operating lease* can be canceled by either the lessee or lessor at any time after due notice has been given. A commitment by both parties to specified charges for the use of an asset for a definite period is a *financial lease*. Temporary use of vehicles, computers, and furniture typify operating leases. More expensive assets are usually subject to financial leases: real estate, railroad cars, airplanes, and construction equipment.

It is common practice for a firm that needs certain equipment to get a price agreement from the manufacturer or distributor and then find a lessor who will purchase the equipment. A commitment is made to pay to the lessor a specified rental that will assure the lessee use of the equipment for the period needed and provide an adequate return on the lessor's investment. It is also possible for a firm to sell an asset it already owns and lease it back from the buyer. This *sale and leaseback* arrangement allows a firm to obtain cash and still have the use of its asset.

Lease or Buy

A graph of a lease-buy situation is shown in Figure 15.3.

The comparison methods presented in previous chapters are applicable to the lease-or-buy question. The two alternatives can be evaluated according to their PW, EAW, or IRR. An after-tax analysis is usually more revealing. The *buy* alternative should include the loan cost if the asset is purchased with all or partly borrowed funds. The *lease* alternative typically includes the following elements:

$$\text{Cash flow (lease)} = \text{revenues} - \text{operating expenses} - \text{rental costs}$$
$$- (\text{tax rate})(\text{revenues} - \text{expenses} - \text{rentals})$$

Particulars for the calculations depend on the specific lease arrangement, such as who pays maintenance costs and which tax provisions are applicable.

An operating lease is normally considered to be an expense that is deductible from taxable income in the year it occurs. However, leases are subject to several criteria that determine their eligibility as deductions. For instance, a rental agreement cannot qualify as an operating lease if it transfers ownership to the lessee at the end of the lease term, has a bargain purchase option, or the lease term is equal to 75 percent of the estimated life of the new leased asset. As is the case for most tax questions, expert advice is needed when a situation does not fit standard prescriptions (see Review Exercise 4).

An example of a lease-buy analysis that considers inflation is given in Review Exercise 5.

The degree of protection from obsolescence and inflation a lessee achieves by leasing an asset depends on the contract. The lessor is also aware of the risks. A lease appears as a "bargain" to a lessee who views the future differently than the lessor and perceives an advantage. Which way the risk shifts as a result of the lease, if it shifts at all, is a function of forecasting ability.

Example 13.5
Leasing versus Buying

Two other options are available for acquiring the heavy construction equipment described in Example 12.5. It can be bought outright with equity funds or it can be leased. The lease would have an annual rental charge of $20,000 per year with an option to buy the equipment at the end of the eighth year for $5000. Assuming that the option is exercised, we must elect to expense the purchase in year 8.

Compare the two alternatives with the 8-year financing plan in Example 12.5. The previous tax rate of 40 percent, minimum rate of return of 10 percent, and depreciation schedule are applicable to the cash-purchase and lease alternatives. It is assumed that rental expense for the leased asset is deductible from income.

Solution 13.5

Of the three methods of acquiring the $100,000 construction equipment, the least expensive is still the 8-year loan plan. As shown in Table 13.6, a cash purchase using internal funds has an after-tax present worth of $72,794. The actual disbursements for the cash purchase are less than those for the other plans, but tax savings are less too; in effect, the government shares part of the interest cost through its taxing rules.

An 8-year lease has the highest total disbursements and the highest tax savings, but the total discounted cost is greater than that of the loan plan, largely because the tax savings are spread over the length of the lease to reflect the lease payments as operating costs instead of being concentrated in the early years as in an accelerated depreciation schedule.

DECIDING WHEN AND HOW TO CONSIDER INFLATION

Including the effect of inflation is a second-order refinement for economic evaluations; the first-order refinement was including the effect of taxes on basic cash flow. Many proposals do not deserve the additional inflation refinement, because their cash flows are too small, are immune from inflation, escalate in concert with competing proposals, or are so uncertain that fine tuning of cash flow is not justified. Other proposals, especially those subject to a wide range of price escalations among cash flow components, are obvious candidates for inflation-sensitive analyses. An indication of the sensitivity of a decision to the effects of inflation can be obtained by using the *sensitivity analysis* techniques presented in the next chapter.

In an inflation-prone economy the rewards from implementing proposals that reduce rapidly escalating costs are great. High rankings for these proposals rely on inflation being included in their evaluation. It is critically important that inflated cash flows be discounted according to an appropriate inflation adjustment to the minimum acceptable rate of return; a common error is to discount actual dollars at a MARR based on real-dollar values. Actual-dollar analysis is recommended for most applications, owing to its practicality, and an after-tax analysis is recommended because it reveals the effect of cash flow components that are not responsive to inflation.

The tedious mechanics of manually including inflation and tax effects in

TABLE 13.6

Comparison of acquiring the 10-year use of a $100,000 asset with no salvage value by cash purchase and by lease with an option to buy for $5000 after 8 years. Tax savings (2) for the cash purchase result from multiplying the annual depreciation charge (1) by the 40 percent tax rate; the total discounted cash flow (3) is the sum of the present worths of total tax savings plus the purchase price. Tax savings (5) for the lease-buy plan result from the 40 percent tax rate multiplied by the annual lease cost and the expensed purchase. The present worth of the cash flow (6) is the net summation [(5) + (4)] discounted at 10 percent.

End of Year	Cash Purchase			Lease-Buy Option		
	Annual Depreciation Charge (1)	Tax Saving (2)	PW of Cash Flow (3)	Annual Rental Charge (4)	Tax Saving (5)	PW of Cash Flow (6)
0			− $100,000			0
1	$ 10,000	$ 4,000	3,636	− $ 20,000	$ 8,000	− $10,909
2	18,000	7,200	5,950	− 20,000	8,000	− 9,917
3	16,000	6,400	4,808	− 20,000	8,000	− 9,016
4	14,000	5,600	3,825	− 20,000	8,000	− 8,196
5	12,000	4,800	2,980	− 20,000	8,000	− 7,451
6	10,000	4,000	2,258	− 20,000	8,000	− 6,774
7	8,000	3,200	1,642	− 20,000	8,000	− 6,158
8	6,000	2,400	1,120	− 20,000	8,000	− 5,598
				(5,000) *	2,000	− 1,400
9	4,000	1,600	679			
10	2,000	800	308			
Totals	$100,000	$40,000	− $72,794	$165,000	$66,000	− $65,419

*$5,000 cost at the end of year 8 to exercise the buy option for the equipment is treated as an expense in that year.

an analysis is alleviated by the use of computers. Appropriate programs can be readily written because the analysis procedure is systematic. Estimating future cash flows and inflation rates is a more serious difficulty, particularly in tense times when prices spurt ahead spasmodically as a result of shortages, political events, and other causes of economic turbulence. However, these same disturbances strongly underline the importance of considering inflation effects for major economic evaluations.

Review Exercises and Discussions

Exercise 1 The rise in the United States Consumer Price Index from 1966 to 1984 is shown in Table 13.7. What return on investment would be required to have a real rate of return of 7 percent during the 1977 through 1979 period?

Solution 1 The compound rise in the CPI for 3 years starting in 1977 was

TABLE 13.7

Record of United States Consumer Price Index figures

Year	CPI	Year	CPI	Year	CPI	Year	CPI
1966	94.5	1971	121.3	1976	170.5	1981	271.8
1967	100.0	1972	125.3	1977	181.5	1982	288.1
1968	104.2	1973	133.1	1978	195.4	1983	299.0
1969	109.8	1974	147.7	1979	217.4	1984	311.6
1970	116.3	1975	161.2	1980	246.8		

$$170.5(1 + f)^3 = 214.1$$

$$f = \sqrt[3]{\frac{214.1}{170.5}} - 1 = 0.08 \quad \text{or} \quad 8\%$$

Combining the real interest rate and the inflation rate yields an inflation-adjusted rate of return of $(1.07)(1.08) - 1 = 0.1556$, or 15.56 percent.

It has now been suggested that the proposal described in Examples 13.2 and 13.3 can be partially financed by a loan of $60,000. The annual interest charge is 10 percent of the amount borrowed, with the principal to be repaid at the end of the fifth year. Evaluate the proposal with after-tax analyses that (1) ignore inflation and (2) recognize an 8 percent general inflation rate. **Exercise 2**

Given the original BTCF data of $-\$100,000$ at time zero followed by net receipts of $28,000 per year, the inflation-free cash flow is changed by the $60,000 loan to $-\$100,000 + \$60,000 = -\$40,000$ at time zero; by annual interest payments of $\$6000 \times 0.10 = \$6000$ that reduce the net income to $\$28,000 - \$6000 = \$22,000$ during each of the years; and by repayment of the loan at the end of year 5, $-\$60,000$. Under these conditions the before-tax PW at MARR = 15 percent is **Solution 2**

$$\text{PW} = -\$40,000 + \$22,000((P/A, 15, 5) - \$60,000(P/F, 15, 5) = \$3915$$

After-tax inflation-free cash flow is

Year	0	1	2	3	4	5
AFTC	$-\$40,000$	$21,200	$26,000	$22,800	$19,600	$-\$43,600$

where the income of $16,400 in year 5 is offset by the loan repayment of $60,000. At MARR = 15 percent, the PW = $2615, which is less, of course, than the before-tax present worth.

The after-tax analysis that includes inflated cash flow is shown in Table 13.8. The feature of the table is that net revenue reacts to inflation while loan interest and repayment are fixed in real dollars according to the loan agreement, as are depreciation charges fixed by law.

A comparison of the present worth when inflation is included ($9688) to the present

TABLE 13.8

After-tax cash flow based on actual dollars and 8 percent inflation when a $60,000 loan finances part of the proposal described in Examples 13.2 and 13.3

End of Year, N (1)	BTCF, Real Dollar (2)	BTCF, Actual Dollar, f=8% (2)(F/P,8,N)=(3)	Loan and Interest (4)	Depre- ciation (5)	Taxable Income (3)+(4)−(5)=(6)	Taxes at 40% (7)
0	−100,000	−$100,000	$60,000			
1	28,000	30,240	−6,000	$20,000	$4,240	1,696
2	28,000	32,659	−6,000	32,000	−5,341	−2,136*
3	28,000	35,272	−6,000	24,000	5,272	2,209
4	28,000	38,091	−6,000	16,000	16,091	6,436
5	28,000	41,141	−6,000	8,000	27,141	10,856
5			−$60,000			

End of Year, N (1)	ATCF, Actual Dollars (3)+(4)−(7)=(8)	Inflation Factor, (P/F, 8, N) (9)	ATCF, Real Dollars (8)×(9)=(10)	Discount Factor, (P/F, 15, N) (11)	Present Worth (10)×(11)=(12)
0	−$40,000		−$40,000		−$40,000
1	22,544	0.92593	20,874	0.86957	18,147
2	28,795	0.85734	24,687	0.75614	18,667
3	27,163	0.79383	21,563	0.65752	14,178
4	25,655	0.73503	18,857	0.57175	10,781
5	24,285	0.68059	16,528	0.49718	8,217
5	−60,000	0.68059	−40,835	0.49718	−20,302
					$9,688

*Negative taxes are considered to be tax offsets against other income in year 2.

worth previously calculated without considering inflation ($2615) shows how gains to the borrower which resulted from repaying the loan principal and interest with inflated dollars more than overcome losses from the real-dollar depreciation schedule.

It is enlightening to observe the present worths that have resulted from the modifications made to the original data given for the proposal in Example 13.2. They are summarized below and clearly underscore how the worth of a proposal depends on what is included in its evaluation.

Comparison Conditions	PW (No Loan)	PW (with Loan)
Before-tax, no inflation	−$ 6,141	$3915
After-tax, no inflation	−$15,485	$2615
After-tax, 8% inflation	−$20,158	$9688
After-tax, 8% expense inflation and 20% revenue inflation	$ 7,747	

A company has developed its own inflation-factor index to allow closer evaluation of future cash flows. The index was prepared by categorizing costs of production, determining the proportion spent in each category, and forecasting an inflation rate for each category according to local conditions. Then the inflation-factor index was calculated as the weighted average of all the inflation factors. The result is shown below.

Exercise 3

Cost Factors	Budget Proportion	Inflation Rate, %	Weighted Rates
Labor (for the company work force)	0.25	8	2.0
Material 1 (raw materials)	0.10	5	0.5
Material 2 (subassemblies)	0.10	15	1.5
Energy (electricity, fuel, etc.)	0.25	20	5.0
General services (general infla-tion rate)	0.30	10	3.0
		Inflation-factor index	12.0%

A proposal is being evaluated to reduce labor costs and material waste by the purchase of automated equipment. The equipment is expected to be in service for 3 years with no salvage value. ACRS depreciation is used, and the tax rate for the company is 50 percent. Data from the proposal are summarized below.

Annual revenue

Net savings in labor	$ 60,000
Net savings in material	$140,000
First cost	$150,000

Annual costs

Additional energy required	$70,000
Added maintenance expense	$30,000

Evaluate the proposal.

After-tax calculations that include the effects of the unique individual inflation rates are shown in Table 13.9. Each revenue and cost component is projected from real dollars to actual dollars by its exclusive inflation rate (maintenance falls within the general-service

Solution 3

TABLE 13.9

After-tax cash flow using exclusive inflation factors and an inflation-factor index

Year, N	BTCF, Real Dollars	BTCF, Actual Dollars	Depre-ciation	Taxable Income	Taxes at 50%	ATCF, Actual Dollars	ATCF, Real Dollars at $f = 12\%$
0	−$150,000	−$150,000				−$150,000	−$150,000
1	100,000	94,800	$49,500	$45,300	$22,650	72,150	69,420
2	100,000	87,234	67,500	19,734	9,867	77,367	61,676
3	100,000	77,624	33,000	44,624	22,312	55,312	39,370

category, where the inflation rate is 10 percent). Note that the BTCF in actual dollars is lower than that in real dollars because expenses increase at a faster rate than do savings. Actual dollars are converted back to real dollars in the ATCF by applying the inflation-factor index, 12 percent.

The BTCF in actual dollars for annual net revenue in years 1, 2, and 3 is determined from the following formula by employing a different inflation rate for each cash flow component:

$$\text{BTCF(actual dollars)} = \$60,000(F/P, 8\ N) + \$140,000(F/P, 5\ N)$$
$$- \$70,000(F/P, 20,\ N) + \$30,000(F/P, 10,\ N)$$

The proposal is evaluated by the rate-of return method. A preliminary before-tax rate of return *without considering inflation* shows a promising value of

$$(A/P, i, 3) = \frac{\$100,000}{\$150,000} = 0.6667$$

for which i is about 45%. A more realistic after-tax approximation using straight-line depreciation, *but still not considering inflation*, is obtained from an after-tax rate of return calculated as shown below.

$$(A/P, i, 3) = \frac{\$100,000 - \$50,000(0.5)}{\$150,000} = 0.5 \quad \text{and} \quad \text{IRR} = 23.4\%$$

The after-tax rate of return calculated *when inflation effects are included* drops still lower:

$$\text{PW} = -\$150,000 + \$64,420(P/F, i, 1) + \$61,676(P/F, i, 2)$$
$$+ \$39,370(P/F, i, 3) \overset{?}{=} 0 \quad \text{at IRR}$$

At $i = 5$ percent,

$$\text{PW} = -\$150,000 + \$61,352 + \$55,942 + \$34,009 = \$1303$$

At $i = 6$ percent,

$$\text{PW} = -\$150,000 + \$60,774 + \$54,892 + \$33,056 = -\$1278$$
$$\text{IRR} = 5.5\%$$

The reason for the drop is twofold: Depreciation charges are not responsive to inflation and the costs (energy and services) in the proposal increase at a higher inflation rate than do the savings (labor and material).

Exercise 4   An asset can be purchased for $80,000 and will have a 10-year useful life. A suggested lease arrangement is to pay rental charges of $28,000 for the first 4 years and then make annual payments of $1500 for the remaining 6 years. Evaluate the suggestion.

Solution 4   A closer look at the lease is needed. It is probable that the Internal Revenue Service would consider the described "lease" as a disguised purchase aimed at reducing the total tax paid by the two parties. Conditions of the lease suggest that the lessee is actually assuming an ownership position for the asset as a consequence of the large initial payments and minimal subsequent payments. The payment schedule would produce a fine return to the lessor, and the lessee's rental expenses would provide larger and earlier deductions from income than allowed for depreciation of a comparable purchased asset. The lease plan is feasible only if it could be proved that the asset did indeed lose almost its entire value in 4 years.

The services of an asset that costs $25,000 will be needed for 3 years. It will have no salvage value and will receive ACRS depreciation if purchased. It can also be leased for $9400 per year, payable at the beginning of each year. Maintenance and operating costs will be the same for either option. Inflation is expected to be 9 percent annually. Compare leasing to buying with equity funds for an after-tax MARR of 13 percent when the tax rate is 38 percent.

Table 13.10 compares the after-tax cash flows with and without inflation for both alternatives. Neither depreciation nor rental is responsive to the 9 percent inflation rate. Maintenance and operating expenses are not included in the comparison because they affect both alternatives identically. Since no revenue figures are available, the negative present worths represent costs, and lower values are therefore preferred.

Results from Table 13.10 indicate that there is very little difference between the costs of leasing and purchasing when inflation is not a consideration. But when inflation affects the cash flow, the lease becomes less costly and purchasing gets more costly.

PROBLEMS

13.1 Use the Consumer Price Index data shown in Table 13.7 to determine the annual rate of inflation for the following periods:
  (a) 1967 through 1972
  (b) 1967 through 1977
  (c) 1974 through 1984
  (d) 1967 through 1984

13.2 Calculate the decline in purchasing power of the dollar over the following periods:
  (a) 1967 through 1972
  (b) 1972 through 1977
  (c) 1977 through 1982
  (d) 1967 through 1982

13.3 A home purchased in 1984 required an $80,000 loan which was to be repaid in equal monthly payments over a 30-year period. Nominal annual interest on the loan was stated as 12 percent compounded monthly. If the inflation rate is forecast to average 6 percent annually, compounded monthly, determine the equivalent present worth of the real-dollar cash flow.

13.4 Electric energy costs within a specific geographic area are expected to increase at an annual rate of 5 percent over the next 5 years. Use a discount rate of 10 percent to determine the present worth of this actual dollar series of cash flows.

13.5 Over the next 5 years the inflation rate is forecast to be 6 percent compounded annually. A potential investment will pay an interest rate of 10 percent for the first 3 years and 12 percent for the last 2 years. What is the average inflation-free rate of return?

13.6 A $10,000 investment would return a series of $3000 year-end payments over the

TABLE 13.10

Comparison of lease and buy alternatives at 0 and 9 percent inflation rates when the tax rate is 38 percent and MARR = 13 percent. For the cash purchase, ATCF with no inflation is composed of P = $25,000, S = 0, and tax savings equal to 38 percent of the depreciation charges.

When inflation is considered for the cash purchase (top table), tax savings (negative taxes) must be deflated at 9 percent to obtain real dollars; PW at 13 percent = ATCF (actual dollars) × $(P/F, f, N)$ × $(P/F\ i, N)$.

Since the lease cost is not responsive to inflation, the actual cash flow with inflation at 9 percent is equal to the real dollar cash flow without inflation. However, when inflation is included in the analysis, the annual lease payments must be deflated at the 9 percent rate to obtain their value in constant buying power

PW (lease) at 13 percent = (annual rent)(1 − tax rate) $(P/F, f, N)(P/F, i, N)$.

Cash Purchase of an Asset when Inflation = 9%

End of Year, N	BTCF, Actual Dollars	Depreciation	Taxes at 38%	ATCF, Actual Dollars	ATCF, Real Dollars	PW at 13%
0	−$25,000			−$25,000	−$25,000	−$25,000
1	0	$8,250	−$3135	3,135	2,876	2,345
2	0	11,250	−4275	4,275	3,598	2,818
3	0	5,500	−2090	2,090	1,614	1,118
						−$18,518

Cash Purchase of Asset when Inflation = 0%

End of year, N	ATCF, Real Dollars	PW at 13%
0	−$25,000	−$25,000
1	3,135	2,774
2	4,275	3,348
3	2,090	1,448
		−$17,430

Use of an Asset Acquired through Leasing:

End of year, N	When Inflation = 0%			When Inflation = 9%		
	BTCF, Real Dollars	ATCF, Real Dollars	PW at 13%	ATCF, Actual Dollars	ATCF, Real Dollars	PW at 13%
0	−$9400	−$5828	−$ 5,828	−$5828	−$5828	−$ 5,828
1	−9400	−5828	−5,158	−5828	−5347	−4,732
2	−9400	−5828	−4,564	−5828	−4905	−3,842
			−$15,550			−$14,402

next 5 years if no inflation were present. However, an average inflation rate of 6 percent is expected to increase the payments accordingly. If the annual market rate of interest remains at 13 percent, determine the present equivalent worth of the investment.

13.7   The investor in Problem 13.6 has an opportunity to accept a revised proposal in which a single $27,000 lump-sum repayment will be made at the end of the 5 year period. Which plan is preferable?

13.8   Net cash flow from the purchase of an asset for $1000 is expected to be responsive to inflation. The inflation rate is forecast to be 5 percent for the next 3 years. Based on this forecast, the expected cash flow in actual dollars is shown below.

Year	0	1	2	3
Cash flow	− $1000	$400	$600	$500

     (a) What is the combined interest-inflation rate if the organization expects a 10 percent return on investments?
     (b) Using this interest rate, calculate the present worth.
     (c) What cash flow estimates in real dollars would produce the same present worth when discounted at 10 percent?
     (d) Do you feel it is a better practice to estimate future cash flow in then-current actual dollars or in now-current real dollars? Why?

13.9   The general inflation rate is 8 percent, and a company requires a real rate of return of 10 percent. A machine purchased for $10,000 will have no salvage value at the end of its 7-year useful life. It is expected to produce a revenue of $2000 the first year, with annual increases of 20 percent in subsequent years. Operating costs are $1000 in year 1 and will rise in proportion to the general inflation rate.
     (a) Calculate the actual-dollar cash flow, convert it to real dollars, and determine the present worth.
     (b) Show that the solution for Problem 13.9a can also be obtained by discounting the actual-dollar cash flow at i_f.
     (c) Use the special interest rate i^* for a geometric series to calculate a modified revenue and show that the resulting PW at $i = 10$ percent is the same as that determined in Problem 13.9a.

13.10   A proposal has been submitted in real dollars that will produce a revenue of $110,000 annually for the next 10 years. It will require an initial investment of $150,000 in depreciable equipment that will be fully written off in 10 years by straight-line depreciation; no investment credit is applicable. Operating costs will be $52,000 per year. In addition, facilities will be leased for $6000 per year for 5 years; then the lease will be renegotiated for 5 more years at a constant annual charge of $6000(1 + f)^5$. The general inflation rate over the study period is expected to be 6 percent. The firm is subject to a tax rate of 46 percent and requires an after-tax rate of return of 4 percent. What is the present worth of the proposal?

13.11   Assume that a firm has an inflation-adjusted rate of return of 24 percent. The general inflation rate is 10 percent. Explain why the firm's real required rate of return is *not* 14 percent. What is it?

13.12   A $10,000 investment can be made today that will produce savings of $2000 annually for the next 7 years. There is no salvage value involved. Calculate the present worth of the investment at MARR = 10 percent. Show that the same PW results when the real-dollar savings inflate at 8 percent annually. Apply the combined interest-inflation rate to discount the actual-dollar cash flow.

13.13 Sales of a new product are expected to grow at a 12 percent compound rate for the next 3 years. Current sales are $1.2 million. The price of the product should increase at the national inflation rate of 6 percent per year. Production-cost categories and associated expected inflation rates are shown below.

Production-Cost Category	Current Total Cost	Inflation Rate, %
Materials	$200,000/year	9
Energy	150,000/year	12
Labor	150,000/year	8
All other	100,000/year	6

Depreciation charges for assets associated with production of the product amount to $200,000 per year. The company's effective tax rate is 45 percent.

(a) Assuming that marketing, overhead, and other allowable tax deductions associated with the product amount to 50 percent of the production costs, determine the after-tax cash flow without considering inflation (the initial investment is the sum of the depreciation charges).

(b) What is the before-tax rate of return when inflation is ignored?

(c) What is the after-tax rate of return when inflation is ignored?

(d) Determine the after-tax cash flow when it is assumed that price and production costs will inflate at the given unique rates while other costs rise at the inflation-factor index rate based on production costs only.

(e) What is the rate of return for the cash flow pattern in Problem 13.13d?

13.14* The relation of interest and depreciation charges, which are predicated on current values, to future dollar receipts tends to cause the rate of return on investments to be overstated during periods of inflation. As explained by George Terborgh, "The reason is not far to seek: tax deductions for depreciation, for 'basis' in computing terminal gains, and for interest on borrowed capital are *unresponsive* to inflation."

To illustrate the effect of inflation on rate-of-return evaluations, consider the asset described in Table 13.11. Its cost is $6194 and it is expected to produce receipts of $1000 (column 1) in present buying power for 10 years. There is no salvage value. Depreciation and interest charges are based on 30 percent financing at 5 percent interest and straight-line depreciation (column 2). These charges remain the same regardless of the effect of inflation upon receipts. With an inflation rate of 3 percent, the real-dollar receipts in column 1 would grow to the values in column 3 in N years.

(a) Calculate the after-tax rate of return for the asset's purchase based on the real-dollar cash flow (columns 1 and 2). The effective income-tax rate is 50 percent.

(b) Calculate the after-tax rate of return based on actual-dollar cash flow when the tax rate is 50 percent.

---

* Adapted from G. Terborgh, *Effects of Anticipated Inflation on Investment Analysis*, Machinery & Allied Products Institute, Washington, D.C.

TABLE 13.11

Cash flow of an asset purchased for $6194 and partially financed by a $1843 loan at 5 percent. Actual dollars responsive to 3 percent inflation result from real dollars multiplied by $(1.03)^N$.

End of Year, N	Net Receipts in Real Dollars (1)	Depreciation and Interest Charges (2)	Net Receipts in Actual Dollars (3)
0	− $4351	$ 0	− $4351
1	1000	711	1030
2	1000	702	1061
3	1000	693	1093
4	1000	684	1126
5	1000	675	1159
6	1000	665	1194
7	1000	656	1230
8	1000	647	1267
9	1000	638	1305
10	1000	629	1344

(c) Explain the erosion of the real after-tax return in your own words. How could it affect engineering economic analyses?

13.15 A portable drilling rig can be purchased for $85,000 or leased at the rate of $30,000 per year. Operating costs will be the same for either option, except that insurance and maintenance costs of $5000 per year are paid by the leasing organization under the lease agreement; these costs are handled as operating expenses for a purchased rig. If purchased, the rig will have a salvage value of $10,000 at the end of its 5-year useful life, and it will be written off for tax purposes by ACRS depreciation. Financing of the purchase will be a $15,000 immediate down payment and equal installments at the end of each year on a loan which charges interest at the rate of 10 percent, compounded annually. Lease payments are to be made at the beginning of each of the 4 years that the rig will be leased.

    (a) If the well driller uses a 15 percent rate of return for economic analysis and has an effective tax rate of 30 percent, which alternative has the lower present worth of outlays?

    (b) All other conditions remain the same except a 10 percent inflation rate is to be included in the analysis. What present worths for the lease-and-buy alternatives result from this added consideration?

EXTENSIONS

13A   Causes and Consequences of Inflation

Some dictionaries define inflation as "an increase in the amount of currency in circulation, resulting in a relatively sharp and sudden fall in its value and rise in prices; it may be caused by an increase in the volume of paper money issued or of gold mined, or a relative increase in expenditures, as when the supply of goods fails to meet the demand.'' A more succinct description is ''too much money chasing too few goods.''

In tune with this explanation, some economists trace the cause of inflation to more money being poured into the economy than the economy is worth. The real wealth of a nation is the goods and services it produces. Money is merely a convenient symbol of wealth, the amount of which is controlled by the government, and governments often feel impelled to create more money (or credit) to pay for old debts and new social programs. When money is generated at a faster rate than the growth in goods and services, it is subject to the old economic law that the more there is of something, the cheaper it becomes. In the case of money, cheaper means it loses purchasing power.

Other economists spray blames for inflation on

- Increases in producers' costs that are passed along to customers, sometimes with disproportionate escalations that push prices up—called *cost-push* inflation

- Excessive spending power of consumers, sometimes obtained at the expense of savings, that pulls prices up—called *demand-pull* inflation

- Impact of international forces on prices and markets, most notably the escalation of energy prices.

- Unresponsive prices that seldom decline, regardless of market conditions, because wages set by union contracts and prices set by some very large firms almost never fall

- Inflation psychology that leads consumers to ''buy ahead,'' often on easily obtained credit, in the belief that prices will inevitably inflate and loans can be repaid in cheaper dollars

Degree of Inflation

Everyone is affected to some degree by inflation. The effects may be immediate and conspicuous, such as a hike in rent. Other effects are subtle but pervasive, and they influence decisions both on and off the job. As a consequence of an inflation-prone economy:

- The general standard of living declines as savings and investments are eroded; people on fixed incomes suffer the brunt of the decline, but wage earners are afflicted by ''tax creep'' when their inflated incomes push them into higher tax brackets.

- Confidence in the economy declines, with a resulting increase in petty crime, political instability for incumbents, greater unemployment, and spreading discontent.

- Business decisions are distorted by efforts to cope with inflation, partially brought on by lower volume and higher taxes, and the ability of business to compete in foreign markets declines.

How large the foregoing consequences loom depends on the degree of inflation. When inflation is *mild*—annual price increases of 2 to 4 percent—the economy prospers, but the condition is temporary because employers are tempted to seek larger profits during periods of growth and unions commensurately bargain for higher wages. So *moderate* inflation occurs with price escalations of 5 to 9 percent. Then people start purchasing more because they would rather have goods than money that is declining in value. Increased demand pulls prices still higher.

Severe inflation occurs when the annual rate reaches 10 percent or more. During double-digit inflation, prices rise much faster than do wages. People on fixed incomes are hurt badly. Only debtors benefit by being able to repay debts with dollars less valuable than those that were borrowed. Beyond this dangerous level is *hyperinflation*—rapid, uncontrolled inflation that destroys a nation's economy. Here money becomes essentially valueless, as the government prints it excessively to pay expenses, while citizens go to a *barter economy* in which goods and services are exchanged without using currency.

Control of Inflation

The destructive effect of rampaging inflation is well-documented. Less known are effective actions to stem rising prices and to lower rates that are already too high.

Inflation first became serious in the United States in 1965, but few realized how dangerous it was and so strong enough measures to restrain it were not adopted. Among the remedies later attempted were price and wage controls, contraction of the money supply, credit restrictions, reduction in demand by raising taxes, increased demand by reducing taxes, enlarged supply of goods through greater productivity stimulated by investment incentives, and wage-price guidelines backed by political persuasion. Lack of success in applying the remedies has been attributed to inconsistency in applying them, inadequate time allowed for a remedy to become effective, and inattention to other factors while concentrating on only one remedy.

Credit for lowering United States double-digit inflation

in 1979–1981 is given to the monetary policy of the Federal Reserve System. The *Fed,* as the Federal Reserve Board is called, adopted a *tight-money* policy which reduced the money supply by such actions as raising reserve requirements for banks to limit their lending capacity and increasing banks' discount rate, causing them to raise interest rates on customers' loans. Other methods to control inflation include adjusting *fiscal policy*—spending and taxing programs—and impositions of wage and price controls. Each action has its advocates and detractors.

Inflation in Economic Analyses

Causes and corrections for inflation are outside the province of most engineering economists. They should, however, be concerned with the effects of inflation on their economic analyses. As a consequence of inflation, a dollar assumes different values at different times. The uninitiated might overlook the decrease in buying power of future dollars compared with today's dollars, as sometimes happens in reverse when managers nostalgically recall how much less expensive things used to be as they review, and possibly reject, proposals for current expenditures. Yesterday's, today's, and tomorrow's dollars do not have the same value. Treating them the same is like calling 10 meters equal to 10 feet; total units are the same but the sizes of the units are different.

Realizing that dollar values are distorted by inflation is a major advance, but the extent of distortion remains difficult to assess. Only foolish forecasters would stake their repu-

tations on multiyear inflation rate predictions. Lacking reliable forecasts, the accuracy of future cash flow equivalence is suspect. But engineering economists can still show what would happen if inflation became moderate or severe. Such scenarios put competing proposals in perspective.

As inflation elevates cost, it also escalates the value of cost-conscious analysis.

QUESTIONS

13A.1 According to Sullivan and Bontadelli,*

At a time when the most productive use of capital is so important, it would be foolish to ignore the anticipated effects of inflation and take the risk of losing competitive position because of indifference (or ignorance) in this regard. Omitting inflation from engineering economy studies is the same as assuming that the monetary unit is a constant-valued measure of worth. This is clearly out of tune with the present and expected future conditions in the business environment.

Do you agree? Why?

13A.2 During periods of severe inflation consumers tend to spend money as soon as it is received and to use credit for large purchases instead of saving money to buy them later. Businesses also have common reactions to severe inflation.

    (*a*) Lending rates are higher. Why?

    (*b*) Investments that increase rapidly in value are prized. What types of investments are most popular to counter double-digit inflation?

13B   Adjustable-Rate Reaction to Inflation

When the United States economy experienced three consecutive years of double-digit inflation, many financial practices were reexamined. Accountants struggled with rapidly rising values for existing assets. Managers sought tighter controls over inventories. Workers wanted their wages and retirement benefits indexed to price increases to protect buying power. And lenders designed multiple-year loans with adjustable interest rates.

The prime rate swung wildly between 11 and 21 percent during the 5 years displayed in Figure 13.2. It was high from the last half of 1980 until the first half of 1982 to fight the lofty inflation that existed in those years. Not knowing what to expect in the future, lenders in the early 1980s were reluctant to make long-duration loans. To make money

available for home purchases, *adjustable rate mortgages* (ARMs) were introduced.

Conventional fixed-term mortgage rates are supposed to provide a rate of return on capital of about 3 percent, compensation for risk of 2 percent, and protection from mild inflation of 2 to 4 percent. As shown in Figure 13.3, moderate inflation causes an extra add-on of about 5 percent as a buffer against rate fluctuations.

Adjustable rate mortgages usually provide lower monthly payments for the first few years of ownership than would be possible with a fixed rate mortgage. Before long, however,

* W. G. Sullivan and J. A. Bontadelli, "The Industrial Engineer & Inflation," *Industrial Engineering,* March 1980.

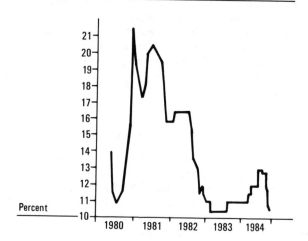

FIGURE 13.2
Five-year history of the prime lend-ing rate, 1980 to 1984.

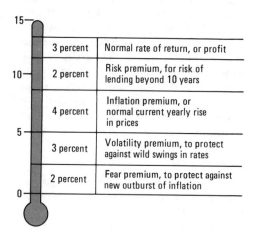

FIGURE 13.3
Interest components that add up to a 14 percent mortgage rate.

payments get bigger if inflation is significant. The amount of increase is tied to change in a respected financial index such as the CPI or yields on United States Treasury securities. Dates on which the adjustments are made are stated in the mortgage contract, often every other year. Sharp rises in interest rates after expiration of the low introductory rate created a new term for ARM holders—payment shock.

QUESTION

13B1 Construct a thermometer similar to Figure 13.3 to explain a mortgage rate high enough to accommodate severe inflation of 12 percent.

CHAPTER 14

SENSITIVITY ANALYSIS

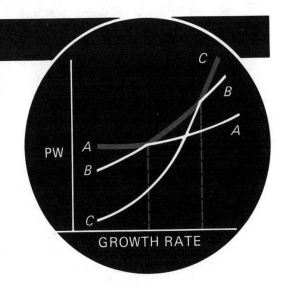

OVERVIEW

Sensitivity analysis provides a second look at an economic evaluation. It questions whether the original estimates adequately represent the future conditions that could affect a proposal if it were implemented. Its purpose is to assist decision makers. By considering the extent to which critical elements can deviate from original estimates before a preference is reversed, either from acceptance to rejection or between competing alternatives, the decision maker gets a better "feel" for the situation—how sensitive it is.

A sensitivity analysis can be performed with PW, EAW, or IRR calculations using after-tax or before-tax cash flows. The sensitivity of any of the elements used in the calculations can be checked. Analyses can be displayed on sensitivity graphs that show the effects of percentage variations for key parameters. The graphs are useful because they consolidate analytical data in a single, easily understood display.

Several formats are suitable for sensitivity studies. Cash flow factors can be investigated individually or in pairs, as in an isoquant. More- or less-favorable estimates can bracket the original estimates to obtain a range of values for a proposal's worth. All the approaches are geared to the question, "What if?"

WHAT IF?

Lurking behind every meaningful decision are "what if" doubts: What if sales differ from forecasts? What if cash flow doesn't follow the planned pattern?

What if a new, far better challenger becomes available? What if inflation is higher than expected? What if shortages disrupt operations? For major decisions the list of "what ifs" is discouragingly long, but decisions still have to be made. To escape inertia caused by nagging doubts, decision makers must first accept the realization that they will seldom, if ever, know all conditions with absolute certainty. Then they can focus on one or more critical factors and investigate what would happen to a proposal as a result of variations in those factors. That is the purpose of sensitivity analysis.

Sensitivity analysis involves repeated computations with different cash flow elements and analysis factors to compare results obtained from these substitutions with results from the original data. If a small change in an element leads to a proportionately greater change in the results, the situation is said to be sensitive to that assumption or variable. In an economic study, the critical point is the level at which an analysis factor causes an economic proposal to change from acceptable to not acceptable, or that reverses a preference between alternatives.

Sensitivity studies are the first step in investigating risk.

An informal sensitivity assessment is natural for most decisions. The amount of analysis time that a decision deserves was examined in Example 1.2. Variability in cash flow estimates and the effects of different values of i, N, P, and S were considered in several chapters. In this chapter, formal procedures for evaluating deviations from basic data are discussed, with emphasis on the value of graphical representations.

Consideration of the sensitivity of assumptions begins at the preproposal stage, where it is decided whether ideas for improvements are worthy of further development, and continues to the proposal presentation stage, where a case is made for the final acceptance or rejection decision. The role of sensitivity analysis for proposal planning, evaluation, and presentation is explored in the following examples.

Example 14.1
What If an Investment Is Made that Increases Fixed Costs?

An engineer has suggested an improvement in a production line that requires a substantial investment. This alteration will increase the fixed cost of producing the products. While it can be shown that the alteration improves quality and reduces variable costs, the question remains whether future sales revenue will provide sufficient additional income to make the investment worthwhile.

Figure 14.1 displays the considerations involved, in the form of a breakeven chart (traditional breakeven analysis and conventional breakeven charts are presented in Chapter 15). The annual cost of the proposed improvement is represented by an increase in the fixed cost of production. This increase raises the total production cost, which is the sum of variable costs and fixed cost. Total cost increases as a function of

annual output volume, as does sales revenue when the output is sold. Potential variations in costs and revenue are shown by shaded areas. If it is reasonable to expect sales revenue to increase enough to offset the added fixed costs, or if the proposal will decrease variable costs enough to pay for itself, or if a combination of greater sales revenue and lower variable costs will produce a large enough positive cash flow, the proposal merits further development.

It is doubtful that a chart would be drawn to enunciate the prospective operation variables in the example. The sensitivity analysis would more likely be a mental evaluation or preliminary calculation to confirm that receipts or savings could plausibly be expected to cover the investment. Then the idea would emerge from the preproposal to the proposal stage.

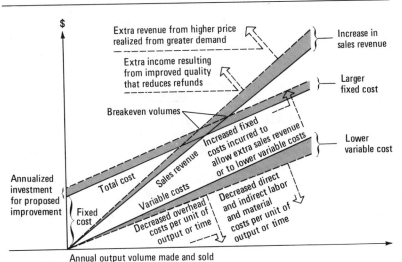

FIGURE 14.1
Possible results of an investment that increases fixed cost to improve quality and reduce variable costs. The critical factor is whether output volume will be sufficient to support the investment.

Example 14.2
What If Demand for Services and Labor Costs Change?

In most service industries, demand fluctuates in response to local or national trends. Inflation, unemployment, size of harvests, and other conditions that cannot be confidently forecasted affect the cash flow of a proposal. In deciding whether to introduce a new labor-intensive service, a company is concerned about the demand for the new service and possible increases in labor costs.

The nature of the service to be offered suggests that demand might exceed the basic estimate by 10 percent or fall short of expectation by 10 percent. Labor costs could increase by 5 percent or 10 percent over original estimates. The effect of these variations is analyzed by calculating the present worth

of the project at each level of anticipated demand and labor costs. Results of the calculations are shown in Table 14.1.

From the table it appears that PW is not very sensitive to labor costs; it is unlikely that a labor-cost estimating error will lead to a "wrong" decision because the proposal's PW does not rise disproportionately to labor costs. However, a 10 percent increase or decrease in demand causes a pronounced change in the present worth, a variation of 17 to 43 percent from the basic data. This sensitivity indicates that more study may be justified to provide assurance that the demand level used in the original data is supportable.

TABLE 14.1

Present worth of a proposal as it is affected by changes in demand and increases in labor costs.

Labor Costs	Demand Level			
	Decrease of 10%	Original Data	Increase of 10%	
Basic data	$250,000	$300,000	$420,000	⎫ Present
Increase of 5%	255,000	310,000	440,000	⎬ worth of
Increase of 10%	265,000	325,000	465,000	⎭ project

Computer-generated graphics are discussed in Extension 14B

Graphs, charts, and tables are often prepared to explain the sensitivity of proposals to key decision makers. The data for the visual aids is generated by substituting different values for the critical variables in formulas used during proposal analysis. The purpose of pictorial presentations is to clarify the considerations that are pertinent to a proposal's acceptance and to present analysis results in a digested but readily understood format.

Example 14.3
What If Profit Increases?

Annual reports from corporations, government reports, news magazines, and other publications that deal with economic statistics use bar graphs, pie charts, and myriad other pictorial devices to display economic relationships. The purpose may be to influence, explain, or educate. When they portray the financial effect of different assumptions, they are agents for sensitivity analysis.

An attention-getting display of how an investment grows as a function of capital structure, operating success, and dividend policy is shown in Figure 14.2. It is a form of trilinear

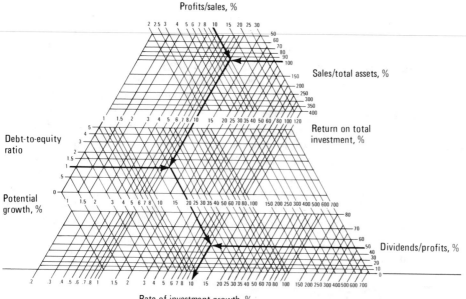

FIGURE 14.2

Trilinear representation of factors affecting investment growth. Bold lines indicate paths based on:

$$10\% \; \frac{\text{profit}}{\text{sales}} \rightarrow 100\% \; \frac{\text{sales}}{\text{assets}} \rightarrow 10\% \text{ return on investment}$$

$$10\% \text{ return on investment} \rightarrow 1.0 \; \frac{\text{debt}}{\text{equity}} \rightarrow 20\% \text{ potential growth}$$

$$20\% \text{ potential growth} \rightarrow 50\% \; \frac{\text{dividends}}{\text{profits}} \rightarrow 10\% \text{ rate of investment growth}$$

chart, a nomograph based on trigometric features of equilateral triangles. It facilitates a sensitivity analysis by providing a range of values for each variable and the paths that relate the variables to each other. In effect, it is a graphical calculator that simplifies an examination of factors that contribute to return on capital.

The top segment of the chart shows return on investment determined from the profit-to-total-asset ratio. Potential growth for the investment is the product of the return given by the top

segment and the debt-to-equity ratio. (A ratio of 1.0 means that total assets are represented equally by debt and ownership funds, which doubles the rate of return when equity is the only concern.) The lower segment shows the rate by which the investment grows as a result of the percent of profits left in the firm after paying dividends. Various combinations of inputs that could produce a given number for a certain parameter can be tested by tracing different paths through the chart.

SENSITIVITY OF A SINGLE PROPOSAL

Assume that a decision is to be made about a business opportunity based on the following estimates and tentative before-tax analysis:

Economic Factors	PW for a 10-Year Study Period at $i = 13\%$
Annual receipts ($35,000)	+$189,917
First cost ($170,000)	−170,000
Salvage value ($20,000)	+5,892
Annual disbursements ($3000)	−16,279
Net PW	+$9,530

The first cost is the most reliably known value in the problem, owing to its immediacy. The other factors could vary considerably over the 10-year period, owing to unforeseeable deviations from anticipated conditions. Even the study period may be inappropriate if the asset's useful life is shorter than 10 years (or longer than 10 years), or if the function it serves does not continue to yield the stated receipts for the full study period. And the 13 percent rate of return might be questioned. Should it be higher to compensate for inflation and the risk of losing invested capital? Or should it be lower?

The sensitivity of an asset's replacement cycle is examined in Extension 14A.

Sensitivity Graph

Such questions are collectively examined by constructing a sensitivity graph as shown in Figure 14.3. Curves are generated by substituting various values for one factor in the PW formula:

$$PW = -(\text{first cost}) + (\text{salvage value})(P/F, i, N) + (\text{receipts} - \text{disbursements})(P/A, i, N)$$

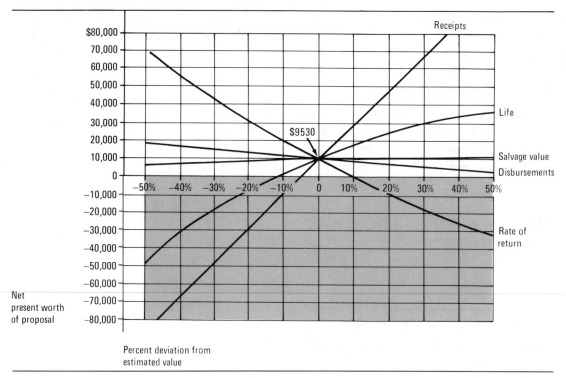

FIGURE 14.3
Sensitivity graph of the effect on a proposal's net PW when factors deviate from their original estimates.

while holding the values of all other factors constant. The abscissa of the graph is the percentage deviation from original values. This dimensionless scale puts all the factors in the same perspective to allow direct comparison of curve shapes. A steeper rising or falling curve indicates greater sensitivity of the proposal's worth to that factor.

Errors of ±50 percent are not uncommon when forecasting the rate of inflation.

The sensitivity graph reveals that deviations of up to 50 percent of the original estimates for salvage value and annual disbursements will not affect the acceptance of the proposal. The other factors—minimum acceptable rate of return, number of years the proposal will stay in effect, and amount of annual receipts—could switch the verdict to the do-nothing alternative if they deviate by only 10 percent from the original estimates. This condition suggests that extra care be given to forecasts of future business conditions that could affect income flow or the continued demand for the asset purchased.

Isoquant

The life of the asset and the size of receipts are not controlled by the analyst, as is the interest rate used in the comparison. The timing and amount of the cash flows are functions of the operating environment. Since these are the two most sensitive factors in the evaluation of the proposal, the limiting combinations for accepting or rejecting the proposal are graphed as an *isoquant* in Figure 14.4. The isoquant forms an *indifference line* which indicates the combinations of the proposal's duration and size of receipts that make the present worth of the proposal neither positive nor negative (indifference condition) when the other factors are unchanged. Thus, a reduction in the life of the asset from 10 to 8 years must be accompanied by an increase in annual receipts for 8 years of at least $36,859 − $35,000 = $1859 for the proposal to be minimally acceptable at $i = 13$ percent:

$$\text{PW} \doteq 0 = -\$170,000 + \$20,000 \, (P/F, 13, 8) + (\$36,859 - \$3000)(P/A, 13, 8)$$
$$0 = -\$170,000 + \$20,000(0.37616) + \$33,859(4.7987)$$

Acceptance-Rejection Zones

A chart divided into acceptance and rejection zones based on more than one parameter of a proposal combines the sensitivity-graph format with an isoquant. Two or three critical parameters can be selected for the sensitivity study. If two are chosen, a formula is written to determine the present or annual worth of the proposal in which one parameter is associated with the x axis of a graph and the other with the y axis. The intent of the formulation is to develop an expression relating values of parameter x and parameter y that generate zero present worth or annual worth. The resulting expression is represented by a line on the sensitivity graph. Percentage changes from the original data that occur on one side of the line maintain a positive worth for the proposal, while changes that fall on the other side cause a negative worth.

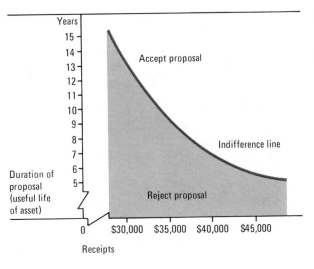

FIGURE 14.4
Isoquant showing combinations of the proposal's life and annual receipts at which the present worth is zero. Any combination that falls above the indifference line indicates that the proposal is acceptable when other factors remain constant.

Indifference lines are usually nonlinear when project life or interest rates are variables.

Example 14.4
Two-Parameter Sensitivity Study

The proposal pictured in Figures 14.3 and 14.4 is to be tested for its sensitivity to annual receipts and disbursements. The intent is to determine how large a joint percentage change in these critical parameters can be sustained without rejecting the proposal; it will be rejected if the equivalent annual worth is negative. Develop a sensitivity graph that identifies the acceptance and rejection zones.

Solution 14.4

Letting x represent a percentage change in receipts and y a percentage change in disbursements, we find that the equivalent annual worth for the proposal is

$$
\begin{aligned}
\text{EAW} = &-\$170{,}000(A/P,\ 13,\ 10) + \$35{,}000(1 + x) \\
&-\$3000(1 + y) + \$20{,}000(A/F,\ 13,\ 10) \\
= &-\$170{,}000(0.18429) + \$35{,}000 + \$35{,}000x \\
&-\$3000 - \$3000y + \$20{,}000(0.05429) \\
= &\ \$1757 + \$35{,}000x - \$3000y
\end{aligned}
$$

where x and y represent percent changes from the original data.

The proposal will be profitable, and therefore acceptable, so long as EAW > 0, or

$$
x > \frac{3000}{35{,}000}\, y - \frac{1757}{35{,}000} \quad \text{so} \quad x > 0.0857y - 0.0502
$$

When this inequality is plotted on a graph with x and y axes scaled in percent, the indifference line separates the chart into

FIGURE 14.5
Sensitivity graph of the percentage change for two factors (receipts and disbursements) that define acceptance and rejection zones for a proposal when all other factors are held constant.

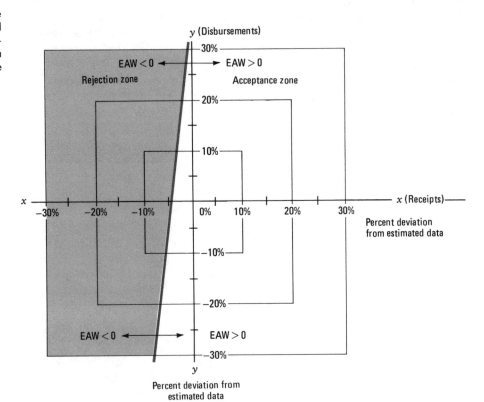

acceptance and rejection zones. The acceptance zone is on one side of the line where EAW >0 and the rejection zone is on the other side. The near-vertical slope of the line indicates that the proposal is highly sensitive to changes in the x factor (receipts), and quite insensitive to the y factor, as is obvious from a comparison of the magnitude of annual receipts relative to annual disbursements.

The center of the sensitivity graph in Figure 14.5 represents the original conditions for the proposal, where EAW = $1757. At x = −5 percent and y = 0,

$$EAW = \$1757 + \$35,000(-0.05) - \$3000(0.0) = \$7$$

At y = −5 percent and x = 0,

$$EAW = \$1757 + \$35,000(0.0) - \$3000(-0.05)$$
$$= \$1907$$

Thus the sensitivity study clearly emphasizes the criticality of an accurate forecast of future receipts.

Range of Estimates

An intensive assessment of a proposal's cash flow is provided by a scenario approach that asks, "What if the future is more or less favorable than originally estimated?" By bracketing the most likely cash flow condition with optimistic and pessimistic scenarios, the analyst exposes the shortcomings and strengths of the proposal. Each element of the proposal is questioned with respect to conditions envisioned for the bleak and promising scenarios. Most elements will probably vary from initial estimates, but some may be stable. The sensitivity of the proposal to each scenario is revealed by comparing the PW, EAW, or IRR of the three sets of data.

The trade-off between data collection costs and the value of more information is explored in Chapter 17.

The range is developed from:

1 *An objective estimate.* The most likely cash flow and the one that would be used if only a single estimate were made
2 *A more-favorable estimate.* An optimistic appraisal based on an advantageous interpretation of future events
3 *A less-favorable estimate.* A pessimistic assessment of the future that adversely affects the cash flow

Neither the more- nor less-favorable estimate is based on an extreme—the best or worst that could conceivably happen. They assess the outcomes of border conditions that are reasonably likely to occur. The three-estimate approach is also applicable to uncertainties besides price; analyses based on ranges of possible levels of activity, such as utilization rates and output quantities, or levels of performance may be more useful than single-estimate evaluations.

Item-by-item estimates are more credible than collective estimates. Additional work is required to develop estimates for smaller cash flow categories, but the resulting summations are more likely to be accurate than are block estimates, and the effort expended makes the analyst better informed about the situation. The cost of extra calculations and estimating is usually insignificant compared with the value of even a small improvement in the decision process.

More sophisticated statistical measures are described in Chapter 16.

Therefore, when price instability casts doubt on future cash flows, or operating levels cannot be anticipated with any assurance, an evaluation encompassing a range of itemized estimates is a sound practice.

Example 14.5
Range of Estimates that Reflect Uncertain Future Prices for a Training Proposal

A proposal has been made to introduce a training program to improve a production process which relies primarily on manual operations. New motions and fixtures for individual operators are expected to lower manufacturing costs from their present level by about $50,000 per year. It is difficult to predict the benefits of the training proposal because the size of the savings depends on the scale of operations and prices, which are functions of the marketplace and general economic conditions.

A range of estimates for possible future conditions is shown. Should the training program be adopted?

Items Estimated	Less-Favorable Estimate	Objective Estimate	More-Favorable Estimate
Additional units produced annually	60,000	75,000	100,000
Price/unit	$2	$3	$3.50
Annual income	$120,000	$225,000	$350,000
Duration of income	5 years	6 years	7 years
Training cost/year	$45,000	$35,000	$30,000
Required period of intensive training	2 years	2 years	1 year
Operating expenses of new process	$90,000	$160,000	$275,000
Investment in consumable supplies	$30,000	$30,000	$30,000

Solution 14.5

A before-tax analysis using a 15 percent required rate of return is conducted by calculating the present worth of the cash flows for the three possible conditions: less favorable (LF), objective estimate (OE), and more favorable (MF).

$$PW(LF) = (\$120,000 - \$90,000)(P/A, 15, 5)$$
$$- \$45,000(P/A, 15, 2) - \$30,000$$
$$= \$30,000(3.3521) - \$45,000(1.6257)$$
$$- \$30,000$$
$$= \$100,563 - \$73,157 - \$30,000 = -\$2594$$

$$PW(OE) = (\$225,000 - \$160,000)(P/A, 15, 6)$$
$$- \$35,000 (P/A, 15, 2) - \$30,000$$
$$= \$65,000(3.7844) - \$35,000(1.6257)$$
$$- \$30,000$$

$$= \$245,986 - \$56,900 - \$30,000 = \$159,086$$

$$PW(MF) = (\$350,000 - \$275,000)(P/A, 15, 7)$$
$$- \$30,000(P/A, 15, 1) - \$30,000$$
$$= \$75,000(4.1604) - \$30,000(0.8696)$$
$$- \$30,000$$
$$= \$312,030 - \$26,087 - \$30,000 = \$255,943$$

Although there is a chance the training program will result in a small loss if things go unfortunately, the opportunity for a very large gain under more favorable conditions makes the proposal very attractive. The three-phase analysis admits the chance of low returns and shows how good returns can be if things turn out favorably.

SENSITIVITY OF ALTERNATIVES

An alternative (A2) has been developed to accomplish the same mission as the proposal (A1) described by the sensitivity relationships in Figure 14.3. Both alternatives are expected to have the same revenue, but proposal *A2* has a lower first cost. However, the annual disbursements for *A2* will increase significantly each year, as shown by the following cash flow estimates:

A sensitivity chart for alternatives being evaluated according to weighted criteria in a priority decision table is shown in Figure 17.15.

Factor	Alternative 1	Alternative 2
Annual receipts	$ 35,000	$ 35,000
First cost	170,000	116,400
Salvage value (year 10)	20,000	0
Annual disbursements	3,000	$3000 the first year and increasing by $2500 each year

On the basis of the given estimates and a study period of 10 years with $i = 13$ percent, both alternatives have about the same PW:

$$PW(A1) = \$9530$$

$$PW(A2) = -\$116,400 + [\$35,000 - \$3000 + (\$2500)(A/G, 13, 10)](P/A, 13, 10)$$

$$= -\$116,400 + [\$32,000 - \$2500(3.5161)](5.4262)$$

$$= \$9540$$

Therefore, a preference for one or the other rests on the interpretation given to the annual cash flows. Since both alternatives are assumed to produce the same receipts, and disbursements can usually be estimated quite accurately, the most questionable feature remaining is the study period. A sensitivity graph for the alternatives' possible useful lives is given in Figure 14.6

The graph indicates that *A2* is preferred to *A1* when the likelihood is that the proposals will have lives of less than 10 years. At $N = 7$, neither proposal

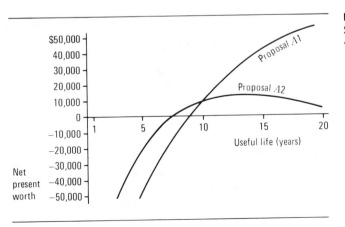

FIGURE 14.6
Sensitivity of two proposals to deviations from their estimated life.

FIGURE 14.7
Sensitivity comparison of present worths of three alternatives that have different growth rates. Alternative A is preferred if the anticipated growth rate is less that G_1, and alternative C is the choice when growth is expected to exceed G_2. Between G_1 and G_2, alternative B has the highest present worth.

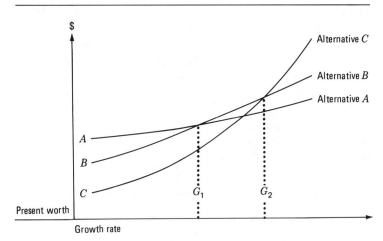

is profitable. Proposal $A1$ offers much larger gains should the life extend beyond 10 years.

The most critical cash flow elements that distinguish one alternative from another are usually quite evident. If one alternative has an exceptionally large salvage value, estimates for N and i should be scrutinized. An after-tax evaluation of proposals where one alternative has a particularly favorable loan arrangement suggests close inspection of interest rates. Any of the elements so picked for further examination can be graphed.

Often a graph will assist in a decision by narrowing the judgments that must be made. For instance, growth rate is frequently a debatable parameter. Growth may be a function of inflation because it affects certain cash flow elements or a function of the rate of service expansion, as in higher production rates needed to satisfy increasing demand for a product. Growth rate is shown on the x axis of Figure 14.7, where the present worths of three alternatives are graphed. Intersections of the plotted lines indicate the range of growth over which each alternative is preferred. Where proposals are presented in this fashion, it is not necessary to substantiate the selection of a certain growth rate for the study; the managers who are responsible for the final selection can use their personal judgment about which alternative best serves their vision of future conditions. Also, the decision makers do not have to fret over a specific growth rate; they need only pick a range in which they expect it to fall.

Sensitivity studies of this type do not require a large database to be quite complex.

Review Exercises and Discussions

Exercise 1   A would-be engineer spent so much time seeing movies while she was enrolled in engineering school that she never passed the differential-equations course. After she dropped out, she inherited $100,000 on her twenty-first birthday (perhaps that is why she did not bother to study much). She greeted this inheritance as a chance to satisfy her passion for flicks by leasing a soon-to-be vacant supermarket and converting it into a CinemaCenter comprising four small theaters in the one building. Based upon what she remembered from the engineering economics course she took before leaving school, she prepared the following cost estimates:

Renovation cost	$80,000
Contingency fund	$10,000
Estimated life of renovations	9 years
Salvage value	0
Annual operating cost (365 days/year)	$62,000
Annual lease expense	$42,000
Other annualized expenses	$16,000
Desired annual profit	$35,000

The CinemaCenter will have 280 deluxe seats in the four projection areas and an elaborate lobby. Income per moviegoer should average $3.35, including net profit from refreshments purchased.

The most questionable estimate involved in this economic evaluation is the number of people who will attend the shows. She has been told that she will be lucky to have a 50 percent utilization rate. What percentage of capacity is necessary to break even (assuming that receipts accumulate to year-end totals and the local cost of capital is 12 percent)? What influencing factors should be considered in forecasting utilization rates?

Solution 1

The percentage of seats occupied by paying customers X needed to break even when the desired profit is not included as a cost is found from

$$0 = (\$3.35)(280)(365)(X) - (\$80,000 + \$10,000)(A/P, 12, 9) - \$62,000$$
$$- \$42,000 - \$16,000$$
$$= \$342,370X - \$90,000(0.18768) - \$120,000$$

and,

$$X = \frac{\$16,891 + \$120,000}{\$342,370} = 0.4 \quad \text{or} \quad 40\%$$

If the estimated 50 percent attendance figure is realistic, the CinemaCenter appears to have a decent chance of surviving, so long as no profit is expected. To achieve the $35,000-per-year profit goal, the utilization rate must exceed 50 percent. Factors to consider in forecasting the chances of bettering 50 percent include the number and success of competitors, demographic data on the number of people in age groups most likely to attend the type of movies to be shown, the average wage and spending habits of the local population.

Exercise 2

Data for Youth Corps Training Program are shown on page 390. The expected duration of the program is 6 years. Both benefits and costs are uncertain. Analyze the desirability of the program from the given data.

Year	Benefit			Cost			(P/F, 10, N)	Present Worth					
								Expected		Minimum		Maximum	
	Min.	Exp.	Max.	Min.	Exp.	Max.		B	C	B	C	B	C
1	0	0	5	10	15	25	0.90909	0	13.6	0	9.1	4.5	22.7
2	5	10	15	10	10	15	0.82645	8.3	8.3	4.1	8.3	12.4	12.4
3	15	20	25	5	5	10	0.75132	15.0	3.8	11.3	3.8	18.8	7.5
4	20	30	30	4	5	8	0.68302	20.5	3.4	13.7	2.7	20.5	5.5
5	10	20	25	3	5	6	0.62092	12.4	3.1	6.2	1.9	15.5	3.7
6	5	15	20	2	5	5	0.56448	8.5	2.8	2.8	1.1	11.3	2.8
Totals								64.7	35.0	38.1	26.9	83.0	54.6

Solution 2   The *expected* benefit-to-cost ratio based on the most likely values is

$$\text{B/C} = \frac{64.7}{35.0} = 1.85 \quad \text{and} \quad \text{B} - \text{C} = 29.7$$

which makes the program acceptable. As is apparent in the table, the B/C ratios under all three conditions are greater than 1.0. However, if maximum cost occurred while the benefit was at a minimum, then

$$\text{B/C(worst condition)} = \frac{38.1}{54.6} = 0.70 \quad \text{and} \quad \text{B} - \text{C} = -16.5$$

The worst combination might be excused from consideration if supporting evidence could be collected to show that the chance is very remote that the highest cost would accompany the lowest benefit.

Exercise 3   Review Exercise 3 of Chapter 13 described a proposal in which the cost factors were subject to individually unique inflation rates. Since forecasting future inflation rates is always an iffy proposition, a sensitivity analysis is appropriate. Based on the data from the original problem (summarized below), develop a sensitivity graph for deviations from initial inflation estimates.

Proposal to Purchase Automated Equipment Designed to
Reduce Labor Costs and Material Waste

Automated equipment
  $P = \$150,000$
  $S = 0$
  $N = 3$ years
Annual revenue
  Net savings in labor = \$60,000; $f = 8\%$
  Net savings in material = \$140,000; $f = 5\%$
Annual costs
  Additional energy required = \$70,000; $f = 20\%$
  Added maintenance expense = \$30,000; $f = 10\%$
Effective tax rate = 50%; ACRS depreciation

    The critical cost factors subject to unique inflation rates are labor, material, and energy. The effects of deviations from the expected inflation rates are determined by

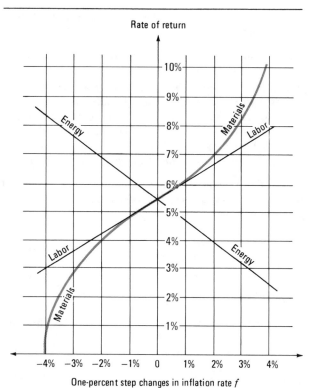

Rate of return

One-percent step changes in inflation rate f

calculating an IRR for each percent change in the inflation rate of one factor while all other conditions are held constant. The curves shown in Figure 14.8 consolidate the results of all the computations.

The intersection of the curves is at the IRR calculated previously for the most likely inflation rates. The most sensitive cost element is material because it has a proportionately much larger dollar value than do the other factors in the proposal; an increase in the inflation rate from 5 to 9 percent almost doubles the rate of return (savings from decreased use of material are worth more as the price of the material goes up). Energy is almost as sensitive, owing to its much higher starting inflation rate, even though it is applied to a factor only half as large as material.

PROBLEMS

14.1   A company has a debt-to-equity ratio of 3:1. Annual sales are 150 percent of total assets and provide a rate of return on total investment of 9 percent. If a growth rate on investment of 25 percent is desired, use the trilinear chart to determine what percent of profit can be paid in dividends and what percent of profit is expected from sales.

14.2   An isoquant is to be developed for program A in Problem 6.16. It will relate the two most sensitive factors; the life of the program and expected volume. Using $i = 8$

percent, $N = 4$, and $P = \$10,000$, determine the dollar value of the specified point on the isoquant.

14.3 A project will have a life of 10 years and is to be evaluated at a social discount rate of 9 percent. It has an initial cost of $1 million, and expected annual costs are $100,000. However, it is possible that these annual costs could consistently increase by $10,000 each year or decrease by $5000 per year. The most likely annual benefits are $350,000, but there is a chance that benefits will decrease by $25,000 per year.

    (a) What is the worst possible B/C that could occur with the given scenario?

    (b) What maximum first cost could be incurred for an acceptable project in which annual cash flows follow the expected (most likely) estimates?

14.4 The purchase of rental property in a neighborhood where real-estate prices are increasing rapidly is being considered. The following estimates have been developed for a preliminary before-tax analysis:

First cost	$50,000
Annual income from rent	$6,000
Annual maintenance	$1,000
Investment period	6 years
Resale value	$60,000
Cost of capital	10%

    (a) Given that the first cost of the property and the investment period are fixed, construct a sensitivity chart showing the effect of changes in all the other elements on the EAW.

    (b) Construct a sensitivity chart for joint variation within a ± 30 percent range of annual income and cost of capital; indicate the acceptance and rejection zones.

14.5 A proposal is described by the following estimates: $P = \$20,000$; $S = 0$; $N = 5$; and net annual receipts $= \$7000$. A rate of return of 20 percent is desired on such proposals. Construct a sensitivity graph of the life, annual receipts, and rate of return for deviations over a range of ± 20 percent. To which element is the decision most sensitive?

14.6 Tennis enthusiasts in a small town want indoor courts to allow them to play tennis during bad weather. Someone suggested that they band together to form a corporation to build and operate a profit-making tennis facility. Two courts with a small lounge could be built for $125,000, not including land. The facility would have a 10-year life and a salvage value of $20,000. Annual operating expenses would be $23,000.

    (a) At a charge of $7 per hour for playing time on a court, how many hours would the courts have to be rented each year for the investors to break even on construction and operating costs, assuming a rate of return of 10 percent?

    (b) Develop a graph of the utilization rate required to yield rates of return between 10 and 25 percent. Assume that the facility will be open 14 hours per day for 320 days each year. Would you recommend the investment? Why?

14.7   Three designs to perform the same function have the cost patterns indicated:
Design 1. Initial cost is $12,000, and annual expenses are uniform at $4000 per year.
Design 2. A low initial cost of $5000 is possible because refinements will be made while the facility is in operation. The first year's operating and refinement cost will be $7500, and it will decrease by $500 each year.
Design 3. Initial cost is $15,000, with annual disbursements that start at $2000 and increase by $1000 each year.

   The principal uncertainty in the evaluation is how long the function will be needed for which the designs were developed. All the designs could perform the function for 12 years, if the function lasts that long, and none of the designs would have any salvage value at any time. Determine the range of life over which different designs would be preferred when the minimum attractive rate of return is 11 percent.

14.8   An aluminum company must decide whether to install a new type of "air float" conveyor for extra thin aluminum sheets or to retain its conventional conveyors. A pilot test reveals that the float conveyor moves the sheets faster and reduces damage, but maintenance costs to keep it operating properly may be high. The amount of savings expected from the new design depends on the quantity of aluminum sheeting produced in the future and the reliability of the new equipment. The engineering department has provided the following estimates:

First cost	$180,000
Economic life	4 years
Annual maintenance expense	$40,000
Annual savings	$100,000

There is no realizable salvage value because the cost of removing the equipment would about equal the scrap value. The company uses a minimum attractive rate of return of 12 percent.

   (a) Because there is still some doubt about the effectiveness of the air float conveyor design, some of the analysts feel that a higher rate of return should be required for the project. Make an isoquant of the maximum first cost that could be incurred to earn rates of return between 5 and 25 percent.

   (b) Assuming the life, maintenance cost, and savings may vary as much as 50 percent on both sides of the given estimate, develop a graph of the effects of individual variations on the net present worth of the project. Do you recommend that the new conveyor system be installed? Why?

14.9   Current mail-sorting operations on one line at a mail distribution center cost about $1 million per year. A newly developed and essentially unproved but promising system for sorting mail by automatic address readers is being considered. Although the devices have been tested and approved in laboratories, there is still doubt about how they will perform in regular service. Questionnaires describing less-than-favorable and more-than-favorable operating conditions were sent to people familiar with the devices and the mail-distribution process. Consensus data from the two scenarios and the most likely estimates are shown on page 394.

Factors Estimated	Pessimistic Estimate	Most Likely Estimate	Optimistic Estimate
First cost, including installation	$2,112,000	$985,000	$915,000
Life, years of full utilization	2 years	2 years	6 years
Annual maintenance and minor repair	$221,000	$81,000	$75,000
Annual operating cost and standbys	$929,000	$714,000	$588,000

Calculate the range of EAC and discuss the results. MARR = 10 percent.

14.10   There is considerable doubt about the need for and performance to be obtained from a new process developed by the R&D department. A decision about launching a small-scale pilot project is being evaluated. Three estimates of possible outcomes of the pilot project are given below.

	Objective Estimate	Less-Favorable Estimate	More-Favorable Estimate
Income/year	$200,000	$150,000	$250,000
Expenses/year	$80,000	$80,000	$90,000
Start-up cost	$300,000	$350,000	$300,000
Life of project	3 years	1 year	4 years
Salvage value	$100,000	$50,000	$100,000

(a) What is the present worth of each possible future when the minimum attractive rate of return is 15 percent?

(b) What other considerations could affect the decision to launch the pilot project? Should the money already invested in research and development be a consideration? Why?

(c) Compare the range-of-estimates method of evaluation with sensitivity analysis.

14.11   Three revenue-producing projects are being considered. Estimates of future returns are uncertain because the projects involve new products. The initial investment is expected to provide adequate production capability for any reasonable project life, and the lives of all projects are considered to be equal. Regardless of useful life, there will be no salvage value.

Project I has a first cost of $200,000 and uniform annual net revenue of $65,000.

Project O has a low initial cost of $100,000 and will return $50,000 the first year, but net revenue will decline each year by an amount 0.5G, where G is a uniform gradient.

Project U has a high initial cost of $250,000 and also returns $50,000 the first year, with the expectation that new revenue will increase by an annual uniform amount G.

Compare the three projects by constructing sensitivity graphs according to the following assumptions, and discuss the results.

(a) Let the project life be 6 years and G = $7500 to test project preference for sensitivity to minimum rates of return up to 25 percent.

(b) Let the project life be 6 years and MARR = 12 percent to test project preference for sensitivity to values of G from 0 to $15,000.

(c) Let MARR = 12 percent and G = $7500 to test project preference for sensitivity to project life.

EXTENSIONS

14A   Sensitivity of Economic Life for Cyclic Replacements

The concept of cyclic replacement was introduced in Extension 8A. It is applicable when an asset continues to be replaced by another asset of the same type. Since the most economical life for the asset is the period of service that minimizes its equivalent annual cost, the replacement period is calculated from

EAC = capital recovery + equivalent annual operating cost

when taxes and inflation are ignored.

It is revealing to observe how sensitive the replacement period is to certain factors used in its calculation. An asset with the characteristics given below is examined.

The asset thus has a purchase price of $15,000 and a potential physical life of 12 years.

Because cyclic replacements are usually associated with a fleet of vehicles or a bank of machines, the typically large investment makes them unmistakable candidates for sensitivity analysis. The nature of the inversely varying cash flows suggests sensitivity to the required rate of return and inflation. A tabular format for calculating the minimum-cost cycle is shown in Table 14.2; it indicates replacement at the end of the fifth year when the MARR is 10 percent and taxes and inflation are not included in the analysis.

By substituting different minimum acceptable rates of

Year, N	0	1	2	3	4	5	6	7	8	9	10	11	12
Salvage value at end of year N	$15,000	$11,000	$8000	$6000	$5000	$4000	$3000	$2500	$2000	$1500	$1,000	$700	$500
Annual operating cost		$3,000	$3500	$4100	$4800	$5600	$6500	$7500	$8600	$9800	$11,100	$12,500	$14,000

TABLE 14.2

Before-tax equivalent annual cost of an asset subject to cyclic replacement at MARR = 10 percent. The minimum-cost replacement cycle is 5 years.

End of year N	0	1	2	3	4	5	6
(1) Salvage value, S	$15,000	$11,000	8000	6000	5000	4000	3000
(2) Operating costs, O		$3000	3500	4100	4800	5600	6500
(3) Decrease in S		$4000	3000	2000	1000	1000	1000
(4) Interest on (1), $i = 10\%$		$1500	1100	800	600	500	400
(5) Capital recovery, (3) + (4)		$5500	4100	2800	1600	1500	1400
(6) Total cost, (2) + (5)		$8500	7600	6900	6400	7100	7900
(7) (P/F, 10, N)		0.90909	0.82645	0.75132	0.68302	0.62092	0.56448
(8) Present worth, (6) × (7)		$7727	6281	5184	4372	4408	4460
(9) ΣPW (cumulative)		$7727	14,008	19,192	23,564	27,972	32,432
(10) (A/P, 10 N)		1.10	0.57619	0.40212	0.31547	0.26380	0.22961
(11) EAC, (9) × (10)		$8500	8071	7718	7434	7379	7447

return in the tabular format, the sensitivity of the replacement cycle to the MARR is revealed. Higher MARRs cause the capital cost to increase but reduce the effect of rising operating costs. The results of repeated calculations with different rates of return provide the data for the sensitivity graph shown in Figure 14.9. The replacement cycle increases from 4 to 7 years as the MARR is raised from 0 to 30 percent, primarily from the effect of higher interest charges on the capital cost of ownership.

When essentially the same procedure is followed to check the sensitivity of other influential factors, the following observations become evident:

• Inflating both capital and operating costs at the same rate does not affect the replacement period in a before-tax analysis; it simply escalates the EAC.

• The replacement period is always longer in an after-tax analysis; for instance, for a tax rate of 50 percent the minimum-cost cycle changes from 5 to 8 years with the data used in Table 14.2 (at MARR = 10 percent).

• There is no change in the replacement interval when different methods of depreciation are used in an after-tax analysis.

• Higher tax rates increase annual costs and tend to lengthen the replacement period only slightly.

• Using borrowed capital for funding lowers the after-tax EAC, depending on the interest paid on the loan, but has little effect on the replacement period. Fifty percent financing from capital borrowed at 15 percent reduces the after-tax EAC by about $160 from full equity funding, but the replacement period is still 8 years.

• A general inflation rate of 10 percent applied to both capital and operating costs in an after-tax analysis at MARR = 10 percent reduces the replacement period to 7 years (as opposed to 8 years without inflation), but further increase in the inflation rate to 40 percent does not change the 7-year period.

FIGURE 14.9
Sensitivity of the cycle for like replacements to changes in the required rate of return. Increasing the MARR increases the replacement cycle. Equivalent annual costs are before-tax amounts.

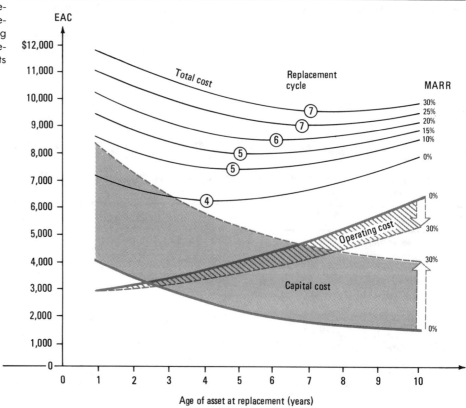

• When operating costs inflate at a 15 percent rate while salvage values escalate at a 10 percent rate and MARR is 10 percent, the replacement period drops to 6 years. Higher inflation rates for operating costs of 20, 25, and 30 percent decrease respective replacement cycles to 5, 4, and 4 years.

• Raising the percentage of borrowed funds makes a purchase more sensitive to inflation rates in after-tax analyses. The respective replacement periods for operating-cost inflation rates of 10, 15, 20, 25, 30, and 40 percent are 7, 6, 5, 4, 4, and 3 years when half the purchase price is borrowed at $i = 15$ percent and MARR $= 10$ percent.

QUESTION

14A.1 You have the responsibility for making a presentation to the capital budgeting committee about future investment plans for the asset described in this extension.

(a) What features would you emphasize? Based on current economic conditions (the present inflation rate and forecasts for inflation at the time of your report), indicate what replacement period you would recommend.

(b) After performing any supplementary calculations that appear necessary, develop a graph, table, or chart that consolidates data about the sensitivity of the proposed cyclic replacement.

14B   Microcomputer Spreadsheets for Sensitivity Analysis

The development of new software packages for microcomputers is encouraging the inclusion of sensitivity analyses as an integral part of many economic studies.

The complexity and number of calculations required for even moderate-sized economic studies often cause limits to be placed on the range of alternatives being considered or in the number of investigations of the effects of variabilities inherent in the estimates being used. Most analyses are still based only on expected values or best-guess estimates with very little time available for the consideration of what-if type questions. The formulation of a basic single-estimate analysis is usually quite time-consuming; and once the primary relations between various parameters are established, there still exists the problem of performing all the necessary calculations.

The advent and later general widespread use of large mainframe computers has provided the economic analyst with a powerful, but very expensive, tool for handling large quantities of calculations at a very high speed. Unfortunately, the high operating costs and a general lack of readily accessible software severely limited most application studies to large corporations, universities, and governmental organizations. The development of multipurpose hand-held calculators in 1974 actually had a much greater impact on the typical engineer. The speed, flexibility, and ready accessibility of these machines provided a quantum leap in speed over the use of slide rules and books of tables. Many of these small units have capabilities comparable to the room-sized computers of previous years.

Microprocessor technology has now advanced to the point where there is widespread availability of desk-top or portable personal computers. With the proliferation of these machines, a number of financial software packages written in languages such as BASIC and FORTRAN have become available. Whereas some of these programs have been developed with the capability of allowing rapid changes in input values, most do not readily facilitate sensitivity studies. The majority require a rigid structure of input data and generally provide only a single answer set per run.

The creation and commercial development of the first electronic spreadsheet—VisiCalc—in 1978 was a significant step in the elimination of some of the difficulties associated with financial analysis on the microcomputer. First of all, the spreadsheet is in a familiar format. As a replacement for the typical columnar pad, calculator, and pencil, it is composed of rows and columns which constitute a grid. Data entry is facilitated through a keyboard and values can easily be corrected, changed, or deleted. Individual data cells can be addressed by either moving a pointer or cursor to the appropriate grid location or by referring to the location's address by a row-column intersection. Although similar to the columnar pad, spreadsheets are considerably larger, with some having over 2000 rows and 256 columns.

The second major advantage in using a spreadsheet is that no formal knowledge of a computer programming language is required. Typical arithmetic operations such as addition, subtraction, etc., are handled by simply typing an appropriate equation in which the addresses of the cells containing relevant data are included as variable names. If it is desirable to perform the same general set of operations a large number of times with different inputs, it is only necessary to alter the numbers contained in the cells. The arithmetic operations will be performed automatically as the new data are entered. Some spreadsheets also have the capability of performing statistical analyses on the results as they are generated. A partial listing of the mathematical,

TABLE 14.3

Functions available in most spreadsheet programs

Financial	Mathematical
Future value	Absolute value
Internal rate of return	Arc cosine
Modified rate of return	Arc sine
Net present value	Arc tangent
Payment	Cosine
	Exponential
Logical	Integer
AND	Logarithm
Error	Modulo
FALSE	Natural logarithm
If . . . Then . . . Else	Pi
IsError	Round
IsNA	Sine
N/A	Square root
NOT	Sum
OR	Tangent
TRUE	Statistical
	Average
Sorting	Count
Alphabetic	Maximum
More than one key	Minimum
Numeric	Sum
Row/Column selection	Standard deviation
	Variance

financial, and statistical functions available in most spreadsheet programs is provided in Table 14.3.

A typical spreadsheet application might be the development of an after-tax analysis similar to that presented earlier in Chapter 11. A spreadsheet (with appropriate labels) could be laid out to represent the data provided in Table 11.1. An example of a typical format is shown in Figure 14.10. In this configuration, the values appearing in the cells representing after-tax cash flows (ATCF) would actually be dependent upon the values entered in other cells. For example:

ACRS Depr. = First Cost * ACRS %
Taxable = BTCF − ACRS Depr.
Tax = Taxable * Tax Rate %
ATCF = BTCF − Tax
    or
ATCF = BTCF − [(BTCF − (First Cost * ACRS%) * Tax Rate%]

Thus any change made in BTCF, First Cost, ACRS%, or Tax Rate % would result in a change being made in ATCF by the program. Another noteworthy feature of this particular spreadsheet is that it has been formulated to calculate the internal rate of return (IRR) of the situation presented. In a matter of seconds after altering one or more of the input variables, the analyst has the IRR of the new cash flows.

In addition to tabular displays, the results of spreadsheet operations may be presented in a variety of graphical forms. Figure 14.11 illustrates two types of charts showing the results obtained in the previous operations.

A more involved spreadsheet analysis could involve the development of a sensitivity analysis similar to that presented in Figure 14.3. The tabular output of this analysis is shown in Figure 14.12.

A computer-generated graphic similar to that provided in Figure 14.3 is shown in Figure 14.13. Whereas the simplifying assumption of a linear relation between data points is not always entirely correct, the relative sensitivity of the figure of merit (PW) is quite obvious and the approximation is more than adequate for most purposes. However, if more accuracy is desired, the spreadsheet can accommodate thousands of additional data points and the integrated graphics program can be used to drive a more precise graphics plotter.

FIGURE 14.10
Spreadsheet representation of data from Table 11.10.

YEAR	Before-Tax Cash Flow	ACRS Depr.	Taxable	Tax	After-Tax Cash Flow
0	($45,000.00)				($45,000.00)
1	$15,700.00	$9,000.00	$6,700.00	$2,814.00	$12,886.00
2	$15,700.00	$14,400.00	$1,300.00	$546.00	$15,154.00
3	$15,700.00	$10,800.00	$4,900.00	$2,058.00	$13,642.00
4	$15,700.00	$7,200.00	$8,500.00	$3,570.00	$12,130.00
5	$15,700.00	$3,600.00	$12,100.00	$5,082.00	$10,618.00

INTERNAL RATE OF RETURN   13.9%

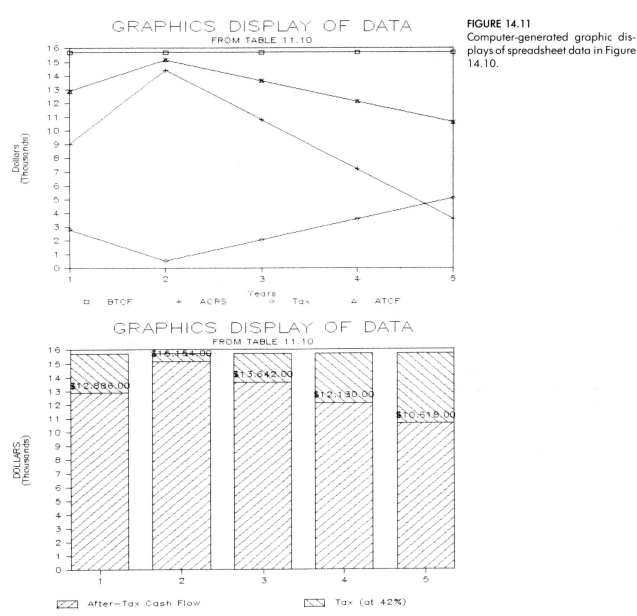

FIGURE 14.11
Computer-generated graphic displays of spreadsheet data in Figure 14.10.

VARIABLE	-40%	-20%	Expected	+20%	+40%
P	$77,532	$43,532	$9,532	($24,468)	($58,468)
S	$7,175	$8,353	$9,532	$10,710	$11,888
R	($64,079)	($28,492)	$9,532	$47,515	$85,499
D	$16,043	$12,787	$9,532	$6,276	$3,020
i	$56,110	$30,728	$9,532	($8,312)	($23,448)
N	($32,472)	($8,916)	$9,532	$23,979	$35,293

FIGURE 14.12
Results obtained from spreadsheet analysis of data on page 299.

FIGURE 14.13
Spreadsheet graphics presented from data provided in Figure 14.12.

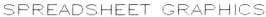

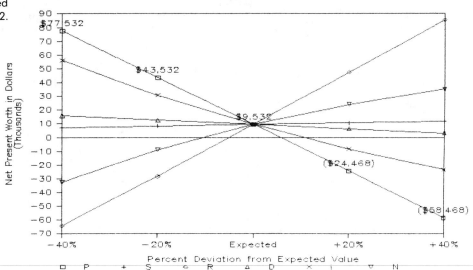

Microcomputer-based spreadsheets are rapidly growing in complexity and capability. The interfaces they provide between powerful arithmetic capabilities and the ability to manipulate and transform large databases are extremely useful to the economic analyst. Their speed and flexibility in scenario development are rapidly proving to be a catalyst to the routine development of sensitivity studies.

SECTION FOUR

ECONOMIC DECISIONS

Over 100 years ago a delightful and prophetic paper appeared in the *Journal of the Statistical Society.** It was written by William Farr, Esq., M.D., D.C.L., F.R.S., and bore the inclusive title "On the Valuation of Railways, Telegraphs, Water Companies, Canals, and other Commercial Concerns, with Prospective, Deferred, Increasing, Decreasing, or Terminating Profits." The following excerpts from that treatise reveal the century-old roots of engineering economic evaluation concepts:

The value of things depends to some extent on their utility. . . . Goods are in general good things: they give life, health, and strength; they yield enjoyment to the highest faculties as well as the lowest wants of human nature. . . . The value of a thing bears no definite or constant relation to its excellence in an aesthetic sense, or to its high place in philosophy. . . . Thus it is found in every case value expressed in money is measured by the mind, and that its price is practically fixed by the concurrence of the seller and the buyer. . . .

As the same thing differs in value at different distances in space, so its value—expressed in money—differs in time. It may be worth 1,000£ if paid for in ready money, but if payable in a year's time it may be worth 905£: due in fourteen year's time it may only be worth 505£ . . .

If there is risk, that is valued by the doctrine of probabilities; thus, if in a lottery there are one thousand prizes of 1£ and forty blanks, the value of one thousand and forty tickets of 1£ will only be 1,000£. The same result, the same depreciation of value, may be caused by risk as is caused by remoteness of payment; and as both risks and profits vary, and their combination produces corresponding effects on the values

*Volume 39, London, September 1876.

involved, several of these combinations are conveniently included in the rates of interest.

The value of annuities at constant risk diminishes with the factor of risk. Thus let an annuity of 4£ in perpetuity be equivalent to 4 percent interest on 100£, then it will be worth twenty-five years' purchase; but if there is a constant risk equivalent to 1/105, it will be worth only twenty years' purchase. The risk rate enters into the dividends of all commercial undertakings; so rate of dividends includes the ordinary profit of capital at no risk, and the insurance premium to cover such normal risk as the concern is exposed to.

The views expressed by Dr. Farr are forerunners of contemporary economic decision making. He recognized the time value of money and the influence of uncertainty on the cost of capital. His words also implied concern for intangible values and risk considerations. These implications are the theme for Section Four.

An investment proposal must satisfy a series of criteria before it is fully accepted. To start with, it must pass the test of technological feasibility—an engineering criterion. Then it must meet the economic criterion of producing a sufficient return on investment, and it is still subject to financial criteria that ration capital to only the most deserving requests for expenditure. Those that survive eventually face the decisive hurdle of management approval. The final review, buttressed by information from previous screenings, ascertains how well a proposal supports future plans of the organization. These plans involve risks and intangible factors which affect a proposal's attractiveness. Although risks and intangible influences cannot be documented as convincingly as can most cash flow elements, they may be conclusive in the final decision on acceptability.

The chapters in this section progress from consideration of known production criteria to recognition of the role of risk in immediate and successive decisions, and conclude with decision-making techniques that accommodate both tangible and intangible elements. The purpose of mastering these refinements to Dr. Farr's concepts is to make engineering economic decision making as complete as possible, to allow better understanding of the opportunities, and to develop evaluation consistency that in the long run will maximize returns.

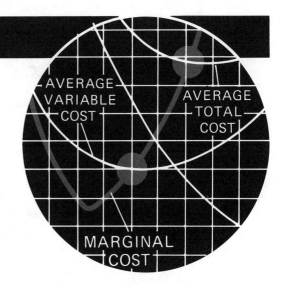

CHAPTER 15

BREAKEVEN ANALYSIS

OVERVIEW

Many economic comparisons are a form of breakeven analysis. The lease-or-buy question from Chapter 13 could be rephrased to ask about what level of service or period of time leasing becomes more expensive than buying. The point where the two alternatives are equal is the *breakeven point*. Most sensitivity studies involve an *indifference* level for a given cash flow element at which two alternatives are equivalent—the breakeven point for the given element. The choice between the two then rests on a judgment about which side of the breakeven point the element will likely register.

In this chapter breakeven analysis is directed to the point at which operations merely break even, neither making nor losing money; changes in operations are evaluated according to their effect on this point. Breakeven analysis, known also as cost-volume-profit analysis, is widely used for financial studies because it is simple and extracts useful insights from a modest amount of data. The studies necessarily include an examination of production costs and operating policies.

Costs which remain relatively constant regardless of the level of a firm's activity are *fixed costs F*. Per-unit *variable costs V* are proportional to *output n*. *Profit Z* is the difference between *total revenue* ($R = nP$, where P = price) and *total cost* ($C = nV + F$). The *breakeven point B* occurs when $Z = 0$. Profit may be increased by raising or lowering P, cutting C, or working with combinations thereof.

In linear breakeven charts, $B = F/(P - V)$, where $P - V$ is the *contribution* per unit. Breakeven charts can be constructed for single or multiple product analysis. *Dumping* is the practice of selling a product at a lower price in a separate market to increase plant utilization.

In a nonlinear breakeven analysis, the average total cost is the sum of ever-declining average fixed cost and the typically saucer-shaped average-variable-cost function. The average-total-cost curve is intersected at its lowest point by the rising marginal-cost curve. Similarly, the point of maximum profit occurs where marginal revenue equals marginal cost. The effects of inflation on breakeven analyses are examined in Extension 15A.

COST AND COMPETITIVENESS

Most organizations strive for profits. They do so through a close scrutiny of their internal operating costs and strict attention to their competitive position. Even intentionally nonprofit organizations must follow the same policy if they are to achieve excellence. The cost-revenue-profit relations are exposed by breaking down a unit of output into its component dollar values.

The rectangular block in Figure 15.1(a) represents a unit of output. This output can be a product, such as an automobile, or it can be a service, such as that of collecting garbage from a subscriber. The unit is divided into three segments which classify the producer's interests. The overall height or price for which it can be sold is a function of the consumer's regard for the item.

Fixed Costs

Costs which remain relatively constant regardless of the level of activity are known as *fixed*, or *indirect*, *costs*. This description implies that the fixed level is maintained whether output is nil or at 100 percent capacity. In some cases this assumption is not valid; fixed costs may tend to increase as output increases, and they can vary with time. However, the change is usually not significant for short-run studies.

Some of an organization's expenditures which can be considered as fixed are shown in Figure 15.1(b). These costs may be thought of as "preparation" expenses. They arise from measures taken to provide the means to produce a product or service. Before painters can paint a house, they must have paint brushes. Whether they paint one house or a dozen with the brushes, the expense has already been incurred and shows as a fixed cost. The painter's insurance and advertisements for work would also be indirect costs.

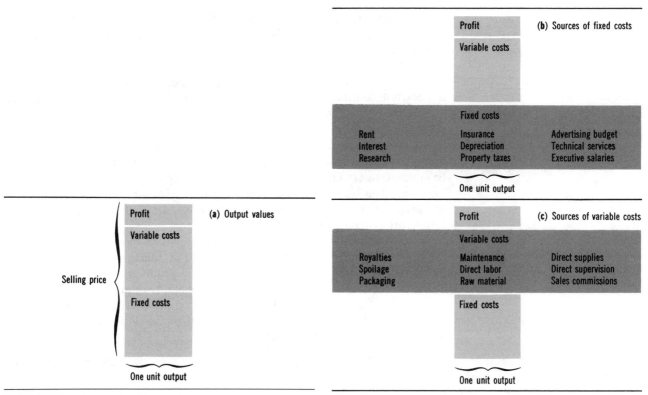

FIGURE 15.1
Unit costs.

Variable Costs

Costs which are generally proportional to output are called *variable*, or *direct*, *costs*. Such costs are relatively easy to determine because they are directly associated with a specific product or service. When there is no output, variable costs are zero. The input material and the time required to make a unit give rise to variable costs. For example, the specific type and quantity of paint painters use in painting a house is a variable cost. The more houses they paint, the more paint they use; the quantity used is a function of their output. In a similar manner, the time they spend painting is a direct cost.

Profit

The dimension of quantity must be included to examine the competitive aspects of profit. A single unit of output is relatively immune to competition. In isolated instances, a fair-sized output distributed in a local area to satisfy a peculiar need is also shielded from competition. However, as output quantity expands, competition is an increasingly apparent factor. Profit is the cause and effect of competitiveness.

A profit (or loss) figure attracts a great amount of attention. It is a handy yardstick of success. Like a thermometer, it measures only the level achieved; it does not control the source it measures. Unlike a thermometer, however, continued low readings may convince the financial temperature takers to eliminate the source.

There are basically three ways to increase profit: (1) increase the selling price, (2) increase the value to increase sales, and (3) decrease the selling price to increase sales. The profit-expansion descriptions are oriented to consumers' interests. The issues become more complicated when we look at them from the producer's viewpoint. Figure 15.2 shows some of the consequences of selling-price manipulations.

The original price-cost-quantity conditions are shown in Figure 15.2(a). *Total revenue* is the product of N units sold at selling price P. *Total cost* is the sum of variable and fixed costs incurred in producing N units. *Profit* is the difference between revenue and total cost (when revenue exceeds costs).

Figure 15.2(b) through (d) shows increased profit. The shaded profit areas of (b), (c), and (d) are equal and are larger than the profit in (a). The dangers and limitations of profit-expansion methods are as follows:

Increased selling price. Competing products or services set an upper limit to price increases. This limit is ultimately controlled by the consumers. Their willingness to pay is a function of the value they expect to receive and their loyalty to a product. Prices higher than competing products or equivalent value

> An old proverb says that there are two fools in every market: one asks too little, one asks too much.

> Conditions of free enterprise are assumed: Similar products or services are available from a number of vendors.

FIGURE 15.2
Methods for expanding profit.

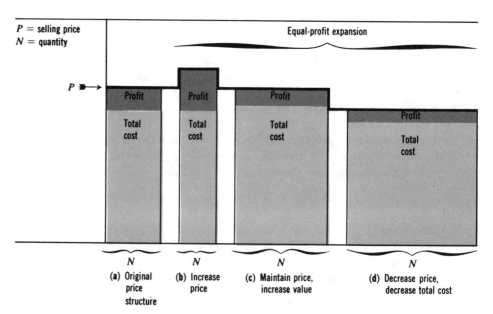

P = selling price
N = quantity

Equal-profit expansion

P ➡→

| Profit | Profit | Profit | Profit |
| Total cost | Total cost | Total cost | Total cost |

N N N N

(a) Original price structure   (b) Increase price   (c) Maintain price, increase value   (d) Decrease price, decrease total cost will reduce the number of units sold. The shrinking share of the market eventually causes a decline in total profit.

Unchanged selling price. One way to increase profit without changing the selling price is to sell more units by increasing the value. The greater value perceived by the consumer can result from better quality, more quantity, or more effective advertising. All these measures increase the total cost of the producer. Higher total cost leads to a lower margin of profit per unit sold. If the market is unstable, a very low profit margin can seriously limit recuperative powers during market fluctuations.

A straightforward means to increase profit while holding prices constant is to reduce total cost. Such a task is the continuous aim of engineers and managers. The obstacle is that it becomes increasingly difficult to make more and more savings in an established operation. At first it is easy. When a product or service is new, it meets a high current demand which compensates for operational inefficiencies. As competition forces the price down, the "fat" is removed from operations. Further effort to reduce costs meets diminishing returns. It is like trying to make a horse run faster. A small whip may help at first, but using ever larger whips fails to force proportional returns in greater speed.

Reduced selling price. New areas of cost reduction are exposed by changing the level of operations or capacity. A greater output often allows new methods to be incorporated. Some of the savings resulting from the new methods are passed on to consumers in the form of a lower selling price. In theory, the decreased price should lead to the sale of more units, which in turn satisfies the conditions for incorporating the new methods.

Limitations are inherent throughout the cost reduction–lower price–increased sales cycle. Cost reductions are limited by minimum levels of quality, maximum levels of expenditure for new equipment, and basic labor or material costs that resist lowering. Reduced prices may be an insufficient incentive to attract enough new sales. However, with reasonable care the cycle rewards the producer and leads to a better standard of living for the consumer.

BREAKEVEN COMPARISONS

After a decision has been made to pursue a venture that promises to be beneficial, a subsequent decision is usually required to select the best tactic to accomplish the venture's purpose. One method of accomplishment seldom stands out as the best possible when all angles are considered. Even if the choice among methods has been narrowed to those that can perform the desired function adequately, it is still likely that different alternatives are better over certain ranges of activity. A breakeven comparison detects the range over which each alternative is preferred. Then the decision maker has only to decide the most likely range of future operations to select the proper method

FIGURE 15.3
Lease-or-buy comparison.

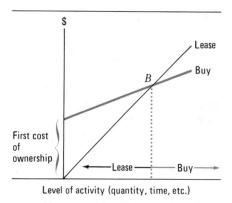

Level of activity (quantity, time, etc.)

to complete the venture. This practice was illustrated for sensitivity analysis in Chapter 14.

A "make-or-buy" decision is frequently encountered. It occurs when an item can be made in-house at a lower variable cost than that of the purchase price for the item from a vendor. The point at which the alternatives are equal depends on the first cost required to begin the in-house production. This situation was illustrated in Figure 1.6 and was used to point out the dangers of suboptimization.

A similar problem arises in a decision to lease or buy an asset. The costs of ownership (purchase price, installation cost, etc.) must be added to operating costs in comparing an owned item to one that is leased. Any fixed-cost investment to lower variable costs with respect to a strictly direct-cost alternative depends on the time period required to pay off the investment from variable-cost savings. One advantage of diagraming a breakeven comparison, as shown in Figure 15.3, has to do with the attention focused on the breakeven point B, which designates the level of activity and implies the time period separating the preference for alternatives. The more conventional practice of comparing costs at a specified level of activity was described in Chapter 13.

BREAKEVEN CHARTS

The best-known breakeven model relates fixed and variable costs to revenue for the purpose of profit planning. The name *breakeven chart* is derived from the concept it depicts, the volume or level at which revenue and total cost of operations exactly break even. At this point, one additional unit made and sold would produce a profit. Until the breakeven point is attained, the producer operates at a loss for the period.

Standard Format

Properties of a typical breakeven chart are displayed in Figure 15.4. The vertical scale shows the revenue and costs in monetary units. The horizontal

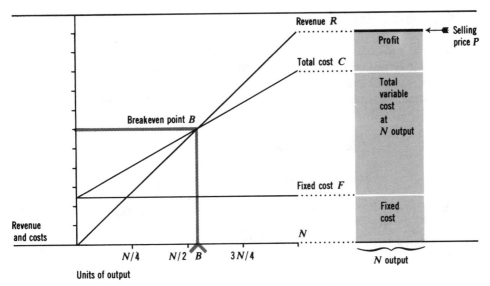

FIGURE 15.4
Standard format for breakeven charts.

scale indicates the volume of activity during the period pictured. The units of volume can be in sales dollars, number of units produced and sold, or output quantity expressed as a percentage of total capacity.

The horizontal line in the chart shows the fixed costs F, which are constant throughout the range of volume. The sloping line originating at the intersection of the fixed-cost line and the vertical axis represents variable costs V plus the fixed costs. For linear relations, variable costs are directly proportional to volume; each additional unit produced adds an identical increment of cost. The sum of variable and fixed costs is the total cost C. The sloping line from the origin of the graph is the revenue line. Revenue R is also assumed to be directly proportional to the number of units produced and sold at price P.

The breakeven point B occurs at the intersection of the total-cost and revenue lines. It thus specifies the dollar volume of sales and the unit volume of output at which an operation neither makes nor loses money. The vertical distance between the revenue line and the total cost line indicates a profit Z to the right of B and a loss to the left.

A block cost diagram used in the discussion of profit expansion is shown alongside the breakeven chart in Figure 15.4. For an output volume of n, the costs, revenue, and profit are the same in both formats. The breakeven chart further indicates the profit or loss expectation at levels of output other than the specific quantity N. This feature helps explain such statements as, "A very low profit margin can seriously limit recuperative powers during market fluctuations." "A very low profit margin" means that the output is barely on the profit side of the breakeven point. An unstable market could easily cause sales to fall below point B and show a loss for the period.

Requisite conditions include:
1 No financial costs
2 Income only from operations
3 P, V, and F constant over time and output
4 All units that are produced are sold

Algebraic Relationships

The graph format is convenient for clarifying or presenting economic relationships. It is possible to obtain quantities for particular conditions by scaling values from the chart. However, the same conditions can be easily quantified by formulas. Calculations generally provide greater accuracy. Using the symbols already defined, we have

$$\text{Revenue/period} = R = nP$$
$$\text{Total cost/period} = C = nV + F$$
$$\text{Gross profit/period} = Z = R - C = n(P - V) - F$$

where n can also be a fraction of total capacity when P and V represent total dollar volume at 100 percent capacity.

At the breakeven point, profit equals zero. To determine the output to simply break even, we have, at B,

$$Z = 0 = R - C = n(P - V) - F$$

and, letting $n = B$,

$$B = \frac{F}{P - V}$$

The term $P - V$ is called *contribution*. It indicates the portion of the selling price that contributes to paying off the fixed cost. At $n = B$ the sum of contributions from B units equals the total fixed cost. The contribution of each unit sold beyond $N = B$ is an increment of profit.

Example 15.1
Breakeven by the Numbers

An airline is evaluating its feeder routes. These routes connect smaller cities to major terminals. They are seldom very profitable themselves, but they feed passengers into the major flights which yield better returns. One feeder route has a maximum capacity of 1000 passengers per month. The contribution from the fare of each passenger is 75 percent of the $120 ticket price. Fixed costs per month are $63,000. Determine the breakeven point and net profit when the effective income tax rate is 40 percent.

Solution 15.1

To find the average percentage of seats that must be sold on each flight to break even, the cost and revenue data could be converted to the graphical breakeven format shown in Figure 15.5. The same information displayed in the breakeven chart is supplied by the following calculations:

Total maximum revenue per month is

$$nP = 1000 \times \$120 = \$120,000$$

Total contribution = $0.75 \times \$120,000 = \$90,000$

$$B \text{ (\% of capacity)} = \frac{F \times 100\%}{\text{contribution}}$$

$$= \frac{\$63,000}{\$90,000} \, 100\% = 70\%$$

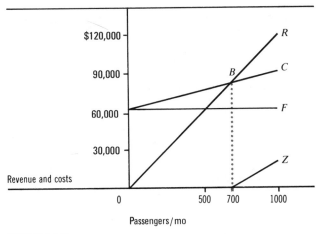

FIGURE 15.5
Breakeven chart for an airline operation.

or

$$B \text{ (passengers)} = \frac{F}{P - V} = \frac{\$63,000}{0.75 \times \$120}$$

$$= 700 \text{ passengers/month}$$

With a tax rate (where t = tax rate) of 40 percent, the net profit at full capacity is

$$
\begin{aligned}
\text{Net profit} &= Z(1 - t) \\
&= (R - C)(1 - t) \\
&= [n(P - V) - F](1 - t) \\
&= [1000(0.75 \times \$120) - \$63,000](0.60) \\
&= \$27,000 \times 0.60 = \$16,200
\end{aligned}
$$

A gross-profit line is shown in the lower-right corner of the chart, starting at B. Net profit is a fraction of Z, which depends on the tax rate for the total earnings of the organization.

LINEAR BREAKEVEN ANALYSIS

A breakeven chart displays cost-volume-profit relations that hold only over a short run. Over the long run, the relationships are altered by internal factors (new products, production facilities, etc.) and external impacts (competition, state of the general economy, etc.). Many of the internal activities that affect long-run changes are initiated from analyses of current cost-volume-profit conditions. Thus, a breakeven analysis is like a medical checkup: the physical examination reveals the current state of health and provides clues about what should be done to become or stay healthy.

Breakeven-Point Alternatives

Any change in costs or selling price affects the breakeven point. We observed the gross effects of profit expansion as a function of selling price. Now we can consider the interaction of revenue, variable costs, and fixed costs in terms of output.

A lower breakeven point is a highly desirable objective. It means that the organization can meet fixed costs at a lower level of output or utilization. A sales level well above the breakeven output is a sign of healthiness. Three methods of lowering the breakeven point are shown in Figure 15.6. The original operating conditions are shown as light lines and are based on the following data:

FIGURE 15.6
Breakeven alternatives.

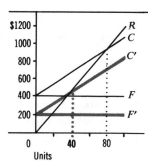

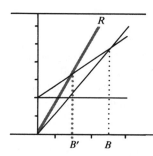

Fixed-cost reduction: By reducing the fixed costs by half, B is halved. Therefore the new fixed cost F' is

$$F' = \frac{\$400}{2} = \$200$$

and the new breakeven output B' becomes

$$B' = \frac{F'}{P - V}$$

$$= \frac{\$200}{(\$12 - \$7)\text{ per unit}} = 40 \text{ units}$$

Variable-cost reduction: Knowing that B' should equal 40 units, we can solve for the associated V' by

$$B' = \frac{F}{P - V'}$$

or $\quad V' = P - \dfrac{F}{B'} = \$12 - \dfrac{\$400}{40}$

$$= \$12 - \$10 = \$2/\text{unit}$$

Selling-price increase: Raising the selling price P' increases the slope of the revenue line and augments the contribution. The contribution required for B' to equal 40 units is

$$\text{Contribution} = \frac{F}{B'} = \frac{\$400}{40}$$

$$= \$10/\text{unit}$$

which leads to

$$P' = V + \text{contribution}$$
$$= \$7 + \$10 = \$17$$

$$V = \$7/\text{unit}$$
$$P = \$12/\text{unit}$$
$$R(\text{at } n = 100 \text{ units}) = \$1200$$
$$C(\text{at } n = 100 \text{ units}) = \$1100$$
$$F = \$400$$
$$B = 80 \text{ units}$$

The bold lines depict the measures necessary to reduce the breakeven point by half, from 80 to 40 units.

Margin of Profit and Dumping

The perspective from which an organization views its breakeven point depends on its particular circumstances at the time of review. A firm struggling to get its products accepted in the market would have concerns different from those of a producer that has already captured a large share of the market. The struggling firm is likely beset with problems of cash availability to meet its fixed costs, whereas the dominant producer probably seeks investments to maintain or expand its sales.

Example 15.2
Making More Money by Raising Costs and Decreasing the Price

A firm produces package waste-disposal units that sell for $35,000 each. Variable costs are $20,000 per unit and fixed costs are $600,000. The plant can produce a maximum of 80 units per year. It is currently operating at 60 percent capacity. The firm is contemplating the effects of reducing the selling price by $2000 per unit, adding a feature to each unit which will increase the variable costs by $1000, and allocating an extra $120,000 per year for advertising. These actions are designed to sell enough additional units to raise plant utilization to 90 percent. Evaluate the alternatives.

Solution 15.2

Under current conditions

$$B = \frac{F}{P - V} = \frac{\$600,000}{\$35,000 - \$20,000} = 40 \text{ units}$$

Since the company now sells $0.60 \times 80 = 48$ units per year, the gross annual profit is

$$Z = \text{units sold beyond } B \times \text{contribution/unit}$$
$$= (48 - 40)\$15,000 = \$120,000$$

The ratio of gross annual profit to fixed costs is

$$\frac{Z}{F} = \frac{\$120,000}{\$600,000} = 0.20$$

and may be thought of as a margin of profit or safety. The same ratio can be obtained by

$$\text{Margin of profit} = \frac{n - B}{B} = \frac{48 - 40}{40} = 0.20$$

where n is the number sold during the period.

Both the current conditions (dotted lines) and the anticipated conditions (solid lines) are displayed in Figure 15.7. As graphed, B increases, as a result of the added expenditures, to

$$B' = \frac{\$600,000 + \$120,000}{\$33,000 - \$21,000} = \frac{\$720,000}{\$12,000} = 60 \text{ units}$$

and the gross profit expected at 90 percent capacity also increases to

$$Z = [(0.90 \times 80) - 60](\$33,000 - \$21,000)$$
$$= 12 \text{ units} \times \$12,000/\text{unit} = \$144,000$$

but the margin of profit remains unchanged at

$$\frac{Z}{F} = \frac{\$144,000}{\$720,000} = 0.20$$

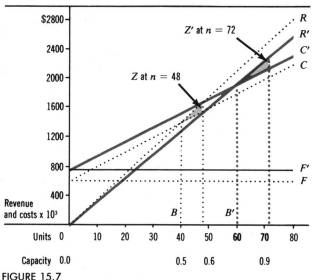

FIGURE 15.7
Original and anticipated conditions for profit improvement from increased plant utilization.

The firm could also follow a course of action in which only one or two of the alternatives are pursued. If the advertising budget is eliminated but price and modifications are retained, the same profit ($144,000) would be obtained at an output of

$$n = \frac{Z + F}{P - V} = \frac{\$144,000 + \$600,000}{\$33,000 - \$21,000}$$

$$= \frac{\$744,000}{\$12,000/\text{unit}} = 62 \text{ units}$$

and

$$B = \frac{\$600,000}{\$12,000/\text{unit}} = 50 \text{ units}$$

which makes the margin of safety $(62 - 50)/50$, or 0.24.

There are, of course, many factors to consider in such a decision. If the market is stable, the margin of profit is less important. Some alternatives are easier to implement than others. Some outcomes are more certain than others.

There is yet another alternative for the firm described in Example 15.2. It would be to sell a portion of the output at a reduced price. This practice is called *dumping*. It can be accomplished by selling to foreign markets at a lower price or by selling the same product at different prices under different names. There are many dangers in this practice, but if it works, profit will increase because of the increased plant utilization.

Figure 15.8 illustrates dumping applied to the original data on waste-disposal-unit sales: 48 units are sold at the regular price ($P = \$35,000$) to account for 60 percent utilization. If 100 percent utilization could be achieved by *dumping* the remaining capacity at a sales price P' of $25,000 per unit, the gross profit would increase from $120,000 to $280,000.

Multiproduct Alternatives

More than one product can be shown on a breakeven chart. Including a whole product line allows the decision maker to evaluate the combined effect of the product mix on plant utilization, revenue, and costs. A slightly different format for the breakeven chart accentuates the effect of multiple products. This type

FIGURE 15.8
Effects of dumping on price, cutting profit, and utilization.

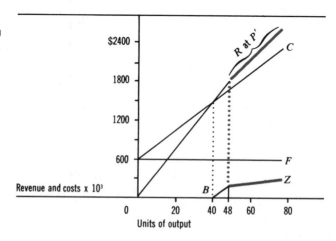

of graph, displayed in Figure 15.9, is called a *multiproduct profit*, or *contribution, chart.*

The chart is constructed by plotting fixed costs as a loss on the vertical axis. The horizontal axis denotes sales revenue. At zero sales the only costs associated with a product are the negative preparation, or fixed, costs. As production and sales develop, each unit sold makes a contribution toward paying off the fixed costs. When enough units have been sold to pay these costs, the breakeven point is reached and the contribution from further sales is profit.

Three products, A, B, and C, are represented in Figure 15.9. The plotted values are based on the following assumptions made for the period shown in the chart:

	Product		
	A	B	C
Selling price per unit	$8	$5	$4
Contribution per unit	$6	$3	$3
Fixed cost	$4000	$7000	$1000
Number of units sold	600	4000	400

Product A is entered in the chart by marking its fixed costs ($4000) on the negative side of the breakeven line at zero sales revenue. The total contribution

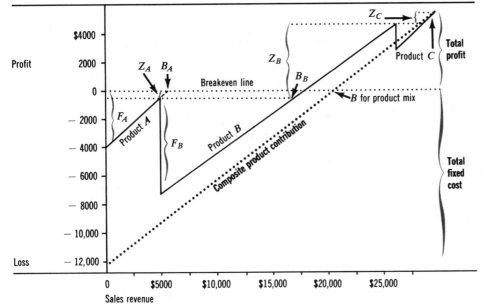

FIGURE 15.9
Multiproduct profit chart.

Another version of a multiproduct
chart based on a cumulative
contribution line is shown in
Figure 15.14.

of A is 600 units times $6 per unit equals $3600. Subtracting the fixed costs from the total contribution leaves a loss of $400 from total sales of $4800.

The remaining products are dealt with in a similar manner. Each fixed cost is entered as a vertical line attached to the highest point on the contribution line for the preceding product. The results are cumulative. The distance that the last contribution segment extends above the breakeven line is the total plant profit. The dotted line extending diagonally across the chart is the composite contribution line. The point at which this line crosses the plant breakeven line establishes the sales volume at which fixed costs are exactly covered.

The value of multiproduct breakeven charts lies in their use for product comparisons. The portion of fixed costs borne by each product is easily observed. A product is preferred when its contribution line is steeper than the composite contribution line. Such considerations are important in decisions to add new products or drop old ones.

NONLINEAR BREAKEVEN ANALYSIS

Cost and revenue functions do not always follow convenient linear patterns. More often than not, realistic cost relationships develop a nonlinear pattern, as typified by Table 15.1. The first four columns in the table relate output n to total fixed cost F, total variable cost TV, and total cost C. The right side of the table shows average and marginal costs derived from the figures tabulated on the left.

Characteristic patterns of average and marginal costs based on Table 15.1 are pictured in Figure 15.10 and discussed on page 417.

TABLE 15.1

Total and average cost data for a firm's operations over a given time period

Total Product, n	Total Fixed Cost, F	Total Variable Cost, TV	Total Cost, C	Average Fixed Cost, F/n	Average Variable Cost, TV/n	Average Total Cost, C/n	Marginal Cost, $\Delta C/\Delta n$
0	$3000	$ 0	$ 3000	—	—	—	
1	3000	700	3700	$3000	$700	$3700	$ 700
2	3000	1300	4300	1500	650	2150	600
3	3000	1800	4800	1000	600	1600	500
4	3000	2400	5400	750	600	1350	600
5	3000	3100	6100	600	620	1220	700
6	3000	3900	6900	500	650	1150	800
7	3000	4900	7900	429	700	1129	1000
8	3000	6200	9200	375	775	1150	1300
9	3000	7800	10,800	333	867	1200	1600

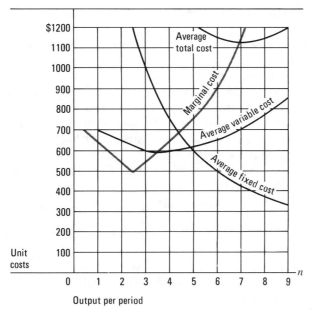

FIGURE 15.10
Relationship of average and marginal costs. Minimum points on the average- and total-cost curves occur where they cross the marginal-cost curve.

Average fixed cost. Since fixed costs are independent of output, their per-unit amount declines as output increases. This feature is recognized when business people speak of "higher sales spreading the overhead."

Average fixed cost $= F/n$

Average variable cost. The typical saucer-shaped average-cost curve declines at first, reaches a minimum, and then increases thereafter. It reflects the law of diminishing returns. Initially, combining variable resources *TV* with fixed resources *F* produces increasing returns, but a point is reached where more and more variable resources must be applied to obtain each additional unit of output. Stated another way, a fixed plant is underemployed when its output is below the minimum average-cost point. As output expands, more complete utilization of the plant's capital equipment will make production more efficient. But continually increasing variable costs will eventually create a condition in which overcrowding and overutilization of equipment impair efficiency.

Average variable cost $= TV/N$

Average total cost. Because average total cost is simply the sum of average fixed and average variable costs, it shows the combined effects of spreading out fixed charges and diminishing returns from variable resources.

Average total cost $= C/n$

Marginal cost. The key to the cost pattern in Figure 15.10 is contained in marginal-cost concepts. Marginal cost is calculated from either *C* or *TV* as the *extra increment of cost required to produce an additional unit of output.* If the last increment of cost is smaller than the average of all previous costs, it pulls the average down. Thus, *average total cost declines until it equals marginal cost. Equivalently, the rising marginal-cost curve also pierces the average-variable cost curve at its minimum point.*

Marginal cost $= \Delta C/\Delta n$

Marginal Revenue and Profit

Both nonlinear revenue and cost schedules may be expressed as formulas. When such equations are available, their analysis is not much more difficult than that of linear models. Since an assumption of linearity makes all monetary increments constant over an extended range of output, nonlinear models call more attention to marginal relationships.

Marginal revenue $= \Delta R / \Delta n$

Marginal revenue is the additional money received from selling one more unit at a specified level of output. For linear revenue functions, the marginal revenue is a constant value P. That is, for each additional unit sold, the total revenue is increased by P dollars. Consequently, a greater output automatically increases the total profit.

When the linear relationship is replaced by an expression such as

$$\text{Selling price} = P = 21{,}000n^{-1/2} \text{ dollars/unit}$$

the price of each unit is not so obvious. Such expressions are examined with differential calculus. For the price function above, the rate of change of revenue with output is

$$\text{Marginal revenue} = \frac{dR}{dn} = \frac{d(nP)}{dn} = \frac{d(21{,}000n^{1/2})}{dn}$$

$$= 10{,}500n^{-1/2}$$

Figure 15.11(a) shows a decelerating revenue rate and linear costs. Decreasing marginal revenue could result from a policy of lowering prices in order to achieve a higher plant utilization. The nonlinear revenue curve fixes two breakeven points. Between these two points the firm operates at a profit. Outside the breakeven points a loss is incurred, as shown in the profit chart, Figure 15.11(b).

The graphs are based on the following data:

$$n = 1 \text{ unit produced and sold/period}$$
$$V = \$1000/\text{unit}$$
$$F = \$100{,}000/\text{period}$$
$$P = \$21{,}000n^{-1/2}/\text{unit}$$

Breakeven analyses that account for inflation in P and v are examined in Extension 15A.

Using the formula for total revenue and the selling-price function, $R = nP = 21{,}000n^{1/2}$, leads to a gross-profit equation of

$$Z = R - C = R - (nV + F)$$
$$= 21{,}000n^{1/2} - 1000n - 100{,}000$$

Knowing that $Z = 0$ at a breakeven point allows the value or values for B to be determined by rearranging the terms for Z:

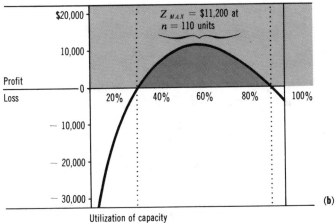

FIGURE 15.11
Nonlinear breakeven charts for decreasing marginal revenue and linear costs.

$$Z = 0 = -10^3n - 10^5 + 21(10^3n^{1/2})$$

or

$$10^3n + 10^5 = 21 \times 10^3n^{1/2}$$

Squaring each side of the equation and dividing by 10^6 results in

$$n^2 + 200n + 10^4 = 441n$$

and by collecting terms,

$$n^2 - 241n + 10^4 = 0$$

For an equation in the quadratic form $Ax^2 + Bx + C = 0$,

$$x = \frac{-B \pm \sqrt{B^2 - 4AC}}{2A}.$$

This expression can be solved with the quadratic formula.

$$n = \frac{241 \pm \sqrt{241^2 - 4 \times 10^4}}{2} = \frac{241 \pm 134}{2} = 53 \text{ or } 188$$

or $\qquad B = 53$ units $\qquad$ and $\qquad B' = 188$ units

The point of maximum profit is especially important when two breakeven points are present. As indicated in Figure 15.11(b), the rate of profit is increasing to the left of the point of maximum profit and decreasing to the right. The rate of change of profit with respect to output is *marginal profit*. At the point of maximum profit, the rate of change (and the marginal profit or slope of the profit line) is zero. Therefore, to find this point, differentiate the profit equation, set the derivative equal to zero, and solve for n:

A graph of marginal profit with respect to marginal revenue and marginal cost is shown in Figure 15.16.

$$\frac{dZ}{dn} = \frac{d(21{,}000n^{1/2} - 1000n - 100{,}000)}{dn} = 0$$

$$= 10{,}500n^{-1/2} - 1000 = 0$$

$$n = \left(\frac{10{,}500}{1000}\right)^2 = 110 \text{ units}$$

Marginal Cost and Average Unit Cost

As production increases, the total cost per unit may also increase, owing to greater maintenance needs, overtime payments to workers, and general inefficiency caused by congestion during stepped-up operation. Under these conditions there is an increasing *marginal cost* (rate of change of total cost with output).

One possible pattern of marginal costs and linear revenue is shown in Figure 15.12. There could be many patterns: two breakeven points (as in Figure 15.11), decreasing marginal costs owing to savings realized from quantity purchases or near-capacity mechanized production, nonlinear functions for both revenue and costs, etc.

The feature points of a breakeven analysis are determined in the manner described previously. To find B, set the profit equation equal to zero and solve for $n = B$. Differentiate the profit equation and set the derivative equal to zero to solve for the output that produces maximum gross profit. In doing so, note that

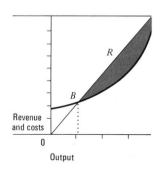

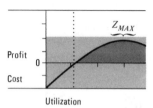

FIGURE 15.12
Breakeven and maximum-profit points for increasing marginal cost and linear revenue.

$$\frac{dZ}{dn} = \frac{d(nP - nV - F)}{dn} = 0$$

$$= \frac{d(nP)}{dn} - \frac{d(nV + F)}{dn} = 0$$

or
$$\frac{d(nP)}{dn} = \frac{d(nV + F)}{dn}$$

Marginal revenue = marginal cost

at the operating level that produces maximum profit, which means that when the change in revenue from one additional sale equals the change in cost of producing one more unit, the point of maximum profit is attained. Graphically, this point is at the output at which the vertical distance between the total-cost curve and the revenue curve is greatest. If marginal revenue and marginal cost were plotted, the output for maximum profit would be at the intersection of the two curves.

The relation of marginal costs and benefits in controllng pollution are explored in Extension 15B.

For linear functions, the average unit cost is

$$\frac{nV + F}{n} = V + \frac{F}{n}$$

where n = a specific output

In this case the average unit cost continually decreases with increasing output. For nonlinear costs this condition is not necessarily true. If the average cost goes through a minimum point and then increases, the slope of the curve will be zero at the output for lowest average unit cost:

$$\frac{d(V + Fn^{-1})}{dn} = 0$$

$$\frac{dV}{dn} - \frac{F}{n^2} = 0$$

$$\frac{dV}{dn} = \frac{F}{n^2}$$

From this equation and the maximum-profit equation it is clear that for nonlinear relations, the point of minimum average unit cost does not necessarily coincide with the maximum-profit point (see Review Exercise 4).

APPLYING BREAKEVEN ANALYSES

A breakeven analysis often tends to oversimplify the decision environment. This is an attribute for presentation purposes and for gross evaluations. It can also be a shortcoming for problems in which detailed measures are needed. A decision to lower the breakeven point for an operation can result from a study of total revenue and costs, but the study alone seldom reveals the in-plant operations that engineers and managers must conduct to implement the decision. The inability to identify tactical procedures is not a defect of a breakeven analysis; it merely indicates that decision makers should be aware of the limitations of the approach in order to apply it appropriately.

The validity of a breakeven chart is directly proportional to the accuracy of the data incorporated in the chart. When several products are lumped together and represented by one line on a chart, there is a distinct possibility that poor performance by one product may go undetected. A firm should have a good cost-accounting system, but data from past performance are not always indicative of future performance. However, examining graphs of previous breakeven conditions calls attention to developing trends in revenues and costs.

Breakeven relationships suggest where engineering efforts can be of most use to an organization. Field or factory-floor engineers can observe the present state of financial affairs and use such observations to guide their cost-control activities. As an engineer's managerial responsibilities increase, the interplay of price, cost, and quantity becomes of greater concern. Then the combined effect of the system's operations and the underlying economic principles merge to steer strategic decisions.

Review Exercises and Discussions

Exercise 1   An engineering consulting firm won a contract to design and supervise construction of a sewage-treatment plant at a remote location. The installation phase will last at most 2 years, and two engineers from the firm will supervise on-site operations. They will need both living accommodations and an office. Three alternatives are available, with the costs shown below.

1   Rent a building with furnished living accommodations and an office: $3000 per month including upkeep and utilities.
2   Buy two furnished trailers to live in and rent an office: The purchase price of a house trailer is $24,000 per trailer (the seller will buy back a used trailer for 40 percent of its purchase price any time within 2 years); trailer upkeep, site rental, and utilities are $200 per trailer per month; and office rental is $800 per month.
3   Buy three trailers: Two house trailers as in alternative 2 and a smaller one to serve as an office, purchased for $16,000 from the same seller.

If all the alternatives provide adequate facilities, which one do you recommend?

Solution 1   Total costs C for the three alternatives are calculated as

$$C(1) = \$3000N \qquad \text{where } N \text{ is the number of months needed}$$
$$C(2) = (2 \times \$24,000)0.6 + [(2 \times \$200) + \$800]N = \$28,800 + \$1200N$$
$$C(3) = [(2 \times \$24,000) + \$16,000]0.6 + (3 \times \$200)N = \$38,400 + \$600N$$

As is apparent in Figure 15.13, $C(1) = C(2)$ at $N = 16$ months; $C(1) = C(3)$ at $N = 16$ months; and $C(2) = C(3)$ at $N = 16$ months. The decision thus narrows to a choice between

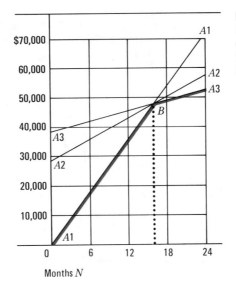

FIGURE 15.13
Breakeven comparison for three alternatives. A2 is dominated by the other two, making the choice between A1 and A3.

alternatives 1 and 3, which, in turn, depends on the engineers' estimate of how long the project will take. If it takes longer than 16 months, 3 is preferred. Otherwise, 1 is less expensive. The convenience offered by the rented building (alternative 1 has upkeep and utilities paid, no trade-in hassles, etc.) would likely sway the decision.

A fertilizer plant is operating at capacity with production of four mixes which have a total sales volume of $2 million. The sales and production-cost figures for mixes W, X, Y, and Z are as shown: **Exercise 2**

	Mix W	Mix X	Mix Y	Mix Z	Totals
Percentage of total sales	10	20	30	40	100
Contribution (percent of P)	45	40	45	35	
Fixed cost charged	$70,000	$180,000	$210,000	$220,000	$680,000
Profit	$20,000	−$20,000	$60,000	$60,000	$120,000

Recognizing the loss incurred with product X, the company is considering dropping the product or replacing it with another mix. If the product is dropped without a replacement, the new sales and cost figures are estimated to develop as follows:

	Mix W	Mix Y	Mix Z	Totals
Percentage of total sales	15	35	50	100 (for R of $1,800,000)
Contribution (percent of P)	45	45	35	
Fixed cost charged	$100,000	$250,000	$290,000	$640,000

Construct a multiproduct profit chart to compare the two alternatives.

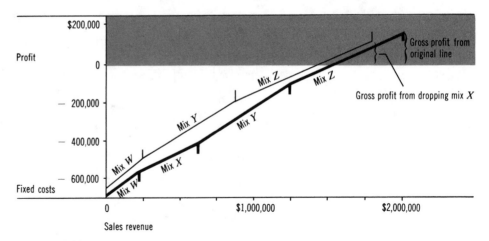

FIGURE 15.14
Multiproduct contribution chart showing reduced profit from the revised mix.

Solution 2   The anticipated and original conditions are displayed in Figure 15.14 in a slightly different form of multiproduct breakeven chart. In this version the first entry is made at zero revenue and the point of maximum fixed cost. Then the contributions of each product are plotted progressively to the right. All other interpretations are the same for the two forms of multiproduct charts.

The new product line without mix X would be less profitable than the former line:

	Mix W	Mix Y	Mix Z	Total
Profit	$21,500	$33,500	$25,000	$80,000

The main reason gross profit falls from $120,000 to $80,000 is that most of the fixed costs carried by mix X did not disappear with its elimination from the product line; depreciation, taxes, engineering services, and other indirect costs are still required to run the batching plant, and their magnitude is not reduced appreciably by dropping mix X. The situation is further aggravated by having the largest sales increase occur for mix Z, which has a lower contribution rate than does the product eliminated.

Exercise 3   A manufacturing company produces several different products that are manufactured on the same equipment but are marketed individually. An audit of one product reveals the data shown in Table 15.2; all figures are corrected to a base year to avoid distortion by inflation or the general economy. The portion of manufacturing overhead allocated to the product as its fixed cost is $400,000.
     Analyze the product's life cycle costs and revenues.

Solution 3   Fixed cost chargeable to the product totals $400,000 and is included in the analysis, as shown in the second column of Table 15.3. Variable costs were high during the start-up period, dropped as production became more efficient, and climbed during the third and fourth years as a result of "relearning" when production runs were not continuous, owing

TABLE 15.2

Life cycle pattern of output, costs, and income

	Cumulative Output, Cost, and Revenue		
Product Life, Age	Output, Units	Variable Cost	Revenue Earned
Year 1	1000	$ 80,000	$ 200,000
	2000	112,000	400,000
	3000	134,000	600,000
Year 2	4000	160,000	780,000
	5000	190,000	930,000
Year 3	6000	252,000	1,050,000
	7000	352,000	1,160,000
Year 4	8000	480,000	1,270,000

TABLE 15.3

Product cost and revenue patterns over a 4-year product life

Output (N units)	Fixed Cost (FC)	Total Variable Cost (TVC)	Total Cost, TC (TC = TVC + FC)	Incremental Marginal Cost (1000 units)	Average Fixed Cost (FC/N)
0	$400,000	0	$400,000	$ 80,000	$400
1000	400,000	$ 80,000	480,000	32,000	200
2000	400,000	112,000	512,000	22,000	133
3000	400,000	134,000	534,000	26,000	100
4000	400,000	160,000	560,000	30,000	80
5000	400,000	190,000	590,000	62,000	67
6000	400,000	252,000	652,000	100,000	37
7000	400,000	352,000	752,000	128,000	50
8000	400,000	480,000	880,000		

Output (N units)	Average Variable Cost (TVC/N)	Average Total Cost (TC/N)	Total Revenue (R)	Incremental Marginal Revenue (1000 units)	Total Production Profit (R − TC)
0	0		0	$200,000	− $280,000
1000	$80	$480	$ 200,000	200,000	− 112,000
2000	56	256	400,000	200,000	66,000
3000	45	178	600,000	180,000	220,000
4000	40	140	780,000	150,000	340,000
5000	38	118	930,000	120,000	398,000
6000	42	109	1,050,000	110,000	408,000
7000	50	107	1,160,000	110,000	390,000
8000	60	110	1,270,000		

to declining sales. Although average fixed cost drops consistently with additional output, the average variable and total cost figures show the effect of increasing production costs.

Marginal costs, collected for increments of 1000 units of output, focus attention on key output levels. The lowest marginal cost occurred in the first year (at 3000 units) when the annual production rate was highest. At about 5000 units, marginal cost equaled the average variable cost at its lowest point. Average total cost bottomed out near 7000 units, where it equaled marginal cost. At about this same level, rising marginal costs met the dropping marginal revenues to signal the point of maximum profit. The decline in marginal revenue, stated in 1000-unit increments, is attributed to pressure from competitors.

Unless variable costs can be slashed or prices increased, the product is a candidate for termination or remodeling. Production factors that will influence the decision include the expected utilization of manufacturing facilities for other products, labor and material availability, and resource requirements (both money and time) for restoring productivity or producing a revamped product.

Exercise 4
A monthly record of operating expenses and revenue for a new manufacturing plant is posted in the manager's office. The purpose is to detect any changes in cost-revenue relationships and to establish the plant's operating pattern.

The graphic results of several months' operations are depicted in Figure 15.15. The pattern appears to indicate that as the output becomes greater, marginal costs increase and marginal revenue decreases. The belief is confirmed by formulas developed for curves that fit the data.

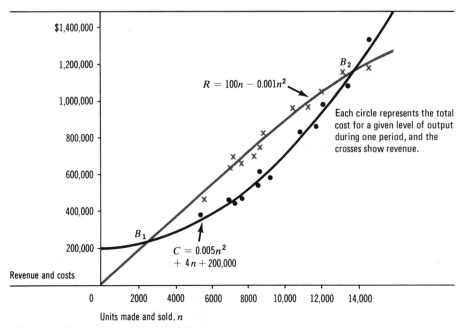

FIGURE 15.15 Production data and fitted curves.

The selling price of finished units varies according to $P = (100 - 0.001n)$ dollars per unit. The price behavior is attributed to lower per-unit quotes given for large orders. Fixed costs are considered reasonable at $200,000 per month. The variable costs, $V = (0.005n + 4)$ dollars per unit, also appear competitive. The plant is designed to produce 12,000 units per month.

Based on the given data, calculate the level of output which produces the greatest profit, the least average unit cost, and the breakeven points. On a graph, show the curves for average unit cost, marginal cost, marginal revenue, and marginal profit.

Maximum profit occurs where marginal revenue equals marginal cost, as shown in Figure 15.16 and the computations below.   **Solution 4**

$$\text{Marginal revenue} = \frac{d(nP)}{dn} = \frac{d(100n - 0.001n^2)}{dn} = 100 - 0.002n$$

$$\text{Marginal cost} = \frac{d(nV + F)}{dn} = \frac{d(0.005n^2 + 4n + 200,000)}{dn} = 0.01n + 4$$

At maximum profit, $100 - 0.002n = 0.01n + 4$

so

$$n = \frac{96}{0.012} = 8000 \text{ units}$$

For an output of 8000 units,

$$Z = R - C = 100n - 0.001n^2 - 0.005n^2 - 4n - 200,000$$
$$= 0.006n^2 + 96n - 200,000$$
$$Z_{8000} = -0.006(8000)^2 + 96(8000) - 200,000 = \$184,000$$

The output for maximum profit is also located by the point at which the marginal profit is zero:

$$\text{Marginal profit} = \frac{dZ}{dn} = \frac{d(-0.006n^2 + 96n - 200,000)}{dn} = 0$$

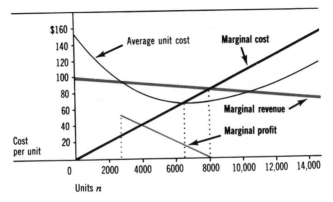

FIGURE 15.16
Average and marginal economic relationships.

$$0 = -0.012n + 96$$

$$n = \frac{96}{0.012} = 8000 \text{ units}$$

The minimum average cost occurs where the output satisfies the relation

$$\frac{dV}{dn} = \frac{F}{n^2}$$

$$\frac{d(0.005n + 4)}{dn} = \frac{200,000}{n^2}$$

$$0.005n^2 = 200,000$$

$$n = 6325$$

At $n = 6325$, the average cost is \$67.20 per unit, as determined from the average-cost formula $(0.005n + 4 + 200,000/n)$, and as shown in Figure 15.16.

The breakeven output is calculated from the gross profit, where $Z = 0$. Thus,

$$Z = 0 = -0.006n^2 + 96n - 200,000$$

$$n = \frac{96 \pm \sqrt{(96)^2 - 4 \times 0.006 \times 200,000}}{2 \times 0.006}$$

$$B = n = \frac{96 \pm 66.4}{0.012} = 2467 \text{ and } 13,533 \text{ units}$$

PROBLEMS

15.1 "Do you mean you would charge me \$78.50 to ride home with you? You're out of your mind! The bus costs only \$51. By air it's just \$66.50. Look at those figures again," implored Gaston.

"Look, Gassie," said Alphonse, "I *know* what it costs me to drive my car. I keep records, complete ones. It's all right here. You look again."

Alphonse handed the paper to Gaston. On it were written the costs of operating a car for 3 years, the length of time Alphonse expected to keep his car before trading it in:

Gasoline (40,000 miles at 7.2¢/mile)	\$2,880
Oil and grease	200
Tires	400
Repairs and maintenance	800
Insurance	900

License fees and property taxes   180

Depreciation in value ($10,000 − $6400)   3,600

Average interest on investment at 10%
interest ($3600/2 + $6400)(0.10)(3)   2,460

         Total cost for 3 years $11,420

Cost per mile = $11,420/40,000 miles = 28.55¢/mile

"OK, Alphie, let's assume your figures are correct. But you haven't used them right. You say the 550-mile trip costs $157. It doesn't. Your car isn't going to lose more value because you drive it an extra 550 miles. Your license fee doesn't go up. Your insurance premium is the same. You still have to pay interest. The only thing this trip will cost you is gas and oil. That's only about 7.7 cents per mile . . . say, $42 for the trip, my share being about $21."

"So you say," retorted Alphonse. "I say I use up my tires. I'm going to have to make repairs sooner. Go hitchhike if you want to be cheap."

"I just might do that. I also thought of catching the plane so I could get home in time for a party Saturday night. However, I will forgo the party and give up a relaxing bus ride to keep you company, if you will be reasonable. Now let's look at that list again, Alphie, and remember that car of yours is going to get older whether I ride in it or not. My contribution is clear profit to you. You've got to go home anyway."

"So have you," replied Alphonse, settling more firmly into his bargaining position.

(a) What is Gaston's fair share of the cost if he rides home with Alphonse? Why?
(b) How do you reconcile the counterarguments that Gaston's share of the trip is pure profit to Alphonse and, conversely, whatever Gaston can save compared with the bus price is pure profit to him?

15.2 Two alternative methods* of crude-oil processing in a producing oil field are available: a manual tank battery (MTB) and an automated tank battery (ATB). Tank batteries are composed of heaters, treaters, storage tanks, and other equipment, which remove salt water and sediment from crude oil prior to its entrance into pipelines for transport to an oil refinery.

Fixed costs for a tank-battery installation include depreciation, taxes, etc. Variable costs that increase in direct proportion to the volume of oil being processed include chemical additives and heating cost. In addition, there are constant costs independent of volume. All these costs are tabulated below.

	ATB	MTB
Annual cost of labor and maintenance	$5485	$7921
Fixed annual cost of depreciation and taxes	4064	2375
Variable cost per barrel of oil processed	0.0114	0.0081

At what level of oil production is the automatic tank battery preferred?

15.3 An import company buys foreign-made sewing machines for $80 per unit. Fixed costs of the operation are $140,000 per year. The sewing machines are sold on commission

* Adopted from E. J. Ferguson and J. E. Shamblin, "Break-Even Analysis," *Journal of Industrial Engineering*, April 1967.

by door-to-door sales representatives who receive 40 percent of the selling price for each machine sold. At what price should the machines be sold to allow the importers to break even on a shipment of 5000 sewing machines?

15.4 A privately owned summer camp for youngsters has the following operating data for a 12-week session:

Charge per camper	$120/week
Variable cost per camper	$80/week
Fixed costs	$48,000/season
Capacity	150 campers

   (a) What is the total number of campers that will allow the camp to just break even?
   (b) What is the profit for the 12-week season if the camp operates at 80 percent capacity?
   (c) What profit would result if the camp stayed open 2 weeks longer and averaged 75 campers paying a reduced rate of $100 per week during the extended season? What might go wrong with the plan?

15.5 A manufacturer of hand-operated power tools can produce a convenient attachment for use with his line of tools at an estimated annual fixed cost of $9000 and variable costs of $3.75 per attachment. He can also buy the complete attachment custom-made for his line of tools at a price of $6 per attachment for the first 5000 and $3.15 for all units purchased beyond the 5000-unit breakoff point. At what increments of unit sales should he purchase or manufacture the attachment? What factors other than cost might influence the decision?

15.6 A lake resort includes 5 three-bedroom cabins which rent for $90 a night, 25 two-bedroom cabins which rent for $70 a night, and 15 single and 15 double rooms in the lodge which rent for $30 and $50 a night, respectively. The cost of cleaning a rental and preparing for new occupants is $20 for a cabin and $12 for a room. Annual fixed costs for operating the resort are $376,000 for the 200-day season. The average stay at the resort is 2 days, and utilization closely follows the proportion of available rooms and cabins.
   (a) What percentage of resort capacity must be rented each night to break even?
   (b) How many rooms and cabins must be rented each day to make an annual profit of $64,000?

15.7 Sales of a desk lamp that wholesales at $4 per lamp have been disappointing. The contribution of each lamp sold is $1.50. It is planned to increase the advertising budget by $0.22 per lamp and reduce the price to spur sales. Twice as much will be allocated for advertising as for price reduction. Current gross profit is $50,000 on sales of 100,000 lamps per year. How many lamps must be sold under the proposed conditions to double the profit?

15.8 A product currently sells for $12. The fixed costs are $4 per unit, and 10,000 units are sold annually for a gross profit of $30,000. A new design will increase variable costs by 20 percent and fixed costs by 10 percent, but sales should increase to 12,000 units per year. What should the selling price be to keep the same profit ($30,000)?

15.9 A boat marina now sells new boats of only one make. The boats are divided into classes by size: *A*, *B*, and *C*. With an average value placed on the accessories sold with each boat, the accounting figures for annual turnover are

Type	Average *P*	Average *V*	Number Sold
A	$ 400	$ 300	300
B	900	500	175
C	2200	1200	100

Adding a new, fancier line of boats, enlarging the display area, and increasing the sales staff could change the cost and sales figures to

Type	Average *P*	Average *V*	Number Sold
A	$ 450	$ 350	425
B	1000	600	200
C	2400	1200	75
D	3600	2000	50

The additional expense of carrying the extra line of boats would double the present $100,000 fixed-cost charge for the original line.

(a) What is the composite contribution for each line?

(b) Should the new line be added?

15.10 A consulting engineer was asked by some of her clients to produce a book on "tilt-up" construction. Her costs for preparation—artwork, typesetting, plates, etc.—came to $12,000. For each 1000 books printed, the variable costs—paper, printing, binding, etc.—will be $4000. She believes that the number of books she will sell depends on the price that she lists for the book. Her estimates show that the following number of books would sell at the prices shown:

Number of Books Sold	Price per Book
2000	$20.00
4000	14.00
10,000	7.00

(a) What price will give her the greatest profit?

(b) An advertising agency claims that a $20,000 promotion plan for the book would double the sales at any of the prices listed above. If the engineer accepts their forecast, what price should she use?

15.11 A plant produces products 1, 2, and 3 at annual rates of 10,000, 7000, and 5000 units, respectively. Product 1 accounts for 30 percent of the plant's revenue, with a 40 percent contribution on its selling price. Product 2 has fixed costs of $35,000 and a contribution rate of $0.5P$, where P is the selling price. The fixed costs for product 3 are

$20,000, while it accounts for 30 percent of the total sales revenue, with the same contribution rate as the composite contribution rate. Total fixed costs for the plant are $80,000, and total sales amount to $250,000.

(a) What is the plant breakeven revenue?

(b) What is the profit for each product?

15.12 Another alternative for the fertilizer batching plant described in Review Exercise 2 is to replace mix X with a new mix XX. With the replacement of mix X with mix XX, the following operating figures are expected:

	Mix W	Mix XX	Mix Y	Mix Z	Totals
Percentage of total sales	10	15	30	45	100 (on $2 million)
Contribution (percent of P)	45	50	45	35	
Fixed cost charged	$70,000	$140,000	$210,000	$275,000	$695,000

(a) Calculate the expected profit from the total product line which includes mixes W, XX, Y, and Z.

(b) A *composite contribution rate*, as shown by the sloping dotted line in Figure 15.9, is the sum of each product's contribution, weighted according to its percentage of total sales. That is,

$$\text{Weighted contribution} = \text{product contribution} \times \frac{\text{product sales}}{\text{total sales}}$$

Then, the composite contribution rate is the sum of the weighted contributions, and the total contribution is the composite contribution rate times total sales, which makes the gross profit equal the total contribution less the fixed cost.

Calculate the composite contribution rate, and use it to reaffirm the gross product computed in Problem 15.12a.

15.13 What are some of the dangers faced by a firm that engages in dumping to increase its plant utilization?

15.14 A small company manufactures rubber matting for the interiors of custom carts. During the past year a revenue of $202,000 from sales was earned with the current costs given below.

Current Operating Costs	
Direct material	$51,000
Direct labor	42,000
Maintenance	11,000
Property taxes and depreciation	17,000
Managerial and sales expenses	35,000

The forecast for the next year is a drastic drop in custom-cart sales, which is expected to limit mat sales to $90,000. There is insufficient time to develop new markets before next year. With a skeleton force for the reduced production, anticipated operating costs are shown below:

Expected Operating Costs	
Direct material	$28,000
Direct labor	23,000
Maintenance	7,000
Property taxes and depreciation	17,000
Managerial and sales expense	35,000

The company can operate at an apparent loss, or mats can be purchased from a large supplier and resold to the custom-cart builders at a price that will meet purchase and handling costs. Either alternative will retain the market for the company until the following year, when sales are expected to be at least equal to last year. Which of the two alternatives is the better course of action?

15.15 Assume that the variable cost in Table 15.1 is composed completely of labor and represents a week's operation of the firm. Suppose that the firm is currently producing 4 units per week with 12 employees (the average wage is $200 per week). The manager recognizes that the firm should operate at a level of 7 or 8 units of output per week to minimize total cost. At a weekly production level of 8 units, variable cost allows an employment level of $6200/$200 per employee $=$ 31 employees at a continued average wage of $200 per week. It thus takes $31 - 12 = 19$ employees to double the output produced by the original work force producing 4 units per week. Or, put differently, the original 12 employees accounted for $4/12 = 0.33$ units of output apiece, whereas the next 19 employees would account for $(8 - 4)/(31 - 12) = 0.21$ output units each. Do the original 12 deserve higher pay than the next 19 employees? Discuss wage policy as a function of marginal cost.

15.16 A company is being organized to produce a new type of one-piece fishing rod and reel to be called the Fish Machine. One million dollars has been budgeted for fixed costs over a 3-year period, including the advertising budget. Sales in the first year are expected to be 20,000 units, and the units will be priced to recover one-half the fixed cost. Second-year sales are expected to be double those of the first year. After all fixed costs have been recovered, the selling price of the Fish Machine may have to be reduced to encourage more sales. In the third year, production will be upped to 50,000 units, and variable costs are expected to rise sharply because production facilities will be operating over their design capacity. The expected total variable cost pattern is given below.

N	TVC	N	TVC	N	TVC
10,000	$320,000	50,000	$1,050,000	90,000	$2,160,000
20,000	520,000	60,000	1,260,000	100,000	2,500,000
30,000	700,000	70,000	1,540,000	110,000	2,860,000
40,000	880,000	80,000	1,840,000		

(a) Prepare a graph of marginal cost, average variable cost, and average total cost.

(b) How much total profit would result if the price could be held at the original level for 3 years?

(c) Assume that the selling price holds at the original level for 2 years but then it must come down to meet competition from larger manufacturers that have entered the market. During the third year, prices must be dropped by 10 percent to sell each additional increment of 10,000 Fish Machines. What is the total profit for 3 years? Show that the maximum profit would be obtained if production were terminated at the output where marginal revenue equals marginal cost.

15.17   A company manufactures industrial clips for assembly work. The marginal revenue has been determined to be

$$\text{Marginal revenue} = 100 - 0.02n$$

where n is the number of clips produced. Variable costs plus fixed costs are calculated by the formula

$$\text{Total cost} = 2n^2 \times 10^{-4} + 10,000$$

Compute the production in clips per year for the following:

(a) Minimum unit cost of sales
(b) Production for maximum profit
(c) Breakeven volume

15.18   Assume that a company can sell all the units it produces, but costs are subject to diminishng returns. Its revenue and cost functions (in thousands of dollars) are

$$R = \frac{3n}{4} \quad \text{and} \quad C = \frac{n^3 - 8n^2 + 25n + 30}{25}$$

(a) Construct a graph of the cost and revenue curves with the breakeven points indicated.
(b) At what output n will profit Z be maximum?

EXTENSIONS

15A   Effects of Inflation on Breakeven Analysis

Moderate inflation rates do not affect breakeven volumes as severely as they affect returns from individual proposals, but they are still influential. One reason why breakeven volumes are less affected is that fixed cost is not responsive to inflation. Since fixed cost is largely composed of depreciation, taxes, and contracts that are typically renegotiated annually (janitorial services, administrative salaries, maintenance agreements, etc.), there is no continuous impact of inflation. In contrast, variable costs and prices are consistently influenced by inflation and tend to rise accordingly.

Effects of escalations on the cost-volume-price relationship can be accounted for by linear or continuously compounded growth. A linear inflation model is conveniently used when inflation rates are moderate and the time taken to reach breakeven volume is short. This model is examined because it is simple yet conceptually complete.

To the symbols already defined for breakeven analyses (P = price/unit, V = variable cost/unit, F = fixed cost/period, n = number of units produced, and B = breakeven volume), the following are added:

$M = P - V$ = contribution margin at beginning of analysis period

N = output or production rate per period (normally 100 percent utilization)

a = inflation rate per period for variable cost V

b = escalation rate per period for price P

t = time ($t = 0$ at start of period, and $t = 1$ at end of period)

Then $n = Nt$ and $B = Nt^*$, where t^* is the time when breakeven occurs.

Knowing that revenue equals total cost at B, we include the effect of inflation on price (Pbt) and variable cost (Vat) at time t in the breakeven relationship as

$$\int_{t=0}^{t^*} (P + Pbt)N \, dt = F + \int_{t=0}^{t^*} (V + Vat)N \, dt$$

By integrating and rearranging terms,

$$\frac{t^*(PbN - VaN)}{2} + t^*(PN - VN) - F = 0$$

Note that if there is no inflation, $a = b = 0$, and the first term in the equation above is eliminated to give

$$t^*(PN - VN) = F$$

and, since $t^* = B/N$,

$$\frac{B}{N}(P - V)N = F \quad \text{or} \quad B = \frac{F}{P - V} = \frac{F}{M}$$

Solving for B in the quadratic equation after substitution for t^* yields

$$B = \frac{-MN \pm \sqrt{M^2N^2 + 2(Pb - Va)NF}}{Pb - Va}$$

As an illustration of the linear inflation model, let $a = 0.5$ percent/month and $b = 1$ percent/month for the feeder airline situation described in Figure 15.5. Given $P = \$120$/unit, $V = \$30$/unit, $F = \$63,000$/month, and $N = 1000$ units (seats available) per month, without inflation.

$$B = \frac{F}{M} = \frac{\$63,000}{\$120 - \$30} = 700 \text{ units/month}$$

When the price of a ticket is going up faster than the variable cost ($b > a$), the breakeven point goes down, as shown by

$$B = \frac{-90(1000) \pm}{\sqrt{(90)^2(1000)^2 + 2[120(0.01) - 30(0.005)](1000)(63,000)}}$$
$$ \overline{120(0.01) - 30(0.005)}$$

$$= \frac{-90,0000 \pm \sqrt{8100 \times 10^6 + 2(1.20 - 0.15)(63 \times 10^6)}}{1.05}$$

$$= \frac{-90,000 \pm 90,732}{1.05} = \frac{732}{1.05} = 697 \text{ units/month}$$

When the percentages are reversed ($a = 1$ percent and $b = 0.5$ percent), prices are still climbing faster than are costs in terms of dollar amounts, and

$$B = \frac{-90,000 + 90,209.75}{0.30} = 699 \text{ units/month}$$

For the given data the change in the breakeven point is slight, mainly because the contribution margin per unit is so large. The effect of inflation is more pronounced when the contribution margin is small and the production rate is high.

The quadratic relationship affords the possibility of two breakeven points. This may occur when M is greater than $\sqrt{2(Va - Pb)/N}$. Whenever Pb is greater than Va, profit continually increases as production increases beyond the breakeven point.

A compound inflation model is developed similarly to the previous model. From

$$\int_{t=0}^{t^*} Pe^{bt} \, dn = F + \int_{t=0}^{t^*} Ve^{at} \, dn$$

the breakeven point B can be determined by integration and numerical solution techniques.[†]

† See D. G. Dhavale and H. G. Wilson, "Breakeven Analysis with Inflationary Cost and Prices," *The Engineering Economist*, Winter 1980.

QUESTION

15A.1 For an annual production rate of 150,000 units, the fixed cost is $200,000 a year. The selling price is $10 per unit, and the contribution margin is $2 per unit.

  (*a*) Solve for *B* when inflation is not considered

  (*b*) Solve for *B* when both *V* and *P* inflate at a 20 percent annual rate by applying the linear inflation model.

  (*c*) What is the breakeven point when prices escalate at 20 percent per year and variable cost inflates at 10 percent? (Use the linear inflation model.)

  (*d*) What is *B* when *a* = 20 percent and *b* = 10 percent, when the linear inflation model is applied?

15B   Engineering, Economics, and Pollution

Life style in highly industrialized nations has been parodied as the "effluent society." The more-developed countries do not have a monopoly on pollution problems. All large population concentrations face the danger of overloading nature's self-regenerating capacity to absorb and recycle wastes. However, pollution is acute and most obvious in industrialized countries because they process more resources per capita and use more advanced technologies that may damage the ecological system on a grand scale.

Alarming examples of pollution practices and frightening forecasts of eventual outcomes from continued contamination are regularly reported in the press. Dangerous chemicals have seeped into aquifers, the sources of drinking water for much of the United States population. Acid rain, caused by smokestack gases reacting with water vapor in clouds, kills fish and wildlife in North America and Europe. How to dispose of radioactive and toxic chemical wastes continues to be a tormenting problem. But progress toward cleansing some parts of our environment is also being reported. The Great Lakes cleanup has brought life back to "dead" Lake Erie. Exhaust fumes from 1980s' automobiles have 90 percent less carbon monoxide and hydrocarbons than did uncontrolled cars in the 1960s. Advances have been made in reducing discharges of untreated wastes into waterways. However, the sweeping drive of the 1970s to clean up the environment appears to be slowing, owing in large part to related engineering and economic difficulties.

Technology to drastically curtail most pollutants is available, but it is expensive. Even if engineering advances can substantially cut the cost of pollution control, the remaining costs must be balanced against other national priorities. What is the cost of a cleaner environment in terms of jobs lost or created, energy consumed, personal inconvenience, and changes in the standard of living—up, down, or simply different? *The present worth of a pollution-control investment is the net cash outlays (−) from public and private sources to prevent a particular form of pollution minus the reduction in the costs (+) created by that form* *of pollution.* Examples of pollution-prevention costs (−) include expenditures of public funds for municipal sewage-treatment plants, expenditures by industrial firms to remove contaminants from waste discharges, and expenditures by individuals for pollution-control devices on automobiles. Reduction in costs (+) to overcome pollution that already exists could include lower expenditures for water-purification systems, medical bills for illnesses due to polluted conditions, and remedies to combat mental and physical discomfort associated with pollution. The costs for pollution prevention are obviously easier to quantify than are the benefits resulting from less pollution.

The relation between pollution costs and the benefits of reduced pollution generally follows the pattern shown in Figure 15.17. The message in this graph centers on the *economically feasible* level *X* of pollution control for a given state of technology and set of economic conditions. Ideally the pollution level would be zero, but the sacrifice in terms of economic disruptions to reach that level is probably prohibitive. However, if the current level of pollution is *Y*, it would be economically sound to spend enough on pollution prevention to lower the level to *X*.

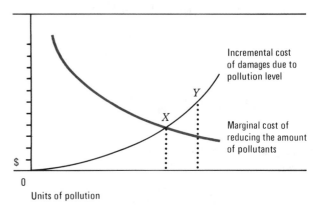

FIGURE 15.17

Relationship of costs and benefits of pollution control.

QUESTIONS

15B.1 Assume that Figure 15.17 represents a sewage-disposal situation in a certain locality. The pollution-prevention costs are expenditures to construct and operate a waste-treatment plant. Pollution damages include downstream water-purification expenses, human health-care costs, and the loss of recreational activities.

(a) How could these costs be determined?

(b) Once the costs are obtained, how should they be used to convince the polluters to allocate resources for prevention when most of the benefits accrue downstream?

(c) Should the downstream beneficiaries contribute to the pollution-prevention costs?

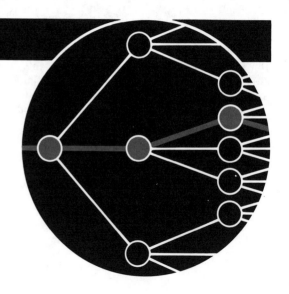

CHAPTER 16

RISK
ANALYSIS

OVERVIEW

Risk is the chance of loss. As applied to engineering economics, risk is the chance that cash flow will fall short of, or exceed, expectation. In previous evaluations we have assumed that cash flow is known with certainty, although sensitivity analysis questioned the effect of cash flow deviations and the cost of capital when risk is considered. In this chapter, risk is a specified dimension for investment decisions. It is quantified as the probabilities of occurrence of different cash flows.

At least some risk is inherent in any economic decision. Inputs and outputs for short-term investment alternatives are usually subject to less variability than are long-term investment proposals. Risk analysis is appropriate when *significant* outcome variations are likely for different future states and *meaningful* probabilities can be assigned to those states.

Probability principles govern risk analyses. Both objective and subjective probability estimates are utilized in economic evaluations. Future states are usually defined by independent probabilities. The widely used *expected-value* measure of preference is a weighted average of outcomes—the sum of the products of outcomes from independent states multiplied by their associated probabilities of occurrence. *Investment-risk profiles* and *acceptable-investment diagrams* are graphic aids that assist in the evaluation of expected values. The *expected value of perfect information* can be calculated to suggest the maximum amount that can be spent to obtain more accurate forecasts of future states. Other criteria such as the *most probable future* and *aspiration level* may influence the final selection of the preferred alternative. Investments to avoid risk are considered in Extension 16B.

RECOGNIZING RISK

Economic analyses in previous chapters were based on the assumption that complete information was available and that any uncertainty connected with a comparison could be tolerated. Thus, a salvage value assigned today is expected to be valid, say, 15 years hence. A positive cash flow of $1000 per year in a comparison model presupposes that exactly $1000 will become available each period on schedule. Unfortunately, real-world conditions do not always follow the models developed to represent them.

Variability is a recognized factor in most engineering and management activities. People are expected to possess individual skills and temperaments and to behave impetuously when exposed to some situations. The properties of materials vary over time. Seemingly identical machines exhibit diverse operating characteristics. Environmental factors are never constant and economic conditions change irregularly. But recognizing variability is much easier than is including its consequences in economic comparisons.

Risk analysis recognizes the existence of variability and accordingly provides ways to select a preferred alternative. Consideration of inflation, sensitivity analysis, and comparisons based on a range of estimates were modifications of assumed certainty suggested earlier to deal with unpredictable future events. In this chapter probability theory and associated management-science tools are introduced. They are applied to discounted cash flows and decision models. Such analyses contribute to a more complete economic evaluation when there are *significant* risks involved that can be represented by the assignment of *meaningful* probabilities.

In the first century Seneca wrote, "Luck never made a man wise." Commensurately, a wise person never disregards luck.

INCLUDING RISK IN ECONOMIC ANALYSES

The problem-solving and decision-making process depicted in the first two figures in this book set the framework for decisions under risk. Every decision of any consequence has overtones of risk. Sometimes the risk is so remote that it can be disregarded as a factor. Even when risk is recognized, it may have to be ignored because the data are insufficient for evaluation or the evaluation requires too much time or money. Such secondary attention to risk factors is neither laziness nor stupidity; it is often a necessity. If every decision were subject to a searching appraisal of risk, management functions would grind along at an intolerably slow pace.

The risk of using data drawn from the past to guide decisions that take place in the future was implied in Figures 1.1 and 1.2.

Definition of the Problem

The investigation leading to a problem definition should give strong hints about the appropriateness of including risk considerations in the evaluation. Objectives that can be satisfied almost immediately are not as subject to change variations that intrude over extended time periods. Narrow objectives bounded by limited means of accomplishment effectively exclude risk by confining the problem and solution to known quantities. Risk enters into the decision situation when current activities are projected into a distant future and the resulting activities are subject to conditional influences.

Collection of Data

Collecting data is a more demanding task when risk effects are to be considered. The first obstacle is the identification of possible alternatives. Both the obvious and the subtle warrant consideration. Creative effort is commonly considered a virtue in developing alternatives and is subordinated to statistical and accounting efforts in assessing outcomes. This is a mistaken belief. Both phases benefit from creativity and both rely on documentation. Moreover, the knowledge acquired in one phase is bound to contribute to the other phases.

In comparing a problem to a tree (Figure 16.1), we might say that the main branches are alternative solutions and the outcomes the secondary branches and twigs. Assessment of these outcomes includes (1) identification of future states or conditions, (2) prediction of the probability of each state, and (3) determination of returns associated with each state. Future states can be anticipated but not controlled. Noncontrollable conditions may include weather, economic or technological developments, political legislation, world affairs, whims of buyers, and so forth. A key question is that of deciding which states are relevant to the problem objectives.

In the problem-tree analogy illustrated in Figure 16.1, note that the roots

> Some analysts believe that there is so much uncertainty in any estimate of future cash flow that evaluating risk is analytical overkill. Other analysts disagree vehemently.

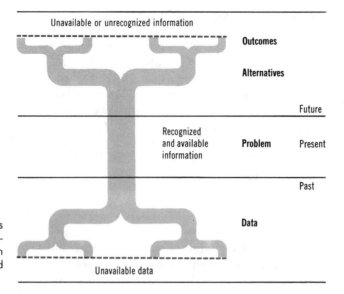

FIGURE 16.1
Problem tree linking data sources to alternative outcomes. The problem results from a current situation which has roots in the past and outcomes in the future.

of the tree or problem lie in the past and constitute the source material. Some information about the outcomes is obtained by investigating the roots. Other information is available from the structure of the limbs, or alternatives. The least-promising alternatives can be pruned out, thereby reducing the effort required for outcome assessments to reasonable limits.

A shortage of pertinent information is a common handicap in risk evaluation. This condition is depicted in Figure 16.1 by the cutoff lines bounding the problem tree. Lack of information may be the result of a scarcity of historical data or an inability to decipher past trends in terms of future outcomes. Failure to identify promising alternatives can be traced to having a lack of data or to simply having overlooked a course of action that should have been apparent. Estimating distant outcomes is at best precarious with good information; without applicable data it becomes an act that is more heroic than rational.

Formulating the Model

A natural reaction to risk is to adopt a conservative stance. Conservatism in economic evaluations often takes the form of making overly pessimistic estimates of future cash flows or requiring an abnormally high rate of return (or an exceptionally short payback period) when there is a significant likelihood that an investment might fail. Both approaches are informal attempts to ration capital to proposals that have better chances of profitability. The weakness of such informal weighting is the subjectivity of the assigned risk penalties. It would be difficult to explain to someone why, for instance, a 5 percent boost in the minimum required rate of return is realistic for one proposal and not for another. For the 5 percent penalty to be meaningful, it should be associated with specific receipts or expenditures that are doubtful. If this is done, only slightly more definitive data are required to utilize a formal model of risk.

Evaluation

The inclusion of risk factors does not automatically increase the accuracy of a study. Probabilities representing risk appear on both sides of the economic balance. Sometimes they can decidedly swing the balance to one side. The important point to realize is that some evaluations are trivial or misleading without risk considerations and others will not benefit from the additional effort required to include risk. The ability to distinguish between the two comes from confidence nourished by familiarity and practice.

Including risk in an analysis can be likened to adding a small flashlight to a key ring; it can be a novelty and a conversational gambit, or it can be used for illumination.

PROBABILITY ESTIMATES

A formal evaluation of risk is feasible when the likelihood of possible futures can be estimated and associated outcomes from alternative courses of action can be identified. The first step is to determine categories of future states that affect the alternatives being compared. If the questionable aspect is expected market conditions, the states might be low sales, average sales, and high sales, defined by specific ranges such as annual sales below $50,000, $50,000 to $125,000, and over $125,000. Finer categories for expected sales could be very low, low, average, high, and very high, with corresponding dollar ranges declared. The breadth of categories appropriate for risk analysis depends mostly on how much the outcomes of alternatives vary between states; when cash flows do not differ much over a wide spectrum of possible future operating conditions, a smaller number of states adequately defines the problem. After the future states are identified and bounded, cash flow outcomes can be estimated by assuming each state, in turn, is sure to occur.

Subjective estimates of future conditions are opinions. They may be wild guesses or judgments based on valid experience. Basing analyses on guesses is a gullible act and basing them on the valid-experience judgments is a venturesome one.

The next step in risk analysis is that of determining the probability that each state will actually occur. The sources may be *objective* or *subjective*. Objective evidence of probability is usually in the form of historical documentation or common experience. It is readily accepted that a coin has two sides and a die has six. A reasonable person would not question the assignment of the probability of 0.50 to the occurrence of a head on the flip of a coin or a probability of $\frac{1}{6}$ for an ace from the role of a die (assuming that both the coin and die were fair or unbiased and that they were tossed or rolled in a fashion that assured an equal chance for all possible outcomes; equivalently, probability data for economic studies are assumed to be impartial and intrinsic). This type of information is *prior* (also known as *a priori*) knowledge.

In practice, prior knowledge of probabilities is seldom possible. Most of the time it is necessary to look at past records of events and use this empirical knowledge as a basis for current probabilities. However, vigilance is required to determine the relation of historical records to the present action; seemingly unrelated events could influence each other. On the other hand, the records of one machine can sometimes be used to predict the performance of a similar machine. Experimental and other measurements that provide *after the fact (a posteriori)* probabilities are necessarily approximate, but they can still be the basis for practical applications.

PROBABILITY CONCEPTS

Betting in poker is a decision under risk. By keeping track of the cards played and relating them to the known distribution of card values in a complete deck, a betting policy can be developed to theoretically maximize returns. However, it takes considerable discipline to abide by the policy consistently and to do it long enough for the laws of probability to have effect. Somewhat the same problems affect economic decisions under risk. Each event has a single outcome and an associated probability of occurrence. It can be discouraging when an undesirable outcome occurs that had odds against it of 99 to 1, but in the long run probability theory promises that if the odds are accurate, the temporary disappointment will be replaced by satisfaction when the same situation is replicated a number of times.

The probability of something is the number of ways it can occur divided by the total number of ways things can turn out.

Events can be either statistically independent or dependent. Statistical *dependence* means that the probability of an outcome is dependent on or influenced by the occurrence of some other event, whereas an *independent* event is not affected by the occurrence of any other event.

Dependent events are discussed in Chapter 17.

The probabilities of mutually exclusive independent events can be added. The probability of drawing a queen from an honest deck of cards is the sum of the probabilities of all four queens in the deck. In functional notation this would appear as

$$P(Q) = P(Q_S) + P(Q_H) + P(Q_D) + P(Q_C)$$

where P is probability, Q means queen, and the subscripts stand for the four suits in the deck. And the numerical values are

$$P(Q) = \frac{1}{52} + \frac{1}{52} + \frac{1}{52} + \frac{1}{52} = \frac{1}{13} \quad \text{or} \quad 0.077$$

Intuitively, the probability of not drawing a queen is 1 minus the probability of drawing a queen:

$$P(\cancel{Q}) = 1 - P(Q) = 1 - \frac{1}{13} = \frac{12}{13} \quad \text{or} \quad 0.923$$

because the outcomes are mutually exclusive and collectively exhaustive (add up to 1.0). Mutually exclusive sets are also additive. The probability that a card will be either a spade or a heart is

$$P(S + H) = P(S) + P(H) = \frac{13}{52} + \frac{13}{52} = \frac{26}{52} \quad \text{or} \quad 0.5$$

Example 16.1
Additive Probabilities

The output of a machine has been classified into three grades: superior (A), passing (B), and failing (C). The items in each class from an output of 1000 items are 214 in A, 692 in B, and 94 in C. If the run from which this sample was taken is considered typical, the probability that the machine will turn out each grade of product is

$$P(A) = 214/1000 = 0.214$$
$$P(B) = 692/1000 = 0.692$$
$$P(C) = 94/1000 = 0.094$$

What is the probability of making at least a passable product?

Solution 16.1

The probability of producing a passable product includes the set of superior and passing grades.

$P(A + B) = P(A) + P(B) = 0.214 + 0.692 = 0.906$

or

$$P(A + B) = P(\cancel{C}) = 1 - P(C)$$
$$= 1 - 0.094$$
$$= 0.906$$

The probability that two or more independent events will occur together or in succession is the product of all the individual probabilities. In terms of the deck of cards, the probability of drawing the queen of hearts twice in a row (provided that the queen is reinserted in the deck and the deck is reshuffled) is

$$P(Q_H Q_H) = P(Q_H)P(Q_H) = \frac{1}{52} \times \frac{1}{52} = \frac{1}{2704}$$

The same reasoning applies to drawing any predesignated cards in a predesignated order or to drawing the same predesignated cards simultaneously from more than one deck.

A probability tree provides a pictorial representation of sequential events. In the probability tree of Figure 16.2 a new deck is used for each draw, and each suit is considered a set. The probability of drawing three hearts in a row is represented by the double line in the probability tree. By formula, this action is equal to

$$P(H_1 H_2 H_3) = P(H_1)P(H_2)P(H_3) = 0.25 \times 0.25 \times 0.25 = 0.015625$$

The same probability is evident for the sequence of drawing two hearts in a row and then drawing any predesignated suit on the third draw.

The probability of not drawing a heart in two consecutive draws is

$$P(\cancel{H})^2 = [1 - P(H)]^2 = (1 - 0.25)^2 = (0.75)^2 = 0.5625$$

It is shown in the tree by the sum of the probabilities in shaded circles which end the paths containing no hearts. Then the probability of at least one heart in two draws is the sum of the probabilities in the light circles and is equal to 0.4375.

Example 16.2
Multiplication of Probabilities

A series of samples from the output of a machine reveals that 4 items out of every 100 sampled are defective. Thus, $P = 0.04$. For an order of 5 items taken directly from the machine without preliminary inspection, what is the probability that the order will be filled without containing a defective item?

Solution 16.2

The probability of no defects $P(0)$ in an order of five is

$$P(0) = (0.04)^0(0.96)^5 = (0.96)^5 = 0.82$$

Further, the probability that the order has not more than one defective is

$$P(1) = 0.82 + 5(0.04)^1(0.96)^4 = 0.82 + 0.17 = 0.99$$

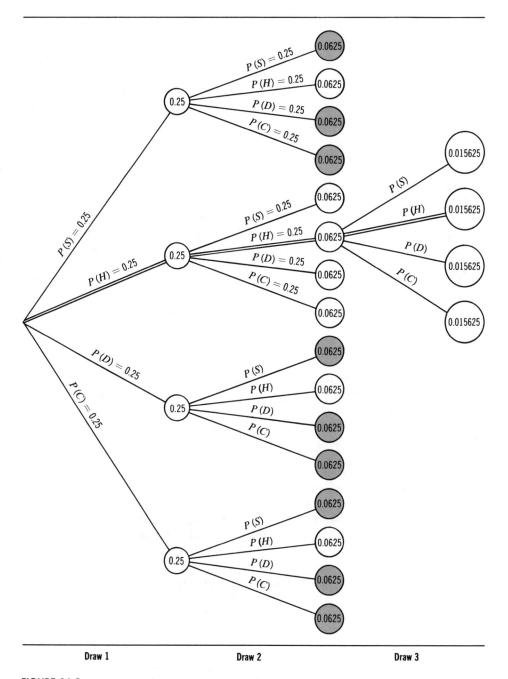

FIGURE 16.2
Probability tree for independent events which registers the suit of a card drawn from a complete deck. The sum of the independent probabilities in each column of circles is 1.0 and represents all possible outcomes. By extension of the tree to include three draws (the incomplete column on the right), the probability of at least two hearts (H) in three draws is 10 × 0.015625 = 0.156.

EXPECTED VALUE

Expected value is a standard measure for economic comparisons involving risk. It incorporates the effect of risk on potential outcomes by means of a weighted average. Outcomes are weighted according to their probability of occurrence, and the sum of the products of all outcomes multiplied by their respective probabilities is the expected value EV:

$$\text{EV}(i) = \sum_j P_j O_{ij}$$

Where P_j is the independent probability of future j, $\Sigma P_j = 1.0$, and O_{ij} is the outcome of alternative i for future j.

As a common example of the expected-value approach, consider the coin-flipping game: A coin is flipped; if it comes up *heads* you win the coin, but when *tails* show you lose. Let the coin be a "fair" dime; then the probability of a head or a tail is, of course, 0.5, and the expected value of winnings from the game is

$$\begin{aligned}
\text{EV(flipping dimes)} &= P(H)(\$0.10) + P(T)(-\$0.10) \\
&= 0.5\,(\$0.10) + 0.5(-\$0.10) \\
&= 0
\end{aligned}$$

The given probabilities of future states are collectively exhaustive (both heads and tails are accounted for, and $\Sigma P = 0.5 + 0.5 = 1.0$); an outcome is associated with each future state (the gain or loss of a dime); and the expected value is the long-term expectation from repeated performances under the same conditions.

A profit-seeking individual or organization prefers a positive expected value, when gains are plus and losses are minus. In comparing alternatives, we find that the one with the highest EV is preferred, other things being equal. For instance, a variant of the coin-flipping game might be to pay $1 for the chance to flip two coins. The payoff could be to win $2 for two heads, $1 for two tails, and nothing when the coins split head and tail. The expected-value calculation,

$$\begin{aligned}
\text{EV} &= P(H_1H_2)(\$2 - \$1) + P(H_1T_2)(-\$1) + P(T_1H_2)(-\$1) + P(T_1T_2)(\$1 - \$1) \\
&= 0.25(\$1) + 0.25(-\$1) + 0.25\,(-\$1) + 0.25(0) \\
&= -\$0.25
\end{aligned}$$

shows that this is not a very good game to get into because the *average* loss per play is 25 cents. But the *actual* outcome for any given gamble is a gain of a dollar, a loss of a dollar, or no change.

The appropriateness of the expected-value model for a once-in-a-lifetime decision is questionable. Other decision criteria such as those presented in

Chapter 17 would enter the verdict. However, truly unique decisions do not occur too often in actual practice; alternatives vary individually with respect to amounts, possible futures, and probabilities, but their evaluation is based on a consistent long-term objective of profit maximization or cost minimization. Since most industries and governments are long-lived and new investment projects are being continually initiated, expected value is a rational measure for most comparisons that recognize risk.

Misuse and misinterpretation of probabilities are surveyed in Extension 16A.

Payoff Tables

A format for organizing and displaying outcomes of alternative courses of action is called a *payoff table*. Each row in the table represents an alternative with its outcomes arranged in columns according to respective future states. In the payoff table below, the present worths of two new products, *A* and *B*, are shown for three states of sales success during a 3-year marketing period.

| Alternative | State of Market Acceptance | | |
	Rejection	Average	Domination
Product A	− $50,000	$200,000	$500,000
Product B	− 200,000	100,000	1,000,000

The initial cost of developing product *A* is $50,000; this amount would be lost if the product were rejected by consumers. If product *A* received average acceptance, the expected net gain would be $200,000; and if it dominated the market, the payoff would be $500,000. Product *B* would cost four times as much as would product *A* to put into production; but if it became a best seller, it would double *A*'s profit. Because production costs for *B* are higher, an average demand would result in only half the payoff expected from *A*.

Some conclusions might be drawn from only the payoffs included in the table. For instance, a loss of $200,000 could be considered disastrous to the company, while a loss of $50,000 would at least be tolerable. With such a severe penalty for failure, alternative *B* would practically be eliminated regardless of the potentially large payoff. However, even more meaningful observations can be made by including the relative likelihood for each outcome. The payoff table below has been modified to incorporate probability factors.

| Product | Outcome and Risk | | |
	$P(R) = 0.1$	$P(A) = 0.6$	$P(D) = 0.3$
A	− $50,000	$200,000	$500,000
B	− 200,000	100,000	1,000,000

The requirement that probabilities of future states add up to 1.0 means that outcomes must be assigned to all possible futures.

It might have been necessary to assign probabilities to each outcome of each alternative, but it is assumed that products *A* and *B* are similar enough to possess the same consumption pattern. In the example both products have a probability of 0.1 of rejection, 0.6 of normal demand, and 0.3 of booming acceptance.

After concluding that the future states are indeed representative, the probability assignments are realistic, and outcomes are estimated as accurately as possible, calculation of the expected values is easy; and preference among alternatives is clearly distinguished. For the comparison between product *A* and product *B*,

$$EV(A) = 0.1(-\$50,000) + 0.6(\$200,000) + 0.3(\$500,000)$$
$$= \$265,000$$
$$EV(B) = 0.1(-\$200,000) + 0.6(\$100,000) + 0.3(\$1,000,000)$$
$$= \$340,000$$

Product *B* is clearly preferred, assuming that all other factors affecting the decision are equal.

Example 16.3 Expected Value of Risky Alternatives

A process line that will continue to be needed for 3 years has annual costs of $310,000, which are expected to remain constant. A novel redesign for the line has been suggested. The new approach will cost $150,000 to install and has a 50 percent chance of cutting annual operating costs to $210,000.

However, there is a probability of 0.25 that annual costs for the redesigned line will increase, from $210,000 for the first year, by $20,000 or $75,000 in each of the next 2 years. If a 12 percent rate of return is required, should the new design be installed?

Solution 16.3

Equivalent annual costs for each future state of the redesigned process line are

$$EAC(\text{at } P = 0.5) = \$150,000(A/P, 12, 3) + \$210,000$$
$$= \$150,000(0.41635) + \$210,000$$
$$= \$272,453$$

$$EAC(\text{at } P = 0.25) = \$150,000(A/P, 12, 3) + \$210,000$$
$$+ \$20,000(A/G, 12, 3)$$
$$= \$272,453 + \$20,000(0.9245)$$
$$= \$290,943$$

$$EAC(\text{at } P = 0.25) = \$150,000(A/P\ 12, 3) + \$210,000$$
$$+ \$75,000(A/G\ 12, 3)$$
$$= \$272,453 + \$75,000(0.9245)$$
$$= \$341,790$$

These costs are inserted in a payoff table, as shown below, where the calculated expected values indicate that the redesign should be attempted even when there is one chance in four of increasing the process-line costs.

	Equivalent Annual Costs			Expected	
Alternative	P(0.5)	P(0.25)	P(0.25)	Value	
No change	$310,000	$310,000	$310,000	$310,000	
Redesign	272,453	290,943	341,790	$294,410	← Preferred

Expected Value of Perfect Information

A decision is at risk whenever the occurrence of the events involved in it are in doubt. The likelihood of outcomes may be known, as in the case when probabilities are available for future states, but which state will actually exist when an alternative is exercised is still unknown. Knowledge of when each state will occur would allow a decision maker to utilize the most profitable alternative on each occasion. This condition of prescience provides *perfect information* for decision making.

In a decision situation represented by a payoff table, the expected result from possessing perfect information is determined by multiplying the best outcome in each column by its probability and then summing the products. For the outcomes in the payoff table of Example 16.3, the *Redesign* alternative has the lowest cost for two future states ($272,453 when $P = 0.5$ and $290,943 when $P = 0.25$) and the *No change* alternative has a lower cost in the other future state ($310,000 at $P = 0.25$). These are the outcomes that the decision maker would choose to have for each future state by exercising the appropriate alternative when it was known which state would occur. Therefore, the *expected result given perfect information* (EV|PI) is

$$EV|PI = \$272,453(0.5) + \$290,943(0.25) + \$310,000(0.25)$$
$$= \$286,462$$

The difference between the expectation with perfect information and the expected value without prior knowledge is called the *expected value of perfect information*. From Example 16.3, where the expected value was computed to be $294,410,

$$\text{Expected value of perfect information} = EV|PI - EV$$
$$= \$286,462 - \$294,410 = -\$7948$$

which is a negative cost that corresponds to a saving of almost $8000. This calculation suggests the maximum investigation costs that could be justified to find out which state will exist each time the decision is due.

> Investments made to reduce risk are discussed in Extension 16B.

> The probabilities of future states happening are not changed by the acquisition of perfect information. They occur in the same proportion in any case; but perfect information reveals the schedule of occurrence.

Example 16.4   Value of Perfect Information

An ambitious entrepreneur has signed a regionally well known rock band to give a concert on the Fourth of July. The concert can be staged in a conveniently located pasture or in a school auditorium. If the weather is sunny, a concert in the pasture could be very profitable, but rain would ruin attendance and cause a loss due to the high fixed costs of preparing the site. Attendance at the auditorium would be largely unaffected by weather, but the maximum capacity is far less than the pasture. Weather records indicate that odds against rain on July 4 are 9 to 1. Using the probabilities given and the estimated returns shown for concerts held at the two locations, calculate the *expected value* and the amount of money that could be paid for *perfect information* about the weather, if it could be purchased.

Alternative	Net Return if There is	
	Rain	No Rain
Auditorium	$24,000	$30,000
Pasture	−27,000	90,000

Solution 16.4

The payoff table indicates that holding the concert in the pasture has the greater expected value.

	Rain (P = 0.1)	No Rain (P = 0.9)	Expected Value
Auditorium	$24,000	$30,000	$29,400
Pasture	−27,000	90,000	78,300 ←Preferred

The shaded outcomes in the payoff table show the highest possible returns for each future state. If the entrepreneur *knew* that it was going to rain on the Fourth of July, the concert

would surely be scheduled in the auditorium to reap a $24,000 profit instead of a loss. Equivalently, assurance of dry weather would lead to a profit of $90,000 from the concert held in the pasture. Since it is dry 9 out of 10 July Fourths, the *expected* result *given* perfect information is

$$\$24,000(0.10) + \$90,000(0.9) = \$83,400$$

and the amount that could be paid for it is $83,400 − $78,300 = $5100, the difference between the returns from a commitment to hold the concert in the pasture and the option to schedule it at the site best suited to the weather. This is the *expected value of perfect information*.

Investment-Risk Profiles

The range-of-estimates approach (Chapter 14) can be modified to reflect risk. Probabilities of occurrence are assigned to each future state in the range. Then all possible combinations of outcomes are collected and their dollar amounts and joint probabilities are calculated. The result is an *investment-risk profile* that reveals the likelihood for an investment to realize various net present-worth returns.

Assume that an investment proposal for income expansion requires an initial outlay of $200,000. The most likely outcome is for after-tax returns to amount to $100,000 per year for 4 years. There is no salvage value and the minimum attractive rate of return is 9 percent. Under the assumption of certainty,

$$PW = -\$200,000 + \$100,000(P/A, 9, 4)$$
$$= -\$200,000 + \$100,000(3.2396) = \$123,960$$

The given probabilities, which tilt toward a shorter life and lower annual returns, foretell an EV less than the value calculated when certainty was assumed.

A risk analysis of the same proposal recognizes that the most likely cash flow has a probability of only 0.5 of occurrence. A pessimistic appraisal of the future concedes returns might amount to only $50,000 per year. Under the most optimistic assessment, returns could total $125,000 per year. The probabilities of pessimistic and optimistic returns materializing are 0.3 and 0.2, respectively. The duration of returns is also questionable; probabilities for 2, 3, 4, and 5 years are, respectively, 0.2, 0.2, 0.5, and 0.1. These data are summarized on the decision-probability tree shown in Figure 16.3.

The joint probability for each outcome is the product of the independent probabilities representing each factor involved in the outcome. These probabilities are entered in the nodes of the probability-decision tree, and the factors are labeled on the lines connecting the nodes. For example, the joint probability of returns of $50,000 per year for 2 years (top of the column in Figure 16.3) is

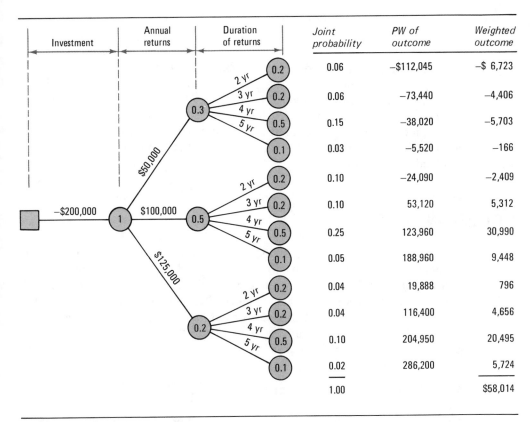

	Annual	Duration	Joint	PW of	Weighted
Investment	returns	of returns	probability	outcome	outcome
			0.06	−$112,045	−$ 6,723
			0.06	−73,440	−4,406
			0.15	−38,020	−5,703
			0.03	−5,520	−166
			0.10	−24,090	−2,409
−$200,000	$100,000		0.10	53,120	5,312
			0.25	123,960	30,990
			0.05	188,960	9,448
			0.04	19,888	796
			0.04	116,400	4,656
			0.10	204,950	20,495
			0.02	286,200	5,724
			1.00		$58,014

FIGURE 16.3
Decision-tree format for an investment-risk profile analysis. Node entries indicate probabilities for the economic factors noted on lines leading to the nodes. Each burst represents
one economic condition subject to risk.

0.06. It results from multiplying the probability of the definite investment (1.0)
by the probabilities of $50,000 returns (0.3) and a 2-year period (0.2).

Present worths of the outcomes are calculated by the usual procedures.
The value listed at the top of the "PW of outcome" column, based on a
required rate of return of 9 percent, is computed as

$$PW(P = 0.06) = -\$200,000 + \$50,000(P/A, 9, 2)$$
$$= -\$200,000 + \$50,000(1.7591)$$
$$= -\$112,045$$

The weighted value of this present worth is the top entry of the last column
on the right in Figure 16.3, and is the product

$$\text{Joint probability}(i) \times \text{PW(outcome } i) = 0.06 \times -\$112,045$$
$$= -\$6,723$$

The remaining outcomes are computed similarly. The sum of these values is the expected value of the proposal, EV = \$58,014. For the given probabilities of occurrence, the expected value is much less than the present worth based on the most likely outcomes. Although the probability estimates are quite likely subject to some error, the exercise of calculating an investment profile contributes to a more complete analysis and a better appreciation of the factors involved in the proposal.

INVESTMENT-RISK DECISION CRITERIA

An investment-risk profile highlights more than only the expected value. When the outcomes and related probabilities are graphed, a proposal's prospects are clearly displayed for capital-budgeting discussions. The data from Figure 16.3 are graphed in Figure 16.4. Lines connect the outcome-risk points to better define the cumulative probability distribution of net present values. The connected points make it quite evident that the proposal has a probability greater than 0.4 of showing a loss.

Acceptable-Investment Diagram

Another graphic display to assist in the evaluation of alternatives subject to risk is called an *acceptable-investment diagram* (AID). Its format is a horizontal axis representing rates of return and a vertical axis scaled to show the probability that an investment will surpass a given rate of return. Criteria for an acceptable investment are blocked off in the chart by setting limits for:

FIGURE 16.4
Tabulated data and graph of an investment-risk profile.

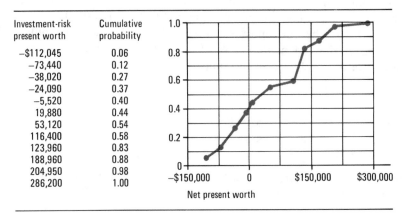

Investment-risk present worth	Cumulative probability
−\$112,045	0.06
−73,440	0.12
−38,020	0.27
−24,090	0.37
−5,520	0.40
19,880	0.44
53,120	0.54
116,400	0.58
123,960	0.83
188,960	0.88
204,950	0.98
286,200	1.00

1 The required probability that an investment's rate of return exceeds a minimum percentage (loss coefficient)
2 The desired probability that an investment's rate of return will exceed an attractive level (payoff coefficient)
3 A line connecting the two coefficients, called an *aspiration level*

Any investment-risk profile that does not intrude on the reject area defined by the listed limits is considered an acceptable investment; its risk-return potential is greater than are the minimum risk-return requirements.

Example 16.5   Application of an AID

An investment of $1000 for 1 year has the possible after-tax returns shown. The probability of each outcome is estimated and the rate of return is apparent, as shown in the table on the right.

The investors seek proposals that provide a probability of 0.95 that they will not lose more than 5 percent and a 0.30 likelihood that their rates of return will be greater than 15 percent. Use an AID to determine the acceptability of the $1000 investment proposal.

Net Return (Outcome)	Rate of Return, %	Probability of Outcome	Probability Investment's RR Will Exceed Rate of Return
−$900	−10	0.05	0.95
1050	5	0.15	0.80
1150	15	0.40	0.40
1300	30	0.30	0.10
1500	50	0.10	0.00

Solution 16.5

The reject area for the acceptable-investment diagram in Figure 16.5 results from lines connecting the loss coefficient horizontally to the ordinate, the loss coefficient diagonally to the payoff coefficient, and the payoff coefficient to the abscissa. The sloping line is the implied aspiration level relating desired returns to allowable risk. The investment's risk-return potential is represented by lines connecting the rates of return associated with their probabilities of occurrence.

Because part of the risk-return curve crosses the reject area, the investment is not acceptable according to the given criteria. Investments passing these criteria can be further evaluated on other merits such as timing of returns, capital availability, and the decision criteria discussed next.

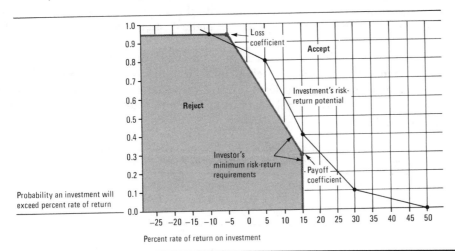

FIGURE 16.5
Acceptable-investment diagram (AID) for evaluating the risk-return potential of a proposal according to the investor's risk-return requirements. The investment proposal shown is unacceptable because it enters the reject region.

TABLE 16.1

Probability of returns from three equal-size, equal-life investments

Alternative	Possible Net Present Worths of Proposals					
	− $1000	0	$1000	$2000	$3000	$4000
A	0	0.11	0.26	0.22	0.02	0.39
B	0.29	0.18	0.07	0	0	0.46
C	0.14	0.10	0.11	0.37	0.28	0

Auxiliary Decision Criteria

The "best-bet" rule optimistically accentuates the highest returns, making it an adventuresome tactic.

Once an investment-risk profile is developed, there are several ways to appraise its desirability. The basic criterion, as already observed, is *maximization of expected returns,* which is determined by expected-value calculations. Other criteria include the *most probable future* and the *aspiration level.* These criteria are illustrated by reference to Table 16.1.

According to the most-probable-future criterion, the investment that has the *greatest return* for the *most probable future* is preferred. The most probable futures for the alternatives in Table 16.1 are

Most Probable Future	PW at Most Probable Future
Alternative B : P = 0.46	$4000
Alternative A : P = 0.39	4000
Alternative C : P = 0.37	2000

They indicate a preference for *B*, because its maximum return, although the same as that of alternative *A*, has a greater probability of occurrence. This criterion thus assumes that the future with the highest probability is *certain* to occur and that the alternative with the highest return for its "certain" future is best.

An *aspiration level* is based on a minimum amount that will satisfy a decision maker. The rationale behind an aspiration level is to achieve assurance that the investment will "at least return a respectable amount" or "surely not lose money." For the sample data from Table 16.1 and an aspiration level of $2000, we have

Alternative	Probability of Return of $2000 or More
C	0.37 + 0.28 = 0.65
A	0.22 + 0.02 + 0.39 = 0.63
B	0.46 = 0.46

Alternative C is preferred because it has a higher cumulative probability of providing $2000 and more. If the aspiration level had been to "at least make some money," alternative A would be selected because it has a probability of $0.26 + 0.22 + 0.02 + 0.39 = 0.89$ of having returns greater than zero, which is larger than that of any other alternative.

Since alternative A would also be selected by the maximum-expected return criterion [because EV(A) = $2320, EV($B$) = $1620, and EV($C$) = $1550], any of the alternatives in Table 16.1 could receive the nod of approval by one criterion or another. This should cause no dismay; it merely demonstrates the importance of a decision maker's perspective, the theme of Chapter 17.

This "safe-bet" rule is associated with a conservative tactic that protects a minimum gain.

Review Exercises and Discussions

The Owlglass Company is considering the manufacture of two mutually exclusive types of sunglass clip-on attachments. One clip-on design is specifically devised for Owl frames only and would not fit the eyeglass frames produced by other manufacturers. The other type is adaptable to competitors' frames but would not fit as securely or be color-coordinated with Owl frames. The investment required to get into the production of either design is $100,000. Estimates of net annual receipts for a 4-year market period are $43,000 for Owl-only clip-ons and $50,000 for general-purpose clip-ons.

Because the "captive" market for clip-ons designed to fit and complement Owl frames appears to have less risk, management decides that the general-purpose clip-on proposal should be evaluated at a risk-adjusted rate of return 8 percent greater than the regular 10 percent required rate. What effect does this decision have on the economic analysis?

Exercise 1

The present worths of the two proposals are calculated using a 10 percent discount for the clip-ons that fit only Owl frames O, and 18 percent for general-purpose clip-ons G:

Solution 1

$$PW(O) = -\$100,000 + \$43,000(P/A, 10, 4)$$
$$= -\$100,000 + \$43,000(3.1698) = \$36,301$$

$$PW(G) = -\$100,000 + \$50,000 (P/A, 18, 4)$$
$$= -\$100,000 + \$50,000(2.6952) = \$34,760$$

The risk-adjusted discount rate switches the preference from general-purpose to Owl-only clip-ons. An engineering economist should question whether the 8 percent addition to the required rate of return fairly represents the perceived risk.

A construction company has determined that it costs $8700 to prepare a bid for a major construction project. It even costs $500 to decide not to submit a bid. The annual profit earned from winning a bid for a major construction project is $400,000. For such projects the competition is severe and the chance of submitting the lowest bid is small. How often must the company's bids win to make the bidding worthwhile?

Exercise 2

Outcomes from the alternatives of bidding and not bidding are shown on page 456 for the future states of winning and losing.

Solution 2

FIGURE 16.6
Graph of two alternatives based on their probability of occurrence. At the indifference point the expected values for the alternatives are the same.

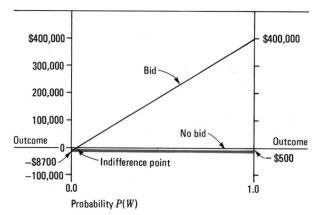

Probability $P(W)$

	Win Bid $P(W)$	Lose Bid $P(L)$
No Bid	$-\$500$	$-\$500$
Bid	$\$400,000$	$-\$8700$

How often the company must win a bid to break even on its bidding can be determined from a graph, as shown in Figure 16.6. A "win" occurs when $P(W) = 1$, and a bid is lost when $P(W) = 0$. As indicated by the no-bid line, the outcome is constant because there is no chance of winning when no bid is submitted. The point where lines representing the bid and no-bid options cross is the *indifference*, or *breakeven*, point.

By letting $P(W)$ and $P(L) = 1 - P(W)$ be the probabilities for the two future states (win and lose), we can calculate the indifference probability from

$$EV(\text{no bid}) = EV(\text{bid})$$
$$P(W)(-\$500) + [1 - P(W)](-\$500) = P(W)(\$400,000) + [1 - P(W)](-\$8700)$$

to yield
$$P(W) = \frac{\$8700 - \$500}{\$400,000 + \$8700} = 0.02$$

which means that the construction company must win over 2 percent of the times it submits bids for new projects. From a slightly different viewpoint, it means that a project is worth bidding on when there is at least a 2 percent chance that the submitted bid will be chosen.

Exercise 3   Owing to the lack of other suitable land, a mill is forced to construct its new settlng ponds for waste disposal along the bank of a river. The stream has no dams to control its flow. Flood-level records for the past 70 years reveal that the river rose above the minimum required height of the settling-pond walls 42 times. Therefore, the economic analysis must include not only construction costs but also the risk of flood damage. Building the walls higher increases the initial investment but lowers the threat of flooding. Engineering calculations for the cost of higher retaining walls in 5-foot increments and the expected damage from floods of different crests are shown in the following table.

River Levels above Minimum Wall Height in Range, Feet	Years the River Crests in Given Range above Wall Height F	Probablity of River Cresting in Range above Height F	Flood Damage when River Crests above Walls	Cost to Construct Walls to Height F
0–5	at $F = 0$:14	14/70 = 0.20	$ 70,000	$135,000
5–10	at $F = 5$:14	14/70 = 0.20	105,000	200,000
10–15	at $F = 10$:8	8/70 = 0.11	150,000	280,000
15–20	at $F = 15$:6	6/70 = 0.09	200,000	370,000
Over 20	at $F = 20$:0	0/70 = 0.0	0	450,000

The settling ponds are expected to be needed for 10 years before the process can be improved sufficiently to eliminate the need for special waste treatment. A rate of return of 10 percent is used by the mill to evaluate federally mandated investments. To what height should the retaining walls be built to minimize annual costs?

The initial cost of the settling ponds is translated to an equivalent annual cost by the capital-recovery factor. For the minimum wall height ($F = 0$), the equivalent annual investment cost is

Solution 3

$$\text{EAC(investment at } F = 0) = \$135,000(A/P, 10, 10) = \$135,000(0.16275) = \$21,971$$

Equivalent annual investment costs for successively higher retaining walls are calculated similarly.

Data arranged for the calculation of expected values of flood damage for alternative wall heights are shown in Table 16.2. The outcomes in the payoff table indicate damages expected from floods exceeding a given wall height by amounts F. Thus, for the minimum wall height, flood waters will cause damage 3 years in every 5 years, on the average, to produce an expected value of

TABLE 16.2

Payoff table (shaded) for five alternative wall heights for settling ponds. Total expected annual costs are the sums of expected flood damages and equivalent annual investment costs.

Wall Height above Minimum, Feet	Probability of River Cresting above Wall Height in Range, Feet				Expected Value	Equivalent Annual Investment	Total Expected Annual Cost
	0–5 $P = 0.20$	5–20 $P = 0.20$	10–15 $P = 0.11$	15–20 $P = 0.09$			
0	$70,000	$105,000	$150,000	$200,000	$69,500	$21,971	$91,471
5	0	70,000	105,000	150,000	39,050	32,550	71,600
10	0	0	70,000	105,000	17,150	45,570	62,720
15	0	0	0	70,000	6,300	60,218	66,518
20	0	0	0	0	0	73,238	73,238

Flood damage from incremental river levels above alternative settling-pond wall heights, F

$$EV(F = 0) = \$70,000(0.20) + \$105,000(0.20) + \$150,000(0.11) + \$200,000(0.09)$$
$$= \$14,000 + \$21,000 + \$16,500 + \$18,000$$
$$= \$69,500$$

which, when added to the equivalent annual investment cost, makes a total expected cost for the minimum-height settling-pond walls of $\$21,971 + \$69,500 = \$91,471$. Expected values for other wall heights are calculated in the same manner, as indicated in the shaded area of Table 16.2. Note that the probabilities do not add up to 1.0 because river levels below $F = 0$ (probability $= 0.40$) have no bearing on the flood-damage expectations.

The settling-pond retaining-wall height that minimizes total annual cost is 10 feet above the minimum level.

Exercise 4   The most likely cash flow for a cost-reduction proposal is for an investment of $4000 to produce after-tax present-worth savings of $1200 per year for 5 years. The net present worth is

$$PW(\text{most likely}) = -\$4000 + \$1200(5) = \$2000$$

Upon further investigation it appears that the initial investment has probabilities of 0.4 of being as high as $5000 and 0.6 of being $3000. The after-tax present worths of annual savings could amount to $2000, $1200, or $800, with respective probabilities of 0.2, 0.3, and 0.5. Determine the investment-risk profile and the expected value of the proposal.

Solution 4   If I and S represent investment and savings levels, respectively, then the pattern of possible future outcomes is shown in the Table. There is a 20 percent chance the proposal will lose money, but the expected value is $2000.

Possible Futures	Net After-Tax Present Worth			Joint Probability	Weighted Outcome
$S(P = 0.2)I(P = 0.6)$	$\$2000(5) - \$3000 =$	$\$7000$		$(0.2)(0.6) = 0.12$	$\$\quad 840$
$S(P = 0.3)I(P = 0.6)$	$1200(5) -$	$3000 =$	3000	$(0.3)(0.6) = 0.18$	540
$S(P = 0.5)I(P = 0.6)$	$800(5) -$	$3000 =$	1000	$(0.5)(0.6) = 0.30$	300
$S(P = 0.2)I(P = 0.4)$	$2000(5) -$	$5000 =$	5000	$(0.2)(0.4) = 0.08$	400
$S(P = 0.3)I(P = 0.4)$	$1200(5) -$	$5000 =$	1000	$(0.3)(0.4) = 0.12$	120
$S(P = 0.5)I(P = 0.4)$	$800(5) -$	$5000 = -1000$		$(0.5)(0.4) = \underline{0.20}$	$\underline{-200}$
				1.00	$\$2000$

PROBLEMS

16.1   Given a fair coin,
(a) Determine the probability of at least one head in three tosses?
(b) Determine the probability of flipping a tail, a head, and a head in that order?
(c) Determine the probability of flipping at most two heads in three tosses?

16.2   In the example used in the chapter, the probability of drawing a queen was $\frac{1}{13}$, and the probability of a heart was $\frac{1}{4}$. The probability of drawing a queen or a heart is called a *union*, because the two sets overlap; they are not mutually exclusive. Therefore, when the probabilities of the two sets are added, the portion that overlaps must be subtracted. By this means we arrive at the formula for the probability of drawing a queen or a heart:

$$P(Q + H) = P(Q) + P(H) - P(QH) = \frac{1}{13} + \frac{1}{4} - \frac{1}{52} = \frac{4}{13}$$

(a) Show the union by a sketch of the overlapping heart and queen sets.
(b) Let three overlapping sets be A, B, and C. Sketch the sets and determine a formula for P(A + B + C).

16.3 A new product introduced by Up-and-Up, Inc. has been so successful that the company is now considering ways to expand production. It can do so by adding facilities or by increasing the utilization of present facilities. The choice depends on forecasts of future market trends for the new product.

The best available probability estimates for future demand are 0.1 to decline slightly from current sales, 0.3 to remain constant, and 0.6 to increase rapidly. If a major expansion of facilities is undertaken now, Up-and-Up should be able to capture most of the new demand before competitors can gear up for production. However, the company would suffer considerable loss from unused capacity if the demand fails to increase or declines. The conservative alternative is to increase utilization of existing facilities from the present 85 percent rate to 100 percent, but there is no way to meet higher levels of potential demand with existing production capacity.

The equivalent annual worths expected to result from each alternative according to each level of future demand are shown in the payoff table below.

	Decline P(D) = 0.1	Constant P(C) = 0.3	Increase P(I) = 0.6
Add new facilities	− $1,800,000	− $50,000	$900,000
Increase utilization	− $100,000	$100,000	$400,000

(a) Calculate the expected values.
(b) Determine what other factors might affect the decision.

16.4 How would you explain the reason for using the expected-value measure to evaluate a unique engineering design project that is unlikely ever to be repeated again in exactly the same form?

16.5 A logging company must decide the most advantageous duration for a paving project. The beginning date of the project has been definitely set. A critical-path analysis has shown that three project durations are feasible. If the paving is completed in 4 months, the basic project cost will be $80,000. A 5-month duration will allow construction savings of $20,000 and it will cost an extra $40,000 over the basic cost to crash the project to 3 months. However, transportation expenses can be cut by $10,000 over the 4-month schedule if the paving is done in 3 months, and an extra transportation expense of $15,000 will be incurred for an extension of the paving time to 5 months.

Since the project must be completed during a period of expected foul weather, the extra expense due to possible weather conditions should also be considered. Weather records indicate that the probabilities for mild rain, heavy rain, and wind and rain are, respectively, 0.3, 0.5, and 0.2. The costs that must be included for these conditions are given in the following table:

Weather conditions	3 Months	4 Months	5 Months
Mild rain	$10,000	$15,000	$ 5,000
Heavy rain	10,000	40,000	60,000
Wind and rain	15,000	55,000	65,000

Which duration has the lowest expected total cost?

16.6 There are several methods available to discover defective welds. A company has investigated two methods. Method 1 costs $0.50 per inspection and detects defects 80 percent of the time. Method 2 costs $2.00 per test but always detects a defective weld. When a defective weld goes undetected, the estimated cost to the company is $30,000 for replacement and other incidental costs. The probability of a defective weld is 0.05. Using the expected-value criterion, determine whether method 1 or 2 should be used, or whether the company is better off with no inspection procedure.

16.7 An investment is being considered that requires $1 million and commits the money for 10 years. During that period it is equally likely that the annual returns from the investment will be $100,000, $150,000, and $200,000. The probability is 0.75 that the salvage value will be $300,000, but there is one chance in four that it will be zero. A minimum rate of return of 10 percent is expected.

(a) Construct an investment-risk profile for the proposal on a chart in which the horizontal axis registers the net PW and the vertical axis is a probability scale ranging from 0 to 1. Draw the curve to show the probability of returns equal to or less than the scaled PWs.

(b) How could the investment-risk profile contribute to the economic evaluation of the million-dollar investment?

16.8 Three mutually exclusive alternatives are described by the data below, which show the probabilities of earning rates of return.

	Rate of Return			
Alternative	−5%	0	10%	20%
A	0.3	0.1	0.2	0.4
B	0.0	0.3	0.5	0.2
C	0.15	0.15	0.4	0.3

(a) Which alternative would be selected using the most-probable-future criterion? Why?

(b) Which alternative would be selected using the expected-value criterion? What is the EV?

(c) Which alternative would be selected if the decision maker had an aspiration level of 10 percent? Why?

(d) Construct an acceptable-investment diagram with a loss coefficient of −5 percent at $P = 0.7$ and a payoff coefficient of 10 percent at $P = 0.3$. Which alternative would be eliminated by the AID?

16.9 A payoff table (in thousands of dollars) is given on the next page for three investments of equal size and duration.

Alternative	Boom (P = 0.3)	So-so (P = 0.5)	Bust (P = 0.2)
A	1000	200	−500
B	300	400	0
C	400	600	−300

Which alternative would you select? Why?

16.10 A cost-saving modification to an existing process is being evaluated. The savings will affect products I and II. The rate of return earned by the investment in the modification depends on how much the process is utilized, which depends on the market conditions for future sales of the two products. Three future states have been identified: *good*, with IRR = 20 percent; *average*, with IRR = 10 percent; and *bad*, with IRR = −5 percent. Since there are two products involved, the maximum possible rate of return under good conditions is 20 percent + 20 percent = 40 percent. The probability of each future for both products is given below.

Product	Future		
	Good	Average	Bad
I	0.20	0.70	0.10
II	0.40	0.30	0.30

(a) What is the expected value of the cost-saving modification?

(b) Draw an investment profile for the proposal on an acceptable investment diagram. The ordinate is the probability that the investment will exceed the percentage return, and the horizontal axis is the rate of return. The investment criteria are to limit losses to 5 chances in 100 of −10 percent return and to be 90 percent sure that the rate of return is at least 20 percent. Should the proposal be accepted? Why?

EXTENSIONS

16A   The Odds Say It Won't Happen, But . . .

Discussions of probabilities conventionally begin with Disraeli's warning: "There are lies, there are damned lies, and there are statistics."

Historians trace probability theory to the gambling casinos of France in the 1650s. Rolling dice was the game of choice in those days. One popular gamble was to place a bet that you could roll a die four times in a row without getting a 6. If a 6 came up in any of the throws, the house won, which it did over half the time. The odds for more complicated bets involving two or more dice captured the fascination of such famous mathematicians as Blaise Pascal and Pierre de Fermat.

Although probabilities still govern gambling, they also underlie the huge insurance industry. An insurance policy is essentially a bet that something bad will happen. As a life-insurance holder, you are betting that you will die sooner than expected, thereby winning the value of the policy for your heirs. The insurance companies counter with actuarial tables for life expectancy by which they set their premiums to recover payments made for early deaths by collecting more from those who live longer. Catastrophe insurers do the same thing with a bit less precision when past exposure is limited. They are "playing the averages."

But *average* can be a tricky statement. Consider the case when a college of engineering boasts that its graduates of 14 years ago now have an average annual salary of $58,426. Since the college is publicizing this handsome figure, one can conclude that the alumni director feels that it will be viewed favorably. Let it also be assumed that the director is honest but pursues numbers that look as good as possible.

A cynic might ask whether the average figure of $58,426 was based on a mean, median, or mode. All three are legitimate averages. If a few alumni earned very large salaries and the sample size was small, the mean could significantly exceed the median. An image-conscious director would then select the mean average.

The cynic could also question the raw data. Perhaps the addresses used in the salary survey represented only class members who had contributed previously to the alumni fund. That creams the list. Perhaps the engineers who are doing well are more likely to return the questionnaire. That purges the list. Those who do reply may be tempted to exaggerate on the high side because they want the college to be proud of them. The featured class may have a notably higher average than that of other classes surveyed. That is selectivity. Perhaps the alumni director practices probabilism. Perhaps the class is not as wealthy as advertised.

Risks of disastrous losses that have very low probabilities of occurring frequently exasperate engineering economists. The "one chance in a million—or a billion" calamity attracts a lot of attention from those who might be affected. Analysts find these dramatic events frustrating to deal with because both the magnitudes of loss and chances of happening are extremely difficult to estimate. Often the events have never happened before, yet protection based on estimates of the extent and likelihood of damage is demanded.

Databases for analyzing the collision of a meteoroid with a spaceship, or accidental creation of a mutant blight that devastates crops, are almost nonexistent. Fault-tree analysis, as described in Extension 17A, is useful in such cases. It attempts to identify all the steps that could lead to a particular disaster and then to estimate the probability of each step occurring. Information about individual events that contribute to a disaster is usually more accessible than are data about the disaster itself. Putting all the pieces together provides an overall estimate with a numerical trail, but the results are still not strictly reliable because of doubts about each piece and the completeness of the whole.

As Disraeli warned, statistical data can be deceptive.

QUESTIONS

16A.1 Calculate the advantage that the house has on a bet that a gambler will not throw a 6 during four consecutive tosses of a fair die.

16A.2 Assume a recent junk-mail delivery contains an announcement that "you may have already won the $1,000,000 jackpot." If this promotion has been mailed to 10 million people, is it worth spending 22 cents for a stamp to mail the entry blank for the contest?

16A.3 The chance of being involved in a traffic accident during a year is about 10 percent and of being hit by lightning is 1 in 2,000,000. But you cannot have one-tenth of an accident nor a hit by 0.000005 of a lightning bolt. You get it all or not at all. Expected value has the same connotation. It is an aggregate figure. EV of a 50-50 chance of winning or losing a dollar is zero. Yet a player never gets nothing for an event. Write a brief explanation for a person who is not familiar with probability theory about why a present worth of $58,014 represents the investment portrayed in Figure 16.3, even though that number is not a possible outcome of the proposal.

16B   Investments to Avoid Risk

For a given set of alternatives there are two ways to modify exposure to risk: (1) alter the probabilities of the applicable future states or (2) make different future states applicable. Both possibilities are usually costly to implement. For instance, the probabilities for future demand of a product could be changed by advertising; the amount of change is typically a function of the amount invested. When probabilities are immune to manipulation, as in the case of weather in a given locality, the odds can be altered by switching operations to a different locality. It may even be possible to effectively omit a future state by eliminating its effect on the alternatives, as would happen to an adverse weather state when an outdoor activity is moved indoors. And there is almost always a recourse to insurance, the conventional reaction to risk that does not reduce the danger but mitigates the damage.

Changing the Odds

An adequate investment can essentially eliminate risk in unique cases. Ski resorts can make provisions for artificial snow in anticipation of a dry winter. Investments are made in massive irrigation projects to eliminate the danger of drought. On a smaller scale, the selection by a city engineer of a storm drain sized to handle the maximum expected runoff is an attempt to nullify risk.

A more typical response in avoiding risk is a compromise. Even if the cause or penalty associated with a risk can conceivably be eliminated, the cost may be prohibitive. Yet the effects of risk can often be limited to a tolerable level by a reasonable expenditure. We dealt with this compromise when we considered benefit-cost ratios for public projects. Similarly, the city engineer could select a storm-drain size which balances the extra costs of larger-diameter sewers against damages that could occur in exceptional storms.

A quality-assurance program is a form of self-protection that attempts to control the probability of error with respect to the cost of increased vigilance. In a production process the cost of defective output is balanced against the cost of a procedure capable of reducing the chance of defects; an inspection plan is often used to guard against inferior quality. Consider the case of a company which received a contract to build 3000 precision instruments of a new type. The terms of the contract strongly suggest an inspection system to avoid the risk of supplying defective instruments; the penalty clause states damages of $300 per faulty unit.

Inspection equipment which would limit undetected faulty units to 0.5 percent could be purchased for $19,000. It would have no value after termination of the 3-year contract. A part-time operator for the testing equipment would have to be paid $7000 per year, and operating costs would amount to $1000 annually. Assuming a desired rate of return of 10 percent, we find that the annual cost of inspection would be

Testing machine:
$19,000(A/P, 10, 3) = $19,000(0.40212)$ $ 7,640

Operator 7,000

Operation 1,000
Total cost of inspection $15,640

In addition to the inspection costs, there is still a penalty for the few defective instruments which pass undetected and adjustment or reworking costs for the detected faulty instruments.

Annual cost of penalty

$$= \frac{3000 \text{ instruments}}{3 \text{ years}} \, 0.005 \times \$300/\text{instrument} = \$1500$$

Annual adjustment, reworking, and scrap costs
$$= X - 0.005 \times 1000 \text{ instruments/year} \times C_{av}/\text{instrument}$$

where X = percentage of defective instruments produced
C_{av} = average cost of reworking a defective instrument to enable it to pass inspection

Then the total cost of the inspection program is the sum of inspection, penalty, and reworking costs. Setting the cost of reworking at C_{av} = $50 per instrument, we have

$$\text{Total cost} = \$15,640 + \$1500 + (X - 0.005) \times 1000 \times \$50$$
$$= \$17,140 + (X - 0.005) \times \$50,000$$

By equating the cost of the inspection program to the penalty cost of no inspections, we get

$$\text{Cost of vigilance} = \text{cost of error}$$
$$\$17,140 + (X - 0.005) \times \$50,000 = X \times 1000 \times \$300$$
$$X = 6.8\%$$

X represents the percentage of defective units produced at which the two costs are equal. If the production capabilities can limit the percentage of defective instruments to 6.8 percent or less, the investment in this inspection program is unwarranted from strictly a cost viewpoint. However, reputation and other intangible considerations should influence the decision. It would also be wise to explore other inspection programs which are not so precise in detecting errors but use less expensive investments.

Spreading the Risk

The dichotomy of risk costs is again apparent in the use of an insurance program which transfers the burden of risk. Paying a premium to an insurer is an investment made to avoid the consequences of a specific disaster. The amount of the premium should reflect the potential magnitude of the disaster in proportion to the probability of its occurrence (plus administrative costs and profit). For a disaster which could cause damages of $100,000 per year with a probability of 0.001, it would be reasonable to expect an annual premium cost of at least $100,000 × 0.001 = $100.

The reason insurance programs are so prevalent is that they spread the cost of protection over a period of time and transfer the risk of a financial calamity to a group better prepared to meet the payment. Simply because the chance of a disaster is one in a thousand, it cannot be stated in exactly what year it will occur or even that it could not occur 2 years in a row. Premiums prorate the disaster shock to the pooled assets of the absorbing insurance company.

A program of self-insurance is followed by some companies for specific risks. Such a policy presumes sufficient resources to cover potential damages and is usually limited to minor risks. Accident liability is a popular area for self-insurance.

Investments may be made to reduce premiums for a group insurance policy or to reduce the probability of risk for a self-insurance policy. Annual investments in a plant safety program are aimed at reducing the probability of accidents. They should result in less damage payment for the self-insured or lower premiums for group insurance.

A feasible amount to spend for a risk-reduction investment can be estimated from insurance rates. Companies which sell insurance normally have a great deal of experience to draw on in setting rates. These rates may be used as estimates of risk for related situations, as described in the following example.

The owners are considering an automatic sprinkling system for a warehouse. The company insuring the warehouse will reduce the annual $12,000 fire-insurance premium by one-third if the system is installed. The owners estimate that opportunity costs incurred as a result of a serious fire, such as relocation expenses, disruption of deliveries, and loss of reputation, would amount to one-half the value of the warehouse. Using the insurance company's evaluation for the extent of risk reduction, we estimate that the expected value of annual savings in opportunity costs from the installation would be $12,000 \times \dfrac{1}{2} \times \dfrac{1}{3} = \$2000$.

If annual taxes and maintenance costs for the sprinklers are $1000, the life of the proposed system is 20 years, and capital is worth 15 percent, a feasible price for the sprinkler system could be determined from the following equality:

$$P(A/P,\ 15,\ 20) = \dfrac{\text{premium}}{\text{reduction}} + \dfrac{\text{opportunity-cost}}{\text{reduction}}$$
$$- \text{ annual taxes and maintenance}$$
$$P = \dfrac{\$12,000 \times \frac{1}{3} + \$2000 - \$1000}{0.15976} = \$31,300$$

QUESTIONS

16B.1 A resident engineer on a construction project has been troubled with parts received from a supplier. Eight percent of the parts have been defective. The scrap and reworking costs for each item average $12. At least 2000 parts will be used over the next 2 years. Two methods have been suggested to reduce the risk of defective parts:

1 Visual inspection by part-time labor would reduce the risk by half. The labor cost would be $500 per year.
2 Only 0.5 percent defective parts would be accepted if gages were purchased for $180 and a worker were trained to use them. The training cost would be $60, and an annual wage for the inspection time would total $700.

Which of the alternatives would be adopted if a 10 percent rate of return is considered acceptable?

16B.2 A private golf course lies in a hollow beside a river. Every other year the river rises to a level that prohibits play. The loss of revenue and damage to the course average $8000 each time the river rises to this level. A new clubhouse has been constructed on higher ground, but flood data indicate a 1/20 chance that the river will reach a level which will inundate the facility. The owners estimate that a flood of this magnitude would cause $40,000 damage to the clubhouse
(a) How much could the course owners afford to pay for a levee that would protect them from the biennial damage? The required rate of return is 8 percent, and the economic life of the levee is expected to be 30 years.
(b) What would be a reasonable annual premium for a full-coverage flood-insurance policy on the clubhouse?
(c) Compare the premium in Problem 16A.2 with a self-insurance policy that the owners could follow of laying aside $\dfrac{1}{20} \times \$40,000 = \$2000$ per year in anticipation of a major flood.

CHAPTER 17

DECISION MAKING

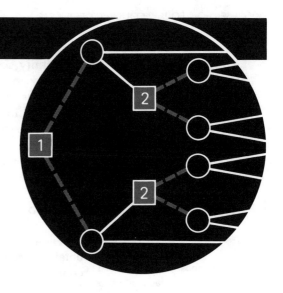

OVERVIEW

Utilizing the expected-value criterion presented in Chapter 16, we analyze decisions involving risk in this chapter with decision trees. The graphic quality of decision trees contributes to comprehensive analysis and cogent communications. Nonrepetitive decisions involving substantial outcomes exposed to risk are well served by the systematic evaluation methodology promoted by decision trees.

Conditional probabilities are the basis for a *discounted-decision-tree* analysis. A graphic tree format displays the effects of successive decisions occurring at timed intervals during a study period. Computations begin with expected values calculated for the most distant decision and roll back to the present by accepting the preferred alternative at each distant decision as a certain outcome for the next closer decision. A completed backward pass discloses the present worth of each immediate course of action. Additional information about a decision is used in *Bayesian analysis* to revise the probability of a future event. The expected value of sampling information suggests how much the additional information might be worth.

Also examined in this chapter are ways to generate and utilize numbers that account for intangible factors that affect decisions. A critical factor is the relative importance of criteria involved in a decision. A *priority decision table* incorporates importance ratings and accommodates mixed dimensions for criteria in a procedure that prioritizes alternative courses of action.

Giving the proper weight to decision factors is one of the toughest problems to contend with in making economic decisions. Rating factors on an *ordinal* scale gives only the order of preference. An *interval* scale shows the relative positions of the factors used. Customized rating forms and utility scaling are used to position the factors on a 1-to-10 interval scale; by one of these means importance factors are developed. Importance ratings are assigned to all the criteria by which alternatives are being evaluated in a priority decision table. Dimensionless numbers are developed by using fractions to convert *ratio*-scaled dimensions to a 10-point scale. The sum of the 10-point scaled criteria ratings multiplied by their respective importance ratings gives a score for measuring alternatives.

DECISION-MAKING DILEMMAS

Even the most straightforward economic decision has a parenthetical question mark that concedes some risk. As President Kennedy said in his 1962 State of the Union Message, "the one unchangeable certainty is that nothing is certain or unchangeable."

Extending economic analyses into the realm of risk exposes the difficulties of obtaining reliable data. Probabilities of future states seldom merit much confidence, and cash flow estimates for some states may be outright precarious. Yet these chancy estimates are made and used in decision models because the advice engendered is valued and could not be obtained otherwise. Even data flawed by uncertainty are prized when the alternative is complete ignorance. A decision-maker's dilemma is that of having to knowingly use data of questionable accuracy and compensate for the deficiencies by sensitivity analysis, or of having to use only data that are demonstrably reliable but which constrict analyses to commonplace situations.

In this chapter probabilities are combined with discounted cash flow principles to produce *discounted decision trees*. These trees provide a structure for analyzing decisions under risk that goes beyond payoff tables to map a sequential decision process.

The chapter also introduces a *priority decision* table. In the same manner that the decision-tree format graphically associates risks with outcomes, a decision table associates importance ratings with decision criteria. Including the relative importance of economic factors inserts a new dimension in the comparison procedures employed in previous chapters. Getting data for the table involves complications that are comparable to probability-estimating difficulties; it is especially difficult to develop ratings for importance and measurements for intangible properties that affect a decision. Subjectivity is involved. The dilemma includes making a choice between using subjective ratings for certain contributing factors, realizing that subjectivity detracts from dependability, or using only purely objective measurements that largely eliminate intangibles from being considered.

Being aware of both dilemmas is a prerequisite to deciding how to decide.

PROBABILITY CONCEPTS

Independent events and their probability relationships were examined in Chapter 16. These relations were applied to evaluate risk-affected situations according to the expected-value criterion. Independent outcomes that could logically be expected for each alternative course of action were displayed in a payoff table or in a decision-tree format (Figure 16.3). In this chapter the same general approach to economic risk is followed, but possible rewards from obtaining additional information are incorporated in the decision tree. In the process, some events become dependent on the occurrence of other events.

Dependent Events

An event is termed *statistically dependent* when its occurrence is affected by the occurrence of one or more other events. For instance, your probability of becoming a millionaire 10 years from now might be 0.9 if you were given $500,000 next year. The event of becoming a millionaire would thus be dependent on the receipt of $500,000. If the probability of receiving $500,000 is 0.01, then your likelihood of being a millionaire in 10 years according to this specific situation is

P(becoming a millionaire *and* receiving $500,000)

$\qquad = P(\text{making a million } given \text{ } \$500,000) \times P(\text{receiving } \$500,000)$

$\qquad = 0.90 \times 0.01 = 0.009$

The millionaire status obviously depends on the likelihood of the occurrence of the $500,000 acquisition.

Dependent relationships are illustrated by the classic "balls in a box" example depicted in Figure 17.1. Two boxes, labeled X and Y, contain black and white balls. Box X contains three white and two black balls; box Y holds one white and four black balls. The probability of drawing a ball of a given color clearly depends on which box is chosen for the draw.

The *marginal probability* of drawing a white ball in this situation is 0.4 [the sum of the probabilities of individual white balls, $P(W) = 1/2 \times 1/5 = 1/10$]. Even though two events may be related, a marginal probability refers to only one of the dependent events. From the probability tree in Figure 17.1, it is clear that the probability of drawing a white ball is affected by the box from which it is drawn. Therefore, the two events (drawing from one of the boxes and drawing a white ball) are related, but the marginal probability of drawing a white ball is still 0.4, because there are 10 balls with an equal probability of selection on the initial draw, and 4 of them are white.

What is the probability that a black ball will be drawn from box Y? This *conditional probability* is expressed symbolically as $P(B|Y)$, where the vertical line is read "given." From inspection of the probability tree, there are four chances in five that a black ball will be selected from box Y. Expressed as an equation, the conditional probability that a black ball will be drawn from box

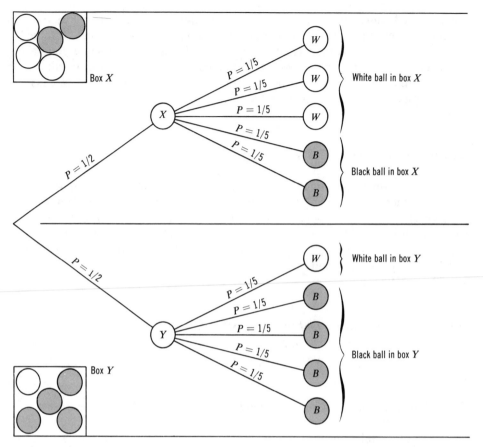

FIGURE 17.1
Population of black and white balls in boxes X and Y. Each burst of probability-labeled lines represents additional available information.

Y is equal to the probability that a black ball will occur in box Y divided by the probability that the draw will be made from box Y. Thus

$$P(B|Y) = \frac{P(BY)}{P(Y)} = \frac{0.4}{0.5} = 0.8$$

Example 17.1
Conditional Probability

The number of defective and acceptable items received in a shipment from two different companies is shown in the adjacent table.

What is the probability of receiving a defective item from company C1?

	Company C1	Company C2	Total
Defective	500	1000	1,500
Acceptable	9,500	4000	13,500
	10,000	5000	15,000

Solution 17.1

The marginal probability of a defective item in the entire shipment is 1500/15,000 = 0.10, but the conditional probability of a defective item, given that it was supplied by company C1, is 0.05. This result could be obtained logically by dividing the number of defective items D in the portion of the shipment supplied by company C1 (500) by the total number received from company C1 (10,000). It could also be calculated by formula:

$$P(D|C1) = \frac{P(DC1)}{P(C1)} = \frac{500/15,000}{10,000/15,000} = \frac{1}{20}$$

The *joint probability* that two dependent events will occur is given by the general equation

$$P(AB) = P(A|B) \times P(B) \quad \text{or} \quad P(B|A) \times P(A)$$

It is readily apparent that this is a restatement of the formula used to calculate conditional probabilities. In the illustration of black and white balls in boxes X and Y, the joint probability that a ball will be black and will come from box Y is

$$P(BY) = P(B|Y) \times P(Y) = 0.8 \times 0.5 = 0.4$$

Similar applications to other joint probabilities for the given situation can be verified by reference to Figure 17.1.

Bayesian Analysis

The opportunity to update and refine probability forecasts by taking advantage of additional information is a powerful analytic tool. Original objective or subjective probability assignments are developed from current knowledge to anticipate possible future outcomes. As time passes, we often have access to new information about the events that we are predicting. The concept of revising prior probability estimates to reflect new data is attributed to Thomas Bayes. His basic formula $P(P|B) = P(AB)/P(B)$ was described in connection with the calculation of conditional probability. The routine followed in finding posterior probabilities is based on this basic formula. The sequence of operations follows this pattern: The equality

Reverend Bayes was an eighteenth-century English clergyman.

$$P(AB) = P(B|A) \times P(A)$$

can be rewritten as

$$P(BA) = P(A|B) \times P(B)$$

which provides the equality

$$P(B|A) \times P(A) = P(A|B) \times P(B)$$

which can be converted to

$$P(B|A) = \frac{P(A|B) \times P(B)}{P(A)}$$

and the marginal probability of A is

$$P(A) = P(A|B_1) \times P(B_1) + \cdots + P(A|B_x) \times P(B_x)$$
$$+ \cdots + P(A|B_n) \times P(B_n)$$
$$= \sum_x P(A|B_x) \times P(B_x)$$

The correct use of this routine allows an event B to be reevaluated when new information concerning the outcome of A becomes available.

The illustration in Figure 17.1, black and white balls in boxes X and Y, will be adapted to illustrate a Bayesian analysis. Assume that the content of each box is known but the identity of the boxes is unknown. This situation is equivalent to knowing the outcome of two possible futures without being certain which future will occur.

In this case we want to identify which box is X and which is Y. Since the two boxes are indistinguishable, there is an equal opportunity that either could be nominated as X or Y.

Event 1   The first event is a random selection of one of the boxes. The prior probability of this event is represented by the probability tree of Figure 17.2.

Event 2   Next a ball is drawn from whichever box was picked in event 1. We will assume that it is a black ball. Since we know the proportion of black to white balls in each box, we can calculate the probability of drawing a black ball, given that it came from a designated box. In box X three of the five balls are white, so the conditional probability for the top branch of an expanded probability tree is 0.6. The likelihood of drawing a ball of a given color from a given box is shown by the joint probability for each branch of the tree. The complete tabulation of joint probabilities (collectively exhaustive) totals to 1.0, but we are interested primarily in the probabilities relating to black balls, because our first draw was black. The sum of the probabilities pertaining to black balls is the marginal probability of drawing a black ball and is shown in the last column of Figure 17.3.

The process of acquiring additional information in this fashion is called sampling.

FIGURE 17.2
First event in the Bayesian-analysis illustration—box selection.

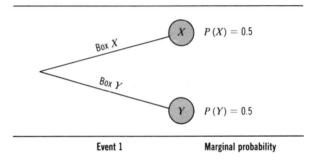

| Event 1 | Marginal probability |

FIGURE 17.3
Box selection and first draw.

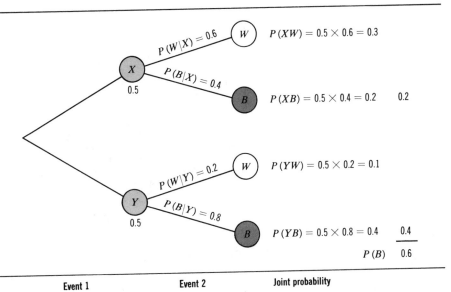

Event 1	Event 2	Joint probability

The posterior probability of identifying the boxes, based on the additional information derived from the draw, is calculated by Bayes basic formula. Box Y is arbitrarily used in the formula to give

$$P(Y|B) = \frac{P(YB)}{P(B)} = \frac{0.4}{0.6} = 0.67$$

Thus the added data have allowed us to revise our probability estimate from 0.5 to 0.67 that the selected box is indeed box Y.

Now assume that another ball is drawn from the same box and is black. **Event 3** Conditional probabilities for this draw are calculated on the basis of the four balls remaining in the box after the first draw. Then the joint probabilities are determined for two successive black draws from either box. These procedures are depicted in the third section of the probability tree in Figure 17.4.

A further revision of the likelihood that the chosen box is box Y now becomes

$$P(Y|B_1B_2) = \frac{P(B_1B_2Y)}{P(B)} = \frac{0.3}{0.35} = 0.857$$

Suppose that the second draw had revealed a white rather than a black ball. This possibility is represented by the dotted lines in the probability tree.

FIGURE 17.4
Box selection and two draws.

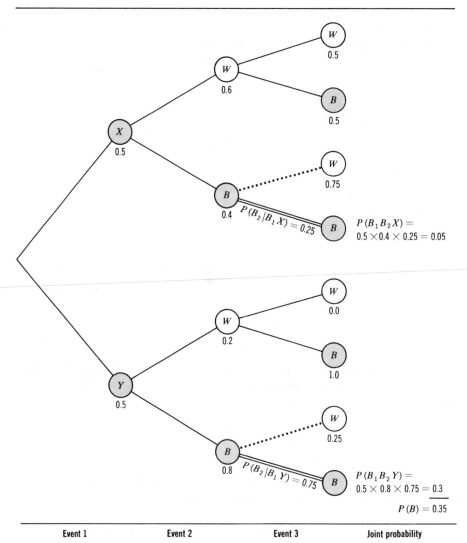

A format for revising the probability that the draws (B_1 and W_2) were from box Y is given in the following table:

Bayesian relationships are fundamental to the development of operations controls and quality assurance; an application is described in Review Exercise 2.

Event 1	Event 2 = B_1	Event 3 = W_2	P(E1E2E3)
$P(X) = 0.5$	$P(B_1|X) = 0.4$	$P(W_2|B_1X) = 0.75$	$0.5 \times 0.4 \times 0.75 = 0.15$
$P(Y) = 0.5$	$P(B_1|Y) = 0.8$	$P(W_2|B_1Y) = 0.25$	$0.5 \times 0.8 \times 0.25 = \underline{0.10}$
			0.25

$$P(Y|B_1W_2) = \frac{0.10}{0.25} = 0.40$$

We would intuitively suspect that a reversed order of the draws (W_1 and B_2) would not alter our revised probability. This suspicion is confirmed by the following values:

Event 1	Event 2 = W_1	Event 3 = B_2	P(E1E2E3)
$P(X) = 0.5$	$P(W_1|X) = 0.6$	$P(B_2|W_1X) = 0.5$	$0.5 \times 0.6 \times 0.5 = 0.15$
$P(Y) = 0.5$	$P(W_1|Y) = 0.2$	$P(B_2|W_1Y) = 1.0$	$0.5 \times 0.2 \times 1.0 = \underline{0.10}$
			0.25

$$P(Y|W_1B_2) = \frac{0.10}{0.25} = 0.40$$

As a final possibility, again assume that the first two balls drawn were black and now a third black ball is picked from the same box. We have obviously been drawing from box Y, because box X originally contained only two black balls. Although formal calculations are unnecessary in this case, the conclusion is verified easily by recognizing that

Event 4

$$P(3 \text{ black-ball draws}|\text{box } X) = 0.4 \times 0.25 \times 0.0 = 0.0$$

and

$$P(3 \text{ black-ball draws}|\text{box } Y) = 0.8 \times 0.75 \times 0.67 = 0.4$$

which makes the marginal probability of three successive black draws $0.0 + 0.4 = 0.4$, so

$$P(\text{box } Y|3 \text{ black-ball draws}) = \frac{0.4 \times 0.5}{0.4 \times 0.5} = 1.0$$

DECISION TREES

The tree structure in Figure 17.4 displays probabilities but does not represent a decision situation. A decision tree has three components:

1 Decision alternatives with associated future events
2 Outcomes for each alternative, given the occurrence of each future event
3 The probability of the occurrence of each event

A convenient set of symbols to graphically portray the decision situation represents decision points as squares and outcomes as circles. Dotted lines between squares and circles symbolize different courses of action, and solid

Samuel Butler noted in his 1912 Notebooks that "There is one thing certain, namely, that we can have nothing certain; therefore it is not certain that we can have nothing certain."

lines from the circles represent the possible consequences of the actions. The decision criterion is the expected value of alternatives at each decision point.

Suppose that it costs $100 to play the game of picking which box is X and Y when the contents of the boxes are known, but outwardly the boxes are identical. Success in naming a box correctly is rewarded by winnings of $180 (net gain of $80 since it costs $100 to play the game). Because the boxes are indistinguishable to the gambler, there is a 50-50 chance of guessing correctly. The expected value from repeatedly playing the game is

$$-\$100 + 0.5(\$180) = -\$10 \text{ per play}$$

A decision-tree representation of the gamble is shown in Figure 17.5.

Now assume that another option has been added to make the game more interesting (and because no one wants to play the original game). A gambler can select one ball from either box for a sampling fee of $25. After the charge is paid and a ball is drawn, the original conditions of the game then apply: an ante of $100 to guess which box is which. The advantage of the sampling draw for the gambler is to gain additional information before making a selection between boxes. The question is whether the additional information provided by the draw can change the expected value of the game to the gambler's favor. The sequential alternatives are displayed in Figure 17.6.

Probabilities for the sampling option are based on the known distribution of black and white balls in boxes X and Y. For instance, once a white ball is drawn, the probability it was drawn from box X is

$$P(X|W) = \frac{P(W|X)P(X)}{P(W|X)P(X) + P(W|Y)P(Y)} = \frac{P(XW)}{P(XW) + P(YW)}$$

$$= \frac{0.6(0.5)}{(0.6)(0.5) + (0.2)(0.5)} = \frac{0.3}{0.3 + 0.1} = 0.75$$

Similarly, the probability of drawing a white ball for the sample when the identity of the boxes is unknown is

FIGURE 17.5
Decision tree for a gamble to select box X or Y when the ante is − $100 and the payoff for correctly selecting the named box is $180.

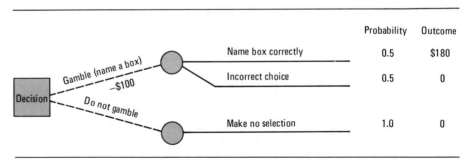

	Probability	Outcome
Name box correctly	0.5	$180
Incorrect choice	0.5	0
Make no selection	1.0	0

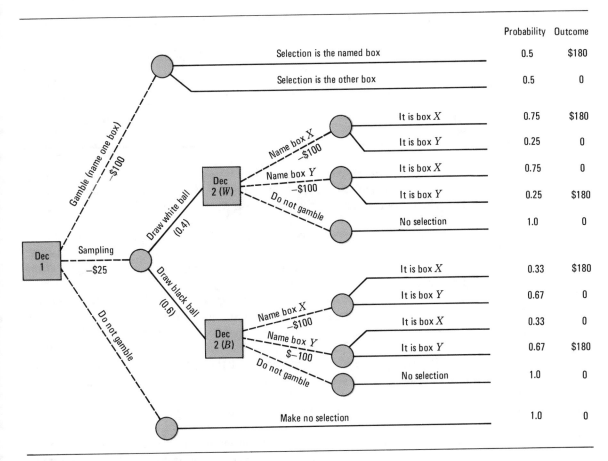

	Probability	Outcome
Selection is the named box	0.5	$180
Selection is the other box	0.5	0
It is box X	0.75	$180
It is box Y	0.25	0
It is box X	0.75	0
It is box Y	0.25	$180
No selection	1.0	0
It is box X	0.33	$180
It is box Y	0.67	0
It is box X	0.33	0
It is box Y	0.67	$180
No selection	1.0	0
Make no selection	1.0	0

FIGURE 17.6
Expanded decision tree for the pick-a-box gamble. Probabilities are based on the boxes described in Figure 17.1. Payoffs are the same as in Figure 17.5, but there is an added option of sampling one ball from a box before deciding to ante $100 for the chance to select a box. The fee for sampling is $25.

$$P(W) = P(W|X)P(X) + P(W|Y)P(Y)$$
$$= (0.6)(0.5) + (0.2)(0.5) = 0.4$$

or, since there are 4 white balls among the 10 in the two boxes,

$$P(W) = 4/10 = 0.4$$

Given the probabilities and the payoffs for each outcome, we calculate the expected value of the sampling option by "rolling back" from decision point 2 to decision point 1. This means that the expected values from point 2 are calculated first and the most profitable alternatives become the outcomes for point 1.

A payoff table for decision point 2(W) (a white ball drawn as the sample) is shown below.

	It Is Box X $P(X\|W) = 0.75$	It Is Box Y $P(Y\|W) = 0.25$	Expected Value
Pick box X	$180 − $100 = $80	−$100	$35
Pick box Y	− $100	$180 − $100 = $80	−$55
Refuse gamble	0	0	0

The preferred alternative is clearly to pick box X when a white ball is drawn. By equivalent calculations, the box should be identified as Y at decision point 2(B) when a black ball is drawn, because the expected value of this choice is

$$EV(Y \text{ given } B) = (\$180 - \$100)(0.67) + -\$100(0.33) = \$20$$

as opposed to naming the box X for an expected value of $-\$40$ or quitting after the draw at no additional gain or loss.

A pruned decision tree is shown in Figure 17.7. It displays the expected values for the original gamble and the outcomes for the sampling alternative. These outcomes are based on selecting the most advantageous alternatives from decision points 2(W) and 2(B). Given the likelihood of drawing a black or white ball for the sample, we find that the expected value given sampling information is

The worths of sampling information and perfect information for this situation are compared in Review Exercise 3.

$$EV|SI = \$35(0.4) + \$20(0.6) - \$25 = \$1$$

The option to buy a sample for $25 thus converts the long-term gain of the gamble from negative to positive, albeit barely profitable. The actual outcome of each round of gambling would be a loss of $25 + $100 = $125 or a gain of $180 − $125 = $55. Over a large number of rounds, the gambler averages a $1 gain per round.

The value of information gained from sampling is the difference between the expected value of the sampling alternative and the *next best alternative*.

FIGURE 17.7
Pruned decision tree showing the expected values for two alternatives and the rolled-back outcomes from decision points 2(W) and 2(B) as determined from Figure 17.6. Expected value of the sampling alternative is $1, making it the preferred course of action for the given data.

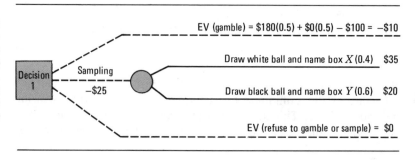

For the "pick-a-box" decision, the alternative of refusing to gamble is compared with the gamble-with-sampling alternative to indicate the

$$\text{Expected value of sampling information (EVSI)} = EV|SI - EV$$
$$= \$1 - \$0 = \$1$$

DISCOUNTED DECISION TREES

Discounted cash flows, probabilities, and expected values are combined in a free format to generate a graphic *discounted decision tree*. Its unique feature is its capacity to display future decision points—times in the future when a decision maker can appraise actual outcomes from earlier decisions to decide if a previously determined course of action should be modified to cope with current conditions.

A discounted decision tree shows decisions separated by time intervals and susceptible to external influencing factors. Branches radiate from an initial decision point to indicate the primary alternatives. Each main branch is divided to show foreseeable outcomes associated with possible future events. The events are then rated with respect to their probable occurrence. When gains can be maximized by introducing new alternatives at a future date, a second decision point is established. A succession of decision points can extend to the limit of forecasting ability. The time value of monetary outcomes is affected by discounting the outcomes to a common point in time.

Formulation of a Discounted Decision Tree

A warehousing problem of a small novelty manufacturing company will serve to illustrate a decision tree for successive decisions. The company is relatively new and has captured a limited segment of the novelty market. It must have additional storage space to meet customer demands and to allow more flexible production scheduling. A primary decision has been made to secure additional inventory storage.

An initial investigation has revealed the availability of only one suitable rental warehouse; it is available only if leased for 10 years. The warehouse has more space than is immediately required, but the company feels that some of the space could be sublet if desired. Estimates solicited from building contractors confirm that the construction of a new warehouse of equivalent size would amount to more than the $23,000 per year lease cost.

Another alternative is to build a small warehouse now and enlarge it if future business activity warrants expansion. The owners feel that in 3 years they will know whether the company's growth will support the addition. In order to evaluate the alternatives, estimates were made of possible business patterns and the likelihood of each. Their optimistic forecasts are shown in Table 17.1.

TABLE 17.1	
Growth patterns for the novelty company	
Growth Pattern	Probability
No increase in activity for 10 years	0.15
No increase for 3 years, but an expanded growth rate during the next 7 years	0.15
Increasing growth for the next 3 years, but no increase during the following 7 years	0.14
Increasing activity for the full 10 years	0.56

The owners place the probability for increased growth during the next 3 years at $0.56 + 0.14 = 0.70$. If the growth materializes, they can use more room than is available in the anticipated small warehouse. Therefore, if they initially decide to build, they must make a decision in 3 years about whether to add to the small warehouse or find other means to obtain extra storage space. At that time the conditional probability that the company will continue to grow is $0.56/0.70 = 0.80$. The spectrum of forecasts and alternatives can be summarized in a decision tree as shown in Figure 17.8.

FIGURE 17.8
Decision tree with alternatives and forecasts.

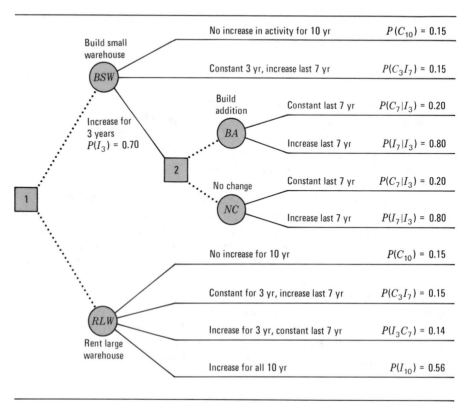

Outcomes

The outcomes for the warehouse proposals are rated according to expected costs. Initial building costs for a small warehouse should be accurate, but the estimated price for an addition is less firm because of possible changes in building conditions at the time of construction. Yearly rental fees for the leased warehouse are a fixed amount. Other annual costs are less certain. Savings, net positive cash flows, could result if the entire capacity is not required for the company's inventory and the extra portion is rented. Conversely, additional costs are incurred when a lack of storage space forces production runs below the economical lot size or causes out-of-stock costs in supplying customers. Estimates of net annual costs for the outcomes of each alternative are tabulated in the decision tree of Figure 17.9.

The cash flow estimates shown in the figure include the first costs associated with each alternative and the annual cost related to the outcomes. Building a small warehouse, *BSW*, has a first cost of $110,000 and a resale value of $40,000 after 10 years; renting a large warehouse, *RLW*, has no initial cost. The second decision point 3 years in the future is a choice between building

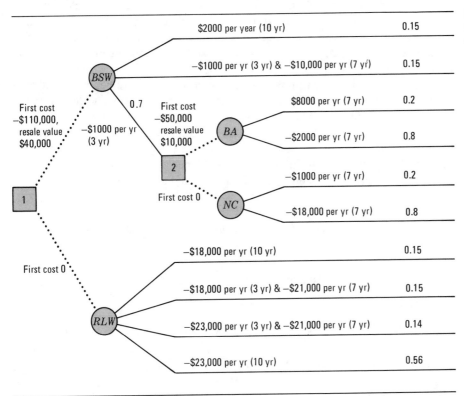

FIGURE 17.9
Costs associated with warehouse alternatives.

First cost and resale value of alternatives

Annual costs

Probability

an addition, *BA*, at a cost of $50,000 with a $10,000 resale value and making no change, *NC*. Net annual returns comprise repairs, taxes, insurance, leasing expense, and opportunity costs, as well as the savings from subletting when possible. The lease for the rental warehouse states an annual charge of $23,000, which includes taxes, insurance, and repairs. If there is no increase in the company's activities, this annual charge can be defrayed by $5000 from subletting extra space. Less rental income is anticipated when the company needs part or all of the space to handle its own increasing activity. The two periods of returns, 3 and 7 years, correspond to the original growth patterns indicated in Figure 17.8.

Evaluation

Two types of calculations are involved in evaluating alternatives. The present values of receipts and expenditures are determined for each outcome and are then weighted according to their probability of occurrence. This procedure amounts to finding the expected value of the present worth of the outcomes. These expected values are compared at a decision point to select the most advantageous alternative.

Comparisons are made in a reverse chronological order. That is, the most distant decision point from time zero is evaluated first. The selected alternative from the first decision then becomes an input to the next decision. The backward pass through successive points is continued until the primary decision is resolved.

For the warehouse example, the discounting procedure begins at decision point 2. At this point in time, 3 years away from the primary decision, the company must decide whether to build an addition to the plant or make no change. The outcomes of each of these two alternatives depend on the level of business activity during the last 7 years of the study period. We will assume that the company uses an interest rate of 12 percent. The present worths of the four outcomes at decision point 2 are calculated as shown.

$$\text{PW}(BA:C_7) = -\$50,000 + \$10,000(P/F, 12, 7) + \$8000(P/A, 12, 7)$$
$$= -\$50,000 + \$10,000(0.45235) + \$8000(4.5637)$$
$$= -\$8966$$

$$\text{PW}(BA:I_7) = -\$50,000 + \$10,000(P/F, 12, 7) + -\$2000(P/A, 12, 7)$$
$$= -\$54,603$$

$$\text{PW}(NC:C_7) = -\$1000(P/A, 12, 7) = -\$1000(4.5637) = -\$4564$$

$$\text{PW}(NC:I_7) = -\$18,000(P/A, 12, 7) = -\$82,142$$

These present worths at decision point 2 are entered in a payoff table as shown

FIGURE 17.10
Payoff table for decision point 2.

	COMPANY GROWTH PATTERNS		
Alternative	*Constant (0.2)*	*Increase (0.8)*	*Expected Value (Costs)*
BA	−$8966	−$54,603	−$45,476
NC	−4564	−82,142	−66,626

in Figure 17.10, where the calculated expected values indicate a preference for the alternative to build an addition, *BA*.

Decision point 1 also has two alternatives. Since we have already considered decision point 2, we will continue on this branch of the decision tree. The initial cost of building is $110,000, with an expected resale value of $40,000 after 10 years. Three outcomes may occur from this course of action. Company activity could increase or remain constant for the entire 10 years, or it could increase the first 3 years and then level off or continue to increase the remaining 7 years. The conditional decision made at point 2 was based on the possible outcomes of the last 7 years, *given* that the activity increased during the first 3 years. This decision becomes an outcome of the primary alternative to build a warehouse. A comparison, assuming end-of-year returns and a 12 percent interest factor, is shown below.

$$PW(BSW{:}C_{10}) = -\$110{,}000 + \$40{,}000(P/F,\ 12,\ 10) + \$2000(P/A,\ 12,\ 10)$$
$$= -\$110{,}000 + \$40{,}000(0.32197) + \$2000(5.6502)$$
$$= -\$85{,}821$$

$$PW(BSW{:}C_{3}I_{7}) = -\$110{,}000 + \$40{,}000(P/F,\ 12,\ 10) + -\$1000(P/A,\ 12,\ 3)$$
$$+ -\$10{,}000(P/A,\ 12,\ 7)(P/F,\ 12,\ 3)$$
$$= -\$110{,}000 + \$40{,}000(0.32197) + -\$1000(2.4018)$$
$$+ -\$10{,}000(4.5637)(0.71178) = -\$132{,}007$$

$$PW(BSW{:}I_{3}) = -\$110{,}000 + \$40{,}000(P/F,\ 12,\ 10) + -\$1000(P/A,\ 12,\ 3)$$
$$+ -\$45{,}476(P/F,\ 12,\ 3) = -\$131{,}892$$

Equivalent calculations applied to the other main fork of the decision tree give the present worths of annual costs associated with leasing a large warehouse.

$$PW(RLW{:}C_{10}) = -\$18{,}000(P/A,\ 12,\ 10) = -\$18{,}000(5.6502) = -\$101{,}704$$

$$PW(RLW{:}C_{3}I_{7}) = -\$18{,}000(P/A,\ 12,\ 3) + -\$21{,}000(P/A,\ 12,\ 7)(P/F,\ 12,\ 3)$$
$$= -\$18{,}000(2.4018) + -\$21{,}000(4.5637)(0.71178)$$
$$= -\$111{,}448$$

$$PW(RLW{:}I_3C_7) = -\$23,000(P/A, 12, 3) + -\$21,000(P/A, 12, 7)(P/F, 12, 3)$$
$$- \$123,457$$

$$PW(RLW{:}I_{10}) = -\$23,000(P/A, 12, 10) = -\$129,955$$

Grouping the outcomes from the two alternatives into a payoff table yields the values in Figure 17.11.

From this analysis it is apparent that the most economical alternative is to sign the lease for renting the large warehouse. The margin of difference is $\$124,998 - \$122,032 = \$2966$.

The novelty company now has a quantitative base for making its decision. It may be satisfied with the information and proceed to sign the lease. It is also possible that some segment of management may not agree on the indicated course of action. There may be a disagreement about estimated growth patterns or cost figures. Assuming that the need for some kind of storage facility is unanimously recognized, we find that there is still another alternative available: The decision can be postponed. This alternative would be procrastination unless it were coupled with a firm desire to obtain more information about the problem. Such information could be provided by further intensive investigation by company personnel or by a study conducted by an independent agency.

The first step in considering the advisability of further research is to determine the worth of additional information. Securing new data costs money, whether the investigation is conducted by company or outside investigators. This additional investment should be exceeded by the expected value of added profits or reduced costs realized from the information.

With *perfect* information the company would *know* which alternative would cost the least. Figure 17.11 revealed that the best alternative would be to build a small warehouse without an addition if it were *known* that the company would experience a constant growth pattern for 10 years. The cost of building ($\$85,821$) is $\$15,883$ less than leasing for this growth pattern. In a similar fashion the best alternative for each state assumed *known* can be determined.

Four future states were identified in the warehouse illustration. It was observed that a small warehouse should be built for a condition of constant

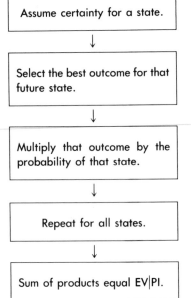

Assume certainty for a state.

↓

Select the best outcome for that future state.

↓

Multiply that outcome by the probability of that state.

↓

Repeat for all states.

↓

Sum of products equal EV|PI.

Flowchart for calculating the expected value given perfect information.

Alternatives	COMPANY GROWTH PATTERNS				
	Constant 10 Years (0.15)	Constant 3 Years, Increase 7 Years (0.15)	Increase 3 Years, Constant 7 Years (0.14)	Increase 10 Years (0.56)	Expected Value (cost)
BSW	−$ 85,821	−$132,007	−$131,892	−$131,892	−$124,998
RLW	−101,704	−111,448	−123,457	−129,955	−122,032

FIGURE 17.11 Payoff table for decision point 1.

growth for 10 years. The best alternative for 3 years of constant growth followed by 7 years of increasing activity is to lease. Figure 17.11 shows that the present worth of renting expense is $111,448 versus building costs of $132,007. The other two states are not so apparent from previous calculations.

For the state of accelerated growth for 3 years followed by 7 years of level activity, the cost for no change in the capacity of the small warehouse was determined to be $4564. Discounting this value to the present and including the initial cost, resale, and annual costs for the first 3 years give

$$PW(BSW{:}I_3C_7) = -\$110,000 + \$40,000(P/F, 12, 10) + -\$1000(P/A, 12, 3)$$
$$+ -\$4564(P/F, 12, 3)$$
$$= -\$102,772$$

which reveals a cost of building that is less than is the lease expense of $123,457. A similar computation for the building cost with 10 years of increasing activity is

$$PW(BSW{:}I_3I_7) = -\$110,000 + \$40,000(P/F, 12, 10) + -\$1000(P/A, 12, 3)$$
$$+ -\$54,603(P/F, 12, 3)$$
$$= -\$138,388$$

which exceeds the leasing cost of $129,955 for the same future state.

The expected value from the preferred alternatives for each state evaluated separately is shown in Table 17.2.

Comparison of the expected value of perfect information with the expected value of the best alternative obtained earlier (Figure 17.10) indicates that the novelty company could afford to pay up to $122,032 - $116,753 = $5279 for a perfect forecast. Securing perfect information is at best a wild hope, but these figures do suggest that significant savings could result from further study of the uncertainties in expected growth patterns.

TABLE 17.2

Expected value of perfect information

State	Preferred Alternative	Cost	Proba-bility	Product
Constant 10 years	BSW	− $85,821	0.15	− $12,873
Constant 3 years, increase 7 years	RLW	− 111,448	0.15	− 16,717
Increase 3 years, constant 7 years	BSW-NC	− 102,772	0.14	− 14,388
Increase 10 years	RLW	− 129,955	0.56	− 72,775
			Expected value	− $116,753

APPLICATION OF DECISION TREES

Nonrepetitive decisions that involve risk and substantial outcomes are good candidates for decision-tree analysis. The tree structure forces an explicit recognition of risk and promotes the comprehensive organization of alternative strategies. The graphic representation of data and competing options improves communication between decision makers and analysts. It also helps in "selling" a decision.

Negative aspects of decision trees stem largely from difficulties in displaying very complex situations. A tree gets too "bushy" to comprehend when branches are proliferated to accommodate complicated relations. Nonquantitative influences may be lost in the quest for data to complete all the branches or ignored because they do not fit conveniently into the tree. Whereas the formalized analysis procedure provides evaluation consistency, it may discourage consideration of interrelations that cannot be definitively expressed. Decision-tree language does not lend itself to intuition.

Extension 17A presents a tree modified to portray the origin of accidents and the data needed to decide whether safety investments are warranted.

Decision-tree concepts are adaptable to diverse managerial and engineering applications. Scratch-pad sketches can be used to "scope out" preliminary designs. Sketches may stimulate the generation of new alternatives through association with displayed strategy. Refinements can be made to the basic tree format to better represent certain problem situations.

DIMENSIONS FOR DECISIONS

Numbers have been called the "language of engineering." Every engineering student is well aware of the emphasis given to mathematics and the sometimes bewildering variety of number manipulations used in solving engineering problems. What tends to be overlooked is the difficulty of acquiring the numbers. "Hard" data, ratio scale numbers for receipts and disbursements, can usually be obtained, although diligent digging may be necessary. "Soft" data, composed of subjective opinions, are normally easy to obtain, but their reliability is questionable. Both hard and soft data are involved in most comparisons.

Many factors that affect a decision have no natural measures. How can you measure the relative *attractiveness* of two designs—by the number of beauty points they possess? If so, how do you define a "beaut"? The evaluation of intangibles is important because the final decision could hinge on the value placed on a factor such as attractiveness.

Jeremy Bentham, in his contentious 1789 book, maintained that managers should try to do "the greatest good for the greatest number." Ever since, the question has been, "How can one measure how good is good?"

No ideal method has yet been devised to quantify intangibles. The choice is to use imperfect methods or ignore the intangible aspects in quantitatively evaluating economic decisions. Where intangibles have little influence, neglect is reasonable. When there are important subjective factors involved, imperfect methods are better than nothing because they at least expose the opinions to formal scrutiny.

Ordinal Scaling

A simple order scale ranks every item in a list in order of preference. It works fine when there are only two alternatives to be subjectively rated for a single criterion. Then it is merely a choice of judging which alternative is more satisfying. However, an order scale does not measure the degree of preference; it is only a listing in preferential order. If three alternatives, X, Y, and Z, are ranked in order 1, 2, and 3, respectively, the ranker expresses preference for X over Y and Y over Z. But there is no clue about how much X is preferred to Y and Z, or why it is preferred.

A separate scale must be established for each merit being ranked, such as more useful than, more convenient than, or easier to understand than. Whether starting from a "most-preferred" level and working down the list or starting from the bottom up, ranking is essentially by pairs.

Assume that there are five alternatives to be ranked for the same criterion: V, W, X, Y, and Z. As shown in Figure 17.12, all possible pairings are listed on the "steps." There will always be one step fewer than the number of items to be compared. The sequence in which the items are listed is purely arbitrary. In the first step, V is paired with W, X, Y, and Z. In the second step, V is dropped, and the next item, W, is compared in turn with the remaining items. The procedure is continued until a single pair remains.

Ranking is accomplished by circling the preferred item in each comparison. In the first step, if V is preferred to W and Z, it is circled for these pairs. Meanwhile, X and Y are circled to show their rank above V. The formal ordering results from a tabulation of the number of times each item is preferred in the complete pairwise comparison. In the example, item Y was circled four times to receive the top rank, and the never-preferred item Z is relegated to the last position.

The tempting violation of ordinal scaling is to read into an ordering a certain mathematical spacing between entries. The simple order of $X > Z > W > Y$ (where $>$ is read "is preferred to") could represent a number set of $100 > 99 > 98 > 2$ or $41 > 10 > 3 > 2$. The lack of specific intervals rules out any arithmetic operations.

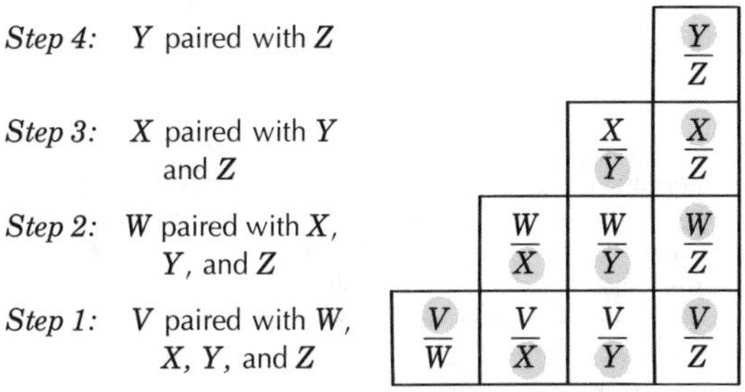

An inconsistent ranking, such as $X > Y > Z > X$, is called *intransitive*. See Review Exercise 5.

Step 4:	Y paired with Z				$\dfrac{Y}{Z}$
Step 3:	X paired with Y and Z			$\dfrac{X}{Y}$	$\dfrac{X}{Z}$
Step 2:	W paired with X, Y, and Z		$\dfrac{W}{X}$	$\dfrac{W}{Y}$	$\dfrac{W}{Z}$
Step 1:	V paired with W, X, Y, and Z	$\dfrac{V}{W}$	$\dfrac{V}{X}$	$\dfrac{V}{Y}$	$\dfrac{V}{Z}$

Alter-native	Number of Circles	Rank
V	2	3
W	1	4
X	3	2
Y	4	1
Z	0	5

FIGURE 17.12
Step-by-step comparison of alternatives to develop an ordinal-scale ranking. Shaded circles indicate the preference among pairs at each step. The tabulation shows that the ordinal scale for the five alternatives is Y over X over V over W over Z.

Interval Scaling

An interval scale is the next improvement from an order scale. This type of scale provides a relative measure of preference in the same way that a thermometer measures relative warmth. An interval scale is a big improvement over ranking, but it still cannot be used like a ratio scale of distance or weight. This limitation stems from the lack of a natural zero. A zero in a ratio scale has a universal meaning; a zero distance or a zero weight means the same thing to everyone. A zero temperature can convey different meanings according to the type of interval scaling employed: Fahrenheit or Celsius. But once this zero value is understood, both temperature scales use standardized units of measurements which allow certain arithmetic operations, such as averaging, to be performed with the scaled values. Two methods for developing an interval scale are described in the following sections.

Rating Forms

A standardized rating form which has written descriptions of each level of desirability is the most commonly used method for rating intangibles. The scales typically run from 0 to 10 with explanations of the attributes expected at each interval. Well-composed rating forms define, in easily understood language, the outcome that qualifies an alternative for each numbered rating.

Personnel ratings are often made on standardized appraisal forms. A satirical form is innocently offered in Figure 17.13. More serious versions are

		EMPLOYEE APPRAISAL GUIDE			
Ranking	*Phenomenal* 4	*Marvelous* 3	*Good* 2	*Not so Good* 1	*Pathetic* 0
Competitiveness	Slays giants	Holds his own against giants	Holds his own against equals	Runs from midgets	Gets caught by midgets
Personal appearance	Could be a professional model	Could model but wouldn't be paid much	Could model as the "before"	Goes unnoticed in a crowd	Panics a crowd if noticed
Leadership	Walks on water consistently	Walks on water in emergencies	Wades through water	Gets caught in hot water	Passes water in emergencies
Intelligence	Knows everything	Knows a lot	Knows enough	Knows nothing	Forgets what he never knew
Communication	Talks to big shots	Talks to little shots	Talks to himself	Argues with himself	Loses those arguments

FIGURE 17.13
A spoof of an employee-ranking system that makes enough sense to be frightening.

widely used. A rater is asked to assign a number or choose a proficiency level that describes the person being rated. The selection of a fitting description fixes a number to a particular criterion of performance. Resulting numbers are collectively taken as a representative measure of stature for the person being ranked.

Rating forms with similar characteristics have been developed to evaluate recurring decision situations. For example, government agencies engaged in research solicit bids from internal and outside investigators for conducting studies. A request for proposals (RFP) contains a statement of the technical requirements of the work and requests bidders to provide cost estimates, time schedules, and proof of competence. The replies are then evaluated by a board according to how well they meet the criteria of acceptance. A typical guideline for assigning numerical ratings for each criterion is given below.

Score				Description
10	9		Very good	Has a high probability of exceeding all the requirements expressed in the RFP for the criterion
8	7	6	Normal	Will most likely meet the minimum requirements and scope of work established in the RFP
5	4	3	Below normal	May fail to meet the stated minimum requirements but is of such a nature that it has correction potential
2	1		Unacceptable	Cannot be expected to meet the stated minimum requirements and is of such a nature that drastic revision is necessary for correction

While using an appraisal form, remember that it is important to keep referring to a mental standard that conforms to each level. In the RFP evaluation, the standards are defined in writing. In personnel rating forms the standards result from experiences with the performance of people who were previously rated in each category. Each rater likely has a different interpretation of what constitutes perfection, based on personal views and past exposures. It is not vital that all raters have the same absolute limits for their interval scale; it is vital that they be consistent in applying their own scale among alternatives. Some raters believe that their consistency is improved by initially giving each criterion a rating of 10 and then subtracting points as the alternatives are compared.

A review of past ratings can improve future ratings. A numerical rating system is only as good as are the rationale exercised in its use. A rater should be prepared to convince a questioner that the judgment was correct. Since intangible judgments are necessarily fragile, they deserve to be handled with care.

Example 17.2
Standardized Scoring

Alternatives are often evaluated by a team of appraisers. Each person evaluates all alternatives using the same scaling technique, and the resulting scores for each alternative are summed to provide a ranking among alternatives. Unless all appraisers have about the same range of scores, bias is introduced. A single score for an alternative that is far outside the range used by the rest of the appraisers can bias the selection toward the alternative favored by one enthusiastic backer. A simple procedure for standardizing the scores reduces this type of biasing.

The procedure described below is used by several organizations to convert each appraiser's ratings to an average value of 500 with a standard deviation of 100. After the scores are standardized, they are averaged for a final evaluation.

To standardize the scores given by each evaluator, calculate:

$$\text{Mean} = \bar{x} = \frac{\Sigma x_i}{N}$$

$$\text{Standard deviation} = s = \sqrt{\frac{\Sigma(x_i - \bar{x})^2}{N - 1}}$$

$$\text{Standardized score} = 500 + 100 \frac{x_i - \bar{x}}{s}$$

where x_i = rating for alternative i
N = number of alternatives

To illustrate the procedure, assume that four people have rated six alternatives, A through F. The raw scores for the alternatives (sum of the criteria ratings multiplied by importance ratings) given by each appraiser are shown in the following table.

	Appraiser			
Alternative	1	2	3	4
A	490	680	610	730
B	515	625	530	610
C	480	640	560	760
D	505	670	500	685
E	460	700	590	750
F	800	600	480	660

The mean score for appraiser 1 is $(490 + 515 + 480 + 505 + 460 + 800)/6 = 542$. The sum of the squares of the differences between each score and the mean is $(490 - 542)^2 + (515 - 542)^2 + (480 - 542)^2 + (505 - 542)^2 + (460 - 542)^2 + (800 - 542)^2 = 81,934$. Then the standard deviation is

$$s = \sqrt{\frac{81,934}{6 - 1}} = 128$$

which makes the standardized score given alternative A by appraiser 1 equal to

$$500 + 100 \frac{490 - 542}{128} = 459$$

Standardized scores for the rest of appraiser 1's ratings are $B = 479$, $C = 452$, $D = 471$, $E = 436$, and $F = 702$.

Utility Scales

Another procedure designed to yield an interval scale is based on utility theory. The top and bottom levels of the scale are mentally fixed by visualizing the perfect outcome of the criterion for a 10 rating and the worst possible outcome for a zero rating. Then the alternative being rated is compared with the extreme examples. The comparison is made like a lottery: The rater selects acceptable odds for a gamble between having a perfect outcome (10) against the worst outcome (0.0) *or* having the certain outcome of the alternative. The mental gymnastics required to conduct this mental lottery are difficult to master, but the scale boundaries for the best and worst outcomes make the ratings comparable for all alternatives.

To further describe utility scaling, assume that graduate schools are being compared. One of the criteria is *prestige,* an attribute with no natural measurements. The first step is to select the most prestigious school imaginable and give it a rating of 10. Next select a school with the least possible prestige for the 0.0 rated outcome. The best and worst limits need not be practical outcomes; that is, it might be impossible to attend the most prestigious institution but it still sets the upper limit.

Once the top and bottom levels of the scale are set, preferably defined by written descriptions, a rater then compares each alternative with the good and bad extremes. If an alternative *is* the best or worst imaginable, as used to define the endpoints, it is so rated. Otherwise, it receives an intermediate rating. This results from mentally positioning each alternative at its conformance level on the scale. For instance, if the three graduate schools under consideration are located at Multiversity, University, and Reversity, they might be assigned to the positions on a "prestige" scale shown in Figure 17.14.

Like a lottery ticket, a midpoint position on the scale means that the rater would trade the alternative for a 50-50 chance of having either the best or worst outcome in its place. The 9 rating for Multiversity means that the rater believes there is only one chance in ten that a more prestigious school could be found. This concept of weighing odds is derived from *utility theory,* which defines procedures for developing an interval scale.

Another perspective for rating is to proportion the scale. A midpoint position on the scale means it is halfway between the best and the worst. This is akin to saying that a 50°C temperature results from mixing equal parts of a liquid from both ends of the 0 and 100° scale. A 4 rating, as shown for Reversity, is equivalent to four parts of the highest mixed with six parts of the lowest. This reasoning may be formalized by developing definitions for each level of the scale somewhat similar to the descriptions in the rating form spoofed in Figure 17.13.

A third approach is to concurrently compare all the alternatives with respect to the best-worst limits. The alternatives are first ranked according to

> Importance ratings that prioritize decision criteria are generally developed according to interval-scale principles.

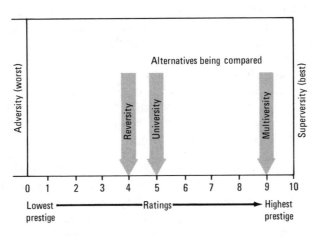

FIGURE 17.14
Interval-scale ratings developed to compare three graduate schools according to their prestige.

how well they satisfy the intangible criterion. They are then "attached" in order to a scale where they can be maneuvered up or down to obtain a spacing that represents their relative value. They are compared in pairs. The highest-ranked alternative is paired with the perception of the highest possible utility. Each remaining alternative is then compared with the next higher alternative to secure its position. This approach is especially effective when the ratings are being developed by a board of judges.

From the foregoing methodology, it should be apparent that putting numbers on intangibles can be frustrating because there is no answer book to check for correct solutions. The purpose of dwelling on techniques for developing number assignments for considerations that resist quantification is that these considerations may be the deciding factor in a close decision. They deserve scrupulous attention. Interval-scaled ratings are also integral parts of a *priority decision table*, which is a widely used decision-making technique.

PRIORITY DECISION TABLE

Significant decisions usually involve objectives that have different dimensions and are of differing importance. To compare weight to distance directly, or to any other dimension in a decision, is cautioned by the old admonition against comparing apples with oranges. Sometimes different dimensions can be converted to the common denominator of dollars. Then a clash to spend minutes to conserve pounds can be resolved by comparing the cost of minutes with the savings from fewer lost pounds. A trade-off between minutes and reputation or appearance is not as easy to reconcile. These difficulties can be overcome by using dimensionless numbers and determining the relative importance of criteria.

Several comparison models based on weighted ratings are available. Multiplicative and exponential models have been proposed. But additive models such as the priority decision table are widely recommended.

A *priority decision table* provides a format for conducting and recording the comparisons that lead to decisions. The table accommodates mixed dimensions and utilizes importance ratings to distill a one-number priority rating from all of the criteria evaluations. This single summary score for each alternative allows competing courses of action to be compared on an equivalent basis. The following six steps outline the comparison procedure.

1 Select independent criteria by which to compare all alternatives.
2 Rate the relative importance of the criteria.
3 Determine whether there is a cutoff score for any criterion that makes an alternative unacceptable, regardless of the scores on the rest of the criteria. If so, state the cutoff ratings.
4 Assign values to the extent that each alternative satisfies each criterion. An alternative is eliminated if any value falls below a criterion cutoff score.
5 Convert any ratio scale dimensions to a 10-point dimensionless scale.
6 For each alternative, multiply its criteria ratings by the respective importance factors, and add all the resulting products. The total scores thus obtained can be compared to determine the most attractive alternative.

Consider the comparison of three prototypes of a new bumper-jack design. Five independent criteria have been selected for evaluation: safety, cost, appearance, weight, and reliability. The relative importance of each criterion is determined as an interval scale rating. With 10 representing the utmost importance, the criteria have the ratings shown in Table 17.3. While the importance is rated, a cutoff score may be recognized that makes an alternative unacceptable. For example, $20 is considered the top limit for a jack's cost, and any alternative that exceeds this cutoff level is eliminated regardless of how well it scores on the other criteria.

Cost, weight, and reliability are ratio scale measurements. Lower cost and weight are preferred, although a higher reliability is desired. All criteria measurements are converted to dimensionless numbers by taking a ratio of each criterion value to the best value available among the alternatives for that criterion. The best value is assumed to have a 10 rating. Ratios used to convert the reliability figures to a 10-point scale are

Most investment decisions have ratio-scaled PW, AW, or IRR as a criterion.

Jack 1	Jack 2	Jack 3
$\dfrac{0.96}{0.96}10 = 10$	$\dfrac{0.81}{0.96}10 = 8.4$	$\dfrac{0.90}{0.96}10 = 9.4$

since 0.96 is the highest reliability rating among the alternatives. The converted cost scores are calculated as

Jack 1	Jack 2	Jack 3
$\dfrac{\$9.95}{\$17.56}10 = 5.7$	$\dfrac{\$9.95}{\$9.95}10 = 10$	$\dfrac{\$9.95}{\$14.47}10 = 6.9$

since $9.95 is the lowest cost for any alternative. Note that none of the costs exceeded the $20 cutoff level, which would have eliminated an alternative from further consideration. Safety and appearance need no conversion because they are already rated on an interval scale with a top score of 10.

TABLE 17.3

Criteria and comparison values for evaluating two bumper-jack designs

	Safety	Appearance	Cost	Weight	Reliability
Jack 1	8	4	$17.56	9.7	0.96
Jack 2	7	9	$ 9.95	6.2	0.81
Jack 3	7	7	$14.47	6.0	0.90
Importance	9	4	8	3	6
Cutoff			<$20		

TABLE 17.4

Typical format by which alternatives are scored according to the sum of the products from dimensionless criterion ratings R multiplied by their relative importance I

Alternatives		Jack 1			Jack 2			Jack 3		
Criteria	I	Observation	Rate	R × I	Observation	Rate	R × I	Observation	Rate	R × I
Safety	9		8	72		7	63		7	63
Appearance	4		4	16		9	36		7	28
Cost (<$20)	8	$17.56	5.7	46	$9.95	10	80	$14.47	6.9	55
Weight, lb	3	9.7	6.2	19	6.2	9.7	29	6.0	10	30
Reliability	6	0.96	10	60	0.81	8.4	50	0.90	9.4	56
Totals		Jack 1		213	Jack 2		258	Jack 3		232

The final step is to multiply the criteria ratings by their respective importance factors and add the resulting products for each alternative. The alternative with the highest total score is the winner. As shown in Table 17.4 jack 2, with a score of 258, is preferred over the other two prototypes.

SENSITIVITY OF DECISIONS

An extremely sensitive and controversial measurement—the value of a human life—is discussed in Extension 17B.

When queried, most decision makers express confidence in their ability but realistically admit that they are not exactly sure how competent they are. No one knows. The reason is that there is seldom a way to determine what the outcome would have been if a different course of action had been taken. Sensitivity analysis attempts to anticipate the consequences of selecting the wrong course. This analysis applied to a priority decision table determines how far the ratings for criteria may vary before the originally preferred alternative is replaced by another contender.

When only two alternatives are compared on the basis of a single criterion, sensitivity is simply a measure of how much that criterion's ratings must change before preference switches. For instance, if price is the sole criterion and alternative A costs $20 versus $10 for alternative B, the decision is relatively insensitive because B would have to double in cost before A became competitive. When the margin of difference is narrower, decision makers should probe into the assumptions upon which the ratings are based.

Narrowly grouped total scores in a priority decision table are a signal for alertness. A sensitivity check is conducted by observing how far a criterion's rating (R) must vary in conjunction with a change in its importance factor (I),

Sensitivity profiles for decision criteria

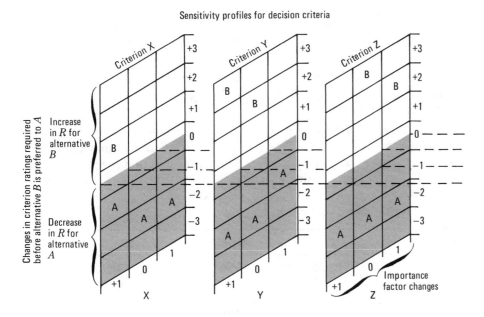

Priority decision table

Alternatives		A			B		
Criteria	Imp.	Observation	R	RxI	Observation	R	RxI
X	8		5	40		9	72
Y	6		8	48		5	30
Z	4		9	36		3	12
Total score				125			114

FIGURE 17.15
Sensitivity profiles for alternatives evaluated by a priority decision table. Each criterion profile displays how much its rating (R) must increase or decrease at different levels of importance (I) before the total priority score, $\Sigma \ (R \times I)$, of alternative B exceeds the score of alternative A, the originally preferred alternative as indicated in the table.

when all other values are held constant, to produce an $R \times I$ product that upsets the previous preference.

To illustrate the sensitivity check, consider the abbreviated priority decision table shown in Figure 17.15. Alternative A is preferred to alternative B by 124 − 114 = 10 points. Criterion X appears to be responsive to variation because it has a high importance factor and a large interval between alternative ratings. With importance (I) held at 8, the total score for alternative B would exceed A's score if the X rating for A dropped by 2 points, as shown below.

Changes for alternative A from the table in Figure 17.15 result from decreasing the rating for alternative A's criterion X by 2.

Alternative A		R × I	Alternative B		R × I
X	(5 − 2) × 8 =	24	X	9 × 8 =	72
Y	8 × 6 =	48	Y	5 × 6 =	30
Z	9 × 4 =	36	Z	3 × 4 =	12
	Total score	108		Total score	114

Alternative B's rating for criterion X cannot be upped enough to make its total score exceed alternative A's total when all other numbers remain unchanged. However, if the importance factor for criterion X is increased by one, $8 + 1 = 9$, then a rating change in X of $+1$ for B or -1 for A causes the preference for alternative A to disappear; these two possibilities are displayed in the following tables.

I is increased by 1 for criterion X. The rating for X is decreased by 1 for alternative A, causing its total score to slide below B's score.

Alternative A		R × I	Alternative B		R × I
X	(5 − 1)(8 + 1) =	36	X	9 × (8 + 1) =	81
Y		48	Y		30
Z		36	Z		12
	Total score	120		Total score	123

I is increased by 1 for criterion X, but this time alternative A's rating for X is held constant and B's is increased by 1, making B's total score higher than A's score.

Alternative A		R × I	Alternative B		R × I
X	5 × (8 + 1)	45	X	(9 + 1)(8 + 1)	90
Y		48	Y		30
Z		36	Z		12
	Total score	129		Total score	132

The intent of the foregoing manipulations is to examine the confidence that can be placed in a decision. When a small change in any rating can reverse the original choice, the ratings should be appraised meticulously. This attentiveness provides ammunition to defend a decision and may furnish ideas about how to conduct a course of action in ways that minimize its weaknesses or questionable features detected during the sensitivity analysis.

Use of the described methods to improve decision making may avoid the situation depicted in Figure 17.16.

Review Exercises and Discussions

Exercise 1

For several years, records were kept of the number of absentees in two offices of a firm. The average number of employees absent from work on each day of the work week at each office are summarized in the following table:

	M	T	W	Th	F	Total
Denver office	40	28	29	32	44	173
Miami office	26	21	20	22	28	117
	66	49	49	54	72	290

FIGURE 17.16

Knowing only that an employee was absent one day, indicate how the absence is related to a day of the week (Friday) and the place of employment (Denver).

Solution 1

The probability that the employee was absent on Friday is $P(F) = 72/290 = 0.248$. The probability that the employee worked in Denver is $P(D) = 173/290 = 0.6$. The probability that an employee who works in Denver was absent on Friday is $P(F|D) = 44/173 = 0.255$.

The probability that an employee was absent on Friday *and* worked in Denver is

$$P(FD) = P(F|D) \times P(D) = 0.255 \times 0.6 = 0.15$$

or

$$P(DF) = P(D|F) \times P(F) = 44/72 \times 72/290 = 0.15$$

Exercise 2

A complaint was received by company C from one of its wholesalers that a certain key part on appliances recently delivered by company C had often failed to operate properly. The company immediately stopped shipment on the remaining appliances made in the same production run as those supplied to the wholesaler. An investigation of similar malfunctions revealed that 70 percent of the time the trouble was caused by poor assembly and 30 percent of the time by inferior materials. It was also determined that the probability of failure from incorrectly assembled parts is 0.40 and the probability of failure from faulty materials is 0.90. If the material is bad, the whole part must be replaced, but if the trouble was caused by improper assembling, the part may be adjusted to perform adequately. The only means of determining the cause of trouble is by destructive testing. A decision was made to test enough parts to determine with a probability of 0.95 that the trouble stemmed from one cause or the other.

If the first five parts tested all failed because of poor materials, what conclusion can be drawn?

Solution 2   As shown by the table below, enough parts have been tested to determine that the probability is 0.96 that the trouble is caused by faulty materials. Since the company planned to take action when it was 95 percent sure of the cause, the five tests provide adequate reason to begin replacement of the defective parts.

Cause	$P(E1)$	$P(\text{failure}\mid\text{cause})$	$P(5\ \text{failures}\mid\text{cause})$	$P(5\ \text{failures})$
Material	0.3	0.90	$(0.9)^5 = 0.59$	0.177
Assembly	0.7	0.40	$(0.4)^5 = 0.01$	0.007
				0.184

$$P(\text{poor material}\mid 5\ \text{failures}) = \frac{0.177}{0.184} = 0.96$$

Exercise 3   Compare the value of sampling information with the value of perfect information for the pick-a-box gamble portrayed in Figure 17.1

Solution 3   The value of sampling information is the amount by which the expected value of the sampling alternative differs from the best expected value among other alternatives. Without sampling, the best alternative is to refuse to gamble: EV(do not gamble) = 0. The previously calculated EV(sampling) = $1 gives a value of $1 − $0 = $1 to the information gained from sampling.

   Perfect information for the pick-a-box choice presumes prior knowledge of the identity of each box. Knowing which box is which eliminates the gamble and is consequently an unlikely proposition. But if it did exist, the "gambler" would pay $100 with complete assurance of winning $180 each time, a net gain of $80 per play. The value of perfect information is then $80 − $0 = $80.

Exercise 4   A firm has identified three potential outcomes for an investment of $1 million. The total return from each investment plus the profit which will occur in less than a year, and the associated probabilities are: $A = \$1,400,000$ with $P(A) = 0.2$; $B = \$1,200,000$ with $P(B) = 0.5$; and $C = \$500,000$ with $P(C) = 0.3$.

   A consultant could be hired to provide additional information. The past record of the consultant in evaluating similar conditions is given in the table below, where A_c, B_c, and C_c represent predictions by the consultant that, respectively, states A, B, and C will occur.

Consultant's Predictions	Occurrence of State			
	A	B	C	
A_c	0.8	0.1	0.1	Probability of the
B_c	0.1	0.9	0.2	consultant's prediction
C_c	0.1	0.0	0.7	given that the state occurs

   a   Construct a decision tree to represent the alternatives, outcomes, and associated probabilities.
   b   What is the value of perfect information, if it could be obtained?
   c   What is the amount that could be paid for additional information based on the consultant's record?

The top branch of the decision tree in Figure 17.17 is the *no-consultant* alternative, which leads to a decision between investing and not investing. From the probabilities and outcomes given in the problem statement for the investment option, the branches are labeled as

Solution 4a

Outcome A = \$1,400,000 − \$1,000,000 = \$400,000   and $P(A) = 0.2$

Outcome B = \$1,200,000 − \$1,000,000 = \$200,000   and $P(B) = 0.5$

Outcome C = \$500,000 − \$1,000,000 = − \$500,000   and $P(C) = 0.3$

If a consultant is hired (the second alternative at decision point 1), outcome A_c, B_c, or C_c could result. Each outcome leads to a second decision about investing. The probability of outcome A *given* the consultant's prediction (A_c) that A will occur is calculated as

FIGURE 17.17

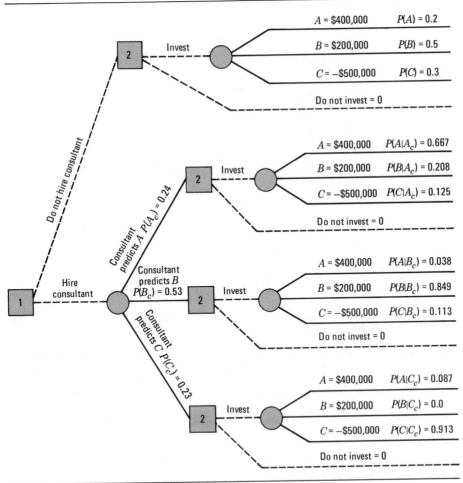

$$P(A|A_c) = \frac{P(A)\,P(A_c|A)}{P(A)\,P(A_c|A) + P(B)\,P(A_c|B) + P(C)\,P(A_c|C)}$$

$$= \frac{P(A)\,P(A_c|A)}{P(A_c)}$$

$$= \frac{(0.2)(0.8)}{(0.2)(0.8) + (0.5)(0.1) + (0.3)(0.1)}$$

$$= \frac{0.16}{0.24} = 0.667$$

The other probabilities in the tree are calculated similarly, as shown below.

$$P(B|A_c) = \frac{(0.5)(0.1)}{0.24} = 0.208$$

$$P(C|A_c) = \frac{(0.3)(0.1)}{0.24} = 0.125$$

$$P(A|B_c) = \frac{(0.2)(0.1)}{(0.2)(0.1) + (0.5)(0.9) + (0.3)(0.2)} = \frac{0.02}{0.53} = 0.038$$

$$P(B|B_c) = \frac{(0.5)(0.9)}{0.53} = 0.849$$

$$P(C|B_c) = \frac{(0.3)(0.2)}{0.53} = 0.113$$

$$P(A|C_c) = \frac{(0.2)(0.1)}{(0.2)(0.1) + (0.5)(0.0) + (0.3)(0.7)} = \frac{0.02}{0.23} = 0.087$$

$$P(B|C_c) = \frac{(0.5)(0.0)}{0.23} = 0.0$$

$$P(C|C_c) = \frac{(0.3)(0.7)}{0.23} = 0.913$$

And, from the above, $P(A_c) = 0.24$, $P(B_c) = 0.53$, and $P(C_c) = 0.23$.

Solution 4b The expected value of perfect information is determined from the payoff table below, which represents the top branch of the tree.

	State A: $P(A) = 0.2$	State B: $P(B) = 0.5$	State C: $P(C) = 0.3$	Expected Value
Invest	$400,000	$200,000	−$500,000	$30,000
Do not invest	0	0	0	0

With perfect information the investment would be made when either state A or state B occurred, but no investment would be made when state C occurred. Therefore,

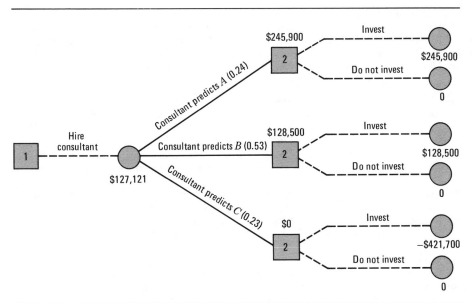

FIGURE 17.18
Decision tree for the hire-consultant alternative. The expected value, exclusive of fees paid to the consultant, is $245,900(0.24) + $128,500(0.53) + $0(0.23) = $127,121.

$$EV|PI = \$400,000(0.2) + \$200,000(0.5) + \$0(0.3)$$
$$= 180,000$$

and the expected value of perfect information is $180,000 − $30,000 = $150,000.

The expected value of additional information from the consultant is calculated by rolling back from decision points 2 to determine the outcome at the node where the three consultant predictions merge. A pruned version of the lower branch of the decision tree is given in Figure 17.18. It shows the expected value at each outcome node and the worth of the preferred alternative at decision points 2. Note that the outcomes do not have to be discounted, because they occur within 1 year.

The expected value from using the consultant's prediction is $127,121, which leads to

Solution 4c

Value of additional information = EV(with added information) − EV(with
                                                                          original data)
$$= \$127,121 − \$30,000 = \$97,121$$

The breakeven point for purchasing additional information is therefore $97,121, which is the maximum amount the consultant should be paid.

Try ranking your three favorite sports. Then stop and ask yourself why you ranked them as you did. For the sake of argument, say the ranking came out skiing, tennis, and jogging. You reason that skiing is more thrilling than tennis, tennis is more fun than jogging, but jogging is handier and cheaper than skiing. How do you get out of this circular reasoning, called *intrasivity*, where X is preferred to Y which is preferred to Z, and Z is preferred to X?

Exercise 5

Solution 5 The exit from circular reasoning is found in more precise objectives. The intangible values of sports could be ranked according to fun per minute of activity, thrills per outing, fellowship enjoyed, satisfaction obtained, etc. Each of these qualities could be the basis for a different ranking episode. They cannot be mixed effectively within one rank. Only one dimension is implicit in a ranking episode. Each quality deserves its own trial.

Exercise 6 Ipso Facto, a thriving independent data-processing company is planning an image-improvement and business-expansion campaign. Three courses of action have been proposed:

1 *Tell-Sell*. Develop a staff to increase personal contacts with old and proposed clients; offer short courses and educational programs on the benefits of modern data-processing methods.
2 *Soft-Sell*. Hire a staff to put out a professional newsletter about data-processing activities; volunteer data-processing services for community and charity projects.
3 *Jell-Sell*. Hire personnel to develop new service areas and offer customized service to potential customers; donate consulting time to charitable organizations.

Only one of the three alternatives can be implemented, owing to budgetary limitations.

The outcomes or returns for each alternative are rated according to desired characteristics, and the importance of each criterion is ranked as shown below, where effectiveness and importance are rated on a 0-to-10 scale; 10 is the top rating.

	Annual Cost	Immediate Effectiveness	Long-Range Effectiveness
Tell-Sell	$250,000	9	7
Soft-Sell	150,000	8	6
Jell-Sell	180,000	6	9
Importance	3	10	6

Using a mixed-rating additive model, determine the overall rankings for the three alternatives.

Solution 6 Since the amount of data is small, it is more convenient to calculate the ratings from an equation than from a matrix. The first expression in each of the following equations converts the annual cost to a dimensionless number; the least expensive alternative is, in effect, given the top rating of 10 by using it as the numerator in each of the cost-criterion ratios. The other criteria already have interval-scale ratings and are consequently multiplied directly by their importance ratings to give them the desired weighting.

Tell-Sell: $\dfrac{\$150,000}{\$250,000}(10)(3) + (9)(10) + (7)(6) = 150$

Soft-Sell: $\dfrac{\$150,000}{\$150,000}(10)(3) + (8)(10) + (6)(6) = 146$

Jell-Sell: $\dfrac{150,000}{\$180,000}(10)(3) + (6)(10) + (9)(6) = 139$

The narrow edge given to Tell-Sell over Soft-Sell suggests that the subjective values used in the comparison model must be appraised carefully. For instance, increasing the importance rating for cost from 3 to 4, with all other figures unchanged, creates the same overall rating for the top two alternatives, 156. Such tests of sensitivity reveal how large a shift in a factor is required to alter the preference from one option to another.

PROBLEMS

17.1 Given the probability that 1 out of every 1000 tire valves is defective, determine the probability that two of the four tires on a car will have defective valves?

17.2 When a machine is properly adjusted, it will produce an acceptable product 9 times in 10. When it is out of adjustment, the probability of an acceptable product is 0.4. The probability of the machine's being adjusted properly is 0.95.
   (a) If the first part tested after an adjustment is not acceptable, what is the probability that the machine was correctly adjusted?
   (b) If the first two parts were acceptable, what is the probability of a correctly adjusted machine?

17.3 A shipment of parts contains 20 items, 8 of which are defective. Two of the items are randomly selected from the shipment and inspected.
   (a) What is the probability that the first one selected is good?
   (b) What is the probability that both are good?
   (c) What is the probability that one is good and one is bad?

17.4 A data-processing firm mails 1000 bimonthly newsletters to present or potential clients. In one issue the firm announced a new service and asked that interested parties write for more information. The firm believed that one of every two replies would come from one of its present customers. From past experience, it is estimated that the probability of a reply from noncustomers is 0.40. On the assumption that the mailing list includes the names of 300 present customers, indicate how many replies can be expected.

17.5 New types of concrete mixes are tested in a laboratory by batching four test cylinders. The probability that a trial batch will yield the specified strength is 0.90 if the mix is properly prepared and tested. Occasionally, about once every 20 times, the trial batch will be improperly handled or the ingredients inaccurately measured. The probability that a poorly prepared mix will yield the specified strength is 0.20. If only one cylinder in a trial batch of four meets the specified strength, what is the probability that the mix was correctly prepared?

17.6 A Louisiana oil operator owns a $5-million oil rig. It costs $75,000 to pull the drills to safety and batten down the rig in anticipation of a bad storm. An uninsured average loss of $400,000 results from a bad storm when no precautionary measures are taken. A weather-forecasting service provides an assessment of the probability of a severe storm. Four out of five times that a severe storm is predicted with a probability of 1.0, it does occur. Only 1 severe storm in 100 arrives unpredicted. Should the rig owner pull the drills when the forecasting service predicts a storm at 1.0?

17.7 A manufacturer has three inspection plans: A, B, and C. The chance that a faulty unit will pass undetected is 2 percent in plan A, 5 percent in plan B, and 10 percent in plan C. The respective inspection costs per unit are $0.35, $0.10, and $0.01. A defective unit going undetected causes opportunity costs of $3.00. The manufacturing process averages 12 percent defectives.

(a) Which inspection plan should be used?

(b) Compare the plan selected in Problem 17.7a with a policy of no inspection.

17.8 Ninety percent of the fruit received at a cannery comes from local growers. The fruit from local sources averages 80 percent grade 1 and 20 percent grade 2. The fruit obtained from other sources averages 40 percent grade 1 and 60 percent grade 2. The markings on a shipment of bins full of fruit were lost. One bin was sampled and from five pieces of fruit inspected, four were of grade 1. What is the probability that the bin came from a local grower?

17.9 How much could be paid for perfect information for the investment decision described in Problem 16.9?

17.10 For the decision situation depicted in Figure 17.19, where capital is valued at 8 percent:

(a) What is the expected profit at decision point 2?

(b) Which alternative should be selected at decision point 1, and what is its expected value?

(c) How much could be paid for perfect information?

17.11 Plans are being developed for the construction of a new school. The city engineer now feels that the probability of growth (G) in the school area is 0.6 as opposed to a stable (S) census probability of 0.4. Two alternative designs are being considered. One is to build a medium-sized school (M) with provisions for adding (A) onto it if needed, and the other is to construct a large (L) facility with the possibility of leasing part of the space to city and county departments if the classroom space is not required.

The study period for the school question is 15 years. It is believed that after 5 years the growth pattern will be evident. If the population is stable for 5 years, there is still a 50 percent chance it will remain stable for the rest of the 15-year period; there is no

FIGURE 17.19

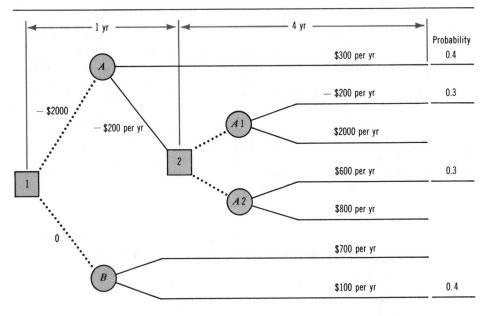

chance that the population will decrease. Given growth during the first 5 years, the probability of continued growth is 0.8, with a corresponding probability of 0.2 for a stable census during the next 10 years.

Additional data are as follows:

Estimated construction costs

Medium-sized school to accommodate stable census	$5,000,000
Addition to medium-sized school to accommodate growth	$4,000,000
Large-sized school to accommodate a growing census	$7,500,000
Remodeling of large school to provide rental space if all the classrooms are not needed	$500,000

Annual costs and income

Maintenance: Medium-sized school	$200,000/year
Large or enlarged school	$350,000/year
Revenue expected from rental space in a large school if all the capacity is not needed for classrooms	$100,000/year
Busing and overcrowding costs if the school is not large enough to accommodate the population after the first 5 years	$500,000/year

The outcomes for the various options are labeled on the partially completed discounted decision tree in Figure 17.20. Nodes are identified by the symbols above, and N means no change. The interest rate for the study is 7 percent.

FIGURE 17.20

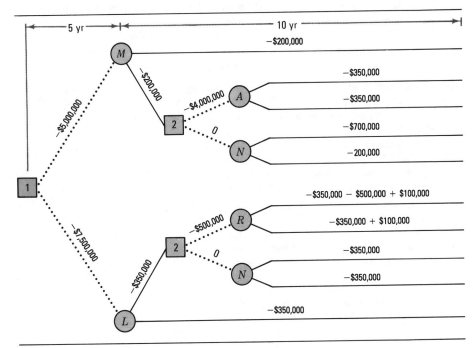

(a) Construct payoff tables for each decision point 2 and determine the preferred alternatives.

(b) Develop a payoff table for decision point 1 and determine the preferred course of action for school construction.

17.12 The editor of a publishing house is deciding whether or not to accept a manuscript. She has already spent $1000 on the development of the manuscript and must now decide whether to

A1 Reject the manuscript and forfeit the $1000
A2 Accept the manuscript without obtaining an expert review
A3 Obtain an expert review at a cost of $800

Data affecting the decision are given in the table below.

Decision and Outcome if Published	Editor-only Evaluation (10 cases)	Evaluation by an Expert (20 Cases)			PW(net returns) per Manuscript if Published
		Bad	Fair	Great	
M0 Do not publish	4	9	0	0	0
M1 Low demand	3	3	2	0	$-$10,000
M2 Good sales	2	1	1	1	$50,000
M3 Best seller	1	0	2	1	$200,000

There are three possible market outcomes if the book is published: M1, M2, and M3. Estimates for the present worth of revenues minus publishing costs are given for each market condition. Previous publishing decisions, made after reviewing, are indicated. The editor has accepted six out of the last ten manuscripts of a similar nature and has obtained one bestseller. The expert has rated 20 books as bad, fair, or great for a fee of $800 and has had the success indicated in the table above.

(a) Construct a decision tree that represents the editor's alternatives. Indicate outcomes and associated probabilities.

(b) Calculate the expected value of sample information.

17.13 A small foundry has had trouble with its old arc furnace. This furnace has been completely depreciated for accounting purposes, but it could currently be sold for $60,000. The immediate alternatives are to overhaul and modify the old machine or to buy a current model which has many desirable features that could not be incorporated in the modification of the old machine. The plans are complicated by the general opinion in the industry that a breakthrough could be made in furnace technology in the near future.

The best estimate the foundry owners can make is that there is a 40 percent chance that a radically improved furnace will be available in about 3 years. If it is developed, the probability that it will make present models noncompetitive is 0.90, and that it will be only a minor improvement is 0.10.

The cost of modifying the old machine is $80,000, and the cost of a new, current-model machine is $250,000. Expected savings and resale values are given in the accompanying table, based on the following three possible future states:

S1 = no technological breakthrough
S2 = furnace developed which provides significant savings
S3 = furnace developed which provides minor savings

Possible Outcomes	Buy New		Modify	
	Savings per Year	Resale at 8 Years	Savings per Year	Resale at 8 Years
S1	$60,000	$80,000	$20,000	$40,000
S2	20,000	20,000	10,000	20,000
S3	30,000	40,000	10,000	30,000

The table is based on a study period and life of 8 years for both furnaces. The sharp decreases in savings and salvage in states 2 and 3 occur because the development of a radically different or even improved furnace would probably cut into the foundry's demand and its general competitive position.

Another alternative exists for the foundry. If the new type of furnace is developed in 3 years, the modified furnace could be sold at that time for $90,000 and the new one purchased for an estimated $450,000. This new furnace would provide a saving of $130,000 per year with a probability of 0.90 and $80,000 per year with a probability of 0.10. It will be worth $200,000, or $150,000 after 5 years with respective probabilities of 0.90 and 0.10.

If a new machine is purchased now, it will be used for 8 years regardless of new developments.

Using a discounted decision tree, determine whether the old furnace should be modified or a new current model should be purchased. Interest is 10 percent.

17.14   A firm has produced a new product which was unusually successful; in order to meet the unexpectedly high demand, it will be necessary to add additional production facilities. The troubling question is whether the high demand will continue, increase, or decrease. Plan A provides a permanent capacity increase and will be more profitable if the demand continues to increase. Plan B is a stopgap measure which can be converted to permanent capacity by a supplementary investment B' after 3 years, when the demand pattern is better known. For a steady or lower demand, plan B is more profitable than is plan A. The estimated future outcomes for an 8-year study period are as indicated in the table.

First 3 years	Last 5 Years	Proba- bility
High	High	0.40
High	Low	0.20
Low	High	0.30
Low	Low	0.10

Initial cost estimates are

Plan A	$1,000,000
Plan B	700,000
Plan B'	450,000

B', the supplementary investment in plan B, will take place after the demand is known for the first 3 years.

Annual income estimates are as follows:

Plan A with a high demand will yield a cash flow of $400,000 per year.
Plan A with a low demand will yield $50,000 per year.
Plan B with a high demand will yield $300,000 per year in the first 3 years and $200,000 annually in the last 5 years.
Plan B combined with B' will yield $400,000 per year with high demand.
Plan B combined with B' will yield $100,000 per year with low demand.
Plan B with a low demand will yield $300,000 per year.

With interest at an annual rate of 8 percent, determine which plan or combination of plans appears most attractive.

17.15 The owners of the novelty company described in the chapter disagree about the solution of their storage problem. There is a minority feeling that more research should be given to the question. One member of the minority group is anxious to do such a study and estimates that it would cost about $5000 and take approximately 6 months to complete. The company could limp through this period with existing inventory facilities. Although the other owners have faith in the person who would make the study, they feel that she is overconservative. Because of this attitude they estimate (1) that if company activity in the next 6 months is very strong, the probability that the study will indicate a continuation of increasing growth will be 0.60, and (2) that if the next 6 months' activity is relatively constant, the chance of a forecast for increasing growth will be 0.10. On the basis of these estimates, an engineer in the company is asked to calculate:

(a) The probability that the study will indicate increasing growth
(b) If the study indicates increasing growth, the probability of continued increase and the chance of leveling off
(c) If the study indicates no increase, the probability of increase and the chance of level activity

Carry out the engineer's assignment, and show the results in a decision-tree format without costs. (*Hint:* You must use the original estimates disclosed in the chapter example in conjunction with those in the problem. The decision tree has three primary alternatives.)

17.16 Engineering Service, a large consulting firm, is considering the acquisition of a computer to lower its project accounting and control costs. It can rent a large computer for $540,000 per year on a noncancellable but renewable 3-year lease. The other alternative is to buy a smaller computer at a cost of $600,000.

There is a probability of 0.70 for a high service demand in the next 3 years. If the demand is large the first 3 years, the probability that it will continue large is 0.60. Expected annual net savings during a period of high demand are $900,000 from a large computer and $510,000 from a smaller model. If demand is low, the large computer will permit a saving of $420,000 per year and the smaller computer's net annual saving will be $330,000. The probability of a continually low demand for the 6-year study period is 0.27.

After 3 years the lease on the large computer could be terminated and a smaller used computer could be purchased for $360,000. Either purchased computer would have a negligible salvage value at the end of the 6-year period. Also, after 3 years the smaller computer could be sold for $360,000 and a large one could be leased for the remaining 3 years for $600,000 per year.

Assume that all receipts and disbursements are end-of-year payments and the acceptable interest rate is 8 percent before taxes. Using a decision tree, determine the most attractive alternative for Engineering Services.

17.17   Based on the raw scores for six alternatives made by four appraisers in Example 17.2,

(a) Calculate the standardized scores for appraisers 2, 3, and 4 and sum the scores for the six alternatives to determine their ranking.

(b) Account for the fact that alternative F would have had the top ranking based on raw scores, but it ranks fifth after the scores have been standardized.

17.18   The most widely used mousetrap is the spring-operated type that uses impact to kill the mouse. This kind has been used for many years and is quite efficient. Both the spring and triggering systems are almost the ultimate in simplicity and economy. However, there are certain disadvantages to this type of trap. It is dangerous because it is not selective. It can kill kittens and puppies and can hurt babies or adults who happen to touch the sensitive bait trigger. It can make quite a mess if the mouse bleeds or is cut in half. Setting the trap and removing the dead victim is not to the liking of many sensitive people.

In an attempt to have the world beat a path to your door by inventing a better mousetrap, assume that the three designs shown on page 508 are your best creations and that you want to select one for a market trial.

(a) List the criteria for evaluating the designs.

(b) Weight the importance of the criteria.

(c) Determine a rating for each outcome of the criteria.

(d) Decide which design is the most promising. Comment on the results of your calculations (sensitivity, confidence, etc.).

17.19   Three designs have been proposed for a new type of can opener. Careful studies have been conducted to evaluate the degree to which each design meets the desired criteria. The characteristics with familiar measurements were determined by design engineers. Opinions of many people were collected to obtain interval-scaled ratings for intangible characteristics. The results are shown below.

Criteria	Design A	Design B	Design C
Cost (minimize)	$3.42	$5.84	$9.88
Cleanability (minimize time)	3.3 min	1.8 min	3.0 min
Reliability (maximize)	0.78	0.91	0.99
Size (minimize)	102 in³	102 in³	320 in³
Appearance (maximize)	6	7	9
Safety (maximize)	7	9	9

(a) Assuming that all criteria are rated equally important, use a priority decision table to determine the preferred design.

(b) Select importance ratings for the criteria using a scale of 1 to 10 and apply the priority decision table to determine the preferred design. Compare the results from Problem 17.19a with calculations based on your importance ratings.

Mousetrap designs for Problem
12.18.

Mousylinder

Operation:

1. Mouse enters cylinder seeking bait.

2. Mouse cannot leave cylinder.

3. Mouse dies from hunger or poisoned bait.

Material as shown

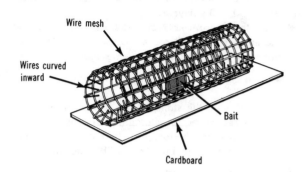

Multicatch Mousetrap

Operation:

1. Mouse goes up ramp.

2. Jumps into recessed section after bait.

3. Trap door sprung by mouse's weight.

4. Mouse caught in well.

5. Slaked lime in well destroys mouse.

Material as shown

Electronic Mouser

Operation:

1. Mouse enters tunnel.

2. Mouse is electrocuted.

Material clear plastic

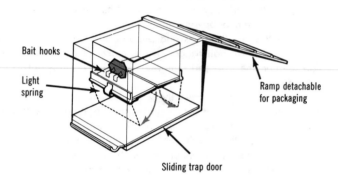

17.20 Prospective sites for a new chemical plant have been narrowed to three locations. The criteria for each alternative and the importance of the criteria are shown on page 509. Higher interval-scaled ratings show a preference. Which site is apparently more attractive? Comment on the sensitivity of the choice.

| Criteria | Alternative | | | Importance |
	Site 1	Site 2	Site 3	
Labor supply	2	8	9	9
Raw materials	3	10	5	8
Transportation	8	7	9	7
Cost of land	$200,000	$700,000	$400,000	5
Building costs	$2,000,000	$3,800,000	$1,800,000	6
Annual taxes and utility costs	$60,000	$120,000	$80,000	7
Climate	7	6	4	5

17.21 Contestants in the "Miss All" competition are judged on beauty, personality, and talent. Beauty is considered twice as important as personality and four times as important as talent. Girls are rated on a simple order scale (1, 2, 3, . . . , n) with 1 being the highest rating. One girl has been judged first in beauty but tenth in talent. She knows that the girl rated second in beauty is sixth in talent.

(a) What is the minimum rating in personality that the girl first in beauty needs to be assured of being selected as "Miss All"?

(b) How could the beauty judging system be improved?

17.22 After a recent visit by OSHA inspectors, a company must install better ventilation and filtering equipment or replace existing machines with newer versions having built-in controls. Costs of two alternatives that will satisfy safety codes are listed below.

	Add Ventilation Equipment to Existing Plant	Install New, Improved Machines
First cost	$80,000	$420,000
Present value of old machines	$150,000	
Economic life, years remaining	4 years	9 years
Salvage value at end of life		$70,000
Added maintenance cost/year	$7,000	
Total annual operating costs	$36,000	$25,000

The company desires a rate of return of 9 percent on investments to improve operations and owing to current cash-availability problems seeks to minimize major capital allocations. Therefore, a temporary priority system has been established to rate the desirability of alternative investments based on need, immediate cash outlays (net first cost), and total present worth, with relative-importance ratings of 10, 10, and 5, respectively. Both alternatives have the same score for need, 10.

(a) Assuming a before-tax economic evaluation for a 9-year study period and using a priority decision table, determine which alternative is preferred?

(b) Discuss the dangers of temporal suboptimization involved in the given importance ratings. What other method can you suggest to include consideration of the conservation of capital in economic evaluations?

17.23 Sites for a new research lab have been narrowed to three localities. The construction cost of the plant will be approximately the same regardless of the location chosen. However, the costs of land and intangible factors largely applicable to personnel recruitment vary considerably from one location to another. Based on the following figures, indicate which site should be selected.

Characteristics	Site 1	Site 2	Site 3	Importance
Availability of technicians	7	10	2	10
Adequacy of subcontractors	5	9	5	8
Proximity to a university	10 miles	40 miles	30 miles	8
Cost of land	$300,000	$400,000	$50,000	6
Recreational potential	7	2	10	4
Climate	6	1	9	2
Transportation	9	10	5	2

17.24 An engineer and an accountant had two things in common. Both saw a potential for computers in production enterprises and both had the same favorite meal—prime rib dinner. The outgrowth of these shared interests was a computer service named Prime Rib Inc. (PRI) that provided customized software for production costing and scheduling. They started the company 3 years ago and now employ 35 people. Business is good and promises to get even better.

At a dinner to commemorate their first 3 years together (serving prime rib, naturally), Ed, the engineer, and Al, the accountant, became engaged in a heated discussion about purchasing a company car. They seldom disagreed on directions that PRI should take to serve customers, but delicate questions about perquisites caused them trouble. In this instance, the question had to do with what type of vehicle the company should buy for their use. They both agreed a vehicle was needed. They disagreed on the type it should be.

Ed: You're thinking too big, Al. We need a limo like a beggar needs a tux. All we require is a nice little economical pickup truck to run errands.

Al: We gotta think big. We're on our way and should let other people know it. A flashy set of wheels will do it. Do you want to meet a customer at the airport in a pickup?

Ed: Maybe that's all we can afford. I figure the price of a fancy sedan is three times that of a pickup. Upkeep and gas will at least be double. That's expensive prestige you want to buy.

Al: It's all a matter of priorities, old buddy. We may have to stretch a bit to meet the sticker price, but we can surely afford the upkeep as we get more accounts. And that "fancy sedan," as you call it, will help us get those accounts. I say prestige should be our top priority in the decision, and my type of car would have at least five times the prestige of yours. You gotta look successful to be successful.

Ed: Okay. I can go along with your rating for prestige, reluctantly. On the same scale, I'd rate cost 8 and upkeep 5, if prestige is 10. I'd also rate utility right next to prestige. We need something versatile, not just a gas hog that impresses people.

Al: I'll accept your numbers, and your argument for utility too, but I also think my big, impressive, luxurious chariot would be great to run errands in. It'll have a trunk big enough to carry our stuff in. I'd rate its utility umpteen times a pickup's.

Ed: No way! We're going in circles. Let's pretend its someone else's problem and solve it the Prime Rib way. I'll set up the decision table.

He then pulled out a pencil, grabbed a napkin, and started scribbling. Assuming Prime Rib Inc. uses a priority decision table for decision analysis, complete the table Ed started. Using the figures given in the dialogue, analyze the situation, paying particular attention to the ratings for utility. Suggest how an interval scale could be set up (describe the 0 and 10 extremes) to determine a rating for utility. Discuss the sensibleness of the decision.

EXTENSIONS

17A Fault Trees*

Various "tree" forms have been developed to represent decision situations found in many disciplines. The shape naturally evolves because decisions typically stem from a central issue and branch out to encompass several contributing factors or alternative courses of action. A graphical display assists in the process of organizing and evaluating data. A *fault tree* is an example of such a structure for analyzing the reliability or other characteristics of a system. It has been applied to the design of spacecraft, quality control, and safety studies. It is illustrated in the following pages as a tool for accident analysis.

Structure

A fault tree is a diagram of relationships or components within a system which collectively describe a specific operation or function. Construction of a tree may reveal the critical, though often hidden, faults of a system. As applied to a safety study, it traces an accident—a fault of a system—to its origins—faulty components within a subsystem. From a list of undesirable events, one is chosen for intense study. This becomes the *head event* for the fault tree.

From the head event, a top-down analysis is conducted by determining primary and secondary events that cause the head event. Factors that contribute to the head event are traced to the smallest subdivisions, called *basic events*. The cascade effect develops by progressively refining groups of causal events through AND and OR relations. Symbols are used to represent factors involved in the analysis, as follows:

1 *Rectangles* identify events that deserve further analysis.

2 *Solid circles* identify basic events that need no further development because sufficient information is known.
3 *Dashed circles* identify events that will not be developed further because there is insufficient data or because the events are relatively inconsequential.
4 A *half circle* identifies an AND gate that indicates two or more inputs to one output. In order for the output event to occur, *all* inputs must occur.
5 A *diamond* identifies an OR gate that indicates multiple possible inputs to one output. In order for the output event to occur, *at least one* of the input events must occur.

Additional symbols or explanatory notes can be used to show special conditions, such as a timing sequence for AND inputs.

An example of a fault-tree representation of one fault in a grinding and polishing operation is shown in Figure 17.21. The head event is eye injury from particles thrown by machines. The first OR gate defines the ways that the head event could occur. For the conditions shown it is assumed that safety records indicate that most eye injuries occur to nonoperating personnel who fail to wear safety glasses; this fault then receives more intense study. Four conditions that must occur before an injury results are shown below the first AND gate. Two of these conditions are further defined by lower gates. Note that the AND gates indicate *what must happen*, whereas the OR gates show *what could be the cause.*

* Adapted from J. L. Riggs, *Production Systems: Planning, Analysis, and Control,* Wiley, New York, 1981.

FIGURE 17.21
Fault-tree analysis of eye injuries caused by particles thrown off during grinding and polishing operations.

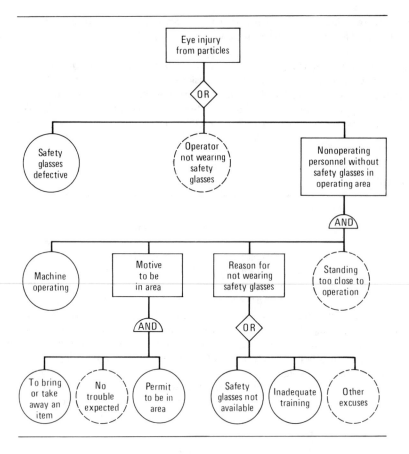

Analysis

Tracing an injury back to its constituent causes gives a safety analyst significant insight, but even more value is realized by determining the cost of the accident and the probabilities of contributing events. The fault-tree structure accommodates this quantitative evaluation. A cost is determined for the occurrence of the head event, and the probability of this occurrence is calculated from the events that lead to the head event. The cost of the head event is multiplied by its probability to provide a *criticality index*. The steps in the procedure are described below.

1 *Head-event cost (negative utility of an accident).* The direct and indirect costs of an accident are included in a measure of the negative utility that results when a head event occurs. Units of measurement can be dollars, lost time, material wasted, or any dimension that indicates the magnitude of loss incurred. The loss should be stated in terms of an appropriate reference unit of production time, such as a million worker-hours. The reference unit is typically scaled to produce a head-event probability that falls between zero and 0.1.

2 *Head-event probabilities.* For some situations it is sufficient to directly measure the probability of the head event. For instance, the probability for the head event "break leg" on a particular ski trail at a resort could be measured by counting the number of broken legs and dividing it by the number of skiers who went down that trail over a period of time; if seven broken legs were recorded in a 2-week period when 19,420 skiers made the difficult run, the probability of a broken leg for that trail is

$$\frac{7 \text{ broken legs}}{19{,}420 \text{ ski runs}} = 0.00036 \quad \text{or} \quad 0.036 \text{ per 100 runs}$$

A more revealing head-event probability is determined from the constituent events in the fault tree. By assigning probabilities to each cause and condition and using these numbers to calculate the resulting head-event probability, we are assured of a more complete study. It is more difficult to collect all these probability factors, of course, but the effort is rewarded by more exact knowledge of the benefits that could result from safety improvements for specific causal events.

Once event probabilities have been assigned, computations proceed from the roots of the fault tree upward to the head event. The following Boolean relations are utilized:

OR *relationship (any of the events cause the subsequent event to occur):*

$$\text{Probability of subsequent event} = P_{\text{OR}} = 1 - \prod_{i=1}^{n}(1 - p_i)$$

where p_i = probability of ith causal event

$\quad\quad n$ = number of parallel events in OR relationship

AND *relationship (all the events must occur before the subsequent event occurs):*

$$\text{Probability of subsequent event} = P_{\text{AND}} = \prod_{i=1}^{n}p_i$$

where n is the number of causal events in the AND series. Sample calculations are shown in Figure 17.22. Note that the probability of a subsequent event is always increased as more events are included in an OR relationship and is always decreased as more events are included in an AND series.

3 *Criticality index.* A measure of the significance of an event that incapacitates the system (human or machine) for normal operations is its *criticality index*—the head-event cost multiplied by its probability of occurrence. For instance, assume that the accident represented by the head event in Figure 17.22 requires first-aid treatment 80 percent of the time and results in temporary disability in 15 percent of the cases and permanent partial disability on 5 percent of the occasions. Costs per incident (including indirect losses from production disruptions and

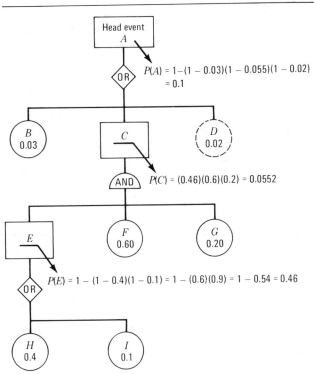

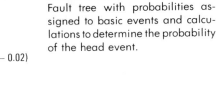

FIGURE 17.22
Fault tree with probabilities assigned to basic events and calculations to determine the probability of the head event.

administrative expenses) for the three classes of accident severity are, respectively, $352, $5014, and $24,480. Then the weighted cost for event A is $352(0.80) + $5014(0.15) + $24,480(0.05) = $2258. Further assuming that the probability of the head event is based on 100,000 worker-hours and that the total annual exposure of the work force to the hazard is 1.2 million hours, we find that the criticality index (CI) has a value of

$$CI = \frac{\$2258}{100,000 \text{ worker-hours}}$$
$$\times \frac{1,200,000 \text{ hours}}{\text{year}} P(\text{head event})$$
$$= \$27,096(0.1) = \$2710/\text{year}$$

4 *Evaluation of fault-correction alternatives.* When money is allocated to improve safety, one or more of the event probabilities in the fault tree should be reduced. Each of the events is a potential alternative for safety improvement. Reducing the probability of any basic event reduces the probability of the head event, which decreases its criticality index. The amount of reduction is a "saving." By comparing the *saving* with the investment required to produce it, we obtain a measure of preference for alternative safety improvements.

Given the accident structure in Figure 17.22 and the associated cost and exposure data, assume that safety improvements have been identified that can reduce the probabilities of events H and F by 0.1. "Savings" resulting from either improvement are calculated in Table 17.5. It is apparent that an investment to reduce the probability of event H causes a greater reduction in the criticality index than does the improvement for event F. If the cost of capital is 12 percent and the study period for the safety proposal is 3 years, the maximum investment P supportable from reduced accident costs due to improved conditions for event H is

$$P = \$217(P/A, 12, 3) = \$217(2.4018) = \$521$$

Other factors that are difficult to quantify, such as improved morale resulting from a safety-improvement expenditure, are likely to influence the investment decision.

QUESTIONS

17A.1 Construct a fault tree in which events F and G must occur prior to event D; D or E prior to C; and B and C prior to head event A. Probabilities for basic events F, G, E, and B are, respectively, 0.4, 0.2, 0.05, and 0.1. These probabilities are based on 1 million operator-hours, and the 3-year

TABLE 17.5

Evaluation of safety improvements for events H and F in Figure 17.22. Effects of reducing the probabilities of events H and F by 0.1 are traced through events E and C to event A. The new criticality index of A is compared with its former value to determine the "savings" provided by the investment.

Calculation Procedure	Safety Improvement to	
	Event H: $P(H) = 0.4 - 0.1 = 0.3$	Event F: $P(F) = 0.6 - 0.1 = 0.5$
Probability of event E	$1 - (1 - 0.3)(1 - 0.1)$ $= 0.37$	$1 - (1 - 0.4)(1 - 0.1)$ $= 0.46$
Probability of event C	$0.37(0.6)(0.2) = 0.0444$	$0.46(0.5)(0.2) = 0.046$
Probability of event A	$1 - (1 - 0.03)$ $(1 - 0.0444)(1 - 0.02)$ $= 0.092$	$1 - (1 - 0.03)$ $(1 - 0.046)(1 - 0.02)$ $= 0.093$
Criticality index of A	$\$27,096(0.092) = \$2493$	$\$27,096(0.093) = \$2520$
Reduction in the criticality index from the proposed safety improvement	$\$2710 - \$2493 = \$217$	$\$2710 - \$2520 = \$190$

production plan calls for 10 million operator-hours annually, during which the injury exposure will continue, unless safety improvements are made.

17A.2 Although the injury rate is not unreasonably high, the severity of each injury is exceptionally great. Previous experience with event *A* indicates that the average loss from each occurrence is \$205,900, including both indirect and direct costs. How large an investment could be supported by the safety saving that results from reducing the probability of event *B* from 0.1 to 0.05? Use a before-tax analysis with a required rate of return of 15 percent.

17B Value of a Human Life

Perhaps the consummate challenge for quantification of intangibles is to assign a number for the value of a human life. Although the challenge may be spiritually repugnant, it exists because many economic decisions unavoidably hinge on the monetary worth placed on human lives: How much can be spent to reduce fatalities associated with a certain stretch of highway, a dangerous occupation, or a particular disease? How large should damages be for a defective design or an irresponsible act that causes a death?

Sir William Petty attempted to place a value on human life in the seventeenth century. His estimate was based on the total earnings of the population and the equivalent total earnings yielded by a capital sum invested at a given interest rate. He divided the capital sum by the total population to obtain a value of 80£ per individual. Since that time many investigators have addressed the question and proposed a variety of answers.*

According to Solomon Huebner, the valuation should be "the capitalization of the value of a human life for the benefit of the dependent household." He suggests the following considerations for determining a numerical value:

1 Individual working capacity should be capitalized in the same way as is business property.
2 Life value is an economic asset that is subject to depreciation of its earning capacity in a manner similar to that of tangible assets.
3 Life has a societal trade-off value that encompasses old-age support, aid to dependents, income taxes, and many other give-and-take relationships.
4 Preservation of life embodies the same principles that apply to the preservation of property; for example, health care is comparable to periodic maintenance of machines, since both are designed to avoid economic losses.

The premises proposed by Huebner have led to four general approaches to human-life valuation: (1) money value

*See references at the end of Extension 17B.

of a person as a wage earner, (2) life insurance as protection for dependents, (3) damage to society resulting from a preventable death, and (4) loss in capital value caused by a person's death. The first two view a person as a "money-making machine," and tabular procedures have been developed that account for the age of the machine with respect to its service life, earnings less consumption, related receivables and payables, and dependencies. The third perspective is often used in benefit-cost analyses and includes the cost to society of losing one of its members who pays taxes, supports dependents, and contributes to the social structure of a nation. The fourth view is sometimes called "human-resources accounting" and is associated with the value of a person to an organization in terms of ongoing activities and replacement cost.

Actual dollar values placed on human life by the valuation methods vary widely, as expected. Amounts are influenced by occupation, productivity record, national standard of living, family relations, age, discount rate used, socioeconomic factors, and a host of other considerations. Consequently, monetary sums assignable to life values among individuals range from a few hundred to millions of dollars, depending on which factors are included, and the amount for one individual may vary by many thousands of dollars according to which valuation philosophy is followed.

In a major liability suit brought against an automobile manufacturer, the defendant sought to prove that the cost of providing special safety features (an additional expense to the public for transportation) would be greater than would be the cost of fatalities and injuries circumvented by the safety features. The value of a human life used in the case, as a set of losses assignable to specific factors in 1971 dollars, is shown in Table 17.6.

The value of a human life is also implied by previous actions taken to avoid fatalities. This surrogate approach makes no attempt to classify the individual factors that contribute to the total value. Instead, it relies on historical perspective to obtain surrogate measures that can be applied to current situations under similar circumstances. For example, assume that \$100,000 has been spent previously in a

TABLE 17.6

Human-life value determined as the summation of fatality costs

Factors	1971 Costs
Future productivity losses	
Direct	$132,000
Indirect	41,300
Medical costs	
Hospital	700
Other	425
Property damage	1,500
Insurance administration	4,700
Legal and court	3,000
Employer losses	1,000
Victim's pain and suffering	10,000
Funeral	900
Assets (lost consumption)	5,000
Miscellaneous accident cost	200
Total	$200,725

given area to reduce the probability of a fatal accident from 0.001 to 0.0001. The consequent value implied for a life is

$$\frac{\$100,000}{0.001 - 0.0001} = \$111,111,111$$

Although this method is not precise in that it recognizes no other contributory factors, it does provide a justifiable reference level from which to estimate or negotiate.

There is obviously no agreement about what value should be placed on a human life, and there probably never will be. Nevertheless, such values will continue to be generated and used in economic justifications because life-and-death considerations are integral to many economic decisions. Including even debatable figures is preferable to ignoring the value of life. The figures at least mitigate the emotional overtones of the subject and encourage objective assessments, which is why the quantification of intangibles is essential to engineering economics.

SECTION FIVE

MONEY VALUE OF TIME

More than 2000 years ago, Theophrastus wrote in Greece, "Time is the most valuable thing a man can spend." The proclamation is still valid. Time may be endless in a theoretical sense, but to each person on earth it is a limited resource. Theophrastus had only 1440 minutes available each day, and that is all we have today. When a resource is limited in supply, it has a money value.

"Time management" is a popular subject. It deals with planning, organizing, and controlling activities to make the most efficient use of personal time. It differs from the management of other scarce resources because of the unique nature of time. Unlike money, time cannot be accumulated, and it can be expended only as it becomes available. The objective is to invest available time to achieve the highest returns possible.

Time is the underlying theme of the five chapters in this section. They generally follow a life cycle sequence for conducting projects:

Obtaining data in Chapter 18

Planning operations in Chapter 19

Performing activities in Chapter 20

Evaluating performance in Chapter 21

Monitoring performance in Chapter 22

This time-oriented flow of subjects also addresses time management from an organizational rather than a personal-use perspective.

Chapter 18, "Accounting and Forecasting," deals with procedures for organizing financial data and methods of analyzing data to anticipate future needs or to predict cash flows.

Chapter 19, "Production Economy," surveys time-zero techniques that search for the most economical way to conduct up-coming and ongoing operations.

Chapter 20, "Resource Management," directly confronts time as a scarce resource that is conserved by diligent scheduling and then relates time to material usage to identify patterns that minimize inventory costs.

Chapter 21, "Productivity and Profitability," explores the relationship between operating effectiveness and financial success, revealing the pervasive effects of productivity on people, capital, and other resources, including time.

Chapter 22, "Productivity Measurement," builds on the basic input/output ratio to develop measures, including difficult-to-quantify factors, that indicate progress from one time period to the next.

The paradox of time has to do with the fact that no one has enough but everyone has all there is. Putting a money value on time gives recognition to its qualities of immediacy and scarcity, just as the time value of money recognizes the effects of immediacy and longevity on capital. Both are measures of preference among alternatives.

CHAPTER 18

ACCOUNTING AND FORECASTING

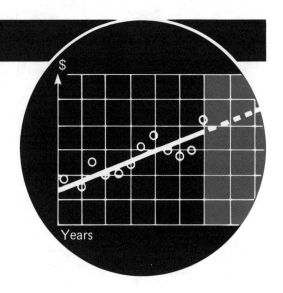

OVERVIEW

Accounting and forecasting methods are bookends for the "money value of time." They are tools used to tell an organization where it has been and where it is going. Accounting measures the past to gauge the present. Forecasting methods examine past and present to estimate the future. This future, of course, will be determined by the present, just as the present has been determined by the past.

Engineering economists need not become accountants to carry out economic evaluations, but they should be aware of the accounting function and how it can contribute to better analyses. An accounting system records financial data. These records indicate the financial condition of an organization, record data from which estimates for future activities can be obtained, and reveal how closely implemented proposals met expectations. An *income statement* shows the sources of revenue and expenses over a period of time. Net earnings from the income statement is an input to the *balance sheet*, where accounts are tabulated for a given date to give values for the basic accounting equation: *assets minus liabilities equal net worth*.

Accounting records are also scorecards that show what actually happened to a proposal after it was accepted and augmented. The use of financial ratios to evaluate the strength of a business is examined in Extension 18A.

Forecasting methods are used to predict future business conditions and technical developments. *Subjective* methods, which rely on personal opinions, are comparatively inexpensive to obtain but may be biased. Historical data are the bases for *line-fitting* and *smoothing techniques* that extrapolate past trends to predict the future. The *least-squares method* fits a straight line to data to form a simple regression model. The choice of alpha in *exponential smoothing* and the number of periods used in a *moving average* determine how sensitive the forecasts are to recent data. *Causal models* attempt to predict future happenings from changes in the causes of events. A *leading indicator* is a statistic that behaves the same but precedes the behavior of the data sought. A *correlation* model examines the closeness of the relation.

PAST, PRESENT, AND FUTURE

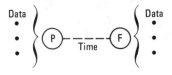

The essence of time value of money is captured by the simple diagram in the margin. Time relates the present P to the future F; data are necessary to establish the value of either endpoint. The present valuation of an asset is an outgrowth of its development or maintenance and its future expectations. Data from accounting records can provide a history of an asset's treatment and an estimate of its current worth. Data generated by forecasting methods suggest what future values will be. These data manifest themselves in engineering economic studies as figures for P, F, A, i, and n. The data may also imply conditions of risk and sensitivity that deserve consideration in a study. Knowing how much confidence to put in historical data and how much faith to put in forecasts are critical inputs for economic decisions.

ACCOUNTING

An accountant and an engineering economist may work with the same financial data but they do so for different purposes. The engineering economist deals with the future to evaluate proposed courses of action. Each proposal is analyzed with respect to its future cash flow and potential for earning a certain rate of return.

THE HARDSHIP OF
ACCOUNTING

Never ask of money spent
Where the spender thinks it went.
Nobody was ever meant
To remember or invent
What he did with every cent.
Robert Frost

The accountant deals with transactions that have already occurred to determine what *was* the return on capital. The accountant does not have an entry for the cost of capital, unless there is an actual disbursement to be made, such as interest due on a loan. The cost of money is implied from the net profit of all operations combined. The views of the accountant and engineering economist coincide when a postaudit is conducted to see how closely actual figures came to figures estimated during the evaluation of a proposal; both are then looking at what happened and how it affected the economic health of the organization.

Balance Sheet

The fundamental accounting equation is

$$\text{Assets} - \text{liabilities} = \text{net worth}$$

Periodically, a *balance sheet* is prepared to show the monetary values of the elements in the accounting equation. This statement is accurate only for the date given on the balance sheet, because specific amounts vary from day to day. It is a measure of financial health in the same way that a report of an annual medical checkup is a measure of an individual's health. Both statements might be different if prepared on a different day, because both are summaries of all contributing factors to the day of preparation. These summary effects make the statements analytically valuable as readings on current conditions and, when compared with previous statements, as an indicator of what to expect in the future.

Extension 18A illustrates the potential of analyzing financial ratios.

An abbreviated balance sheet is shown in Figure 18.1. The left side tabulates the assets that total to an amount equal to the sum of the liabilities and net worth on the right side. The *assets* fall into three categories that represent the resources of the organization:

Current assets. Assets that will become cash within 1 year (materials, credit due, etc.)

Fixed assets. Assets that cannot be readily converted to cash (mortgages, vehicles, buildings, etc.)

Other assets. Assets that are intangible or difficult to classify (prepaid insurance, patents, etc.)

Liabilities. Obligations of known amounts owed to creditors. They are usually listed in order of due dates; *current liabilities* are payable within 1 year and *long-term liabilities* are not due for over a year. The *net worth* (of stockholders' equity) represents the current worth of investments in the organization. *Retained earnings* are profits from past operations which belong to the stockholders and have been retained within the organization for reinvestment.

FIGURE 18.1
Balance sheet showing an organization's major accounts as of a specific date.

Balance Sheet June 30, 19xx			
ASSETS		**LIABILITIES**	
Current assets		*Current liabilities*	
Cash	$ 5,400	Accounts payable	$35,200
Accounts receivable	9,000	Bank loans	9,000
Raw-material inventory..	11,000	Accrued taxes	4,800
Work in progress	10,000	*Long-term liabilities*	
Finished products	16,600	Bonds (due in 12 years).	15,000
Fixed assets			$64,000
Land	37,000	NET WORTH	
Buildings and equipment.	83,000	Capital stock	$100,000
Other assets		Retained earnings	24,000
Prepaid services	16,000		$124,000
	$188,000		$188,000

522   SECTION FIVE / MONEY VALUE OF TIME

Fixed assets are usually shown at their current value as determined by the difference between what was paid for the assets minus accumulated depreciation. If the buildings and equipment in Figure 18.1 had originally cost $100,000 and had since been depreciated by $17,000, the entry in the balance sheet could have taken the form:

Fixed assets

Land....................................		$37,000
Buildings and equipment	$100,000	
Less: reserve for depreciation........	17,000	
		$83,000

There are many other variations and different headings for accounts in balance sheets. Since all the accounts are summaries, supplementary accounting records are necessary to observe the cash flow.

Example 18.1
The Balance of a Business

Sam Sales is an engineering sales representative for an air-conditioning manufacturer. He is paid a commission on the amount of commercial air-conditioning equipment that he sells. As he traveled within his assigned territory calling on prospective customers, he became fascinated by the variety of novelties displayed near the cash registers at restaurants and stores.

One day Sam saw a person filling up one of the novelty displays and started talking to him. Before they were done talking, Sam agreed to purchase the novelty distributorship. The seller wanted to go to a different climate; and Sam figured that he could combine the novelty business with his other travels.

The distributorship, priced at $23,000, included an exclusive franchise for Freaky postcards, $14,200 in inventory stored in a warehouse where rent was prepaid for 6 months at $200 per month, and accounts receivable of $3100 from ongoing sales. Sam agreed to take over the accounts payable of $1900 and accrued taxes of $300 from the seller.

Sam met the $23,000 selling price by putting up $5000 from his savings, borrowing $8000 on a 2-year bank loan, and getting the former owner to take an unsecured note promising to pay the remainder over 3 years. Then Sam withdrew $10,000 more from his savings for *working capital* to handle day-to-day cash transactions for his venture, now called "Sales' Gimmicks." Construct a balance sheet for the new company as of the current date, April 1, 19xx.

Solution 18.1
Sam Sales' beginning balance sheet is shown in Figure 18.2.

Income Statement

A statement of the earnings of an organization over a stated period of time is called an *income statement* or *profit-and-loss statement*. The net profit or loss it discloses represents the difference between revenue from the sale of goods or services and the sum of all the operating expenses, including taxes. The net profit may be held as retained earnings or distributed to the owners. An

FIGURE 18.2

SALES' GIMMICKS
Balance Sheet
April 1, 19x0

ASSETS		LIABILITIES	
Cash	$10,000	Accounts payable	$ 1,900
Accounts receivable	3,100	Accrued taxes	300
Inventory	14,200	Bank loan	8,000
Prepaid rent	1,200	Note due seller	10,000
Franchise	4,500*	NET WORTH	
Total assets	$33,000	Sam Sales' equity	$12,800†
		Total liability and equity	$33,000

*The value of the franchise is the amount included in the purchase price for existing customer goodwill and the exclusive territorial rights to distribute Freaky postcards.

†Sam's equity is derived from the $15,000 withdrawn from savings minus the amount of accounts and taxes payable ($1900 + $300).

abbreviated income statement is shown in Figure 18.3. Its development precedes the balance sheet, because the bottom-line figure, net income after tax, is an entry in the balance sheet under net worth. Both the income statement and the balance sheet are needed to understand an organization's financial position—and then only the more prominent features are revealed.

Cost Accounting

The major accounts displayed in the balance sheet and income statement are composed of many smaller accounts, each of which is, in turn, composed of more detailed items. For instance, the cost of goods sold in the income

Income Statement
Year Ended June 30, 19xx

Income		
Product sales	$247,500	
Fees for services	16,500	$264,000
Expenses		
Cost of goods sold	$119,000	
Administrative	82,000	
Depreciation	9,200	
Utilities	6,000	
Other (insurance, rent, etc.)	7,800	$224,000
Income before income tax		$40,000
Federal income tax (at 40%)		16,000
Net income after tax (to retained earnings)		$24,000

FIGURE 18.3
Income statement for the same organization during the same time period represented by the balance sheet in Figure 18.1.

statement includes direct labor, direct material, and factory overhead. More detailing breaks down the costs of direct labor by craft and work assignment, and the cost of material by type, use, and timing. This type of information is in the accounting system and is recorded because the summarizing balance sheets and income statements are built up from the elemental cost data. The collection of these costs for producing a product or service is called *cost accounting*.

Cost estimating is examined also in Chapters 12 and 19.

The cost-of-goods-sold category is particularly useful to an engineering economist because it contains data that can be extracted and interpreted to estimate costs for proposals that alter the production process. The cost data about the current process are applicable to the do-nothing or defending alternative, and *predictive cost accounting* estimates the cost of future operations of the same nature from past performance.

Example 18.2
A Summary Statement of Profit or Loss

After Sam Sales bought the novelty business, he laid out a year-long operating plan in the form of a monthly budget. He knew that his postcards and other novelties would enjoy their highest sales during the summer. He also realized that he would not have time to service all his outlets during the busy season and continue with his primary job in engineering sales; and he was not nearly confident enough in his Gimmicks Company to rely on it alone for a living.

He could hire helpers part time in the summer as route servicers, but he was worried about his cash flow position. With only $10,000 for working capital, he could not afford to build up his inventory levels too high or he would run out

of cash to pay the bills. The results of juggling amounts between cost categories and the time periods are shown in the monthly budget accounts in Table 18.1. The plan anticipates a decline in the postcard inventory during the winter to stock up on novelties. During the summer the monthly receipts would pay for the additional help and the cash balance should not drop below $4000.

Develop an income statement for the year by assuming that the actual receipts and disbursements came out exactly as budgeted in Table 18.1 for the first year's operation of Sales' Gimmicks. The tax rate is 20 percent on the $18,500 income, and the after-tax total is held as retained earnings.

Solution 18.2

The income statement for the first year's operation of Sales' Gimmicks Company, based on the budget shown in Table 18.1, is in Figure 18.4. The income and expense categories summarize the expected effects of operations during the 1-year period analyzed. Sam did fairly well; the return on his initial investment was about 100 percent in 1 year.

FORECASTING

Estimating is more of a science than an art when conditions are stable, at which time standard costs and catalog prices are satisfactory data. But nothing is static for long. Changes in the economic environment may be as gradual as wind erosion in the desert or as sudden as a tornado. Economic estimators are particularly sensitive to a sudden shift, for it may mean the difference between fortune and fiasco. The prediction of a shift is a work of art.

TABLE 18.1

Monthly budget for cash receipts and disbursements for Sales' Gimmicks Company. Loan expense includes $2200 in interest and $9000 toward repayment. Before-tax income is $28,500 − $10,000 = $18,500.

	Sales' Gimmicks Company Monthly Budget					
	Jan.	Feb.	March	April	May	June
Sales						
Postcards	$6000	$5,000	$6,000	$6,000	$ 8,000	$9,000
Other novelties		500	1,000	1,500	2,000	3,000
Cash received	$6000	$5,500	$7,000	$7,500	$10,000	$12,000
Disbursements						
Postcards	1000		2,000	4,000	6,000	12,000
Other novelties	3000	1,000	1,000	500	500	1,000
Route servicers' salaries	600	600	600	600	1,400	2,100
Travel expense	1000	1,000	1,000	1,000	1,000	1,300
Office salaries	450	450	450	450	450	700
Office expense	100	100	100	150	150	200
Other expense	180	180	180	220	220	280
Loan repayment				1,000		
Cash disbursements	$6330	$3,330	$5,330	$7,920	$9,720	$17,580
Gain (or loss) from operations	(330)	2,170	1,670	(420)	280	(5,580)
Cash balance ($10,000)	$9670	$11,840	$13,510	$13,090	$13,370	$ 7,790
	July	Aug.	Sept.	Oct.	Nov.	Dec.
Sales						
Postcards	$10,000	$13,000	$11,000	$7,000	$5,000	$ 6,000
Other novelties	4,000	5,500	4,500	2,000	2,000	4,000
Cash received	$14,000	$18,500	$15,500	$9,000	$7,000	$10,000
Disbursements						
Postcards	6,000	4,000	2,000	2,000		1,000
Other novelties	2,000	1,000	1,000	1,000	1,000	200
Route servicers' salaries	2,100	2,100	1,950	650	650	650
Travel expense	1,300	1,300	1,100	600	600	800
Office salaries	700	700	700	450	450	1,050
Office expense	200	200	200	200	200	300
Other expense	640	340	280	220	1,040	220
Loan repayment	4,700			1,000		4,500
Cash disbursements	$17,640	$ 9,640	$ 7,230	$ 6,120	$ 3,940	$ 8,720
Gain (or loss) from operations	(3,640)	8,860	8,270	2,880	3,060	1,280
Cash balance ($10,000)	$ 4,150	$13,010	$21,280	$24,160	$27,220	$28,500

FIGURE 18.4

SALES' GIMMICKS COMPANY
Year Ending April 1, 19x1

Income

Net Sales: Postcards	$92,000	
Other gimmicks	30,000	
Total net sales		$122,000
Cost of goods sold: Postcards	40,000	
Other gimmicks	13,200	
Total cost of goods sold		53,200
Gross profit on sales		$68,800

Expenses

Route servicers' salaries	$14,000	
Travel expenses	12,000	
Office salaries and expenses	9,100	
Other expenses (storage, pilferage, etc.)	4,000	
Loan repayment ($2200 interest + $9000) ...	11,200	
Total expenses		$50,300
Net operating profit........................		$18,500
Income tax		3,700
Net profit (to retained earnings)		$14,800

Forecasting attempts to predict future events by the best means possible. The simplest form of forecasting is that of a mental voyage into the future to guess what might occur. Experienced estimators can do amazingly well, but seat-of-the-pants forecasting promises a rude jolt to the inexperienced. More disciplined forecasting methods utilize statistical analysis and other formal procedures to draw conclusions about future events from past events. Regardless of the methods used (see Figure 18.5), the basic issues to predict for economic comparisons are the amount and timing of resource requirements and the associated cash flow streams.

Forecasting is predominately concerned with future economic conditions. *Business forecasting* comprises the prediction of market demands, conversion costs, availability of labor, and the like. Nearly every major engineering decision is affected by business forecasts because engineering projects inevitably are launched to satisfy future demands. Both the pace and the magnitude of engineering expenditures are linked to economic conditions, and business activity is the barometer of these conditions.

Prediction of new engineering developments and scientific advances, assumed to be independent of business activity, is called *technical forecasting*. It is concerned with activities that are largely under the control of an organization or an association sharing common interests. For example, technical forecasts have been developed for advances in computer-aided technology and laser developments. These forecasts could be used by engineering economists to estimate the economic life and usage of assets in investment proposals for computer and laser applications.

FIGURE 18.5
Forecasting methods.

Attempts to seek a competitive edge from a glimpse into the future are nothing new. Rulers in the Middle Ages had their favorite crystal gazers and fortune tellers. Today astrologers and palm readers have believers, including some business people and engineers. Other organizational leaders rely on market surveys, product questionnaires, and quantitatively trained forecasting experts to anticipate levels of future activity. Yet no forecasting technique is infallible.

There are many ways to develop predictions. For convenience in getting an overview of the methods utilized, they can be grouped into the following three categories:

Subjective methods, based on opinions

Historical methods, based on past performance

Causal methods, based on underlying causes of events

Since more precise predictions are customarily more expensive to develop, it would be financial foolishness to allocate the same estimating time and cost to small and large investment evaluations. Similarly, only limited forecasting effort need be devoted to comparisons of alternatives that are identically affected by future conditions, and many comparisons do fit this category. An indication of the relation between forecasting methods and costs is given in Figure 18.6.

Even the overwhelming amount of timely data about current conditions and advanced computer analyses now available cannot ensure accurate forecasts. Ask any weather forecaster.

FIGURE 18.6
The method of forecasting must be justified by its economic feasibility. More sophisticated techniques should produce more accurate predictions. Since the total cost includes both the expense of making a forecast and the cost of damage resulting from inaccurate forecasts, more funds should be allowed for predictions upon which larger investments will be judged.

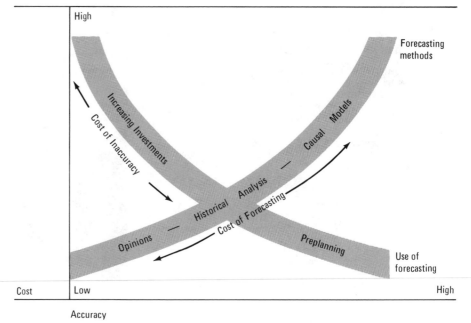

FORECASTS BASED ON OPINIONS

When an estimator does not have access to a staff of forecasting specialists, subjective predictions may be the most feasible forecasting method. A prime source is the opinion of the forecaster. This judgment has a better chance for accuracy if it is grounded in facts. An abundance of free information is available to aid estimators. General data about national economic health, pricing indexes, consumer spending trends, etc., are offered in magazines, newspapers, and publications from trade associations and government agencies. Even specific data and item-by-item forecasts about particular industries are available from various publishers and consultants.

The danger in basing forecasts on opinions is that individuals may let their personal feelings take command over what they know are facts. As an example of well-meant but biased advice, consider the case of a young entrepreneur who went to an unfamiliar area some years ago to invest in a coin-operated car wash. He sought advice on the investment and was put in touch with a person who had considerable experience in the automotive field; he had been a mechanic, an automotive service manager, and the operator of a taxi company. The entrepreneur asked him, "What do you think of the possibilities for success of coin-operated car washes in this area?" Now this person had a particular aversion to any mechanical device for cleaning the finish of a car. He simply did not believe that it was good for the finish to use pressure hoses, mechanical scrubbers, spray waxes, etc. He did believe in hand washing with

Cynics say that any new idea evokes one of three reactions: (1) It is impossible—don't waste my time; (2) it is possible, but it's not worth doing; (3) I said it was a good idea all along.

sponge and chamois and in hand waxing and polishing. As a result, his answer was "Those coin-op car laundries are a flash in the pan. They'll never have any success around here."

It so happened that the facts indicated the opposite. Sales of new cars, personal incomes, traffic counts, and population growth curves all would have said yes to our entrepreneur's question. That area was one in which salt was used on the roads in winter. This fact alone leads many people to wash their cars frequently in coin-operated washes. Now there are several hundred of these car-wash units in the area. The sad part of the tale is that all these coin-ops are now operated by other people. The prediction our entrepreneur got was based upon subjective opinion rather than objective data.

Personal opinions are not necessarily bad sources of advice. In many instances they are the only source of advice. Even when more objective sources can be called upon, opinions may offer valuable insights or provide extra assurance about a prediction. Many companies routinely seek opinions from people who buy or sell their products. Five formal methods of obtaining subjective information are described below.

A person who has actually bought a product is the most logical source of information about that product's sales appeal. Questions to the buyers are often part of the product's guarantee. Follow-up questionnaires are frequently mailed to the purchasers of a large item, such as an automobile. Sometimes surveys are made of potential rather than actual consumers. Replies from this audience must be interpreted carefully, because consumers' tastes change very rapidly and their intentions or desires may be far removed from what they actually do.

1 Consumer survey

An experienced sales force is in a position to observe both the actions of suppliers and the behavior of consumers. They can give warnings about changes in buying trends and the activities of competitors. Sales engineers are particularly well qualified to suggest design changes to improve a product's acceptance. The optimism or pessimism of individuals in the sales force can be balanced by averaging the predictions made by several sales representatives and sales managers.

2 Opinions of sales representatives and distributors

Many forecasts used by executives are made by executives. The effect of individual biases is reduced by generating forecasts as a group effort. The mixture of interests and experience that ensures a good cross section of estimates makes a consensus difficult to obtain.

3 Executive views

The most elaborate means of gathering consumer opinions is to sell a new product on a trial basis. The trial is designed to be a carefully controlled experiment. A small market area is selected to represent, as realistically as possible, the whole market that the product will compete in. Care is taken to avoid obstructing the competition and to be sure that the results are timely.

4 Market trials

Since the cost of these experiments is high, market trials are limited to significant product promotions.

5 Delphi method   A systematic routine for combining opinions into a reasoned consensus has the prophetic name *Delphi*. The technique was developed at the Rand Corporation and has gained fame in technical forecasting of future scientific developments. The procedure followed is to solicit and collate opinions about a certain subject from experts and feed back digested appraisals to narrow the differences among opinions until a near agreement is obtained.

A carefully prepared questionnaire is delivered to a panel composed of experts from professional specialties that pertain to the forecasting problem. The survey can be conducted in a group meeting or by mail. Each questionnaire solicits written opinions about specific topics and requests supporting reasons for the opinions. These reasons are summarized by the Delphi moderator to assure anonymity of responses and then returned for consideration by the whole panel. The process is continued until the exchanged arguments and transfer of knowledge forges a consensus prediction. The end product may be a time-scaled "map" charting the nature of future technological developments. Advocates of the Delphi method claim that the anonymity of written responses preserves the desirable features of a committee of specialists while reducing the "bandwagon" and dominant-personality effects that unduly sway group opinions.

FORECASTS BASED ON HISTORICAL RECORDS

Basing forecasts on historical evidence relieves most of the uneasiness connected with relying on personal opinions. Historical data are simply facts. It is up to the forecaster to make intrepretations from them. A forecaster can, of course, introduce personal bias into the intrepretation, but this bias is open to exposure because the facts from which the forecast was drawn are still available for reexamination. In addition to the factual attractiveness of historical data, two adages bear out the value of past performance in estimating future performance: "History repeats itself," and "we learn from experience."

Line Fitting

The most direct way to observe historical patterns is to plot data on a time scale, a natural approach for engineering-minded analysts. A quarterly sales record for 10 years is plotted in Figure 18.7. The purpose of graphing is to expose a recurring shape or pattern that gives a clue about where the next point or series of points should fall. The assumption that existing patterns will continue into the future is more likely to be correct over a short rather than a long estimating horizon, unless the patterns are unusually stable.

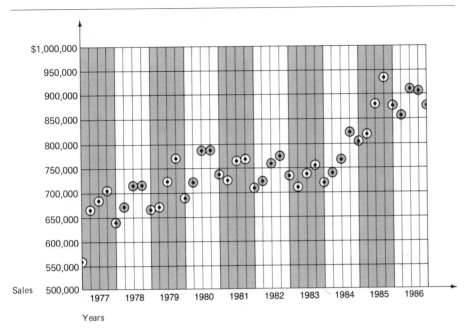

FIGURE 18.7
Ten-year record of sales, where each dot represents quarterly sales.

The scattered sales-level dots from Figure 18.7 are repeated in Figure 18.8, with the addition of a *trend line*. This straight line is constructed to show the *long-term trend* in sales. It can be sketched freehand or fitted mathematically

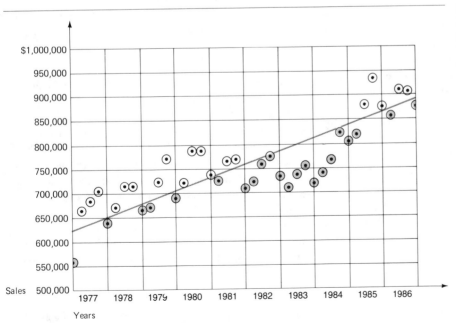

FIGURE 18.8
Long-term trend line imposed on the sales data from Figure 18.7. The trend line shows the general direction of sales during the 1976–1986 decade.

by use of regression techniques. By either method, roughly half the dots should be above the line and half below. The best possible fit exists when the sum of the squared distances from each dot to the line is smaller than that sum for any other line that might be drawn. A routine to quantify this condition is appropriately called the *least-squares method* and is explained in Example 18.3. By this method, the formula for the trend line shown is

$$Sales = \$625,000 + \$26,500X$$

where X is the number of years in the future from the end of the base year 1976 for which an estimate is sought. For instance, the sales level indicated by the trend line for year 1986 ($X = 1986 - 1976 = 10$) is

$$Sales_{1986} = 625,000 + (26,500)(10)$$
$$= 625,000 + 265,000 = \$890,000$$

Similarly, the forecast for 1987 would be

$$Sales_{1987} = 625,000 + (26,500)(11) = \$916,500$$

Example 18.3
Line Fitting with the Least-Squares Method

Whenever plotted data points appear to follow a straight line, the least-squares method can be used to determine the line of best fit. This line is the one that comes closest to touching all the data points. Calculations provide the equation of the line for which the sum of the squares of the vertical distances between the actual values and the line is at a minimum. A further property of the line is that the sum of the same vertical distances equals zero.

A straight line is defined by the equation $Y = a + bX$. For our purposes, Y is a forecast value at a point in time, and X is measured in increments such as years from a base point. The objective is to determine a, which is the value of Y at the base point, and b, which is the slope of the line.

Two equations are employed to determine a and b. The first is obtained by multiplying the straight-line equation by the coefficient of a and then summing the terms. With the coefficient of a equal to 1 and N as the number of data points, the equation becomes

$$\Sigma Y = Na + b\Sigma X$$

The second equation is developed in a similar manner. The coefficient of b is X. After multiplying each term by X and summing all the terms, we get

$$\Sigma XY = a\Sigma X + b\Sigma X^2$$

The two equations thus obtained are called *normal equations*.

The four sums required to solve the normal equations (ΣY, ΣX, ΣXY, and ΣX^2) are obtained from a tabular approach. The calculations can be simplified by carefully selecting the base point. Because X equals the number of periods from the base point, selecting a midpoint in the time series as the base makes ΣX equal to zero. The smaller numbers resulting from a centered base point also make other required products and sums easier to handle. After the four sums are obtained, they are substituted in the normal equations and the values of a and b are calculated. Then these values are substituted into the straight-line equation to complete the simple regression forecasting model:

$$Y_F = a + bX$$

where Y_F = the forecast value X periods in the future

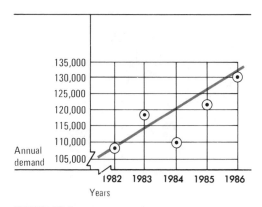

FIGURE 18.9
Sample data for forecasting examples.

QUARTERLY AND ANNUAL DEMAND					
Year	1982	1983	1984	1985	1986
Quarter 1	19,000	28,000	27,000	30,000	32,000
Quarter 2	37,000	42,000	36,000	43,000	44,000
Quarter 3	30,000	31,000	28,000	29,000	32,000
Quarter 4	22,000	18,000	19,000	20,000	22,000
Annual demand	108,000	119,000	110,000	122,000	130,000

To illustrate simple linear regression, the linear relationship between an independent and dependent variable, the least-squares method will be applied to the data in Figure 18.9. A forecasting equation is sought to relate demand to time.

A straight sloping line appears to be a reasonable fit for the data in Figure 18.9. To make the calculations easier, a centered base point is used; 1984 is the midpoint of the historical data and the date selected for $X = 0$. A tabular format with Y as demand in 1000-unit increments is set up to obtain the sums required to solve the normal equations:

Year	Y	X	X^2	XY	
1982	108	−2	4	−216	
1983	119	−1	1	−119	
1984	110	0	0	0	←Base point
1985	122	+1	1	122	
1986	130	+2	4	260	
Sums	589	0	10	47	

The sums are substituted into the normal equations, with $N = 5$, as

$$589 = 5a + b(0)$$
$$47 = a(0) + b(10)$$

and solved to yield $a = 589/5 = 117.8$, or 117,800 units, and $b = 47/10 = 4.7$, or 4700 units.

A forecasting equation is developed by substituting values of a and b into the straight-line equation:

$$Y_F = 117,800 + 4700X$$

The forecast for 1987, which is 3 years away from the 1984 base point for X, is given by substituting 3 for X in the forecasting equation:

$$\text{Demand}_{1987} = 117,800 + 4700(3) = 131,900 \text{ units}$$

Historical data may also reveal *cyclic patterns* within the long-term trend. Two types of wavelike variations are noticeable in Figure 18.10. The large wave shows a business cycle of some sort. It may represent repeating fluctuations in local economic conditions, repeating changes in styles or consumer tastes, recurring fads and buying habits, or a combination of these and other factors.

The other cyclic pattern is attributable to seasonal variation. The insert in Figure 18.10 displays the data collected in respective quarters of a year. It

FIGURE 18.10

Business cycle and seasonal variation patterns imposed on the sales data from Figure 18.7.

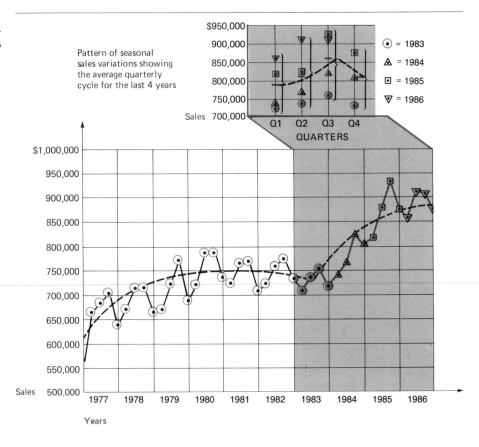

is evident that higher sales can be expected in the middle two quarters of each year. Many products that depend on a particular season (ski equipment, swimming wear, gardening supplies, holiday treats, etc.) exhibit a repetitive sales pattern each year.

The purpose of distilling different patterns from a collection of data is to improve the forecast. A long-term trend gives a valid prediction of the general region in which future sales should fall. (This trend is modified by cyclic influences when they appear.) The longer-term cycles are difficult to identify, but seasonal cycles are quite apparent and predictable. The forecast for 1987 from the given sales data would probably be only slightly higher than the current (1986) sales, with the highest demand concentrated in the middle two quarters. A forecast for 1988 and 1989 would likely predict a sales pattern about the same as 1987, owing to the long-run cyclic influence. Less confidence can be placed in the forecasts as they are pushed farther into the future, because new factors could enter to disrupt the traditional patterns.

Soviet economist Nikolai Kondratieff proposed in 1920 that capitalist countries would have economic crises about every 50 years. His long-wave theory is roughly substantiated by depressed conditions in the 1840s, 1890s, and 1930s.

Exponential Smoothing

A line-fit forecast gives equal weight to all the historical data included in the regression model. It may be advisable to give more weight to recent occurrences. This is accomplished in the exponential-smoothing model by selecting a weighting factor α that gives more or less prominence to recent happenings. The exponential-smoothing formula is very convenient to apply, since it has only three inputs:

1 LD = latest demand, the most recent data on actual accomplishments of the subject being forecast. LD for a sales forecast would be the sales made during the current period.

2 PF = previous forecast, the forecast produced for the current period by applying the exponential-smoothing formula. The forecast made with the LD for the current period will be the PF for the next forecasting period.

3 α = alpha, the smoothing constant, with a value between 0.0 and 1.0. The value selected for α determines how much emphasis is put on the most recent data.

These factors are combined as shown below to yield a forecast NF for the next period:

$$\text{Next forecast} = NF = \alpha(LD) + (1 - \alpha)(PF)$$

This expression can be rewritten to show that the new forecast is simply the previous forecast corrected by adding a percentage α reflecting how much the old forecast missed the actual demand:

$$NF = PF + \alpha(LD - PF)$$

As an example, let alpha have a value of 0.4, sales for June be 4200 units, and the forecast made in May for the sales expected in June be 4500 units. Applying the two versions of the exponential-smoothing formula shows

$$
\begin{aligned}
NF_{July} &= \alpha(LD_{June}) + (1 - \alpha)(PF_{June}) \\
&= 0.4(4200) + (1 - 0.4)(4500) \\
&= 1680 + 2700 = 4380 \text{ units}
\end{aligned}
$$

and

$$
\begin{aligned}
NF_{July} &= PF_{June} + \alpha(LD_{June} - PF_{June}) \\
&= 4500 + 0.4(4200 - 4500) \\
&= 4500 - 120 = 4380 \text{ units}
\end{aligned}
$$

It is apparent from these equations that the value used for alpha significantly affects the forecast. If alpha is set equal to zero, the original forecast never changes, no matter how much the actual demand varies from it. At the other extreme, when $\alpha = 1.0$, the next forecast always equals the last demand experienced. A projection that the next period will be the same as the last period is extensively used for short-term scheduling and is known as *persistence forecasting*. Between the extremes, α values closer to zero produce more stable predictions that may not detect current trends, and values closer to 1.0 closely track actual demands with forecasts that may fluctuate erratically.

The best way to choose alpha is to apply different values to historical data to see which one would have provided the most useful forecasts in the past. A simulation exercise to evaluate alpha values is shown in Figure 18.11. As expected, forecasts made with higher alphas are more responsive to current demand and consequently follow the actual demand pattern more closely. Other formulas using more involved smoothing functions can yield even better tracking, but the simple record-keeping, the easy computations, and the often commendably accurate basic exponential-smoothing formula make it attractive for tactical planning.

In sporting circles, persistence forecasting is called "sticking with a winner." The current champ is thus usually the betting favorite.

More elaborate versions of tactical planning are incorporated in the Box-Jenkins technique; see G. E. P. Box and G. M. Jenkins, *Time Series Analysis, Forecasting, and Control*, Holden-Day, San Francisco, 1970.

Moving Average

Another popular forecasting method predicts the demand for the next period from the average demand experienced during several recent periods. This method is appropriately called the *moving average*. It is quite similar to exponential smoothing in that both methods "smooth," or "average out," the fluctuations in past demands to produce a forecast. The main difference between the two methods is that the choice of alpha in exponential smoothing determines how sensitive this forecast is to the most recent demands, whereas a moving-average forecast gives equal weight to each demand period included in the forecast.

A 3-month moving-average forecast for the next month would simply be the average demand experienced for the last 3 months. For example, the 3-month moving-average forecast for June would be calculated as

$$\text{NF}_{\text{June}} = \frac{\text{March demand} + \text{April demand} + \text{May demand}}{3}$$

In the same pattern, the forecast for July would result from averaging the demands for June, May, and April. The most recent block of actual demands is used in each successive forecast.

The choice of how many periods to include in the moving average involves the same considerations used to select an alpha value for exponential smoothing. A forecast based on very few periods is more sensitive to the latest events, as is an alpha value near 1.0. An average derived from data taken over several

FORECASTS DEVELOPED FROM IDENTICAL DATA WITH ALPHA VALUES OF 0.2, 0.4, AND 0.6

Date	Actual Demand	Forecast at $\alpha = 0.2$	Absolute error	Forecast at $\alpha = 0.4$	Absolute error	Forecast at $\alpha = 0.6$	Absolute error
Oct. 8	180						
15	164						
22	162	177	15	174	12	170	8
29	151	174	23	169	18	165	14
Nov. 5	143	169	26	162	19	157	14
12	158	164	8	154	10	148	10
19	154	163	9	156	2	154	0
26	160	161	1	158	7	154	6
Dec. 3	172	161	11	159	16	158	14
10	178	163	15	164	16	166	12
17	188	166	22	170	18	173	15
24	174	170	4	177	4	182	8
31	172	171	1	175	4	177	5
Jan. 7		171	—	174	—	174	—
Average error:			12.3		11.5		9.6

FIGURE 18.11

Comparison of forecasts provided by the exponential-smoothing method using different alphas. To start the forecasting process, the first prediction is approximated by letting it equal the first demand. Thus, the forecast for October 22, when $\alpha = 0.2$, was calculated as $NF_{Oct.\ 22} = 0.2(164) + 0.8(180) = 177$. Subsequent forecasts then include the estimate for the actual demand in each weekly period. The graph demonstrates how the lag between forecast and actual demand fluctuations is decreased by larger alpha values.

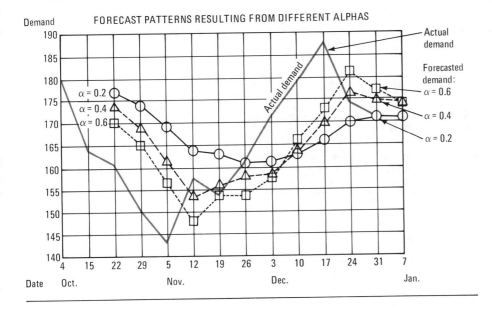

FORECAST PATTERNS RESULTING FROM DIFFERENT ALPHAS

periods will better reduce fluctuations caused by random events, but it may be too stable to detect current trends. As was the case for exponential smoothing, the best way to decide how much smoothing is desirable is to experiment with past demands, trying different spans for the moving average and selecting the one that would have predicted previous demands most accurately.

A *seasonal index* referenced to a moving average improves the forecast unless the demand pattern is relatively constant. An index value is calculated by dividing the actual demand by the centered moving average for that period. A more reliable index is obtained by averaging several index values for common time periods. The forecast is thereby the product of the most recent centered moving average for a period and the index value for that period. The procedure for developing and using a seasonal index is demonstrated in Example 18.4.

Example 18.4
Moving-Average Forecast Applied to the Quarterly Demand for the Sample Data in Figure 18.9

With the demand by quarters given in Figure 18.9, a four-quarter moving average will be used to forecast the demand for the next two quarters, $Q1_{1987}$ and $Q2_{1987}$. The four-period average is chosen because it smooths out seasonal variations over a full year.

The first moving average is one-fourth the total sales for 1982; it represents a point in time between the end of the second quarter and the start of the third quarter. The second

moving average is the sum of the last three quarters of 1982 and the first quarter of 1983 divided by 4. This value is associated with the end of $Q3_{1982}$ and the start of $Q4_{1982}$. An average of these two numbers gives a moving average centered at $Q3_{1982}$. This procedure for finding the midpoint value for all quarters is continued to obtain the four-period and centered moving averages in Table 18.2.

The last column in Table 18.2 is the seasonal index for

TABLE 18.2

Computation of four-period moving averages, centered moving averages, and quarterly seasonal index for the sample data from Figure 18.9

Year	Quarter	Sales; 1000s of Units	Four-Period Moving Average	Centered Moving Average	Seasonal Index
1982	Q1	19.0			
	Q2	37.0			
	Q3	30.0	27.0	28.1	1.07
	Q4	22.0	29.2	29.8	0.74
1983	Q1	28.0	30.5	30.6	0.91
	Q2	42.0	30.7	30.2	1.39
	Q3	31.0	29.7	29.6	1.04
	Q4	18.0	29.5	28.7	0.63
1984	Q1	27.0	28.0	27.6	0.98
	Q2	36.0	27.3	27.4	1.32
	Q3	28.0	27.5	27.9	1.00
	Q4	19.0	28.3	29.2	0.66
1985	Q1	30.0	30.0	30.1	1.00
	Q2	43.0	30.3	30.4	1.42
	Q3	29.0	30.5	30.7	0.94
	Q4	20.0	31.0	31.1	0.64
1986	Q1	32.0	31.2	31.6	1.01
	Q2	44.0	32.0	32.2	1.37
	Q3	32.0	32.5		
	Q4	22.0			

TABLE 18.3

Calculation of an adjusted seasonal index

Year	Q1	Q2	Q3	Q4
1982			1.07	0.74
1983	0.91	1.39	1.04	0.63
1984	0.98	1.32	1.00	0.66
1985	1.00	1.42	0.94	0.64
1986	1.01	1.37		
Totals	3.90	5.50	4.05	2.67
Average seasonal index	0.975	1.375	1.0125	0.6675
Adjusted seasonal index	*0.97*	*1.37*	*1.00*	*0.66*

each quarter. It is obtained by dividing the actual sales for a quarter by the centered moving average for that quarter. A better estimate of a quarter's index is obtained by averaging all the values available (see Table 18.3).

Before the average seasonal index is applied, two checks should be made:

1 The average of the periodic indexes should total to 1.0. In the example, the average is

$$\frac{0.9750 + 1.3750 + 1.0125 + 0.6675}{4} = \frac{4.03}{4} = 1.0075$$

Therefore, the indexes must be adjusted or else the quarterly forecasts will exceed the implied annual forecast by 0.75 percent.

2 Attention should be given to any obvious trends in a quarterly index. In Table 18.3, Q1 appears to be increasing and Q3 has a distinct downward trend.

The adjusted index reflects the considerations above. A fraction (4.00/4.03) of each average index was taken and the results were rounded off with respect to the trend in Q3. Thus, calculations provide the forecasting framework, but finishing touches are supplied by judgment.

The final step is to make the forecast. It results from taking the product of the most recent centered moving average and its respective seasonal index. Forecasts for the first two quarters of 1987 are

$$Q1_{1987} = 31.6 \times 0.97 = 30.7 \quad \text{or} \quad 30,700 \text{ units}$$
$$Q2_{1987} = 32.2 \times 1.37 = 44.1 \quad \text{or} \quad 44,100 \text{ units}$$

Extensions of Historical-Basis Forecasting Methods

Only the perimeter has been explored in this survey of methods of forecasting from historical data. Simple regression, a model relating a dependent to an independent variable, can yield fits of many shapes besides straight lines. Curvilinear regression based on the equation $Y_F = aX^b$ could be used to develop certain curve-shaped fits. Hyperbolic and polynomial fits are also possible. Multiple regression, either linear or nonlinear, is used when the forecast depends on the relation of several variables.

An indication of how much faith can be put in a forecast results from calculating statistical control limits for regression lines. Such calculations can also provide prediction limits. For instance, a statement could be made that there are only 5 chances in 100 that the accumulated demand will exceed a certain value at a certain time in the future.

The main value of regression models as compared with weighted-average techniques (exponential smoothing and moving average) is that the regression line can be extended several periods beyond the present. This advantage is gained at the expense of extra computational effort. But this burden can be shifted to computers; canned programs that handle huge regression models are widely available.

FORECASTS BASED ON CAUSE

Very reliable predictions are possible if the causes for future demand can be identified and measured. The causal approach to forecasting has produced sophisticated models that attempt to include all the relevant causal relationships; in the most elaborate models there may be hundreds of quantified relationships. But causal models can also be quite simple. Perhaps knowledge of a single cause is sufficient for predicting, as in the case of a college bookstore which orders books based on knowledge of the number of students admitted to the school for next year. The manager of the bookstore could apply a factor based on past history to the number admitted to forecast the demand by courses in the forthcoming terms.

Leading indicators are frequently used in business forecasting. These indicators are statistics compiled and published by many sources: *The Wall Street Journal, Business Week,* United States government reports on almost everything, local government reports on economic activity, university publications, etc. If an indicator can be found that consistently predicts what demand to expect in, say, 6 months, current activities can be planned accordingly. It is even more comfortable to have two or three leading indicators pointing in the same direction. The problem is to find the accurate indicators.

Correlation Model

A search for a leading indicator starts with assumptions about the causes of demand. A house builder could hypothesize that housing demand depends on population growth, prices for houses, personal incomes, demolition of old homes, housing starts by competitors, and availability of financing. After data about the possible indicators have been collected, correlation studies are conducted to see if they are indeed related to housing demand. Ideally, an indicator will be identified that has a pattern similar to the historical demand but that leads it by several periods. For example, it might be discovered that housing demand is correlated with interest rates: As rates climb, demand declines; but the change in demand follows a change in interest rates by 6 months. In this case, a builder could observe today what is happening to interest rates and use the observation as a leading indicator to predict with some confidence what the demand for houses will be in 6 months.

A correlation study examines the degree of relation between variables. The variables are represented in mathematical expressions, such as the forecasting formula developed by the least-squares method in Example 18.3. Because not all the data points coincide with the fitted line, part of the relationship among variables remains unexplained. This dispersion of data points about regression lines is characterized by three sums-of-squares:

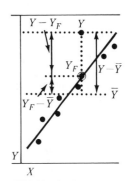

FIGURE 18.12
Deviations of a data point.

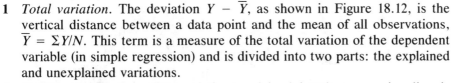

$$\Sigma(Y - \overline{Y})^2 = \Sigma(Y_F - \overline{Y})^2 + \Sigma(Y - Y_F)^2$$

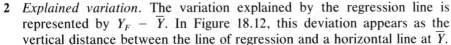

| Total variation | Explained variation | Unexplained variation |

1   *Total variation.* The deviation $Y - \overline{Y}$, as shown in Figure 18.12, is the vertical distance between a data point and the mean of all observations, $\overline{Y} = \Sigma Y/N$. This term is a measure of the total variation of the dependent variable (in simple regression) and is divided into two parts: the explained and unexplained variations.

2   *Explained variation.* The variation explained by the regression line is represented by $Y_F - \overline{Y}$. In Figure 18.12, this deviation appears as the vertical distance between the line of regression and a horizontal line at $\overline{Y}$.

3   *Unexplained variation.* The sum-of-squares of the unexplained variation expresses the error between the forecast and actual values, $Y - Y_F$. Each deviation appears as the vertical distance from a data point Y to the regression line.

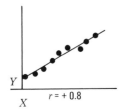

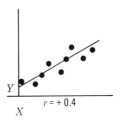

The ratio of the sum-of-squares of the unexplained variation to the sum-of-squares of the total variation, $\Sigma(Y - Y_F)^2/\Sigma(Y - \overline{Y})^2$, measures the proportion of the total variation that is not explained by the regression. Therefore,

$$1 - \frac{\Sigma(Y - Y_F)^2}{\Sigma(Y - \overline{Y})^2}$$

measures the proportion of the total variation explained by the regression line, and the square root of this expression is the *coefficient of correlation, r:*

$$r = \sqrt{1 - \frac{\Sigma(Y - Y_F)^2}{\Sigma(Y - \overline{Y})^2}}$$

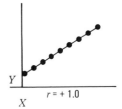

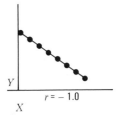

FIGURE 18.13
Correlation patterns.

The value under the radical can never be greater than 1 or less than 0. However, because the radical has both positive and negative roots, the value of r is between $+1$ and -1. The plus or minus is indicative only of the slope of the regression line, as depicted in Figure 18.13. When $r = +1$, all the data points fall on an upward-sloping regression line. When r is between $+1$ and

FIGURE 18.14

Visually apparent correlation between 3-month T-bill interest rates,
housing starts, and softwood lumber consumption. The scale for T
bills is inverted to highlight the
close negative correlation. As T-bill
rates go up, housing starts and
lumber consumption go down. The
T-bill trend leads housing and lumber trends by a few months.

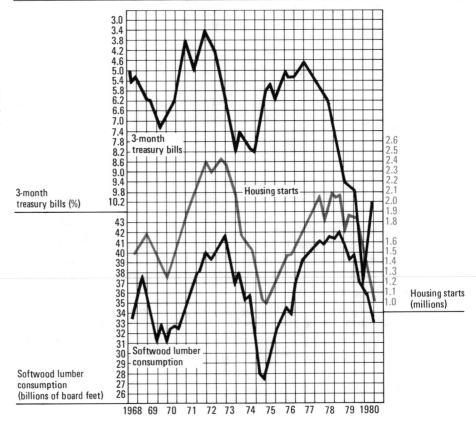

0, the regression line still slopes upward, but data points fall on either side of
the line. The closer they cluster around the line, the closer r approaches 1.

When sufficient data are available, r is calculated from the sum-of-squares
expression or, more directly, from the formula

$$r = \frac{N\Sigma XY - (\Sigma X)(\Sigma Y)}{\sqrt{N\Sigma X^2 - (\Sigma X)^2}\,\sqrt{N\Sigma Y^2 - (\Sigma Y)^2}}$$

A graphical display of actual correlation between a leading economic
indicator and production demand is shown in Figure 18.14. It indicates a
negative correlation between the interest rates paid for 3-month Treasury bills
(T bills) and housing starts; the scale for T-bill interest is inverted to better
portray the relationship. This relationship is attributed to higher T-bill rates
attracting money that would otherwise have gone for home mortgages and,
concurrently, forcing mortgage rates up. Both conditions hurt the demand for
housing. The close correlation between housing starts and softwood lumber

consumption is visually obvious. The total pattern suggests that interest rates for T bills provide a few months' warning of what lumber sales to expect.

Econometric Models

Econometrics is a discipline concerned with the measurement and definition of economic systems. It utilizes statistical and programming methods to develop models based on the quantitative aspects of system behavior. The models vary from the simple regression methods previously described to very elegant and rich representations of large systems involving hundreds of variables.

Difficulties and disconcertations encountered in forecasting future economic conditions are discussed in Extension 18B.

The more sophisticated economic models are based on causal relationships. They have been developed for different engineering systems, most notably for transportation. The initial effort in developing a causal model is considerable, and after the causal forecast is accepted, the model must be continually

---

Example 18.5
Development of a Leading-Indicator Forecast by Regression and Correlation Models

The Slam-Bang Company manufactures automatic nail drivers as modern replacements for the conventional hammer. The company sensibly reasoned that the sales of nail drivers should be related to the total amount spent on building construction. If a relationship does exist, published government and construction-industry forecasts of anticipated building levels can be used as a sales indicator. First a check was made to see that the building-level forecasts were relatively accurate. Next the national figures were broken down to conform to the Slam-Bang marketing areas. Then the records of monthly building volume and nail-driver sales for corresponding months were collected. The resulting data are tabulated in Table 18.4, where the construction volume is in \$100-million units and product sales are in \$10,000 units. The column headings correspond to the values needed for the calculation of r and a linear-regression equation.

From this reasonably large sample, the coefficient of correlation was calculated as

$$r = \frac{N\Sigma XY - (\Sigma X)(\Sigma Y)}{\sqrt{N\Sigma X^2 - (\Sigma X)^2}\sqrt{N\Sigma Y^2 - (\Sigma Y)^2}}$$

$$= \frac{20(519.5) - (50.4)(202.4)}{\sqrt{20(130.5) - (50.4)^2}\sqrt{20(2076.62) - (202.4)^2}}$$

$$= \frac{10,390 - 10,201}{\sqrt{2610 - 2540.16}\sqrt{41,532.4 - 40,965.7}}$$

$$= \frac{189}{(8.36)(23.81)} = 0.95$$

Although the correlation is not exact, it is definitely worthy of consideration for prediction purposes. With the sums developed in Table 18.4, the least-squares method is applied to obtain a forecasting equation. The normal equations are

$$\Sigma Y = Na + b\Sigma X \qquad 202.4 = 20a + b(50.4)$$
$$\Sigma XY = a\Sigma X + b\Sigma X^2 \qquad 519.5 = a(50.4) + b(130.5)$$

They are solved for a and b to obtain

$$a = 3.37 \text{ or } \$33,700 \qquad b = 2.68 \text{ or } \$26,800$$

The forecasting equation is then

$$Y_F = \$33,700 + \$26,800X$$

Thus, a projected construction volume of \$275,000,000 for the next month would suggest a nail-driver sales volume of

$$Y_F = \$33,700 + \$26,800(2.75) = \$107,400$$

TABLE 18.4

Correlation data for sales of Slam-Bang nail drivers and construction volume. The last three columns are the computations required to calculate r and Y_F.

Nail-Driver Sales $\times 10^{-4}$, Y	Construction Volume $\times 10^{-8}$, X	Y^2	X^2	XY
7.1	1.8	50.51	3.24	12.78
9.9	2.3	89.01	5.29	22.78
9.0	1.9	81.00	3.61	17.10
10.4	2.6	108.16	6.76	27.04
11.1	3.1	123.21	9.61	34.41
10.9	2.8	118.81	7.84	30.52
10.5	2.9	110.25	8.41	30.45
9.8	2.4	96.04	5.76	23.52
11.1	2.8	123.21	7.84	31.08
10.2	2.5	104.04	6.25	25.50
9.7	2.3	94.09	5.29	22.31
10.9	2.8	118.81	7.84	30.52
8.8	2.1	77.44	4.41	18.48
8.6	1.9	73.98	3.61	16.34
12.3	3.2	151.29	10.24	39.36
11.4	3.0	129.96	9.00	34.20
11.2	2.8	125.44	7.84	31.36
10.2	2.6	104.04	6.76	26.52
10.7	2.7	114.49	7.29	28.89
8.6	1.9	73.96	3.61	16.34
202.4	50.4	2076.62	130.50	519.50

monitored to confirm that the modeled relationships are still valid. Causes of demand can change, and the cause of the causes can change.

Causal models are almost never constructed simply to generate better data for specific economic comparisons. When the models are already available, they should certainly be utilized in the comparisons, but a decision would have to be of "make-or-break" proportions to provoke the development of an original model. However, causal forecasting is more apt than is historical forecasting to predict a turning point in demand, a point at which growth switches to decay or the reverse, and even a hint of a turning point is vital information to the estimator. Reputations and fortunes are made or lost by alternatives selected in anticipation of a shift from the existing trend.

Review Exercises and Discussions

Exercise 1   A rough indication is needed to determine whether additional trucks will be required to transport raw materials during the coming year. If the preliminary indication is positive,

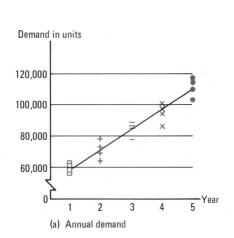

(a) Annual demand

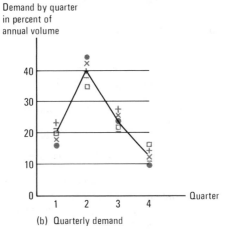

(b) Quarterly demand

FIGURE 18.15
Demand data with forecasting trends made by inspection.

an economic study will be made to see how many and what kinds of trucks should be added to the company fleet.

Annual demands for the past 5 years have been plotted and free-hand trend lines drawn, as shown in Figure 18.15(a). The "eyeballed" trend line appears to start at 60,000 units and increase by 12,500 units per year.

A breakdown of annual demand by percentage per quarter is shown in Figure 18.15(b). A simple forecast by quarters would result from an arithmetic average of all the data. This average value would be of little use for the trucking problem because fleet size is a function of the *maximum* transportation needs any time during the year. A visual inspection is made to reveal that second-quarter demand accounts for about 41 percent of the year's total and is therefore the critical figure.

Statistics on national demand over the past 30 years suggest that the current business cycle is on an upswing. Therefore, a judgment value of +8 percent, or 1.08 times the trend-line forecast, is selected to represent the effect of this cycle.

Using these unsophisticated forecasting methods, discuss what forecast seems reasonable for the maximum demand that will occur during the next year.

Solution 1

"Unsophisticated" is not a synonym for worthless. The eyeball approach delivers a reasonably close forecast at a minimum cost. Since a "guesstimate" does not depend on a formal procedure, it can put more weight on recent data or other factors that the forecaster feels are especially pertinent. From the given data, maximum demand should occur during the second quarter. The forecast results from multiplying the trend value by the cyclic correction and the quarterly percentage:

$$\text{Demand}_{\text{2d quarter}} = [60,000 + 12,500(5)](1.08)(0.41) = 54,243$$

Exercise 2

Develop a forecast by the exponential-smoothing method for the demand in 1987 based on the sample data in Figure 18.9. Let $\alpha = 0.3$. To develop the PF value, start the process with a forecast for 1984 based on the demand in 1982 as PF and the demand in 1983 as LD:

$$NF_{1984} = 108 + 0.3(119 - 108) = 111.3 \quad \text{or} \quad 111{,}300 \text{ units}$$

and $\quad NF_{1985} = 111.3 + 0.3(110 - 111.3) = 110.9 \quad \text{or} \quad 110{,}900 \text{ units}$

Continue the procedure to obtain NF_{1987}. Compare this forecast with the one obtained using the least-squares regression model in Example 18.3.

Solution 2   The sequence of calculations leading to the 1987 forecast is

$$NF_{1986} = 110.9 + 0.3(122 - 110.9) = 114.2 \quad \text{or} \quad 114{,}200 \text{ units}$$

$$NF_{1987} = 114.2 + 0.3(130 - 114.2) = 118.9 \quad \text{or} \quad 118{,}900 \text{ units}$$

Compared with the simple regression model in which $NF_{1987} = 131{,}900$ units, the exponential-smoothing model is more pessimistic. This is a result of the selection of 0.3 for the value of α, which puts more weight on the older data. Changing α to 0.8 increases the forecast for 1987 to 128,000.

PROBLEMS

18.1   What use can be made of general-accounting and cost-accounting records by an engineering economist? What precautions should be exercised in this usage?

18.2   Compare the features that are desirable in a depreciation method with the causes of declining value of an asset. What does this comparison suggest about the accounting concept of depreciation versus the considerations involved in an economic analysis?

18.3   Construct a balance sheet from the accounts given in Table 18.5. What are the dollar values of the terms in the basic accounting equation?

18.4   Prepare a cost-of-goods-sold schedule from the amounts given in Table 18.5. Include materials, labor, and factory overhead. What was the cost of goods sold in the period from July 1 to December 31?

18.5   Prepare an income statement for the 6-month period from the data in Table 18.5. What is the net profit if the accrued-taxes account represents the taxes owed for the period?

18.6   A cost index is a dimensionless number that relates a certain type of cost at a given time to that cost during a reference year. That is,

$$C_p = C_r \frac{I_p}{I_r}$$

where C_p = present cost, dollars
C_r = cost during reference period
I_p = present value of the cost index
I_r = index value during reference period

TABLE 18.5

The basic data in Table 18.5, about the In-Out Universal Corporation, are to be utilized in Problems 18.3 to 18.5 and Extension 18A. They show the condition of I-O-U on December 31 and the transactions that occurred during the prior 6 months.

Transactions for In-Out Universal during the 6 Months prior to December 31, 19xx	
Category	Amount
Administration expenses	$ 30,000
Depreciation charged	30,000
Direct labor costs	70,000
Factory overhead charged	35,000
Finished goods (July 1, 19xx)	75,000
Interest and debt payments	40,000
Inventory, materials (July 1, 19xx)	15,000
Materials purchased (July 1 to Dec. 31)....	115,000
Sales: Cash...........................	200,000
Credit..........................	200,000

State of Accounts for In-Out Universal on December 31, 19xx	
Account	Balance
Accounts payable	$ 50,000
Accounts receivable....................	80,000
Accrued taxes.........................	20,000
Buildings (net value)....................	205,000
Cash on hand	40,000
Dividends payable	20,000
Equipment (net value)..................	180,000
Finished goods........................	30,000
Land value	115,000
Long-term mortgages	390,000
Material inventory.....................	40,000
Notes payable	60,000
Stockholders' equity....................	150,000

(a) Cost estimates are being developed to justify the construction of a warehouse. A structure of similar lay-up design was built 6 years ago at a unit price of $14.40 per square foot ($155 per square meter) of wall when the cost index was 116. The index now is 170. What would the construction cost be today for a warehouse with 60,000 square feet (5574 square meters) of wall?

(b) According to present plans, the warehouse construction will not get underway until 3 years from now. If the cost index continues to increase at its present rate, what construction bid can be expected for the warehouse?

18.7 Persistence predictions may seem too naïve to be classified as a forecasting method, but they can be surprisingly successful under some conditions. Comment on the following observations:

(a) In sporting events the usual forecast is for the current champion to win again.

(b) A decision not to make a decision is a persistence prediction.

18.8 What approximate mathematical relationship between the α used in exponential smoothing and the number of periods used in a moving average makes both methods equally responsive to changes in historical data?

18.9 Sales figures for two products first marketed 6 months ago are as shown:

Month	Product 1	Product 2
January	$110,000	$ 54,000
February	102,000	63,000
March	95,000	80,000
April	85,000	98,000
May	78,000	112,000
June	70,000	133,000

(a) What July forecast for each product is obtained by using a 6-month moving average? How do you explain these forecasts with respect to the dissimilar sales patterns?

(b) What forecast would you make for each product for July?

18.10 The least-squares method can be applied to fit a curve to data when the forecasting equation is $Y_F = ab^X$, in which Y changes at a constant rate b each period for X periods. When this equation is translated from the exponential form to its logarithmic form,

$$\log Y = \log a + X \log b$$

normal equations can be set up as

$$\Sigma \log Y = N \log a + \Sigma X(\log b)$$
$$\Sigma(X \log Y) = \Sigma X(\log a) + \Sigma X^2(\log b)$$

to allow the tabular approach of the least-squares method in determining the values of a and b. When the base point is selected to make $\Sigma X = 0$, the solution reduces to

$$\log a = \frac{\Sigma \log Y}{N} \quad \text{and} \quad \log b = \frac{\Sigma(X \log Y)}{\Sigma X^2}$$

Set up a table with column headings for Y, X, X^2, $\log Y$, and $X \log Y$, and use the annual-demand data from Figure 18.9 to determine a curve-fitting forecasting formula. Compare the forecast for 1987 with the straight-line forecast.

18.11 Given the following data:

Year	1	2	3	4	5	6	7	8
Demand	90	100	107	113	123	136	144	155

(a) Plot the data and establish a forecast for year 9 by observation.

(b) Use the least-squares method to develop a forecasting formula. What is the forecast for year 9?

(c) Determine what the forecast is for year 9 by the exponential-smoothing method when $\alpha = 0.25$?

18.12 Quarterly unit demands for a product are:

Year	Winter	Spring	Summer	Fall
1	81	64	73	83
2	80	70	84	74
3	86	59	71	73
4	98	72	74	64
5	106	68	75	60

(a) Using a four-period moving average, determine a seasonal adjusted index and establish a forecast for each quarter of next year.

(b) Use line-fitting methods to determine a forecast for each period of next year.

18.13   Assume that you have the franchise to sell refreshments for theatrical performances given at the college playhouse. Besides the usual soft drinks and popcorn, you offer homemade tarts. Since the tarts must be baked within a day of when they are to be sold and the market for stale tarts is quite limited, predicting sales is very important. The sales record for the first 10 performances is tabulated below. Each performance was a sellout, and all future performances are expected to play to capacity crowds.

Date	Jan. 5	6	12	13	26	27	Feb. 2	3	16	17
Tarts sold	120	120	120	144	191	171	172	186	153	145

In reviewing the sales, you recall that you sold all the tarts you baked for the first four performances. Then you increased production for the remaining performances to 16 dozen (192) and started having leftovers. For the last two performances you barely covered costs because so many tarts remained unsold. Now the question is how many to bake for the next performance.

(a) What forecasts for tart demand would have been made for the last four performances by the exponential-smoothing formula with $\alpha = 0.4$? What is the forecast for the next performance?

(b) Repeat Problem 18.13a using a 3-month moving average.

(c) Determine the forecast for tart sales for the next performance when alpha and the number of moving-average periods are doubled ($\alpha = 0.8$ and $N = 6$)?

(d) What forecast do you recommend? What additional information might improve your prediction?

18.14   A patented new product was introduced 9 years ago. Sales since its introduction are:

Years since introduction	1	2	3	4	5	6	7	8	9
Annual unit sales	2023	2102	2009	2768	3291	3881	4622	5494	5557

The original plant that produces the product is now operating at 100 percent capacity. A second plant, to be built immediately, must have the capacity to meet the additional demand expected for the next 5 years. What should be the design capacity of the new plant?

18.15   Suggest possible leading indicators of readily available data for the following products and services:

(a) Industrial production in a developing country

(b) Enrollment in a private business school specializing in data-processing training

(c) The number of convicts for whom space must be provided in penitentiaries in less populous states

(d) Demand for prepared baby foods

18.16   Private-aircraft sales in a three-state marketing area and the number of students enrolling each year to take flying lessons are shown in the table on page 550.

Year	Aircraft Sales	Student Starts
1	300	4100
2	400	4600
3	350	4800
4	450	4300
5	500	5800
6	400	5600
7	600	6000
8	550	6300
9	600	6800
10	650	7200

(a) Using only the past history of aircraft sales, estimate the number of sales to expect in year 11 and year 12.

(b) Calculate the coefficient of correlation to determine if student starts are a leading indicator for aircraft sales. Assume that the sales pattern lags the starts pattern by 2 years.

(c) Develop a forecasting formula for sales based on student starts. Develop the equation

$$\text{Sales}_{\text{year } t} = a + b(\text{starts}_{\text{year } t-2})$$

and use it to find the expected sales in years 11 and 12 ($t = 11, 12$) from the student starts in years 9 and 10 ($t - 2$).

EXTENSIONS

18A Analysis of Financial Statements

A sound financial position is the bedrock upon which successful enterprises are built. Accounting practices measure the financial conditions. Accounting statements are the reports of the measurement. Interpretation of the reported figures is the responsibility of the analyst. Engineering economists may become involved in financial-statement analyses through their professional responsibilities to evaluate dealings with other companies, such as a subcontractor's capacity to meet contractual commitments or a supplier's ability to continue follow-up services, and in personal investing to evaluate the operating performance of a company.

The relation of accounts to the return on investment is broached in Figure 18.16. The left leg of the progression shows the accounts that constitute *total investment*, where *permanent investment* represents land, buildings, equipment, etc. *Sales* divided by total investment is the *turnover*, or number of times per year the capital is used; it suggests how hard the investment is being worked. This turnover times the earnings as a percent of sales gives the *return on investment*. For earnings on sales of 3 percent, as determined by subtracting the value of sales from the cost of sales and then dividing this amount by total sales, and for a turnover of 4 times per year, the annual return on investment is 12 percent.

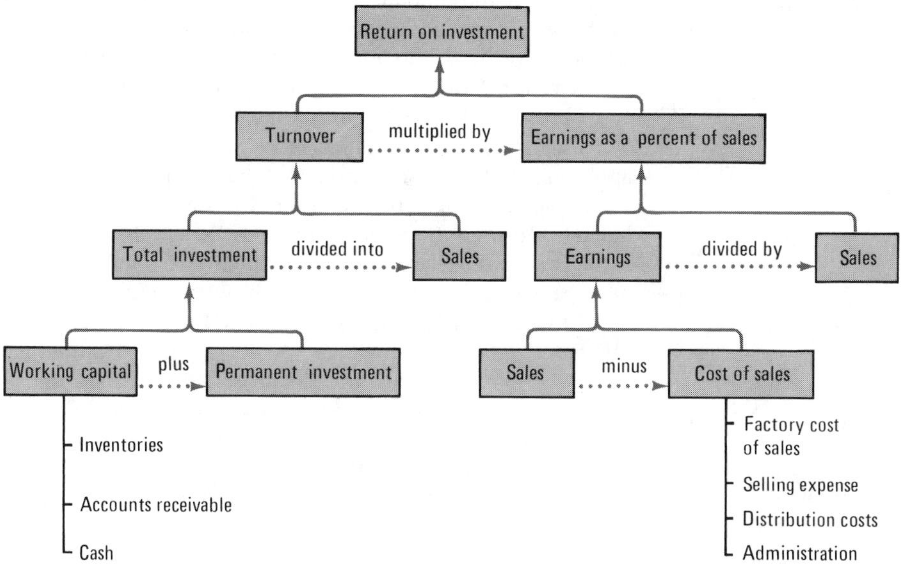

FIGURE 18.16
Relation of factors affecting return on investment.

Certain figures from accounting records have been found to be good barometric readings of a firm's future performance. A half dozen of the more prominent ratios employed to analyze financial conditions are described below and are applied to the data in Figures 18.1 and 18.3.

1 The *current ratio* gives an indication of a company's liquidity—its ability to meet current financial obligations. The current ratio applied to the balance-sheet data given in Figure 18.1 is

$$\text{Current ratio} = \frac{\text{current assets}}{\text{current liabilities}} = \frac{\$52,000}{\$49,000} = 1.06$$

which is below the level of 2:1 or 3:1 that is generally considered satisfactory. A lower ratio suggests that the company may have a cash flow problem in meeting current obligations.

2 The *acid-test ratio* is a more rigorous evaluation of liquidity made by subtracting inventories and other prepaid expenses from current assets to obtain *quick assets*—assets readily convertible into dollars. With a 1:1 ratio deemed reasonable, the firm represented in Figure 18.1 appears to have a fairly serious liquidity problem:

$$\text{Acid-test ratio} = \frac{\text{current assets} - \text{inventories}}{\text{current liabilities}}$$

$$= \frac{\$52,000 - 37,600}{\$49,000} = 0.29$$

3 The *equity ratio* indicates the dependency of a company upon its creditors for working capital:

$$\text{Equity ratio} = \frac{\text{total capital}}{\text{total liabilities and capital}}$$

Based on the data from Figure 18.1 and the assumption that retained earnings are actually owned by the stockholders rather than the corporation,

$$\text{Equity ratio} = \frac{\$124,000}{\$188,000} = \frac{\$100,000 + \$24,000}{\$64,000 + \$124,000} = 0.66$$

which is below the generally preferred lower limit of 0.75.

4 The *operating ratio* shows the percentage of every dollar received from net sales, which is needed to meet the cost of production and operations:

$$\text{Operating ratio} = \frac{\text{total expenses}}{\text{net sales}} = \frac{\$224,000}{\$247,500} = 0.905$$

A lower ratio is preferred by management for obvious reasons. The relatively high ratio for the firm represented in the income statement of Figure 18.3 suggests that cost-reduction or sales-promotion efforts should be considered.

5 *Inventory turnover* is a ratio used to evaluate the passage of goods through a firm's operations:

$$\text{Inventory turnover ratio} = \frac{\text{net sales}}{\text{inventory}}$$

$$= \frac{\$247,500}{\$37,600} = 6.6$$

where the total inventory given in the balance sheet is assumed to be the average level for the year. With typical turnover ratios of 5 to 10, the given firm appears to have a reasonable balance of inventory on hand. Desirable ratios vary widely for industries with different operating requirements.

6 A *net-profit ratio* compares the profit after taxes with sales, total assets, or stockholders' equity and is usually expressed as a percentage:

$$\text{Net-profit ratio} = \frac{\text{profit after taxes}}{\text{sales}}$$

or

$$\frac{\text{profit after taxes}}{\text{total assets}}$$

or

$$\frac{\text{profit after taxes}}{\text{stockholders' equity}}$$

$$\text{Net-profit ratio} = \frac{\$16,000}{\$247,000} = 6.5\%$$

or

$$\frac{\$16,000}{\$188,000} = 8.5\%$$

or

$$\frac{\$16,000}{\$124,000} = 12.9\%$$

With sales in the denominator, the ratio indicates the volume of business needed to earn a certain profit. When the denominator is total assets or equity, the ratio is indicative of the return on investment. The sample company appears to be providing adequate but not handsome returns.

Financial ratios are much more meaningful when they are compared with industrial averages and past ratios. Industrial averages are published by several sources to indicate norms for specific industries. A graph that plots changes in ratios with respect to time is an indicator of management effectiveness, especially when compared with an equivalent graph of industry norms. However, the ratios offer only general guidelines. A deeper investigation is needed if it is important to fully understand a company's operations. For instance, a major research effort could absorb funds that would otherwise be displayed as profit. The resulting low net-profit ratio would make the firm appear weak, whereas in reality it might have developed a strong position for future profits.

QUESTIONS

18A.1 Consolidated balance sheets and income statements for two companies engaged in similar operations are presented in Figures 18.17 and 18.18. Compare the two companies as potential investments or with respect to their financial health as potential suppliers to your company. Discuss the meaning of each ratio in terms of managerial effectiveness and the financial positions of the companies.

18A.2 Utilize the basic data in Table 18.5 to determine the values of the ratios listed below for the In-Out Universal Corporation. Industry norms are also listed. Compare the calculated ratios with the industry norms in appraising the economic health of I-O-U. Assume that the income-statement values are doubled for a full year's operation and that 10 workers are employed.

Ratio	Industry Norms
Current	2.0
Acid test	1.0
Equity	0.8
Net profit/equity	15%
Net profit/worker	$2000
Accounts-receivable turnover: $\dfrac{\text{Credit sales}}{\text{Average accounts receivable}}$	8.0

	Income Statement Year Ended December 31, 19xx	
	Salt Co.	*Pepper Co.*
Net sales	$220,000	$126,800
Cost of goods sold	108,000	55,000
Gross profit on sales	$112,000	$ 71,800
Operating expenses		
Maintenance and repairs	$ 6,000	$ 4,000
Depreciation	15,200	12,000
Bad-debt expense	3,600	3,200
Selling expense	37,000	32,000
Administrative and general . . .	12,600	11,000
Total operating expense	$74,000	$62,200
Net operating profit	$37,600	$9,600
Other income	3,000	1,600
Other expense	17,000	8,000
Net income	$23,600	$3,200

FIGURE 18.17   Income statements for Salt and Pepper Companies.

18B   Predictable Perils of Economic Forecasting

The burdens of forecasting are many. Everyone knows that there is no infallible seer, but decision makers continue to base their actions on predictions. Failures are blamed on the predictors. Success is seldom shared with the seers. Even when forecasters achieve a series of successful predictions, they realize the transitory nature of their successes—and their reputations.

Economic forecasters acquired considerable renown in the 1960s and early 1970s. Their reputation was earned by cogent analyses and sound advice. The stable economy of those years deserves some of the credit too; it was easy and accurate to predict steady growth. The more-turbulent 1980s have seen reputations falter because economic advice has often been misleading and explanations of events inconsistent. For instance, nearly all forecasters failed to call the 1981–1982 United States recession, predicted a recovery that never materialized, and then did not foresee the magnitude of the strongest business expansion to occur in more than 30 years. Comparably, no ready rationale explained why the

United States public's saving rate remained almost constant at 6 percent of disposable income when economists said it would climb after interest rates on savings accounts were raised, tax-free retirement accounts were allowed, and big income tax cuts were legislated.

Not unexpectedly, worst results were recorded for long-range, large-scale forecasts. Even the Paris-based Organization for Economic Cooperation and Development, arguably the world's premier forecasting body, missed on predictions of the early 1980's energy needs, unemployment rates, inflation rates, industrial growth, and other economic megatrends. Better results can be expected from 6- to 12-months forecasts with a limited scope, such as the demand for a class of machines by a defined industry segment within a certain region during the next year. Engineering economists are fortunate that narrowly focused forecasts meet most of their needs.

Any economic forecast can be foiled by political shocks, erratic currency values, and impulsive shifts in consumer

FIGURE 18.18
Balance sheets for Salt and Pepper Companies.

Balance Sheet
December 31, 19xx

	Salt Co.	Pepper Co.
Current assets		
Cash	$ 60,000	$ 30,000
Marketable securities	12,000	16,000
Accounts receivable	19,000	25,400
Inventories	80,000	48,000
Total current assets	$171,000	$119,400
Fixed assets		
Investments......................	$ 32,000	$ 40,000
Land, buildings, equipment (net)	346,000	251,200
Total fixed assets	$378,000	$291,200
Prepaid expenses	10,400	3,000
Total assets	$559,400	$413,600
Current liabilities		
Notes payable	$ 7,000	$16,000
Accounts payable	32,000	31,200
Other current liabilities	58,000	48,000
Total current liabilities	$97,000	$95,200
Fixed liabilities		
Mortgage payable (due in 10 years) ...	$111,600	$180,000
Total liabilities	$208,600	$275,200
Net worth		
Preferred stock (6% cumulative)	$ 84,000	$ 40,000
Common stock....................	120,000	68,000
Appropiated surplus	48,000	21,600
Unappropriated surplus	98,800	8,800
Total net worth	$350,800	$138,400
Total liabilities and net worth	$559,400	$413,600

behavior. Such surprises cannot be fully anticipated in mathematical forecasting models. Even rich and sensitive causal models fail to predict turning points when basic conditions suddenly lose stability. Experts say that more attention should be given to such noneconomic factors as social, cultural, and political issues. They also suggest that business forecasts longer than 12 months should indicate what will transpire when specific conditions emerge, such as demand for a product when inflation or interest rates reach a certain level, rather than make an overall prediction.

Regardless of the abuse directed at "crystal balling," government and industry will still require economic projections. Well-reasoned assumptions will still guide cash flow

estimates. Analysts must therefore point out the risks and decision makers should consider contingency courses of action for outcomes that fail to materialize.

QUESTIONS

18B.1 Laboratory techniques are often used to investigate economic theories and to explore market behavior. Although most studies rely on human beings, experiments conducted with rats and pigeons have investigated the relative value of money versus leisure time and tested the effects of a negative income tax (Does "free" aid reduce the incentive to be self-supporting?). Companies commonly use employees and local groups to evaluate new products. Design an experiment to determine a cash value for "peace of mind" claimed as a benefit for a public project to provide better street lighting in a city. What reservations about the experiment might be raised by critics?

18B.2 Technical forecasts are generally conceded to be more reliable than are economic forecasts. Yet Admiral William Leahy told President Truman in 1945, "The (atomic) bomb will never go off, and I speak as an expert on explosives." Thomas Edison once said, "There is no plea which will justify the use of high tension and alternating currents, either in a scientific or a commercial sense." Admitting that even experts can be wrong, discuss how technical advice should be obtained for cash flows of futuristic proposals.

CHAPTER 19

PRODUCTION ECONOMY

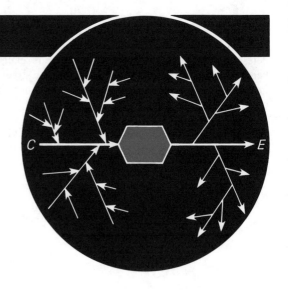

OVERVIEW

Broadly speaking, production is the process by which resources are combined to perform a service or produce a product. The resources are people, equipment, energy, facilities, and materials. *All are driven by capital.* Economy of operations is an immediate concern of engineers and managers because it directly affects the bottom line where profit versus loss is most conspicuous.

Day-to-day operating decisions are less glamorous than capital-commitment questions. Daily decisions individually have minor consequences, but collectively they determine the success of production. Because factors involved in production are so varied, no single analysis technique dominates economy studies, unless it is common sense. While there are sophisticated models for certain applications, many of the methods used to conserve time and resources are expressions of simple logic. The methods presented in this chapter involve no formal consideration of risk and require no discounting. Thus they provide simply a taste of economizing opportunities; they are appetizers for the more extensive applications of engineering and economics to production.

Ways to economize are as diverse as they are numerous. *Value engineering* involves a systematic search for ways to increase value by discovering less expensive means to fulfill necessary functions. Examples of production improvement programs include *value analysis* to control costs, *operator-machine* charts to minimize idle-time costs, and *personnel training* to reduce operating costs. In evaluating such programs, comparisons should be based on equivalent outcomes to avoid *false economy.*

A preferred level for the quality of goods and services balances the cost of attaining that quality against the value placed on the quality by consumers. Process capabilities, product design, and inspection practices are considered in minimizing the total cost of quality assurance.

ECONOMIZING

Engineering economy begins with a state of mind, a reference frame that seeks more efficient ways to utilize resources: materials, machines, processes, and people. Good intentions to conserve resources have a much better chance of succeeding when the mission is guided by proven procedures. Fresh sets of procedures are regularly unveiled. The "new" programs may simply recast old concepts in rejuvenated language, but they add vigor to the quest for economy.

Several programs and methods to encourage economy are described in this chapter. All are drawn from the same tool bin of cost-conscious concepts. Many suggestions could be classified as common sense, although they rely on uncommon ability to employ them effectively.

Value-improvement programs focus on short-run actions to better operating performance. They emphasize motivation to seek out improvements and propose ways to hasten accomplishments. They lead eventually to the question of which improvements deserve implementation. Will the new way detract from traditional values as it reduces cost? Is the new way good enough, or should further improvement be sought? How high is the cost of high quality? What are the causes and effects of the new way of doing things?

Economizing techniques in this chapter feature thrifty and efficient operations—conservation of capital—rather than capital investment.

VALUE ENGINEERING

The design of a new product, process, or service offers an exceptional opportunity to economize. It is a chance to start anew, build better from experience, rectify old mistakes with refinements, utilize the latest technological developments, and break away from stagnating traditions or customs. The search for and selection of new means to reduce cost and improve value during the design phase is the forte of *value engineering* (VE). The concepts involved grew from government-sponsored cost-prevention campaigns in the 1940s to become a recognized discipline with a national organization called Society of American Value Engineering, cleverly initialed SAVE.

Value engineering is an organized effort to identify and eliminate unnecessary costs without sacrificing quality or reliability. Some larger companies have slots in their tables of organization for value-engineering staffs. In other companies the VE function is less formally performed by engineers in research and development, design groups, and plant-engineering departments. The innovative approaches forwarded by value engineering are equally applicable to service organizations and individual economizing.

Types of Economic Value	
Use-value	The properties or qualities which accomplish a use, work, or service
Esteem value	The properties of an object which make its ownership desirable
Cost value	The sum of labor, material, overhead, and other costs required to produce economic value
Exchange value	The qualities of an object that make it possible to procure other items in its place

Reasons for Unnecessary Costs	
Incomplete information	Failure to gather all the facts
Lack of an idea	Failure to explore all possible ways of performing a service or making a product
Honest wrong beliefs	Decisions made on what is believed to be true, and not on the facts
Habits and attitudes	Habits which take us where we were yesterday and attitudes which tend to keep us there

FIGURE 19.1 Economic values and unnecessary costs.

Value

The methodology of value engineering is intended to (1) identify the function of a product or service, (2) establish a value for that function, and (3) develop a means to provide value. Value is established by comparison, and by no other means. It is not assumed to be an automatic ingredient of any design. Since many of today's products and systems are designed by teams, design groups cannot always devote detailed attention to all subassemblies. This leads to suboptimization and subassemblies designated by standard or rote practices. The end result may be a decrease in economic value and an increase in unnecessary costs. Figure 19.1 lists four types of economic value and four reasons for unnecessary costs.

Value engineering emphasizes *use* value. The value of anything can be no more than the value of the function it performs. Unnecessary costs arise when there is a less-expensive method that performs the same function without downgrading quality.

Function

A value-engineering study starts with the identification of an item's primary function by means of a verb and a noun. According to this technique, the primary function of a table would be to "support weight," and that of a shipping container would be to "provide protection." Any secondary functions are also identified; for example, a shipping container might also "attract sales." All subassemblies, such as a zip opener for a container, should contribute to the value of either the primary or secondary function.

Next, a value is assigned to the function by comparing it with something else that will do the same task. Areas of costs are evaluated to determine where current quality standards or expenses are excessive. A search is made for alternative methods that preserve the desired quality. For instance, a manufacturer had a contract for engines ranging from 10 to 20 horsepower.

Creativity is a valuable talent for value engineering. Methods for increasing innovative practices were discussed in Extension 1A.

Under the terms of the contract the engines were to be packed in individual crates. The requirement appeared to be unnecessarily costly. After a study of the problem, a proposal was submitted to the buyer to use returnable steel racks in lieu of the crates and to cover the six engines on each rack with a plastic dust cover. The proposal was accepted, and significant savings resulted.

If the noun in the function description has a measurable parameter, such as weight or length, a mathematical evaluation is feasible. The key to comparison is an assignment of dollar values to the parameters of equivalent quality. When the comparison looks favorable, there is still the problem of implementation. As a rule, any change will meet resistance. The effort required to produce a usable idea often pales into insignificance when compared with the effort required to sell it. In a sound study, objections are anticipated and means to overcome them are included.

Another example of value engineering concerns a lock on a small compartment door that appeared unnecessarily expensive. The door was on the outside of a newly designed truck cab and led to a tool compartment. The original design called for a tool-actuated lock which cost $16.54. Its functional description was "secures door." The search for alternatives revealed a spring lock that served the purpose equally well and cost $6.22. A further search led to a clip fastener that cost only $0.84. Finally, the whole assembly was questioned and it was determined that there was no need even to have a door, because the tool-storage compartment was readily accessible from inside the cab. The study led to a total saving of $52.84 per unit.

The cost of a VE study must be deducted from the savings derived from the study. In order to receive maximum return for the effort expended, studies should be concentrated first on items for which the largest expenditures are planned. There may be many candidates. A selection is made by examining the factors involved in a study by means of the formula

$$\frac{\text{Potential savings}}{\text{Study and implementation costs}} \times \frac{\text{probability of}}{\text{implementation}} = \text{rating factor}$$

Items with the highest ratings are given first attention, and studies with lower cost-reduction potential are made when time permits.

Example 19.1
T-Chart Comparison of Design Alternatives

Unusual solutions to usual problems are characteristic of value engineering. Techniques such as brainstorming and synectics are employed to enhance engineering creativity. A brainstorming session could very well generate a hundred different ways to perform a certain function. Some of the ideas are probably so wild and impractical that they can be easily discarded. Other ideas are filtered through a screening process to test their feasibility. One method of screening is to compare each new alternative with the existing or conventional method by means of a T-chart. The T-chart consists of a list of criteria that an acceptable solution should satisfy and a place to mark whether the new design is better or worse than another design in reference to each criterion. A sample T-chart is shown on page 560.

Design X: Present Design	Better	Worse
Initial cost	✔	
Operating cost		✔
Reliability		✔
Appearance	✔	
Comfort		✔

Check marks, as used in the sample T-chart, are usually adequate to eliminate less-desirable ideas in a preliminary screening. Design *X* in the T-chart above would probably be rejected because its lower first cost and more attractive appearance are outweighed by the lower reliability, less comfort, and higher operating costs of the existing design. A finer T-chart screen requires that percentages replace check marks in the comparison.

Develop a T-chart to evaluate the following suggestion for a company-sponsored plan to conserve energy and assist employees commuting to and from work.

The company will purchase 12-passenger deluxe vans and recruit employees to drive them. The fees paid by the first 10 passengers will be allocated to gasoline, upkeep, and payment for the van over a 5-year period. The driver will get to keep the fees paid daily by the eleventh and twelfth passengers and have free use of the van at night and on weekends. The company will assist in signing people up for this *Vanpool* service and will provide reserved parking adjacent to the plant. There is no public transportation available at the plant location.

Solution 19.1

The Vanpool plan has been exceptionally well received in actual practice. Initial reports indicate that the vans are popular because they are comfortable (air-conditioned, carpeted, stereo-equipped, etc.) and the passengers enjoy the compan-ionship (some drivers even installed a bar in the back to serve coffee in the morning and cocktails on the way home). The T-chart comparison of commuting by individual cars against Vanpool commuting might appear as follows.

Vanpool Plan versus Private Cars	Better	Worse
Conservation of energy (fuel and materials)	✔	
Cost of commuting (to commuters)	✔	
Flexibility of transportation (emergencies)		✔
Condition of commuters arriving for work	✔	
Convenience (to commuters)		✔
Parking facilities (adequacy and upkeep)	✔	

VALUE-IMPROVEMENT PROGRAMS

The cause-and-effect diagram introduced in Extension 19A is useful for detecting and examining value-improvement opportunities.

Any operation can be improved. What was thought of as being the ultimate in efficiency a few years ago is now commonplace or outdated. Further improvements are not always dramatic leaps ahead, but multiple small refinements can gradually fashion a breakthrough. Greater value might result from a minor design change that makes a product more useful to the consumer or easier to build by the producer. A small change in the layout of the workplace can make operations more convenient or safer for the operator. A larger investment in robots can reduce the exposure of humans to dangerous or distasteful work.

The criterion of acceptability of all these engineered improvements is that the value gained from each change must be greater than the cost incurred.

Materials

The control of material costs is becoming more significant as higher rates of inflation boost prices from one period to the next and resources become scarce owing to either temporary shortages or dwindling world supplies. The potentials for material savings are manifold; the challenge is to discover those that retain the required output quality but lower cost.

Minimum-cost inventory policies are discussed later in this chapter. These policies are important but constitute only one aspect of material cost control. *Value analysis* is a name given to studies that determine how appropriate purchased materials are for the functions they are to perform. The analysis is very similar to value engineering, because it seeks answers to questions such as: Could a less-expensive substitute be used that would still perform the required task?

Engineers are conventionally responsible for specifying materials for products. Sometimes they are guilty of *overdesign*—specifying closer tolerances, stronger construction, or costlier materials than are actually needed.

Weight, strength, machineability, and appearance are typical selection alternatives for materials. For instance, a building contractor comparing galvanized-steel gutters with aluminum gutters faces a material-selection decision. Both materials are adequately strong and rust- or corrosion-resistant. The desired box-type gutter is available in both materials. Appearance is less important because the gutters will be painted. The main differences have to do with weight and price. Weight affects installation and shipping costs. On-site costs for a 10-foot (3.05-meter) trough section would amount to:

	Basic Cost	Weight, lb	Difference in Cost due to Weight	Cost Comparison
Aluminum	$13.20	4 (1.81 kg)		$13.20
Steel	11.40	7.5 (3.40 kg)	$1.14	12.54

This cost comparison could change abruptly if the gutters had to be shipped a greater distance. The 3.5-pound (1.59-kilogram) difference per section would cause a major cost contrast. A building site in Alaska could change the outcome ratings to:

	Basic Cost		Shipping Cost		Installation Cost		Total-Cost Comparison
Aluminum	$13.20	+	$1.92	+	$2.46	=	$17.58
Steel	11.40	+	3.57	+	3.42	=	18.39

Machines

To get the maximum value from a machine, it must be used for the proper purpose and be kept in operation. Idle machines, like idle materials, represent capital that is not earning any returns. And machines, like people, have seemingly individual characteristics. Some machines do more precise work than do others. Some take a long period to start or set up. Others have frequent breakdowns and are unreliable. Operating costs vary widely. Because of these diverse characteristics, the assignment of work to different machines can be frustrating.

Most machine considerations involve time: machine running time, setup time, checking time, and operators' times.

Ways to minimize idle time of resources have long been of interest, probably because an idle resource is such an obvious exemplification of waste. Solution methods range from queuing theory (Chapter 23) to block diagrams. *Operator-machine charts* fit the latter category. These charts are used to display the relation between operator and machine time when one operator or a crew operates a machine or a battery of machines. The operations required to keep the machines producing form a repetitive cycle. Each cycle is represented in a chart by a schedule of operations for the operators and the machines.

Queuing theory is a discipline for mathematically analyzing waiting time in queues. It is presented in Extension 23A

Assume that a large number of semiautomatic machines producing identical products are serviced by operators who follow the repetitive operating sequence given below.

Operation	Standard Time, Minutes
Load machine	1.2
Remove finished unit	0.4
Inspect finished unit	2.3
Package finished unit	3.1
Walk to next machine	0.3

One unit of output is produced during each machine cycle of 17.4 minutes. As shown in Figure 19.2, the proportion of idle time between operator and machine varies with the number of machines serviced by an operator.

The most economical operator-machine ratio is determined from the relative costs of idle time. If the machine burden rate (maintenance, depreciation, operating cost, etc.) is $30 per hour and the labor rate for an operator is $12.60 per hour, the minimum cost per unit is calculated as

Minimum machine cycle time = loading-unloading time + running time
$$= 1.6 + 17.4 = 19 \text{ min}$$
Minimum operator cycle time = loading-unloading time + handling and travel time
$$= 1.6 + 5.7 = 7.3 \text{ min}$$

which can be used to roughly fix the solution area:

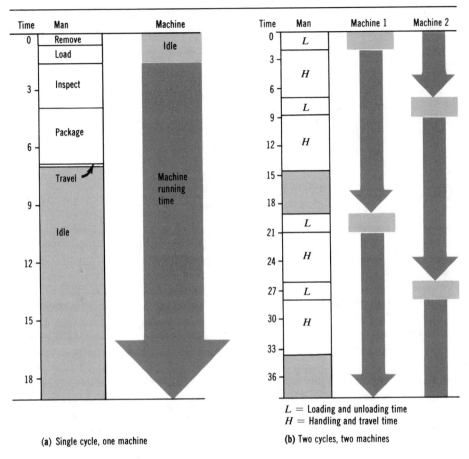

FIGURE 19.2
Operator-machine charts for one
and two machines serviced by one
operator.

L = Loading and unloading time
H = Handling and travel time

(a) Single cycle, one machine

(b) Two cycles, two machines

$$\text{Approximate number of machines} = \frac{19}{7.3} = 2.6$$

If one operator services two machines, the cycle time is set by the machine's running time, 19 minutes. The unit cost is

Labor: $(19 \text{ min}) \dfrac{\$12.60}{60 \text{ min}} = \$ \ 3.99$

Burden: $19(2) \dfrac{\$30.00}{60 \text{ min}} = \$19.00$

Total cost per cycle = $\$22.99$

Unit cost = $\dfrac{\$22.99}{2} = \$11.50$

If one operator services three machines, the cycle time is set by the required operator's time, $3 \times 7.3 = 21.9$ minutes. The unit cost is

Labor: $\qquad (3 \times 7.3) \dfrac{\$12.60}{60} = \$\ 4.60$

Burden: $\qquad (21.9)(3) \dfrac{\$30.00}{60} = \$32.85$

Unit-cost evaluations are similar to average-cost analyses (discussed in Chapter 15) applied to individual assets or operations.

$$\text{Total cost per cycle} = \$37.45$$

$$\text{Unit cost} = \frac{\$37.45}{3} = \$12.48$$

As is apparent in the preceding example, *unit cost* is the periodic total cost divided by the units of output. Unit cost is a convenient measuring scale by which to compare different machines, methods, and services.

Example 19.2
A Robot or Not, That Is the Question

Recent availability of more sophisticated machines makes possible the substitution of automatic equipment for human workers in an expanding variety of situations. Both white- and blue-collar workers are subject to replacement by steel-collar workers. Machines have obvious advantages for certain applications. They can work in a hostile environment, continue repetitious tasks with consistent quality for 24 hours a day, and exceed humans in most motor-performance capabilities plus some mental tasks. But humans are more versatile, creative, and generally less expensive.

Many emotional issues cloud decisions to employ robots or other machines in place of humans. A straightforward economic comparison equates the cost of doing a task with humans to machine cost for the same work output. The output must be equivalent, of course, or economic compensation must be allotted for faster, better, or more accomplishments. A complete tabulation of costs over a machine's life must include maintenance, insurance, taxes, purchase and installation price,

operating costs, and other expenses. Human costs include direct and indirect wages, training, special facilities, and many incidental expenses associated with operating foibles. More difficult to determine are social costs for people whose jobs have been lost to computer-assisted automatons.

An evaluation of proposals to increase output by either adding workers or raising capital investment is locally less sensitive than is a proposal to replace workers with machines. Consider the data in Table 19.1. The firm manufactures 1000 units per month but intends to double the output to meet rising market demand. Current monthly income is $100,000 and labor cost is $30,000. Production can be doubled by doubling the number of workers (proposal W) or by purchasing robots and automated material-handling equipment that has a monthly capital recovery cost of $60,000 (proposal R). If the operation is automated, no additional workers will be hired.

Evaluate the proposals.

Solution 19.2

If the proposals are compared on the basis of unit cost, the current $70,000/1000 units = $70/unit is lower than proposal R's $150,000/2000 units = $75/unit but higher than proposal W's $60/unit. However, both proposals provide a larger monthly profit than does the current operation:

$$\begin{aligned} \text{Profit (current)} &= \$100{,}000 - \$70{,}000 \\ &= \$30{,}000 \\ \text{Profit (proposal } W) &= \$200{,}000 - \$120{,}000 \\ &= \$80{,}000 \end{aligned}$$

TABLE 19.1			

Monthly revenue and cost breakdown for the current operation and two options for increasing capacity

	Current	Proposal W	Proposal R
Sales volume (units)	1,000	2,000	2,000
Total revenue (P = $100/unit)	$100,000	$200,000	$200,000
Total labor cost	30,000	60,000	30,000
Material and other variable costs	20,000	40,000	40,000
Overhead cost	20,000	20,000	20,000
Capital recovery cost			60,000
Total cost	$70,000	$120,000	$150,000

$$\text{Profit (proposal } R) = \$200,000 - \$150,000$$
$$= \$50,000$$

Therefore, if there is no limit on the number of qualified workers that can be hired, proposal W is preferred. Oppositely,

if a labor shortage precludes implementation of proposal W, proposal R should be accepted because it yields greater profit than does the present manufacturing operation, even though the unit cost will rise.

People

The most-conspicuous, best-studied, yet least-comprehensible resource in any system is the human resource. Materials and machines obey the rules of physics and are predictable. People are not. Since the behavior of large groups of people is more predictable than is individual behavior, collective responses to particular conditions allow the formulation of macroeconomic "laws." Deviations become more pronounced as the group gets smaller. The response by any given individual to an economic incentive may fall anywhere within the spectrum of possibilities. Because engineering economists deal with both individuals and groups under shifting economic circumstances, they should never be bored by lack of variety.

Just as individual opinions vary, capabilities also vary. Unit costing can measure what an operator has done before, but it may not be accurate for tomorrow's performance. Uniformity and consistency are encouraged by training and motivation programs.

Improvement programs are packaged under many names and are backed by a varied host of supporters. Most programs do spur improved performance, at least for a short time, in the way a pep talk arouses go-get'em emotions before a contest. Some programs rely on researched principles of information feedback on past performance to improve future performance, and others focus on changing the environment to affect worker behavior. Regardless of

Improvements that result from longer experience on tasks or projects tend to follow a predictable pattern, called a learning curve, that is discussed in Extension 19B.

FIGURE 19.3
General trend of costs and savings caused by an improvement program.

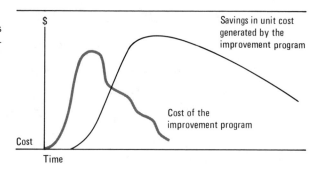

the methods used, the value of the program ultimately rests on the difference between the output before the program was instituted and the output afterward.

Figure 19.3 indicates the general pattern of costs for conducting a performance-improvement program and savings owed to it. Savings generated by training or exposure usually lag the presentation expenses but drop off more slowly.

Variations in the cost-savings pattern for value-improvement programs make economic analyses difficult, because measurements taken at different periods exhibit different saving-to-cost ratios as do ratios of total savings to total costs over different time intervals. A yardstick of effectiveness for programs to improve personnel practices is how long it takes for the savings resulting from a program to equal its cost. This payback rating can be used to compare trials of different programs to determine which ones deserve continuation. In selecting which programs to try initially, an adaptation of the formula used for selecting value-engineering studies is useful:

$$\text{Trial improvement program rating factor} = \frac{\text{Potential savings per period}}{\text{Cost of improvement program per period}} \times \begin{array}{l} \text{probability of success} \\ \text{within the trial period} \end{array}$$

FALSE ECONOMY

Decision makers can outsmart themselves. In a rush to get the best deal, they can be distracted from the main issue by the attraction of a "bargain." A practice of making comparisons strictly on the basis of immediate cost will inevitably lead to some long-run disappointments. Comparisons limited to input factors are not complete unless the outputs resulting from all the alternatives are essentially equivalent.

Wage payments demonstrate two sides of comparability. The rewards for extreme proficiency are most apparent in the "star" ratings of the entertainment industry. Famous people receive great remuneration from motion pictures, television, recordings, book writing, and athletics. Stars deserve extravagant rewards if their presence is repaid by proportional increases in revenue. A major-league baseball player could expect to earn at least a million dollars a year if his batting average were .400 and his presence might increase the club's ticket sales by an amount equal to his big salary. Yet the star cannot play the game alone. Should the rest of the players on the team be compensated in proportion to their proficiency measured against the star's? If so, a .100 hitter would earn (.100/.400)($1,000,000) = $250,000. That a .100 hitter cannot even stay in the league is evidence that comparisons based on proficiency alone are inadequate. Yet it warrants consideration along with other factors.

Consider a similar example from manufacturing. A worker receives an hourly wage of $12, and the indirect costs of employment are 50 percent of wages. The worker operates a machine with a burden rate of $42 per hour and averages 10 pieces of output per hour. The resulting unit cost is

$$\text{Unit cost} = \frac{\$12 + (0.5 \times \$12) + \$42}{10} = \$6.00/\text{piece}$$

Another worker uses the same type of machine but produces only 7 pieces per hour. An equivalent pay-scale calculation shows that this worker's wage W should be such that

$$\frac{W + 0.5W + \$42}{7} = \$6/\text{piece}$$

or

$$W = \frac{\$42 - \$42}{1.5} = \$0/\text{hour}$$

which means that on a relative basis, the second worker deserves no pay.

While this example exaggerates the equivalent-pay concept, it does highlight the problem of proper rewards for greater productivity. Standardized hourly wage contracts, step-graded annual salaries, guaranteed annual wages, minimum wages, and other payment plans that are not tied to output are deemed socially beneficial by custom, unions, and Congress. But they severely restrict incentives for productivity improvement. However, when wage adjustments for proficiency are possible, it is always difficult to measure the *value added* by a specific operator because all workers depend on an extensive support system which has to function correctly to sustain any output.

Measurement of work performance and utilization of resources, and their relationship to profitability, are discussed in Chapter 21.

Example 19.3
When Per-Unit Profit Is Overwhelmed by Overhead

A small company makes only two products, A and B. Each product requires 4 hours to manufacture and only one unit of each is produced per day. The cost-accounting data shown in the table indicates a breakeven condition in which labor cost is divided equally between the products, and the overhead charges are allocated in proportion to the selling price.

Could total profit be increased by producing only product A?

	Product A (per unit)	Product B (per unit)
Selling	$2000	$3000
Material and variable cost	400	1200
Labor cost ($1400 per day)	700	700
Overhead cost ($2000/day)	800	1200
Profit per unit	$100	−$100

Solution 19.3

An operating loss would occur if two units of product A were produced every day:

$$\text{Profit (two units of product } A/\text{day)} = 2(\$2000) - 2(\$400)$$
$$- \$1400 - \$2000$$
$$= -\$200$$

The reason for the apparent switch in profitability is that overhead cost is proportioned to selling price rather than a more appropriate production input.

A better indication of profitability is obtained by calculating the gross margin that would result from doubling the production of either product: gross margin = marginal profit − fixed cost.

$$\text{Gross margin (product } A) = 2(\$2000 - \$400) - \$3400$$
$$= -\$200$$
$$\text{Gross margin (product } B) = 2(\$3000 - \$1200) - \$3400$$
$$= \$200$$

This confirms that per-unit profit falsely suggests exclusive production of product A. Instead, manufacturing only product B would increase daily profit to $200 from its present net zero.

MINIMUM-COST OPERATIONS

Engineers are continually confronted with design problems of a type suggested by Figure 19.4: If you vary a certain parameter, the change causes one characteristic to improve and simultaneously causes another characteristic to get worse. It is the engineering equivalent of the adage that there are two sides to every question. This chapter deals with the question of what design level produces the minimum total cost when some of the operating costs vary directly and others inversely to changes in the design variable.

Five factors are considered in a minimum-cost analysis. Their relationship is evident in Figure 19.4 and is expressed by the formula

$$C = Ax + \frac{B}{x} + K$$

A multitude of real-world conditions are represented by this minimum-cost model. The design variable can have dimensions such as money, length, weight,

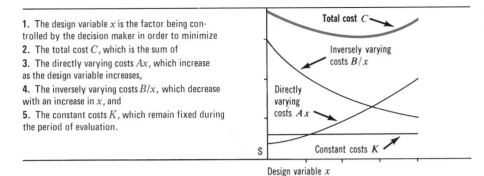

1. The design variable x is the factor being controlled by the decision maker in order to minimize
2. The total cost C, which is the sum of
3. The directly varying costs Ax, which increase as the design variable increases,
4. The inversely varying costs B/x, which decrease with an increase in x, and
5. The constant costs K, which remain fixed during the period of evaluation.

FIGURE 19.4
Minimum-cost factors.

number of parts, and time, or combined units such as speed. Design problems naturally get more complex when several design variables are involved. Only the easiest case of one variable affected by two opposing cost functions will be considered here. However, the relatively simple case covers many problems commonly met in engineering practice, and it sets the framework for handling more complicated situations.

There are two ways to calculate the minimum-cost point. Both can be supplemented by graphs, which are less precise for a solution but which excel for explanation and discussion purposes.

Classic examples of inversely varying costs are Lord Kelvin's optimum wire-size formula (Problem 19.16) and inventory lot size models described in Chapter 20.

Tabular Approach

A tabular method is appropriate for a design variable that changes in incremental steps. Cumulative direct, inverse, and total costs are tabulated for consecutive changes in the design variable. The tabulation ends when a minimum point is evident in the total-cost column. When the cost patterns consistently rise or fall as a function of the design variable, like the curves in Figure 19.4, the minimum-cost level is identified as the point preceding the first upturn in total cost. If the directly or inversely varying costs behave erratically along their general trend line, it is necessary to check several points beyond the first minimum.

Formula Approach

Equations can be used to find the minimum-cost point when the direct cost and inverse cost are proportional to the design variable. The total-cost expression, $C = Ax + B/x + K$, is subjected to the same mathematical manipulations used in nonlinear breakeven analysis. Since the minimum point on a continuous curve occurs where the slope of the curve is zero, the total-cost formula is differentiated to find the slope, and the derivative is set equal to zero:

$$\frac{dC}{dx} = \frac{d(Ax + Bx^{-1} + K)}{dx} = 0 \qquad A, B, \text{and } K \text{ are constants}$$

$$= A - \frac{B}{x^2} = 0$$

which results in

$$x = \sqrt{\frac{B}{A}}$$

at the point where total cost is lowest. At this point the directly varying costs are equal to the inversely varying costs, as is apparent when the minimum-cost value of X is substituted into the total-cost equation:

$$C_{\min} = AX' + \frac{B}{X'} + K \qquad X' = \text{minimum-cost point} = \sqrt{\frac{B}{A}}$$

$$= A\sqrt{\frac{B}{A}} + \frac{B}{\sqrt{B/A}} + K$$

$$= \sqrt{AB} + \sqrt{AB} + K$$

VIGILANCE VERSUS ERRORS

Most individuals and organizations take pride in the quality of their output. Higher quality tends to be automatically equated with superior performance, which is assumed to produce more profit. This relation usually holds true, but there are occasions when the cost of higher quality is not repaid from the receipts earned by the achievement.

Both service and manufacturing quality start with a process capable of producing to the design specifications and continue with a quality-assurance program that ascertains if standards are being met. The initial decision concerning specifications is based on what customers supposedly want. It is moderated by the realization of what the supplier can provide. These two considerations can take the shapes shown in Figure 19.5, in which the customer's perceived value of a service or product levels off at higher degrees of refinement while the cost of providing refinements continues to increase.

A supplier's calibrated-to-profit perspective focuses on the difference between perceived value and the cost-refinement capabilities of the process. Suppliers with high refinement capabilities might choose to advertise their highly refined offerings with the intent of educating customers to appreciate

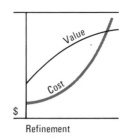

FIGURE 19.5
Product value and cost as a function of refinement.

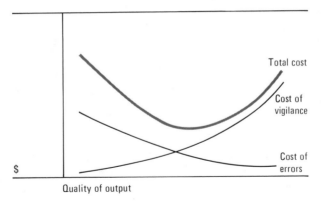

FIGURE 19.6
Relation of vigilance to errors in
quality assurance.

them. Wine, automobile, and clothing advertisements abound with examples
of persuasions to buy high quality at a high price.

Another view of value-cost relations centers on cost. For a given degree
of refinement there are many contributing factors affecting cost. One of the
most critical is the expense of quality assurance. A minimum-cost quality-
assurance program balances the cost of maintaining a desired level of quality
(vigilance) against the cost of failure (errors). In most production situations
the cost of vigilance varies directly with the excellence of output, and the cost
of errors varies inversely, as shown in Figure 19.6.

Cost of Inferior Quality

Failure to meet quality standards creates internal costs for the production
process and external costs associated with the output. Internal costs include
labor, material, and overhead charges on scrapped products; the expense of
activities to rework products to make them acceptable; disruptions in the
production process caused by quality deficiencies; and the engineering time to
correct deficiencies. External costs are due mainly to loss of consumer
confidence and the complaint or field services designed to placate customers.

Enormous attention has been focused on quality in this decade as
companies attempt to implement *just-in-time production* (also called Kanban
and zero-inventory production). The goal is to remove all in-process inventory
by having parts and materials arrive at a work station just when they are
needed. Not only are storage costs slashed, but less floor space is needed for
the same level of production and the number of employees can be reduced
because there is less material handling. Extremely high quality is critical to
just-in-time production. With no stockpiles to cushion delays and compensate
for defective workmanship, arriving parts and material of poor quality immediately hurt the production process.

Intense competition in some
industries has concurrently driven
costs down and quality up, often
to the surprise of the competitors
and much to the pleasure of
consumers.

Example 19.4
Damage from Defectives Depends on Demand

A manufacturer makes only one kind of product, producing 20,000 units per month, 20 percent of which are defective. Each acceptable unit sells for $20; it requires $8 of direct material and other variable costs. Salaries, indirect expense, and depreciation charges per unit are, respectively, $3.00, $1.60, and $2.40. Defective products are discarded with no net scrap value. Ignoring taxes, determine what loss is caused by inferior quality.

Solution 19.4

The cost of defective units depends on how many products can be sold. If monthly demand for good units is only 16,000, which is met by manufacturing 20,000 units with the expectation that 4000 of them will be defective, the cost of inferior quality is only the variable cost per unit of defective production:

If the plant could sell all it produced, the loss would be $20 for each defective unit produced:

Cost when demand exceeds output
$$= \$20/\text{unit} \times 4000 \text{ units} = \$80,000$$

Cost when good output equals demand
$$= \$8/\text{unit} \times 4000 \text{ units} = \$32,000$$

Cost of Prevention

Errors can be prevented throughout the production process. The first opportunity is in the design of the product. Designers must beware of tolerances and standards that are unnecessarily restrictive; tight tolerances introduce more chances for error, and unrealistic standards can raise enforcement costs prohibitively. Clever engineering designs may even reduce the cost of production errors by making their correction and prevention easier.

Deviations from design specifications are detected by inspection. Materials purchased for the production process are inspected before they are accepted; in-process inspections confirm quality during the production process; and output is inspected before it is released to customers. Higher inspection costs normally provide a greater degree of protection from the effects of defective output, but the marginal cost of detection rises sharply as the percentage of undetected errors approaches zero. A point is usually reached where further improvement is not worthwhile. For instance, if 98 percent viability of grass seeds is considered acceptable, decreasing the proportion of nonviable seeds below 2 percent by inspection would be uneconomical unless the higher quality is rewarded with a higher price.

For occasional situations in which exceptional reliability is demanded, quality is pursued at any cost, e.g., space vehicles and medical supplies, such as an artificial heart.

Quality-control and design engineers are thus exposed to economic questions of vigilance versus errors. Would it be better to set a standard that can be held within stated tolerances 90 percent of the time at a lower cost or accept a higher cost to have 95 percent meet the standard? Should a higher-cost process be used to consistently produce to a certain degree of excellence or should a lower-cost process be coupled with intensive inspection to produce the same degree of quality? Answers are straightforward when complete information is available, but costs often involve human value systems and performances that are difficult to verify and subject to abrupt changes.

Example 19.5
Expectations from Inspections

A manufacturing plant makes 600 Wambos each day to get 400 good ones. Labor costs are $3 per good unit of output, and daily overhead is $1200. A Wambo goes through two stages of processing which require an equal amount of labor. Inspection currently takes place after the second stage. The 200 defective units are melted and returned to the first stage as a raw-material input at approximately the same cost as other raw materials.

A new inspection plan is proposed to add an intermediate inspection point at the end of the first stage to detect defectives before they are processed through the second stage. The plan will add $150 per day to overhead costs and is expected to reject one-sixth of the flow (another one-sixth will be rejected at the final inspection). If plant capacity is limited to the present level of material input, should the plan be implemented?

Solution 19.5

Under present conditions, material for 600 Wambos is processed daily, 400 units from new raw material and 200 units from reject material. The new inspection plan will not improve the capability of the process, but it lowers labor costs because less labor is spent on rejected units. The present flow pattern of raw-material sources, process stages, and output is:

Since labor costs are evenly divided between the two stages, $400 \times \$3/2 = \$600$ is spent at each stage, or $1 per unit processed. If the output is maintained at 400 units per day, the input X when one-sixth of the flow is rejected at each of two stages is

$$X \frac{5}{6} \frac{5}{6} = 400 \quad \text{and} \quad X = 576 \text{ units}$$

which means only $5/6 (576) = 480$ go through the second stage. At $1 per unit, the labor savings are $400 \times \$3 -$ $(576 + 480)\$1 = \$144$ per day. Under this arrangement the plan would be turned down.

But if the flow were kept at the maximum level of 600, $600 \times 1/6 = 100$ units would be rejected at the first inspection and $(600 - 100) 1/6 = 83$ at the second. Therefore the raw material input would equal the output at $600 - 100 - 83 = 417$ units per day. In this pattern labor costs at $1 per unit processed decline by $100, but the 17 units of increased output valued at $3 per unit would contribute $17 \times \$3 = \$51$ to make total savings of $\$100 + \$51 = \$151$. The cost of the extra inspection is thus recovered, barely.

Review Exercises and Discussions

Two types of machines with the capacities and costs shown in Table 19.2 are capable of producing a certain kind of unit. The annual demand for these units is 45,000. If all   **Exercise 1**

TABLE 19.2

	Type S	Type L
Depreciation, taxes, insurance/year	$7500	$16,800
Labor rate of operator/hour	$10.50	$11.20
Power consumption/hour	$0.30	$0.50
Production rate/hour, finished units	10	15

production is done during the normal working year of 250 eight-hour days and the machines are not needed for any other production, what type of machine or combination of types should be used?

Solution 1   Unit costs for the three ways to meet the annual demand for 45,000 units with the two types of machines are calculated as

$$\text{Unit cost (three } S\text{-type machines)} = \frac{(\$10.80)(3)}{10 \times 3} + \frac{(\$7500)(3)}{45,000} = \$1.58/\text{unit}$$

$$\text{Unit cost (two } L\text{-type machines)} = \frac{(\$11.70)(2)}{15 \times 2} + \frac{(\$16,800(2)}{45,000} = \$1.53/\text{unit}$$

The unit cost using one S-type machine plus one L-type machine is determined by first checking which machine has the lowest unit cost when used to capacity:

$$\text{Unit cost}_{S\text{-type}} = \frac{\$10.80}{10} + \frac{\$7,500}{10 \times 8 \times 250} = \$1.455/\text{unit}$$

$$\text{Unit cost}_{L\text{-type}} = \frac{\$11.70}{15} + \frac{\$16,800}{15 \times 8 \times 250} = \$1.34/\text{unit}$$

Thus, the L-type machine should produce 30,000 of the 45,000 units required and the S-type machine the remainder to give

$$\text{Average unit cost (one machine of each type)} = \frac{1}{2}\left(\frac{\$10.80}{10} + \frac{\$7500}{15,000} + \$1.34\right)$$

$$= \$1.46/\text{unit}$$

which is the lowest-cost alternative.

Exercise 2   New improvement programs for quality and safety are being considered.

A "Down with Defects" promotion would have a big kickoff rally with banners and speeches, where workers would pledge how much they expect to reduce defective output during the 6-month program. Individual pledgers who successfully achieve their defect-reduction goals would receive publicity and rewards. Potential savings from defect reduction are $1 million per year, but it is estimated that there is only a 50 percent chance of reaping half the potential. The promotion would cost $100,000.

An "Avoid Accidents" program would involve hiring another safety engineer and increasing the safety budget. New classes would be developed to explain unsafe acts associated with each type of job and all the workplaces would be visited to detect unsafe conditions. A better safety record would reduce insurance costs and avoid work disruptions to save $265,000 annually, but slower production due to workers' anxiety about being careful might reduce output by $25,000 per year. The accident-reduction program would cost $60,000 per year, and the probability of success is 80 percent.

Assuming that the estimates are not overly optimistic, indicate what you would recommend.

Using the program-selection formula to develop rating factors for the two proposals, we get

Solution 2

Down with Defects:
$$\frac{\$1,000,000 \times 0.5}{\$100,000} 0.5 = 2.5$$

Avoid Accidents:
$$\frac{\$265,000 - \$25,000}{\$60,000} 0.8 = 3.2$$

which indicates a preference for the safety program even though the potential savings are over twice as great for the defect-reduction program. The lower presentation cost and higher probability of success suggest that the safety program should be done first. It is usually not as productive to have more than one major improvement program underway at one time.

Operator 1 has an hourly wage of $12 and produces 50 parts each hour on a machine with a $28-per-hour burden rate. The raw material in each part costs $1.00 and if spoiled, its scrap value is $0.18. This operator has a spoilage rate of 10 percent.

Exercise 3

Operator 2 does the identical job for the same pay and has the same spoilage rate but works slower and always detects a ruined part by the time the operation that produces it is 50 percent complete.

How many parts per hour should operator 2 produce to have the same unit cost as operator 1?

The equation for unit cost is

Solution 3

$$\text{Unit cost} = \frac{W}{N} + \frac{M}{N} + \text{SR} \frac{\text{RM} - \text{SV} + (W/N + M/N)K}{100\% - \text{SR}}$$

where W = wage = $12/hour
M = machine burden = $28/hour
SV = scrap value = $0.18/part
K = percent of operation completed when scrap is detected
  = 100% for operator 1 and 50% for operator 2
SR = scrap rate = 10%
RM = raw materials = $1.00/part
N = units/hour = 50 for operator 1

The unit cost for operator 1 is

$$\frac{12}{50} + \frac{28}{50} + 0.10 \frac{1.00 - 0.18 + (12/50 + 28/50)(1.0)}{1.0 - 0.10} = 0.80 + \frac{0.10}{0.90}(0.82 + 0.80)$$

$$\text{Unit cost} = \$0.80 + \$0.18 = \$0.98$$

Then, for operator 2 to have the same unit cost,

$$\$0.98 = \frac{12}{N} + \frac{28}{N} + 0.10\frac{1.00 - 0.18 + (12/N + 28/N)(0.5)}{1.0 - 0.10}$$

$$= \frac{\$40}{N} + \frac{0.10}{0.90}\left[0.82 + \left(\frac{\$40}{N}\right)(0.5)\right]$$

or
$$N \doteq \frac{42.222}{0.889} \doteq 47.5 \text{ parts}$$

which is about 97 percent of the output of operator 1.

Exercise 4 Assume that the plant described in Example 19.4 plans to cut its product defective rate in half, from 20 percent to 10 percent. How much could be spent for this improvement?

Solution 4 The investment justified by the quality improvement depends on the demand for the product. Recalling that 20,000 units were manufactured per month to produce 16,000 good units [16,000/(1 − 0.2) = 20,000], we find that the number required to get the same 16,000 good units when the defective rate is 10 percent is

$$16,000/(1 - 0.1) = 17,778$$

which reduces the required production volume by 2222 units.

If only 16,000 products can be sold, the $8 variable cost is saved for each of the 2222 units: $8(2222 units) = $17,776. This is the amount that could be spent per month for the improvement when monthly demand is limited to 16,000 units.

When the plant can sell all it produces and its capacity is 20,000 units, the improved defective rate reduces the number of defective products to 20,000(0.2 − 0.1) = 2000 units from 4000 units. The plant therefore gains $20(2000) = $40,000 per month. This is the amount that could be spent for the improvement when 18,000 products can be sold.

Exercise 5 A temporary supply of running water is needed at a construction site situated high above a river. The water is free but must be pumped to the site. Larger hose sizes decrease the pumping cost but are initially more expensive. An adequate supply of water can be provided by a 3-, 4-, or 5-inch-diameter (7.62-, 10.16-, or 12.70-centimeter-diameter) hose. Net costs for these sizes (first cost minus resale value) are $460, $590, and $770. Pumping costs during the 6-month project are estimated at $440 for a 3-inch-diameter hose, 20 percent less for a 4-inch size, and half as much for the 5-inch size. What size hose should be purchased?

Solution 5 A typical tabular approach lists the design variable (in this case the hose size) and respective costs as shown below.

Design Variable (Hose Diameter)	Directly Varying Cost (Hose Purchase)	Inversely Varying Cost (Pumping Cost)	Total Cost	
3 in (7.62 cm)	$460	$440	$900	←minimum cost
4 in (10.16 cm)	590	352	942	
5 in (12.70 cm)	770	220	990	

The smallest hose size exhibits the lowest total cost.

PROBLEMS

19.1 Value engineers are expected to think positively in order to encourage the generation of new ideas for improvements. Negative attitudes prevail among people who are satisfied with present conditions and get disturbed when someone suggests change. Some customary comments that discourage changes are listed below. What response to each one could a positive-thinking value engineer make so that the countercomment would make the negative thinker think again? For instance, the negative comment that "We're making a profit now, why change?" could elicit a VE response of "That's no reason not to make more."

    (a) Why change it? It has been done this way for years.
    (b) The customer will never agree to it!
    (c) It costs too much.
    (d) It's against company policy!
    (e) Plastics are too weak.

19.2 A tie clasp issued to all new recruits in a branch of the military is chrome-plated, has a pivoted clasp, and contains a replica of the insignia of the military branch in the center. The description of the working function of the clasp could be "hold parts," where the parts are the tie and shirt. A secondary (or perhaps the basic) function is "create impression." Show how the creation of substitute means and the method of evaluation depend on the functional description of the item.

    (a) If the present tie clasp costs $2.93, determine a less-expensive alternative based on the hold parts function.
    (b) How does the alternative from Problem 19.2a satisfy the function "create impression"?

19.3 Four items have been singled out by the communications department as possibilities for value-engineering studies. The data accumulated for these items are shown in the table below.

Item	Estimated Yearly Savings	Estimated Study and Implementation Cost	Probability of Implementation
1	$14,000	$4000	0.9
2	6,500	1000	0.6
3	41,000	9000	0.7
4	3,300	600	0.8

    (a) According to the VE rating expression, which alternative should be selected?
    (b) What are some other considerations that are not included in the data that could influence the decision?

19.4 An appliance store has received a franchise to distribute a new type of recorder. Plans are being made to advertise the recorder by a direct-mail campaign of at least 14,000 flyers. A mimeograph machine is available at the store. If it is used, the flyers can be produced for $0.04 apiece. However, an artist will have to be employed to make the line drawings of the recorder and to do the lettering. The art charge is $390. Franked envelopes can be purchased for $200 per thousand.

Another alternative is to have the flyers printed. Plates are supplied by the manufacturer, but printing costs will still amount to $1230. A slight change in the format will allow the flyers to be sent without envelopes. The alterations will cost $120, and mailing expense will be $0.12 per flyer.

A suitable and equivalent advertisement will result from either method. The addressing costs will be the same for both alternatives. Which one has the lowest unit cost?

19.5   An operator earns $12.50 per hour running a plastic molding machine that has an hourly burden rate of $13.00. The production rate is 16 units per hour, of which 10 percent are rejected. Rejection takes place during a midprocess inspection when the operation is 60 percent complete. Raw-material costs are $1.80 per piece and there is no scrap value.

A plan is being considered under which the operator will receive a $2 per hour pay increase to operate two machines. Average production should be 14 units per hour from each machine, but the spoilage rate will double and the operator will not have time for the midprocess inspection.

(a) What is the unit cost for each situation?
(b) What is the cost of producing a rejected unit when two machines are being operated?

19.6   Develop a graph of unit costs versus units produced according to the data and conditions given in Review Exercise 1. How does the graph refute the traditional belief that "the more you produce, the less expensive it is per unit"?

19.7   The 4-day 40-hour workweek has been proposed as an alternative to improve worker motivation and morale. It has been tried in several industries and many government agencies with mixed success. Develop a T-chart to compare the traditional 5-day week with a 4-day pattern, both for a 40-hour workweek in a manufacturing industry. Select criteria that are important to employees and employers and evaluate the alternative by checking whether the 4-day pattern is better or worse from the standpoint of operating effectiveness for the firm, realizing that satisfied employees are usually more productive.

19.8   An operator earning $9 per hour runs a machine with a burden rate of $14.40 per hour. What is the unit cost when one operator runs two and three machines? The activity times, in minutes, for producing one piece are:

Insert piece in machine	0.60
Remove finished piece	0.30
Inspect piece	0.50
File burr and set aside	0.20
Walk to next machine	0.05
Machine running time	3.95

19.9   Two operators, each paid $11.00 per hour, operate two semiautomatic machines. Both operators must work together at all times. The element times, in minutes, for the work cycle are as follows:

Load	1.2
Run	3.4

Unload   0.7

Inspect prior to loading 0.4 (the part is on the worktable during inspection)

Inspect after completion 0.8 (done while the machine is running)

The machines are located next to each other, and transportation times are included in the above elements. If material costs $31.60 per part, direct machine cost is $48 per hour, and overhead is calculated to be 25 percent of labor, materials, and machine costs, determine the cost of each completed part.

19.10 Plans are underway to introduce a new machine to increase output. There are two alternatives, machines A and B. Acquisition cost of the former is $30,000 and for the latter is $50,000. Operating costs are paid at the end of each month in proportion to the number of units produced. Fixed costs are paid at the end of each half year. Assume that fixed costs are not required for zero units of production, the useful life of either machine is 4 years, and the cost of capital is 1 percent compounded monthly. Specific costs for each machine are shown below.

	Machine A	Machine B
Operating cost (per unit)	70 cents	50 cents
Fixed cost (every 6 months)	$800	$1300

  (a) Find the breakeven production level at which machines A and B are equally attractive.
  (b) If the two machines have already been purchased, determine which machine should be used at different levels of production for greatest economy.

19.11 In a company in which the annual average labor cost per employee is $30,000, a plan is being considered that would reduce the workforce by 10. Other revenues and expenses are not expected to change when the plan becomes effective. Assuming that the useful life of required equipment is 7 years and that it has an annual maintenance factor of 3 percent (added to the capital recovery cost), determine how much investment is justified for the plan when taxes are not considered. MARR = 10 percent.

19.12 An electronic specialty company employs 60 people to fabricate miniature resistor assemblies. The workers average 2000 hours of work per year, with an average production of 16.2 assemblies per hour, of which 6 percent are defective. The top third of this group produces 19.1 assemblies per hour with only 3.4 percent defective. The cost to the company to rework each defective unit is $0.30.

  All the work is paid at a straight piecework rate of $0.25 per assembly. Each employee uses a scope and special equipment which has an annual fixed cost of $612. Indirect costs and supervision amount to $740 per year per worker.

  For the same total production, how much could be spent per year to select, train, and motivate the workers so that all of them could produce at the level of the top 20? (*Hint:* The original top 20 are part of the new smaller assembly crew).

19.13 Three products, X, Y, and Z, are being produced at the rate of one unit per day, each requiring one-third of a workday to manufacture. Cost figures are shown in the

	Product X	Product Y	Product Z
Selling price	$400	$600	$1000
Labor cost ($240/day)	80	80	80
Raw material cost and other variable costs	200	380	740
Overhead cost ($400/day)	71.8	118	210.2
Profit	$48.2	$22	− $30.2

table above. Overhead cost of $400 per day is proportionately allocated among the products based on the total raw materials and labor costs of each product. Labor cost is proportional to the hours worked on each product. Assuming that three units of any of the products can be sold, indicate what product mix would be most profitable.

19.14 A concrete contractor uses different sizes of crews for commercial and residential work. A typical residential job requires 3 hours on site and ½ hour travel each way, with a minimum crew of two, a finisher and a laborer. A finisher is paid $14 per hour and laborers get $8 per hour.

Adding another pair (finisher and laborer) to the minimum crew cuts the working time required for a typical job by 60 percent. Adding only one laborer to the minimum crew reduces the time by 30 percent and adding two laborers cuts it by 50 percent. Because of the sequence of work and space limitations, the maximum crew size is four.

Transportation costs are unchanged by the size of the crew. An overhead charge of $40 per hour for equipment is also independent of crew size and is levied for both working and travel time. What makeup of crew will provide the most economical performance?

19.15 For a rough analysis of different bridge designs, it is assumed that costs are proportional to the span length X between piers. As the span length increases, a greater amount of steel is required to support the superstructure. However, fewer piers are required when span lengths are longer. The cost of each span for a certain bridge design is

$$C_s = 50X^2 + 5000X - 100,000$$

and the cost for a pier or an abutment is given by

$$C_p = 200,000 + 1000X$$

(a) Use the tabular approach to find the minimum cost for a bridge crossing a 300-foot (91.44-meter) bay.
(b) Use the formula approach to find the number of spans that minimize the cost for bridging a 300-foot (91.44-meter) bay.

19.16 Lord Kelvin developed a classic minimum-cost analysis for the most economical wire size to conduct electricity. He proposed that the investment cost in wire diameter (directly varying cost) should be equal to the energy loss due to wire resistance (inversely varying cost). The energy loss in watts is equal to I^2R, where I is the current to be conducted and R is the electric resistance which is inversely proportional to the cross-sectional area of the wire. Using the following symbols for the cost factors, determine the formula for the wire size which produces the minimum total cost:

A = cross-sectional area of the wire
I = current conducted
R = resistance of the wire
C_c = rate charged for the use of invested capital, percent
C_t = rate charged to cover taxes and insurance, percent
C_A = cost of wire
C_I = cost of energy

19.17 Of 1000 units produced daily, 10 percent are defective. Per-unit selling price and variable cost are $10 and $2, respectively. Fixed cost per day is $5500. There is no salvage value for a defective unit. More units can be sold than can be manufactured under present conditions. Daily output could be increased to 1200 units if machine operating speeds were increased, but experience indicates that the reject rate for defective units will increase to 15 percent if the speedup is instituted. Determine the gain or loss in profit if the operating speeds are increased.

19.18 A factory has excess capacity for its single machine-made consumer product. The wholesale price of a carton of the products is $400. Labor, indirect manufacturing, and administrative cost per carton are $60, $40, and $60, respectively; all three costs are considered to be fixed at the given levels. Only material costs and other variable costs of production are proportional to output; these costs are $160 per carton.

One day a part broke in a critical machine. It was replaced in a few minutes, but the machine had produced 100 cartons of defective units before being stopped. How much loss could have been prevented if the breakdown had been avoided by a better preventive maintenance program?

19.19 During a fruit harvest a group of orchard growers estimate that they can reduce the number of culls sent to the packing plant by hiring field inspectors. Their crop averages 16 percent culls without inspection on 80 tons picked per day. They expect that for every two inspectors hired the percentage of culls would be cut in half (that is, two inspectors, 8 percent culls; four inspectors, 4 percent culls; etc.). An inspector would be paid $50 per day. The growers will save $0.80 per ton for each percent the culls are reduced. How many inspectors should be hired?

19.20 Where products are inspected during a production process depends on cost and the nature of the process. Inspections are logically scheduled in front of operations that are costly, irreversible, or masking. A study of an assembly process reveals the following rejection percentages and cumulative manufacturing costs for five consecutive operations:

	Operation				
	1	2	3	4	5
Percentage rejected	5.1	0.6	4.8	3.2	1.2
Cumulative production process costs	$2.75	$3.98	$4.27	$4.85	$5.12

The standard unit of inspection is 100 products, and inspectors are paid $4 per hour. Inspection times in minutes per product are, in order of operation, 1.2, 0.6, 0.5, 2.0, and 1.8.

(a) If two inspection stations are to be established, where in the sequence of operations should they be located? Why?

(b) If three inspection stations are allowed, where should they be placed? Why?

EXTENSIONS

19A Cause-and-Effect Analysis

The toughest part of an economic analysis is often getting started. Inertia is a physical law that seems to have a counterpart in mental motion. Once mental activity gets under way and directed, things happen. It is comparatively easy to keep them happening after that first movement, even when there is an uncomfortable suspicion that not enough is known about the situation. As the solving process progresses, the uneasiness may be accompanied by frustrations in correlating facts and drawing meaningful interpretations from data. Better organization of data alleviates some of the difficulties by highlighting pertinent information. Cause-and-effect (C&E) diagrams assist in defining a problem by categorizing related factors to make their influence more observable.

Most improvement studies are precipitated by an unsatisfactory economic condition. An early phase of a study is data collection and organization. For this phase, a C&E diagram provides a systematic procedure for classifying and displaying data. The display may stimulate creative proposals for correcting difficulties. It also suggests the criteria by which the proposals should be evaluated and the factors that influence cash flow. Although a C&E analysis is superfluous for simple familiar problems, it can be extremely useful when a group is assigned to work together on a problem and when the relation of factors that affect a complex problem must be explained to others.

Structure of a C&E diagram

A C&E diagram is a portrait of a problem. It serves much the same purpose as does the black-box concept in engineering studies. A black-box diagram, as shown in Figure 19.7, represents an undefined transformation process that converts known inputs into specified outputs. Examination of the inputs and outputs is supposed to expose the workings of the black-box conversion process. In a similar fashion, also displayed in Figure 19.7, a C&E diagram elaborates on a problem (black box) through the identification of its causes (inputs) and effects (outputs).

The cause-and-effect concept is especially fitting for economic studies. In most situations the basic problem is known. The question is not how does a mechanism work to convert inputs to outputs but which inputs (causes) are needed to solve the problem or accomplish the mission to produce the wanted outputs (effects). Specific alternatives are *not* identified in the initial investigation. The characteristics associated with the problem are identified and noted on the diagram. Then the monetary values corresponding to each alternative's reaction to the problem characteristics are examined.

The mechanics of constructing a C&E diagram are elementary, but the mental effort preceding the drawing is demanding. The first step is to develop a pithy statement of the problem, situation, or objective. This is entered in a six-sided box. Below the box is a boundary line that sets a limit (such as when the solution is to be implemented) that denotes when the causes take effect. Labeled ribs are then drawn to or from the cause-and-effect spine (bold arrows in Figure 19.7) to specify the main causes and effects. Short definitions in ovals at the end of the rib arrows identify these main factors. Finally, smaller arrows identifying subfactors run to or from the main ribs and, in turn, sub-subfactor arrows can lead to or from these to provide greater detail. The construction procedure is illustrated in the following example.

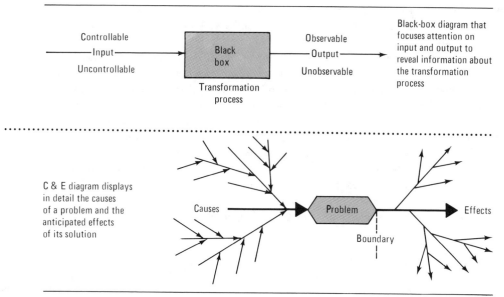

FIGURE 19.7
Structure and purpose of a C&E diagram compared with the traditional black-box analysis approach.

Example 19.6
C&E Diagram Developed to Examine the Practicality of Adopting a 4-Day Workweek

An engineering economist in the public works department of a large city is investigating the relative costs and benefits to expect from replacing the conventional 5-day workweek with a 4-day workweek. Both workweeks constitute 40 hours; the lengths of the workdays differ. There is no wage differential.

The main categories of factors involved in "adopt a 40-hour 4-day workweek" are shown by the labeled ribs in Figure 19.8. These titles serve the same function as do headlines or chapter headings, attention-getters for related data. The

FIGURE 19.8
Main factors affecting the adoption of a 40-hour 4-day workweek are shown as ribs joined to the spinal cause-and-effect arrows. Brief descriptions of the prime considerations are enclosed in ovals for easy recognition.

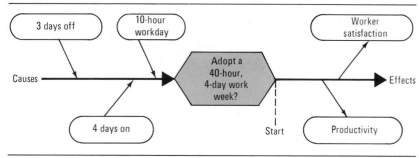

boundary line below the problem description is labeled "start" because this is a feasibility study, not an action plan.

Specific conditions of interest are next listed on the diagram through arrows attached to the titled ribs. Either an individual or a group can discuss the main factors to pinpoint the considerations that are most important to a decision. Elaboration of successively more detailed factors creates the branching effect of arrows shown in Figure 19.9. This preliminary sketch can be circulated among affected persons to solicit additional considerations.

After the comprehensive structure has been refined, numerical values for appropriate factors are estimated to evaluate specific alternatives. Although the same considerations are involved, the advantages and disadvantages of adopting a 4-day workweek are quite different for distinct departments or segments of the work force. The police and fire departments would benefit from overlapping shifts, made more convenient by the 10-hour days, during their busiest hours. More hours would be available each day during the early part of the week for sanitation workers, when they are most needed. Equipment utilization would be better for some types of work and worse for others, as would several other factors.

Once the receipts and disbursements are collected, the evaluation continues according to the most suitable comparison method. In the analysis of a problem that has numerous intangible factors, such as "motivation" and "attitudes" in the 4-day-workweek question, a mixed-rating comparison model may be most appropriate.

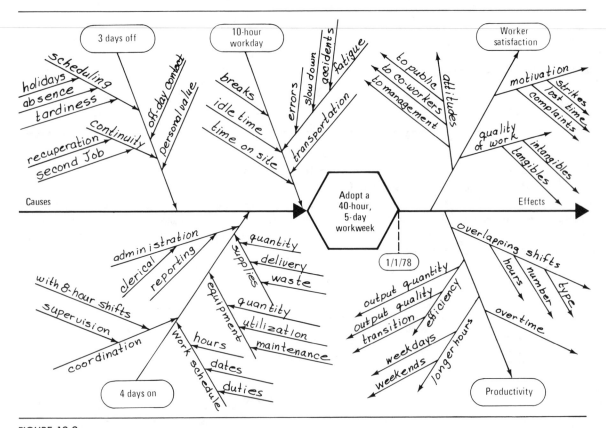

FIGURE 19.9

C&E sketch of considerations involved in switching to a 4-day workweek. Additional subfactors can be added during the investigation to determine alternative courses of action. After alternatives have been itemized, measurements are inserted in a refined diagram to represent specific proposals.

C&E Applications

A cause-and-effect analysis is conducted to explore a situation. The visual impact tends to strengthen an investigation by suggesting additional considerations associated with the factors already recorded. Familiarity gained by the diagraming exercise reveals what data are needed to compare alternative courses of action.

There are no rigid rules for the construction of a C&E diagram. The finished figure can be a sketch on a napkin or a colorful poster for presentation purposes. Innumerable modifications can be made to suit the situation.

A one-sided diagram, either the cause side or the effect side, is appropriate when the input or output of a situation is predetermined and unalterable. A study featuring the cause side is called an *Ishikawa diagram* after its originator, Kaoru Ishikawa of Tokyo University, who devised this form of analysis in 1953 for quality-control studies. In this version the quality goal is known and the contributors to quality are sought. A similar condition exists when, for instance, the desired product output is known and the question is whether a current machine or a new one would achieve that output at lower cost. As shown in Figure 19.10, the cost categories are identified but no specific cost figures are given for individual machines. These categories reveal what data must be collected to compare the defender with the challengers in a replacement study.

A C&E variation introduced by Ryuji Fukuda, based on his work at Sumitomo Electric Corporation, is called CEDAC—Cause and Effect Diagram with the Addition of Cards. It has the same skeleton as an Ishikawa diagram, but the single words and brief phrases on the arrows are replaced in a CEDAC by cards that hold a short sentence. The sentences are supposed to express more completely what is known about the factors, and the cards can be changed easily when a diagram needs revising. A CEDAC posted in a conspicuous location in a factory or office encourages employees to add their ideas and comments by pinning more cards on the diagram.

It is sometimes easier to visualize a process by considering factors in an accustomed sequence. Clusters of arrows in a C&E diagram for a material-handling study could group related factors in a sequence that follows the physical flow of material: *purchase→receive→store→ship* on the cause side and *inspect→display→sell* on the effect side. Other sequential scales could be time, organizational hierarchy, level of technological development, etc. Any routine that stimulates a more complete diagram increases the chance that the needed data will be available for a thorough evaluation. Occasionally, the diagraming operation exposes a new alternative or divulges a solution that might otherwise have been overlooked, as described in Example 19.7.

FIGURE 19.10
One-sided diagram displaying the factors to be considered in a machine-replacement study. It is assumed that any machine selected will produce the required output (effect).

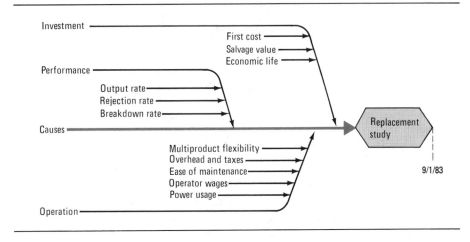

Example 19.7
A Search for Alternatives

The State Department of Vocational Training sent an industrial engineer to a local sheltered workshop to help overcome a production problem. The shop serves as a rehabilitation center for handicapped workers. It provides training and work for

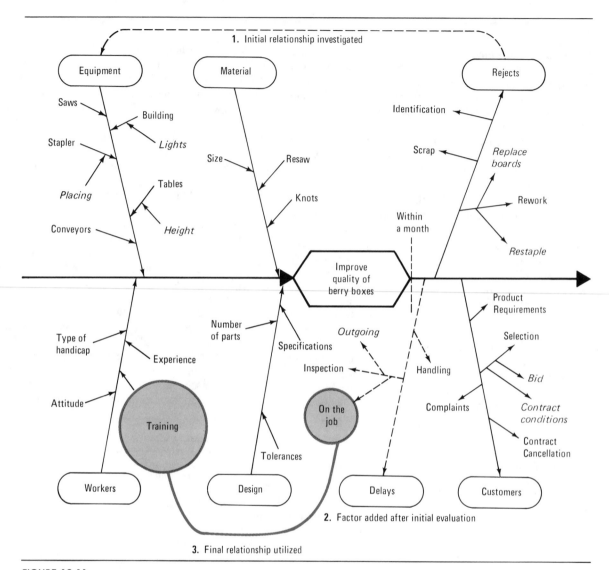

FIGURE 19.11
Initial and revised C&E diagram representing the production of berry boxes in a rehabilitation center. The steps in the solution are numbered in the order of their development.

the handicapped in assembly operations and in the production of wood products such as pallets, packing cases, wishing wells, and picnic tables. The work is purposely designed to emphasize hand labor and the processes consequently employ minimal mechanization.

A quality-control problem occurred when the center began work on a contract to produce berry boxes. The customer threatened to cancel the contract if sturdier boxes were not produced within a month. The first analysis of the problem

identified the causes and effects shown in Figure 19.11. An obvious solution appeared to be to add an inspection station for outgoing boxes to ensure the use of good wood and adequate fastening. Then it was noticed that attention was being focused strictly on effects, not causes. Maybe it would be better to have the workers inspect as they produced (note 1 at the top of the diagram).

As a result of this internal-inspection idea, attention was given to the need for better physical facilities, brighter lights, lower work tables for wheelchair workers, jigs for more accurate stapling, etc. These and other considerations led to the addition of a new effect, *delays* (shown in Figure 19.11 by note 2 and dashed arrows). To minimize delays, a suggestion was made to improve the motion patterns of workers. Finally, the subject of training was considered, and it was then recognized that training was currently directed to general skills, but little effort was made to tell individual workers what was expected of them and what particularly should be watched in satisfying each production contract.

Thus, the berry-box problem was solved simply by informing the workers what was needed and how to do it with existing facilities. The answer is indicated by the connected and circled factors in the lower portion of the diagram. No investment funds were needed. Perhaps the answer may appear patently obvious, but how many times have sophisticated solutions been installed where a simple change could have produced the same results at lower cost and less loss of dignity?

QUESTIONS

19A.1 The State Highway Department is planning to construct several highway rest stops. The purpose of such stops is to improve highway safety by encouraging motorists to get out of their cars and relax for a few minutes during long trips. The locations and appearance of the rest stops largely determine how much they are used. Develop a C&E diagram to show what data are required to compare alternative sites and designs for rest stops. Some of the main causes are *location, natural features, construction*, and *design*; some main effects are *utilization, upkeep*, and *safety*.

19A.2 Many innovative ideas have been proposed to solve the problems of intracity transportation. They range from moving sidewalks to individualized jitney service. A poll to determine what consumers believe are the most important characteristics of a carrier-based city transportation service revealed the following attributes:

1 Arrive and leave on schedule
2 Have a seat for the whole trip
3 No transfers during trip
4 Low fares, comparable with the cost of using a car
5 Shelters at pickup points
6 Available at all hours
7 Use direct routes to shorten travel time
8 Easy way to pay fares
9 Space provided for packages
10 Adjustable seats provided
11 Clean vehicle interiors
12 Adjustable air, light, and sound
13 Coffee, refreshments, and reading material on board
14 Stylish vehicle exterior

Develop a C&E diagram that contains the key features by which any intracity transportation proposal could be evaluated. Include the units of measurements by which each entry in the diagram would be measured.

19A.3 After an audit revealed that fuel costs had increased 30 percent in recent months, the building supervisor took steps to correct the situation, figuring that significant savings would be made by lowering the temperature settings while buildings were unoccupied. Each building had a master steam control which night janitors were told to set at the new lower level. However, savings failed to reach expectations. An investigation revealed that the janitors were using the lower settings *except* while they were cleaning the building; then the setting for a whole building was put at the regular working-day level even though only a few janitors occupied the building.

The supervisor changed the temperature controls. The new controls could be altered only by using a special tool. One person was assigned to reset the heating level each morning and evening. Before long, a grievance was filed against the new practice by the union, claiming that the janitors could not be expected to work in cold buildings.

(a) Draw a cause-and-effect diagram to identify and organize the factors affecting this problem.
(b) Circle the most influential factors and suggest a solution based on them.

19B   Learning Curves

Some types of repetitive operations exhibit decreasing cost per unit as the production run continues. The declining costs are attributable to experience gained by individual operators and by engineering-management support teams. Studies have shown that about 15 percent of the total time or effort reduction comes from improved operator performance, 35 percent from industrial-engineering cost-reduction activities, and the rest from redesigns or process improvements by product engineering. The total effect of the experience is called *learning*

Learning curves were first identified by the airframe industry in the 1940s. Since then they have been found applicable in many high-cost, low-volume industries, such as shipbuilding, computers and electronics, machine tools, and building or plant construction. In cost estimating, the existence of learning can have a major effect. Bids or price quotes should be lower on successive orders of major items that benefit from learning. Make-or-buy decisions can be affected by the learning rate in the manufacturing process.

An 80 percent learning curve is a fairly representative figure for many large-scale repetitive projects. This learning rate means that there will be a 20 percent reduction in worker-hours per unit between *doubled* units. That is, if it took 100,000 hours to produce unit 1, it would take 100,000(100 percent − 20 percent) = 100,000(0.8) = 80,000 hours to produce unit 2. Then unit 4 would take 80,000(0.8) = 64,000 hours; unit 8 (double unit 4) would take 64,000(0.8) = 51,200 hours; and so on. This diminishing marginal-improvement pattern reveals the characteristics of a process subject to learning:

1 The time to complete each unit is less than the time to complete the previous unit.
2 Unit times decrease at a decreasing rate.

Note that learning rates usually refer to operational times rather than costs. These times are normally proportional to direct labor costs, but other production costs such as materials and indirect labor are likely paced by other factors.

The relation between direct labor hours and the number of units produced is

$$Y_N = KN^X$$

where Y_N = effort per production unit to produce the Nth unit, as in (direct labor hours)/unit
K = effort to produce the first unit, in dimensions compatible with Y_N
N = unit number: 1, 2, etc.
X = improvement function = [log (learning rate)]/ log 2, as in (log 0.8)/0.3010

Table 19.3 shows values of the improvement function x as it is related to various learning rates. For the product that took 100,000 hours for the first unit, the time to produce the eighth unit when the learning rate is 0.8 is calculated as

$$Y_8 = 100,000(8)^{\log 0.8/\log 2} = 100,000(8)^{-0.322}$$
$$= 51,200 \text{ hours}$$

Individual and cumulative times to produce 10 units at this improvement function are given in Table 19.4.

When it appears that learning is taking place in a production run, the rate can be calculated from any two unit times, Y_i and Y_j. Suppose that the only data available about a process are that the fifth unit took 59,560 hours to produce

TABLE 19.3

Learning rates and associated improvement function (x) values

Learning Rate	Improvement Function
1.0 (no learning)	0
0.95	−0.075
0.90	−0.152
0.85	−0.234
0.80	−0.322
0.75	−0.415
0.70	−0.515
0.65	−0.621
0.60	−0.737
0.55	−0.861
0.50	−1.000

TABLE 19.4

Unit, cumulative, and cumulative average time or effort required to produce 10 units of a product when the first unit took 100,000 hours and the learning rate is 80 percent (improvement function = -0.322)

Unit Number, N	Unit Production Effort, Y_N	Cumulative Production Effort	Cumulative Average Production Effort
1	100,000 = K	100,000	100,000
2	80,000	180,000	90,000
3	70,210	250,210	83,403
4	64,000	314,210	78,553
5	59,560	373,770	74,754
6	56,170	429,940	71,657
7	53,450	483,390	69,056
8	51,200	534,590	66,824
9	49,290	583,880	64,876
10	47,650	631,530	63,153

and the ninth unit took 49,290 hours. Then $Y_5 = K(5)^x$ is divided by $Y_9 = K(9)^x$ to obtain

$$\frac{Y_i}{Y_j} = \frac{K(N_i)^x}{K(N_j)^x} \quad \text{or} \quad \frac{59,560}{49,290} = \left(\frac{5}{9}\right)^x$$

The logs of both sides of the equation are taken as

$$\log \frac{Y_i}{Y_j} = x \log \frac{N_i}{N_j}$$

from which

$$x = \frac{\log Y_i - \log Y_j}{\log N_i - \log N_j} = \frac{\log 59,560 - \log 49,290}{\log 5 - \log 9}$$

$$= \frac{4.7750 - 4.6928}{0.6990 - 0.9542} = \frac{0.0822}{-0.2552} = -0.322$$

The learning rate is then determined from

$$\log (\text{learning rate}) = x \log 2$$
$$= -0.322(0.301) = 9.9031 - 10$$

where the antilog of both sides reveals that the learning rate

is 80 percent, as expected from the values in Table 19.4.

QUESTIONS

19B.1 A condominium complex is to be composed of five identical housing units. Direct labor costs for the first unit are $531,000. It is believed that costs for the remaining units will follow a 90 percent learning curve. What is the estimated total direct labor cost for the condominium complex?

19B.2 A contractor is supplying 20 custom rigs to your organization. The ninth rig required 59,800 hours of direct labor, and the eighteenth rig required 50,850 hours. Work is now under way on the twentieth rig, and you find that you will need one additional rig beyond the contracted 20. How many direct labor hours should be anticipated for the construction of the twenty-first rig? (*Hint:* Determine x, then K, then Y_{21}.)

19B.3 Units produced under proposal *A* are expected to have a 90 percent learning rate with the first unit costing $12,000. The first unit produced under proposal *B* will cost $18,000, but production is expected to have an 80 percent learning rate. What would be the cost for the fifteenth unit produced by each proposal? Assume that costs are for direct labor only.

CHAPTER 20

RESOURCE MANAGEMENT

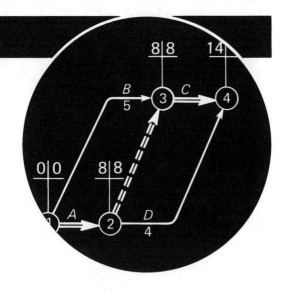

OVERVIEW

No management functions better epitomize concern for the money value of time than do *scheduling* and *inventory management*. A schedule is an ordered list of events or a timetable of activities for accomplishing something. It must be specific because it not only lays out individual assignments, it also coordinates resource utilization to accomplish the mission of the project. Similarly, inventory management relates material usage rates to the holding cost incurred to have a supply available when needed. The minimum cost pattern reveals the point in time when a designated amount of material should be ordered.

Network scheduling is a graphical approach to the sequencing and coordination of activities necessary to complete a project economically and on time. The most renowned version of network scheduling is the critical path method (CPM). A related version, called the program evaluation and review technique (PERT), is discussed in Extension 20B.

The first step in a CPM application is to break the project down into its component operations to form a complete list of essential activities, as explained in Extension 20A. A network is drawn to conventions, where arrows representing activities connect nodes showing the sequence of events. *Dummy* arrows are included to allow distinctive nodal numbering for computer applications and to show certain event restrictions. *Boundary times* are calculated for all network activities to determine the float available for noncritical activities and the chain of activities that sets the total project duration—the *critical path*.

Time-cost trade-offs can be conducted with network activities to determine the least-cost measures to reduce the project duration. PERT/

COST is a systematic accounting method that tracks cost and time deviations for the project's anticipated schedule.

Computer programs for CPM applications are widely available. Manual calculations are feasible for smaller problems, but electronic assistance is almost a necessity for larger ones. In network analyses, a computer can perform the boundary-time calculations and generate bar charts for easier checks of resource assignments.

Basic inventory costs include order or set-up costs O that decrease with larger order quantities Q, and holding costs H that increase with increases in Q. When demand and delivery times are constant, the economic order quantity (EOQ) is calculated by $Q = \sqrt{2OD/H}$, where D is annual demand. The use of quantity discounts is evaluated by comparing the lowest total inventory cost C possible at each price level P, where

$$C = \frac{OD}{Q} + \frac{HQ}{2} + DP$$

When items are consumed as they are produced, the economic production quantity (EPQ) formula is

$$Q = \sqrt{\frac{2OD}{H(1 - D/M)}} \qquad \text{where } M = \text{annual manufacturing rate}$$

Material management is changing rapidly as a consequence of electronic assistance. Computerized material requirements planning (MRP) has replaced many lot-size inventory systems. But the simple ABC classification system still guides the management of both small- and large-scale supply inventories.

CRITICAL PATH SCHEDULING AND INVENTORY MANAGEMENT

Most engineers work with network management diagrams at some time in their careers—as developers who plan the work, as analysts who refine the workings, or as managers who commit the resources to work the plan.

A network depicts the sequence of activities necessary to complete a project. Segments of a project are represented by lines connected together to show the interrelation of operations and resources. When a duration is associated with each segment, the model shows the time orientation of the total project and its internal operations. This information is used to coordinate the application of resources. The most celebrated versions of network scheduling are the *critical path method* and the *program evaluation and review technique*.

PERT's origin in R&D explains its probabilistic approach that recognizes scheduling risks, while CPM's origin in manufacturing recognizes the deterministic nature of production schedules.

Few if any new management tools have received such wide acclaim so rapidly as have CPM and PERT. The U.S. Navy developed PERT in 1958 to coordinate research and development work. It was applied to the Polaris ballistic missile program with significant success. In the same year a similar network scheduling tool was developed by the DuPont Company for industrial projects. It was called the critical path method in deference to the path of critical activities that control the project's duration. Critical path scheduling is a management control tool for defining, integrating, and analyzing what must be done to complete a project in an economical way and on time.

Economic lot-size formulas are management tools for analyzing material-usage factors that must be satisfied to manufacture products in an economical way and on time. The computations seek a minimum cost point from a combination of directly varying, indirectly varying, and constant costs, as displayed in Figure 19.4. Costs vary with respect to the quantity of materials ordered at one time or manufactured in one production run. Thus cost and time are unavoidably related.

Material management is much more than a collection of inventory formulas, of course. Engineering economists may engage material management from the investment aspect to justify facilities to store and move material, and from the operating aspect to evaluate expenditures for acquiring, stocking, handling, and monitoring material flow.

ARROW NETWORKS

A network representation of a small maintenance construction project is shown in Figure 20.1. Five activities, indicated by labeled arrows, are needed to finish the project. The normal duration for each activity is shown below its arrow. The arrows also designate the order in which the activities must be completed. For instance, the placement of arrow 1, 2 (node numbers at the beginning and end of the arrow) shows that old machines must be removed before maintenance (activity 2,4) can begin or site preparation (activity 2,3) can commence. Further, activities 2,3 and 1,3 must be completed before the new machines can be installed (activity 3,4).

The double-lined arrows indicate the *critical path*—the longest chain of

FIGURE 20.1
Arrow network for the installation of new machines and concurrent maintenance of associated equipment.

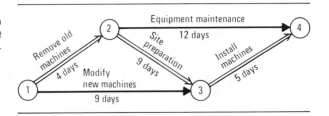

activities through the network. The sum of the durations of activities on the critical path gives the total project duration, $4 + 9 + 5 = 18$ days.

Although the little project in Figure 20.1 oversimplifies the planning process, it embodies the basics of network analysis. Activities are identified. Their relation to each other is displayed graphically. Times are computed to determine a schedule. Then the schedule is analyzed with respect to the time, money, and physical resources required for the activities.

NETWORK CONVENTIONS

Every analysis tool seems to require special conventions. CPM is no exception. The basic symbols used in arrow networks are displayed in Figure 20.2. The activity description or a letter symbol is written above the activity arrow. An arrow can be drawn at any angle and with any length because arrow networks are not drawn to a time scale; the duration of the activity is noted by a number below the arrow. The circle nodes at both ends of the activity arrow represent the beginning and end events of the activity. The numbers in the circles identify the activity for computer programs; activity A would be known as activity 1,2. Note that the end node of an activity is the beginning node of the next one.

A dummy is a dashed arrow used to show a precedence relation among activities. A "logic" dummy is needed when two or more activities share a common postrequisite but at least one of the activities also has a unique postrequisite. In Figure 20.2, activity A has the same postrequisite (C) as does activity B, but it also has a unique postrequisite in activity D. The dummy shows that event 2 must be completed before event 3 can begin or, equivalently, that activity A must be completed before activity C is begun. The arrowhead on the dummy shows the direction of the one-way restriction.

Another use of dummy arrows is displayed in Figure 20.3. Since some computer programs for CPM calculations require that each activity have a unique pair of nodal numbers, a problem occurs when two or more activities

The development of activity descriptions, times, and restrictions is discussed in Extension 20A.

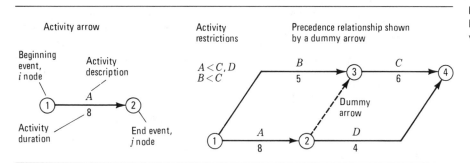

Activity arrow

Beginning event, i node
Activity description

A
8

End event, j node

Activity duration

Activity restrictions

$A < C, D$
$B < C$

Precedence relationship shown by a dummy arrow

B
5

C
6

Dummy arrow

A
8

D
4

FIGURE 20.2
Basic symbols used in arrow networks.

FIGURE 20.3
Conventions used for arrow networks in this text.

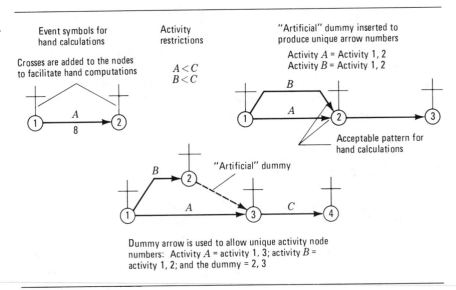

have the same postrequisite. As shown in the top network of Figure 20.3, activities A and B have the same node numbers because both begin and end with the same circles. This condition is acceptable for hand calculations, but a dummy should be inserted to form unique activity numbering for computer applications. In the lower network, a dummy (2, 3) has been inserted between activities B and C. This "artificial" dummy maintains the same restrictions but forces unique numbering for A and B; A becomes activity 1,3 and B becomes activity 1,2 for computer coding. All the CPM networks in this text will use artificial dummies, even when calculations are to be performed manually. The event crosses illustrated in Figure 20.3, however, will be added to the nodes only when hand calculations are shown.

Example 20.1
Network Drawing

Information required to construct a network for a prototype-development project is as follows.

The arrow network which conforms to the activity list in the table is given in Figure 20.4. The diagram faithfully portrays the logic and information from the activity list. Note that a dashed dummy arrow is used to show a one-way restriction. It is treated as an activity arrow with zero duration and indicates that activity C must be completed before activity E can begin.

Activity Description	Symbol	Duration, Weeks	Restrictions
Design prototype	A	6	$A < B,C$
Obtain materials	B	2	$B < D$
Order parts	C	2	$C < E,F$
Manufacture parts	D	3	$D < E$
Assemble component no. 1	E	2	$E < G$
Assemble component no. 2	F	4	$F < G$
Assemble prototype	G	1	

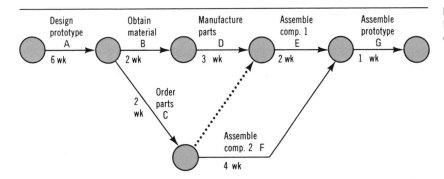

FIGURE 20.4
Network diagram for the development of a prototype.

THE CRITICAL PATH

After estimated activity performance times are secured, each duration is entered beside the appropriate arrow in the network. The network is then ready for boundary-time calculations. These calculations can be performed either by hand or with computerized assistance to provide time relationships suitable for activity scheduling. The most useful boundary times are described below.

Earliest start (ES). The earliest time an activity can begin when all preceding activities are completed as rapidly as possible

Latest start (LS). The latest time an activity can be initiated without delaying the minimum project completion time

Earliest finish (EF). The sum of ES and the activity's duration D

Latest finish (LF). The LS added to the duration D

Total float (TF). The amount of surplus time or leeway allowed in scheduling activities to avoid interference with any activity on the network critical path; the slack between the earliest and latest start times (LS − ES = TF)

Some CPM programs report exotic forms of float, such as "free" and "safety" float, that have managerial implications. They are not very useful.

Identification of the critical path is a by-product of boundary-time calculations. A critical activity has no leeway in scheduling and, consequently, zero total float.

The ES times are calculated manually by making a forward pass through the network, adding each activity duration in turn to the ES of the prerequisite activity. When a merge is encountered, the largest ES + D of the merging activities is the limiting ES for all activities bursting from the event. Dummies are treated exactly the same as other activities. Each limiting ES is recorded on the left bar of the event markers to keep track of the cumulative entries. The cross at the *i* node of an activity indicates the ES for that activity.

Earliest start times for the network in Figure 20.2 are shown in Figure 20.5. The circled values below and at the head end of activity arrows are EF

FIGURE 20.5
Earliest start- and finish-time calculations.

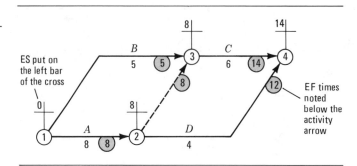

times. Observe that the largest EF at a merge (node where arrows converge) is also shown on that node's cross.

Latest start and finish times are calculated by a *backward pass* through the network. The latest start for all activities terminating at the last event are computed by subtracting each activity's duration from the project completion time (the largest EF in the network). Then the smallest LS of any activity leading from an event is entered on the right bar of the event's cross. This value then becomes the latest finish for the next round of calculations. The procedure is described below with reference to Figure 20.6.

1 The largest EF from the forward pass is entered on both bars of the last cross, node 4. This is the minimum project duration, 14.
2 Latest starts for C and D, calculated as $LS_C = 14 - 6 = 8$ and $LS_D = 14 - 4 = 10$, are entered near the origin of their respective activity arrows.
3 At node 3, the LS of C is entered on the cross because it is the only activity originating at the node. This entry is the LF for all activities (and the dummy) terminating at node 3.
4 At node 2 the smallest LS (8 from the dummy) is entered on the cross.
5 Latest start times for activities A and B are determined by subtracting their

FIGURE 20.6
Latest start- and finish-time calculations.

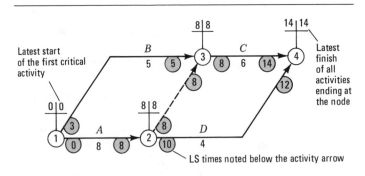

durations from their termination nodes as $LS_A = 8 - 8 = 0$ and $LS_B = 8 - 5 = 3$.

6 At node 1 the entry is the smallest LS from A or B; $LS_A = 0$. Unless a mistake has been made in the backward pass, the initial project event will have the same entries on both bars of the cross.

Total float is a measure of the leeway or slack available in scheduling activities. As is apparent in the sample networks, some activities have no leeway; they must start and finish at the given times if the project is to be completed at its calculated minimum duration. An activity with LS larger than ES has leeway. If $LS - ES = 2$, an activity can begin two time units later than its earliest start, or it can take up to two time units longer to complete than anticipated, without delaying the project completion. Total float is evident for activities B ($LS_B - ES_B = 3 - 0 = 3$) and D ($10 - 8 = 2$) in Figure 20.6. The same values could be obtained from the network by using the latest start and finish times: The total float of B is $TF_B = 8 - 5 = 3$ and $TF_D = 14 - 12 = 2$.

The chain of activities stretching from the first to last event with no float is known as the *critical path*. There may be several parallel critical paths in a project. In Figure 20.6 the critical path goes from node 1 to 2 to 3 to 4. It is easily identified by crosses with the same entries on both bars and is often marked by double lines to give added emphasis. A delay in starting or completing any of the critical activities will cause a corresponding delay in the project duration unless corrective action is taken.

Example 20.2
Network Boundary Times

Calculate the five boundary times for the network given in Figure 20.4.

Solution 20.2

As diagramed in Figure 20.7, the project start is set by a zero on the left bar of the first cross. Then $0 + 6 = 6$ is the EF for the first activity (entered below the arrowhead of activity A), and this value in turn is entered on the left bar of the cross at node 2 to give the ES for activities B and C. Note that dummy 4,5 is treated as an activity. The calculation of EF and ES values continues as a forward pass through the network until the cross on the last node is filled. This last entry is the earliest completion of the project and the earliest start for a following project that relies on the same resources.

A backward pass, right to left, through the network allows the calculation of *latest start* (LS) and *latest finish* (LF) times. These latest-time calculations are the reverse of ES and EF calculations: An activity's duration is subtracted from the entry on the right cross of its arrowhead node and is then entered below the activity's arrow at its tail end. For example, in Figure 20.7 at node 7, $LS_G = 14 - 1 = 13$ is entered below the tail of arrow G and on the right bar of the cross at node 6, because 13 is the latest time activities E and F can be completed in order not to delay the project completion time of 14. At node 2, $LS_B = 8 - 2 = 6$, and $LS_C = 9 - 2 = 7$ are shown as tail-end numbers below outgoing arrows B and C, and the smaller LS of the two, 6, is entered on the cross as the latest finish of activity A.

After the start and finish times are inscribed on the network, they can be collected in a boundary timetable as

FIGURE 20.7
Network calculations for ES, EF, LS, and LF times for the project described in Table 20.1.

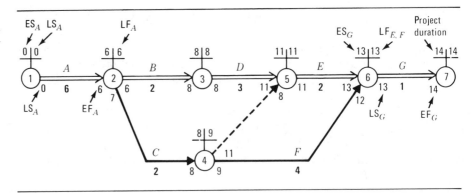

shown in Table 20.1. The critical activities that constitute the critical path are those with no float, TF = 0.

Both network construction and calculations get easier with practice because they are largely mechanical. The measure of merit in the network analysis is the validity of the input data and what is done with the boundary time output to produce a better project plan.

TIME-COST TRADE-OFFS

Time management, as a discipline to improve the utilization of personal time, is explored in Extension 20C.

In *The Way to Wealth* (July 7, 1757), Benjamin Franklin wrote, "Dost thou love life, then do not squander time, for that's the stuff life is made of." He would likely have relished the thought of buying time. In an informal way, everyone buys time occasionally: hiring someone to do a task that frees you to do something else, or purchasing a machine that cuts the time required to do something. On a larger scale, capital intensive industries make maximum use of high-investment assets by working them 24 hours per day, or a rush order is completed early by having crews work overtime. In all these cases,

TABLE 20.1

Boundary timetable developed from network in Figure 20.7

Activity	Duration	ES	LS	EF	LF	TF
A	6	0	0	6	6	0
B	2	6	6	8	8	0
C	2	6	7	8	9	1
D	3	8	8	11	11	0
E	2	11	11	13	13	0
F	4	8	9	12	13	1
G	1	13	13	14	14	0

an extra premium is paid to conserve time. Deciding whether to pay the premium is a time-cost trade-off problem.

Most opportunities to buy time are settled by simply comparing the incremental cost of gaining a time unit against the incremental profit earned from acquiring that time unit. Equivalently, a decision to sell personal time rests on a value judgment for alternative uses of time. When there are time-cost trade-off options, the one providing the most net gain is selected. But identifying that most profitable option gets complicated when activities have to be completed in a certain sequence and the resources involved have different prices for time.

Project Costs

The performance cost of an activity is the sum of direct-labor, material, and equipment charges required to complete it. When a separate agency such as a consulting firm or subcontractor undertakes the entire activity, the cost is the amount paid to that agency. Thus, activity time-cost relations take a variety of forms. Several of the more common configurations are shown in Figure 20.8.

Costs which increase with longer project durations are composed of overhead expenses associated with the project and outage or penalty costs caused by disruption to ongoing operations or by the condition that there is no facility available when it is needed.

For the project depicted in Figure 20.1, inversely varying activity costs are assumed to be linearly related to time, as indicated in Table 20.2. The cutting cost in the last column of the table is calculated for each activity with the formula

$$\frac{\text{Crash cost} - \text{normal cost}}{\text{Normal duration} - \text{crash duration}} = \text{cutting cost/day}$$

From Table 20.2, the cutting cost of activity 2,4 is

$$\frac{\$1530 - \$1200}{12 - 9} = \frac{\$330}{3 \text{ days}} = \$110/\text{day}$$

Directly varying costs are $220 per day and represent the loss in output from the machines not being available for production during the project duration.

Trade-Off Analysis

At first glance it might appear that the duration of all activities in the project should be cut to the minimum because all the per-day cutting costs are less than the $220 daily cost of lost production. This impression is incorrect,

FIGURE 20.8
Variations of activity cost patterns.

Linear increase of cost with decreasing time: a job that can be efficiently implemented by increasing resources

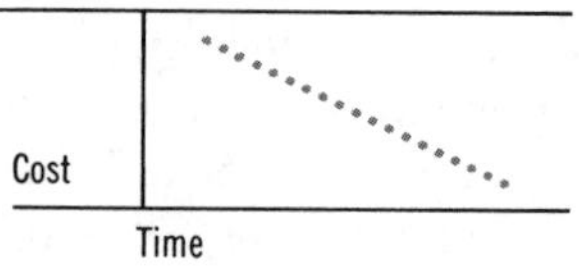

Constant cost which does not vary with time: a subcontracted job with an established minimum time

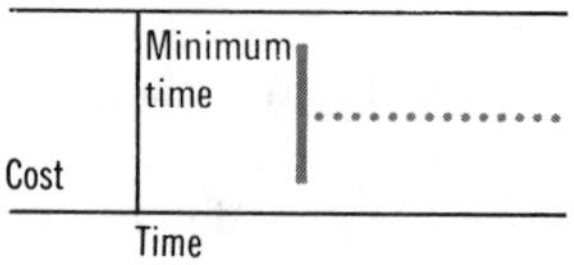

A distinct step increase of direct cost at a point in time: a job where a major cost rise occurs if the duration is to be shortened, due to an incremental resource charge

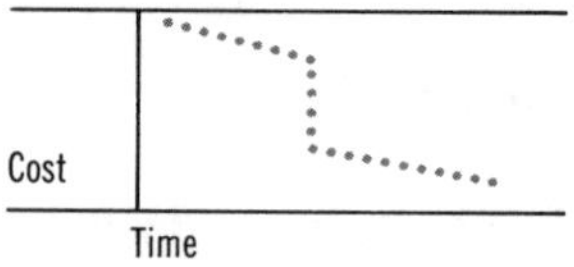

A concave cost-time relationship: a job where some limitation prevents an increasing resource application from showing a proportionate return. This very common pattern is represented by a piecewise linear approximation

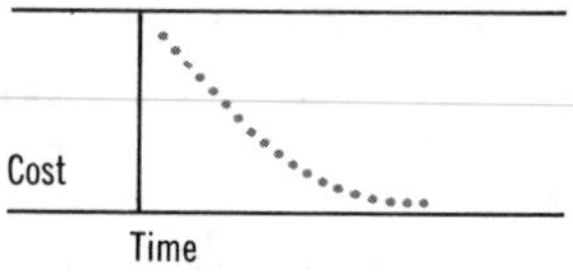

A noncontinuous time-cost relationship: an activity, such as delivery time, where there is only a normal time cost and a crash time cost

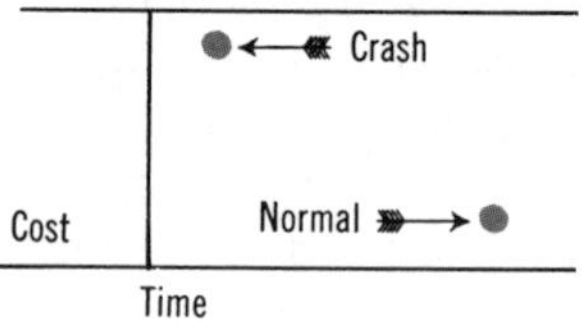

TABLE 20.2

Per-day costs of reducing normal activity durations to their minimum (crash) times

Activity	Normal		Crash		Cutting Cost, $/day
	Day	Cost	Day	Cost	
1,2	4	$1200	3	$1400	200
1,3	9	1800	7	2100	150
2,3	9	1440	5	2080	160
2,4	12	1200	9	1530	110
3,4	5	1650	5	1650	
Totals		$7290		$8760	

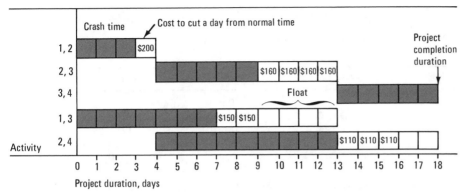

because sequencing requirements dictate that certain activities should be done concurrently. The result is that the cost to cut a day from the project when concurrent activities are being performed is the sum of cutting costs for all activities going on at the same time.

The limiting time restrictions are shown on a bar graph in Figure 20.9. The solid portion of each bar is the activity's minimum duration. Costs to cut an activity from its normal to its crash duration are shown in segments to the right of the solid portion. The empty segments on the extreme right of some bars indicate "float," or leeway in scheduling. For instance, activity 1,3 can begin when the project starts and will not interfere with the start of any other activity so long as it is completed in 13 days (activity 3,4 must start at the end of the thirteenth day to avoid delaying the project's 18-day completion time). Since activity 1,3 normally takes only 9 days to complete, it has $13 - 9 = 4$ days of float. The three critical activities with no float are shown at the top of the bar chart and are connected to highlight the critical path.

From the bar chart, it is possible to develop various project durations and total costs. Since a critical activity must be shortened on each cut, activity 2,3 is the one to cut first. This cut increases the project cost by $160 and shortens the total duration to 17 days and reduces the float available for activities 1,3 and 2,4 by 1 day. Cutting continues until the cost of the next cut exceeds the savings from directly varying costs:

16-day schedule. Activity 2,3 is cut again at a cost of $160 to reduce its duration to 7 days. This cut allows activity 3,4 to start at the end of the eleventh day and eliminates the float available for activity 2,4, making it critical.

15-day schedule. The least expensive cut now available is in activity 1,2 at a cost of $200. This cut allows activities 2,3 and 2,4 to start after the third day and activity 3,4 to begin after 10 days. The only remaining noncritical activity is 1,3 with 1 day of float.

TABLE 20.3

Minimum cost schedule for the project conditions graphed in Figure 20.9.

Activity	Start	End	Cost
1,2	Day 0	Day 3	$1400
1,3	Day 0 or 1	Day 9 or 10	1800
2,3	Day 4	Day 10	1760
2,4	Day 4	Day 15	1200
3,4	Day 11	Day 15	1650
		Total inversely varying cost =	$7810
	Total directly varying cost = 15 days × $220/day =		3300
		Total project cost =	$11,110

14-day schedule. The next cut requires that activities 2,3 and 2,4 be shortened at a cost of $160 + $110 = $270. Since this cut costs more than the savings it produces, 15 days is the lowest-cost project duration. Times and costs for the 15-day duration are shown in Table 20.3.

In summary, 3 days were cut from the original 18-day project duration given in Figure 20.1, at a cost increase of $160 + $160 + $200 = $520. The shortened schedule reduces directly varying costs by $220/day × 3 days = $660, effecting a saving of $660 − $520 = $140.

Application of Network Cost Analysis

The preceding example demonstrated that time-cost trade-offs are clumsy to conduct by hand, even for a tiny project. Computer programs are available to relieve manual calculations. Inputs to a program include the sequential ordering of activities obtainable from an arrow network and time-cost data as typified by Table 20.2. Most programs will handle only linear cost relationships, as described by the top cost-time pattern in Figure 20.8.

Minimum-cost evaluations have not been used as much as was anticipated when the procedure was introduced in the early 1960s. A major reason is the lack of historical accounting data from which to obtain time-cost relationships for individual activities. Accounting records are traditionally maintained for line items, not collections of cost categories associated with different activities needed to perform a maintenance or construction project.

"Crashing" is possibly the only salvation for a project that has fallen seriously behind schedule and faces stiff penalty costs for late completion. The project engineers are exhorted to find ways to capture time. Each remaining activity is examined to find out if it can be shortened and if so, how and at what cost. In this situation, manual time-cost trade-off calculations are feasible because only a relatively small portion of the project's activities need be

Crash schedules vividly demonstrate the importance of early planning to avoid a finish-at-any-cost rush to complete a project. Crashing is usually even more costly in practice than it is on paper because disruptions and inefficiencies take an added toll beyond overtime and overuse.

investigated, and their cutting costs can be estimated accurately with respect to current conditions. The advantage of the systematic CPS cost-analysis method is that it limits time-reduction efforts to the activities where they are most profitable and thereby avoids the panic-button approach of trying to buy time wholesale.

PERT/COST

Network-based cost control for projects is a natural outgrowth of network-based project planning. In PERT/COST the accounting units are ''work packages'' composed of individual activities or groups of related activities found on the network schedule for a project. The rationale for this approach is that managers responsible for the completion of segments of a project should also bear responsibility for their costs.

The basic advantage of cost accounting by work packages is that cost overruns are easier to detect and corrective actions are more readily initiated. If traditional accounting is used, higher than expected expenditures in a department might be caused by any number of activities in which that department is engaged. When costs are measured and reported by activities, the source of the overrun is immediately apparent.

Several disadvantages of PERT/COST have limited its wider acceptance. Besides the obvious physical problem of converting an existing accounting system to a new form, there are theoretical questions about how to allocate overhead charges. Should all activities be charged for legal costs incurred on behalf of the entire project? If so, how should such costs be prorated? If overhead expenses are spread among activities, users may find themselves inundated with details, perhaps more than they can handle and certainly more than they wish to pass along to their customers or contracting agencies.

Once the difficulties of implementing PERT/COST have been overcome, the amount and speed of information feedback it fosters can lead to better project control. An example of the graphic reports which result from a PERT/COST application is shown in Figure 20.10. The solid line (line A) in the top section of the display shows the planned cost for the project. The timetable of expense is obtained by adding cumulatively the cost for all activities scheduled to be worked on during each weekly or monthly period. The resulting line is the budgeted cost schedule against which actual performance is compared.

The other two lines in the top section are a record of actual performance. The dashed line (line B) shows the amount of money so far expended. The dotted line (line C) indicates the value of work thus far completed, as measured on the basis of the original cost estimates. In the example given, there is an overrun in costs. Note that the actual-cost line B is running above the work-value line C.

Construction and interpretation of PERT networks are explained in Extension 20C. The most prominent feature of PERT is its statistical base, which accommodates novel projects such as construction of a new plant or production of any first-of-a-kind product.

FIGURE 20.10
Comparison of actual project costs and work completed with the budgeted schedule. A cost budget is prepared by estimating the expenses for all activities drawn in a project network. Cumulative costs for the budget are determined b' adding the costs of individual activities scheduled to be worked on in each time period. Inspectors then estimate the actual cost and amount of work done during each period as the project progresses.

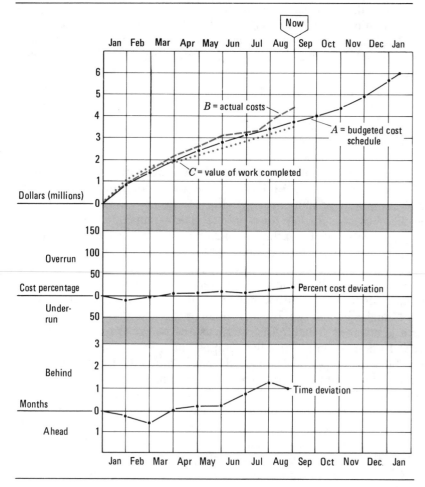

An indication of the seriousness of an overrun is given by the curves in the lower portion of the display. The cost-deviation line is based on the formula

$$\text{Percent cost deviation} = \frac{\text{actual cost} - \text{value of work completed}}{\text{value of work completed}} \times 100$$

where a negative value indicates an underrun. The graph at the bottom displays current completion progress compared with progress anticipated in the original budget. The cost values of two curves (line A and line C) are converted to a time value by measuring the *horizontal* distance between curve C and the point with the same dollar value on curve A. When curve C is below A, the project is behind schedule.

The manager of the project displayed in Figure 20.10 faces a serious situation. The current cost overrun is about 29 percent and work progress is about a month behind expectation. In addition, neither excesses in cost or time appear to be diminishing as the project nears completion. Both the magnitude of deviations and their trends guide the manager's decision about the time to act and the options to exercise.

When cost overruns get excessive and delays approach the danger point, a manager seeks the specific sources of overages. Cost curves are developed for major subdivisions of the project, such as progress on different components, testing facilities, fabrication output, site preparation, or basic operations. When the troublesome subdivisions are detected, more details are sought from reports on appropriate work packages and related activities. This information should reveal whether the overages are the result of general problems or of isolated instances of poor performance.

After the sources responsible for the overages are known, the project manager must still decide whether to exert resources to catch up with the original time and cost schedule or to negotiate a new schedule. Both options are loaded with pitfalls. A catch-up program can be treacherously expensive because resources tend to be expended inefficiently owing to crowded conditions, makeshift arrangements, and hurried planning. Often it is impossible to get the needed resources for a crash program even if it seems the desirable option. Then the project must be extended. The extra time may afford an opportunity to regain control over costs in order to approach the original total project budget. But the time extension could have hidden long-run penalties from loss of reputation and the ire of customers who had counted on the promised delivery schedule. Thus, neither a hasty catch-up race nor a contrived stretchout is a very satisfying solution, but either one is better than lack of awareness of the project difficulties until the full failure becomes evident at the end.

> The theme of numerous books written about project management is the money value of time. For instance, interest costs during a construction period are a major project expense.

INVENTORY ANALYSIS

Every organization needs supplies in order to operate. But stored supplies (accumulated inventory) are an idle resource. That the resource is idle does not mean that it is serving no purpose. It is available if needed. It serves as an insurance policy to protect the organization from breakdowns, delays, unexpected surges in demand, and other disturbances that could upset ongoing activities. Insurance is not free. The idle resource can be damaged or become obsolete before it serves any purpose. The objective is again to secure an economic balance between the cost of loss and the cost of preventing it.

Engineering activities are affected by inventory policies in a number of ways. Designers should be aware of current stock levels and delivery time for items that must be ordered. Engineers in production are concerned with supply of materials to the production process and the storage of finished products. Sales engineers must know which of their products are immediately available to buyers and which have delivery delays. Engineers themselves are occasionally an "idle resource" when organizations hire more engineers than are immediately needed for operations in order to be prepared for the award of a large contract.

Inventory analysis is a classic minimum-cost evaluation. One of the first formulas was developed by an engineer, F. W. Harris, in 1915. Since that time, expecially in the past two decades, an immense volume of literature has been written about inventory problems and solution methods. The content ranges from crude one-line rules to elaborate computer models. So much emphasis is owed to the universality of inventory usage; supplies are consumed in all kinds of organizations from family units to government agencies, and from one-person companies to multinational conglomerates.

Basic inventory cost relationships are explored in this section. The discussion is limited to short-run current conditions and all data are assumed to be known accurately. These restrictions are eased in Chapter 23, where analyses recognize the uncertainties of future events.

Under conditions of either assumed certainty or risk, two critical and related questions are how often should supplies be replenished and what amount should be replenished each time.

INVENTORY FACTORS

This condition demonstrates the Pareto principle: A few activities in a group of activities always account for the major share of the resources used or gained. (See the section on ABC analysis in this chapter.)

An organization orders, receives, stores, and uses a great variety of supplies, but relatively few classes of supplies usually account for most of the inventory costs. Therefore, analytical evaluations are applied only to key classes in which close control will promote the most savings. After an inventory item has been selected for analysis, the following information is needed:

D = *annual demand or usage for the item.*
M = *annual manufacturing rate for the organization producing the item.*
O = *order, or set-up, cost to procure the item.*

Order costs to obtain supplies from a vendor include the fixed cost of maintaining an order department and the variable costs of preparing and executing purchase requisitions. Set-up cost accounts for the expenses incurred in physically preparing for a production run to produce the item and includes the overhead costs for shop orders, scheduling, and expediting.

Order and set-up costs remain relatively constant regardless of order size.

H = *holding cost for keeping each item in storage 1 year.* Costs originating from many sources are consolidated under the heading of holding cost. In general, holding cost exhibits a fixed charge for a very low inventory level and then varies directly with additional quantities stored. Costs collected under H include operating and ownership expenses for a storage facility; handling charges composed of wages and equipment costs; protection and insurance for stored items; taxes on the items stored and the storage facilities; and damage, pilferage, or obsolescence of inventory during the storage period.

P = *price of the item.* The value of an item is its unit purchase price if it is obtained from an outside supplier, or its unit production cost if it is produced internally. The amount invested in an item being manufactured is a function of its degree of completion as it passes through the production process. The price per piece for outside purchases can vary as a function of quantity discounts for larger orders.

i = *interest rate for money invested in inventory.* The amount of money invested in an item is an amount of capital not available for other purposes. If the money were invested in a bank, a return on the investment would be expected. Therefore, a charge to inventory expense is made to account for this unreceived return from money represented by stored items. The interest rate charged, i, is the percentage return expected from other investments and is applied as an annual capital cost based on the price of the item, iP.

The interest rate charged is usually less than the MARR because risk is insignificant.

Design Variable

The number of items procured for each inventory replenishment is the design variable denoted as Q. If all the items needed for an entire year are procured at one time, then $Q = D$. When inventory is replenished more than once a year,

$$Q = \frac{D}{N}$$

where N is the number of procurement periods per year, or $N = D/Q$.

Directly Varying Costs

The cost H of holding items in storage and the interest on money invested in stored items are directly proportional to the inventory level. Immediately after a procurement is made, there are Q units in storage. If a constant usage rate is assumed and all the items in storage are depleted before another procurement is received, the average inventory stored is $Q/2$.

Then the total annual holding cost is

$$\text{Holding cost} = H\frac{Q}{2}$$

And the annual interest charge is the value of the average inventory, $PQ/2$, times the interest rate i, or

$$\text{Interest charge} = \frac{iPQ}{2}$$

which makes the total carrying cost

$$\text{Carrying cost} = (H + iP)\frac{Q}{2}$$

Inversely Varying Costs

The cost O associated with placing an order or preparing to produce an order occurs once each procurement period. Total annual preparation cost is then the order cost for each procurement period times the total number of periods per year N, or

$$\text{Procurement cost} = \frac{OD}{Q}$$

Constant Costs

If P is also assumed constant throughout a year, the total annual purchase cost is

$$\text{Purchase cost} = DP$$

Total Cost

The sum of the costs above is the total inventory cost C, or

$$C = \frac{OD}{Q} + \frac{(H + iP)Q}{2} + DP$$

Total cost = procurement cost + carrying cost + purchase cost

The total-cost formula can now be adapted to represent different real-world inventory situations. The terms in the formula are changed to fit specific inventory patterns, but the enduring objective is to minimize C by calculating the optimal quantity Q.

ECONOMIC ORDER QUANTITY

An organization ordering items from an internal or external source can usually minimize its costs by adhering to the economic order quantity Q. The calculations for Q are based on the assumptions and inventory pattern shown in Figure 20.11.

Under the stated assumptions, the previously encountered minimum-cost formula appropriately designates the economic order quantity (EOQ) as

$$\text{Design variable} = Q = \sqrt{\frac{\text{inversely varying costs}}{\text{directly varying costs}}}$$

$$= \sqrt{\frac{2OD}{H + iP}}$$

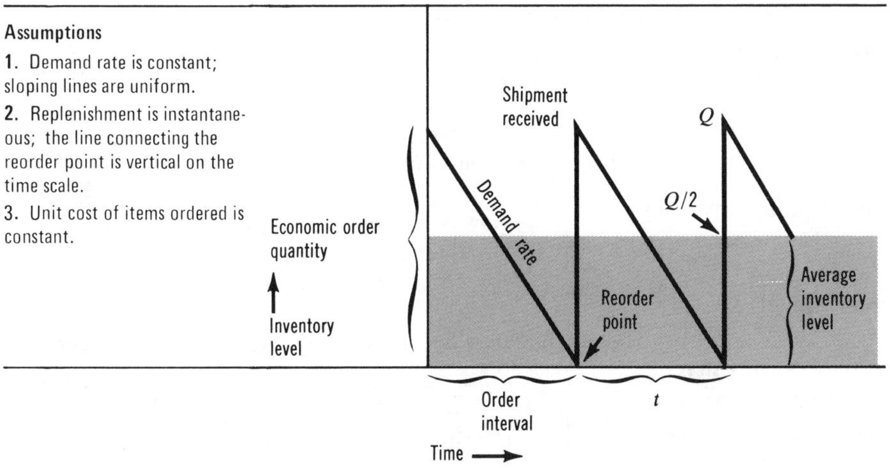
Assumptions
1. Demand rate is constant; sloping lines are uniform.
2. Replenishment is instantaneous; the line connecting the reorder point is vertical on the time scale.
3. Unit cost of items ordered is constant.
Economic order quantity
Inventory level
Shipment received
Q
Demand rate
Q/2
Reorder point
Average inventory level
Order interval
t
Time

FIGURE 20.11
Inventory usage pattern and controlling assumptions.

When the interest charge is included as part of the total holding cost H, as is often the case, the order-size formula reduces to

$$Q = \sqrt{\frac{2OD}{H}}$$

and related inventory values based upon Q are

$$\text{Numbers of orders per year} = N = \frac{D}{Q} = \sqrt{\frac{HD}{2O}}$$

$$\text{Order interval} = t = \frac{\text{working time per year}}{N}$$

$$\text{Total annual cost} = C = \frac{OD}{\sqrt{2OD/H}} + \frac{H}{2}\sqrt{\frac{2OD}{H}} + DP$$

$$= \sqrt{2ODH} + DP$$

Example 20.3
EOQ Based on Maximum Inventory Level

A retail paint distributor with several outlets in a large city annually sells 8000 gallons (30,280 liters) of Do-or-Dye stain. The costs of determining the order amount of each shade of stain, preparing the order forms, and transporting the product amount to $300. The firm has its own warehouse, which is used to store only its own merchandise. Warehousing and handling costs are prorated at $0.42 per gallon ($0.11 per liter) per year. The average price paid to the stain manufacturer is $6.30 per gallon ($1.664 per liter). Insurance and interest charges are 12 percent. What is the economic order quantity for Do-or-Dye stain, and how often should orders be placed? (*Hint:* Warehousing costs are based on the maximum level of inventory held because warehousing and handling expenses are fixed costs and remain the same whether the warehouse is full or empty.)

Solution 20.3

Relating the given costs to formula symbols, we have

D = 8000 gal/year (30,280 l/year)

O = $300/order

H = $0.42/gal/year ($0.11/l/year) on maximum inventory

P = $6.30/gal($1.664/l) wholesale price

i = 12% of P on the average inventory level

The statement of the problem indicates that the storage space allowed for paint and stain is not used for any other purpose. This means that space is provided for the maximum number of gallons received in an order. As the stain is sold, less space is occupied, but the total cost of having the space available is constant. Therefore, the directly varying costs are

$$\text{Carrying costs} = HQ + \frac{iPQ}{2} = \frac{(2H + iP)Q}{2}$$

Then the economic-lot-size formula is

$$Q = \sqrt{\frac{2OD}{2H + iP}}$$

and substituting figures into the EOQ expression, we get

$$Q = \sqrt{\frac{2 \times \$300 \times 8000}{2(\$0.42) + 0.12(\$6.30)}} = \sqrt{\frac{\$4,800,000}{\$1.596}}$$

$$= 1734 \text{ gal/order (6556 l/order)}$$

The number of orders per year is $N = 8000/1734 = 4.6$ orders per year, which makes the order interval in weeks $t = $ (weeks/year)/$N = 52/4.6$, or about 11 weeks between orders.

QUANTITY DISCOUNTS

Price discounts are offered by some suppliers to encourage larger orders. "Cheaper by the dozen" is a household slogan which reflects availability of discounted prices for large-volume purchases. Lower freight rates and handling charges for bigger shipments create the same effect as do price discounts. These benefits must be measured against the incremental increase in carrying costs required to accommodate purchases larger than the economic order quantity.

Total-cost curves for a commodity subject to a price discount are shown in Figure 20.12. The top curve C is the annual inventory cost at price P for different order quantities. The lower curve C' shows annual inventory cost based on a lower price P'. However, price P' is applicable only for order size Q' or larger, as is indicated by the solid portion of curve C' to the right of Q'. Because C' is lower at Q' than C is at its minimum point Q, the minimum-cost ordering policy is to order Q' units each time replenishments are needed. Note that if the cutoff quantity for price P' were at Q'' on curve C', it would be less costly to order quantity Q at price P.

The procedure outlined in Figure 20.13 discloses the minimum-cost order quantity for most common quantity-discount patterns. It is applicable to a commodity with several price breaks that would be graphed as a family of curves for different prices similar to the two-curve family in Figure 20.12.

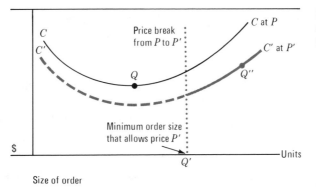

FIGURE 20.12
Minimum-cost order quantities for purchase-price discounts.

To amplify the quantity-discount procedure, consider the costs shown in the margin for a product with a large annual demand.

For the lowest unit price ($P3 = \$0.45$), the order quantity is (step 1)

$$Q_{P3} = \sqrt{\frac{2 \times 25 \times 400,000}{0.06 + 0.2(0.45)}} = 11,547 \text{ units}$$

which is less than the 25,000 order size needed to qualify for the lowest unit price. The annual total cost at the $0.45 price break is (step 2)

$D = 400,000$ units/year
$O = \$25$/order
$H = \$0.06 + 0.2P$/unit, based on average inventory
$P1 = \$0.50$/unit on orders up to 9999
$P2 = \$0.47$/unit on orders from 10,000 to 24,999
$P3 = \$0.45$/unit on orders greater than 24,999

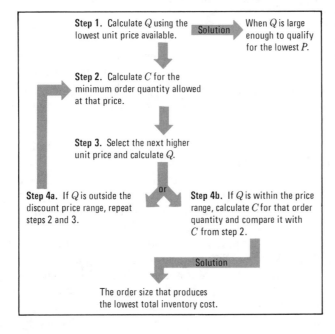

FIGURE 20.13
Flowchart for order-size calculations when quantity discounts are allowed.

$$C_{P3} = \frac{(25)(400,000)}{25,000} + [0.06 + 0.2(0.45)]\frac{25,000}{2} + 400,000(0.45)$$

$$= 400 + 1875 + 180,000 = \$182,275$$

The next higher price is $P2 = \$0.47$, and Q at this price is (step 3)

$$Q_{P2} = \sqrt{\frac{2 \times 25 \times 400,000}{0.06 + 0.2(0.47)}} = 11,396 \text{ units}$$

This time the order size is within the price discount range. The total cost is then calculated as (step 4b)

$$C_{P2} = \frac{25 \times 400,000}{11,396} + [0.06 + 0.2(0.47)]\frac{11,396}{2} + 400,000(0.47)$$

$$= 877.50 + 877.50 + 188,000 = \$189,755$$

The solution is to select the lowest-total-cost order size: Order 25,000 units at $0.45 per unit. In this example the solution was obtained by passing once through the steps outlined in Figure 20.13. If $P2$ had been less than 10,000 units, it would have been necessary to pass again through steps 2 and 3 using the next higher price, $P1$. That the EOQ at $P1$ equal to $0.50 does not yield a lower total inventory cost is demonstrated by

$$C_{P1} = \sqrt{2ODH} + DP = \sqrt{2 \times 25 \times 40,000[0.06 + 0.2(0.50)]} + 400,000(0.50)$$

$$= 1789 + 200,000 = \$201,789$$

Thus, the largest price break provides the lowest cost, but it should be remembered that bigger orders always increase the dangers associated with carrying higher inventory levels. The degree of risk is gauged by the stability of past demand, the resale value of stock, and market trends.

ECONOMIC PRODUCTION QUANTITY

An economic production quantity (EPQ) is associated with a manufacturing environment, whereas the economic order quantity (EOQ) is more common in purchasing situations. The difference between the two quantities is a result of delivery time. EOQ calculations are based on an assumption of instantaneous delivery time. EPQ calculations realistically show that inventory is gradually built over a period of time, as illustrated in Figure 20.14.

If the manufacturing rate M exactly equals the usage rate D, the items will be used as fast as they are produced. If M is greater than D, inventory is accumulated at the daily rate of $(M - D)$ divided by the number of operating

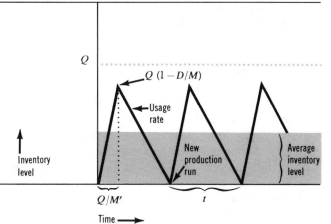

Assumptions

1. Demand rate is constant.

2. Production rate is constant and equal to or greater than the usage rate.

3. Order cost includes the cost of setting up production lines and procuring material for the production run.

4. Unit costs are constant.

FIGURE 20.14
Inventory accumulation-usage pattern and controlling assumptions.

days per year. The inventory level reaches a maximum after Q/M' days, where M' is the daily production rate. This level is equal to $(M - D)Q/M$, and the average inventory is half this amount, or $(Q/2)(1 - D/M)$. Thus, the total cost for a production situation is

$$C = \frac{OD}{Q} + \frac{HQ}{2}\left(1 - \frac{D}{M}\right) + DP$$

and the economic production quantity is

$$Q = \sqrt{\frac{2OD}{H(1 - D/M)}}$$

INVENTORY PRACTICES

The methods and formulas presented in this chapter to determine minimum-cost inventory policies are indicative of what can be done. They should not be used indiscriminately because each formula was based on specific assumptions. Only the tip of the inventory iceberg was exposed. A multitude of adaptations has been developed to fit different situations. For instance, more realistic inventory models include the cost of running out of stock and the effects of variable delivery times for ordered materials.

Inventory control involves much more than simply the ordering policy. *Safety stock*, a reserve supply beyond the amount consumed by average usage, is usually maintained to cushion the consequences of running out of needed items. This reserve obviously increases the carrying cost, but the buffer it provides may be worth the extra holding expense when demands or delivery

Risk analysis applied to inventory policy is discussed in Chapter 23

Unit costs for materials can be accounted for in a number of ways. *Original cost* assumes that the oldest material is used first [called FIFO (first in, first out)]. *Last cost* assumes that the most recently acquired material is used first [called LIFO (last in, first out)]. *Replacement cost* assumes that the next arrival will cost the same as the last one [called NIFO (next in, first out)].

times are uncertain. Even when economic order quantities and acceptable safety-stock levels are known, the administration of material controls is a demanding and critical task. Records must be maintained to know when stock levels have dropped to a reorder point, the condition of reserve supplies must be checked, and materials must be physically handled to make room for new deliveries while sustaining access to existing stores.

Fundamental departures from conventional order-point techniques have evolved in recent years. *Material requirements planning* (MRP) is a form of material management based on the requirements of the end product. That is, orders for components are based on known demands for finished products in MRP systems instead of historical data to estimate future needs for parts, as is done in statistical (order-point) inventory control. MRP is a classic example of production advances afforded by computers; huge amounts of data are involved in MRP applications to coordinate the demand for components required for the end product, subcomponents needed to produce components, sub-subcomponents needed for subcomponents, etc. Tracing and updating material requirements in conformance with the laddered needs of the master production schedule would have been completely impractical in precomputer days, but MRP is now a preferred alternative to order-point techniques for many manufacturing applications.

MRP applications typically occur in larger manufacturing organizations where products are progressively refined or assembled.

At the other end of the spectrum from computer-driven MRP is the Japanese *Kanban* system. It is known as the "just-in-time" technique and relies on manual monitoring of inventory movements to have material arrive exactly when it is needed in the production process. Between MRP and Kanban are other inventory-management practices designed to serve special functions or to take advantage of particular physical facilities. Some warehousing systems are automated to mechanically fill orders on command and to place orders for replenishment automatically according to programed decision rules. Such adaptations typify the quest for greater production economy.

ABC INVENTORY CLASSIFICATION SYSTEM

Not all inventory situations require the sophisticated controls associated with MRP and Kanban. Small companies may need only a system as simple as the *two-bin plan* in which the supply of an item is physically divided into two parts, one large and one small. The smaller portion is not touched until the large stock has been consumed. When the reserve supply is tapped, a replenishment order is called for. A more general and widely used method of inventory management is called the ABC classification system. It utilizes the *Pareto principle*.

Few professionals have achieved the lasting fame of Vilfredo Pareto. And he did it in three professions: engineering, economics, and sociology. In 1869, Pareto graduated with a degree in engineering from the Polytechnic Institute of Turin, Italy. For 13 years he worked as an engineer and manager in railways and iron works. Then he turned to economics, holding the chair of political economy at the University of Lausanne in Switzerland. Still later he switched to sociology and brought to this field his rational methodology.

While studying the distribution of wealth in different nations, Pareto observed that the greater the personal fortune, the smaller the number who possess it. Plotted as a histogram with wealth increasing to the right on the horizontal axis, the tallest bar representing the population with the lowest income appears on the left. Heights of the bars decrease for greater wealth. A chart that exhibits this pattern is known as a Pareto diagram.

Companies and federal installations routinely stock thousands and even tens of thousands of different items in their warehouses. The same amount of attention is not deserved for every type of item. The Pareto principle is applicable here because a few items in an inventory of materials account for most of the dollar volume. Selection of the items for concentrated attention is guided by the assignment of A, B, and C classes which follow the Pareto distribution.

The relative importance of each inventory item is measured by how much the total value of purchases is for that item each year. This measure is termed the *activity level*. It shows the cash flow per period. An item that costs 10 cents per unit with annual purchases of 1 million units has an activity level of $100,000. It is considered more important than an item which costs $200 per unit with only 100 purchases annually.

It is expected that 10 to 20 percent of the total number of items in stock constitute 60 to 80 percent of the total annual value of purchases. These are classified as A items. At the other extreme, about 50 to 60 percent of the inventory items constitute only 5 to 10 percent of the value and are classified as C items. Items between the A and C ranges are in the B class. Exceptions may be made for highly critical but low-activity-level items, but the typical distribution of classifications follows the curve in Figure 20.15.

A Pareto diagram is a vertical bar chart in which classes of data are arranged in descending order from left to right according to the frequency of their observation or magnitude of their effect. See example below.

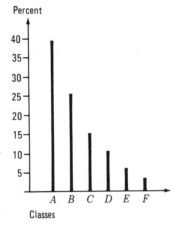

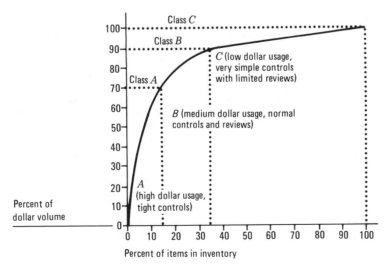

FIGURE 20.15
Distribution of inventory items in ABC classes.

Besides being used for inventory management, Pareto analysis has been used in the investigation of accidents, crime, absenteeism, and quality assurance. "First things first" is the message.

To allow more attention for A items, C items in storage are left almost unattended under the theory that whatever problems arise in the C class will be relatively minor. Total inventory management costs are expected to be lower by committing extra capital to have larger safety stocks for little-used C items than to keep records accurate enough to avoid C-class stockouts. The opposite is true for A items. The tightest possible controls are exercised for A items through close supervision of deliveries, frequent audits of records, and strict policing of ordering policies.

Review Exercises and Discussions

Exercise 1   Given the data below for a research and development project, construct an arrow network and boundary timetable. [See Extension 20B for an explanation of PERT time estimates calculated according to the formula $t_e = (a + 4m + b)/6$.]

Research and Development Project

Activity	Postrequisites	Time Estimates (Weeks)			
		a	m	b	t_e
A (1,2)	A < B, C, D	4	5	12	6
B (2,4)	B < E	1	1	7	2
C (2,3)	C < E, F	2	3	4	3
D (2,6)	D < H	3	3	3	3
E (4,5)	E < G	3	4	5	4
F (3,5)	F < G	1	7	7	6
G (5,6)	G < H	1	2	3	2
H (6,7)		3	3	9	4

Solution 1   The arrow network, complete with start- and finish-time calculations, is shown with a boundary timetable in Figure 20.16.

Exercise 2   Continue to reduce the duration of the project charted in Figure 20.9. Calculate the cost of the minimum duration and show the resulting schedule on a bar chart.

Solution 2   Two additional cuts can be made to reduce the project from the described 15-day duration to a minimum 13-day duration. The first cut takes a day off activities 2,3 and 2,4 at a cost of $160 + $110 = $270. The next cut involves activities 2,3, 2,4, and 1,3 at a cost of $160 + $110 + $150 = $420. No further cuts are possible according to the conditions stated in Table 20.2. The minimum-duration schedule and associated costs are shown in Figure 20.17, where all activities are now critical. The project cost for a minimum duration is greater than the minimum-cost schedule ($7810 + 15 × $220 = $11,110) by $250.

Exercise 3   One department of a plant produces switch plates which are used in the assembly department of the same plant. Switch plates can be produced at the rate of 4000 units

FIGURE 20.16
PERT applied to a research and development project.

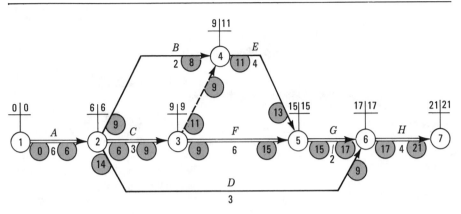

PERT BOUNDARY TIMETABLE

| Activity | Description | EARLIEST | | LATEST | | Total Float | Critical |
		Start	Finish	Start	Finish		
1,2	Complete design	0	6	0	6	0	*
2,4	Layout, mechanical	6	8	9	11	3	
2,3	Layout, electrical	6	9	6	9	0	*
2,6	Train operators	6	9	14	17	8	
3,4	Dummy	9	9	11	11	2	
3,5	Build controls	9	15	9	15	0	*
4,5	Fabricate frame	9	13	11	15	2	
5,6	Assemble prototype	15	17	15	17	0	*
6,7	Test prototype	17	21	17	21	0	*

per day. The assembly department needs only 182,000 per year. The cost of authorizing switch-plate production and setting up the machines is $300 for each run. The plant operates 260 days per year. The annual cost of holding one unit in storage is $0.10. How many switch-plate production runs should be scheduled each year?

The EPQ for the given data is calculated as follows:

Solution 3

$$D = 182,000 \text{ units/year}$$

$$O = \$300/\text{production run}$$

$$H = \$0.10/\text{unit/year}$$

$$M = 4000 \times 260 = 1,040,000 \text{ units/year}$$

$$Q = \sqrt{\frac{2 \times 300 \times 182,000}{0.10 \times [1 - (182,000/1,040,000)]}} = 36,400 \text{ switch plates/run}$$

$$N = \frac{182,000}{36,400} = 5 \text{ runs/year}$$

FIGURE 20.17
Minimum-duration bar graph for
the project begun in Figure 20.9.

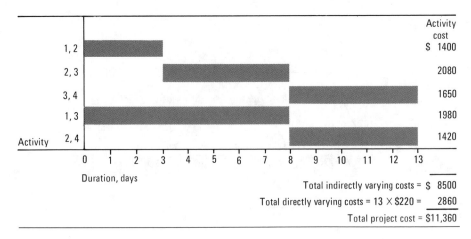

Total indirectly varying costs = $ 8500
Total directly varying costs = 13 × $220 = 2860
Total project cost = $11,360

PROBLEMS

20.1 Construct an arrow-network segment for each set of restrictions given below:

(a)	(b)	(c)	(d)
A < C	A < D	A < C	A < C,D
B < C	B < D,E	B < D,E	B < C,D
C < E	C < E	C < D	C < E
D < E	D < F	D < F	D < E
	E < F	E < F	

20.2 Construct an arrow network to conform to the relationships below. Note that the activities are not in alphabetical order and that there are redundant restrictions.

(a)		(b)	
Activity	Must Precede	Activity	Must Precede
A	C	K	X
B	A,C,F	R	X
D	C,F	D	A,T,N,R
E	A,C	L	X
		N	B,R,X
		T	L
		A	R,L
		B	C,X

20.3a   Based on the following restriction list, construct a network.

Activity	Duration	Restriction
A	3	A < E,F
B	5	B < C,D,E,F
C	3	C < H
D	4	D < G,H
E	6	E < G,H
F	7	
G	4	
H	5	

(b) From the given durations, calculate the boundary times. What is the length of the critical path?

(c) If the ES for the first event is 23, determine the LF for the last event.

20.4   The list below gives nodal-numbered activities and their durations. Each activity is described by an *i* and *j* node. What is the minimum time in which the project can be completed?

Activity	1,2	1,5	2,3	2,4	2,5	3,4	4,6	5,6
Duration	5	4	3	4	Dummy	Dummy	2	7

20.5   The activities and restrictions for a fabricating project are given below. Construct an arrow network and develop a boundary timetable for the project.

Activity	Description	Duration	Restrictions
A	Make parts list	2	A < B,C
B	Prepare routings	3	B < D
C	Order and procure materials	4	C < E,F
D	Make schedule	3	D < E,F
E	Process parts for subassembly 1	3	E < G
F	Process parts for subassembly 2	2	F < H
G	Assemble subassembly 1	2	G < I
H	Assemble subassembly 2	4	H < I
I	Final assembly	1	

20.6   Given the following activity list, time estimates, and restrictions, draw an arrow network and complete a boundary timetable.

Activity	Duration	Postrequisites	Activity	Duration	Postrequisites
A	2	B	H	6	M
B	4	G,F,L	I	11	K
C	3	E,I,J	J	7	K
D	4	I,J	K	12	M
E	7	G,F,L	L	9	K
F	16	M	M	5	
G	6	H			

20.7 Plans are being made to remodel a shipping and receiving area. The network for the project is shown below. Also shown are earliest and latest start times, normal and crash activity durations, and normal and crash costs. The remodeling project has a $1100 fixed burden for overhead costs and supervision charges plus a $100/day directly varying cost that results because extra help and additional storage space are needed while the old area is being torn up.

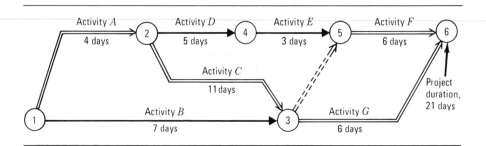

Activity	Normal Times				Crash Times	
	Days	ES	LS	Cost	Days	Cost
A (1,2)	4	0	0	$200	3	$250
B (1,3)	7	0	8	300	5	450
C (2,3)	11	4	4	800	8	950
D (2,4)	5	4	7	100	2	180
E (4,5)	3	9	12	300	3	300
F (5,6)	6	15	15	180	4	240
G (3,6)	6	15	15	560	4	800

(a) Develop a bar chart similar to that of Figure 20.9 to show the sequence, float, and cutting costs of project activities.

(b) Determine the duration and total cost for the project that minimize the sum of directly and inversely varying costs.

20.8 A closer look at the activity duration costs for the project described by the network for Problem 20.7 leads to a conclusion that all the activity costs do not linearly increase from normal to crash amounts. A revised cost schedule is shown on page 621, where an intermediate duration and associated cost are given for each activity.

	Normal		Intermediate		Crash	
Activity	Days	Cost	Days	Cost	Days	Cost
A	4	$ 200	4	$ 200	3	$ 250
B	7	300	6	350	5	450
C	11	800	10	870	8	950
D	5	100	3	140	2	180
E	3	300	3	300	3	300
F	6	180	5	210	4	240
G	6	560	5	600	4	800
Totals		$2440		$2630		$3170

Assuming the given costs vary linearly between each pair of the three designated values and that the directly varying costs given in Problem 20.7 are still accurate, determine the minimum-cost project duration and its total cost.

20.9 Construct a network based on the data below and determine:
    (a) The normal duration of the project
    (b) The minimum crash duration
    (c) The minimum cost to cut the project by 1 day, 2 days, and 3 days (indicating which activities are cut and their cost)

		Activity Duration		Cost to Cut Activity by 1 Day
Activity	Postrequisites	Crash	Normal	
V	—	1	1	No cut possible
F	V,M,O	6	6	No cut possible
B	F	6	7	$30
D	S	3	5	$50/day
M	—	2	3	$40
S	V,M,O	5	6	$70
A	F	2	4	$25/day
O	—	3	4	$85
C	D	2	2	No cut possible
K	A,D	3	4	$80

20.10 Given the project history in Figure 20.18, which shows the budgeted cost schedule, actual cost, and value of work completed,
    (a) Determine the approximate cost deviation and time deviation at time "now."
    (b) Analyze the project performance with respect to a scheduled completion date of December 30.

20.11 Depending on their responsibilities, different engineers in a production organization would likely have different opinions about the most advantageous quantity of a certain item to produce in each production run. State whether the following individuals would probably prefer longer or shorter production runs and what reasons they would give to support their contentions:

FIGURE 20.18
Time and cost performance records.

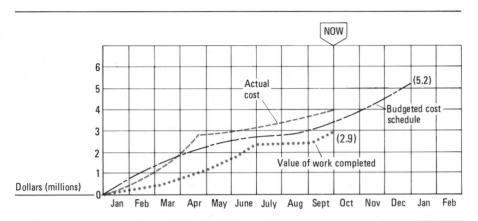

(a) Sales engineer concerned with customer relations
(b) Material-handling engineer concerned with warehousing
(c) Design engineer concerned with purchasing
(d) Engineering economist concerned with capital budgeting
(e) Production engineer concerned with the production line

20.12 A company buys $100,000 worth of a certain material each year. Order costs are 1 percent of the amount of each order and carrying costs are 10 percent of the average inventory. If annual order and carrying costs are equal, how many weeks of supply should be ordered at one time?

20.13 A type of raw material is used at the rate of 80,000 pounds (36,288 kilograms) per year. The current price is $1.00 per pound ($2.205 per kilogram). Storage costs are estimated at $0.09 per pound ($0.198 per kilogram) per year on the maximum inventory value. The cost of placing an order is $10 and the lead time is negligible. What is the EOQ?

20.14 "I propose that we apply EOQ calculations to all our inventory procurements," said the machine shop's all-purpose engineer. "At the present time we have one all-purpose purchasing agent who operates in a very informal manner, so informal that we don't even find out how much it costs us to place an order. We know he gets paid $32,000 a year, occupies an office that we figure costs $4000 a year, has a telephone bill averaging $1000 per year, and incidental expenses of about $250 per month. Since our small company has no major material requirements, he just circulates around the building to see what supplies are needed. Then he phones the orders to local merchants and writes the checks himself to make all the payments. Apparently he places about 2500 orders each year, ranging in size from a few dollars to $600 or $700. That comes out to about $16 ordering cost per order. We can probably cut the number of orders down to 1500, which means a savings of (2500 − 1500 orders) $16/order = $16,000/year."

Discuss three attributes of the present system that should receive consideration before changes are made to formalize the inventory policy.

20.15 A wholesaler forecasts annual sales of 200,000 units for one product. Order costs are $75 per order. Holding costs of $0.04 per unit per year and interest charges of 12 percent are based on the average inventory level. The cost to the wholesaler is $0.80 per unit acquired from the factory.

    (a) What is the EOQ?

    (b) What is the annual inventory cost?

    (c) What is the time between orders based on 250 working days per year?

20.16 A heavy equipment manufacturing company produces components which are later fabricated into finished products. One product is produced at the rate of 6 units per day. A component used in this product can be manufactured at the rate of 30 units per day and has annual carrying costs per unit of $1.50 based on the average inventory level. Administrative and set-up costs total $240 per order. If the company operates every day of the year, what is the value of the EPQ?

20.17 Determine the economic production quantity for the following conditions:

Set-up costs per production run	$80
Variable production costs per unit produced	$0.30
Percentage charge for interest and insurance	25
Selling price per finished unit	$0.60
Manufacturing rate per year, units	110,000
Usage rate per year, units	20,000

20.18 A manufacturer of children's toys uses approximately 200,000 nuts and bolts of one size each year. The nuts and bolts are purchased in standard quantities of 5000 at a cost of $50 per package. The storage costs of $0.001 per year to hold one bolt and matching nut are based on the maximum expected inventory level. Interest and insurance charges are 13 percent of the average value of inventory on hand. Typical order costs for preparing a purchase order, mailing, receiving, inspecting, and transporting are $25 for each purchase. The company operates 250 days per year.

    (a) What is the minimum-cost order quantity?

    (b) What is the time interval between orders?

20.19 Determine the economic order size for the following conditions, where n refers to the number of units, and holding costs are based on the average inventory size.

$$O = \$10 + \$0.01n$$

$$H = \$2n + \frac{0.01P}{n^2}$$

$$P = \$40/\text{unit}$$

$$D = 800 \text{ units/year}$$

20.20 The Hour Glass Girdle Works has been offered a 2 percent discount if it purchases its annual material requirements in equal quarterly amounts. The costs are given on page 624. Note that the carrying cost is based on maximum amount in stock at any time.

Ordering cost = $30/order

Carrying cost = 10% of maximum inventory level

Annual demand = $270,000

Should the girdle works use the discount?

20.21   Four thousand tons (3,628,800 kilograms) of raw material are used each year. Order costs are $20 per order; carrying costs are $8 per ton ($0.0088 per kilogram) based on the maximum storage requirement. The supplier has offered to reduce the $40-per-ton ($0.044-per-kilogram) price by 5 percent if the minimum order is 500 tons (453,600 kilograms) and by another 5 percent if the minimum order is 1000 tons (907,200 kilograms). The capacity of the present storage facility is 250 tons (226,800 kilograms). Any increase in capacity will increase the carrying costs in direct ratio. Show the calculations and assumptions needed to prove whether or not the quantity discount should be utilized.

20.22   Determine the economic lot size under the following conditions:

Annual usage rate is 3000 units.

Cost of acquisition is $8 per order.

Interest, insurance, and damage charges are 12 percent of the value of the average inventory.

Annual warehousing cost is $0.08 per unit, based on the maximum inventory level of 500; if the inventory level exceeds the 500 limit, an additional warehouse must be leased to make the cost $0.12 per unit for the maximum number of units stored.

The purchase price of $3 per unit is subject to a

4 percent discount on orders of 200 or over

7 percent discount on orders of 400 or over

9 percent discount on orders of 500 or over

10 percent discount on orders of 600 or over

EXTENSIONS

20A   Network Activities

One of the tougher tasks in constructing either a PERT or a CPM network is breaking a project into its component activities. There are few rules to guide decomposition. An early step is to peg the beginning and end points of the total project. This sounds easy, but it can be tricky. For instance, suppose that the project is to paint a room. When does the project begin? Does it begin when the paint is selected, when an order is placed for painting supplies, when the paint brush first touches a surface, and so forth? Likewise, how do you tell when the job is finished? After all the paint is on? After it dries? After clean-up? After passing inspection? A good scheduler must identify the distinct acts which mark the boundaries of the project.

Not only must the project have a distinct beginning and end, but all activities throughout the project must also have distinct beginning and end events. The room-painting project mentioned above might be only part of the renovation of an entire office building. The following considerations might influence the activity divisions:

1 Work done by each agency, subcontractor, or department should be separated into individual activities.
2 Improvement of administrative control comes from activity breakdowns that recognize special crafts, such as salespeople, carpenters, department heads, plumbers, police, and engineers.
3 Structural elements such as digging a foundation, laying floors, or constructing walls are natural activity divisions.
4 When a certain resource is limited as it might be with one-of-a-kind machines or a special technician, work requiring that resource should be separated from the other activities.

The object of activity designation is to separate the entire project into easily distinguishable steps, each based on separate tasks with measurable outcomes.

Activity Times

Time allocations for completing an activity are generally based on expectations of normal working conditions and the usual standards of performance. However, if it can be anticipated in advance that unusual obstacles must be overcome, estimated times should reflect the situation. For example, if construction work must be done outdoors in a northern locality, more time must be allowed to complete the activity during the winter than during the summer. Similarly, more time should be scheduled when shortages of experienced workers occur, when hazardous weather conditions develop, or when unfamiliar project requirements are forecast.

Activity Restrictions

Another phase of scheduling involves determining the sequence in which activities are to be done. Sequencing is usually easier than is setting activity descriptions or estimating times. We tend naturally to think of activities in a more or less logical order of occurrence. Therefore, as each activity is considered, its location relative to other activities in the project can be set by determining what activities must precede it and follow it. Some schedulers prefer to work backward from the end event because they feel it is easier to think of work which must precede an activity than to think of work that must follow it. Others prefer to build from the first event, listing in order each following or postrequisite activity. Either way is acceptable as long as it results in the final production of accurate sequencing.

Example 20.4
Planning a Survey

Even small projects may benefit from network analysis, if for no other reason than that systemized procedures lead to more thorough planning. Suppose that city officials wish to know public opinion concerning some industrial development involved in an urban-renewal project. Those in charge of the project have decided that it will start with the approval of the survey and will end with the analysis and distribution of the results. Between the beginning and end events a huge number of activities could be inserted if a finely detailed schedule were needed. To avoid becoming bogged down with minute details, most schedulers first make a skeletal activity list, which includes only major functions. After putting this skeletal list in order of

occurrence, schedulers can subdivide activities to whatever level of detail they may need.

Figure 20.19 is a worksheet for the attitude-survey project. Each activity is described by a short phrase, usually a verb and a noun. Symbols are assigned to activities to reduce writing time for later notations, such as the postrequisite listing. A postrequisite activity is one that can begin immediately after but not before the preceding activity is completed. A comment column is used in the illustration to record the events that bracket each activity. Comments could also include responsibility or crew assignments and required resources.

FIGURE 20.19
Worksheet for activities involved in an attitude-survey project. All estimated times are in units of 8-hour workdays. To better define each activity, the beginning event (B) and the end event (E) are described in the comment column. All activities could be subdivided to allow more detailed descriptions. For example, the activity "analyze results" could include the subactivities "form analysis team," "tabulate data," "apply statistical tests," "interpret test results," "write preliminary report," "review report," "write final report," "approve report," and "submit report." The appropriate degree of detail depends on the intended use of the network.

Activity Description	Estimated Time	Activity Symbol		Post-requisite	Comments
Approve survey	10	A	<	B, C, D	(B) Council meeting (July 11) (E) Signed approval
Hire canvassers	5	B	<	E	(B) Advertise for workers (E) Select personnel
Design questionnaire	8	C	<	E, F, G	(B) Hire public-relations firm (E) Questionnaire approved
Select households	3	D	<	H	(B) Get city directory (E) Selected addresses listed
Train canvasers	5	E	<	H	(B) Training plan received from public relations firm (E) Training completed
Publicize survey	4	F	<	H	(B) Contact newspapers (E) Notices printed
Print questionnaire	3	G	<	H	(B) Deliver form to printer (E) Receive all copies
Conduct survey	7	H	<	I	(B) Start canvassers on routes (E) Completed questionnaires
Analyze results	12	I			(B) Questionnaires to analysts (E) Receive analysis report

QUESTIONS

20A.1 Draw the network for the survey project represented in Figure 20.19. Avoid crisscrossing arrows and number the nodes chronologically, starting from 1 for the first node.

20A.2 Determine the critical path for the network in the solution to Question 20A.1.

20A.3 Construct a boundary time table from the data in the solution to Question 20A.2. List according to the activity nodal numbers and include activity descriptions, earliest and latest start and finish times, and the total float.

20A.4 Assume that you must interview workers who build specialized assemblies. You are to make a network diagram of the sequence of operations to record what must be done if the experienced workers are not available when a special assembly is needed in the future. There are eight steps in the assembly process, identified as A to H. To appreciate the difficulties of gathering data, read the following paragraph only once (very slowly) or have someone read it to you and then organize the information to enable construction of the network. What suggestions can you make for improving the extraction of information from interviews?

"OK, here's how we make it," said the lead worker. "First I start with A. I have to finish A before Ed and Bill can work on F and H. You can't do G or B before H is done. F also comes before G and C. And B has to come after D. C is the last thing we do but we can't start it until G and B are done. Oh yeah, we have to get E from the shipping department in order to start C. There's another thing too; Ed usually does G and B, but he can't do them at the same time."

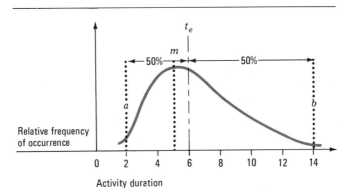

FIGURE 20.20
Relationship of a, b, and m to t_e.

20B Program Evaluation and Review Technique (PERT)

A statistical approach to network analysis characterizes the program evaluation and review technique (PERT). It is appropriate for unfamiliar or experimental projects. In such projects it is difficult to agree on one most likely time estimate for each activity because there is little or no precedent for the operations involved. In research and development work, only very naive or adventurous managers will commit themselves to a definite completion date when they are unsure of what they have to do or how they must do it.

Much of the hesitancy concerning time estimations is relieved by allowing a range of estimates. Instead of a commitment to a single duration estimate, a most likely time m is bracketed between an optimistic a and a pessimistic b estimate of performance time. The most likely time is the duration that would occur most often if the activity were repeated many times under the same conditions. The optimistic and pessimistic estimates are the outside limits of completion time when everything goes either all right or all wrong.

It is assumed that there is a very small probability that an actual completion time will fall outside the range a to b and that the proportion of durations within the range follows a beta distribution. If an activity were performed many times and the relative frequency of completion times were plotted, a distribution curve as shown in Figure 20.20 would result. The mean point dividing the area under the curve into two equal parts is called the expected time t_e; it is calculated as a weighted average of the three time estimates according to the formula

$$t_e = \frac{a + 4m + b}{6}$$

where $a \leq m \leq b$. After the expected times are computed for each activity, they are treated the same as are single time estimates in determining the critical path and boundary times, as illustrated in Figure 20.21.

Each activity in a PERT network also has a variance associated with its completion time. This variance measures the dispersion of possible durations and is calculated from the formula

$$\sigma_{ij}^2 = \left(\frac{b - a}{6}\right)^2$$

where σ_{ij}^2 is the variance of an individual activity with pessimistic and optimistic performance time estimates of b and a, respectively.

When $b = a$, the duration of an activity is well known and the variance is consequently zero. A wide variation in the outside limits of estimated times produces a large variance and indicates less confidence in estimating how long it will actually take to complete the activity. Sample calculations of activity variances are provided in Figure 20.21.

Having a variance for each activity offers the interesting potential of calculating the probability that a certain time scheduled for an event will be met. The scheduled start SS is called a *milestone*; it represents the time of the planned accomplishment of an event of particular managerial importance. This significant date may originate from the need to

FIGURE 20.21
Expected times, variances, and critical-path calculations.

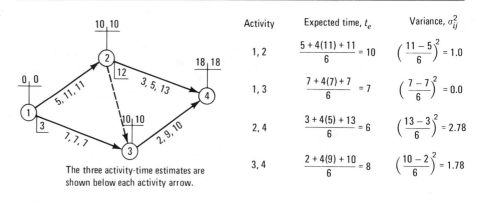

The three activity-time estimates are shown below each activity arrow.

Activity	Expected time, t_e	Variance, σ_{ij}^2
1, 2	$\dfrac{5 + 4(11) + 11}{6} = 10$	$\left(\dfrac{11 - 5}{6}\right)^2 = 1.0$
1, 3	$\dfrac{7 + 4(7) + 7}{6} = 7$	$\left(\dfrac{7 - 7}{6}\right)^2 = 0.0$
2, 4	$\dfrac{3 + 4(5) + 13}{6} = 6$	$\left(\dfrac{13 - 3}{6}\right)^2 = 2.78$
3, 4	$\dfrac{2 + 4(9) + 10}{6} = 8$	$\left(\dfrac{10 - 2}{6}\right)^2 = 1.78$

complete a vital activity by a given date so that other portions of the project can commence, or it may result from contractual obligations outside the project.

A milestone date is compared with the ES of an event. The ES evolves from a cumulative total of activity expected durations to the milestone event. Similarly, the variance associated with that event is the sum of the variances of the activities along the path determining the milestone ES. The difference between the SS and the ES is the numerator in the equation

$$Z = \frac{SS - ES}{\sqrt{\Sigma\sigma_{ES}^2}}$$

where Z is the probability factor from which the relative likelihood is derived, and $\Sigma\sigma_{ES}^2$ is the sum of the individual variances of the activities used to calculate ES.

The actual completion times for the milestone event are assumed to be normally distributed about ES. Therefore, the probability factor is a function of the normal-curve areas given in Appendix F. A Z value of zero means that the probability of completing the event on the scheduled date is 0.50. With reference to Figure 20.21, the Z factor for an SS of 17 for event 4 is

$$Z_4 = \frac{SS - ES}{\sqrt{\sigma_{1,2}^2 + \sigma_{3,4}^2}} = \frac{17 - 18}{\sqrt{1 + 1.78}} = \frac{-1}{1.67} = -0.6$$

which indicates that the SS (17) is 0.6 standard deviations less than the ES (18). This Z value corresponds to a probability of 0.2743. Therefore, if time "now" is zero and the time units are days, we could expect to complete the project depicted in Figure 20.21 on or before 17 days with a likelihood of 0.27. This probability is applicable without

any expediting measures applied to the project. When an extra day beyond the expected 18-day project duration is allowed, the probability of meeting the milestone (19) is 0.7257:

$$Z_4 = \frac{19 - 18}{\sqrt{1 + 1.78}} = \frac{1}{1.67} = 0.6$$

A back-door approach to milestones would set the SS at a date with a desired probability of completion. A manager might thus promise a finished product on a date that is two standard deviations beyond the ES of the last event, to give a 0.977 probability of meeting the date. Any probability could be incorporated in the formula $SS = ES + Z(\sqrt{\sigma_{ES}^2})$ to provide the desired safety margin.

QUESTIONS

20B.1 Based on the following restriction list and the three time estimates for each activity, construct a boundary timetable.

Activity	Time Estimates			Postrequisites
	a	m	b	
A	3	5	7	
J	1	2	9	M
U	5	7	9	
K	3	3	3	G
M	9	9	9	A
C	4	6	14	A,U,T
T	1	1	13	D
G	1	1	7	T
D	3	3	21	

20B.2 With the data shown below for the activities required to launch the promotion of a new project:

(a) Calculate the expected time for each activity.

(b) Complete a boundary timetable.

(c) Calculate the activity variances, and determine the probability of completing the project in 12 weeks.

(d) Calculate the probability that the advertising will be "released" by the time of its LF.

Activity	Time Estimates			Postrequisites
	a	m	b	
Develop train-ing plans	2	6	10	Conduct training course
Select trainees	3	4	5	Conduct training course
Draft brochure	1	3	4	Conduct training course Print brochure, Prepare advertising
Conduct train-ing course	1	1	1	
Deliver sample products	3	4	4	
Print brochure	4	5	6	Distribute brochure
Prepare adver-tising	2	5	7	Release advertising
Release adver-tising	1	1	1	
Distribute bro-chure	2	2	3	

20B.3 For the network at the bottom of this page, answer the following questions:

(a) What is the most likely duration for the project? What is its probability of occurrence?

(b) Using basic PERT assumptions, calculate the duration for which there is a 92 percent likelihood that the project will be done by that time or sooner.

(c) What is the probability that activity I can begin after 16 days?

20B.4 In a PERT network, a noncritical path terminating at a critical event will occasionally cause the merge node to have a lower completion probability than that derived from the critical activity ending at the same event. This happens when activities on the noncritical path have very high variances, as compared with those on the critical path. Therefore, whenever a scheduled start (SS) calculation is being conducted, it is advisable to check suspicious non-critical paths as well as the critical path to determine which chain has the limiting (lowest) probability. For the network on page 630, calculate the limiting probability and indicate the path from which it was taken, for each of the following conditions:

(a) To complete event 7 by day 25

(b) To complete event 7 by day 28

(c) To complete event 5 by day 16

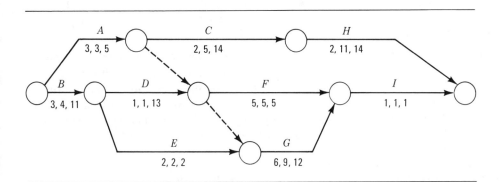

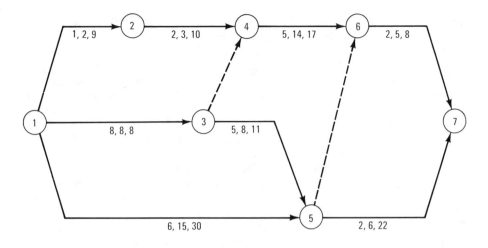

20C Time Management, or Use It or Lose It

Many studies have been conducted about how managers and engineers spend their time. The constant is that an 8-hour day for all of them is the same length. Variables include lengthening the standard workday by stealing minutes or hours from other pursuits, adjusting to uncontrollable conditions during the workday, and having better management of controllable activities. Comparisons show with remarkable consistency, irrespective of the nationality or type of industry, that the biggest wasters of professional time are:

• Telephone and drop-in-visitor interruptions

• Meetings and impromptu socializing

• Crises and responses to shifting priorities

• The necessity of doing things that should be done by others or not done at all

• Inability to say "no" to special requests

• Indecision and procrastination in routine affairs

• Receipt and composition of unnecessary or unorganized reports

• Confused objectives, priorities, or responsibilities

There is no order of importance for this list because different professions and positions are exposed to different time consumers, but the first three are nearly universal.

If, as Peter Drucker wrote, the difference between efficiency and effectiveness is *doing things right* and *doing the right things right,* then the rule for improving time-utilization effectiveness is the elimination of nonessential time-consumers to afford more time for essential activities—Pareto's principle applied to time management. Time- and money-minded engineers can lead the assault on productivity-inhibiting time dissipators by the following actions:

Make people aware of the cost of wasted time. From 20 to 40 percent productivity improvement is not an unrealistic expectation to have as a result of time-conservation efforts.

Detect duplication of effort. One company found that its auditors were checking freight invoices while it was paying a verifying agency to do the same thing. It is engineering management's duty to dig out such duplications.

Get rid of built-in time wasters. Nothing is more frustrating than the hurry-up-and-wait syndrome. Every queue in an organization represents an opportunity to save time. Studies indicate that clerical workers lose about 18 percent of their workday waiting for work.

Be a time watcher. Observe how time is parceled out all day and ration it to the best uses. This means prioritizing activities. It also means not wasting other people's time.

Prepare a schedule. Everyone knows that the most important things should be acted upon before mundane matters get attention. The trick is to divide the important things into small goals that can be accomplished one after the other: divide and conquer.

Know how to conduct a meeting. Posting the agenda in advance, starting promptly, maintaining the pace, and summarizing at the end are respected practices for conducting group meetings. What may be forgotten is that essentially the same preparations improve one-on-one meetings. The tremendous potential for managers to manage meeting time is demonstrated in Example 20.5.

Example 20.5
How Managers Manage their Time

Comparisons of the way managers allocate their time are interesting but not very definitive, because individual styles and duties vary so much. A little spice is added by comparing times from two cultures. A 1980 survey conducted by the Japanese Management Association and another by IBM to study time management by middle- to upper-level managers produced the following results:

Management activities*	Percentage of Time Spent on Each Activity by	
	Japanese managers	American managers
Attending scheduled meetings	24.4	13.1
Conferring with others	21.6	25.2
Reading	15.7	16.6
Writing	16.3	20.0
Traveling outside headquarters	8.5	13.1
Planning or scheduling	5.7	4.7
Other	7.8	7.2

*Each activity category includes related operations. For example, the 20 percent allocation for "writing" by American managers includes actual writing (9.8 percent), dictating (5.9 percent), proofreading (1.8 percent), calculating (2.3 percent), and copying (0.2 percent).

Japanese managers, as compared with American managers, spend more of their time attending meetings, as may be expected in a culture that values consensus. American managers, in contrast, spend the largest chunk of their time participating in unscheduled meetings, over half of which are by telephone. The popular conception that managers devote a major share of their time to thinking and planning is shattered by the low percentages; 5.7 percent in Japan and 4.7 percent in the United States. The Japanese survey reported that an average working day for managers lasted 10 hours and 9 minutes.

CHAPTER 21

PRODUCTIVITY AND PROFITABILITY

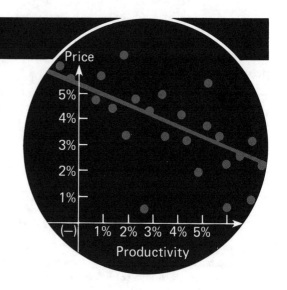

OVERVIEW

One dictionary defines *productivity* as "the quality or state of being productive." Furthermore, *productive*, in an economic sense, means "the creation of goods and services." This accounts for the output of a system but only implies the existence of inputs. A more appropriate definition for engineering economists is

> Productivity is the measure of how specified resources are managed to accomplish timely objectives stated in terms of quantity and quality.

The general nature of this definition suggests the diversity of productivity measures and the breadth of its influence. It affects prices, employment, and utilization of physical resources. More-productive organizations are more likely to be profitable organizations. Price escalation is slower for products from industrial sectors that have achieved greater productivity growth, thereby helping to lower national inflation rates.

The most prominent factors affecting productivity growth are capital, government, technology, and the workforce. Capital investments fund the facilities. Governments control the economic environment. Technological advances provide better methods, machines, and materials. Effective performance by employees converts the other factors into goods and services. Productivity gains or losses at the national level are determined by measurements of output and input taken by the U.S. Bureau of Labor Statistics.

Paths to productivity improvement include better management practices, superior physical facilities, and more employee involvement. All the paths draw on concepts and techniques from many disciplines, conspicuously including the principles of engineering economics.

CONCEPTS OF PRODUCTIVITY

Productivity is a subject that has reached headline proportions in recent years. Yet it is often misunderstood. It can be defined simply as a ratio of output to input, or more completely as the relation between output of goods and services and the quantities of inputs—labor, capital, and natural resources—that represent the performance of a production system. The concept of productivity is as applicable to a division of a firm or the whole firm as it is to an entire industry or the economy of a whole nation.

Productivity is more than a measure of production. An increase in production means that the volume of output has risen, but it does not necessarily mean that productivity has increased. Quantity produced is only the numerator of the productivity ratio. The denominator contains the inputs required for the given output. If added output is attained at the expense of a disproportionate increase in input, productivity declines even as production climbs. Awareness of such relations becomes more important as the need to conserve resources becomes more critical.

Profitability and *productivity* are closely related, although they differ conceptually. Greater productivity usually supports higher profits, but not always. Highly efficient production of products that cannot be sold is futile efficiency. Conversely, even incompetent production of prized products can yield high returns. Profitability is thus influenced by market pricing effects, accounting adjustments, and other financial actions unrelated to the proficiency of production performance.

Many conventional economic analyses are, in effect, productivity analyses. When a rate of return is calculated for the savings expected from an investment in productive assets, a productivity type of ratio is used: savings (output)/ capital (input). Similarly, an economic evaluation to select a preferred alternative is a productivity evaluation if all the proposals are designed to maintain or increase the level of productivity. Other economic comparisons may ignore productivity aspects, as in rationing capital among independent proposals according to their yield on investment. Then a supplementary productivity analysis is needed if productivity is a policy consideration.

Beyond the well-known statement that productivity means "working smarter, not harder" are other popular slogans such as
- Doing more with less
- Working wisely
- Getting it all together
- Knowing you've done your best
- Striving, yet at ease
- Performing effectively
- Making tomorrow better than today

Example 21.1
Profits, Prices, and Productivity

A highly productive process is not always a highly profitable process, although there is usually a positive correlation with success. Exceptions occur when shortages or an inflationary economy masks inefficiencies, as when any price increase tends to be accepted and when the output of an efficient process is unwanted regardless of attractive price and quality. The conventional relation between changes in profitability, price contribution, and productivity are shown in Figure 21.1.

The center column shows how profit variations are driven by changes in revenue and cost. Top and bottom rows, respectively, link changes in product quantity and price to revenue, and changes in resource quantity and price to cost. These are familiar relationships.

Less familiar are the outside columns that indicate that productivity change is a function of changes in production quantities and resource consumption, and that price contribution depends on changes in the price of products and resources. Productivity is thus the ratio of output products to input

FIGURE 21.1
Interrelations of the prices and quantities of resources and products to profit. (*Adapted from John Parsons, "Productivity Profits and Prices," National Productivity Institute, Pretoria, S.A.*)

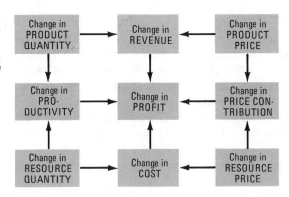

resources. Price contribution measures how well resource price increases are compensated by higher prices for products sold. When a product's price goes up faster than the cost of resources to produce it, profit probably climbs, even without a productivity increase; this condition is inflationary pricing.

The same relationships can be observed in formulas. Starting from the traditional equation of profitability,

$$\text{Profitability} = \frac{\text{revenues}}{\text{expenses}}$$

which can be expanded to

$$\text{Profitability} = \frac{\text{output quantity} \times \text{unit price}}{\text{input quantity} \times \text{unit cost}}$$

that converts to

$$\text{Profitability} = \frac{\text{output}}{\text{input}} \times \frac{\text{price}}{\text{cost}}$$
$$= \text{productivity} \times \text{contribution factor}$$

The last term in the equation above is also called a *price recovery factor* to characterize an ability to pass input price increases to the customer as a higher output price.

Consideration of the formulations above integrates operating efficiency with financial effectiveness, a natural perspective for engineering economists.

PRODUCTIVITY PERSPECTIVES

Productivity is a puzzling subject as well as a huge subject, encompassing the entire economy and affecting everyone, yet each person can draw a meaning to fit his or her own situation.

To generalists—legislators, educators, corporate executives—productivity is a web of interacting influences radiating from their favored nexus. Its strands are taxes, government policies, capital, access to resources, adequacy of labor and management, competitors, and the flow of goods and services.

Special-interest groups have their own perspectives. Union and industrial leaders often position themselves on opposite sides of a productivity issue. Conservationists and consumerists will direct attention to other facets of the same issue. Parties and issues can be further fractured by divergent views from regional or political factions. Possible permutations are endless.

Many of these entanglements are excluded when productivity is viewed from the reference point of a single organization, be it a factory, government agency, or service unit. Cause-and-effect relations are more visible, and more urgent. Managers coordinate input with output to meet schedules. They buy technology within budgetary limits. They search for efficiencies. They perceive productivity improvements as a duty of their office but less imperative than is the drive for profitability. This perception can be devastatingly short-sighted.

The classic example is an aging production base caused in part by corporations maintaining a reputation-building dividend rate on their stock at the expense of modernization. Old plants, outdated machines, and low R&D budgets are blamed for part of the declining trend in United States productivity.

Individual workers, concerned with their immediate occupations, have a still narrower view. To them, productivity means having sharp tools and materials available, knowing what to expect and what is expected of them, and working in a supportive environment. These anticipations apply to workers at all levels, but fulfillment depends on individual situations. A typist and a janitor may share equal motivation yet suggest entirely different paths toward productivity improvement.

Varying views should be expected and cultivated. They make the pursuit of productivity always challenging. There is no all-purpose solution, no guru to point out the path to a proven answer. There is, however, a wealth of productive practices that can be screened to reveal those that conform to each perspective.

The popular quick-fix of throwing money at a problem gives it quick recognition and may even solve it, but the chances of lasting gains are slim.

Example 21.2
Farce, Fable, or Foretoken?

The following passage was entered in the Congressional Record on January 31, 1980 (p. E296) by the Honorable J. William Stanton, representative from Ohio:

This is the story of the U.S. pencil industry. Remember, we are looking back from our vantage point of 1990. It's strange to think that, back in 1979, just anyone could use a pencil any way they wanted to.

You see, it all started when the Occupational Safety and Health Act carcinogen policy went into effect. The graphite in the pencil lead always contained a residue of crystalline silica. And there was at least one animal test and one in-vitro test indicating that crystalline silica produced tumors, so the material became regulated as carcinogen. There was no alternative for pencils, so exposure had to come down almost to zero.

The Environmental Protection Agency, acting under the Clean Air and Clean Water Acts, required drastic reductions in emissions and effluents. The control technology was quite expensive, and only the largest manufacturers could afford it. This caused a flurry of antitrust suits in the early '80s when there were only three pencil makers left in the country. One of the three was split into smaller companies, but they soon went out of business since they were unable to afford increasingly stringent workplace and pollution control requirements. Then foreign pencil manufacturers began to

threaten to dominate the pencil market, and our government, in an abrupt about-face, allowed a merger of the two remaining companies.

The Consumer Product Safety Commission then became concerned with what the newspaper headlines were calling the "pencil problem." Rubber erasers could be chewed off and choke small children. The sharp points of pencils could also be dangerous. There were residual solvents in the paint used on pencils, and pencil-chewing seemed to be a more widespread habit than anyone had realized. Printing a legend on each pencil that said: "This Pencil Could Be Hazardous to Your Health" did not seem to affect consumer pencil habits, a Harvard study indicated. In fact, the study found additional potentially harmful uses, such as stirring coffee. This led FDA to declare that harmful substances could be dissolved out of the pencil into the coffee, and thus pencils violated food additive laws.

Trying to salvage its business, the pencil company began making pencils without paint, without erasers, and with only soft leads so they would not hold a sharp point. But consumers were outraged, and sales declined.

Then someone invented a machine that could measure

crystalline silica below the parts-per-trillion level, and work-place, air emission, water effluent and waste disposal regulations required that the best practicable technology be used to reach this low level. The pencil company was threatened with financial ruin because of the large sums needed to purchase new control equipment. There were those who wanted to ban pencils entirely under the Toxic Substances Control Act, but the government decided that pencils were necessary, particularly since they were used to write new regulations. Besides, the Senators from the state where the pencil company was located declared that pencils were as American as baseball.

So the government bailed out the pencil company with a large guaranteed loan. But, of course, that was only a temporary measure, and to protect the pencil business, the government eventually nationalized it.

It is comforting to know that, after all, society is being protected against a danger that was so obvious we didn't even notice it for many, many years. There are still those who complain about paying $17 for a pencil, but you really can't put a price tag on health or safety.

The tongue-in-cheek example above parodies the evolution and spillover effects of government regulations. Public and industry opinions seem to reside uneasily between wanting the security that regulations promise and fearing the excesses that are associated with bureaucratic interference. Productivity measures could be the intermediary perspective for evaluating opposing postures. Then compliance inputs would be compared with output benefits of regulations. This would force policy makers to weigh alternative regulatory objectives, applying budget discipline to regulations, and industry managers to weigh their perceived costs against long-run benefits bestowed by regulations.

OBSTACLES AND OPPORTUNITIES

C&E diagrams were introduced in Extension 19A.

The statement in the hexagon is the central issue defined by the causes and effects.

Productivity is an index of economic welfare. More goods and services per worker-hour are produced when productivity increases. More output per input means that more product value can be purchased from the same paycheck. This translates into a rising standard of living.

A cause-and-effect representation of the factors that contribute to productivity and the results of productivity growth or decay are diagramed in Figure 21.2. The four most apparent contributors are listed on the main ribs slanting from the left to the subject box at the center of the diagram: capital, government, technology, and workforce. Specific issues associated with each of the four causes are noted by arrows on the ribs. Effects to the right of the subject box are treated similarly. Further refinements could be made to every arrow. For instance, the *composition* arrow on the workforce rib could be supplemented by subarrows for age, sex, status, skills, and location. The wealth of possible entries demonstrates the complexity of the subject.

The four prime causes are examined below with respect to their economic influence.

Capital

"The name of the game is business, and the score is kept with money" was the maxim for money management mentioned in the introduction to Section Three. A revised version attuned to today's concerns would be "The name of the game is prosperity, and the score is kept by productivity." Or the two

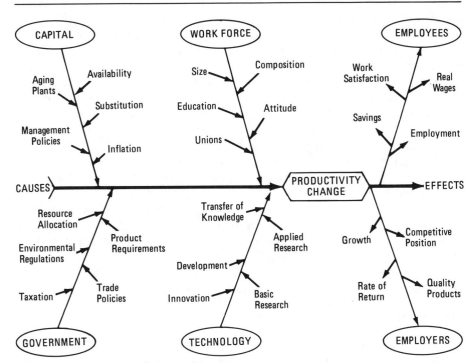

FIGURE 21.2
Causes and effects of productivity. Prominent factors that *cause* productivity change and the resulting *effects* are displayed in a C&E diagram.

could be combined: "The name of prosperity is productivity, and you score with capital."

Newer factories and equipment take advantage of the latest technology to produce more with less input. They conserve energy and often yield higher-quality goods. Their products cater to current needs. A natural response is therefore to call for more capital investment. Aside from the question of where the money should be directed, an equally intricate question is where it would come from. A larger share of the gross national product (GNP) devoted to investment shrinks the share for individual consumption and government programs. If consumers increased their savings to propel capital formation, there would be less money to spend on discretionary goods. Similarly, federal budgets would be squeezed by any tax rebates granted to companies to increase their cash accumulation, causing fewer dollars for social programs unless taxes were raised elsewhere. Every pro has a con.

Americans have never been enthusiastic savers. Since 1948 the average household has saved about 6 percent of its annual disposable income. In Europe the percentage is close to 14 and in Japan it is about 20.

Government

Ask who is to blame for any nation's productivity ills and you find that government is on everyone's list. Then ask for the cure and you get government again. Any action that the government takes is at once acclaimed and criticized, depending on whose pocket is fingered.

Consequences of questionable government interventions were hypothesized in Example 21.2.

The taxation arm of government affects capital formation. That same muscle subsidizes certain industries at the expense of others, distorting the marketplace. Trade agreements, tariffs, and informal import limits similarly provide an umbrella from competition that dissuades pursuit of productivity by the protected industries.

Energy disturbances of the past decade illustrate the dependence of industrial productivity on broad government policies. Ignoring the role of international politics in the energy "crises," federal containment policies shook the roots of the production base. The focus of investment and innovation shifted away from labor efficiency toward energy efficiency. Labor was substituted for more expensive energy, often to the detriment of productivity. Money spent for energy conservation was money unavailable for productivity in labor and materials.

Social regulations which require capital investments to correct situations deemed unfair may also detract from productivity by drawing money from more productive investments.

Economic regulations, covering matters such as price setting and entry into an industry, have been around a long time. Business has learned to live with them. However, recent deregulation of transportation and major antitrust suits have strong productivity implications. Where inefficiencies have hidden behind the shield of limited access, allowing the franchise holders to escape competition, deregulation will contribute to productivity growth.

Technology

Capital investment and government policies are intertwined with tendrils from technology. Although the significance of technological innovation in productivity progress is unquestionably proclaimed, solid evidence is scarce. Indirect measures correlate productivity growth with spending for research and development, number of patents awarded, and imports or exports in high-technology industries.

Strong incentives for increased R&D efforts were included in the 1981 tax act.

Following the national preoccupation with science in the 1960s, highlighted by humanity's first footprints on the moon, government inflation-adjusted spending on basic research has declined. Industrial R&D spending shifted away from basic research and toward more applied research until 1978, but threats from foreign competition based on fundamental science discoveries have since reversed the trend.

The same conclusion about the value of R&D can be drawn from the performance of R&D-intensive industries.

Agriculture is a sterling example of R&D's contribution to productivity. American farmers lead the world in productivity. Since 1950, farm labor has enjoyed a spectacular average productivity increase per year of about 6 percent. The gain parallels the growth of agricultural research. A Cooperative Extension Service, established in 1914, meshes scientists advancing knowledge with engineers inventing technology with farmers producing food. The well-documented result is an annual average rate of return of about 50 percent on the investment in agricultural research.

Workforce

Labor is the cornerstone of productivity. All other considerations—investments in the production base, government taxing policies and regulations, advances in technology—support the performance of the workforce. And that performance determines the nation's productivity.

Logic dictates that employees' attitudes toward their work must reflect their performance. Logic also suggests that employers can improve the attitudes of employees. Yet there is little solid documentation about the extent that performance is affected by attitude or the actual source of attitudes. Theories are plentiful, but which one should be followed to correct sloppy craftsmanship, idleness, or outright sabotage? The decision resides with management. Productivity is at stake. Evidence points toward attitude improvement, not by temporarily exhorting employees to excel but by treating them as collaborators in a mutually rewarding exercise of betterment that includes satisfying work conditions and stimulating work.

A copying machine has replaced the forge as a symbol of America's growth. For over half a century heavy industry has been under the magnifying glass of industrial engineers who have been probing for each tiny extra motion that can be eliminated to gain efficiency. Comparable scrutiny of white-collar operations is not yet underway, although engineers and management analysts are eagerly snipping at red tape. The direction employment has taken, away from higher-productivity factory work toward lower-productivity office work, is another partial explanation of United States productivity losses in the past decade and is a beacon for the productivity push of this decade.

Analysis of the capital-labor-productivity relation is explored in Extension 21A.

Nonfarm civilian employment shifted to the following proportions between 1960 and 1980: Blue-collar jobs dropped from 39.7 to 32.6 percent, white-collar jobs rose from 47.1 to 53.7 percent, and service jobs went from 13.2 to 13.7 percent.

Example 21.3 Productivity and Inflation

A significant effect of national productivity gains is the dampening action they exert on inflation. As a rule, industries with above-average productivity have shown, either with or without pressure from competition, smaller price increases for their products. The relation is not perfect, as indicated in Figure 21.3, but it is apparent that rate of change in output per employee hour closely correlates with inflation rates. A commendable example is the price of radio and television sets, which declined annually by more than 1 percent on the average while productivity was increasing by almost 4 percent per year over 20 years. At the other extreme is bituminous coal mining,

which became less productive while prices soared at an annual average rate of more than 10 percent.

Many factors besides productivity affect the general inflation rate, unfortunately. Fiscal policy is a major influence. Political decisions at home and abroad, resource shortages, cartels, and consumers' spending philosophy also contribute to price escalations. However, the rate of productivity growth is the underlying governor. Whenever wage increases exceed gains in production efficiencies, the goods produced from labor and capital have become more expensive.

NATIONAL MEASURES OF PRODUCTIVITY

The first measures of productivity for the United States were attempted in the 1880s by the Bureau of Labor. By 1940 a division was created within the Bureau of Labor Statistics (BLS) to compile regular reports of estimated output per labor hour in various sectors of the United States economy. An inclusive productivity measure of the real output of the nation's private business sector per unweighted hour for all workers is now published quarterly and annually.

Productivity in the private business sector is a closely watched statistic.

FIGURE 21.3
Price escalation versus productivity growth during the period from 1960 through 1979, as compiled by the U.S. Bureau of Labor Statistics.

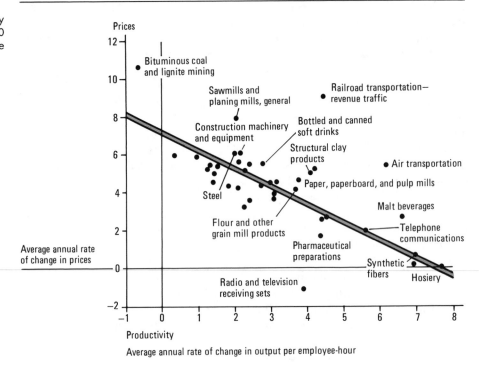

A chart of United States productivity rates since 1968 is given in Figure 22.1.

For the 20 years preceding 1968, United States productivity growth was consistently close to 3.2 percent per year. Then it became irregular and conspicuously lower. In some years the rate was even negative. This poor performance created unprecedented concern because of its potential effect on international trade and domestic living standards.

Measurement Method

The output of the private business sector is the gross national product, with certain exceptions. It includes the market value of goods and services produced by labor and capital that generate observable income transactions and imputed values for certain resources without observable market transactions, such as food produced and consumed on farms. Excluded from the GNP are domestic workers employed in households, government employees, and employees of not-for-profit organizations. These and a few other deductions from GNP are made because output measures are inadequate or inappropriate.

For specifics, see *Measurement and Interpretation of Productivity*, National Academy of Sciences, Washincton, D.C., 1979.

Data for the numerator of the private business sector productivity ratio come from many sources. Besides the data collected by government agencies, trade associations and large companies provide information. Figures from these sources are in current dollars. They are deflated to constant dollars (real dollars at a base year) according to applicable price indexes.

The denominator is the unweighted sum of the hours of all workers who produced the output measured in the numerator of the private business sector ratio. The labor input is usually calculated by multiplying the number of workers by the average hours worked per person. Data are obtained primarily from current employment statistics and current population surveys. Unlike many other productivity ratios, no weighting is given for degree of skill, educational attainment, wages, or any other means of differentiation among workers.

Productivity measures for corporations are presented in Chapter 22.

Additional Government Measures of Productivity

The Bureau of Labor Statistics measures the productivity of several divisions within the private business sector (farming, mining, construction, communications, manufacturing, utilities, finance, etc.). Of these, only the manufacturing-division measures are regularly published; this division has consistently outperformed the aggregate productivity rate of the total private sector. Procedures for measuring industry divisions are essentially the same as those described previously.

The lowest level of breakdown for BLS measures is by individual industries. For example, detailed industries within the *mining* subsector include

Iron mining—crude ore

Iron mining—usable ore

Copper mining—crude ore

Copper mining—recoverable metal

Coal mining

Bituminous coal and lignite mining

Nonmetallic minerals

Crushed and broken stone

Measures for such detailed divisions show which industries are leading or lagging overall productivity trends and allow an individual company to compare its productivity record with that of its industry.

Since 1973, BLS has prepared productivity measures for selected federal government activities. Inasmuch as a final output from the federal government is difficult to observe, much less measure, attention is focused on the quantifiable output of work done by selected agencies. A sample of agency activities and associated output indicators is given in Table 21.1. From such indicators, composite indexes are developed for 20 functional areas within the federal government—for example, audit of operations, communication, library services, medical services, postal service, and records management. Each productivity ratio is the composite output over unweighted employee years.

A systematic method to utilize performance ratios for evaluating white-collar and service jobs is introduced in Chapter 22. It is called the objectives matrix.

TABLE 21.1

Output indicators used in the BLS federal government productivity program

Activity	Output indicator
Dining-facilities operations	Number of meals served
Flight training	Student-years trained
Invoices and travel processing	Number of invoices and travel claims processed
Background checks on persons assuming sensitive positions	Number of cases closed
Consumer complaints	Number of complaints processed
Patent-application examinations	Applications completed

The focus of the program to measure productivity in the federal government is that of improving efficiency. Consequently, the issue of what is the final (total) product loses significance. When the amount of labor expended to provide a service is related to output indicators for that service, an agency has a continuing record of its performance efficiency. Indicators are more useful when they measure smaller activities. For instance, it would be preferable to categorize inspections by type rather than to simply list the number of inspections completed.

Example 21.4 Productivity Information Reporting by BLS

1984 marked the centennial anniversary of collection and publication of data about United States business activities by the government. Currently, BLS is the agency that collects, analyzes, and distributes information about productivity and production costs. These data are tracked closely by both private and public analysts as indicators of future business trends and the relative health of various sectors of the economy.

Table 21.2 is a sample of the data published quarterly by BLS. It features the business sector. Its output is equal to gross national product, less the rest-of-the-world sector, general government, output of paid employees of private households and nonprofit institutions, rental value of owner-occupied dwellings, and the statistical discrepancy in computing the national income accounts. Corresponding exclusions are also made in labor inputs. Business output has accounted for about 77 percent of gross national product in recent years.

The productivity and associated cost measures describe the relation between output in real terms and the labor time involved in its production. They show the changes from period to period in the amount of goods and services produced per hour. Although these measures relate output to hours of all persons engaged in a sector, they do not measure the specific contributions of labor, capital, or any other factor of production. They reflect the joint effects of many influences, including changes in technology; capital investment; level of output; utilization of capacity, energy, and materials; the organization of production; managerial skill; and the characteristics and effort of the work force.

PATHS TO PRODUCTIVITY IMPROVEMENT

One of the most intriguing things about productivity is that there are so many ways to improve it, ranging from hard technology to soft behaviorals and all the mutations between. It is truly an interdisciplinary subject, but this immense

TABLE 21.2

Business sector productivity, hourly compensation, unit labor cost and prices, seasonally adjusted. Productivity changes in other sectors during the first quarter of 1984, when the total business sector was 4.1 percent, were 3.5 percent in the nonfarm business sector, 2.9 percent in total manufacturing, 5.3 percent in durable goods manufacturing, −0.6 percent in nondurable goods manufacturing, and 2.0 percent in nonfinancial corporations.

Year and Quarter	Output per Hour	Output	Total Hours	Real Compensation per Hour*	Unit Labor Cost	Unit Nonlabor Payments†
			Indexes 1977 = 100			
1983						
I	102.4	107.1	104.6	99.3	156.9	140.9
II	103.9	110.4	106.2	99.1	156.0	145.7
III	104.2	112.4	107.9	99.0	156.9	147.6
IV	105.3	114.5	108.8	99.5	157.9	149.9
Annual	103.9	111.1	107.0	99.2	156.9	146.1
1984						
I	106.3	117.8	110.7	99.8	158.6	151.9
			Percent Change from Previous Quarter at Annual Rate			
1984						
I	4.1	11.8	7.3	0.9	1.7	5.7
			Percent Change from Corresponding Quarter of Previous Year			
1984						
I	3.9	10.0	5.9	0.5	1.1	7.9

* Wages and salaries of the employees plus employers' contributions for social insurance and private benefit plans adjusted for changes in the Consumer Price Index.
† Nonlabor payments include profits, depreciation, interest, rental income, and indirect taxes.

lode of lore causes confusion. Where are the nuggets? Are they in the veins of engineering, psychology, business administration, politics, or what? "In all of them" is the unsatisfying answer. There is simply no single pocket of wisdom from which to extract all the right methods to treat productivity.

Three fertile fields for productivity-oriented engineering economists are introduced in the section: managerial practices, physical facilities, and employee involvement. Together they delineate the broad path to productivity improvement.

Management Practices

To manage in a business sense is to direct, conduct, and administer. In most cases, managers do not actually produce goods and services that appear in a productivity measure. They nonetheless account for a big chunk of labor cost while their direct contribution to output is tenuous. They do not do; they facilitate doing. As a result, the work scene is seemingly bifurcated into "management" and "workers," them and us. Bosses, staff, and everyone else in the front office are often collectively regarded as a distinct social group with special interests and characteristic economic views. This demarcation castigates management by inference that only the workers work while managers ride along extracting an unjust share of labor's reward. Such misgivings are often vocalized during a drive to increase productivity when workers ask, "What are *they* [managers] doing? What's their contribution?"

As one small example of a managerial practice that could enhance productivity, consider how capital-budgeting procedures could be modified to give priority to expenditures that boost productivity. It has already been observed that medium-range proposals cluster around cost-reduction, product-enrichment, and regulation-mandated requests. Required outlays to satisfy government regulations receive top priority, because ignoring them could shut down operations or result in fines. Cost-reducing proposals are attractive because results can be accurately predicted and successess can directly boost profit. Product-improvement proposals are a bit riskier than are the others, but such improvements are needed to protect earnings. Many of the proposals in these categories also favorably affect productivity. Most investment selection processes, however, do not award special priority to proposals that contribute to technological innovation and higher work-life quality. Potential adaptations that favor productivity-oriented proposals include the following:

The "merger mania" of the early 1980s absorbed cash reserves and executive attention, making both unavailable for operational improvements.

- A percentage of capital funds available each year for internal investment could be set aside for productivity projects. Within this mandated allocation, preference among the productivity proposals could be determined by ranking proposals according to either their expected rate of return or their status in a vote by a productivity council.

- Larger organizations usually have a preprinted request-for-expenditure form that is filled out for each proposed project. One of the categories on the work sheet should be a required explanation of what effect the project will have on productivity and, if pertinent, on the quality of working life. The factors listed in this category should then receive special attention by the committee that selects the proposals to be funded.

- A "seed" fund could be established to encourage workers to develop their ideas for improving products or processes. It would provide time and materials for them to experiment with their inspirations. Unceremonious application forms coupled with ceremonious acceptance announcements increase participation. One company boarded off a small section of a shop, put a lock on the door, which was opened by a very large key, and issued keys to workers who had been awarded "research contracts" to work in the "think tank annex."

- A version of the seed fund, called a "sapling" fund, can be established for staff members who are regularly assigned to R&D or product development. It allows them to spend time on offbeat projects that do not conform to the edict that marketability supersedes makeability. The intent of a sapling fund is to relieve some of the market-driven pressure to work only on products that already have consumer acceptance.

Physical Facilities

At the same time that people revel in the achievements of technology, they distrust the consequences. In general, managers look toward new hardware for part of their future productivity growth and workers have misgivings about the future effects of labor-saving machines. Neither group fully appreciates the ramifications of the rapid movement toward smart machines and computer-aided processes because no one can accurately forecast the eventual directions that the movement will take.

In search of greater productivity, organizations can build, buy, or borrow technology. Building from scratch is an expensive and lengthy undertaking, but its reward could be exclusive ownership of a new device or even the founding of a new industry. Buying ready-made technology means upgrading production facilities by acquiring modern machines and structures from vendors and possibly modifying them to perform new functions or do designed functions better. Technology is borrowed by taking something that has been designed for another purpose and adapting it to solve a different problem, often at the suggestion of people who are faced with the problem.

Experts say that it takes 20 to 25 years before a new technology is assimilated by industry. Computers are of age. Machine-paced automation is also a mature manufacturing medium. The following technologies are still in the adolescent range:

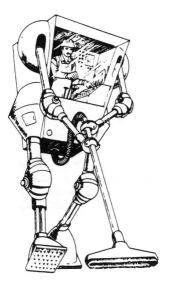

The race to embrace robots could conceivably produce this absurd automaton—the ultimate janitor.

- *Lasers.* Already engaged in cutting steel, welding metals, and performing delicate eye surgery, lasers will also be used in hardening metals, making precise measurements, reading video disks, and performing many data-oriented operations.

- *Fiber optics.* Future applications of fiber optics technology will expand to many areas beyond its proven worth in telephonic communications. Even now it is being applied to information processing and for quality control inspection of a wide variety of products.

- *Energy technology.* Fuel scares in the 1970s spawned a deluge of energy research projects that created exotic products that are now ready for commercial application. Changing sunbeams into electric energy (solar photo voltaics) is a proven process waiting for a means of low-cost production. Strange-shaped, more-efficient windmills in "wind farms" are generating electricity. Fuel cells that produce electricity through electrochemistry show promise. The tongue-tying technology named magneto-hydrodynamics sends gases from superheated fossil fuels through a magnetic field to create electric current.

Justification procedures for investments in high-tech facilities is considered in Extension 21B.

Whether technology is built, bought, or borrowed, its acquisition involves capital risk and transfer of knowledge. Trying anything new is risky and exposes managers to the money auditors who, according to managerial lore, enter the fray after the battle is lost to bayonet the wounded. Chances of escaping a bayonet are improved by fielding knowledgeable recruits. However, it is still possible to win the battle to implement technology and lose the war to improve productivity. Engineering economists should be aware of what is required to win both engagements.

Employee Involvement

Whereas the rapid deployment of technological advances can aptly be called revolutionary, the trend toward recognition of employee power is strictly evolutionary. Management theorists have been expounding on the merits of a "motivated" work force ever since the word was coined and have sundered the mechanics of motivation to compile a bewildering mixture of explanations and suggestions. Bits and pieces of behavioral methodology have steadily found their way into employee-relation practices, sometimes building trust that produced lasting results but more often arousing grand prospects that soon clashed with the realities of entrenched beliefs that equated worker participation with management giveaways.

Well-developed worker participation plans in Europe include *industrial democracy* and *self-management*. These are formal collective programs with government-legislated union-industry power sharing.

Even a cursory look at a firm's financial figures shows why workers are worth thinking about. A typical manufacturing firm finds each dollar of revenue roughly distributed in the following pattern:

Compensation to employees	43¢
Paid to suppliers of material and services	45
Debt retirement and asset replacement	7
Federal, state, and local taxes	3
Return to shareholders	2

Putting the spotlight on employee compensation reveals that a large chunk of income is parceled out within the walls of the company as wages and fringes. Less attention is often given to the validity of these payroll outlays than is given to bills from outside suppliers. Closer regard to possible returns from payroll dollars might disclose how an active employment involvement program could generate savings comparable to the better equipment investment proposals.

The most direct method of encouraging worker participation is to offer monetary incentives for suggestions and improved performance. Incentive pay packages come in all shapes and sizes, ranging from the old standby of pay per piece produced to executive bonuses. Recent programs often carry the name *productivity sharing* and take the form of a company-sponsored plan for group activities, such as regular meetings of work units organized by a trained facilitator and led by an appointed team leader. The most recognizable United States examples are *quality circles* and *quality-of-work-life* programs. They are designed to expand the responsibilities and influence of employees by delegating authority downward to allow wider participation in planning and decision making. It is assumed that people want to participate in work-related activities, are willing to work toward common objectives, and will improve productivity as they do so.

See Example 21.5, wherein three of the best-known incentive-wage plans are described.

Membership in a quality circle is typically made up of six to ten people who are engaged in similar work.

A pragmatic consideration of the involvement question suggests that greater employee participation may be the most cost-effective way to secure productivity gains; it relies more on an investment of management dedication than do investments in hardware. It also offers a humanitarian bonus of improvements for the work lives of all contributors to productivity.

Example 21.5
Scalon, Rucker, and Improshare Plans for Incentive Wages

The first and simplest version of productivity sharing is straight piece-rate compensation—direct payment for the amount of work done. Many mutations have evolved as a response to criticism and a need to fit new situations. A base-pay level with incentive premiums paid for above-normal output ensures a minimum wage. "Measured daywork" is a widely used version that gives a worker a standard hourly base rate, which is bumped up or down when actual performance is compared with the established standards.

Wage incentive plans for individuals and small groups are generally opposed by unions for pitting employees against each other and having standards of questionable accuracy. From the standpoint of productivity, workers often resist the introduction of new equipment and work methods because of the possible impact on their earnings and the incentives that reward only greater output provoke workers to waste materials, squander energy, and abuse machines. Other incentive plans are more suitable for indirect workers, employees in service agencies, and operators in highly automated processes.

A precise relation between productivity gains and worker performance can be obtained by mathematically defining the size of the gain and proportionate shares. Although formulas vary considerably, their similarities include provisions for frequent bonus payments, production rather than sales-based

measurement of gains, and emphasis on employee involvement. The three most frequently cited plans are described below.

Scanlon Plan

Since Joseph *Scanlon* conceived his plan in the 1930s, it has undergone numerous modifications, but its basic concepts of employee involvement and recognition have remained intact. Employees participate through a formalized suggestion system that operates at two levels: (1) monthly meetings of a production committee composed of elected employee representatives to review suggestions, and (2) a higher-level screening committee composed of representatives from the production committee and management, meeting monthly to discuss company operations.

Recognition for increased productivity comes from a monthly bonus based on a single ratio:

$$\text{Base ratio} = \frac{\text{payroll costs}}{\text{value of production}}$$

If the historical base ratio is 0.20 and the employees' share is 75 percent, a production value (sales plus or minus inventory) of $1 million leads to an allowed labor cost of

$$\$1{,}000{,}000 \times 0.20 = \$200{,}000$$

If actual labor cost during the month was $160,000, the employee bonus pool would be 75 percent of the difference between allowed and actual cost:

$$(\$200,000 - \$160,000) \times 0.75 = \$30,000$$

A portion of the bonus is usually withheld to cover any deficit months, and the accumulated leftovers are paid as a year-end jackpot.

Rucker plan

The *Rucker* plan operates similarly through a suggestion system and Rucker committees to improve labor-management communications. The bonus calculation is based on a historical relation between labor cost and value added. For example, if the Rucker standard ratio is 0.60 and

$$\text{Net sales} - \text{purchases} = \$1,000,000 - \$400,000 = \$600,000$$

then the allowable payroll is $600,000 \times 0.60 = \$360,000$.

Any month that the actual labor cost is less than 60 percent of the production value, a bonus is earned. The formula encourages employees to save on materials and supplies because they share the gains.

Improshare plan

Mitchell Fein is the architect of *Improshare*—Improving Productivity through Sharing. The plan compares the number of work-hours needed to produce a certain quantity of units in the current period with the number of work-hours needed to produce the same quantity of units during a base period. Both direct and indirect work-hours are included to allow both production and support workers to share the gains achieved together. For example, assume that 100,000 work-hours were consumed in producing 10,000 units in the base period. If it took only 95,000 hours to produce the same 10,000 units in the current period, the productivity gain is 5000 hours. This gain is split 50-50 between the company and all the workers in the plant. When the gain is facilitated by capital investment, the sharing proportion is changed to reflect the new equipment's contribution, perhaps 80 to 20 in favor of the company.

PROBLEMS

21.1 Name and discuss the implementation considerations of the five changes in the numerator and denominator of the basic productivity equation (productivity = output/input) that indicate a productivity increase. (*Hint*: The most obvious change is to increase the numerator while maintaining the same input level, which is accomplished by more efficient utilization of input resources.)

21.2 Popular slogans associated with productivity are listed in the margin on page 633. Associate each slogan with one of the following professions and name a ratio that would uniquely measure the productivity of a person in that profession. (*Hint*: "Working wisely" could be associated with the teaching profession and a measure of productivity for a teacher would be *number of students* per *full-time teacher*.)

(a) Engineer
(b) Optimistic worker
(c) Philosopher
(d) Psychologist

(e) Scholar
(f) Production manager
(g) Economist

21.3 Figure 21.4 is a graphic representation of the price contribution and productivity relationship described in Example 21.1. The vertical and horizontal scales are productivity and price contribution, respectively. The diagonal line connects all points where productivity

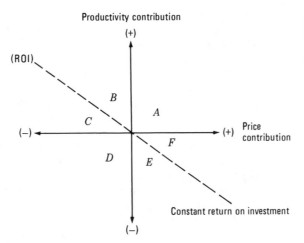

and price contributions are equal and opposite. Along this line the return on investment is constant because there is no change in profit for invested capital.

For each segment of the chart named below, indicate what happens to ROI when that condition occurs in an organization. Also state what advice you would give to a company that found itself in such a position. (*Hint*: Quandrant A—Happyday—is the ideal condition. The organization has high (+) productivity and prices are strong. It should continue whatever practices it has followed to achieve that position and institute new policies to position itself still higher and farther to the right in the Happyday quandrant.)

(A) Happyday   (D) Doomsday
(B) Start worrying   (E) Try harder
(C) Worry faster   (F) Keep trying

21.4 Several obstacles and detours on the three paths to productivity improvement (management practices, physical facilities, and employee involvement) have been blamed for the poor productivity performance in the United States during the 1970s and early 1980s. For each of the possible inhibitors to productivity listed below, discuss how it is being relieved currently or how you believe it could be reduced in the future.

(a) Management decisions motivated by short-term gains rather than directed toward the long-term prosperity of the firm.

(b) Decline in the work ethic among young employees.

(c) Government regulations which mandate investments to improve environmental or social conditions that siphon money from investments that could boost productivity. (Consider both sides of this situation: what happens when social responsibilities are neglected and when social expenditures are so high that industries become unproductive and noncompetitive.)

(d) Lack of concern about designing and building higher-quality products.

(e) Inadequate support for basic research.

(f) Influx of untrained or less-trained workers into the work force as a result of high birthrates from 1946 to 1964 and a larger proportion of females seeking employment during the 1970s.

(g) Low rate of saving by Americans that has limited the amount of capital available for borrowing for industrial purposes.

(h) Tax policies that provide little incentive for modernizing production facilities and investing in technologically advanced manufacturing processes.

(i) Confrontational relation between management and unions which restricts cooperation in productivity-improving projects.

(j) Management resistance to employee participation plans that give workers more voice in job design and decisions about working conditions.

(k) Work rules that enforce inefficient use of resources, especially labor.

(l) Shift of employment from manufacturing, where productivity is high, to service industries that generally have lower productivity.

(m) Lack of government support for ailing industries and an unwillingness on the part of the government to protect new industries from foreign competition.

(n) Reluctance of employees at all levels to have wage increases tied to measured increases in productivity.

(o) General lack of awareness about the negative impact of declining productivity on the standard of living.

21.5 Productivity awareness is considered to be an important factor for improving productivity. Much like Adam Smith's invisible hand, pursuing productivity improvements as a selfish goal, only to better one's own economic health, collectively has a widespread benefit to everyone through improved utilization of resources.

Construct a C&E diagram to explore what could be done to increase workers' awareness of the value of high productivity to their own and everyone else's prosperity. (*Hint:* See Extension 19A.) Use the following main ribs and complete the diagram with factor arrows to or from all ribs.

Causes. Students, universities, governments, workers, corporations, professional associations. (For each of these headings, indicate categories of activities or activists that could publicize and promote the awareness issue.)

Effects. Publications, nonprint media, programs, contacts. (For each of these headings that represent the results of the identified causes, indicate on arrows the specific channels of information flow.)

21.6 Wages based directly on measured performance are often resisted by workers, including managers.

(a) What disadvantages are associated with the following classes of incentive plans?
  (1) Piece-rate wage plans (e.g., paying a specified amount for completing a specific unit of work, such as $1.36 per piece correctly assembled.)
  (2) Bonuses based on performance (e.g., annual amount paid above regular wages for accomplishing a given level of preset goals.)
  (3) Formula incentive plans (e.g., see Example 21.5).
  (4) Profit sharing (e.g., percentage of gross profit distributed to employees according to an agreed-upon schedule.)

(b) What productivity advantages are associated with each of the plans given in Question 21.6a?

EXTENSIONS

21A   Analysis of National Productivity Measures

Measurement of productivity change is the first step in understanding national productivity. The second step is analysis. The purpose of analysis is to prepare directions for productivity improvement. Because the dimensions are vague and factor interactions are so intricate, measurements are only approximations. This complicates analysis. The complications, in turn, obfuscate efforts to improve productivity. However, much can be done with even less-than-perfect measures.

Labor Productivity and Economics

There is growing evidence that output per employee is a significant factor in the explanation of general price trends.* At the national level, changes in the output/labor ratio affect the supply of goods and services available. They can have a significant impact on the general level of prices for any given monetary or fiscal policy. Improving labor productivity alone cannot eliminate inflation, of course, but it can relieve some of the inflationary pressures.

Industries that have enjoyed a steady increase in output per worker have generally contained or even lowered their price per product; witness the computer and electronics industries. In most cases, these industries have not raised wages disproportionately. Instead, they have reduced prices or, during inflationary periods, have had lower-than-average increases, owing in part to competitive pressure from new firms attracted by the favorable cost prospects. Some industries have voluntarily reduced prices in accordance with the classical economy theory that a larger volume of output sold at lower prices can increase profit.

Experiences such as those just cited tend to refute fears that gains in productivity cause higher unemployment. It is true that the isolated impact of higher labor productivity is normally a reduction in employment per unit of output. Lower employment results only when output quantity is constant. However, the indirect effects of the productivity gain will likely swamp the direct effects. Favorable pricing should increase the demand, depending on the coefficient of price elasticity for the particular commodity. In the long run, records show that employment in industries that have superior labor productivity usually increases faster than does the industrial average.

A better understanding of the relation between productivity practices, effects on prices, impact on inflation, sharing of gains, and effects on employment is needed at the national level to guide policy making. Comparable understanding is needed at the firm level to allocate resources and design programs to foster productivity.

Labor and Capital Productivity

Empirical evidence supports the contention that capital investment enhances labor productivity. Good equipment and facilities logically contribute to more output per worker. Newer capital is more productive than is old because technological improvements are normally built into new capital goods. The impact of technology is not as readily measured as are dollars invested and labor hours, but it undoubtedly is significant; it is sometimes characterized as "knowledge productivity."

Extensive research has been devoted to assessing the relative contribution of productivity to economic growth.† Capital-labor ratios have received a lot of attention. A particularly frustrating aspect has been *technical change*, a phrase used to describe a residual measure intended to represent changes in the efficiency with which capital and labor are employed.

A simplified approach to explain the relation of productivity inputs begins with a production function

$$Q = L^\alpha K^\beta A$$

where Q = aggregate output
   L = labor input
   K = capital input
   α, β = respectively, share of labor and capital
      $(\alpha + \beta = 1)$
   A = residual representing technological change

*E. Denison, *Accounting for Slower Economic Growth*, Brookings Institute, Washington, D.C., 1979.
†S. Maital, and N. M. Meltz, *Lagging Productivity Growth*, Ballinger, Cambridge, Mass., 1980.

Taking logarithms and differentiating give the above expression the form

$$\frac{\Delta Q}{Q} = \alpha \frac{\Delta L}{L} + \beta \frac{\Delta K}{K} + \frac{\Delta A}{A}$$

For convenience, this can be rewritten as

$$R_A = R_Q - (\alpha R_L + \beta R_K)$$

where R denotes rate of growth and all subscripts and symbols retain their previous definitions; for instance, R_A = rate of growth of technology, and R_K = rate of growth of the capital input. When conditions of competition and profit maximization are assumed, α is equal to the proportion of income due to labor, as β is the proportion due to capital. From this relationship, the technical-change residual can be calculated from knowledge of the rate of change of the other factors.

As an example of "growth accounting," suppose that the following values of R are known: R_Q = 4 percent, R_L = 3 percent, and R_K = 2.5 percent. When α = 0.6 and β = 0.4, the following figures are inferred:

Rate of growth of output R_Q	4.0 percent
Contribution of labor ($\alpha R_L = 0.6 \times 3\%$)	1.8 percent
Contribution of capital ($\beta R_K = 0.4 \times 2.5\%$)	1.0 percent
Contribution of technology and other factors ($R_A = R_q - \alpha R_L - \beta R_K$)	1.2 percent

The 1.2 percent residual is thus a measure of contributors to output growth that cannot be accounted for by the number of workers and machines in the economy. By itself it does not reveal what is needed to promote greater gains in productivity, but continued analysis should eventually expose the workings of forces that control the growth.

Example 21.6 The Vital Role of Engineering and Technology in Productivity Growth

According to John Kendrick, the most important source of productivity growth is the advance in technological knowledge when applied to the ways and means of production through cost-reducing innovations. Informal inventive activities of managers and workers produce a myriad of small technological improvements that cumulatively spark total progress. Formal R&D programs also contribute. Data are scarce for the informal innovations, but R&D programs are known to have been about 3 percent of the GNP in the mid-1960s, and to have declined to about 2.2 percent of GNP in 1978.*

The prominence of the *knowledge* input to productivity

growth is apparent in Table 21.3.† According to Edward Denison, national output in the United States grew by an average annual rate of 3.65 percent during the 1948–1973 period. Labor and capital inputs together accounted for 2.13 percent of the growth. The "residual" input is the difference between the growth rate of aggregate output and the traditional factor input: 3.65 − 2.13 = 1.52 percent. Of this amount, he attributes 1.10 percentage points to advances in technology and its application. When one recognizes the unwieldiness of labor and capital inputs to expeditious change, the message in this productivity analysis is unmistakable.

21B   Justification of High-Tech Investments

The economics of computer-integrated manufacturing (CIM) is becoming more complex as technology advances. A growing set of initials is entering the vocabulary of analysts who evaluate high-tech investments. Among the better-known technologies are computer-aided process planning (CAPP), computer-assisted design (CAD), computer-aided

manufacturing (CAM), group technology (GT), automatic storage/retrieval systems (AS/RS), manufacturing resource planning (MRP II), computer-aided engineering (CAE), flexible manufacturing systems (FMS), computer-aided testing (CAT), and robotics, which needs no initials.

Some of these computer-driven systems are extremely

---

*J. W. Kendrick, *The Formation and Stocks of Total Capital,* National Bureau of Economic Research, New York, 1976.
†Adapted from E. F. Denison, *Accounting for Slower Economic Growth,* The Brookings Institution, Washington, D.C. 1979.

TABLE 21.3

Sources of growth in real gross product of the United States, 1948–1973

Sources	Average Annual Percentage Rates of Change[a]
Real gross product	3.65
Real product per unit of labor	2.13
Capital labor substitution	0.71
Changes in quality at hand	0.00
Improvements from resource reallocations	0.29
Net government impact	−0.03
Volume changes from economics of scale	0.32
Advances in knowledge	1.10
Other factors	−0.16

[a]Tentative percentage-point contributions for the 1973–1978 period given in order of the categories above, starting from the real gross product, are 2.4, 1.1, 0.3, −0.2, 0.3, −0.2, −0.1, 0.8 and −0.11.

expensive, whereas others are less costly than are traditional methods for certain applications. All have the potential to yield very large jumps in productivity. They are usually more productive in terms of labor than manual or conventional automatic machines, offer more consistent and higher levels of quality, reduce scrap, shorten product lead times and manufacturing cycle times, lower inventory levels at both raw-material and in-process stages, allow more flexibility in reacting to demand fluctuations, and provide safer, more comfortable working conditions. Most of the benefits are quantifiable, but data may be difficult to obtain.

When conventional investment-justification thinking is applied to automated systems, full credit may not be awarded to them. The major omission is the *synergy* that results when several subsystems, such as those listed above, are tied together into one overall system. When components of a system are evaluated individually, only those savings can be counted that were achieved by fitting the new component into the existing layout, which may not have the capacity to fully utilize the automated features of the addition. This process fails to recognize that the cost advantage of a fully integrated system of automated components is greater than the sum of the benefits of its individual parts.

Other benefits of automated equipment may also be neglected because they are difficult to measure. The need for less management attention, more accurate controls, greater reliability, and vastly increased flexibility of CIM components are difficult to determine, especially before the system is installed.

Considerations involved in justifying the acquisition of high-tech equipment can be illustrated by an evaluation of an automatic material-handling system. Traditionally, AC/RS have been justified by using savings in labor, inventory, and building costs as offsets against the project purchase price and operating cost of the equipment. Savings generally are grouped into two categories, hard and soft savings. Hard savings are those that are readily provable and measureable, leaning heavily on material-handling labor and building-occupancy cost. Other savings areas, principally indirect labor and inventory savings, are usually considered to be soft savings and are sometimes excluded from formal justification calculations.

The key obstacle to including indirect labor, inventory savings, damage control, response time, flexibility, and other soft benefits can be summarized in one word—*credibility*. First, confusion exists about the relation of system functions to the overall benefits that can be achieved by integrating AC/RS with automated production. Then, since benefits tend to be concentrated in areas that historically have not been well measured, people charged with preparing justifications do not know how or where to gather information. Finally, even if the savings are quantified, top management often does not believe that the savings will materialize, because of past failures to realize promised benefits from computerized equipment.

Because they are attuned to technological considerations and systems thinking, engineering economists tend to realize the significance of soft benefits and are prepared to search for them. The challenge is to be able to substantiate the soft savings to the satisfaction of the proposal reviewers.

QUESTIONS

21B.1 Firms regularly invest in segments of their business without a formal financial justification. Examples include new personnel hires, changing grades of raw materials, internal company activities (picnics and executive retreats, etc.), and promotional events. Justification is based almost entirely on faith that the derived benefits are worthwhile regardless of the expense and trouble generated.

Consider money spent to maintain a company "image" through advertising, the funding of community projects, charity contributions, and other public relations projects that are probably not subjected to justification requirements comparable to equipment-acquisition proposals. Could an investment in a major component of a high-technology system be accepted by similar reasoning when it fails to meet the MARR? Explain why.

21B.2 Bela Gold suggests the development of a portfolio of projects whose net return meets the MARR hurdle rather than an attempt to justify each project by itself. ("CAM Sets New Rules for Production," *Harvard Business Review*, November–December 1982.) By including system synergy and strong projects in the portfolio, weaker projects are, in effect, subsidized.

Discuss the pros and cons of Gold's approach.

CHAPTER 22

PRODUCTIVITY MEASUREMENT

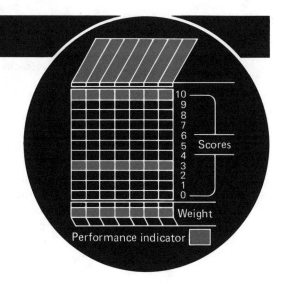

OVERVIEW

The effectiveness of production planning and economy is manifested in the productivity of an organization. The purposes of measuring productivity are to determine how efficiently resources are being utilized and to detect ways to improve operating performance. A creed for productivity is embodied in the following procedure:

To improve productivity, you must manage.

To manage effectively, you must control.

To control consistently, you must measure.

To measure validly, you must define.

To define precisely, you must quantify.

Broadly stated, productivity measurement includes all measures that relate one or more inputs to outputs—usually in the form of a ratio. Measurement is complicated by diverse outputs, multifactor inputs, mixes within inputs, and varying values for prices, costs, and quality caused by market factors and business cycles. Several measurement models have been developed to avoid certain complications and to emphasize different productivity goals.

Four types of productivity ratios are:

1 *National productivity measures* (industry divisions or sectors). Relate real output to labor input for major economic units, as discussed in Chapter 21
2 *Total productivity index* (firm or division). Relates the output to aggregate input associated with all or part of an organization

3 *Partial productivity index* (any level of aggregation). Relates output to one or more classes of inputs, as in output per labor-hour or output per labor plus capital inputs

4 *Work unit productivity indicators* (cohesive groupings of employees). Relate performance measures to the mission shared by unit members

An *objectives matrix* (OMAX) provides a format for calculating a comprehensive productivity indicator for any size organizational unit, although it is most widely applied to operational work units. It can be used for monitoring professional and service work performance, as well as manufacturing performance.

The matrix is constructed by identifying the *productivity criteria* that contribute to work performance, determining eleven levels of accomplishment for each criterion that correspond to a 0-to-10-point scale, in which level 10 represents a goal for future performance, and assigning weights to all criteria to indicate their relative importance. People in the work unit that is being measured should ideally be involved in identifying criteria, setting work objectives (level 10), and determining the scoring scale. Managers are responsible for weighting the criteria.

Productivity is measured in OMAX by determining the current level of performance and converting it to a score on the 0-to-10 scale. The *performance indicator* for the period monitored is calculated by summing the products of each criterion's score multiplied by its importance weight. A *productivity index* is the rate of change of a work unit's performance indicator from one measurement period to the next.

MEASURES OF PRODUCTIVE PERFORMANCE

The purpose of measuring productivity is to improve it. Measures contribute to improvement by calling attention to accomplishments or deficiencies, allowing comparisons between similar organizations, and providing a scale by which to assess progress. A creed for productivity measurement could be

To improve productivity, you must manage.

To manage effectively, you must control.

To control consistently, you must measure.

To measure validly, you must define.

To define precisely, you must quantify.

A broad spectrum of measurements has been exposed in previous chapters: dollars for cash flow analyses, time estimating for scheduling, product demand for forecasting, probabilities for decision making, numbers for accounting, product dimensions for quality control, and many more. Engineering economists rely mostly on monetary figures, but they utilize other dimensions as needed to extend and improve their analyses. Productivity measurement affords another perspective for economic evaluation of resources. It focuses on efficient utilization. Productivity studies suggest alternatives for improvement, instead of comparing alternative investment proposals for improvement, as occurs in typical engineering economic studies. Thus the two perspectives complement each other, both being contributors to the decision-making role of engineering economists portrayed in Chapter 1.

On page 2, engineering-economic decisions were said to embody (1) identification of alternative uses for limited resources and (2) analysis of data to determine a preference.

This chapter concentrates on productivity measurement as a means to promote productivity, not to raise quality, conserve time, increase profit, or reduce costs, although these may be outgrowths of productivity improvement, but to credit the type of work performance and behavior that efficiently utilizes resources. This credit flows to the providers of productivity who gain recognition from measurements that acknowledge their contribution. An ideal measurement system should therefore be equally applicable to skill-based and knowledge-based performance, and should motivate conscientiousness in both.

Necessary features of any measurement system that involves people are shown in the margin. They emphasize trustworthiness and procedural practicality. Based on the mission of productivity monitoring, the attributes listed in Table 22.1 are desirable for a measurement system that effectively audits workers' performance and resource utilization as it concurrently encourages improvement.

Requirements of an acceptable measurement system:
Feasible. Needed data are available
Consistent. Repeatable procedures and results
Accurate. Measures what is sought
Fair. Accepted as unbiased; trusted
Understandable. Clear methods and results
Timely. Provides feedback promptly
Economical. Value exceeds cost

TABLE 22.1

Desirable characteristics of a productivity measurement system that motivates as it monitors

1 *Achievement is directly attributable to its source.* Reciprocally, the people who do good things should be credited with their accomplishment.

2 *Only performance that contributes to the organization's goals deserves recognition.* Little or no credit is bestowed on activities unrelated to the primary thrust of production or service, regardless of superficial attractiveness or convenience. People tend to work toward what is measured, so what is measured should be what will do the most for productivity. If resource conservation is a critical issue, for instance, its importance should be stressed to those who control it.

3 *Objective rather than subjective measures are employed wherever possible.* Numbers should accurately reflect contribution and be so recognized by both those who grade and those who are graded. Simple imperfect measures are preferable to complex closer-to-perfect measures, if the sophisticated assessment is not understandable.

4 *Quality is clearly tied to quantity of output, as is initiative to dependability of performance.* Trade-off proportions are defined as specifically as possible, as in time versus cost versus quantity versus quality.

(*Continued on page 568.*)

TABLE 22.1 (continued)

5 *The same measurement procedure and rationale of performance expectations are applicable to all workers.* "Workers" include managers and staff as well as hourly employees. Uniformity enhances acceptability, given it is rational.

6 *Motivation is stimulated by measurement scaled to achievable objectives.* Indicated standards of performance should be an incentive that implies no ceiling on accomplishment. Unreasonably high expectations discourage any drive for improvement, embarrasing both the drivers and the driven. Alternatively, easy-to-reach objectives offer little challenge. Every criterion of performance should expose a specific pathway to improvement.

7 *All measured attributes of performance are controllable by the performers.* Corrective action to improve deficiencies cannot be initiated by individuals or work units if they have no control over inhibiting factors.

8 *Measurements are perceived as nonthreatening.* People should be rated with respect to an accepted profile of competency. It is natural for productivity purposes to measure the performance of work units instead of individuals. Unit measures are generally not as personally threatening and tend to encourage cooperation toward common goals. Scores should not be used to compare or pit one team against another, but a single score should audit a team's contribution.

9 *Individuals or units participate in determining performance criteria, establishing the scoring system, and monitoring scores.* Involvement in the formulation of a measurement system almost guarantees that it is understood and greatly increases the chance that it will be willingly endorsed. However, participation is fostered at the expense of customizing the measurement process to fit participants. The cost can be high, but it will be rewarded generously if it closes the feedback loop.

10 *The measurement system is practical.* Data to fuel the system should be obtainable without excessive effort, and the mechanism for generating ratings should operate smoothly with minimal administrative burden. It should produce consistent readings for analysis purposes. Its procedures should be flexible enough to accommodate modifications in response to changing conditions, yet its uniformity should facilitate stable productivity reporting. Its value must exceed its cost of implementation.

No measurement system will completely satisfy the 10 ambitious attributes in Table 22.1. The familiar ratio of output over input for productivity has historically centered on the value of goods and services produced per labor hour, which only vaguely suggests what could be done to raise the ratio. Adding other traditional inputs, such as capital, energy, and materials, still directs attention only toward the broad factors that affect productivity. Extensive analysis of the ingredients that collectively constitute the output/input ratio is required before credit or blame for the result can be resolved. Nonetheless, national and firm-level measurements are valuable because they lead to a single number, equivalent to a return-on-investment figure, that scores performance. That number can signal alarm. It can be a target. It provides historical comparisons. A productivity index is a statement of progress in the quest for resource utilization efficiency.

Example 22.1
Relation of Unit Labor Cost to Manufacturing Productivity

The two charts in Figure 22.1 appear to be inversely related; when productivity increases, unit cost decreases during the same time period. And since labor cost is a critical contributor to inflation, it too tends to decrease when productivity climbs. In general, these pictured relations are valid in practice but there are other factors to consider.

Reports about productivity usually feature the manufacturing sector, which is normally higher than are all other sectors except farming. For instance, from 1979 through 1984, United States manufacturing productivity increased at an annual average rate of just over 2.5 percent, whereas nonfarm, nonmanufacturing productivity (mining, construction, communications, utilities, and services) rose less than half a percent.

Thus the high correlation with unit labor cost is reserved for manufacturing productivity.

Sources of productivity changes are not evident from the charted percentages. In 1984, United States manufacturing productivity rose because the gain in output exceeded the increase in hours worked. During the same period two countries in Europe recorded higher productivity gains than did the United States even though their output declined; the decline was accompanied by an even larger drop in labor hours.

Comparison of unit labor cost between countries can be easily misinterpreted. In 1983 unit labor costs fell in the United States, Japan, and Germany and rose less than 1 percent in Canada, 2 percent in the United Kingdom, over 7 percent in

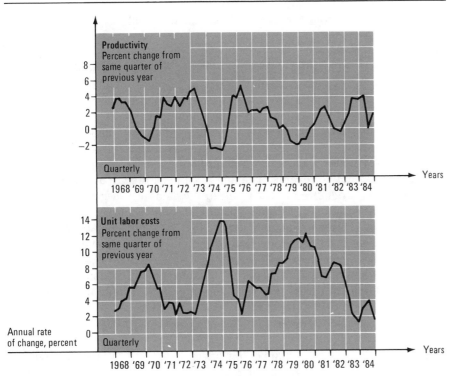

FIGURE 22.1
Patterns of annual rate changes in productivity and unit cost that reveal an inverse relationship.

France, and 16 percent in Italy. When measured in United States dollars, however, unit labor costs rose only 3 percent in Italy and declined substantially in the other European countries because of the rapid appreciation of the United States dollar. In contrast, Japanese unit labor cost, measured in United States dollars, increased over 2 percent, because the average value of the Japanese yen rose in 1983.

Predicting the future from the hills and valleys of Figure 22.1 is hopeless. Both productivity and unit labor cost are variables that reflect the joint effects of many influences, including new technology, capital investment, capacity utilization, energy use, and managerial skills, as well as the skills and efforts of the work force. However, knowledge of the current effect of these factors is a major step toward managing them more effectively.

Measures of Output

National measures of productivity were discussed in Chapter 21. In this chapter, firm-level and work-unit productivity measurements are featured. All are outgrowths from the same origin: output/input. Since they have shared ancestry, several precautions about the content of the numerator and denominator are generally applicable.

The selection of a weighting method is an early decision in any productivity measurement. When physical output is homogeneous, or nearly so, a pure quantity measure is sufficient. This is the case for production of a raw material of essentially constant quality, such as corn, ice, or one grade of coal. When quality varies, as in a combination of anthracite and bituminous coal, weighting is needed. Price is the most common weight.

The use of a general price index as a deflator is demonstrated in Example 22.2.

Additional cautions for output measurement include:

1 Goods and services in the numerator of a productivity ratio must be those produced by the inputs included in the denominator. Excluding an output produced by an included input causes a spurious decline in productivity whenever that input is increased, and vice versa. Equivalently, a reverse effect is caused by including an unearned output.
2 Since many different models, sizes, and product lines may be included in the total output of an organization, care must be given to the grouping of products. Greater accuracy is obtained from smaller groupings, provided that representative revenue data and price deflators are available.
3 Changes in product mix may result in the sale of a discontinued product during the period when inputs are directed toward the introduction of a new model. This situation may require that the reporting period be divided into intervals, one for production of the old model and the other for the new model.
4 Output is limited to the number of units produced, *not* units sold, during the measurement period. Partially completed units should be counted with the finished products. Their value is determined as

$$\text{Value of partial units} = \begin{pmatrix} \text{number of work-} \\ \text{in-process units} \end{pmatrix} \begin{pmatrix} \text{average} \\ \text{percent} \\ \text{completed} \end{pmatrix} \begin{pmatrix} \text{base-period} \\ \text{price per unit} \end{pmatrix}$$

Measures of Input

Measurements of inputs share most of the considerations given for the measurement of output. Additional features are discussed in this section for three classes of tangible inputs: labor, capital, and intermediate.

The common measure of labor is the number of hours expended by all people engaged in production of the output. National measures of productivity treat labor hours as a homogeneous input that can be simply added together. Most industrial measures of labor differentiate and weight the hours according to some pattern of labor composition.

Labor Input

     Several reasons support the belief that weighting provides a preferable measure of labor input. It is argued that different kinds of labor make different contributions to output. For instance, differentiating between skilled and unskilled labor is supported by reasoning that more skilled labor represents a larger labor input per hour of work. Weighting by a skill-related dimension (years of experience, training, etc.) also accommodates recognition of a change in labor mix or wages.

When labor is weighted by wage rates, a firm could drastically reduce the number of employees and still not significantly decrease its labor input if it laid off only the lowest-paid employees.

     The composition of the labor input reflects the intent of the analysis. A conventional limitation is to include only direct labor. A more comprehensive input includes managers, support personnel (clerical and administrative), professionals (essentially independent contributors), and workers. When the numerator of a productivity index constitutes the total output of a firm, it is logical that the labor input should include all the employees.

Physical capital is conceptually analogous to labor, sometimes called *human capital*, but there is less agreement on how it should be measured and weighted. The intent is to measure the efficiency with which capital assets are used. The concept varies somewhat from traditional accounting measures which are more concerned with the profitability of the total investment than its "productiveness." The monitoring difference between human and physical capital is the difficulty in tracking the depletion of dollar investments and establishing their earning rates. Four basic methods of capital productivity accounting are in use, with numerous variant forms proposed to overcome perceived inequities.

Capital Input

     An estimation of capital stock based on asset service life is followed in many nations. It inventories the years of service that an asset possesses productive capacity and the value of that service each year. At purchase, an asset's annual service contribution is estimated to establish a schedule of depreciation based on actual use, *not* tax life. The perpetual inventory is then a summation of the worth of different vintages of capital assets available in a given year. Each year the portion of an asset's worth that has been "consumed" is dropped from an updated inventory.

1 *Perpetual capital inventory*

     To illustrate perpetual capital inventory, let us assume that a new business purchases a machine for $100,000 with the expectation that it will provide adequate service for 4 years and then be retired at no additional cost or revenue. We will further assume that the machine becomes less efficient with longer ownership, its end-of-year values decreasing to: year 1, $60,000; year 2, $35,000; year 3, $15,000; year 4, $0. According to this schedule, the capital input during the first year is $100,000. In the second year of ownership, $60,000 of the investment capital is left, $40,000 having been expended to produce the first year's output. Next, we assume that a second identical machine is purchased one year after the first one at a cost of $110,000. If productivity is measured according to a base year taken as the year the business began, all capital inputs are deflated to that date. Therefore, the $60,000 value for the first machine in year 2 is already stated in base-year worth, but the second machine's price must be deflated to its base-year real-dollar value. If the price index during the second year went up from 213 to 238, the total capital input in year 2 for both machines would be

$60,000	(remaining value of 3 more years of service from machine one)
$+98,445$	$= 110,000 \times \dfrac{213}{238}$ (first cost of machine 2, deflated by the price index to the same real-dollar value as machine 1)
$\overline{\$158,445}$	(total capital, in base-year value, that contributed to production of the firm's output during its second year of operation)

The perpetual inventory thus tracks the amount of investment funds employed each year in operations.

2 Annual capital cost

An assumption of an average life for all assets is open to second guessing. Other questionable assumptions for capital input include adjustments for declining efficiency, proper percentages to use for capital return, and changes in the utilization of capital assets.

A conceptually different and simpler approach is to charge a consumption plus return percentage against total real-dollar capital stock to obtain the annual capital input. Instead of considering the aggregate investment accounted for in a perpetual capital inventory, only the decrease in total capital attributable to operations is counted. Calculation is simplified by assuming an average life for all physical assets. An assumption of 20 years means that 5 percent of the assets are "consumed" each year; a 15-year average service life implies that each asset annually loses 6.67 percent of its value. In addition to the charge for depreciation, a charge is also made for the cost of using the capital. In theory, capital earnings are held to be equivalent to wages paid for labor. If a rate of 13 percent is deemed to be a reasonable return for investors and a 15-year asset life is assumed, annual capital input is calculated by multiplying the total investment by 19.67 percent, that is, 13 percent + 6.67 percent.

Working capital may also be included. Sometimes the value of inventory is lumped together with the amount of cash, accounts receivable, and notes receivable. This total is deflated by an appropriate price index to obtain its real-dollar base-year value and is then multiplied by the selected rate-of-return percentage to obtain the working-capital portion of the capital input.

As an example of the annual capital cost method, let total investment this year be $10 million in physical assets and $1 million in working capital. Assume that the price index deflator for the current year is 0.83, a 15-year average asset life is applicable, and a realistic *rate of return* is 13 percent. Then the capital input is

$$\$10,000,000 \times (0.0667 + 0.13) = \$1,967,000$$
$$1,000,000 \times 0.83 \times 0.13 = \underline{107,900}$$
$$\text{Annual capital cost (this year)} \quad \$2,074,900$$

Next assume that another year has passed in which inflation has pulled down the price index deflator to 0.76, working capital has increased to $1.2 million, and another $500,000 has been invested in capital assets. The value of physical assets is the preceding year's real-dollar investment total minus the amount depreciated last year, plus the current year's additional real-dollar investment:

$$\$10,000,000 - \$10,000,000(0.0667) + \$500,000(0.76) = \$9,713,000$$

Capital input for the next year is thus

$$\$9,713,000 \times (0.0667 + 0.13) = \$1,910,547$$
$$1,200,000 \times 0.76 \times 0.13 = \underline{118,560}$$
$$\text{Annual capital cost (next year)} \quad \$2,029,107$$

The sum is considered to be the capital input for the second year, down slightly from the preceding year.

The rental-price approach visualizes all capital assets as being leased, including working capital. A rental value could theoretically be determined by going to the marketplace to see how much it would cost to rent each item used in production. But for many productive assets, there is no rental agency. Who rents pot lines for producing aluminum or almost any piece of specialized equipment? The alternative is to calculate a "rental" value by deflating the purchase price of an asset to its base-year value and then annualizing its cost over its expected life. The cost includes a profit for the leasor, which is equivalent to the rate of return expected by an investor. **3** *Rental price*

For an individual asset, the calculation of its annual rental cost is quick, but collectively the procedure is tedious. After deflating an asset's purchase price, its base-year value is multiplied by the capital recovery factor. If the real-dollar price of an asset is $100,000, its life is 10 years, and the expected rate of return is 15 percent, its annual "rental" cost is

$$\$100,000 \, (A/P, 15, 10) = \$100,000(0.19925)$$
$$= \$19,925$$

The sum of the annual rental costs for all assets employed is the capital input for that year.

4 *Book value* The simplest way to estimate capital input is to use book values taken directly from accounting records. A total capital investment is then the book value of the complete company, and capital cost is simply the sum of all depreciation charges in a year; either figure could be used as the capital input, depending on the analyst's preference for aggregate or annual figures. This simplicity is gained at the expense of accuracy. Book values are less likely to represent the physical consumption of assets than they are to be write-offs taken to maximize tax advantages for depreciation. The value of capital is distorted by eliminating base-year adjustments. Nonetheless, the convenience of book values allows a rough fix on capital input that might otherwise be forfeited to data-generating difficulties.

Intermediate Inputs All other inputs besides labor and capital are classified as intermediate inputs—commodities consumed in the process of producing other goods. Examples are raw materials, semifinished goods that require further fabrication, fuel, and services provided by one firm to another. The most-used partial indexes are for materials and energy.

Alternative ways to include intermediate goods in a productivity ratio are illustrated in Exercise 5.

To measure intermediate consumption in real terms, the current year's nominal values are deflated to their base-year equivalents by an appropriate price index. The only conceptual issue involves the basic productivity index. In a value-added model the intermediate inputs are subtracted from gross output to determine the output numerator. In a total-output model, intermediate inputs are added to the other inputs in the denominator.

COMPANY MEASURES OF PRODUCTIVITY

It could be said that financial accounting records where you have been, productivity analysis reveals where you are, and economic analysis suggests where you could be in the future. All three measures of performance are needed to manage a company. Profitability is the cardinal indicator of success, but it is not the only indicator. Profits are affected by factors external to and beyond the control of the firm. Shifts in demand, inflation, or resource availability can wipe out profit, even when a company is highly productive. Measurements of productivity are based on internal operations controlled by company managers and therefore indicate the efficiency of that management.

What managers learn from measuring productivity depends on the design and extent of their study. Calculations of a *total-productivity index* for a firm provide a comparison of company performance from one time period to the next and allow interfirm comparisons. Measures can also be made for units within a firm, to contrast efficiency among organizational elements. *Partial-productivity indexes* result from relating company output to only one input factor or an exclusive combination of inputs. These partial indexes are useful in tracing the utilization patterns of particular resources, such as labor or energy. A variety of *productivity indicators*, output/input ratios of selected financial or technical data, can be developed to contrast operating efficiencies from one period to the next; for example, ratios used in the foundry industry include

$$\frac{\text{Cost of fuel}}{\text{Weight of metal charged}} \quad \text{and} \quad \frac{\text{Net weight of all good castings}}{\text{Weight of metal charged}}$$

In practice, an analyst will usually use two or more measures of productivity to examine a firm's performance. Partial and total productivity are mutually complementary measures because they rely on the same data and help explain the source of performance changes. Productivity indicators are particularly useful in evaluating the performance of small operating units, such as a single department or a team of workers, engaged in distinguishable production.

General methods of calculating these productivity indexes and indicators are described in this section. Actual formulations may vary from the general models, in recognition of different needs and conditions among individual firms.

Total-Productivity Index

Total productivity is the ratio of real product to an aggregate of all the relevent production inputs. In effect, it expresses the efficiency of an entire organization as a single figure. This figure can be used to compare efficiency from one year to the next because all the data are indexed to a base year to eliminate the effects of inflation.

Many models have been developed to calculate the total-productivity index. They differ conceptually with respect to what outputs and inputs should be included and the proper way to state them. From the safe start in which

$$\text{Total productivity} = \frac{\text{output}}{\text{input}} \quad \text{or} \quad \text{TP} = \frac{O}{I}$$

the ratio can be expanded to

$$\text{TP} = \frac{\Sigma O}{\Sigma I}$$

where the summations refer to productivity measures aggregated from different levels of an organization or values that represent distinct groupings of outputs and inputs that collectively constitute the factors of production for the entire organization. One popular form of the expansion is

$$\text{TP} = \frac{\text{finished plus partial products}}{\text{labor + capital + energy + materials + other inputs}}$$

Surveys indicate that about 1 percent of United States companies that measure productivity use total productivity measures, 19 percent use partial productivity measures, and 80 percent use other nonstandard productivity indicators. (D. J. Sumanth, *Productivity Engineering and Management*, McGraw-Hill, New York, 1984.)

A slightly different version utilizes weights to combine outputs and inputs as

$$TP = \frac{\sum_i w_i O_i}{\sum_i w_i I_i}$$

where i factors are included and weighted according to their relative contribution, w. In practice, the weights $w_1 w_2, \ldots, w_n$ show the proportional contributions of respective outputs and inputs to a production process.

A single productivity measure is about as useful as one shoe. At least two consecutive measurements are needed to reveal a productivity change. To make the two ratios equivalently representative of actual operations, unusual conditions such as a temporary plant closing must be accounted for, and prices must be inflated or deflated to convert them to constant base-year dollars. The base-year reference is often the first year in which a productivity measurement is compiled. The reference year may be updated when a distinctly different product mix becomes evident, new basic models are introduced, a new accounting system is installed, or abnormal developments seriously disturb operations.

Although statements of yearly productivity gains or losses predominate, firms may calculate monthly or seasonal indexes. Regardless of the reporting interval, actual dollar values for outputs and inputs should be translated to constant base-year dollars. This is commonly accomplished by deflating actual dollars according to one of the established national price indexes. As discussed in Chapter 13, several indexes are available. Which one to use depends on the location, product, labor composition, and nature of the firm. The frequently used U.S. Consumer Price Index (CPI) is charted in Figure 13.1. Given CPI values of 299.0 and 311.6 for 1983 and 1984, respectively, a 1984 output of $100 would be deflated to a 1983 base-year equivalent as $100(299.0/311.6) = $95.96.

One of many possible total-productivity models is demonstrated in the following example.

Example 22.2
Calculation of a Total-Productivity Index

The total-productivity model can be simply stated as

$$TP = \frac{O_1 + O_2 + O_3}{I_1 + I_2 + I_3 + I_4 + I_5}$$

but the individual outputs and inputs are not so easily determined. Assume that the representative price index has risen 13 percent between the 2 years for which productivity is being measured, 1987 and 1988. This makes the 1988 general price deflator equal to $100/113 = 0.885$.

Individual output and input data are described and developed below.

O_1 = sales (net sales billed for the measured year)

If the sales for 1988 amount to $133,333,000 and they are to be compared with 1987 sales, the 13 percent inflation is removed by applying the 0.885 deflator. Thus 1988 sales in constant 1987 dollars are $133,333,000(0.885) = $118,000,000.

O_2 = inventory change (variation in the value of inventory levels between the beginning and end of the measured year)

Inventory includes work in process and finished goods. Each category of inventory could have its own deflation factor. Assuming that a deflator of 0.85 is applicable for all material categories and that the total value of all inventories on hand increased by $3.5 million during the year, we find that the amount of inventory change has a constant dollar value of $3,500,000(0.85) = $2,975,000.

O_3 = plant (value added internally to the organization)

The capabilities of an organization are often increased by the output of its employees in ways that increase the worth of the production system but do not directly increase income through sales. Maintenance, R&D, and company-produced machinery are examples. The value contributed by plant improvement can be measured as the internal cost or market value but in either case must be deflated to constant dollars. Assume that the value of the plant increased by $2.2 million (constant dollars) in 1988.

I_1 = labor (wages and benefits paid to employees or set aside in reserve accounts)

Annual compensation for labor includes hourly wages, salaries, overtime, vacation and sick pay, insurance, profit sharing, social-security tax, and bonuses. Assume that after deflating by the Consumer Price Index, labor costs for 1988 were $35 million.

I_2 = material (raw materials and purchased components consumed in production)

Assume that after applying appropriate material deflators, the net consumption for 1988 was $61 million.

I_3 = services (externally purchased energy and supplies plus rented equipment and contracted labor)

Annual expenditures are deflated to constant dollars: 1988 services equal $2,500,000.

I_4 = depreciation (decrease in capital value of the assets that provide production capability)

Since depreciation charges are not responsive to current price trends, they are included at their current value and are assumed to be $3.2 million in 1988.

I_5 = investment (cost to investors for fixed and working capital in the production system)

Fixed capital includes land, buildings, machinery and equipment, and deferred charges. Working capital includes cash, notes, accounts receivable, inventories, and prepaid expenses. The investment input is then the rate of return that should be earned by all the capital assets. If we assume a constant value of 15 percent for the annual cost of capital to the organization, total capital assets of $58 million, and a price deflator of 0.885, the constant-dollar investment input for 1988 was $58,000,000(0.15)(0.885) = $7,700,000.

From these output and input values, total productivity for 1988 is calculated as

$$TP(1988) = \frac{\$118,000,000 + \$2,975,000 + \$2,200,000}{\$35,000,000 + \$61,000,000 + \$2,500,000 + \$ 3,200,000 + \$7,700,000}$$

$$= \frac{\$123,175,000}{\$109,400,000} = 1.126$$

This figure has more meaning when it is compared with the index for the same organization from the preceding year. Assuming that the total productivity measure for 1987 was 1.073,

$$\text{Productivity index (1988)} = \frac{1.126}{1.073} = 1.0494$$

Productivity increase (1987 to 1988) = 0.0494

or = 4.94%

Partial-Productivity Indexes

A partial-productivity measure results from relating output to only one input or a combination of inputs. The most common measure is output per labor hour. The combination of labor and capital, called *factor productivity*, is also a broadly applied partial-productivity measure because it focuses on controllable inputs; the cost of materials is considered to be a throughput. A partial-productivity measure is most valuable where one or more inputs are clearly the dominant factors of production.

Partial-productivity indexing is burdened with the same conceptual difficulties as is total-productivity indexing. A critical one is the weighting of outputs and inputs. Consider the indexes that could be developed from the data given in Table 22.2.

The total quantity of units produced can be compared with the labor hours consumed in each year:

$$\begin{aligned}\text{Unweighted partial} \atop \text{productivity index} &= \frac{(\text{units produced in year 1})/(\text{labor hours in year 1})}{(\text{units produced in year 0})/(\text{labor hours in year 0})} \\ &= \frac{8000/19,000}{7000/16,500} = \frac{0.421}{0.424} = 0.99\end{aligned}$$

According to this physical-accounting index, productivity declined by about 1 percent. Being based on summary product-unit and labor-hour counts, it fails to register changes in the product and labor mix.

An index more sensitive to production economics is developed by including prices and costs. Working again with the given unit and labor totals, but

---

TABLE 22.2

Base-year and current-year output and labor-input data

	Model A	Model B	Models A and B		Rate 1	Rate 2	Rates 1 and 2
			Base Year (0) Production Data				
Units produced	5000	2000	7000 (total)	Labor hours/units	9000	7500	16,500 (total)
Price/unit	$40	$70	$55 (average)	Labor cost/hour	$9	$15	$12 (average)
			Current Year (1) Production Data				
Units produced	6000	2000	8000 (total)	Labor hours/unit	11,500	7500	19,000 (total)
Price/unit	$60	$80	$70 (average)	Labor cost/hour	$10	$16	$13 (average)

weighting them according to their average dollar values, the following expression results:

$$\text{Average-weighted partial-productivity index} = \frac{Q_1 P_{av_1}/L_1 C_{av_1}}{Q_0 P_{av_0}/L_0 C_{av_0}}$$

$$= \frac{Q_1 P_{av_1}/Q_0 P_{av_0}}{L_1 C_{av_1}/L_0 C_{av_0}}$$

where Q and L represent units and labor hours, respectively, and subscripts av_0 and av_1 indicate average values in the base year and current year. Substituting numbers from Table 22.2 produces

$$\text{Average-weighted partial-productivity index} = \frac{8000(\$70)/7000(\$55)}{19,000(\$13)/16,500(\$12)}$$

$$= \frac{\$560,000/\$385,000}{\$247,000/\$198,000}$$

$$= \frac{1.4545}{1.2475} = 1.17$$

which suggests that productivity rose 17 percent. Note that output value increased by over 45 percent at the expense of a 25 percent increase in input cost. This condition could have resulted from improvements in the product line that made the output more valuable without a corresponding rise in production costs. However, the use of average prices and costs may yield misleading outcomes.

A slightly different version of the preceding formula is known as the *Edgeworth* ratio. It uses the average price or cost as the weighting factor. As applied to the index numerator, the Edgeworth expression for output is

$$\text{Edgeworth ratio} = \frac{\Sigma Q_1(P_0 + P_n)/2}{\Sigma Q_0(P_0 + P_n)/2}$$

For the previously used prices, $(P_0 + P_1)/2 = (\$55 + 70)/2 = \$62.50$, and for the same output quantities

$$\text{Edgeworth ratio} = \frac{8000(\$62.50)}{7000(\$62.50)} = 1.14$$

The *Laspeyres* ratio produces the same figure from the formula

$$\text{Laspeyres ratio} = \frac{\Sigma Q_1 P_0}{\Sigma Q_0 P_0} = \frac{8000(\$55)}{7000(\$55)} = 1.14$$

Still another respected calculation method is the *Paasche* ratio:

$$\text{Paasche ratio} = \frac{\Sigma Q_1 P_1}{\Sigma Q_1 P_0} = \frac{8000(\$70)}{8000(\$55)} = 1.27$$

The Paasche ratio differs from the previous two because it emphasizes current prices. In effect, the Edgeworth and Laspeyres ratios are quantity-weighted whereas the Paasche ratio is price-weighted.

When the Edgeworth and Laspeyres ratios are applied to both the numerator and denominator (output units and labor hours) with the given data, the resulting index is the same as the previously calculated unweighted partial productivity index, 0.99. The Paashe ratio produces a partial-productivity index of 1.17 when applied to both output and input.

Each model has its advocates. The recommended model is indicated below. It adjusts for quality changes in both output and input and utilizes base-period weighting to assist in trend analysis, as shown by

$$\frac{\text{Partial-productivity}}{\text{index (PPI)}} = \frac{[(\text{constant-price output})/(\text{constant-cost input})]_{\text{current period}}}{[(\text{constant-price output})/(\text{constant-cost input})]_{\text{base period}}}$$

This may be written as

$$\text{PPI} = \frac{[(\Sigma O_1 P_0)/(\Sigma I_1 C_0)]}{[(\Sigma O_0 P_0)/(\Sigma I_0 C_0)]} = \frac{[\Sigma O_1 P_0)/(\Sigma O_0 P_0)]}{[(\Sigma I_1 C_0)/(\Sigma I_0 C_0)]}$$

where the subscripts 0 and 1 indicate, respectively, the base and current periods; O and I indicate, respectively, output and input quantities; and P and C indicate, respectively, output price and input cost.

The ratios above can be recognized as Laspeyres indexes. They avoid the use of deflators by weighting both output and input quantities according to base-year dollar values. As applied to the data from Table 22.2, the basic PPI model can accommodate changes in the output mix (A and B) and labor rates (1 and 2) as shown below:

$$\text{PPI} = \frac{[6000(\$40) + 2000(\$70)]/[5000(\$40) + 2000(\$70)]}{[11,500(\$9) + 7500(\$15)]/[9000(\$9) + 7500(\$15)]}$$

$$= \frac{\$380,000/\$340,000}{\$216,000/\$193,500} = \frac{1.11}{1.11} = 1.00$$

According to the recommended partial-productivity formula, there has been no productivity gain attributable to the labor input.

The sample application above dealt with only one input, labor. Capital, materials, and energy are other candidates for partial-productivity indexing.

A total-productivity index results from a Laspeyres index of complete output in the numerator divided by a summation of Laspeyres indexes for all inputs.

Two complementary indexes have been developed to extract more information from data collected for a partial-productivity measurement. These variations of the basic PPI analyze operations from different vantage points. They contrast prices and costs which occurred in the base period with those in the current period and therefore make the indexes sensitive to economic changes.

A *price-recovery index* reveals to what extent the firm has been able to absorb increases in the costs of inputs by its pricing policy:

$$\text{Price-recovery index} = \frac{\Sigma O_1 P_1 / \Sigma O_1 P_0}{\Sigma I_1 C_1 / \Sigma I_1 C_0}$$

This expression is a Paasche index in which current output O_1 and input I_1 are weighted by base (P_0 and C_0) and current (P_1 and C_1) prices and costs, respectively; P_1 and C_1 are *not* deflated values.

A *cost-effectiveness index* contrasts current output and input performance with that in the base period:

$$\text{Cost-effectiveness index} = \frac{\Sigma O_1 P_1 / \Sigma O_0 P_0}{\Sigma I_1 C_1 / \Sigma I_0 C_0}$$

> The price-recovery and cost-effectiveness indexes quantify the relationships described in Example 21.1.

This is perhaps the most intuitively logical way to measure productivity—directly compare current value with base values. However, as a basic index it would be deceptive because it contains no inflation correction; P_1 and C_1 are the *actual* price and cost. As a complementary index it summarizes the effects of quantity and economic-value changes between the two periods measured.

The relation between the basic partial-productivity index and the price-recovery and cost-effectiveness indexes is shown by

$$\text{PPI} = \frac{\text{cost-effectiveness index}}{\text{price-recovery index}}$$

$$= \frac{(\Sigma O_1 P_1 / \Sigma O_0 P_0)/(\Sigma O_1 P_1 / \Sigma O_1 P_0)}{(\Sigma I_1 C_1 / \Sigma I_0 C_0)/(\Sigma I_1 C_1 / \Sigma I_1 C_0)} = \frac{\Sigma O_1 P_0 / \Sigma O_0 P_0}{\Sigma I_1 C_0 / \Sigma I_0 C_0}$$

To illustrate the nature of the complementary indexes, they are applied below to the data of Table 22.2 which produced the basic PPI $= 1.0$.

$$\text{Price-recovery index} = \frac{[6000(\$60) + 2000(\$80)]/[6000(\$40) + 2000(\$70)]}{[11,500(\$10) + 7500(\$16)]/[11,500(\$9) + 7500(\$15)]}$$

$$= \frac{1.368}{1.088} = 1.26$$

$$\text{Cost-effectiveness index} = \frac{[6000(\$60) + 2000(\$80)]/[5000(\$40) + 2000(\$70)]}{[11,500(\$10) + 7500(\$16)]/[9000(\$9) + 7500(\$15)]}$$

$$= \frac{1.529}{1.214} = 1.26$$

As expected, PPI = 1.26/1.26 = 1.0. This result indicates that the increase in labor costs, due mainly to a change in labor mix and quantity, is compensated for by a corresponding increase in the value of output, primarily as a result of producing more units of the model that commanded much higher prices.

Example 22.3
Total- and Partial-Productivity Indexes Developed by Deflating Actual Dollars to Constant Dollars

Still another version of productivity indexing is demonstrated in Table 22.3. It utilizes data that are readily available from accounting records. Five categories are assumed to collectively represent a firm's productivity posture. Four of the classifications are deflated by their associated price indexes to provide comparable year-to-year figures. The other category, depreciation, is not deflated, because it is the sum of the year's depreciation charges, based on the productive life of assets, which is assumed to be nonresponsive to current price changes.

Output and input figures in constant dollars indexed to a

TABLE 22.3

Breakdown of input and output data from the annual corporate financial statement used for productivity analyses. Current dollars are converted to the 1986 base-year values as (current dollars × 100)/(price index) = constant dollars. 1988 operations are compared with 1987 by the productivity rise as [1988 − 1987 constant dollars] (100%)/(1987 constant dollars). Note that total productivity increased at a lesser rate than did other productivity ratios, a condition caused mostly by the steep rise in material costs.

	Productivity Analysis of Annual Operations						
	1987			1988			Productivity Rise (1987 to 1988)
	Current Dollars (in 1000s)	Price Index (1986 = 100)	Constant (1986) Dollars	Current Dollars (in 1000s)	Price Index (1986 = 100)	Constant (1986) Dollars	
1. Net sales	$1831	108	$1695	$2293	122	$1880	
2. Labor	295	107	276	352	120	293	
3. Materials	880	105	838	1161	131	886	
4. Services	301	106	284	365	118	309	
5. Depreciation			96			122	
6. Total inputs (2 + 3 + 4 + 5)			1494			1610	
7. Net output (1 − 3 − 4)			573			685	
8. Factor input (2 + 5)			372			415	
9. Labor productivity (7/2)			2.08			2.34	12.5%
10. Factor productivity (7/8)			1.54			1.65	7.1%
11. Total productivity (1/6)			1.13			1.17	3.5%

base year are collected in various ratios to furnish insights about the organization's performance:

Labor productivity (9) relates *net* output (sales minus material and service expenses) to total labor costs to measure labor efficiency.

Factor productivity (10) is a measure of the efficiency of the combined labor and capital (represented by depreciation charges) input. It is also based on *net* output.

Total productivity (11) is the ratio of total output (sales) to all inputs.

Although index ratios are not shown, the percent differences between the productivity measures for the 2 years are given in the right-hand column of the shaded section of Table 22.3.

Other partial-productivity indexes could be resolved from the given data or more detailed breakouts, such as ratios of net output to raw materials or purchased parts. These indexes provide insights about operating performance that are not in financial statements, especially when trends are analyzed.

PRODUCTIVITY INDICATORS

Financial ratios representing asset dispositions have long been favored by investment analysts. They provide handy reference points from which to assess the security of investments and by which alternative investments can be compared. These measures of financial inflows and outflows, as discussed in Extension 18A, are prominent in annual reports of corporate performance. To better understand the sources of these financial performance standards, it is enlightening to relate them to their production constituents.

Physical flows can be interspersed within traditional financial flows to obtain firm-level productivity indicators. They focus on the operations that produce revenue and consume capital. Attention to the indicators could channel cash flow to the betterment of total investment. The familiar ratio for the return on investment is progressively factored to reveal contributing elements as follows:*

$$\frac{\text{Profit}}{\text{Total investment}} = \frac{\text{profit}}{\text{output}} \times \frac{\text{output}}{\text{total investment}}$$

but

$$\frac{\text{Profit}}{\text{Output}} = \underset{\text{(weighted prices)}}{\frac{\text{product value}}{\text{output}}} - \underset{\text{(weighted costs)}}{\frac{\text{total costs}}{\text{output}}}$$

where

$$\frac{\text{Total costs}}{\text{Output}} = \frac{\text{wages}}{\text{output}} + \frac{\text{material cost}}{\text{output}} + \frac{\text{other costs}}{\text{output}}$$

*Adapted from S. Eilon, B. Gold, and J. Soesan, *Applied Productivity Analysis for Industry*, Pergamon, Oxford, England, 1976.

and

$$\frac{\text{Output}}{\text{Total investment}} = \underbrace{\frac{\text{output}}{\text{capacity}}}_{\substack{\text{(utilization} \\ \text{rate)}}} \times \underbrace{\frac{\text{capacity}}{\text{fixed investment}}}_{\substack{\text{(productivity} \\ \text{of fixed} \\ \text{investment)}}} \times \underbrace{\frac{\text{fixed investment}}{\text{total investment}}}_{\substack{\text{(internal allocation} \\ \text{of capital)}}}$$

so

$$\frac{\text{Profit}}{\text{Total investment}} = \left(\frac{\text{product value}}{\text{output}} - \frac{\text{wages}}{\text{output}} - \frac{\text{material cost}}{\text{output}} - \frac{\text{other costs}}{\text{output}} \right)$$
$$\times \frac{\text{output}}{\text{capacity}} \times \frac{\text{capacity}}{\text{fixed investment}} \times \frac{\text{fixed investment}}{\text{total investment}}$$

The constituent productivity indicators could easily be converted to partial-productivity indexes by linking ratios from sequential periods, and indeed they constitute most of the elements previously mentioned for a total-productivity index. However, productivity indicators are designed to serve a different purpose. They provide a perspicuous reading on short-term performance in critical areas of operations. They are managerial control measures.

Two of the most advantageous applications of productivity indicators are for interfirm comparisons and as work-unit performance indicators.

Interfirm Comparisons

Since its founding in 1959, the Centre for Inter-firm Comparison in London has conducted comparisons in 91 different United Kingdom industries, trades, and services.

Several nations have productivity centers that advise companies about how to improve their productivity. A service by some centers is a program that allows a firm to compare its performance with other competing companies that produce similar products by similar techniques. The comparisons use information generally available in a company's records and transform it into ratios that indicate how well resources are being managed.

Starting from financial ratios, technical ratios are developed that rely on operating data such as output per labor-hour, machine throughput per hour, and theft cost per unit. A few of the many possible ratios are shown in Figure 22.2 Each isolates an area in which inefficiencies may become evident as a result of comparison with other companies. For instance, a higher than median ratio in*

Ratio 4. Indicates disproportionately large manufacturing expense

Ratio 9. Suggests excessive spoilage, lack of standardization, or inefficient purchasing

*Adapted from *Interfirm Comparison*, Australian Department of Productivity, Canberra, 1979.

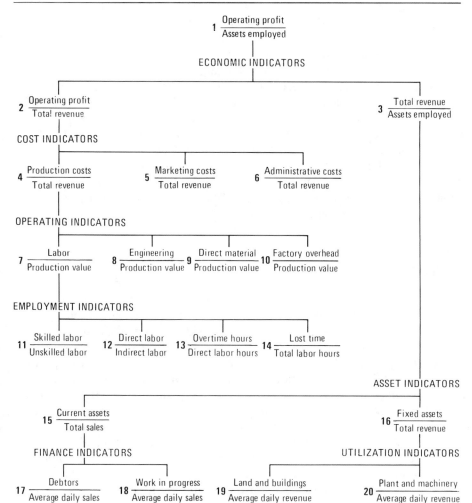

FIGURE 22.2
Typical ratios that can be used as productivity indicators for interfirm or intrafirm comparisons of financial and operating efficiencies.

Ratio 10. Indicates higher-than-average energy costs, maintenance, or level of depreciation

Ratio 14. Suggests the possibility of unpleasant working conditions, poor staff relations, unattractive factory location, or personality problems

Ratio 18. Indicates problems in work flow, plant layout, length of the production cycle, or number of ''rush'' orders received

Ratio 20. Indicates ownership of more capacity or more expensive plant and machinery than other companies have, which should be compensated for by significantly lower labor cost, ratio 7

Productivity indicators are useful for intrafirm as well as interfirm comparisons. A single company with several organizational divisions benefits from a ratio-incited information exchange within its operating units. The ratios indicate areas of excellence in different divisions. Divisions that excel in one area can share their efficiency techniques, particularly with divisions where significant inefficiencies were detected.

The compilation of productivity indicators for internal analysis by a single company may be worthwhile. Instead of comparing ratio values with other firms or divisions, historical values are compared. This version of trend analysis promotes an awareness that encourages managers to investigate incipient troubles before they get aggravated.

Work-Unit Productivity Indicators

Productivity-improvement effort is directed toward the people who actually produce the output by direct measurement of their performance. Productivity indicators differ from standards set for accomplishing a given amount of work within a given time. Team indicators measure characteristics of an operation or project being done by a group of people. The composition of the groups may cross traditional organizational boundaries. The criterion for inclusion is that all members of the group are engaged in the production of a definable end product or accomplishment—maintenance of certain production facilities, operation of a warehouse, an administrative function such as accounting, fabrication of a subassembly, the issuing of licenses, completion of one stage of production, etc. The size of the group naturally depends on the situation.

To be effective, productivity indicators must be readily understandable, easily measured, competently administered, and acceptable by those being measured. Most employees are apprehensive about having their output formally and critically evaluated. Being group- instead of individual-oriented, team measurements are usually considered less threatening. However, they still must be reasonable indicators of performance quality.

No single parameter can capture the gist of productivity. Because productivity results from a complex interaction of a number of parameters, many of which are interrelated, productivity indicators must be selected carefully. Omitting a significant indicator, or including an unimportant one, lowers the credibility of the whole measurement exercise. Examples of indicators applied to research and development activities, an exceedingly difficult area to measure, are presented in Example 22.4.

The worth of measuring team productivity is ultimately decided by the management action it provokes. Benign indifference undermines its value as surely as does disciplinary overreaction.

Example 22.4
Productivity Indicators for R&D

Hughes Aircraft Company sponsored a 5-year study of the productivity of research and development.* According to the study, "efficiency, effectiveness, and *value* are the key factors related to productivity. Work can be efficient but highly ineffective and of little or no value; conversely, work can be effective and valuable but grossly inefficient."

Productivity in R&D is an elusive subject because it involves so many variables: individual abilities and work habits, vague standards of performance, erratic schedules, changing goals, etc. However, some basic tenets were identified to foster productivity improvement:

- Improvements in any organization are "there for the asking," but few take advantage of them.

- There are underutilized resources in every organization and individual.

- Higher productivity results from greater involvement by management.

- Seemingly small individual improvements add up to significantly increased productivity and this can frequently have a large impact on profits.

Quantitative productivity indicators suggested by the study include the following:

1 Profits generated per R&D dollar spent
2 Dollar value of proposals won versus dollars spent to secure them
3 Number of errors detected per square foot of drawing
4 Number of changes per drawing per year
5 Ratio of staff personnel to line personnel
6 Secretarial support ratios—managerial, scientific, and engineering
7 Absentee rate
8 Employee voluntary turnover rate

It was generally agreed that productivity evaluations should be conducted at frequent intervals by responsible line managers.

THE OBJECTIVES MATRIX

Most of the desirable characteristics for a productivity measurement system listed in Table 22.1 are accounted for in the *objectives matrix* (OMAX), a performance measurement method that uses productivity indicators and a weighting procedure to develop a total performance indicator.

In the ideal method, practicality forbids expensive data collection and elaborate monitoring procedures. Its need for accountability suggests that *surrogate measures* be substituted for actual counts of output. For instance, the output of "knowledge workers" is difficult even to define, and much more so to measure in concrete values. Their output is therefore represented by activities that foster the production of goods and services instead of physical measures of output. This personalizing of measurement also forfeits consistency. When individuals or teams are evaluated by surrogates that represent their unique contribution to productivity, the scores obtained cannot be compared on a one-to-one basis with other individuals and groups working in different situations. With the exception of these concessions to practicality,

Performance indicators are most readily accepted in the service and government sectors, where few direct-product measurements are possible.

* *R & D Productivity*, Hughes Aircraft Co., Culver City, Calif., 1978.

FIGURE 22.3
Basic structure of an objectives matrix. Circled numbers indicate sections of the matrix that are explained in Table 22.4.

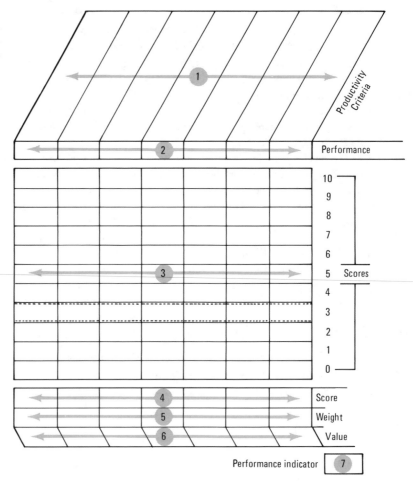

the *objectives matrix* approach to performance measurement meets most of the listed attributes for effective monitoring.

By definition, a matrix is "that within which something originates, takes form." As applied to performance measurement, an objectives matrix is a compact format within which performance goals originate and from which a profile of accomplishment takes form. The structure of the matrix is shown in Figure 22.3 and the different sections, as identified by circled numbers, are described in Table 22.4.

The objectives matrix was created in the mid-1970s by J.L. Riggs. The given format evolved during applications at the Oregon Productivity Center.

Productivity Criteria

Most employees know how their activities affect the productivity of their work unit. They realize which of the functions they perform support the organization's output, as opposed to those which are incidental to production or are simply

TABLE 22.4

Features of the objectives matrix that yield a performance indicator for an individual or a group of workers. The seven features correspond to the sections of the matrix shown in Figure 22.3.

Each named portion of the matrix, starting from the top section, has the following inputs which are keyed to the circled numbers in Figure 22.3:

① *Productivity criteria.* Activities and behaviors that promote productivity in specific jobs or work positions are characterized by ratios, such as output per hour, scrap per 100 units, and lost time per month. These are the characteristics of performance that collectively delineate a productive individual or group, similar to a job description but oriented to measurable acts. The ratios are entered as headings for the columns of the matrix.

② *Performance.* Measurements of performance over a period of time are entered in this line for all the criteria. These are actual accomplishments registered as a single number derived from the ratio developed for each criterion. Data to complete the ratios are obtained from production, accounting, and personal records, plus information provided by supervisors.

③ *Scales.* The body of the matrix is composed of levels of achievement for all the productivity criteria. There are eleven levels, ranging from unsatisfactory performance at level 0 to a realistic objective for superior accomplishment at level 10. When an objectives matrix is initiated, the prevailing level of performance is considered to be at level 3. Each higher score is achieved by passing a hurdle defined by the ratios for each performance criterion, such as passing from level 3 to level 4 by increasing output from 3.1 units per hour to 3.3, or by decreasing lost-time hours from 16.8 hours per month at level 3 to 16.2 hours at level 4. Level 10 scores for these two criteria might be 7.4 units per hour and 13.1 hours per month, both being strong but acceptable challenges for improvement. The *scales* are ideally developed cooperatively with the work unit that is to be measured.

④ *Score.* In the line immediately below the body of the matrix, the performance number, registered in the *performance line* immediately above the body, is converted to a *score*. This is done by comparing the recorded performance number with the hurdle scores. If the hurdle at level 4 is 3.3 units per hour and is 3.5 at level 5, output at a rate of 3.4 units per hour earns a score of 4, because the hurdle at level 4 has been surpassed, whereas the requirement for a score of 5 has not yet been attained.

⑤ *Weight.* Importance ratings are attached to all the criteria to indicate their relative impact on the total organization's productivity objectives. These ratings are developed by a management group above the level of the individuals being monitored. The sum of individual criterion weights totals to 100, which corresponds to 100 percent of the performance being evaluated. For instance, if the weight for output per hour is 30 and there are five more criteria, 70 points would be distributed among the remaining five ratios.

⑥ *Value.* The score for a period multiplied by the weight for the criterion in which the score was recorded is the *value* of performance for that period. In the case of an output-per-hour performance of 3.4, which earns a score of 4 when the output criterion has a weight of 30, the value is $4 \times 30 = 120$.

(*Continued on page 680.*)

TABLE 22.4 (continued)

⑦ *Performance indicator.* The sum of the values (weighted scores) is the performance indicator for the period. The initial indicator would have a value of 300, because all the criteria rate a score of 3 for the level of achievement at the start of the monitoring process. Thereafter, productivity improvement is measured by increases in the indicator. A climb from 300 to 321 would constitute a productivity gain of $(321 - 300)/300 = 0.07$, or 7 percent for the period. This is the *productivity index.*

customary behavior. These significant functions are the criteria of productive performance.

Different groups of workers share a set of work characteristics which distinguish their contribution to the organization's productivity. What contributes most in an office may not be as important in a machine shop. Activities of consequence to productivity differ between supervisors and their subordinates. Yet a few general categories encompass most of the factors. The seven broad classifications given in Figure 22.4 encompass most of the productivity criteria pertinent to nonmanagerial positions.

Actual entries within the headings shown in Figure 22.5 take the form of ratios. Two or three ratios may be required to fully account for one category, such as quality, or a category may be ignored if it does not affect productivity in a given situation. Since most white-collar and knowledge-worker positions do not have readily distinguishable outputs, their ratios are measurable modes of behavior that are known to affect their collective output. Typical ratios for various categories applicable to different work situations are given in Table 22.5.

Performance Scores

Performance scales in the body of the objectives matrix run from 0 to 10. There are thus 11 levels of accomplishment for each criterion; a single criterion occupies a column that stretches from the top to bottom of the matrix. Levels of accomplishment extend across the body of the matrix, as indicated by the rows marked from 0 through 10. Assignment of results expected at each level

FIGURE 22.4
General categories which define a person's or group's contribution to productivity at the level of their work unit. Comparable categories with more specific designators might be, respectively from left to right, *Units/hour, Defects/unit, Late orders/ total orders, Rejects/100 units, Equipment downtime/operating hours, Accident rate* and *Turnover rate.*

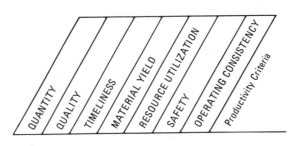

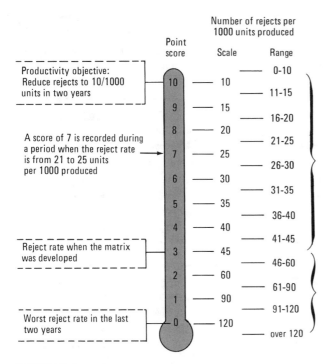

Number of rejects per
1000 units produced

Point
score   Scale   Range

Productivity objective:
Reduce rejects to 10/1000
units in two years — 10 — 10   — 0-10

9 — 15   — 11-15

8 — 20   — 16-20

A score of 7 is recorded during
a period when the reject rate — 7 — 25   — 21-25
is from 21 to 25 units
per 1000 produced   6 — 30   — 26-30

5 — 35   — 31-35

4 — 40   — 36-40

Reject rate when the matrix
was developed — 3 — 45   — 41-45

2 — 60   — 46-60

1 — 90   — 61-90

Worst reject rate in the last
two years — 0 — 120   — 91-120

— over 120

In this example, a linear scale was calculated by dividing the difference between the base reject rate at level 3, and the objective rate at level 10 (45 − 10 = 35) by the number of steps between levels 10 and 3 (7) to obtain the number of rejects at each step (35/7 = 5).

A nonlinear scale was used for steps between 3 and 0 to accommodate a range of reject rates between that recorded during the base period and the previous worst-performance period.

FIGURE 22.5
A 10-point thermometer represents the scoring levels for a quality-of-output criterion. Associated with each level is a performance scale. For this criterion, a score of 3 is earned when the number of defects falls within the range of 40 to 36, whereas a smaller number (fewer defects) shows improvement.

is the crucial part of scaling, because the results set specific hurdles that reflect accomplishment of a work unit's productivity objectives. The scale is anchored by designated numbers at three levels:

- *Level 0.* The lowest level recorded for the criterion ratio over a recent period of time, say, the last year, in which normal operating conditions existed; nominally the worst ratio reading that might be expected.

- *Level 3.* Operating results indicative of performance proficiency at the time the rating scale was established; current ratio reading at the time measuring is initiated.

- *Level 10.* A realistic estimate of results that can be attained in the foreseeable future, say, 2 years, with essentially the same resources that are now available; a stimulating productivity objective.

The objectives matrix gets its name from the level 10 goal-setting process. The two most critical lines in the matrix are the score 10 line and the weight line.

Level 0 and 3 are clearly defined benchmarks. Level 10 is the challenge. An overly optimistic objective may later prove to be discouraging by its unattainability and a conservative goal may inhibit motivation if it is too easily achieved.

TABLE 22.5

Examples of ratios frequently encountered in manufacturing activities, the government and service sectors, and work groups in any organization.

Ratios Associated with Manufacturing

Number of units produced / Labor hours	Hours of rework required / Units produced
Equivalent vehicles unloaded / Labor hours	Weight handled or loaded / Labor hours
Number of defects / Total units produced	Machine operating hours / Total possible machine hours
Pounds of waste / Total pounds processed	Finished product weight / Raw material weight

Ratios Associated with Services

Number of customer complaints / Orders delivered	Number of data entry errors / Total data lines written
Errors made on policies / Number of new policies	Number of drawings / Total drafting hours
Number of pages completed / Employee day	Number of late deliveries / Total number of deliveries
Number of customers / Number of employees	Project planned cost / Project actual cost
Investigations completed / Investigator hours	Department expenses / Total hours worked

Ratios Associated with Work Groups

Hours missed from irregular absences / Total possible work hours	Overtime hours / Regular hours
Number of appointments missed / Total appointments	Number of quits / Average group size
Hours lost to accidents / Total paid hours	Number of calls handled / Total operator hours
Number of orders processed / Number of department hours	Volume of case loads / Employee days

Equal intervals of ratio results are commonly assigned to each score on the scale, as shown for levels 3 through 10 in Figure 22.5. These seemingly linear proportions are actually increasingly larger percentages of improvement for each upward step in the scale when better performance is indicated by lower numbers. To progress from the lowest ratio result that meets a score of 8 (20 rejects per 1000 units) to gain a score of 9 (15 rejects) is a $(20-15)/20 \times$

100 percent = 25 percent improvement, as opposed to a $(45 - 40)/45 \times 100$ percent = 11.1 percent improvement required to go from a score of 3 to a score of 4. Linear or nonlinear intervals (where every scoring interval increases the performance range by a constant percentage) can be utilized, and a single criterion may even use both, as demonstrated by the nonlinear ranges below level 3 in the thermometer. The issue of concern is less about the structure of the scale than about how well it is understood by those whose performance it registers.

For any criterion, the scaling may differ between work units according to the size of the teams and the nature of their work. Even for identical units, scales may vary according to their objectives. Divisions of the scale are of no great consequence because it is the rating earned from one period to the next that measures gains or losses in each performance criterion. The important consideration is that work-unit members agree that the scales are meaningful and fair.

A continuous scale is more appropriate for some criteria. A graph is the easiest way to relate earned points to performance measures when the measures extend over a wide range. As shown in Figure 22.6, the range of losses resulting from defects stretches from $0 to $3000 for a given quantity of production. Losses are determined from assigning a cost to each class of defect. The cost-weighted reject tally is translated to performance points by a reading taken for the curve which was constructed by the team. For the scale in Figure 22.6, a composite loss of $1600 yields a point rating of 7. This score could have come from 34 class 1 rejects at $15 each, 72 class 2 rejects at $10 each, and 110 class 3 rejects at $3 each, for a total loss of $1600 in the production of 1000 units.

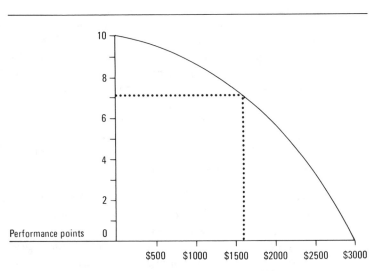

FIGURE 22.6
Graphical point scale for a performance criterion that has no easily distinguishable steps.

Cost-weighted loss from rejects per 1000 units produced

FIGURE 22.7
Calculation of the performance indicator. The sum of score × weight = value for all the criteria is the performance indicator which is entered in the box.

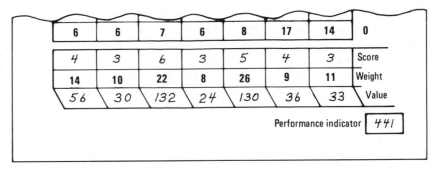

Performance Scores, Weights, Values, and Indicator

All the criteria of productive performance do not have equal effect on the overall productivity of the work unit. Assigned weights, 100 points distributed among the criteria, reflect management's perceived contribution of each criterion to the total organization's productivity objectives.

Weight assignment is not a trivial undertaking. It provides an opportunity to direct attention to activities that have the greatest potential for improving productivity. For instance, if reducing the amount of wasted raw material is a critical problem, the "waste-reduction" criterion would be weighted heavily. A weight for waste reduction twice as large as "output per hour" encourages material conservation, perhaps at the expense of lowering output to avoid scrap. Ambitious groups naturally concentrate on the criteria that tender the most recognition.

The final phase ties together criteria scores and weights to determine a performance index. Data for ratios are collected periodically, once a week, month, or quarter depending on the use of the monitoring system. Results are entered on the *performance* line of the matrix and translated to scores according to the "thermometer" of each criterion. Scores are entered in the *score* line and are multiplied by the weight immediately below each score to complete the *value* line. The procedure is demonstrated in Figure 22.7.

The sum of the numbers in the *value* row is entered in the nearby box. This is the *performance indicator* for the OMAX application. It is a single number that represents the composite performance of the work unit or organization being monitored. A *total* performance indicator results when several work-unit indicators are combined.

Example 22.5 Determining an OMAX Performance Indicator

There are as many ways to develop a monitoring system as there are reasons for establishing one. Measurements may be needed to provide continuity, correct operating deficiencies, improve supervision and management, justify wages, or co-ordinate operations. The purpose often establishes who directs the monitoring and the procedure that is used. As an example of a typical application of an objectives matrix, consider the one shown in Figure 22.8. It represents an assembly operation

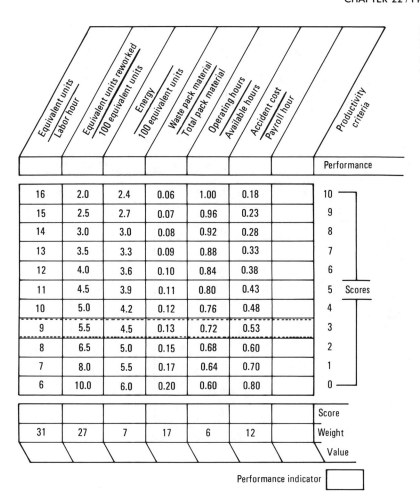

FIGURE 22.8
Productivity criteria, performance scales, and criteria-importance weights for an assembly operation.

in which operators connect and solder parts, test completed units, rework as necessary, and package sets of completed products.

Six criteria were identified for the assembly operation. Each was structured as a ratio. "Equivalent" units of output were defined to account for the variety of units assembled on the line. Equivalency was based on the time required to complete each type of unit rather than the commercial value of units, because time is controllable by assemblers. Correspondingly, accident cost was determined to be a more comprehensive measure of safety than was the number of accidents; a formula for cost was developed accordingly.

The critical performance objectives for a score of 10 were

assigned after consultation with assembly-line workers. Following an explanation of the intent and procedures of measuring, operators were asked to set reasonable goals for improvement over the next 2 years. They were assured that whatever assistance they needed to achieve the objectives would be afforded by the productivity-improvement program. Then the managers met to assign weights to the criteria. Both economic and production factors were considered in distributing 100 points to each work-unit's matrix. After the weights were aligned with organizational resources and objectives, the managers returned to the work units to explain the total matrix and how it would be used.

A scored matrix completed for the first reporting period

FIGURE 22.9
Calculation of the first-period performance indicator for the assembly operation.

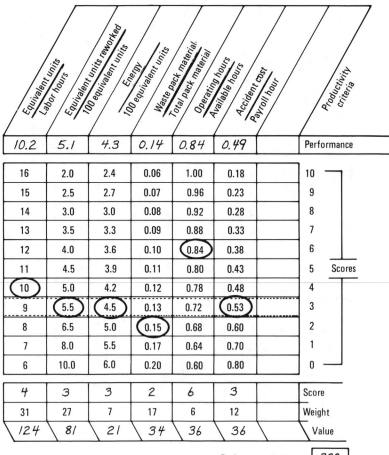

Equivalent units / Labor hours	Equivalent units reworked / 100 equivalent units	Energy / 100 equivalent units	Waste pack material / Total pack material	Operating hours / Available hours	Accident cost / Payroll hour		Productivity criteria
10.2	5.1	4.3	0.14	0.84	0.49		Performance
16	2.0	2.4	0.06	1.00	0.18		10
15	2.5	2.7	0.07	0.96	0.23		9
14	3.0	3.0	0.08	0.92	0.28		8
13	3.5	3.3	0.09	0.88	0.33		7
12	4.0	3.6	0.10	(0.84)	0.38		6
11	4.5	3.9	0.11	0.80	0.43		5
(10)	5.0	4.2	0.12	0.78	0.48		4
9	(5.5)	(4.5)	0.13	0.72	(0.53)		3
8	6.5	5.0	(0.15)	0.68	0.60		2
7	8.0	5.5	0.17	0.64	0.70		1
6	10.0	6.0	0.20	0.60	0.80		0
4	3	3	2	6	3		Score
31	27	7	17	6	12		Weight
124	81	21	34	36	36		Value

Scores (bracket spanning scores 0–10)

Performance indicator | 332

is shown in Figure 22.9. Circled scores correspond to the ratio results for the period. Criterion values calculated by multiplying each score by its weight are summed to determine the composite indicator of the group's total effort. In this case, the score of 332 indicates a productivity gain of about 11 percent for the reporting period, because the initial performance level was assigned a score of 3. That is, no improvement during the first reporting period would have yielded a performance indicator of 300.

Productivity Index from Performance Indicators

A series of performance scores must be collected before they become truly useful. The first one is simply a score. What counts is the rate of change from one period to the next. The productivity index for a period is calculated as

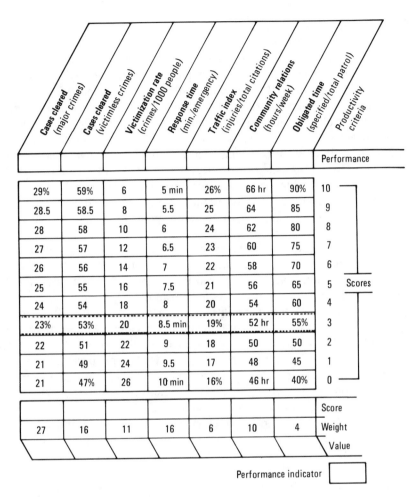

FIGURE 22.10
Objectives matrix for a small-town police department. Note the use of percentages and single dimensions for criteria.

$$\text{Productivity index} = \frac{\begin{array}{c}\text{performance during this period}\\ \text{minus the previous period's indicator}\end{array}}{\text{previous period's performance indicator}} \times 100\%$$

From the data in Example 22.5, where the first performance indicator showed a gain of 32 points over the base level of 300, which is the starting point for an OMAX application, the improvement is displayed as

$$\text{Productivity index} = \frac{332 - 300}{300} \times 100\% = 10.7\%$$

The perspective obtainable by continuous indexing is destroyed by changes in the rating scale or weights. Maximum usefulness results from correctly setting up the matrix the first time and continuing to apply it for several periods.

OMAX Implementations

The value of using a *productivity audit* to assist the installation of a measurement system is discussed in Extension 22A.

An objectives matrix is more than simply a scorecard. Although the format has tic-tac-toe simplicity, its application recipe includes ingredients of acclaimed management practices. Its development follows time-management advice to identify goals, assign priorities, construct to-do lists, and track progress. Its versatility is also a valued feature. A weighted indicator method is perhaps most attractive for nonmanufacturing activities, where measurement is generally more difficult. Where engineered standards are available, the ratio of *actual time/standard* usually has a heavy weight. When standards are unavailable, the criteria that represent accomplishment of service functions are identified and weighted according to their impact on related functions. Examples of objective matrices applied to service organizations are shown in Figures 22.10 and 22.11.

Although the examples provided have illustrated group monitoring, the objectives matrix is easily applied to the appraisal of an individual person. Performance criteria are drawn from the major duties and responsibilities of the position. Since the purpose is less to monitor productivity than to guide development and judge capability, criteria are keyed to effectiveness and inferential evaluations of performance. Measurement scales are consequently most likely to be subjective. Importance weights serve the same purpose regardless of where the matrix is applied. An inexperienced new-hire is expected to receive an index of about 200 initially and to work up to a summary value of around 700 within a reasonable time. Indexes above 800 are reserved for a few truly innovative and outstanding persons.

OMAX applications have been most successful at the work-unit level where group measurements are perceived as less threatening than evaluations of individuals and cooperation within work units to achieve performance objectives is encouraged.

Given that there is no perfect measurement system and no cure-all for productivity ills, engineering economists may still be called upon to analyze the productivity of operations and to suggest ways to improve them. The objectives matrix provides a tool that combines analytical objectivity with managerial practicality to systematically motivate and monitor productive performance.

Review Exercises and Discussions

Exercise 1 The following data describe the labor productivity for a product:

	Period 1		Period 2	
	Quantity, O_1	Price, P_1	Quantity, O_2	Price, P_2
Product	1000 units	$100/unit	1200 units	$125/unit
	Quantity, I_1	Cost, C_1	Quantity, I_2	Cost, C_2
$L(1)$	5000 hours	$8/hour	5200 hours	$9/hour
$L(2)$	1500 hours	$12/hour	1000 hours	$14/hour

FIGURE 22.11
Objectives matrix for financial operations. Note the emphasis on quality. Two quality-oriented criteria account for 50 percent of the importance weights.

Loans serviced / Department hours	Late computer reports / 22 days	Manual coupons / Payments received	Number of policy errors / New policies	Number of inquiries / Loans serviced	Investor complaints / Commercial loans	Hours missed / Department hours	Scores
							Performance
120	0	36	0.25	0.09	0.8	1.2	10
111	9	32	0.50	0.16	1.5	2.2	9
102	18	28	0.75	0.24	2.3	3.3	8
93	27	25	1.0	0.32	3.1	4.4	7
84	36	22	2.0	0.40	3.9	5.5	6
75	45	19	4.0	0.48	4.7	6.6	5
66	54	16	7.0	0.56	5.5	7.7	4
57	83	13	11.0	0.64	6.3	8.8	3
48	72	10	16.0	0.72	7.1	9.9	2
39	81	7	22.0	0.80	7.9	10.2	1
30	90	4	29.0	0.90	8.7	11.0	0
							Score
15	5	10	20	15	30	5	Weight
							Value

Performance indicator [ ]

Calculate the partial-productivity index (labor) for the given data. Then compute the price-recovery and cost-effectiveness indexes. Interpret the results from all three indexes.

Solution 1

The change in labor mix and costs in relation to the variations in the output quantity and price is analyzed according to the following ratios.

$$\text{Labor productivity index} = \frac{1200(\$100)/1000(\$100)}{[5200(\$8) + 1000(\$12)]/[5000(\$8) + 1500(\$12)]}$$

$$= \frac{1.200}{0.924} = 1.3 \quad \text{or a 30\% increase}$$

$$\text{Price-recovery index} = \frac{1200(\$125)/1200(\$100)}{[5200(\$9) + 1000(\$14)]/[5200(\$8) + 1000(\$12)]}$$

$$= \frac{1.250}{1.134} = 1.1 \quad \text{or a 10\% increase}$$

$$\text{Cost-effectiveness index} = \frac{1200(\$125)/1000(\$100)}{[5200(\$9) + 1000(\$14)]/[5000(\$8) + 1500(\$12)]}$$

$$= \frac{1.500}{1.048} = 1.43 \quad \text{or a 43\% increase}$$

The indexes indicate that sales revenue is increasing faster than are costs and that the productivity increase is more significant than is the product price increase:

$$1.43 = 1.3 \times 1.1.$$

Exercise 2   Additional information has been collected for the small firm described by the data in Review Exercise 1. For the previously given total output, the other inputs shown below were utilized in the period measured.

Capital   Increased from a base-year amount of $78,000 to $83,600. Base-year capital cost is 10 percent.

Materials   Quantity used in the base year was 840 units at an average cost of $12 per unit. The quantity increased to 1020 units in the next year.

Energy   3000 units of energy were used in the base year at a composite price of $0.30 per unit. In the next year 3400 units were used.

What are the partial- and total-productivity indexes for this firm?

Solution 2

$$\text{Capital productivity index} = \frac{1200(\$100)/1000(\$100)}{\$83,600(0.10)/\$78,000(0.10)} = \frac{1.200}{1.072} = 1.12$$

$$\text{Material productivity index} = \frac{1.200}{1020(\$12)/840(\$12)} = \frac{1.200}{1.214} = 0.99$$

$$\text{Energy productivity index} = \frac{1.200}{3400(\$0.30)/3000(\$0.30)} = \frac{1.200}{1.133} = 1.06$$

Total productivity is the output divided by the sum of all the inputs:

$$\text{Total-productivity index} = \frac{1200(\$100)/1000(\$100)}{\begin{array}{c} 5200(\$8) + 1000(\$12) + \$83,600(0.10) + 1020(\$12) + 3400(\$0.30) \\ \hline 5000(\$8) + 1500(\$12) + \$78,000(0.10) + \ \ 840(\$12) + 3000(\$0.30) \end{array}}$$

$$= \frac{\$120,000/\$100,000}{\$75,220/\$76,780} = \frac{1.200}{0.980} = 1.22$$

The productivity figures indicate that most of the productivity gains are owed to labor and that material usage should be examined.

Exercise 3   Many firms have established standard labor hours for manufacturing. These standards provide a direct measurement of labor productivity from the relationship

$$\text{Labor productivity ratio} = \frac{\text{standard labor hours for given output}}{\text{actual number of hours for given output}}$$

Three products constitute the output for a small manufacturing firm. Standard hours for products 1, 2, and 3 are 7.4, 5.6, and 3.2, respectively. During one month, output of the three products was 80, 100, and 200 units produced with 1820 labor hours, and during the next month was 80, 120, and 240 units of products 1, 2, and 3, respectively. If total labor hours in the second month were 1970, what was the productivity gain?

Solution 3

The labor productivity index compares successive months as

$$\text{Monthly labor productivity index} = \frac{[80(7.4) + 120(5.6) + 240(3.2)]/1970}{[80(7.4) + 100(5.6) + 200(3.2)]/1820}$$

$$= \frac{1.031}{0.985} = 1.05$$

which indicates a 5 percent productivity gain for the month but only a 3 percent increase over the standard rate.

Three years of performance data for a company are shown in the table below. Also indicated are the consumers (CPI) and producers (PPI) price indexes for each year. The CPI is used to deflate sales and the PPI deflates inputs. The company expects a base-rate return on capital of 9 percent. Calculate the value-added and total-factor productivity indexes for each year.

Exercise 4

Annual Company Performance Data and Price Indexes

	Year		
	0	1	2
CPI	191	205	230
PPI	285	313	340
Sales output	$23,000,000	$28,000,000	$31,000,000
Labor input	11,000,000	13,000,000	16,000,000
Tangible assets	42,000,000	43,000,000	48,000,000
Intermediate inputs	8,000,000	9,500,000	12,000,000

Sales are deflated to base-year equivalence by the CPI index as follows: sales (year 1) × 191/205 and sales (year 2) × 191/230. Labor and intermediate inputs from years 1 and 2 are converted to constant year 0 dollars by the respective PPI deflators, 285/313 = 0.91 and 285/340 = 0.84. The capital input each year is the sum of constant-valued tangible assets multiplied by the 9 percent rate of return; in the base year, capital input is $42,000,000 × 0.09 = $3,780,000. In years 1 and 2 the value of invested capital is translated to its base-year value by a CPI deflator; in year 1 the capital input is $43,000,000 (0.93)(0.09) = $3,599,100.

Solution 4

The value-added index, shown in the lower section of Table 22.6, is obviously more sensitive to changes than is the total-productivity index, as indicated by the wider swing in the percentage change in productivity during the 3 years. Including intermediate inputs

TABLE 22.6			

Calculations of total and value-added productivity indexes.

	Year 0	Year 1	Year 2
Sales (output)*	23.00	26.09	25.74
Labor input*	11.00	11.83	13.44
Capital input*	3.78	3.60	3.59
Intermediate input*	8.00	8.65	10.08
Total input*	22.78	24.08	27.11
Output/(total input)	1.01	1.08	0.95
Annual change		7%	−12%
Output − intermediate input*	15.00	17.44	15.66
Factor input*	14.78	15.43	17.03
$\dfrac{\text{Output} - \text{intermediate input}}{\text{Factor input}}$	1.01	1.13	0.92
Annual change		12%	−19%

*Values in millions of dollars

in the denominator, with total sales in the numerator, dampens the magnitude of change from year to year.

Exercise 5   The conceptional issue about where to include intermediate goods and services in calculating a productivity measure is more cosmetic than cardinal. Intermediates can be either added to the numerator of a productivity ratio by considering them to be ordinary inputs or subtracted from the numerator of the ratio by assuming them to be "pass-throughs" that neither affect nor are affected by the production process. Consider the following data given in inflation-adjusted dollars.

	Last Year	This Year
Gross output	$21,000,000	$23,000,000
Labor input	4,000,000	5,000,000
Capital input	6,000,000	6,000,000
Intermediate goods	10,000,000	11,000,000

Calculate the total productivity and factor productivity indexes for the process.

Solution 5   Based on the given data, the measures of total productivity (TP) and factor productivity (FP) are

$$TP_{\text{(last year)}} = \frac{\$21,000,000}{\$4,000,000 + \$6,000,000 + \$10,000,000} = 1.050$$

$$TP_{\text{(this year)}} = \frac{\$23,000,000}{\$5,000,000 + \$6,000,000 + \$11,000,000} = 1.045$$

$$FP_{\text{(last year)}} = \frac{\$21,000,000 - \$10,000,000}{\$\ 4,000,000 + \$\ 6,000,000} = 1.100$$

$$FP_{\text{(this year)}} = \frac{\$23,000,000 - \$11,000,000}{\$\ 5,000,000 + \$\ 6,000,000} = 1.091$$

from which the productivity index for this year is calculated as

$$\text{Total productivity index (TP)} = \frac{TP_{\text{(this year)}}}{TP_{\text{(last year)}}} = \frac{1.045}{1.050} = 0.995 \quad \text{or} -0.5\%$$

$$\text{Factor productivity index (FP)} = \frac{FP_{\text{(this year)}}}{FP_{\text{(last year)}}} = \frac{1.091}{1.100} = 0.992 \quad \text{or} -0.8\%$$

Both indexes show the same trend, declining productivity. The role of labor and capital receives greater emphasis in the FP index, because only the value added by the factor inputs is considered in the firm's output. Swings in a productivity index are generally larger when a value-added ratio is evaluated; proportionately higher intermediate inputs have a damping effect on fluctuations in total productivity measurement.

PROBLEMS

22.1 Complete the following table to make the sales data equivalent through the use of quantity and price indexes. Entries in the bottom line (comparable sales) should be equal; the 1982 sales are corrected for price and the 1983 sales are corrected for volume.

	1982	1983
Sales	$2,000,000	$4,800,000
Price per unit	$80	_____
Units sold	_____	40,000
Price index	100	_____
Quantity index	100	_____
Comparable sales	_____	_____

22.2 Production of a product starts in the fabrication department and finishes in the assembly department. Last year 20,000 units were processed through the two departments and this year the total is 22,080. Standard labor hours per unit in fabrication and assembly are 0.47 and 0.5, respectively. Actual labor hours last year were 9640 in fabrication and 9920 in assembly; corresponding actual labor hours this year are 10,064 and 10,820.
   (a) What are the partial productivity indexes for each department?
   (b) What is the labor productivity for the product?

22.3 Charley's Chickenpluckers buys chickens, processes them, and sells the cut meat to wholesalers. From the data shown at the top of page 694 about Charley's production last period and this period, determine the following:
   (a) The quantity productivity index.
   (b) The price-weighted productivity index.
   (c) The conclusions that can be drawn from the two indexes.

	Last Period		This Period	
	$/pound	Pounds	$/pound	Pounds
Input	1.39	22,000	1.24	24,200
Output	1.65	17,500	1.73	19,400

22.4 Toyco is a small labor-intensive producer of toys. During the last month of 1987, the president of the company, encouraged by recent marketing success, acquired the assets of a small competitor that had been forced into bankruptcy. This doubled the production capacity and consequently doubled the work force. The resulting financial statistics, shown below, caused the president to proclaim the acquisition a complete success. Assume that the data are given in constant dollars.

	1987	1988
Net sales	$520,000	$1,040,000
Cost of goods sold:		
Material	100,000	200,000
Labor	300,000	620,000
Other expense	50,000	100,000
Depreciation	10,000	20,000
Total	460,000	940,000
Gross income	60,000	100,000
Tax (50%)	30,000	50,000
Net income	30,000	50,000

(a) Calculate the productivity ratios applicable to Toyco for 1987 and 1988 by using the method given in Example 22.2. Comment on the productivity changes from one year to the next.

(b) Was the president's claim justified? Why or why not?

(c) What suggestion might you make to improve Toyco's productivity?

22.5 Last year, as a result of competition and inflation, Zorzaz Corp. barely made a profit. Near the end of the year the corporation had an opportunity to purchase rights to a patent that appeared likely to lower the cost of production and increase the amount of sales in future years. A contract was made to give the patent owner 1 percent of net sales by Zorzaz for the next 5 years. Use of the patented process in equivalent manufacturing situations has reduced labor costs by 10 percent with no additional capital expenditures for equipment. Financial data for the 2 years in constant dollars are given below.

	Last Year	This Year
Net sales	$3,000,000	$3,300,000
Cost of goods sold:		
Material	1,000,000	1,050,000
Labor	1,400,000	1,500,000
Patent royalty		33,000
Other expense	400,000	410,000
Depreciation	150,000	150,000
Total	2,950,000	3,143,000
Gross income	50,000	157,000

(a) Calculate the productivity ratios using the model from Example 22.2 for both years and interpret the results.

(b) What logic should guide the distribution of productivity gains expected in the future from the application of the patented process?

22.6 (a) Calculate the base-period labor productivity index for a mine that last year produced 1 million tons of coal valued at $50 per ton and 1½ million tons valued at $80 per ton. To do this, 1.5 million labor-hours at $12 hour and 2.5 million labor-hours at $16 per hour were expended.

(b) If prices and costs both increase by 10 percent next year while the quantity and proportion of labor-hours remain constant, how much must the output of the mine increase to produce a 5 percent productivity gain for the year? The proportion of the two types of coal remains unchanged.

22.7 Two products, J and K, are the output of a manufacturing company. Quantities sold and price per unit for 2 years of operation are shown below. Also shown are input hours and cost per hour for two labor classes, L1 and L2.

	Last Year		This Year	
	Quantity	$/Unit	Quantity	$/Unit
Product J	1,000 units	$200	1800 units	$240
Product K	2,000 units	150	1500 units	150
Labor L1	5,000 hours	8.50	7500 hours	8.50
Labor L2	10,000 hours	8.50	9000 hours	9.00

(a) What is the labor productivity each year based *only* on quantity? Explain the weakness of this ratio.

(b) Calculate the labor productivity index.

(c) Calculate the labor price-recovery index.

(d) Calculate the labor cost-effectiveness index.

(e) Compare the three indexes, indicating the meaning of each and what the three together suggest about the operations of the company.

22.8 Quantities and costs for three additional inputs that are used in the production of products J and K from Problem 22.7 are given below.

	Last Year		This Year	
	Quantity	$/Unit	Quantity	$/Unit
Energy	300,000 units	$0.09	280,000 units	$0.15
Material	100,000 units	$2.00	120,000 units	$2.10
Capital	$3,000,000	0.12*	$3,200,000	0.13*

* Percentage cost for invested capital.

(a) Calculate the energy productivity index.

(b) Calculate the energy price-recovery index.

(c) Calculate the energy cost-effectiveness index.

(d) Compare the energy indexes to the labor indexes calculated in Problem 22.7. What conclusions can be drawn?

(e) Calculate the material productivity index. Compare it with the energy factor-productivity index and discuss the implications.

(f) Calculate the capital productivity index and discuss its meaning to the company.

(g) Calculate the total-productivity index using output for products J and K and inputs of labor, energy, material, and capital.

22.9 Your organization has the productivity pattern, corrected to a base-year index, shown in the following table. Assume that net profit increases at a commensurate rate with total productivity gains each year.

	1987	Change, %	1988	Change, %	1989
Net profit (1)	100	15	115	13	130
Labor inputs (2)	50	0	50	14	57
Capital input (3)	40	25	50	4	52
Labor productivity (1 ÷ 2)	2.0	15	2.3	0	2.3
Capital productivity (1 ÷ 3)	2.5	−9	2.3	9	2.5
Total productivity [1 ÷ (2 + 3)]	1.11	4	1.15	4	1.19

(a) Is the increase in total productivity from 1987 to 1988, at no increase in labor input, sufficient reason to grant a wage increase to workers? Why?

(b) If the decrease in capital productivity in 1987 resulted from a mandatory capital investment to conform to a new law restricting pollution, should this social benefit be recognized differently in the productivity measure? Why?

(c) Who should benefit from the 4 percent gain in 1989? What percentage, if any, of the total profit for 1989 should be distributed to each of the following parties: (1) consumers, (2) workers, (3) owners or stockholders and (4) the organization, as retained earnings? Why?

22.10 The owner-manager of a custom bakery faces a delicate but pleasant problem of how to distribute rewards. A year ago the bakery was close to going out of business because of competition from bakery-product sections built into the new shopping centers in the suburbs. Fewer people visited the downtown site of the bakery and those that did were purchasing less. The turnaround in business originated at a meeting of all employees called by the manager last year.

The purpose of the meeting was to urge the employees to further reduce costs or face the possibility of having the shop close. As the meeting progressed and the employees became aware of how serious the situation was, they started offering many ideas that the manager had never considered before. Suggestions ranged from proposing an easier way to make butter-flake rolls to that of opening a coffee-roll sidewalk bar and a discount booth inside the shop for day-old products.

The manager shrewdly recognized the productivity-improvement potential of the suggestions and gave them backing. Highlights of the resulting year's activities compared with the previous year are listed below, along with price-index data.

• Total sales increased by 60 percent (including a 5 percent general increase owing to inflation in the overall economy) to $480,000.

- Cost of materials used increased by 50 percent (including a 10 percent industrywide rise in material expenses) to $150,000.

- Labor costs increased by 10 percent (the same rate as the industrywide wage average) to $110,000.

- Annual depreciation charges jumped by $15,000 (owing to investments made to facilitate innovations) to $65,000.

- Other expenses increased by 40 percent (including inflation) to $140,000.

The increased output with essentially the same number of employee-hours worked was a result of improved methods instituted by the bakers and partially the new equipment. Sales increased more than did material costs because waste was reduced.

  (a) Calculate the percentage increase in different productivity measures to determine the effect of revised operations in the bakery. Interpret the significance of the different ratios.
  (b) How do you think the manager should distribute the benefits from the productivity rise in order to sustain the gain?

22.11 After several weeks of planning and indoctrination, a company initiated a unit productivity improvement program. Teams were formed to represent natural divisions of the production process according to the nature of their value-added contributions. Organizational structures were established to provide unit training and to respond to team suggestions.

  During weekly meetings the teams developed their own objectives and planned ways to achieve their goals. Concurrently, performance indicators were identified, and rating scales were established. Then management's "productivity council" proposed a weighting plan for each team. An abbreviated version of one team's performance indicator table is shown in Figure 22.12.

  (a) Based on the data in Figure 22.12, construct an objectives matrix using the standard format.
  (b) The unit that developed the performance efficiency indicators in Figure 22.12 has measured its performance recently with the following results:

Delivery. Average days late per delivery = 1.2

Facilities. 11 hours downtime per 168 operating hours = 0.07

Material. 61 pounds salvaged for 748 finished units = 8.2

Operations. 748 units completed

Personnel. 15 overtime hours used

Quality. Rework required on 88 of 748 units = 0.12

  Determine the performance indicator for the period.
  (c) Assume that the performance indicator calculated in Problem 22.11b is based on the most recent measurements and that the measurements of operations taken 6 months previously by the same unit produced a performance indicator of 695. What is the productivity index for the period?

PERFORMANCE EFFICIENCY INDICATORS

Point Scale	Delivery Days late / Total orders	Facilities Hours downtime / Line operating hours	Material Pounds salvaged / 100 units	Operations Completed units / Period	Personnel Overtime hours / Period	Quality Rework / Units Produced	Point Scale
10	0	0	9	800	0	0	10
9	0–0.2	0–0.02	8.9–8.8	799–780	1–8	0–0.10	9
8	0.2–0.5	0.03–0.04	8.7–8.6	799–755	9–16	0.11–0.20	8
7	0.5–1	0.05–0.06	8.5–8.3	756–725	17–24	0.21–0.30	7
6	1–2	0.07–0.10	8.2–7.9	726–690	25–36	0.31–0.40	6
5	2–3	0.11–0.13	7.7–7.4	691–650	37–48	0.41–0.50	5
4	3–4	0.14–0.17	7.3–6.8	651–610	49–60	0.51–0.60	4
3	4–5	0.18–0.21	6.7–6.1	611–570	61–84	0.61–0.70	3
2	5–6	0.22–0.25	6.0–5.3	571–530	85–108	0.71–0.80	2
1	6–7	0.26–0.30	5.2–4.4	531–490	109–132	0.81–0.90	1
0	Over 7	Over 0.30	Under 4.4	Under 490	Over 132	Over 0.90	0
Effectiveness weights	0.05	0.10	0.05	0.30	0.20	0.30	

FIGURE 22.12 Performance indicator table for one team. Columns list the points earned for each level of accomplishment for six performance-efficiency criteria. The weighting scale for each criterion is shown in the bottom row.

22.12 Performance records for the seven criteria used in the objectives matrix shown in Figure 22.11 are listed below. These results were obtained at the end of the first two evaluation periods for the financial operations.

	Ratio Results	
Criteria	First 6 months	Next 6 Months
Loans serviced/department hours	63	67
Late computer reports/22 days	60	55
Manual coupons/payments received	14	12
Policy errors/number of new policies	6.8	3.9
Number of inquiries/loans serviced	0.73	0.88
Investor complaints/commercial loans	5.4	5.6
Hours missed/department hours	7.0	6.6

(a) Calculate the performance indicators for both evaluation periods.

(b) Calculate the productivity indexes for each of the periods and the productivity index for the first year.

(c) Consider the performance in each criterion. Which criterion might be considered "out of control" on a control chart. Does any overall pattern of progress appear to be developing?

22.13 A company with 200 employees produces over 100 different types of envelopes and mail containers. In an attempt to increase productivity, the management instituted a sharing program based on a company-wide objectives matrix, as shown in Figure 22.13.

Five criteria were selected for the total plant monitoring:

1. *Quantity*—Number of units produced divided by the total number of straight-time and overtime hours for the entire work force.

2. *Quality*—All credits given to customers during the period as the result of plant quality problems or errors, based on a production level of 4 million units per week.

3. *Materials*—Equivalent pounds of wasted raw materials, mostly papers of comparable value, per 1000 units produced.

4. *Safety*—Number of days lost during the period for 200 employees owing to industrial accidents.

5. *Safety*—Number of doctor-related accidents during the period times 200,000 divided by total hours worked.

Goals were set and scale levels developed to measure improvement above existing performance, designated as level 3. Dollar values for cost savings resulting from a change in each scoring level of all criteria were calculated. Based on these values, the company agreed to split the savings 50-50 with employees, paid as a bonus every 6 months.

Employees endorsed the plan wholeheartedly and were rewarded with significant bonuses when productivity improved by 30 percent during the first year; quantity, quality, and materials increased by a point each, while the other criteria held steady at level 3.

FIGURE 22.13
Productivity sharing plan displayed on an objectives matrix format.

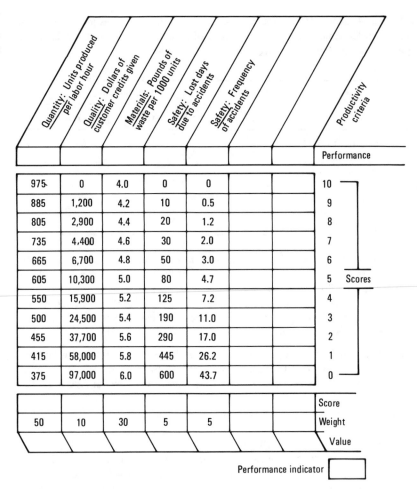

Quantity: Units produced per labor hour	Quality: Dollars of customer credits given	Materials: Pounds of waste per 1000 units	Safety: Lost days due to accidents	Safety: Frequency of accidents			Productivity criteria
							Performance
975	0	4.0	0	0			10
885	1,200	4.2	10	0.5			9
805	2,900	4.4	20	1.2			8
735	4,400	4.6	30	2.0			7
665	6,700	4.8	50	3.0			6
605	10,300	5.0	80	4.7			5 Scores
550	15,900	5.2	125	7.2			4
500	24,500	5.4	190	11.0			3
455	37,700	5.6	290	17.0			2
415	58,000	5.8	445	26.2			1
375	97,000	6.0	600	43.7			0
							Score
50	10	30	5	5			Weight
							Value

Performance indicator ☐

During the next 6-month period, orders were heavy and the enthusiastic employees boosted quantity to 610 units per hour, but quality costs rose to $39,270 in credits and waste increased to 5.35 pounds per 1000 units. Safety did not change.

(a) What was the rate of change of productivity during the period?

(b) Management began to worry about the effect of declining quality on the reputation of their products. Perhaps the penalty for lower quality was not large enough. Should they consider changing the matrix? How might it be changed? What effect might a change have on employee morale? What do you suggest?

22.14   Jonesy is the only manager in the Fancy Fence Company. He is also the owner. There are 12 employees: a bookkeeper/secretary, estimator, and 10 fence builders. All the fences are constructed of cedar, but some of them are very fancy; there are 19 basic designs. By far the largest cost of operations is for materials. Next comes labor and then travel.

Most orders are received by phone. The estimator calls on a prospective customer and makes a bid to do the job or if a standard fence is ordered, a bid is quoted over the phone. The estimator is very good. Jonesy does the work scheduling, material ordering, and general administration.

Jonesy has scheduled a company meeting for next Friday at two o'clock. He plans to set up an objectives matrix to monitor performance. He is also thinking about giving rewards for improvements that are justified by productivity increases. Before going into the meeting, he wants to prepare notes and a tentative matrix. He asks your help. You agree.

   (a) Prepare at least five ratios that could be used as criteria of performance. Consider how output can be measured when standard hours per foot of fence are not known and the times to build fences of different designs vary greatly.
   (b) Place weights on the criteria you have developed. Provide notes on why certain criteria are accorded more importance than others.
   (c) What types of recognition for improved productivity do you suggest? Why? Should the estimator and secretary be included? What level of recognition should trigger a reward?

EXTENSION

22A Purpose of a Productivity Audit

A logical way to begin any drive to improve an organization's productivity is to learn as much as possible about current process inefficiencies and unproductive practices. The people who participate in the process and perform the practices are the prime information source. A systematic procedure for acquiring this information is called a *productivity audit*. Most audits involve questionnaires and follow-up interviews. They may be supplemented by group discussions, inspections by internal staffs or outside consultants, and managerial task forces. The common intent of all the associated activities is to identify specific opportunities for bettering productivity.

A productivity audit is not an employee attitude survey. An attitude survey is usually sponsored by the personnel or human-resources department to find out how employees feel about their wages, supervision, fellow employees, work situation, and other factors that contribute to their feeling of job satisfaction; they are often asked to compare present conditions with previously experienced conditions and to those found in other organizations. Although a productivity audit may also seek employee perceptions about human relations and employment benefits that affect morale, and consequently may affect personal productivity, the main issues are work conditions, methods, facilities, and environment. An audit concentrates on factors that influence the quantity, timeliness, and quality of output, whether that output is a manufactured product or a service provided by a public or private organization. The immediate nature of an audit is apparent from the sample of typical questions listed below.

- *What are the most serious problems that inhibit production*—absenteeism, turnover, accidents, slowdowns, equipment downtime, reject rates, poor quality of supplies or raw material, etc.?

- *What characteristics of operations contribute to lower than expected performance*—inattention, horseplay, lack of enthusiasm, make-work, no respect for quality, poor work flow, tardiness, wrong assignments, too many delays and disruptions, etc.?

- *What opinions do employees have about working conditions*—adequacy of facilities (crowding, lighting, ventilation, noise, etc.), condition of equipment, sufficient conveniences (cafeteria, parking, in-plant transportation, restrooms, lockers, etc.), availability of modern machines and tools, etc.?

- *What apprehensions do employees have about the organization*—paternalism, nepotism, misemployment, restrictive procedures, too much paperwork, unfair performance appraisals and promotion policies, provincialism, weak internal communications, etc.?

- *What is the prevailing perception of management*—willingness to listen, effectiveness of structuring assignments, adequacy of supervision in terms of coordination and direction, interest in employee concerns, keeping promises, willingness to share information about future plans, etc.?

- *What is the extent of management's commitment to improving operations*—willingness to share responsibilities with workers and to revise procedures, dedication to the concept of employee involvement, provisions for financial backing of worker projects, recognition of the value of better training and skill development, support of technologically advanced processes and encouragement of technological innovations by the work force, etc.?

Based on the answers to such questions, an organization should comprehend current strengths and weaknesses, be able to set reasonable objectives, and be in a position to decide how to move from here to where it must go. However, some managers perceive a productivity audit as a blunt instrument that hits only weaknesses in the organization without rewarding strengths. It is indeed directed toward detection of problems because that is the direction of opportunity. Strengths appear only as areas of stable accomplishment. This is the nature of an audit and must be accepted as such, since its purpose is to promote change.

An audit sets the stage for introducing productivity measurement and any other program that fosters productivity, by increasing employee awareness of the relation between productivity and prosperity, especially when management vigorously disseminates the results. A wealth of suggestions for improvement can be harvested when workers believe that their contributions will be acted upon and appreciated. Direct feedback from a productivity audit may also reveal promising investment opportunities of the type that yield high immediate returns, and analyses of the data may suggest long-range improvements that strengthen both productivity and profitability.

SECTION SIX

ENCOUNTERS WITH RISK AND UNCERTAINTY

It has been said that the only thing certain is uncertainty. This adept observation applies to all our activities. We can respond by ignoring the uncertainty of future events, assuming that they will happen as planned and adapting to changes if they do not, or we can try to anticipate contingencies through selection of the most promising course of action. Which response we follow depends largely on the consequences, although there are gamblers who habitually dare the future and risk avoiders who worry incessantly over trivialities and do nothing. A realistic response to economic risk is to admit that cash flows are subject to change, question the potential effects of change with respect to consequences, and use insights from the analysis to guide decisions.

Informal responses to risk take the form of arbitrary increases in the discount rate and contraction of payback periods. The vexing problem with these responses is the size of the adjustment. More elaborate risk-recognition methods include sensitivity analysis and minimum–most-likely–maximum estimates. Formal risk analysis involves probabilities of future events as incorporated in expected value, forecasting, PERT, and other techniques.

When probabilities are available to weight the possible occurrence of future events, a condition of risk is assumed. A condition of uncertainty is admitted when frequency distributions of outcomes are unknown. Previously described techniques for handling risk and uncertainty are enlarged upon in this section. Models for risk analysis in production operations (Chapter 23) and financial decisions (Chapter 24) along with techniques for decision making under complete uncertainty (Chapter 25) are presented.

The meaning of risk or uncertainty is more personal when viewed from the perspective of a poker player. If a poker player unfairly uses marked cards, a condition of absolute certainty exists in terms of knowing the outcome of each hand. This is akin to returns promised by a government bond. A gambler who fairly relies on good card sense, being alert and trading on available knowledge, plays under assumed certainty, knowing that information is not complete but believing enough is known to make a decision. Risk analysis is employed when a poker player keeps meticulous track of cards played and relates these data to the known distribution of card values in the complete deck. A case of uncertainty would exist for a bizarre game in which players bet without looking at the cards dealt to them. In this case the poker player faces an array of possible outcomes guided only by a personal betting policy.

The poker player's options are magnified in investment decisions. The dubiety is when to utilize powerful but expensive risk models. They are powerful because they can provide prudent advice and costly because they consume time and money to develop and apply.

Many sophisticated methods have been suggested for evaluating risk in capital investments.* In assessing the applicability of probabilistic models it should be remembered that the probabilities on which they operate are seldom objectively verifiable. They are generally subjective in the same way individual cash flow estimates rely on personal judgment. Thus the models do not eliminate risk in a decision. Instead they refine an analysis by disclosing expectations from specific assumptions about the probabilities assigned. The assumptions can be improved by getting

- *More data.* Better estimates of future conditions and their effects

- *More reliable information.* Reduction of biases and inaccuracies

- *More analytical experience.* Familiarity with similar decisions that avoid misapplication of models and misinterpretation of results

Beyond this preparation is the sage gambler's strategy—hedge your bets. For engineering economists with sufficient resources this means diversifying investments to spread the risk, selecting proposals for which a condition that causes a poor outcome in one investment creates a good outcome in another.

*C. D. Zinn, W. G. Lesso, and B. Motazed, "A Probabilistic Approach to Risk Analysis in Capital Investment Projects," *The Engineering Economist*, Summer, 1978.

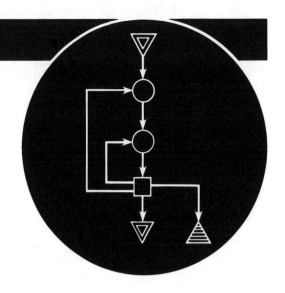

CHAPTER 23

RISK ANALYSIS IN PRODUCTION OPERATIONS

OVERVIEW

The outcome of any economic decision is usually subject to an environment of uncontrollable influence. The longer the planning horizon, the greater the exposure to chance events. The time to consider risk is before making a commitment to a course of action. Perhaps a portion of the potential profit should be traded for partial immunity to risk. Mathematical models of the riskiness of operations assist decision makers by suggesting the impact of chance events on economic outcomes.

Minimum-cost models for short-term operations can be adapted to include risk. The basic inventory patterns presented in Chapter 20 are subject to variations in *lead times* and *demand rates*. These variations are analyzed by calculating expected values, as introduced in Chapter 17. The most economical *safety-stock* size is determined from the expected value of opportunity costs and holding costs for different ordering alternatives. *Maintenance-replacement* policies are determined from the expected failure patterns and the sum of the costs resulting from equipment failures and replacement operations. Group item replacement plans are based on the premise that it is sometimes more economical to replace large groups of items in mass rather than to wait for individual failures.

The total production cost of an item must include the expenses incurred for maintaining quality. The risk of producing substandard or defective products must be compared to the cost of *quality assurance* activities. The form and location of inspection operations in the production process are always important in determining the total cost per unit produced.

Costs are also incurred when delays are encountered in production operations or customer service activities. The costs of reducing or eliminating these delays must be balanced by the cost of the delay. Equipment and personnel may stand idle while awaiting service or customers may leave the system. The expected values of these costs may be obtained through the application of *queueing theory*.

RISKS IN INVENTORY POLICY

The analysis methods for determining minimum-cost operating levels in Chapter 20 are based on assumptions of certainty. Inputs to the basic formula employed in the analyses, $C_{min} = AX' + B/X' + K$, are assumed to be known and unchanged in future states. Such assumptions allow direct evaluation of the basic properties of the problem at the sacrifice of some realism. The appropriateness of developing and applying more rigorous minimum-cost models is subject to the same benefits and drawbacks discussed for developing more sophisticated discounted cash flow models. Sample applications that include risk in the analysis of inventory policies are presented on the following pages.

Inventory Policy for Perishable Products

Some firms handle merchandise which has negligible utility if it is not sold almost immediately. Products in this category include newspapers, printed programs for special events, fresh produce, and other perishable commodities. Such items commonly have a high markup. The large difference between wholesale cost and retail price is due to the risk a vendor faces in stocking the item. The vendor faces obsolescence costs on the one hand and opportunity costs on the other.

In most cases a retailer cannot forecast exactly what the demand will be on a given day. The loss incurred from any unsold items left after their limited selling period must be balanced against the loss in profit caused by unfilled orders. With records of past sales and the belief that the pattern will not change appreciably in the future, the retailer can calculate an order size which will permit a maximum profit over an extended period of time.

Consider the case of a vendor who has an exclusive franchise to sell programs at a municipal coliseum. During the basketball season he orders programs in increments of 500. Using his records from the previous two seasons, he has developed the sales forecast shown in Table 23.1. It will be used to determine an ordering policy for typical games. Special attractions would require a separate forecast, because they have a different pattern from that of regular games.

The vendor purchases the programs for $0.15 each and sells them for $0.50. From his markup of $0.35 he must pay the commissions and fixed expenses. As a simplifying measure, we will call the $0.35 markup his profit.

His forecast does not tell him how many programs will be sold at a regular

TABLE 23.1

Program sales

Sales per Game	Number of Games	Probability of Programs Being Sold
2500	4	0.10
3000	6	0.15
3500	10	0.25
4000	16	0.40
4500	4	0.10
	40	1.0

game, but it does tell him, for instance, that there is a 50 percent chance that he will sell at least 3500 copies. It also indicates his profit outlook. If he orders 4000 programs but sells only 3500, his profit will be $3500 \times \$0.35 - 500 \times \$0.15 = \$1150$. In an extreme case of misjudgment, he could order 4500 and sell 2500. This calamity would result in a minimum profit of $2500 \times \$0.35 - 2000 \times \$0.15 = \$575$. A maximum profit of $1575 would result from ordering and selling 4500 programs.

Between $575 and $1575 is a range of possible profits. These are conditional profits. That is, each is the profit that will result from a possible demand, given that a certain number of programs has been ordered. The range of possible profits is determined by the range of possible sales. There is no need to extend the range, because all past performances are included. All the conditional profits are tabulated below.

| Possible Order Supply | Possible Demand | | | | |
	2500	3000	3500	4000	4500
2500	$875	$ 875	$ 875	$ 875	$ 875
3000	800	1050	1050	1050	1050
3500	725	975	1225	1225	1225
4000	650	900	1150	1400	1400
4500	575	825	1075	1325	1575

The vendor's profit status can be further refined by analyzing his sales history in combination with his conditional profits. This combination takes the form of a payoff table. The future states are the possible sales levels. Associated with each state is the probability of its occurrence, derived from sales records. The alternatives are the various order sizes and the outcomes are the conditional profits.

The vendor can expect to make the most profit by ordering 4000 programs each time. By following this policy he will average $1187.50 profit for each

TABLE 23.2

Expected profit from program sales

	State of Demand					
Alternative (Order Size)	2500 (0.10)	3000 (0.15)	3500 (0.25)	4000 (0.40)	4500 (0.10)	Expected Profit
2500	$875	$ 875	$ 875	$ 875	$ 875	$ 875.00
3000	800	1050	1050	1050	1050	1025.00
3500	725	975	1225	1225	1225	1137.50
4000	650	900	1150	1400	1400	1187.50
4500	575	825	1075	1325	1575	1137.50

game, provided that the actual demand follows the pattern of past sales. He is not guaranteed this amount for a given game, but any other quantity will leave him with less *average* profit.

After studying the calculations reported in Table 23.2, the vendor might ponder the possibility of improving his operation. Perhaps he could use weather forecasts to anticipate the size of the crowd at a game. Maybe he could devise a program which would retain its appeal for postgame sales. The amount he could afford to pay to develop improvements is the difference between the expected profit calculated in Table 23.2 and the profit he could make if he possessed perfect information. Table 23.3 shows that the maximum possible profit is $1268.75. Therefore, he could afford $1268.75 − $1187.50 = $81.25 *per game* to develop means to avoid accumulating obsolete programs.

Safety Stock to Counteract Risk

The classic sawtooth pattern of inventory usage given in Figure 20.11 rests on assumptions that the demand rate is constant and replenishments are received

TABLE 23.3

Maximum profit with perfect information

Program Sales	Preferred Alternative	Profit	Probability of Demand	Expected Profit from Perfect Data
2500	2500	$ 875	0.10	$ 87.50
3000	3000	1050	0.15	157.50
3500	3500	1225	0.25	306.25
4000	4000	1400	0.40	560.00
4500	4500	1575	0.10	157.50
Maximum profit				$1268.75

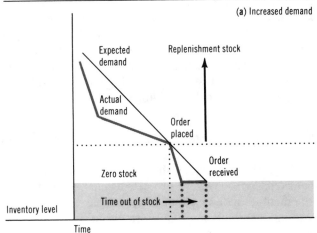

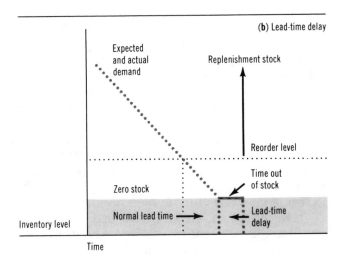

FIGURE 23.1
Irregularities in the classic inventory-usage pattern (see Figure 20.11) that contribute to stockouts. The time out of stock would be reduced by holding a reserve supply represented by the shaded area.

when ordered. Many practical realities frustrate these assumptions. Causes of disruptions to the regular flow of inventory are shown in Figure 23.1. Receiving an order a week later than anticipated may cause an interruption in production or a loss of sales. An unexpected increase in demand may cause a stockout of merchandise or supplies. Such risks are relieved by carrying a reserve inventory, or *safety stock.**

Safety stock acts as a buffer between anticipated and actual inventory levels. In Figure 23.1, safety stock would appear as a plus (+) inventory in

* A procedure for considering the interaction between EOQ and safety-stock is presented in T. M. Whitin, *The Theory of Inventory Management*. Princeton University Press, Princeton, NJ, 1957.

the shaded section to lower the zero stock level, serving thereby to eliminate or reduce the out-of-stock time. Reducing stockouts reduces opportunity costs, but it increases holding costs. A very large safety stock could completely eliminate the change of stockouts, and, correspondingly, it would substantially raise insurance, damage, interest, and other holding costs. The problem is to determine a safety-stock level which will provide adequate protection from stockouts at reasonable holding costs.

<table>
<tr><td colspan="2">**TABLE 23.4**</td></tr>
<tr><td colspan="2">Probabilities of different lead times</td></tr>
<tr><td>Lead Time, Days</td><td>Relative Frequency</td></tr>
<tr><td>6</td><td>0.00</td></tr>
<tr><td>7</td><td>0.04</td></tr>
<tr><td>8</td><td>0.08</td></tr>
<tr><td>9</td><td>0.38</td></tr>
<tr><td>10</td><td>0.24</td></tr>
<tr><td>11</td><td>0.12</td></tr>
<tr><td>12</td><td>0.09</td></tr>
<tr><td>13</td><td>0.03</td></tr>
<tr><td>14</td><td>0.02</td></tr>
<tr><td></td><td>1.00</td></tr>
</table>

Minimum-Cost Inventory Model for Variable Lead Times

Placing orders for replenishments well in advance of the time stock levels are expected to drop to zero protects against stockouts by creating a safety stock. One method of determining a reasonable ordering policy when lead times vary is to modify the basic EOQ formula to include *opportunity costs*, the loss in revenue (or increase in costs) from not having stock available when needed. To illustrate the problem of irregular replenishment periods, consider a manufacturing plant which uses special chemicals in its finishing department. The chemicals must be stored in a controlled environment. Because the chemicals deteriorate with age, the producer must make a new batch for each delivery. This leads to a considerable variation between the time an order is placed and the time at which it is received. The distribution of lead times is shown in Table 23.4. The minimum delivery time is 7 days. Cumulating the lead times shows there is a probability of 0.5 that a delivery will take at least 10 days.

Many factors in the problem are known with relative certainty. Continuous production entails an annual demand for 1150 cylinders used at a constant rate through the year. Each cylinder costs $40, in addition to the order and inspection costs of $55 per order for a typical order size. Holding costs are 25 percent of the value of the average inventory in storage.

If we disregard stockout costs temporarily, we can calculate an order size based on known conditions for costs and demand by the economic lot-size formula developed for conditions of assumed certainty:

Economic-lot-size equations were first developed by F. W. Harris in 1915.

$$Q = \sqrt{\frac{2OD}{H}} = \sqrt{\frac{2 \times 55 \times 1150}{10}} = 113 \text{ cylinders}$$

where O = order costs = $55/order
D = annual usage rate = 1150 cylinders
H = holding costs = $0.25 \times \$40 = \$10$/cylinder-year

Since Q and D are known, the number of orders per year is

$$\text{Orders/year} = \frac{D}{Q} = \frac{1150}{113} \doteq 10.2$$

Assuming 230 working days per year, we find that the time between orders is

$$\text{Order interval} = \frac{230}{10.2} \doteq 23 \text{ days}$$

and the daily usage rate is

$$D_{daily} = \frac{1150}{230} = 5 \text{ cylinders/day}$$

Now the effect of the lead-time distribution on the basic inventory plan can be determined. For a fixed lead time of 7 days the company would place an order whenever the inventory fell to $7 \times 5 = 35$ cylinders. Under conditions of certainty there would be no chance of late delivery and therefore no danger of running out of chemicals. However, from the table of lead times it is apparent that a 7-day lead time occurs with a probability of only 0.04. This indicates that the chances of running out of stock for at least 1 day are 96 out of 100 if an order is not placed until the inventory level reaches 35. Ordering when the stock level is higher than 35 reduces the chance of a stockout but adds the costs of holding a safety stock. Both holding and opportunity costs must be included in the economic-lot-size calculations to achieve a minimum-cost inventory policy.

Figure 23.2 shows the effect of each ordering alternative (amount of lead time provided). The upper portion indicates the shortage that will occur for each day of delay from the lead time provided. The lower portion shows the incremental size of safety stock which will be accumulated by providing more lead time than is required. If orders are based on a 7-day lead time, there will never be old stock on hand when a new order arrives. At the other extreme, an order policy based on a lead time of 14 days assures the company that it will never run out of stock, although as many as 35 cylinders could be in storage when a new supply is delivered.

FIGURE 23.2
Excess or deficient cylinder supply.

Lead Time Provided	LEAD TIME REQUIRED, WEEKS							
	7	8	9	10	11	12	13	14
	Number of cylinders not supplied							
7		−5	−10	−15	−20	−25	−30	−35
8	5		−5	−10	−15	−20	−25	−30
9	10	5		−5	−10	−15	−20	−25
10	15	10	5		−5	−10	−15	−20
11	20	15	10	5		−5	−10	−15
12	25	20	15	10	5		−5	−10
13	30	25	20	15	10	5		−5
14	35	30	25	20	15	10	5	
	Number of cylinders stored							

Lead Time Provided	LEAD TIME REQUIRED, WEEKS								EXPECTED VALUE	
	7 (0.04)	8 (0.08)	9 (0.38)	10 (0.24)	11 (0.12)	12 (0.09)	13 (0.03)	14 (0.02)	HC	OC
7	$ 0	$200	$400	$600	$800	$1000	$1200	$1400	$ 0	$562
8	50	0	200	400	600	800	1000	1200	2	370
9	100	50	0	200	400	600	800	1000	8	194
10	150	100	50	0	200	400	600	800	33	94
11	200	150	100	50	0	200	400	600	70	42
12	250	200	150	100	50	0	200	400	113	14
13	300	250	200	150	100	50	0	200	161	4
14	350	300	250	200	150	100	50	0	210	0

FIGURE 23.3   Expected cost of lead-time alternatives.

The opportunity cost of running out of stock is estimated to be $40 per cylinder. This cost accrues from the disruption of production. The manufacturing plant would sustain a loss of 5 × $40 = $200 per day for late deliveries based on a daily usage of five cylinders. By multiplying the stock deficiencies in Figure 23.2 by $40 and the excess stock by the $10-per-cylinder holding costs, the payoff table of Figure 23.3 is generated.

In Figure 23.3 the alternatives are the lead times provided, and the future states are lead times that could occur. The outcomes result from two types of costs: holding costs and opportunity costs. These two costs affect the ordering policy in different ways. Opportunity costs occur only when an order takes longer to arrive than the lead time that was allowed. Since this cost is associated with the order interval, it can be treated as an ordering expense or a set-up cost. Holding costs are annual expenses and reflect the extra inventory or safety stock that is held in storage.

The expected-value column on the right of the table has two divisions. One applies to holding costs for an alternative and the other to the opportunity costs. Both are calculated from pertinent outcome costs. When the lead time provided coincides with the actual lead time required, the outcome is a zero cost. Obviously, one zero outcome must occur for each alternative. The costs to the left of the zero outcome arise from holding costs, while those to the right are opportunity costs. Each type is multiplied by the likelihood of its occurrence, and the sum of the products is entered in the appropriate expected-value column. Thus, a 7-day lead time shows no holding cost, and a 14-day lead time has no opportunity costs. For a lead-time alternative of 11 days,

$$EV(HC_{11}) = \$200 \times 0.04 + \$150 \times 0.08 + \$100 \times 0.38 + \$50 \times 0.24$$
$$= \$8 + \$12 + \$38 + \$12 = \$70$$

where HC = holding costs of safety stock

and

$$EV(OC_{11}) = \$200 \times 0.09 + \$400 \times 0.03 + \$600 \times 0.02$$
$$= \$18 + \$12 + \$12 = \$42$$

where OC = opportunity costs

The expected values for the other alternatives are obtained in a similar manner.

Each lead-time alternative provides a different economic lot size because of different opportunity costs. By adding the expected value of the opportunity costs to the other costs, we can apply the same economic-lot-size formula used previously. For a 7-day lead time,

$$Q_7 = \sqrt{\frac{2(O + OC)D}{H}} = \sqrt{\frac{2(55 + 562)1150}{10}} = 377$$

Lead Time Provided	Economic Lot Size
7	377
8	314
9	239
10	186
11	150
12	126
13	117
14	113

All the lot sizes are shown in the accompanying table.

The final step is to determine which lot size will allow the lowest total cost. If the purchase price of cylinders is not subject to quantity discounts, price breaks will not affect the ordering policy. Then the total cost becomes the sum of annual ordering costs and holding costs. The ordering costs include both the cost of placing an order and the opportunity cost. Holding costs are composed of the storage expense due to the order size, H, and the additional storage expense incurred from providing a safety margin for delivery delays, HC. This latter cost is the expected value of the holding cost for each lead-time alternative in Figure 23.3. Total costs are calculated from the formula

$$C_{LT} = \frac{(O + OC)D}{Q} + \frac{HQ}{2} + HC$$

This yields the total cost for a 7-day lead time as

$$C_7 = \frac{(55 + 562)(1150)}{377} + \frac{10 \times 377}{2} + 0 = \$3767$$

and for an 8-day lead time as

$$C_8 = \frac{(55 + 370)(1150)}{314} + \frac{10 \times 314}{2} + 2 = \$3130$$

Lead Time	Total Cost
7	$3767
8	3130
9	2402
10	1884
11	1564
12	1373
13	1326
14	1335

From the complete tabulation of costs shown in the total-cost table, it is clear that the most economical policy is to allow a 13-day lead time. An order

TABLE 23.5

Sensitivity to opportunity costs

Opportunity Cost	Lead Time	Total Cost
$10	12	$1272
20	13	1305
30	13	1314
40	13	1326

should be placed whenever the stock level declines to $13 \times 5 = 80$ cylinders. By following this policy, the company should expect to run out of chemicals twice in every 100 order periods.

Opportunity costs are considered more difficult to estimate than are holding costs. It is interesting to observe the *sensitivity* of the lead-time calculations to different opportunity-cost estimates. Total annual costs for the cylinder-inventory problem with four different opportunity costs are shown in Table 23.5. All other factors in the problem are unchanged. It is apparent that estimates of the same general magnitude will not change the results significantly. Consequently, there should not be too much concern if opportunity costs cannot be estimated with the desired degree of accuracy.

Consideration of Variable Demand Rates

Variations in demand or usage rates produce situations similar to those produced by variable lead times. If the lead time is fixed, the risk of a stockout is dependent on the actual demand exceeding the anticipated demand. Holding a safety stock reduces the risk. The problem then becomes one of selecting a safety-stock size which balances the expense of holding the safety stock against the cost of a stock shortage.

Consider a maintenance section which receives deliveries of spare parts from a distributing firm. Deliveries are made regularly a week after an order is placed. The ordering policy has been to reorder class 2 parts whenever the inventory level dips to 200 parts. A series of stockouts has indicated that the policy should be reviewed. It was decided to retain the same lot size but to carry a safety stock for certain parts. Part no. 323 has the usage rate depicted by the histogram in Figure 23.4. It shows the relative frequency of demand for a 1-week period. Annual holding costs for the part are $2, and the stock is replenished about six times a year. Opportunity costs, estimated to be $1 per week, are the result of special ordering procedures and inconveniences caused by the part's being required but unavailable.

Before seeking a solution to this inventory situation, let us inspect the conditions carefully. Here we have a fixed order size and a fixed lead time. A

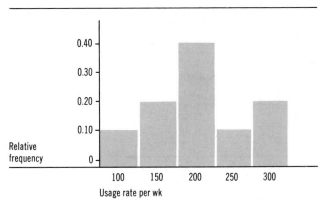

FIGURE 23.4
Histogram of demand for part no. 323.

stockout can occur only if the demand increases during the lead time. The solution space has been limited to the investigation of an appropriate safety stock for certain parts. These conditions considerably simplify the calculations. Without a fixed order size, we would follow the same general approach to a solution as that used for variable lead times. A zero safety stock would be associated with a 200-part reorder point. This is equivalent to decreasing the reorder point, and it would be a reasonable alternative if storage space is limited or if negligible inconvenience results from stockouts. In turn, altering the reorder point would affect the order quantity by increasing costs associated with the order interval. But with the fixed order interval, only the effect of adding a buffer stock to the accepted reorder level needs to be considered.

A reorder level of 200 parts will lead to a stockout 30 percent of the time. Holding an extra 50 parts would limit running out of parts to a probability of 0.20, and a 100-part cushion would completely eliminate the chances of a stockout if the demand pattern remains unchanged. The expected value of the opportunity costs for each of these alternatives is shown in Figure 23.5. Each outcome is the product of the cost of a shortage and the number of times a shortage would occur in a year.

The total cost of each of the alternatives is the sum of the opportunity costs and the holding costs. No additional holding costs are incurred for a

FIGURE 23.5
Expected value of opportunity costs.

| Parts Provided | Reorder Point | Safety Stock | SAFETY STOCK REQUIRED | | Expected Value OC |
			50 (0.10)	100 (0.20)	
200	200	0	$50 \times \$1 \times 6 = \$300$	$100 \times \$1 \times 6 = \$600$	$150
250	250	50	0	$50 \times \$1 \times 6 = \$300$	60
300	300	100	0	0	0

TABLE 23.6

Total-cost table for part no. 323

Safety Stock Provided	Opportunity Costs	Holding Costs	Total Cost
0	$150	$ 0	$150
50	60	100	160
100	0	200	200

zero safety stock. For each safety-stock increment of 50 parts, the annual holding cost is $50 \times \$2 = \$100$. From the totals in Table 23.6, it appears that the original ordering policy was sound and the reorder point should remain at 200 for part no. 323.

Far more sophisticated models are available for including risk in the evaluation of inventory policies. For instance, the case in which both demand and lead times vary complicates the issue considerably and requires a more extensive analysis. *Simulation* is often used for such situations because it can closely duplicate actual conditions without the development of a specific optimization model. At the other extreme from elegant mathematical formulations are rules of thumb and SOPs that have evolved from long experience with inventory problems. How deeply to pursue reality in risk analysis is akin to limiting routine responses to operational problems, as described in Figure 1.3; the depth depends on the cost of operating errors and the prospect of eliminating errors by the analysis.

PROBABILISTIC REPRESENTATION OF FAILURE PATTERNS

Equipment may fail for a wide variety of reasons, ranging from erosive wear to operator carelessness. In many cases it is easy to determine direct causes for wear-related failures whereas in others the cause may remain undetected. Although the causes are not always determinable, a significant proportion of failure patterns may be adequately represented by a small number of statistical distributions. The normal and beta distributions were introduced earlier, in Chapter 20. We will now look at the uniform, Poisson, and exponential distributions.

Uniform Distributions

The uniform, or rectangular, distribution may be in either discrete or continuous form. It represents situations in which all possible outcomes have an equally likely probability of occurring or in which the probability of an occurrence of an event is constant throughout an extended period of time. Perhaps the most

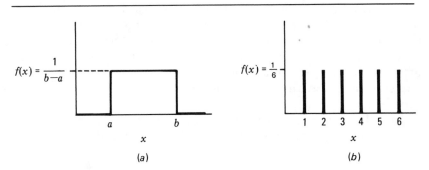

FIGURE 23.6
Examples of continuous and discrete uniform distributions. (a) Continuous distribution; (b) discrete distribution

widely recognizable example of a discrete uniform distribution is the rolling of a single die. In this situation the points shown on each face have a probability of occurrence equal to one-sixth on each roll of the die.

The continuous and discrete cases of the uniform distribution are illustrated in Figure 23.6. In the continuous case the probability density function (pdf) of the distribution is

$$f(x) = \frac{1}{b - a} \qquad \text{for } a \leqslant x \leqslant b$$

$$= 0 \qquad \text{elsewhere.}$$

The mean and variance of the continuous case are, respectively,

$$\mu_x = \frac{(a + b)}{2} \qquad \sigma_x^2 = \frac{(b - a)^2}{12}$$

For the discrete case, the pdf is $f(x) = 1/(d + 1)$, where d equals the difference between the maximum and minimum values. In the case of a die, d is equal to $6 - 1$, or 5, thus $f(x) = 1/6$. The mean and variance of the discrete case are, respectively,

$$\mu_x = \frac{(a + b)}{2} = a + \frac{d}{2} \qquad \sigma_x^2 = \frac{d(d + 2)}{12}$$

The uniform distribution may be used to represent emergency maintenance or replacement activities for components which fail at a constant rate over an extended period of time. It is also useful in representing inventory levels for items undergoing constant demand.

Poisson Distribution

The Poisson distribution occurs often in both natural and industrial processes. In the class of discrete distributions it plays a role as powerful as does the

The sum of the points obtained from rolling a pair of dice is represented by the triangular distribution.

Other useful discrete distributions include the binomial, geometric, hypergeometric, and Pascal.

normal distribution in the continuous class. It is particularly useful in situations wherein the probability that an event occurs is proportional to the length of time, area, or volume involved in a study. Practical applications in which the Poisson distribution is appropriate include such varied examples as the number of visible defects on a roll of newsprint, the number of busy circuits on a telephone microwave channel, or the number of arrivals per hour at a gasoline service station.

In a Poisson process with an average occurrence rate of λ, the probability of having exactly x occurrences is given by the expression

$$p(x; \lambda) = \frac{e^{-\lambda}\lambda^x}{x!}$$

As an example, consider the situation where an average of two defective solder joints are detected on each circuit board inspected. In this case $\lambda = 2.0$ and the resulting Poisson distribution would be calculated as

$$p(x; \lambda) = \frac{e^{-2.0}2.0^x}{x!} \qquad \text{for } x \geqslant 0$$

The results of these calculations are shown in Table 23.7.

A unique feature of the Poisson distribution is that the variance is equal to the mean. Thus

$$\mu_x = \sigma_x^2 = \lambda$$

Complete tables for evaluating Poisson probabilities are available in many statistics texts. However, owing to the simple mathematical form of the pdf, the values may be easily obtained by direct calculation.

TABLE 23.7

Poisson distribution representing defects in circuit-board manufacturing

Number of Defects x_k	Probability $p(x_k)$	Cumulative Probability $P(x_k)$
0	0.135	0.135
1	0.271	0.406
2	0.271	0.677
3	0.180	0.857
4	0.090	0.947
5 or more	0.053	1.000

Exponential Distribution

The exponential distribution is frequently used in both reliability and queueing (waiting-line) studies. It is a continuous distribution which is directly related to the Poisson. Whereas the Poisson distribution may represent the number of occurrences of an event per unit of time, the exponential distribution will represent the amount of time between occurrences. If we expect a process to produce five defects per hour with a Poisson distribution, we would thus expect the average time between defects to be 12 minutes with an exponential distribution. If the mean of the Poisson process is λ, the mean of the exponential distribution is $1/\lambda$, or μ. The pdf of the exponential distribution is given by

Other useful continuous distributions include the gamma, log-normal, and Cauchy.

$$f(x) = \frac{1}{\mu} e^{-x/\mu} \qquad \text{for } 0 < x < \infty$$

The variance of the exponential distribution is μ^2.

The form of the exponential distribution shown above is generally referred to as the negative exponential distribution. In addition to being used to describe interarrival times, it is also useful in predicting failures in electronic systems and representing service times in queueing models. The cumulative distribution function (cdf) of the exponential is

$$F(x) = 1 - e^{-x/\mu}$$

The development of a queueing analysis is presented in Extension 23A.

In our example of an average of five defects per hour, the probability of having less than 10 minutes (1/6 hour) between two defects could be calculated as

$$p(x \leq 10) = 1 - e^{-10/12} = 0.565$$

Time between failures on many mechanical and electric systems is often modeled using this approach.

Distributions for Modeling Reliability

Some of the most important applications of probability distributions are in the areas of modeling the reliability and maintenance patterns of industrial systems. The number of hours or years or operation cycles of a system is a random variable the value of which depends on many factors. The development of an appropriate probabilistic model of failure-rate patterns throughout system life is often critical in making economic decisions involving system maintenance.

Figure 23.7 is a representation of the failure rate with respect to operating time. For many electronic and mechanical systems the failure (or hazard) rate is uniform during the majority of the operating, or service, life of the equipment and is therefore independent of time. However, during the early life of

FIGURE 23.7
A typical failure-rate (hazard) curve for an electronic component.

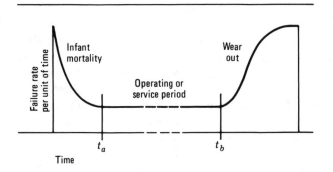

"Reliability is the probability of a device performing its purpose adequately for the specified period of time under the operating conditions encountered"—Electronics Industries Association.

equipment, or what is often called the infant period, the failure rate is high. It is also high near the end of the life, when wear-out and failure due to deterioration take place. In this period the failure rate is so high that it is extremely difficult to maintain the equipment and it must be replaced. This type of hazard function is appropriate for many systems and individual components; it is often referred to as the "bathtub curve" owing to the unique shape.

From the discussion in the previous section it follows that time to failure is often exponentially distributed. Even in cases where the time-to-failure distribution for a component is not exponential over its entire life, it is during the operating period. For example, an electronic component whose hazard function fits that represented in Figure 23.7 would usually be installed only after it had been burned in until time t_a. It would then be replaced at time t_b by another component which had also been burned in. The time to failure for the components in service could thus be represented by the exponential distribution because the hazard function is constant with an expected value of μ failures per time unit.

The Weibull and Erlang distributions are also useful in time-to-failure and reliability studies.

The use of the exponential distribution to represent time between failures has widespread justification on strictly empirical grounds. Many studies on resistors, capacitors, motors, and entire control systems support the validity of the theory. However, the engineer should take considerable care before making such an assumption for a particular system. The validity of such assumptions should always be confirmed before proceeding with an economic study concerning maintenance or reliability projects.

RISKS IN MAINTENANCE-REPLACEMENT POLICY

Maintenance costs were influencing factors in long-term replacement decisions made under assumed certainty, and they are equally significant in short-term replacement decisions which include risk. The short-term effect of wear often follows a consistent pattern. After the pattern has been identified, a user has the alternative of dealing with each item as it fails or following a policy of collective remedial action before failure. Preventive maintenance is a well-known example of the latter alternative.

Maintenance policies are extremely varied in practice. They depend on the characteristics of the items, available maintenance facilities, and relevant costs. We will consider a basic model which may be adapted to fit the circumstances of a particular situation. It includes the factors common to most situations: probable life pattern, alternative courses of preventive action, and outcome costs for the alternatives. The objective is to find the interval at which preventive action is most suitable and then compare this with the alternative of remedial action to decide which is more economical.

To illustrate a maintenance policy, we will assume that a factory has 30 similar machines which exhibit the probability distribution of failures shown in the accompanying table. The cost of remedial action after a breakdown averages $100, and the cost of providing preventive maintenance is $30 per machine.

Months after Mainte- nance	Probability of Failure
1	0.2
2	0.1
3	0.1
4	0.2
5	0.4

The total cost of a preventive-maintenance program is the sum of servicing expense for all the machines each maintenance period (30 machines × $30 per machine = $900) and the cost of breakdowns occurring between services. For a *monthly* preventive-maintenance policy, PM1, the cost is $900 plus $100 for each breakdown expected in the first month after servicing. This amounts to

$$PM1 = \$900 + (\$100 \times 30 \times 0.20) = \$1500/month$$

More data concerning replacement periods is available from the Machinery and Allied Products Institute (MAPI)

A *bimonthly* policy must again include the $900 basic group-servicing cost and the cost of individual breakdowns. In the first month of the period, 30 × 0.20 = 6 machines are expected to break down. During the second month, 30 × 0.10 = 3 machines serviced at the regular period are likely to break down, and 20 percent of the individual breakdowns treated during the first month will *again* fail. These calculations are depicted in Figure 23.8, where a modified expected-value table is utilized.

The first alternative in the table represents a monthly preventive-mainte-nance policy. The expected value of this alternative is six machines treated each month, which agrees with previous calculations. The second line of outcomes depicts the failures of the second month. The total number of

PM Periods	INDIVIDUAL FAILURES DURING MONTH		EXPECTED VALUE	
	1 (0.2)	2 (0.1)	Individual	Cumulative
1	30	0	6	6
2	6	30	4.2	10.2

FIGURE 23.8
Expected failures for two PM periods.

FIGURE 23.9
Expected failures for three PM periods.

	INDIVIDUAL FAILURES DURING MONTH			EXPECTED VALUE	
PM Periods	1 (0.20)	2 (0.10)	3 (0.10)	Individual	Cumulative
1	30	0	0	6	6
2	6	30	0	4.2	10.2
3	4.2	6	30	4.44	14.64

expected breakdowns in 2 months is the cumulative value for both months (6 + 4.2 = 10.2 machines). The cost of a bimonthly policy is

$$PM2 = \$900 + (\$100 \times 10.2) = \$1920$$

which makes the cost per month equal to $1920/2 = $960.

A policy of servicing all the machines every 3 months would lead to the expected failure record shown in Figure 23.9. Now a pattern can be observed. Some of the original group of 30 machines continue to break down each month, and some of those repaired fail again individually. The repairs made each month start a new cycle which must follow the failure distribution. Thus, the expected number of breakdowns from one month is always the first outcome (0.20 probability of failure) for the next month. Similarly, the expected value 2 months ago is the second outcome for the current month, and the cumulative value is the total number of expected breakdowns for a cycle of so many months. A 3-month cycle would have a total cost of

$$PM3 = \$900 + (\$100 \times 14.64) = \$2364$$

or a monthly cost of $2364/3 = $788.

Figure 23.10 can be converted to costs and expanded to include all the necessary calculations for the cost of every preventive-maintenance alternative. The outcomes are converted from units of machines to dollar values by

	INDIVIDUAL FAILURE COSTS DURING MONTH					EXPECTED VALUE				
PM Period	1 (0.20)	2 (0.10)	3 (0.10)	4 (0.20)	5 (0.40)	Individual	Cumulative	PM Cost	Total Cost	Monthly Cost
1	$3000	$ 0	$ 0	$ 0	$ 0	$ 600.00	$ 600	$900	$1500	$1500
2	600	3000	0	0	0	420.00	1020	900	1920	960
3	420	600	3000	0	0	444.00	1464	900	2364	788
4	444	420	600	3000	0	791.00	2255	900	3155	789
5	791	444	420	600	3000	1564.60	3819	900	4719	944

FIGURE 23.10 Expected cost of preventive-maintenance alternatives.

multiplying the number of machines by the individual remedial cost ($100). Columns are added to account for the cyclic group PM costs, the sum of individual and group costs, and the prorated monthly costs. The completed table in Figure 23.10 displays the same recurring pattern of outcomes described previously.

The last step is to determine the costs associated with a policy of performing no preventive maintenance; machines are serviced whenever they break down. The expected period between breakdowns is calculated from the original failure distribution. The expected period is

1 month $\times$ 0.2 + 2 months $\times$ 0.1 + 3 months $\times$ 0.1
$$+ \text{ 4 months } \times \text{ 0.2 } + \text{ 5 months } \times \text{ 0.4}$$
= 3.5 months between breakdowns

Then, with the cost of servicing individual breakdowns pegged at $100 per breakdown, a remedial-action policy costs

$$\frac{30 \text{ machines} \times \$100/\text{machine service}}{3.5 \text{ months/service}} = \$857/\text{month}$$

A comparison of the $857 monthly cost for the remedial-action alternative with the minimum monthly cost of a preventive-maintenance policy from Figure 23.10 ($788) indicates that the latter alternative is preferable. Both the 3- and 4-month preventive-maintenance periods show a lower expected cost than that for dealing with machines only after they fail.

This approach can be applied to a variety of situations. The items being evaluated could fail completely, such as electric light bulbs or electronic tubes. Failures may represent personnel who are no longer available owing to transfers or retirement, and replacement could be the recruiting or training policy. There are also numerous modifications which fit special maintenance and repair situations.

RISK IN QUALITY ASSURANCE

The economic aspects of quality-assurance programs have been introduced earlier, in Chapters 17 and 19. Manufacturing quality starts with a process capable of producing design specifications and continues with a quality-assurance program that ascertains if standards are being met. The manufacturer's perspective focuses on the difference between the value of a quality product and the degree of cost-refinement necessary to produce at a high level of quality. For any given level of refinement there are many contributing factors affecting total manufacturing cost per unit. One of the most critical is the expense of a quality-assurance program. A minimum-cost quality-assurance program balances the cost of maintaining a high level of quality against the cost of shipping an inferior product.

Placement of Inspection Stations

A basic knowledge of matrix algebra is necessary for an appreciation of this section.

The placement of quality inspection stations can have a pronounced effect on the cost of production. Value is added to a product at each state of the production process. If not detected at an early stage, units which were improperly processed early in the operation will continue to have expensive production operations performed with no value being added. A basic model for determining the economic number and location of inspection points will be developed in this section.

Problem Formulation and Solution

The basic methodology for this technique was originally developed by A. A. Markov around 1900.

The formulation of a model to represent the characteristics of a series of production operations is most easily handled through the application of the Markov process theory. First-order Markov processes can be used to model physical or economic systems with the following properties:

1   There is a finite set of outcomes for each operation represented in the model.
2   The probability of the outcome from an operation depends only on the input to that operation.
3   The probabilities are constant over time.

Each individual outcome is called a state. In a process the outcomes from each operation would generally be that the product would either be scrapped, reworked, or passed on to the next operation.

Figure 23.11 represents a very simple production line consisting of two operations, two inspection stations and two absorption states—finished goods and scrap.

States 1 through 4 represent operations in the production line and the values of p_{ij} represent the proportion of work from state i which moves to state j. For instance p_{21} represents the probability that a part leaving inspection will be returned to operation A for rework while p_{23} represents the probability that a part proceeds on to operation B and p_{26} represents the probability of a part's being scrapped. Naturally, the probabilities represented by p_{21}, p_{23}, and p_{26} must sum to 1.0.

The entire production operation can be easily represented in matrix form as

$$
\begin{array}{c|cccc|cc}
 & S_1 & S_2 & S_3 & S_4 & S_5 & S_6 \\
\hline
S_1 & 0 & p_{21} & 0 & 0 & 0 & 0 \\
S_2 & p_{21} & 0 & p_{23} & 0 & 0 & p_{26} \\
S_3 & 0 & 0 & 0 & p_{34} & 0 & 0 \\
S_4 & p_{41} & 0 & p_{43} & 0 & p_{45} & p_{46} \\
\hline
S_5 & 0 & 0 & 0 & 0 & 1 & 0 \\
S_6 & 0 & 0 & 0 & 0 & 0 & 1 \\
\end{array}
$$

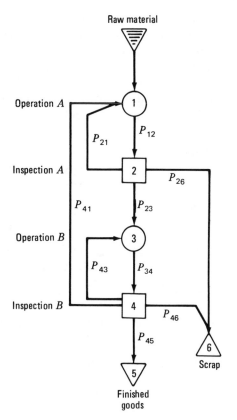

FIGURE 23.11
Flowchart for production line with inspections and rework loops.

In general form the matrix may be described as a set of four submatrices such that

$$P = \begin{vmatrix} Q & | & R \\ \hline 0 & | & I \end{vmatrix}$$

Taken individually, they contain the following information about the process being modeled:

Q is a square matrix containing all the transition probabilities of going from any nonabsorbing state to any other nonabsorbing state (production operations and inspections).

R is a matrix containing all the probabilities of going from any nonabsorbing state to an absorbing state (finished goods or scrap).

I is an identity matrix representing the probability of staying in an absorbing state.

O is a matrix representing the probabilities of escaping an absorbing state. Naturally, it is always equal to zero.

An introductory discussion of Markov processes is available in *Operations Research* by J.E. Shamblin and G.T. Stevens (McGraw-Hill, New York, 1974.)

Since the matrix Q represents the probabilities of going from one nonabsorbing state to another nonabsorbing state in exactly one step, Q^2 gives the probabilities of performing the same transition in exactly two steps. Looking at the total matrix P, we find that if we are interested in how long a part remains in the process or in which absorbing state it finally enters, we must determine the expected number of times it will be in each nonabsorbing state i. This value can be calculated as follows:

$$
\begin{aligned}
E\,(\text{times in } i) = & \;1 \times P\,(\text{being in } i \text{ at start}) \\
& + 1 \times P\,(\text{being in } i \text{ after one transition}) \\
& + 1 \times P\,(\text{being in } i \text{ after two transition}) \\
& + \cdots \\
= & \;1Q^0 + Q^1 + Q^2 + Q^3 \cdots
\end{aligned}
$$

Since the values of p_{ij} in the Q matrix are all less than 1, Q^n approaches 0 as n becomes large. The series of matrices can now be represented by a variation of the algebraic power series.

$$
1 + x + x^2 + x^3 + x^4 \cdots = \frac{1}{1-x} \qquad \text{for } \left| x \right| < 1.0
$$

Thus for the matrix Q,

$$
Q^0 + Q^1 + Q^2 + Q^3 \cdots = (I - Q)^{-1}
$$

With this relationship, the matrix $(I - Q)^{-1}$ provides the expected number of times the process will be in each nonabsorbing state before absorptions. For the purpose of analyzing the production problem we will be most interested in inspecting the first row of the $(I - Q)^{-1}$ matrix. This row indicates the expected number of times that a part passes through each operation before being placed in finished goods or scrap. The importance of this information in determining production cost per unit will be shown in Example 23.1.

Another factor of interest in the analysis is the determination of what total proportion of the units entering production will eventually end up as finished goods. If it were not possible for units to recycle through the system several times as rework, this question could easily be answered. However, as more potential return loops from inspections are created, the complexity of a direct probabilistic solution grows rapidly. Fortunately, the Markov model provides an easy direct solution.

In order to solve the absorption probability problem, it is necessary to recall the probabilities of transitioning from a nonabsorbing state to an absorbing state which are represented by submatrix R. However, not all absorbing states are accessible from all nonabsorbing states. Before reaching a specific absorbing state j, the process which starts in state i may pass through some other nonabsorbing state k. Thus while the probability of going from i to j in one step is R, the probability of going from i to j in two steps must include an intermediate stop at k. The resulting probability for this two-step process is therefore QR, and the probability of a three-step process is Q^2R. It can now be seen that the probability of transitioning from stage i to stage j in exactly n steps is equal to $Q^{n-1}R$.

The probability of going from i to j in an infinite number of steps is thus the sum of the terms in a converging series.

$$\begin{aligned} P(i \text{ to } j) &= R + QR + Q^2R + Q^3R + \cdots \\ &= IR + QR + Q^2R + Q^3R + \cdots \\ &= (I - Q)^{-1}R \end{aligned}$$

Performing this series of operations of the submatrices representing the production problem results in a matrix of absorption probabilities A. This matrix takes the form

$$(I - Q)^{-1} R = A = \begin{array}{c} \\ S_1 \\ S_2 \\ S_3 \\ S_4 \end{array} \begin{array}{|cc|} S_5 & S_6 \\ \hline a_{15} & a_{16} \\ a_{25} & a_{26} \\ a_{35} & a_{36} \\ a_{45} & a_{46} \end{array}$$

where a_{ij} represents the probability that a part which is in stage i will eventually be absorbed in stage j (either finished goods or scrap). This result is particularly important when considering that all parts enter the process at stage one. Thus a_{15} represents the net percentage of good production after some reworks have been included and a_{16} represents the net percentage of scrap produced.

This type of Markov process model can be used to represent a wide variety of industrial and sociological processes. Although this form of analysis is particularly useful in the study of situations involving feedback loops, straight-flow processes such as waiting-line behavior at service stations can also easily be represented.

EXAMPLE 23.1
Production Process Analysis

A small production line has four workstations where various operations are performed on semiautomatic equipment. The machine operator at each station discards obviously defective units after removing the parts from the production equipment. However, the current quality level is quite low and a final inspection is necessary before shipping. The final inspection equipment is also used to determine if a part may be salvaged by rework, and, if so, where the part should be reintroduced into the production line. Determine the overall quality level of the production area and the total cost per unit shipped using the data provided in Figure 23.12.

FIGURE 23.12
Flowchart of production line
modeled in Example 23.1.

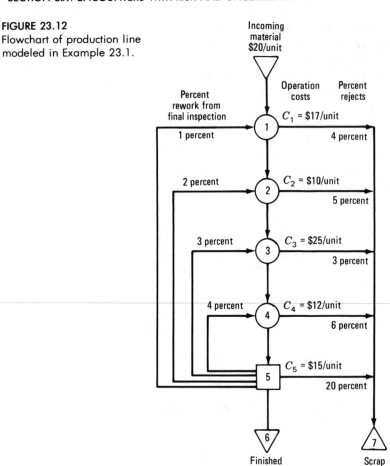

Solution 23.1

Formulating the problem as a Markov process yields the following transition matrix:

$$
P = \begin{array}{c|ccccc|cc}
 & S_1 & S_2 & S_3 & S_4 & S_5 & S_6 & S_7 \\
\hline
S_1 & 0 & 0.96 & 0 & 0 & 0 & 0 & 0.04 \\
S_2 & 0 & 0 & 0.95 & 0 & 0 & 0 & 0.05 \\
S_3 & 0 & 0 & 0 & 0.97 & 0 & 0 & 0.03 \\
S_4 & 0 & 0 & 0 & 0 & 0.94 & 0 & 0.06 \\
S_5 & 0.01 & 0.02 & 0.03 & 0.04 & 0 & 0.70 & 0.20 \\
\hline
S_6 & 0 & 0 & 0 & 0 & 0 & 1 & 0 \\
S_7 & 0 & 0 & 0 & 0 & 0 & 0 & 1
\end{array}
$$

Thus $(I - Q)^{-1}$ yields the following matrix of expected stages before absorbtion:

$$
(I - Q)^{-1} = \begin{array}{c|ccccc}
 & S_1 & S_2 & S_3 & S_4 & S_5 \\
\hline
S_1 & 1.01 & 0.99 & 0.97 & 0.97 & 0.91 \\
S_2 & 0.01 & 1.03 & 1.01 & 1.01 & 0.95 \\
S_3 & 0.01 & 0.03 & 1.06 & 1.07 & 1.00 \\
S_4 & 0.01 & 0.03 & 0.06 & 1.10 & 1.03 \\
S_5 & 0.01 & 0.03 & 0.06 & 0.11 & 1.10
\end{array}
$$

Inspection of the first row of this matrix yields the expected number of times which a part entering the system at S_1 will pass through each process point in the production line. Total expected production and inspection costs for each item entering the system can thus be calculated as

$$E(\text{TPC}) = \$17(1.01) + \$10(0.99) + \$25(0.97)$$
$$+ \$12(0.97) + \$15(0.91) = \$76.61$$

However, we do not yet know what percentage of parts which enter the system actually emerge as good production units. In order to obtain this figure, it is necessary to determine the A matrix, where $A = (I - N)^{-1} R$.

This calculation yields

$$A = \begin{array}{c} \\ S_1 \\ S_2 \\ S_3 \\ S_4 \\ S_5 \end{array} \begin{array}{cc} S_6 & S_7 \\ \hline 0.64 & 0.36 \\ 0.67 & 0.33 \\ 0.70 & 0.30 \\ 0.72 & 0.28 \\ 0.77 & 0.23 \end{array}$$

The value of 0.64 for a_{16} indicates that 64 percent of the entering parts are finally absorbed in the stage representing acceptable quality units.

Thus the total production and inspection cost for each good unit produced is calculated as

$$\frac{(\text{Raw material cost} + \text{production expense})/\text{unit}}{\text{Percentage of acceptable production}}$$

or

$$\frac{(\$20.00 + \$76.61)}{0.64} = \$150.95/\text{unit}$$

Review Exercises and Discussions

The demand for a specific-type voltage regulator averages three per day at an automobile parts distributor. The units are ordered in prepackaged lots of 60 when stock reaches a minimum level of 10 items. Assuming the demand for regulators fits a Poisson distribution and 3 days are required for an order to arrive, determine the probability that the stock of regulators will be exhausted before the replacement shipment arrives.

Exercise 1

The expected demand for regulators over a 3-day period would equal $\lambda_1 + \lambda_2 + \lambda_3 = 3 + 3 + 3 = 9$. Since the Poisson assumption is based on each day's demand being independent, the total demand over the 3-day period also fits the Poisson distribution. Thus the probability of exhausting the stock of regulators is equal to one minus the probability of having a demand of 10 units or less. Thus,

Solution 1

$$P(11 \text{ or more}) = 1 - P(10 \text{ or less})$$

$$= 1 - \sum_{x=0}^{x=10} \frac{2^x e^{-2}}{x!}$$

$$= 1 - 0.458 = 0.542$$

An electronic system is composed of four subsystems connected in series. Thus if one subsystem fails, the entire system fails. Each of the subsystems has an expected mean

Exercise 2

time between failures of 1000 hours, and the failure rates are exponentially distributed. Determine the probability that the entire system will operate for at least 500 hours without failing.

Solution 2   The average failure of a single subsystem is equal to 0.001 per hour. The probability of operating less than 500 hours is equal to

$$P(x < 500) = 1 - e^{-(0.001)\,(500)}$$

$$= 0.3935$$

Thus the probability of a single subsystem's operating longer than 500 hours is equal to 0.6065. For the entire system to successfully operate over this period, each subsystem must survive. The probability of this occurring is calculated as

$$P(\text{system will operate over 500 hours}) = P_1 \times P_2 \times P_3 \times P_4 = (0.6065)^4 = 0.1353$$

Exercise 3   In order to ensure the reliability of a total system, component subsystems are sometimes connected in parallel. Thus, if one of the parallel systems fails, the system will continue to operate with the remaining component. Assume that a single component has a reliability of 0.90 of successfully meeting operational requirements throughout a year. The total equivalent annual cost of the component is $10,000 and the cost of having a system failure during the year is $150,000. Should an additional component be installed in parallel with the first?

Solution 3   The expected total equivalent annual cost of having one component in the system can be calculated as

$$E(C) = \$10,000 + \$150,000 \times (\text{probability of failure})$$

$$= \$10,000 + 150,000\,(0.10)$$

$$= \$25,000$$

If two components are connected in parallel and are assumed to have independent failure rates, the probability of at least one system operating is calculated as

$$P(\text{at least one system works}) = 1 - P(\text{both systems fail})$$

$$= 1 - (0.10 \times 0.10)$$

$$= 0.99$$

The expected annual cost of this system is therefore

$$E(C2) = (2 \times \$10,000) + \$150,000\,(0.01)$$

$$= \$21,500$$

The parallel system should be installed.

Exercise 4   Assume that the production process analyzed in Example 23.1 is such that no rework is possible, but with the exception of the percentage of rejected parts at final inspection, all other values remain as stated. The final inspection will now reject 25 percent of the parts reaching that point. Determine the total production and inspection cost per acceptable unit produced.

The modified production system can still be easily modeled as a Markov process. However, since no feedback loops are involved, a direct probabilistic analysis is also possible.

Solution 4

Operation	Cost	Proportion Raw Material Processed	Expected Cost
Raw material	$20.00	1.000	$20.00
1	17.00	1.000	17.00
2	10.00	$1.000 \times 0.9600 = 0.9600$	9.60
3	25.00	$0.9600 \times 0.9500 = 0.9120$	22.80
4	12.00	$0.9120 \times 0.9700 = 0.8846$	10.62
5	15.00	$0.8846 \times 0.9400 = 0.8315$	12.47
			$92.49

The expected total cost of operations on every item entering the system is $92.49. However only $0.96 \times 0.95 \times 0.97 \times 0.94 \times 0.75 = 0.6237$ of those entering as raw material exit as acceptable production. The total expected cost per good unit produced is therefore

$$\$92.49/0.6237 = \$148.29$$

A large number of replaceable purifier units are required in a production process. The units cost $8 apiece and have the following failure pattern:

Exercise 5

	Week						
	1	2	3	4	5	6	7
Proportion of units that fail during week	0.30	0.05	0.05	0.10	0.20	0.25	0.05

Total charges for replacing a single unit that fails during operation are $50 for maintenance and downtime. The cost of replacing one unit during a blanket replacement over a weekend is $10 for labor plus the cost of the unit.
a Determine the expected cost per week of a replace-as-fail policy.
b The current policy is to replace all units on a 4-week cycle, with individual replacements made as needed. Consideration is being given to a new 5-week blanket replacement cycle. Is this a better policy?

a The expected replacement time for individual units is

Solution 5

$$0.30(1) + 0.05(2) + 0.05(3) + 0.10(4) + 0.20(5) + 0.25(6) + 0.05(7) = 3.8 \text{ weeks}$$

This expectation yields

$$\text{Individual replacement cost} = \frac{\$50 + \$8}{3.8 \text{ weeks}} = \$15.26/\text{unit/week}$$

b The pattern of individual replacements during blanket replacement cycles, for a reference base of 100 units, is

Cycle	Replacements during Week per 100 Units	
Week 1	100(0.30)	= 30
Week 2	100(0.05) + 30(0.30)	= 14
Week 3	100(0.05) + 30(0.05) + 14(0.30)	= 10.7
Week 4	100(0.10) + 30(0.05) + 14(0.05) + 10.7(0.30)	= 15.4
Week 5	100(0.20) + 30(0.10) + 14(0.05) + 10.7(0.05) + 15.4(0.30)	= 28.86

During a 4-week cycle, individual replacements per 100 units will be 30 + 14 + 10.7 + 15.4 = 70.10 units. Including the group replacement cost per 100 units of 100($18) = $1800,

$$\text{Total cost per unit(4-week cycle)} = \frac{70.10(\$58) + \$1800}{100\text{-unit cycle}} \times \frac{\text{cycle}}{4\text{ weeks}}$$

$$= \frac{\$14.66/\text{unit}}{\text{week}}$$

By a similar process,

$$\text{Total cost per unit (5-week cycle)} = \frac{(70.10 + 28.86)(\$58) + \$1800}{100\text{-unit cycle}} \times \frac{\text{cycle}}{5\text{ weeks}}$$

$$= \frac{\$15.08/\text{unit}}{\text{week}}$$

The present 4-week blanket replacement policy is therefore preferable.

PROBLEMS

Delivery Lead Time, Weeks	Prob-ability
2	0.1
3	0.2
4	0.3
5	0.3
6	0.1

Data for Problem 23.1.

23.1 A motorcycle dealer has kept a record of the lead time for delivery of new bikes from the factory. He estimates that the opportunity cost of lost sales averages $60 for each motorcycle not on display when it might have been. His sales are relatively uniform at seven per week. Ordering costs average $20 per order and holding costs are $120 per year for each bike in stock. What ordering policy should the dealer use?

23.2 Compute the reorder point for the following situation:

Average lead time	5 days
Holding costs/unit/year	$2
Optimum order interval	2 months
Cost of stockouts/order interval	$20
Average demand rate	15 units/day

23.3 A small plant uses wood chips as a raw material for one of its products. Chips are delivered in units of railway cars. The average demand is four cars per month, but the demand varies according to the accompanying table.

Since special unloading equipment is rented when a delivery arrives, it is desirable to have large orders. Space is available to store any size of order, but stored inventory is subject to weather damage. The costs associated with wood-chip inventory are

Holding costs	$60/year for each unit
Order costs	$300/order (includes rental of unloading equipment)
Opportunity costs	$80/unit (includes altering production schedules)
Lead time	1 month (4 weeks)

Demand during Lead Time	Probability
60	0.05
65	0.05
70	0.15
75	0.40
80	0.20
85	0.10
90	0.05

Data for Problem 23.2.

What safety stock should be carried?

23.4 The Able Company places an order for 300 gallons of paint (the EOQ when opportunity costs are not considered) 25 times each year. The average lead time is 7 calendar days. The usage rate is 20 gallons per day, 60 percent of the time. The probabilities of 15-, 25-, and 30-gallon usage are, respectively, 0.10, 0.20, and 0.10. If the order costs are $30 per order and the cost of being out of stock is $20 per gallon, what safety stock should be carried? Develop an expected-cost table and determine the total cost for each safety-stock alternative.

Demand	Probability
6	0.10
5	0.20
4	0.50
3	0.20

Data for Problem 23.3.

23.5 Because of a shortage of wood chips, the lead time for the small plant's orders (Problem 23.3) can no longer be considered fixed. The estimated future lead times are described by the distribution shown in the accompanying table. The costs and demand rate from Problem 23.3 are still applicable. Determine the order size and the inventory level at which this size order should be placed.

23.6 Demand for a specified part from a tool crib fits a Poisson distribution with a mean of 5.0 units per day.

(a) Determine the probability that demand in any single day is between four and six units, inclusive.

(b) Determine the expected total demand and the variance of the demand for a 5-day week.

(c) Determine the probability that demand in any randomly selected week will be between 20 and 30 units, inclusive.

(d) Use the normal distribution to determine an approximate value for the probability of weekly demand being between 20 and 30 units, inclusive. Compare the result to the value obtained in Problem 23.6c.

Lead Time, Weeks	Probability
1	0.1
2	0.1
3	0.3
4	0.5

Data for Problem 23.5.

23.7 Service times at a truck weigh station are exponentially distributed with a mean of 4.0 minutes. Determine:

(a) The probability that a truck can pass through the station in less than 3.0 minutes.

(b) The probability that more than 6.0 minutes are required to pass through the station.

(c) The variance of the time required to pass through the station.

23.8 A container is filled with a large number of poker chips each having an indicated value between 0.00 and 1.00. Assuming that the values are uniformly distributed within the stated range, determine the expected sum and the variance of the sum obtained by adding the points shown on 12 randomly selected chips. What distribution would you expect these sums to approximate if a large number of samples of size 12 were selected?

23.9 It has been suggested to a data-processing firm that it adopt a policy of periodically replacing all the tubes in certain pieces of equipment. A given type of tube is known to have the mortality distribution shown in the table. There are approximately 1000 tubes of this type in all the combined equipment. The cost of replacing the tubes on an individual basis is estimated to be $1.00 per tube, and the cost of a group-replacement policy averages $0.30 per tube. Compare the costs of preventive versus remedial replacement.

Tube Failure during Week	Probability of Failure
1	0.3
2	0.1
3	0.1
4	0.2
5	0.3

Data for Problem 23.9.

23.10 A vending machine operator has machines in 40 locations. There is an equal probability each day during a 10-day period that the machines in one location will be emptied. After 10 days all the machines will be empty. The cost to individually replenish the machines at one location (travel and working time) is $18. The loss in profit from idle machines in one location is $10 per day. Replenishments may be made individually as requested when the machines in one location are empty, or they may be made all at one time with a total cost to service the 40 locations of $250. What is the lowest-cost replenishment policy?

23.11 An assembly line has 30 identical machines. The pattern of breakdowns is shown in the accompanying table. Breakdowns can usually be fixed in a short period of time, but the disruption of the production line creates considerable expense. One way to eliminate this disruption is to provide standby machines. The daily cost of keeping a standby machine is estimated to be $12. The cost of an out-of-order machine is $150 per day. How many standby machines should be provided?

Number of Machines Out of Order at One Time	Probability
0	0.5
1	0.2
2	0.1
3	0.1
4	0.1

Data for Problem 23.11.

23.12 Operators working in a "clean room" environment use magnifying equipment that has a breakdown pattern that follows an arithmetic progression: In 50 percent of the shifts there are no equipment failures, in 25 percent of the shifts one unit fails, in 12.5 percent two fail, in 6.25 percent three fail, etc. The 30 operators now have five standby machines. The cost of each standby is $40 per shift; the cost of production and servicing for a down machine averages $300 per shift. There is also a lost-time cost of $30 to get the replacement machine in position when a breakdown occurs. Determine whether the present number of standby machines is the optimum number.

23.13 Another approach to the replacement problem described in Review Exercise 5 is being considered. All new units will be pretested on a test bench to minimize the high failure rate during the first week of operation. Surviving units will then be used in the production process. If the pretest cost is $3 per unit, compare the total cost of the plan with the cost of individual replacements.

23.14 The Covell Company uses a large number of sacrificial catalysts to protect specialized production equipment. The units cost only $10.00 each but are subject to sudden failure, and emergency replacement costs amount to $50.00 in labor and lost production. Analysis of production records has revealed the information shown in the table concerning failure patterns.

(a) Determine the expected weekly cost per unit of a replace-as-they-fail policy.
(b) The company currently uses a blanket replacement policy on the weekend following a 4-week cycle. Labor cost involved in replacing one unit during weekend shutdown is $9.00. Management feels that this cycle should be extended to 5 weeks. What do you recommend?

Week	Proportion of Units Failing during Week
1	0.25
2	0.05
3	0.05
4	0.15
5	0.20
6	0.25
7	0.05

Data for Problem 23.14.

(c) A suggestion has been made to place all units on a test bench for 1 week to minimize the initial high failure rate. The surviving units would then be placed in the production line. The test cost is $2.00 per unit for the equipment and personnel involved, and the company will use a replace-as-they-fail policy. Compare the cost of this plan to that obtained in Problem 23.14a.

23.15 Nitpick Industries uses a large number of electronic flyash mashers to control pollution from the various stochastic nitheaters installed throughout the plant. These units cost $200 each and must be replaced immediately upon failure if production operations are to be sustained. Replacement costs total $500 per unit if maintenance occurs during scheduled production hours, and $50 per unit if it occurs during weekend scheduled downtime. A study of flyash equipment over the past year reveals the equipment failure pattern, as shown in the table

(a) What is the cost per week of a replace-as-they-fail policy?

(b) What would be the optimum blanket replacement policy and how would the cost of such a policy compare with that in Problem 23.15a?

(c) It would cost $40 per unit in extra labor and equipment expense to place all new units on a burn-in stand for 1 week before placing the survivors on production-line equipment. Using a replace-as-they-fail policy, determine if this would result in an improvement over the policy in Problem 23.15a?

Week of Operation	Failure Rate, %
1	20
2–10	Less than 1
11	10
12	40
13	30

Data for Problem 23.15.

23.16 Given the following data concerning the transitional probabilities, determine the expected production cost of the process in Figure 23.11. The per-unit cost of each operation is $20.00 and each inspection costs $4.00.

$$p_{21} = 0.20, \ p_{26} = 0.08, \ p_{41} = 0.20, \ p_{43} = 0.10, \ p_{46} = 0.15$$

23.17 A proposed modification of the production line in Example 23.1 involves the placement of semiautomated inspection points immediately following each operation. The equivalent annual cost of each inspection would total $5.00 per unit and the transitional probabilities within the process would be changed. Assuming that the new inspection stations provide 100 percent accuracy, determine the total production cost per unit of the modified system in Figure P23.17.

EXTENSION

23A The Risk of Waiting

Whenever waiting lines form in an industrial system there is reason to investigate the situation. A line, or queue, means congestion and lost time. The cost of congestion may be directly observable, as when trucks are waiting to be loaded, or the costs may be more subtle, as when a potential customer leaves a line that has formed at a sales counter. But there are also costs associated with relieving congestion. Providing more services—more loading crews or sales clerks—eliminates the congestion at the risk of creating excess service capacity. Surplus capacity merely transfers the cost of idle time from the calling population to the service facility. The objective of waiting-line, or queueing, analyses is to minimize the sum of congestion and service costs.

Two primary methods are available to reduce total costs: schedule and control the flow of arrivals into the system and/or provide the correct service capacity. The preferred approach depends upon whether the input flow or service facilities (or both) can be suitably altered. Such problems are analyzed according to a group of mathematical relations collected under the title of *queueing theory*.

In order to utilize queueing theory to investigate a waiting-time situation, the primary operating characteristics of the system must be known. The arriving unit that requires some service to be performed is called the *customer*. Customers may include people, parts to be machined or inspected, documents, or telephone inquiries. This input to the system will come from a source or *population* which may be either finite or infinite in size. As a general rule of thumb, the assumption of an infinite population will be valid when the presence of one customer in the system does not significantly alter the probability of another arrival. The queue of customers awaiting service may also be finite or infinite according to the number of calling units it can contain. If the queue is full, arriving customers will be turned away, thus altering the effective arrival rate of new units into the actual service station.

We will restrict our analyses to a limited subset of the wide field of queueing studies. Unless otherwise noted, the following standard terminology and notation will be used:

1 *Arrival rate.* Arrivals per unit of time will follow a Poisson distribution. The mean arrival rate is referred to as λ.
2 *Service rate.* The rate at which one service channel can perform the required customer service. Service times are generally assumed to follow an exponential distribution. The service rate is therefore Poisson with a mean rate of μ.
3 *Queue discipline.* Arrivals into the system are served on a first-in, first-out (FIFO) basis. No special priority or emergency provisions are included.

4 *Customer discipline.* Arrivals are not affected by the length of the queue when they arrive. This means a long queue will not discourage arrivals, and disgruntled arrivals will not leave the queue after spending time in the waiting line.
5 *Service channels.* The customers arriving in a system with multiple service channels will form a single queue in a general waiting area and then proceed to the next available service station in a FIFO manner. A multiple service channel system is shown in Figure 23.13.

The basic queueing model is that of a single channel queue serving an infinite population. The basic parameters of the system are λ = arrival rate, units/time period, and μ = service rate, units/time period. The probability that the system is idle at any time, P_0, is simply $1 - \lambda/\mu$, and the probability of there being a finite number of customers, n, in the system at a randomly selected time is $P_n = (\lambda/\mu)^n P_0$. Using this relation, we can develop equations for describing the steady-state performance of the system for all situations where $(\lambda/\mu) < 1.0$. Among the more common equations are the following:

L = Expected number of customers in the system

$$= \frac{\mu}{\mu - \lambda}$$

L_q = Expected number of customers waiting in the queue

$$= \frac{\lambda^2}{\mu(\mu - \lambda)}$$

W = Expected time a customer spends in the system

$$= \frac{1}{\mu - \lambda}$$

W_q = Expected time a customer spends in the queue

$$= \frac{\lambda}{\mu(\mu - \lambda)}$$

$P(W > t)$ = Probability that a customer spends longer than t time units in the system

$$= e^{-u(1 - \lambda/\mu)t} \qquad \text{for all } t > 0.0$$

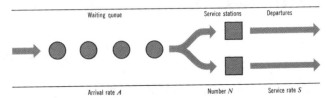

FIGURE 23.13 Waiting-line (queue) characteristics.

Example 23.2
Single-Service Facility

A tool crib is maintained for machine operators to obtain cutting tools, spare parts, and miscellaneous stock. The total labor cost of an operator is $24.00 per hour and it is estimated that machine idle time and lost production amount to $50.00 per hour while the operator is away from the production floor. Operators call at the crib at an average rate of eight per hour and it takes the crib operator an average of 5 minutes to look up the stock record, locate and pull the material, and record the transaction. Cost of maintaining the crib and contents is $60.00 per hour and total hourly labor cost for the operator is $20.00

With these facts we can investigate several aspects of the system. The utilization rate for the crib is

$$\rho = \frac{\lambda}{\mu} = \frac{8/\text{hour}}{12/\text{hour}} = 0.67$$

And the average number of operators in the system is

$$L = \frac{\lambda}{\mu - \lambda} = \frac{8}{12 - 8} = 2.0 \text{ operators}$$

More importantly, the total time an operator spends in the system is

$$W = \frac{1}{\mu - \lambda} = \frac{1}{12 - 8} = \frac{1}{4} \text{hour}$$

Total cost of the present system would therefore be calculated as

Cost of operator and cost production $+$ Cost of tool crib operation

$$(0.25 \text{ hour}) (8 \text{ operators/hour}) (\$50.00 + \$24.00)$$

$$+ (\$60.00 + \$20.00) = \$228.00/\text{hour}$$

Since the operators are obviously spending a significant amount of time waiting for service, a proposal has been made to install a computerized location directory and bar-code checkout system. Estimated total equivalent hourly cost for the new system is $25.00 and the average service time will be reduced to 3 minutes. Is the proposed system economically justifiable?

Installation of the new equipment would result in the following average values for system parameters

$$\rho = \frac{8}{20} = 0.40 \qquad L = \frac{8}{20 - 8} = 0.667$$

$$W = \frac{1}{20 - 8} = 0.083 \text{ hour}$$

The new total hourly cost would be calculated as

$$(0.0833 \text{ hour}) (8 \text{ operators/hour}) (\$50.00 + \$24.00)$$
$$+ (\$60.00 + \$20.00 + \$25.00) = \$154.33/\text{hour}$$

The proposed system would result in a net savings of $73.67 per hour.

The equations presented above for the single-service-channel case can be extended to multiple-channel systems. We can use these equations without going into their derivations if we recognize the basic limitations on their applications. The model presented below is limited to infinite population, infinite queue-length situations, where Poisson arrival rates and exponentially distributed service rates are present.

With multiple-service channels, s, the general equation for determining the probability of an empty system is

$$P_0 = \frac{1}{\sum\limits_{n=0}^{n=s-1} \left[\frac{1}{n!} \left(\frac{\lambda}{\mu} \right)^n \right] + \frac{1}{s!} \left(\frac{\lambda}{\mu} \right)^s \frac{s\mu}{s\mu - \lambda}}$$

The expected time in the system can then be calculated as

$$W = \frac{\mu(\lambda/\mu)^s P_0}{(s-1)!\,(s\mu - \lambda)^2} + \frac{1}{\mu}$$

$$W_q = W - \frac{1}{\mu}$$
$$L_q = W_q \lambda$$
$$L = L_q + \frac{\lambda}{\mu}$$

Fortunately, it is possible to calculate the other parameters through the application of the following relationships:

Example 23.3
Two-Channel Facility

In Example 23.2, the comparison was made for a single-channel system with and without a computer system. We will now investigate a third alternative—hiring another clerk rather than installing the computer. It will be necessary to assume that the second clerk will have the same hourly cost as the first and that there will be no interference with both employees occupying the crib space.

The probability that there are no customers in the system is equal to

$$P_0 = \frac{1}{\left[\frac{1}{0!}\left(\frac{8}{12}\right)^0 + \frac{1}{1!}\left(\frac{8}{12}\right)^1\right] + \frac{1}{2!}\left(\frac{8}{12}\right)^2\left(\frac{2\times12}{2\times12-8}\right)} = 0.50$$

And the expected total time a customer spends in the system is

$$W = \frac{8\,(8/12)^2\,(0.50)}{1\,(24-8)^2} + \frac{1}{12} = 0.0937$$

The new total hourly cost would be

(0.0937 hour) (8 operators/hour) ($50.00 + $24.00) + ($60.00 + $20.00(2)) = $155.49/hour

With the two most economically attractive alternatives relatively equal, the final decision would require the inclusion of non-monetary factors and possibly the development of a sensitivity analysis.

Summary

Additional equations are available to model a wide range of queueing situations. Finite populations, finite queues, variable and constant arrival and service rates, and priority systems have all been studied in great detail. These equations are readily available in many texts, but care must be taken to ensure that the assumptions made in developing the equations pertain to the actual system. Whenever such models are used, it is necessary to give particular attention to the most sensitive characteristics. In queueing models, the total system is highly influenced by small changes in arrival and service rates.

QUESTIONS

23A.1 An automobile dealer has used cars which need reconditioning before they can be sold. They have a Poisson arrival rate which averages four per week. The dealer is going to hire one of two people to clean up the cars. The first is capable of reconditioning cars at the rate of seven per week but requires a wage of $300 per week. The alternative is to hire a less efficient worker, for $195 per week, who is capable of reconditioning only five cars per week. The dealer estimates the cost of holding a car off the used-car lot at $135 per week. Which worker should be hired? Assume that the cars are reconditioned at an exponential rate.

23A.2 Machines break down in a factory at an average rate of four per day. Breakdowns occur in a manner which closely follows a Poisson time distribution. An idle machine costs the company $30 per day. The current method of servicing follows an exponential repair rate averaging six per day. The cost of this facility is $60 per day. By purchasing diagnostic equipment for $12,000, the capacity of this facility will be doubled while still following an exponential repair-time distribution. The diagnostic equipment will last 10 years and the operating costs will be $35 per day. A minimum rate of return of 12 percent is expected on all new investments. There are 250 working days per year. Should the equipment be purchased?

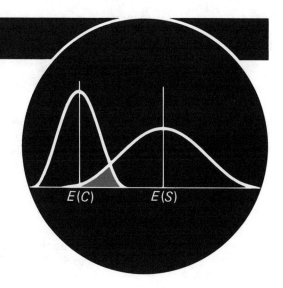

CHAPTER 24

RISKS IN FINANCIAL ANALYSIS

E(C)   E(S)

OVERVIEW

The true future worth of currently available alternatives is known with only a partial degree of assurance. This risk concerning future cash flows makes economic decision making one of the most challenging tasks faced by managers throughout public and private organizations. The development of techniques for determining the expected value of various proposals was discussed in Chapter 16. Measures of the variation of outcomes complement the expected-value criterion. For instance, if two proposals have the same expected value, the one with outcomes having less variation from the mean is considered to have less risk. In many cases, a risk-adverse manager may well select an option with a lower expected value if it involves less risk.

The *standard deviation* of a cash flow is often considered the primary measure of risk. The variance of future cash flows can readily be converted to appropriate measures of risk in terms of net present worth. Although there is sometimes considerable subjectivity involved in the development of cash flow distributions, descriptive information provided by statistical analysis of risk furnishes valuable insights.

A distribution representing the variation of the present worth of an alternative is generally the result of aggregating a number of future cash flows and their respective probability density functions. Some of these cash flows may be independent from all other factors and others may be highly correlated. The evaluation of many representative problems may be accomplished through direct analytical methods, whereas others may require *simulation*-based studies. The *Monte Carlo* technique is widely used for computerized simulation of financial problems.

EXPECTED VALUES AND VARIANCES

Consider the outcomes from two investment proposals depicted in Figure 24.1. Both have the same expected value, but the range of estimated outcomes indicates that the forecaster is more confident that proposal A will return close to $1000. The three estimates constitute a *discrete distribution*. Histograms for proposals A and B would both show a central bar at $1000 on the horizontal axis with a height indicating a probability of occurrence of 0.6 on the vertical axis, but the equal-height ($P = 0.2$) bars representing the other two outcomes for proposal A would be much closer together than the corresponding bars for proposal B.

A useful quantitative measure of variability can be obtained from calculating the variance and standard deviation of the population. By definition the variance, var (X) or σ_x^2, is equal to $E(X^2) - E(X)^2$, where $E(X^2)$ and $E(X)$ are the expected values of X^2 and X, respectively. For discrete distributions, this expression can be restated as

$$\text{var } (X) = \Sigma \, [p(x_i) \cdot (x_i - \bar{x})^2]$$

or

$$[\Sigma p(x_i) \, (x_i^2)] - \bar{x}^2$$

where $p(x_i)$ is the probability of occurrence of the value represented by x_i and $\bar{x}$ is the mean of the population. As applied to proposal A from Figure 24.1,

$$\text{var } (A) = [0.2(900)^2 + 0.6(1000)^2 + 0.2(1000)^2] - 1000^2$$
$$= 4000$$

and the standard deviation σ is equal to $\sqrt{\text{var } (A)} = \$63.25$.

The visually obvious greater variability of proposal B is confirmed as

$$\text{var } (B) = [0.2(400)^2 + 0.6(1000)^2 + 0.2(1600)^2] - 1000^2$$
$$= 144{,}000$$

and $\sigma_B = \$379.47$.

If the three states of demand in Figure 24.1 were visually expanded to include all possible demand levels represented by their respective variances,

FIGURE 24.1
Variability of two proposals that have the same expected values.

	DEMAND			
	$P(high) = 0.2$	$P(average) = 0.6$	$P(low) = 0.2$	*Expected Value*
Proposal A	$900	$1000	$1100	$1000
Proposal B	400	1000	1600	1000

the result would be a continuous distribution for each proposal. Figure 24.2 illustrates the two proposals with their respective variances in continuous form.

Probability in a continuous distribution is given by the area under the distribution curve, and the expected value can be computed by integration. For the normal distribution, a table of probability values catalogued by standard deviations from the mean (Z values) is provided in Appendix F. The mean of a symmetrical distribution is the expected value of the variable represented by the horizontal axis.

Once a well-behaved theoretical distribution is assumed to adequately represent outcome probabilities, an estimate of the standard deviation is necessary. It can be estimated directly or derived from an assessment of the risk associated with a certain outcome. For instance, assuming that the normal distribution is appropriate and has a mean of $1000, we can estimate the standard deviation forthrightly as $100. A mental check on the soundness of this estimate is made by comparing assessments of outcomes with their probabilities in a normal distribution with $\mu = \$1000$ or $\sigma = \$100$. From the relationship

The two most popular techniques for evaluating goodness-of-fit are the chi-squared and Kolmogorov-Smirnov tests.

$$P(O < x) = P\left(Z < \frac{x - \mu}{\sigma}\right)$$

where O is a normally distributed outcome, x is a selected outcome value with a subjective estimate of the probability of its occurrence, and Z is the standard normal deviate; an assessment that there is only a 10 percent chance that the outcome will be less than $875 is checked against the estimate of $\sigma = \$100$ as

The 0.106 probability is obtained by entering the table in Appendix F at $Z = 1.25$. At this point the area under the curve as measured outward from the mean is 0.3944. Since Z is negative in the formula, the left tail of the curve is indicated. Its area, which is the probability that $Z < -1.25$, is $0.5 - 0.3944 = 0.1056$.

$$P(O < \$875) = P\left(Z < \frac{\$875 - \$1000}{\$100}\right)$$
$$= P\left(Z < \frac{-125}{100}\right) = P(Z < -1.25)$$
$$= 0.106$$

The result indicates that the distribution conforms closely to the stated belief

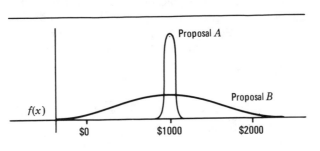

$f(x)$

$0 $1000 $2000

Proposal A

Proposal B

FIGURE 24.2
Comparison of variances for proposals A and B.

that the probability of an outcome of less than \$875 is 0.1 when the mean is \$1000 and the standard deviation is \$100.

By similar computations, a standard deviation can be obtained from such an assessment as "there is a 20 percent chance that the outcome will exceed the expected value of \$1000 by \$200 or more." The mathematical counterpart of this statement is

$$P(O \geq \$1200) = 0.2 = P\left(Z \geq \frac{\$1200 - \$1000}{\sigma}\right)$$
$$= P(0.84 \geq 200/\sigma)$$

which leads to an estimate that
$$\sigma = \$200/0.84 = \$238$$

Comparable relationships for the beta distribution are given by the equations associated with Figure 20.20, in which EV is substituted for T_e.

RELATIONSHIPS OF FINANCIAL RISK

Simply calculating the variance of outcomes from a proposal does not provide conclusive evidence for acceptance or rejection. The intangible considerations discussed in Extension 10C are influential. Three additional considerations deserve attention: (1) elapsed time before the outcome will occur, (2) magnitude of variance with respect to size of investment, and (3) relation of a proposal to other investments already made by the organization.

1 Riskiness over Time
It is intuitively logical that estimates made for a period several years in the future are not as reliable as are those made for the near future. This concept is portrayed by the outcome distributions in Figure 24.3. All three distributions have the same expected value, but the standard deviation increases with extension of the planning horizon. An approximation of the increase is given by the product of the earliest standard deviation and the square root of the number of periods to which the planning is extended: $\sigma_n \doteq \sigma_0\sqrt{n}$, where $n =$ time periods from time zero.

2 Coefficient of Variation
A standard deviation can be a misleading indicator of risk when alternatives differ in size. Let a third proposal C be added to the alternatives given in Figure 24.1. For the same future states, proposal C has expected outcomes of \$980,000, \$1,000,000, and \$1,020,000, with respective probabilities of 0.2, 0.6, and 0.2. The expected value is obviously \$1 million and the standard deviation is

$$\sigma_C = \sqrt{0.2(980,000 - 1,000,000)^2 + 0.6(0)^2 + 0.2(1,020,000 - 1,000,000)^2}$$
$$= \sqrt{160,000,000} = \$12,649$$

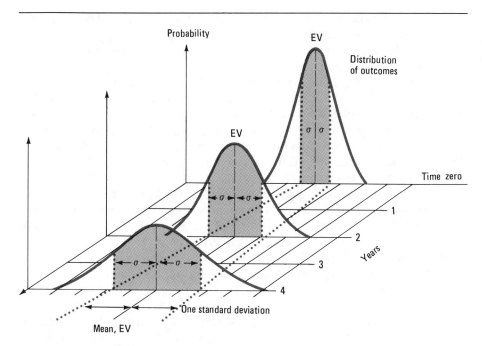

A direct comparison of σ_C with σ_A = \$63 and σ_B = \$379 might be taken as an indication that proposal C is riskier because the standard deviation is much larger. The erroneous impression is erased by calculating the *coefficient of variation*.

The customary way to handle this situation of scale is to divide the standard deviation by the mean expectation EV to obtain

$$\text{Coefficient of variation} = \frac{\sigma}{\text{EV}}$$

for proposal C,

$$\text{Coefficient of variation} = \frac{\$12,649}{\$1,000,000} = 0.0126$$

Comparing this value with the coefficient of variation for proposal A (\$63/\$1000 = 0.06) and proposal B (\$379/\$1000 = 0.379) makes it apparent that proposal C is subject to less outcome variability.

The notion of spreading risk among financial assets has prospered as a construct of *portfolio theory*.* The logical extension of this theory to capital budgeting

3 Asset-Deployment Policy

---

*H. Markowitz, "Portfolio Selection," *Journal of Finance,* March, 1952.

suggests that individual proposals for capital expenditures should be compared not only against each other but also with returns from previous investments. That is, the impact of an investment proposal on overall return earned from total assets should be considered.

It is generally desirable for a firm to avoid large swings between negative and positive cash flows. A steady flow averts operating disturbances caused by temporary financial woes and allows consistent funding for long-range programs. A proposal that generates income during the same periods when income generation is high from other assets tends to amplify cash flow surges. Conversely, a proposal that produces income while other assets drain income is one that steadies cash flow.

The correspondence between a proposal's cash flow and cash flow from other investments is measured by *correlation analysis,* as described in Chapter 18. A positive *coefficient of correlation r* of 1.0 means that a proposal's cash flow coincides with the total cash flow pattern. The reverse is true when $r = -1.0$, as shown in Figure 18.13. The degree of correlation is related to asset deployment by the following guidelines, assuming that a proposal does dominate total investment:

1 For a given coefficient of variation, a proposal with cash flow that correlates negatively with overall cash flow is preferred. Risk is reduced by smoothing fluctuations through diversification.
2 Uncorrelated proposals ($r = 0$) are preferred to those having cash flows positively correlated with total flow.
3 The more closely a proposal's cash flow moves with total cash flow, the less risk is reduced. A proposal with perfect positive correlation ($r = 1.0$) produces no diversification of cash flow and therefore contributes nothing to risk reduction.

Most proposals are positively correlated with a firm's total cash flow because they generally complement the firm's basic operations, products, and markets. The purpose of considering cash flow correlation is to give prominence to proposals which may reduce fluctuations, seasonal or cyclic. Formulas based on coefficients of variation and correlation have been developed to rank investments.* Unless diversification benefits are overwhelmingly important, an informal policy favoring proposals that promote more uniform total cash flow is usually sufficient.

Example 24.1
Risk and Cash Flow Correlation

A modest-sized firm is considering two investment proposals that would have significant effect on the firm's total cash flow. Reducing the variability of this flow lessens the chance of bankruptcy and lowers financing costs during adverse states of the economy. Evaluate the cash flow stability of the proposals described in Figure 24.4.

*C. W. Haley and L. D. Schall, *The Theory of Financial Decisions,* McGraw-Hill, New York, 1973.

State of the Economy	Proba-bility	Cash Flow Outcome	Firm's Other Cash Flow	Firm's Total Cash Flow
		PROPOSAL I		
Boom	0.2	$20	$200	$220
Normal	0.6	30	160	190
Bust	0.2	40	120	160
		PROPOSAL II		
Boom	0.2	40	200	240
Normal	0.6	30	160	190
Bust	0.2	20	120	140

FIGURE 24.4
Cash flows for individual proposals and for the total firm.

Solution 24.1

Both proposals have the same expected value and variance, but the cash flow of proposal I is negatively correlated with the firm's other cash flows. As a result, adopting proposal I would decrease the standard deviation of total cash flow to

$$\sigma_{\substack{\text{total} \\ \text{cash flow}}} = \sqrt{0.2(220 - 190)^2 + 0.6(190 - 190)^2 + 0.2(160 - 190)^2}$$
$$= \$18.97$$

If proposal II were accepted, the standard deviation would be

$$\sigma_{\substack{\text{total} \\ \text{cash flow}}} = \sqrt{0.2(240 - 190)^2 + 0.6(190 - 190)^2 + 0.2(140 - 190)^2}$$
$$= \$31.62$$

By adding the outcome from proposal I to the firm's other cash flow, the range of variation from boom to bust conditions is reduced from $200 - $120 = $80 to $220 - $160 = $60. Proposal I is plainly preferable.

VARIANCE OF COMPOSITE CASH FLOWS

Through the use of appropriate statistical relationships, different cash flow patterns can be accommodated in risk analysis. For instance, to determine the variance of a random variable which is the sum of other random variables, we must recall the original definition of variance and then expand the derivative process to cover the new situation. Thus if

$$X = A + B$$

then the variance of X is given by var $(X) = E(X^2) - E(X)^2$, which can be expanded to

$$\text{var }(X) = E[(A + B)^2] - E(A + B)^2$$

which then yields

$$\text{var}(X) = E(A^2) - E(A)^2 + E(B^2) - E(B)^2 + 2[E(AB) - E(A)E(B)]$$

Since $E(AB) - E(A)E(B)$ defines the covariance of the random variables A and B, it is evident that

$$\text{var } (X) = \text{var } (A) + \text{var } (B) + 2 \text{ cov } (AB)$$

If the random variables A and B are independent, then

$$\text{var } (X) = \text{var } (A + B) = \text{var } (A) + \text{var } (B)$$

A similar expansion for $X = A - B$ shows that for this case var (X) is also equal to var (A) + var (B) when A and B are independent random variables.

The equation for the variance of the product of two independent random variables can also be obtained through a similar series of operations. This expression stated in terms of the mean μ and variance σ^2 of the component distributions is

$$\sigma_{AB}{}^2 = \mu_A{}^2 \sigma_B{}^2 + \mu_B{}^2 \sigma_A{}^2 + \sigma_B{}^2 \sigma_A{}^2$$

There is no general formulation for the variance of a product of two random variables in which covariance exists between the variables.

Allowing the random variable A in the multiplicative expression to be a constant with $\sigma_A{}^2 = 0$ reveals that the variance of a random variable multiplied by a constant k is simply $k^2\sigma^2$. Thus

$$\text{var } (kB) = k^2\sigma_B{}^2$$

Example 24.2
Product of Two Random Variables

Items being inspected after several production operations are found to have either zero, one, two, or three appearance defects with respective probabilities of 0.70, 0.15, 0.10, and 0.05. Cost to repair each defect depends upon the degree of the flaw. Light paint work repairs which constitute 50 percent of the work cost only $5.00; moderate refinishing which costs $10.00 per operation accounts for 30 percent of the work; and major material repairs cost $30.00. Assuming that the severity of the defects is independent of their number, determine the expected value and the standard deviation of the total repair cost per unit.

Solution 24.2

The expected number of defects per unit is equal to

$$E(N) = 0.70(0) + 0.15(1) + 0.10(2) + 0.05(3)$$
$$= 0.50$$

with a variance of

$$\text{var } (N) = [0.70(0)^2 + 0.15(1)^2 + 0.10(2)^2$$
$$+ 0.05(3)^2] - 0.50^2$$
$$= 0.75$$

The expected cost of repair is calculated as

$$E(C) = 0.5 (\$5.00) + 0.3 (\$10.00) + 0.2 (\$20.00)$$
$$= \$9.50$$

with a variance of

$$\text{var}(N) = [0.5 (\$5.00)^2 + 0.3 (\$10.00)^2$$
$$+ (0.2) (\$20.00)^2] - (\$9.50)^2$$
$$= 32.25$$

Number of Repairs	0	1	2	3
Probability	0.70	0.15	0.10	0.05

FIGURE 24.5
Joint distribution of probabilities and costs for repair of production defects.

Cost	Probability				
$5.00		$0	$5.00	$10.00	$15.00
	0.50	0.350	0.075	0.050	0.025
$10.00		0	10.00	20.00	30.00
	0.30	0.210	0.045	0.030	0.015
$20.00		0	20.00	30.00	60.00
	0.20	0.140	0.030	0.020	0.010

The expected total repair cost per unit is therefore

$$E(TRC) = [E(N)] \cdot [E(C)] = \mu_N \cdot \mu_C$$
$$= (0.50 \text{ repairs/unit}) (\$9.50/\text{repair})$$
$$= \$4.75$$

Calculation of the variance of expected total repair costs may be represented by the data shown in Figure 24.5. Use of the tabular information makes it possible to verify the calculation of the expected value as

$$E(TRC) = \$[0.350(0) + 0.075(5) + 0.050(10)$$
$$+ 0.025(15) + 0.210(0) + 0.045(0)$$
$$+ 0.030(20) + 0.015(30) + 0.140(0)$$
$$+ 0.030(20) + 0.020(40) + 0.010(60)]$$
$$= \$4.75$$

and the variance as

$$var(TRC) = [0.350(0)^2 + 0.075(5)^2 + 0.050(10)^2$$
$$+ 0.025(15)^2 + 0.210(0)^2 + 0.045(10)^3$$
$$+ 0.030(20)^2 + 0.015(30)^2 + 0.140(0)^2$$
$$+ 0.020(40)^2 + 0.010(60)^2] - 4.75^2$$
$$= 99.9375$$

However, since both the number and the cost of repairs are independent random variables, the variance may be calculated through the use of the equation

$$var(AB) = \mu_A^2 \sigma_B^2 + \mu_B^2 \sigma_A^2 + \sigma_A^2 \sigma_B^2$$

Thus

$$var(TRC) = (0.50)^2 (32.25) + (9.50)^2 (0.75)$$
$$+ (0.75) (32.25)$$
$$= 99.9375$$

And the standard deviation equals 9.99.

Example 24.3
The Distribution of Differences

A small firm is analyzing a potentially significant methods change in one production area. The proposal calls for an equivalent annual expenditure of $112,000 with an annual savings of $130,000 expected to result from improved operational efficiency. However, there is a significant amount of variation in the estimates of both cost and expected savings. The standard deviation of the equivalent annual cost is only $8000; but owing to difficulties in predicting the performance of an untried system, the potential savings have an estimated standard deviation of $22,000. Assuming that both variables are normally distributed and independent, determine the probability that installation of the production change will actually result in a loss.

Solution 24.3

The distributions representing the estimated costs and savings resulting from the proposal are shown in Figure 24.6(a). The shaded area indicating overlap of costs and savings represents the area in which a loss could occur.

FIGURE 24.6
A graphic representation of the difference between two distributions and the distribution of the differences.

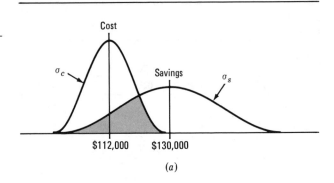

(a)

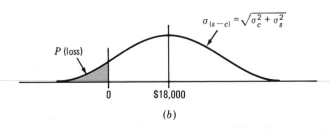

(b)

The normal distribution in Figure 24.6(b) indicates the distribution of differences between costs and savings. In this example, the difference would be the net savings resulting from implementation of the proposal.

$$E(\text{net savings}) = E(\text{savings}) - E(\text{cost})$$
$$= \$130,000 - \$112,000$$
$$= \$18,000$$

The variance of the net savings would be calculated as

$$\text{var(net savings)} = \text{var(savings)} + \text{var(costs)}$$
$$= (\$22,000)^2 + (\$8000)^2$$
$$= 548,000,000$$

and the standard deviation would be

$$\text{SD(net savings)} = \sqrt{\text{var(net savings)}} = \$23,409$$

The probability of the project implementation resulting in a loss is calculated as

$$P(\text{net savings} < 0) = P\left(Z < \frac{\$0 - \$18,000}{\$23,409}\right)$$
$$= P(Z < -0.769)$$
$$= 0.22$$

Thus there is a 22 percent chance that the investment will result in a net loss to the company.

PROBABILISTIC BREAKEVEN ANALYSIS

As long as the assumption of independence between random variables can be maintained, complex situations involving numerous variables can easily be analyzed. An extension of the basic breakeven problem presented in Chapter 15 provides an excellent example of introducing risk analysis into a previously deterministic solution.

Net profit from a typical production-sales operation may be calculated as

$$\text{Net profit} = \left[\begin{array}{c} \text{sales} \\ \text{volume} \end{array} \left(\begin{array}{c} \text{sales} \\ \text{price} \end{array} - \begin{array}{c} \text{variable} \\ \text{production} \\ \text{cost} \end{array} \right) \right] - \begin{array}{c} \text{fixed} \\ \text{cost} \end{array}$$

This expression may more realistically be stated as a series of expected values, thus

$$E(NP) = E(S)\,[E(P) - E(V)] - E(F)$$

If it is possible to obtain reasonable estimates of the variances for V, S, P, and F, then the variance of the expected net profit may be determined and the amount of risk associated with the proposal investigated.

The first step in such an analysis is the determination of the variance of the contribution $K = P - V$.

$$\begin{aligned} \text{var}(K) = \text{var}(P - V) &= \text{var}(P) + \text{var}(V) \\ &= \sigma_p^2 + \sigma_v^2 \end{aligned}$$

The variance of the sales volume times the contribution may be calculated as

$$\begin{aligned} \text{var}(SK) &= E(S)^2\,\text{var}(K) + E(K)^2\,\text{var}(S) + \text{var}(S)\,\text{var}(K) \\ &= \mu_s^2\sigma_k^2 + \mu_k^2\sigma_s^2 + \sigma_s^2\sigma_k^2 \end{aligned}$$

Substituting the original variance of k back into this expression yields

$$\text{var}[S(P - V)] = \mu_s^2(\sigma_p^2 + \sigma_v^2) + (\mu_p - \mu_v)^2\sigma_s^2 + (\sigma_p^2 + \sigma_v^2)\sigma_s^2$$

Since the fixed cost F is subtracted from $S(P - V)$, the variance of expected net profit NP can now be expressed as

$$\text{var}(NP) = \mu_s^2(\sigma_p^2 + \sigma_v^2) + (\mu_p - \mu_v)^2\sigma_s^2 + (\sigma_p^2 + \sigma_v^2)\sigma_s^2 + \sigma_f^2$$

RISK IN AGGREGATED CASH FLOW

Under conditions of certainty the present worth of a proposal is obtained routinely by discounting the cash flow stream to time zero. There are other occasions when the indefinite nature of receipts and disbursements must be acknowledged. If the mean and variance of cash flow in each period can be realistically estimated, the distribution of possible present-worth values can be determined.

The appropriate PW risk model depends on the relation of periodic cash flows. For instance, there may be a correlation between cash flows such that a high flow in year 1 leads to high flows in the following years; in this case there is a positive *covariance* of the two flows, since the flow in the second year tends to exceed its mean when the first-year flow exceeds its mean. Hillier* suggests that it is unreasonable to expect accurate estimates of covariance. He proposes models in which cash flows are considered to be independent of one another, perfectly correlated, or mixed (part are independent and part perfectly correlated). For practicality, the following discussion is limited to the case of independent cash flows for a single investment.

Starting from the familiar formula for present worth in which Xs are random variables representing cash flows,

$$\text{PW} = X_0 + \frac{X_1}{(1 + i)^1} + \cdots + \frac{X_N}{(1 + i)^N}$$

we take the expectation E from both sides of the equation to obtain

$$E(\text{PW}) = \mu_0 + \frac{\mu_1}{(1 + i)^1} + \cdots + \frac{\mu_N}{(1 + i)^N}$$

where μ_n are the means of the distributions of cash flow during n periods. This expression indicates that the expected present worth of a series of cash flows is the sum of the expected values of individual cash flows.

By invoking the *central limit theorem,* which credits a normal distribution to the sum of independently distributed random variables as the number of terms in the summation increases, we can consider PW normally distributed with a mean of $E(\text{PW})$ and a variance of $V(\text{PW})$. The variance is given by the formula

$$V(\text{PW}) = \sigma_0^2 + \frac{\sigma_1^2}{(1 + i)^2} + \cdots + \frac{\sigma_N^2}{(1 + i)^{2N}}$$

where σ_n^2 are the variances of the periodic cash flows. This formula is derived from relationships which state that:

1 The variance of a constant, $(1 + i)^{-N}$, times a random variable, μ_N, is the constant squared, $(1 + i)^{-2N}$, times the variance of the random variable, σ_N^2.
2 The variance of a sum, PW, is the sum of the variances, $V(\text{PW})$.

To illustrate the foregoing relationships, suppose that a company plans to invest \$2000 in a productive asset. Net returns are believed to be independent normally distributed random variables, each with a mean of \$1000 and a standard deviation of \$200. This belief implies probabilities of 0.68 and 0.95

*F. S. Hiller, "The Derivation of Probabilistic Information for the Evaluation of Risky Investments," *Management Science*, April, 1963.

that annual cash flow ranges, respectively, from \$800 to \$1200 ($\mu \pm \sigma$) and from \$600 to \$1400 ($\mu \pm 2\sigma$). If the cash flow stream continues for 3 years, MARR = 10 percent, and there is no salvage value, then

$$E(PW) = -\$2000 + \frac{\$1000}{(1 + 0.1)^1} + \frac{\$1000}{(1 + 0.1)^2} + \frac{\$1000}{(1 + 0.1)^3}$$

$$= -\$2000 + \$909.09 + \$826.45 + \$751.32 = \$487$$

which equals PW under the assumption of certainty when mean values are considered to be "known" cash flows:

$$PW = -\$2000 + \$1000(P/A, 10, 3) = \$487$$

Although the investment yields a positive PW, its acceptance is not assured when risk is considered. The degree of risk is measured by the variance:

$$V(PW) = 0 + \frac{(\$200)^2}{(1 + 0.1)^2} + \frac{(\$200)^2}{(1 + 0.1)^4} + \frac{\$200)^2}{(1 + 0.1)^6}$$

$$= 0 + \$33,058 + \$27,321 + \$22,580 = \$82,959$$

in which the first term, σ_0^2, is zero because the initial cash investment is assumed known with certainty. The probability distribution of PW is defined by the standard deviation,

$$\sigma_{PW} = \sqrt{V(PW)} = \sqrt{\$82,959} = \$288$$

and the assumption that PW is a normally distributed random variable. From Z values in Appendix F, the probability of having a loss from the investment is calculated as

$$P(PW < 0) = P\left(Z < \frac{0 - E(PW)}{\sqrt{V(PW)}}\right)$$

$$= P\left(Z < \frac{0 - \$487}{\$288}\right) = P(Z < -1.69)$$

$$= 0.046$$

Thus there is close to a 5 percent chance that the present worth of the investment will be negative.

Example 24.4
Variances That Increase with Time

Upon reconsidering the cash flow from the \$2000 investment examined above, we find it reasonable to expect the variances of future cash flows to be less sure. In recognition of the risk-time relation displayed in Figure 24.9, estimates of the cash flow deviations in years 2 and 3 are enlarged from \$200 to, respectively, \$300 and \$400. What effect does this admission of less certainty have on the probability that the investment's PW will be negative?

Solution 24.4

The expected value of the cash flow is unchanged at $487 when the minimum acceptable rate of return stays at 10 percent. However, the variance increases to

$$V(PW) = 0 + (\$200)^2(P/F, 10,2) + (\$300)^2(P/F, 10, 4)$$
$$+ (\$400)^2(P/F, 10, 6)$$
$$= 0 + \$33,058 + \$61,472 + \$90,317$$
$$= \$184,847$$

Based on a standard deviation of $\sqrt{V(PW)} = \$430$,

$$P(PW < 0) = P\left(Z < \frac{0 - \$487}{\$430}\right) = 0.129$$

indicates that the probability of a loss is almost tripled as a result of attributing larger variances to more distant cash flows.

SIMULATION

To simulate, in a general sense, means to feign or to assume the outward appearance of something without being the real thing. In an engineering economics sense, simulation is used to feign a real system in order to observe and learn from the behavior of the replica. This departure from reality has several advantages over that of observing the real system. It is usually easier, less disruptive, far less expensive, and possibly more illuminating to confine ones attention to characteristics of particular interest.

The earliest publications on simulation were by W. S. Gosset, alias "Student," in 1908.

The use of simulation techniques is increasing rapidly, owing to the wider availability of computers along with the greater sophistication of today's analysts. Computer simulation is an effective way to deal with complex economic relations without suffering the penalties of real trial-and-error experiences. Some problems do not yield conveniently to ready-made solution methods and models, such as inventory problems in which both lead times and demand rates are variable. In such cases, when no brief, handy models are available, problems can be "run" to see the numerical effect of different alternatives rather than solved by analysis. However, a run does not guarantee an optimum solution for the given data as does an optimization model.

Monte Carlo Simulation

Monte Carlo is the colorful name given to a simulation technique in which random numbers are generated to select events from a probability distribution of occurrences. The name is derived from possible random-number generators: a flipped coin, a tossed die, a cut of a deck of cards, or even a roulette wheel.

Many hand-held calculators have the capability to generate pseudorandom numbers.

However, the most-used generator is a random-number table, as displayed in Figure 24.7. The groups of numbers follow no pattern or special order; they are randomly distributed. The main concern a user should have is to avoid imposing a pattern by repeatedly using the same set in a consistent order. The figures can be read in any manner desired—by rows or columns, diagonally, up or down, etc.

32867	53017	22661	39610	03796	43693	18752
43111	28325	82319	65589	66048	04944	61691
38947	60207	70667	39843	60607	63948	49197
71684	74859	76501	93456	95714	87291	19436
15606	13543	09621	68301	69817	39143	64893
82244	67549	76491	09761	74484	91307	64222
55847	56155	42878	23708	97999	40131	52360
94095	95970	07826	25991	37584	56966	68623
11751	69469	25521	44097	07511	88976	30122
69902	08995	27821	11758	64989	61902	32131

FIGURE 24.7
Random numbers.

For an example of simulation, the pie charts in Figure 24.8 may be assumed to represent the relative frequency of lead times and usage rates for a commodity. The relative size of the slices in the "pies" corresponds to the chance occurrence of each increment of lead time or demand. Thus, possible lead times are 8, 9, 10, and 11 days, occurring with respective relative frequencies of 0.20, 0.20, 0.30, and 0.30.

The simulation of stock movement is conducted by randomly selecting both a lead time and a usage rate. This could be done by mounting spinners on the two pies. A spin on the lead-time pie might show 9 days, and the associated spin on the usage pie could point to 30 units. The two spins are combined to indicate one level of total demand that could occur after an order has been placed.

A random-number table would commonly be used in place of spinners. In a group of 10 numbers, any digit from 0 to 9 is equally likely to occur. By letting each digit be equal to 10 percent, we can assign digits to each increment

See Extension 24A for another example of a simulation application.

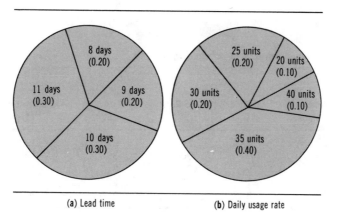

FIGURE 24.8
Relative frequency of lead times and usage rates.

(a) Lead time   (b) Daily usage rate

of lead time and usage according to its likelihood of occurrence. An arbitrary assignment might be

Lead Time		Usage Rate per Day	
Days	Digits	Units	Digits
8	0 and 1	20	0
9	2 and 3	25	1 and 2
10	4, 5, and 6	30	3 and 4
11	7, 8, and 9	35	5, 6, 7, and 8
		40	9

Each pair of numbers in the random-number table could then represent a lead-time duration and the number of units used per day during that duration. Using the first two numbers in the first column of Figure 24.7, we find that the inventory pattern would take the shape shown in Table 24.1. The reorder level is based on an average lead time of 10 days and an average usage rate of 30 units per day. A typical order is placed when the stock on hand falls to $10 \times 30 = 300$ units.

If the simulation procedure started in Table 24.1 is continued, a distribution can be determined for expected stock levels at the time of delivery. From this distribution an inventory policy can be developed which establishes a minimum-cost balance between holding and opportunity costs. The computations would be similar to those followed for a variable-lead-time inventory policy. A more direct tactic would be to set a tolerable limit for stockouts per year and hold a safety stock which conforms to this limit.

It is logical that the greater the number of trials, the more closely the simulation will correspond to the actual inventory pattern. Computers are almost always used for simulation because actual distributions usually cover a far greater range than do those used in the example, and a relatively large number of trials is required to give reliable information.

TABLE 24.1

Simulation of five reordering periods

Random Number	Lead Time (1)	Usage Rate (2)	Demand during Lead Time [(1) × (2)] (3)	Average Reorder Level (4)	Stock on Hand when Order Arrives [(4) − (3)] (5)
32	9	25	225	300	+75
43	10	30	300	300	0
38	9	35	315	300	−15
71	11	25	275	300	+25
15	8	35	280	300	+20

Cash Flow Simulation

The computational effort required to analyze risk is apparent from the section on cash flow aggregation. For more extensive applications computerized simulation is a sensible alternative. Sufficient trial outcomes can be obtained at a reasonable cost to reveal the approximate distribution of the returns. Since most of the distributions for cash flow components are subjective forecasts, the approximation by simulation forfeits negligible reality. Therefore, a decision to use cash flow simulation rests on the cost and time of direct analysis versus the availability and expense of computer simulation; the alternative of manual simulation is seldom feasible because of the exhausting tediousness of cumbersome iterations.

The general procedure for the simulation of an investment's return is shown in Figure 24.9. Inputs may be any of the cash flow factors for which distributions are obtainable. Random numbers are generated to establish quantitative values for all the variable inputs in each trial run. These simulated inputs are then combined with other known factors according to relationships written into the computer program. Each trial then produces one outcome for the characteristics of interest (PW, IRR, AW, or a factor such as asset life N). When enough outcomes have been accumulated, a pattern emerges that suggests the distribution and expected value.

A graphic representation of the sources of data for the statistical trials is given in Figure 24.10. The discrete distribution in Figure 24.10(a) is an asset's life; $P(N = 1) = 0.2$, $P(N = 2) = 0.3$, $P(N = 3) = 0.4$, and $P(N = 4) = 0.1$. A comparable chart for normally distributed random variables is shown in Figure 24.10(b); the mean of the distribution is zero, and deviations from the mean are expressed in standard deviations. A manual simulation application using these distributions is presented in Example 24.5.

General-purpose computer languages such as FORTRAN and PL/1 can be used to implement simulation. Several special simulation languages have been developed for greater programming efficiency.

It is important to validate any simulation model. Both the basic model and its internal logic, including program debugging, should be verified. Verification starts with a check of model assumptions against the purpose of the study and extends to checking the output against reality. Checking includes the sufficiency of the number of trials. An easy way to determine the stability of a solution is to plot the average outcome (vertical axis) against the number of trials; when the line connecting average outcomes is essentially horizontal, the range of error for the solution is narrow.

Specialized computer languages for simulation studies include DYNAMO, GPSS, GASP, GERT, SLAM, and SIMSCRIPT.

Example 24.5
Manual Monte Carlo

Construction at an exploration site has a mean cost of $2 million and a standard deviation of $500,000. Costs are assumed to be normally distributed. The project may last from 1 to 4 years with the distribution shown in Figure 24.10(a). If the before-tax rate of return is 20 percent and no salvage value is expected, what is the annual capital-recovery charge for the project?

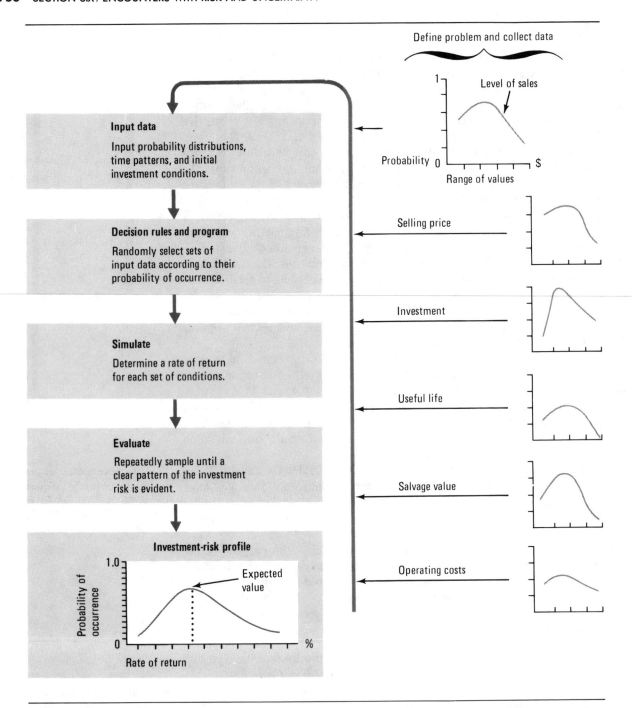

FIGURE 24.9
Flowchart for simulation procedures to evaluate an investment subject to risk.

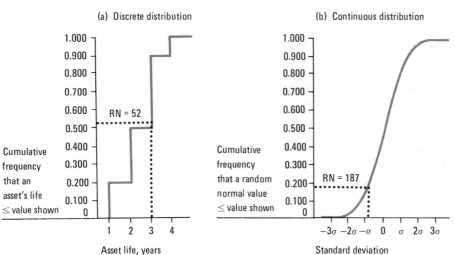

(a) Discrete distribution

(b) Continuous distribution

Asset life, years

Standard deviation

Solution 24.5

Assuming that the construction costs and life of the project are statistically independent, we can calculate the equivalent annual cost (EAC) as

$$\text{EAC} = \text{FC}(A/P, 20, N)$$

where first cost FC and life N are simulated in each trial. First cost is estimated by associating a random number with a normal deviate from Figure 24.10(b). For instance, a three-digit random number would correspond to each number between 0.000 and 0.999 on the vertical axis: A random number of 500 is associated with zero deviation and produces FC = $2,000,000; a random number of 834 matches 1σ, which means that FC = $2,000,000 + $500,000 = $2,500,000. By an equivalent procedure, random numbers of 500 and 834 lead, respectively, to lives of 2 and 3 years in Figure 24.10(a). The values so generated are substituted in the equation above to determine EAC for one trial.

Ten trials are shown in Table 24.2. Random numbers for FC are taken from the first three digits of the right-hand column of Figure 24.7. The last two digits in each random number in the same column are used to determine N. The top number in the column (18752) provides 187 as the entry shown in Figure 24.10(b), which leads to a standard deviation of -0.89. (A more exact reading can be taken from Appendix F, where 187 corresponds to the Z value for a proportion of $0.500 - 0.187 = 0.313$.) The last digits in the same random number are 52 and indicate a life of 3 years, as shown in Figure 24.10(a). The equivalent annual cost for the first trial is

$$\begin{aligned}\text{EAC}_{\text{trial 1}} &= [\$2,000,000 - 0.89(\$500,000)]\,(A/P, 20, 3) \\ &= \$1,555,000(0.47473) = \$738,205\end{aligned}$$

The estimate of EAC based on the very limited 10-trial sample is the average value of the outcomes,

$$\$10,368,622/10 = \$1,036,862$$

TABLE 24.2

Ten simulations of first cost and life to determine the equivalent annual cost of a project

	First Cost			Project Life			Equivalent Annual Cost = FC(A/P, 20, N)
Random Number	Normal Deviate ND	FC = $2,000,000 + ND($500,000)		Random Number	Life N	(A/P, 20, N)	
187	−0.89	$1,555,000		52	3	0.47473	$ 738,205
616	0.30	2,150,000		91	4	0.38629	830,524
491	−0.02	1,990,000		97	4	0.38629	768,717
194	−0.86	1,570,000		36	2	0.65455	1,027,644
648	0.38	2,190,000		93	4	0.38629	845,975
642	0.37	2,185,000		22	2	0.65455	1,430,192
523	0.06	2,030,000		60	3	0.47473	963,702
686	0.49	2,245,000		23	2	0.65455	1,469,465
301	−0.52	1,740,000		22	2	0.65455	1,138,917
321	−0.47	1,765,000		31	2	0.65455	1,155,281
Total							$10,368,622

Review Exercises and Discussions

Exercise 1   Given the anticipated outcome distributions for proposals M and W, analyze the risk.

	Probability of Outcomes				
	$O_1 = -11$	$O_2 = -3$	$O_3 = 5$	$O_4 = 13$	$O_5 = 21$
Proposal M	0.2	0.2	0.2	0.2	0.2
Proposal W	0.0	0.4	0.3	0.2	0.1

Solution 1

$$E(M) = 0.2(-11) + 0.2(-3) + 0.2(5) + 0.2(13) + 0.2(21)$$
$$= 5$$

$$E(W) = 0.4(-3) + 0.3(5) + 0.2(13) + 0.1(21)$$
$$= 5$$

$$\sigma_M = \sqrt{0.2(-11 - 5)^2 + 0.2(-3 - 5)^2 + 0.2(5 - 5)^2 + 0.2(13 - 5)^2 + 0.2(21 - 5)^2}$$
$$= \sqrt{128} = 11.31$$

$$\sigma_W = \sqrt{0.4(-3 - 5)^2 + 0.3(5 - 5)^2 + 0.2(13 - 5)^2 + 0.1(21 - 5)^2}$$
$$= \sqrt{64} = 8$$

Both proposals have the same expected value, but proposal W is preferred because it has lower risk, owing to less variability of outcomes.

Exercise 2

Year N	3	4	5	6
Probability P(N)	0.1	0.4	0.3	0.2

A proposed investment of $9000 will produce annual revenue of $4000. The investment risk is how long the revenue stream will continue. The given estimate of probabilities for the investment's duration appears reasonable.

Regardless of N, there will be no salvage value, and a before-tax rate of return of 25 percent is required.

a   What are the expected present worth of the investment and its variance?

b   Determine the probability of a positive present worth.

a   The expected value of the investment is the sum of the four possible cash flow outcomes, weighted according to their probabilities of occurrence: **Solution 2**

$$PW(N = 3) = -\$9000 + \$4000(P/A, 25, 3) = -\$1192$$
$$PW(N = 4) = -\$9000 + \$4000(P/A, 25, 4) = \$446$$
$$PW(N = 5) = -\$9000 + \$4000(P/A, 25, 5) = \$1757$$
$$PW(N = 6) = -\$9000 + \$4000(P/A, 25, 6) = \$2806$$
$$E(PW) = -\$1192(0.1) + \$446(0.4) + \$1757(0.3) + \$2806(0.2)$$
$$= \$1148$$

The variance is calculated from the relationship

$$V(PW) = E(PW^2) - [E(PW)]^2$$

for which the $E(PW)$ has already been determined. Then

$$E(PW)^2 = (-\$1192)^2(0.1) + (\$446)^2(0.4) + (\$1757)^2(0.3) + (\$2806)^2(0.2)$$
$$= \$2,722,495$$

and

$$V(PW) = \$2,272,495 - (\$1148)^2 = \$1,404,591$$

b   The probability of a positive present worth, based on the discrete distribution of project duration, is the sum of the probabilities of outcomes that have a positive value. From part a, where PWs were calculated for each N,

$$P(PW > 0) = P(N = 4) + P(N = 5) + P(N = 6)$$
$$= 0.4 + 0.3 + 0.2 = 0.9$$

The most likely outcome for an investment is $1000. The most optimistic and pessimistic   **Exercise 3**
estimates are, respectively, $600 and $2000. What are the expected value and variance of the proposal if outcomes are represented by the beta distribution?

Using the formulas from Chapter 20, where the symbols for estimates are a = optimistic,   **Solution 3**
m = most likely, and b = pessimistic, we estimate that the approximate values for the mean EV and variance σ^2 are

$$EV = \frac{a + 4m + b}{6} = \frac{\$600 + 4(\$1000) + \$2000}{6} = \$1100$$

$$\sigma^2 = \left(\frac{b - a}{6}\right)^2 = \left(\frac{\$2000 - \$600}{6}\right)^2 = 54,443$$

Exercise 4 Analyze the proposals below and select the most promising one based on coefficients of variation and correlation.

Proposal	Coefficient of Variation	Correlation with Other Cash Flows
P1	0.8	0.4
P2	0.4	0.2
P3	0.3	−0.2
P4	0.3	0.0
P5	0.5	0.2

Solution 4 P1 is the riskiest proposal because it has the highest correlation with the firm's other cash flows *and* the highest coefficient of variation. P3 offers the least risk because of its negative correlation and low variation. The order of preference from least to most risk is P3, P4, P2, P5, and P1.

Exercise 5 A new cost-reduction proposal is expected to have annual expenses of $20,000 with a standard deviation of $3000, and it will likely save $24,000 per year with a standard deviation of $4000. The proposed operation will be in effect for 3 years, and a rate of return of 20 percent before taxes is required. Determine the probability that implementation of the proposal will actually result in an overall loss and the probability that the PW of net savings will exceed $10,000.

Solution 5 The expected value of the present worth of savings and costs is

$$E(\text{PW}) = (\$24{,}000 - \$20{,}000)(P/A, 20, 3)$$
$$= \$4000(2.1064)$$
$$= \$8426$$

The variance is calculated from the relation

$$\sigma^2_{\text{savings} - \text{costs}} = \sigma^2_{\text{savings}} + \sigma^2_{\text{costs}}$$

to obtain

$$V(\text{PW}) = (\$3000)^2(P/F, 20, 2) + (\$3000)^2(P/F, 20, 4) + (\$3000)^2(P/F, 20, 6)$$
$$+ (\$4000)^2(P/F, 20, 2) + (\$4000)^2(P/F, 20, 4) + (\$4000)^2(P/F, 20, 6)$$
$$= \$25{,}000{,}000(0.69445 + 0.48225 + 0.3349)$$
$$= \$37{,}790{,}000$$

from which

$$\sigma = \sqrt{V(\text{PW})} = \$6147$$

Assuming that the PW is normally distributed, we find that

$$P(\text{loss}) = P\left(Z < \frac{0 - 8426}{6147}\right) = P(Z < -1.37)$$
$$= 0.0853$$

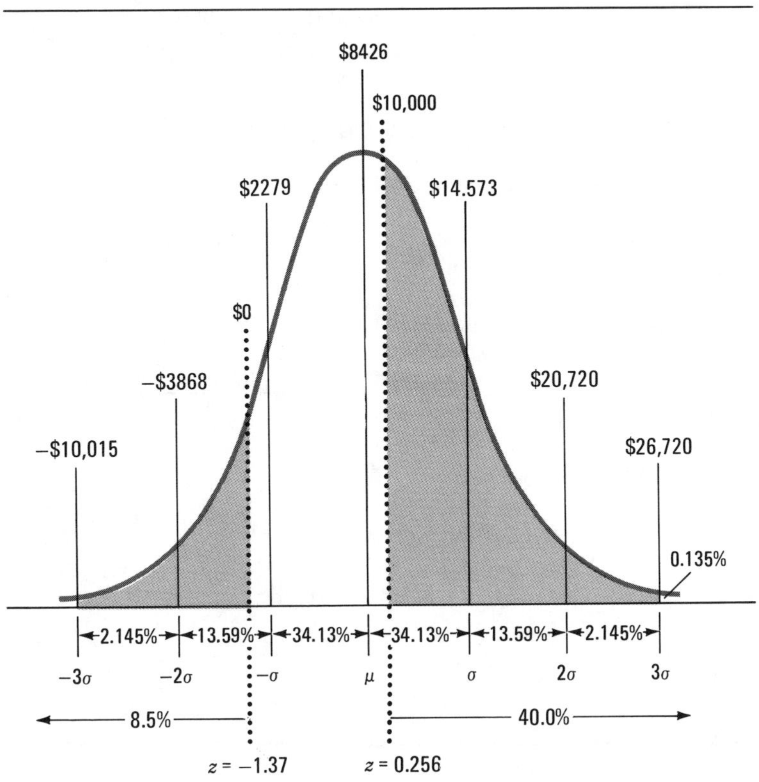

FIGURE 24.11
Normal distribution of present worth (for Review Exercise 5) with a mean of $8426 and a standard deviation of $6147.

and

$$P(\text{PW} > \$10,000) = P\left(Z > \frac{10,000 - 8426}{6147}\right) = P(Z > 0.256)$$
$$= 0.40$$

These probabilities are represented graphically by the shaded areas under the normal distribution curve shown in Figure 24.11

Exercise 6

A new product is expected to be marketable for 5 years. The first cost of necessary production equipment is estimated at $40,000 with a standard deviation of $2000; it will have no salvage value when the product is discontinued. The contribution per unit produced, exclusive of capital-recovery costs, has a predicted mean of $3 and standard deviation of $1.50. The before-tax MARR is 30 percent.

a What is the annual breakeven volume?

b Determine the mean and standard deviation of the present worth at a production level of 10,000 units per year.

Solution 6

a The expected values of first cost FC and contribution C are used to determine the breakeven point B as

$$B = \frac{FC}{PW_C} = \frac{\$40,000}{\$3(P/A, 30, 5)} = \frac{\$40,000}{\$3(2.4355)} = 5475 \text{ units}$$

b $E(PW) = -\$40,000 + \$3(10,000)(P/A, 30, 5) = \$33,069$

The previously applied formula for calculating the variance,

$$V(PW) = \sigma_0^2 + \sigma_n^2 \sum_{n=1}^{N} (1 + i)^{-2n}$$

is made computationally easier when cash flow is uniform by utilizing the relation

$$\sum_{n=1}^{N} (1 + i)^{-2n} = \frac{(P/A, i, 2N)}{2 + i}$$

to obtain

$$V(PW) = (2000)^2 + \frac{(1.50 \times 10,000)^2(P/A, 30, 10)}{2 + 0.30}$$

$$= 4 \times 10^6 + \frac{(225 \times 10^6)(3.0915)}{2.3}$$

$$= 306.43 \times 10^6$$

and $\qquad\qquad \sigma_{PW} = \sqrt{306.43 \times 10^6} = \$17,505$

PROBLEMS

24.1   An asset has a first cost of $50,000 and a salvage value which is dependent upon how long it remains in service. For service periods of 4, 5, 6, and 7 years, the respective estimated salvage values are $20,000, $15,000, $12,000, and $10,000. Given that all the service periods are equally likely, determine the mean and standard deviation of the asset's present worth using an interest rate of 15 percent.

24.2   An asset is expected to produce a net annual operating profit of either $18,000, $24,000, or $30,000 per year during the 5 years it remains in service. Assuming that the profit amounts are independent from year to year and that each value has an equally likely probability of occurrence, determine the mean and standard deviation of the net present worth using an interest rate of 15 percent.

24.3   Determine the expected value and the standard deviation of the PW for the situation described in Problem 24.2 if the annual profits are uniformly distributed in a continuous distribution between $18,000 and $30,000.

24.4   The six proposals described in the table on page 763 are under consideration for funding.

   (a) Compute the coefficients of variation and rank the proposals accordingly.
   (b) Rank the investments according to their correlation with other cash flow within the firm.

Investment Proposal	Expected Value	Standard Deviation	Correlation with the Firm's Other Cash Flow
A	$ 700,000	$400,000	0.4
B	200,000	0	0.0
C	1,000,000	400,000	0.6
D	600,000	100,000	−0.3
E	400,000	100,000	0.1
F	500,000	250,000	−0.2

(c) Comparing the lists developed in Problems 24.4a and b, indicate how you would rank the proposals in terms of risk. Where there is no clear choice, indicate the reasoning that you would use to make your selection.

24.5 Given the following estimates of a project's cash flows, in which the flows are assumed to be independent of each other, determine the probability that the present worth will be positive and hence desirable at an interest rate of 20 percent.

End of Year	Expected Value of Cash Flow	Standard Deviation
0	− $20,000	0
1	8,000	$1500
2	8,000	2000
3	8,000	2500
4	8,000	3000

24.6 An investment proposal requires an immediate payment of $50,000 and is expected to provide an annual return of $18,000 for each of the next 6 years. However, the positive cash flow which is independent from year to year has a standard deviation of $3000.

(a) Determine the expected value and the variance of the PW for this investment using an interest rate of 20 percent.

(b) Determine the probability that the investment will result in a rate of return below a MARR of 20 percent if the annual cash flows are expected to be normally distributed.

24.7 Given the cash flows shown in the table below, determine the probability that the investment will provide a positive net present worth if interest is charged at 15 percent.

End of Year	Expected Value	Standard Deviation
0	− $30,000	0
1	10,000	$1,000
2	9,000	1,200
3	8,000	1,400
4	7,000	1,600
5	6,000	1,800

24.8 An equipment supplier states that there is a 90 percent probability that annual maintenance costs on a new machine will be between $4800 and $6000 over a 5-year service life. Using a before-tax interest rate of 20 percent, determine the expected value and the standard deviation of the present worth of these costs.
(a) Assuming that the costs are normally distributed.
(b) Assuming that the costs are uniformly distributed.

24.9 Estimates for the construction of a new dock and breakwater indicate a need for 100,000 cubic yards of fill material. The average total cost for placing the material has been estimated at $2.50 per cubic yard with a standard deviation of $0.20. Given that the estimators are 90 percent confident that the amount of fill material required will be between 85,000 and 115,000 cubic yards, determine the mean and variance of the total cost of the filling operation.

24.10 A project is estimated to require an investment of $25,000 and have an annual net cash flow of $16,000 with zero salvage value. The life is estimated to be 1 year, 5 years, or 10 years with respective probabilities of 0.1, 0.5, and 0.4. If the minimum acceptable rate of return is 15 percent, what are the expected value and variance of the net annual worth?

24.11 Project Perplexity is estimated to require an investment of $25,000 and to have no salvage value. Expected revenue outcomes at the end of each potential duration are shown below. The MARR is 15 percent.

Life	Probability	Probability		
		0.3	0.5	0.2
3 years	0.25	$30,000	$35,000	$40,000
5 years	0.40	50,000	50,000	50,000
8 years	0.35	75,000	85,000	100,000

(a) Plot a histogram showing the probabilities of different present worths.
(b) Calculate $E(PW)$.
(c) Calculate $V(PW)$.
(d) What is the probability that the present worth will be positive?

24.12 A proposed acquisition of material-handling equipment will likely have annual operating and capital-recovery costs of $12,000 with a standard deviation of $2000. Estimated gross savings from the use of this equipment are $15,000 per year with a standard deviation of $4000. Assuming that the figures are given in discounted constant dollars, determine the probability that in actuality a loss will result from the installation of the equipment.

24.13 An initial investment of $120 in a productive asset results in annual receipts of $50 until production is terminated. The before-tax required rate of return is 20 percent.
(a) If the salvage value is zero at all times and the probabilities of receipts continuing for 3, 4, and 5 years are, respectively, 0.5, 0.4, and 0.1, what is the expected present worth?
(b) What is the probability of a loss for the conditions given in Problem 24.13a?

(c) If the salvage value has an equal chance of being -30 percent or $+30$ percent of the first cost in any year and the distribution of asset life is unchanged from Problem 24.13a, what is the expected present worth?

(d) What is the probability of a negative present worth for the conditions in Problem 24.13c?

(e) If the life of the asset is certain to be 4 years but the annual receipts have a normal distribution with a standard deviation of $20, what is the expected present worth when the salvage value is still zero?

(f) What is the probability of a loss for the conditions in Problem 24.13e?

(g) If the initial investment may vary with a standard deviation of $40, what is the probability of a loss when $N = 4$, $A = \$50$, and $S = 0$ are known with certainty?

24.14 Annual sales of a new product are expected to average 10,000 units with a standard deviation of 1500 units. The sales price of the product has been set by market conditions at $29.95 and total variable manufacturing costs have been estimated at $23.00 with a standard deviation of $1.10. Fixed manufacturing costs are expected to average $20,000 with a standard deviation of $2000.

(a) Determine the mean and the variance of the net annual profit for this item.

(b) Research and development costs for the product totaled $80,000. Assuming that it will remain on the market for 5 years, determine the mean and the standard deviation of the net present worth using a before-tax interest rate of 30 percent.

24.15 The local school board is considering the installation of a large trash compacting system to reduce the number of trips necessary to pick up refuse at the high school. The compactor will be purchased on a bid basis, but it is expected to have a first cost of $65,000 with a possible standard deviation of $3000. The equipment is expected to last 6 years and no significant salvage value is projected. Operating expenses have been estimated at $12,000 per year with a 90 percent probability that they will be between $10,000 and $14,000. The gross savings should average $35,000 annually, but this estimate is thought to have a standard deviation of approximately $5000.

(a) Determine the mean and the variance of the net present worth of this proposal using an interest rate of 20 percent.

(b) Determine the probability that a net loss results from the installation of the compactor.

24.16 Bureaucratic Resources Incorporated (BRI) is considering the feasibility of bidding on a contract to haul excess red tape from the state capitol to a nearby power plant where it will be burned. The contract will run for 6 years and will require that BRI purchase a new truck. Initial cost of the vehicle is expected to be $60,000 with a standard deviation of $2000 and the salvage value is estimated at $10,000 with a standard deviation of $3000. Indirect costs of the operation are expected to be uniformly distributed between $8000 and $14,000 per year and direct costs are estimated at $3.00 per ton with a standard deviation of $0.20. BRI expects the legislature to annually produce 5000 tons of excess red tape with an estimated variance of 2500 tons.

(a) Using a before-tax interest rate of 20 percent, determine the mean and variance of the present worth of this project.

(b) Determine the sales price per ton that BRI must obtain for the red tape if the company is to be 90 percent confident that the trucking project will make a profit.

Quarts	Probability
0	0.4
1	0.3
2	0.2
3	0.1

24.17 Harry's Hotshot Service Station sells both gasoline and diesel fuel. Customers arrive at a Poisson rate of 6.0 per hour for gasoline and 2.4 per hour for diesel. On the average, gasoline customers purchase 15 gallons with a standard deviation of 4 gallons and diesel customers purchase 18 gallons with a standard deviation of 5 gallons. Customers also occasionally purchase one or more quarts of oil. This demand fits the distribution as shown in the table. Gasoline currently sells for $1.20 per gallon, diesel for $1.05 per gallon and oil for $1.50 per quart at Harry's. Develop the appropriate Monte Carlo simulation model and simulate 1 hour of sales at the station.

24.18 A machinery supply and service company advertises that orders received by 8:00 A.M. will be delivered that day or the customer will have to pay only half price for the order. The amount of an average order is $372. The number of orders received before the deadline varies according to the following pattern:

Orders per day	16	17	18	19	20	21	22	23
Probability	0.05	0.10	0.10	0.20	0.25	0.15	0.10	0.05

Each order is delivered by van. The fixed cost of a van is $30 for a normal 8-hour working day. Van variable costs depend on the length of time required to deliver an order. The daily average order-time distribution and the associated variable costs are as shown:

Average Hours/Order during 1 Day	Variable Cost	Probability
½	$ 3.50	0.40
1	7.00	0.30
1½	10.50	0.20
2	14.00	0.10

Use simulation to determine the approximate number of vans that the company should operate.

24.19 A new type of automatic assembly equipment costs $100,000 and may be in use for either 4, 6, or 8 years, depending upon the development of new markets for the product line being serviced and technological changes in the industry. The salvage value of the equipment will be either $40,000, $20,000, or $10,000, depending on how long the unit is retained in service. Maintenance costs are expected to be $25,000 for the first year and $30,000 for the second year, and they will continue to increase at a rate of $5000 per year for each year of use. Annual cost reductions resulting from the installation of the equipment are unknown at this time but are expected to be closely approximated by a normally distributed random variable with a mean of $35,000 and a standard deviation of $10,000. If the probabilities with the three potential service periods are equal, determine by Monte Carlo simulation the distribution of a present worth resulting from the installation of this equipment. Use a before-tax rate of return of 20 percent.

24.20   The estimated first cost of a piece of equipment is $100,000. However, there is some doubt about this figure, because several bids are being received. The estimated standard deviation of the bids is $10,000 and their dispersion is assumed to be normally distributed.

Annual expenses associated with the equipment are also normally distributed as follows:

Maintenance. $\mu_M = \$5000$ and $\sigma_M = \$700$

Number of breakdowns per year. $\mu_B = 10$ and $\sigma_B = 2$

Cost per breakdown. $\mu_C = \$900$ and $\sigma_C = \$200$

In addition, direct labor expenses are expected to be uniformly distributed between $4000 and $7000 per year.

If $N = 5$ and MARR $= 10$ percent (disregarding taxes), determine by Monte Carlo simulation the approximate annual income required to make this a feasible purchase.

EXTENSION

Extension 24A   Simulation of Waiting Lines

Simulation methods are appropriate means of evaluating the effect of waiting time. This general approach can be used regardless of the distribution of arrivals or service times. The techniques involved are the same as those described for inventory simulation. In a waiting-time application we have the objective of determining a service policy which will minimize the total cost of idleness within the system.

Machine breakdowns in a factory could be considered as arrivals to a repair (servicing) facility. Breakdowns are randomly distributed, with an average arrival rate of 40 minutes. The cost of an idle machine awaiting repair is estimated at $24 per hour. A mechanic hired at a wage of $10 per hour can repair machines according to the distribution of service times shown in Table 24.3. The service rate for two mechanics working together is also shown. Should one or two mechanics be employed?

One way of simulating random arrivals with random numbers is to select one digit to represent a breakdown. This digit will average one appearance in a group of 10 random numbers. Then, for this example, each group of 10 digits represents 40 minutes of factory time. The number of times the selected digit appears in each group of 10 numbers indicates how many breakdowns occurred in that 40-minute

TABLE 24.3

Distribution of service times

One Mechanic at $10/Hour		Two Mechanics at $20/Hour	
Time, Minutes	Probability	Time, Minutes	Probability
10	0.10	5	0.10
20	0.10	10	0.10
30	0.30	15	0.20
40	0.30	20	0.20
50	0.20	30	0.30
		35	0.10
34 minutes average		21 minutes average	

period. The number of breakdowns will vary from zero in a period to the remotely possible maximum of 10.

A more definitive arrival simulation could be used if more information were available. A detailed probability distribution of periods between breakdowns would specify the breakdown pattern more precisely. In this example we are assuming that at most 10 breakdowns could possibly

TABLE 24.4

Waiting-time simulation for one mechanic

Work Period, Minutes	Random Numbers	Arrival Number	Breakdown Time	Random Number	Service Period	Service Begins	Service Ends	Queue (Machine Waiting Time)
0–40	3961003796	1	20	1	20	20	40	0
		2	40	8	50	40	90	0
40–80	6658966048	3	80	7	40	90	130	90 − 80 = 10
80–120	3984360607	4	100	5	40	130	170	130 − 100 = 30
		5	120	2	30	170	200	170 − 120 = 50
120–160	9345695714	...	...	...	...	...	...	...
160–200	6830169817	6	200	6	40	200	240	0

occur in a 40-minute period and we must arbitrarily set the pattern within the period for multiple breakdowns. We will assume that:

If two occur, one is at the beginning of minute 21 and the other is at the end of the period.

If three occur, one will be at the start of the period, one at the end, and one at the beginning of minute 21.

If more occur, they will follow a similar periodic arrangement.

Service-time simulation is accomplished by assigning digits in proportion to the frequency of each repair time. For the distribution of service times with one mechanic, we could let digit 0 = 10 minutes, 1 = 20 minutes, 2, 3, and 4 = 30 minutes, 5, 6, and 7 = 40 minutes, and 8 and 9 = 50 minutes.

With digit 0 = an arrival, a table of waiting times for a single mechanic could appear as shown in Table 24.4. The random numbers are from Figure 24.7, columns 4 and 5.

The tabulations in Table 24.4 are only a beginning. Many more samples would be required before legitimate conclusions could be drawn, but the same procedures would be followed. For each arrival (breakdown) a service time is simulated. The difference between the time a breakdown occurs and the time a service mechanic can begin repairs is the waiting time per machine. The total waiting time divided by the number of machines repaired is the average waiting time per machine. In an 8-hour day (480 minutes) the average number of breakdowns is

$$\frac{480 \text{ minutes/day}}{40 \text{ minutes/machine}} = 12 \text{ machines/day}$$

and the total daily cost of waiting time plus servicing is calculated from

Repair costs per day = number of mechanics × daily wage
+ average number of breakdowns per day
× average waiting time (hours) per breakdown
× hourly cost of idle machines

For the simulated data from Table 24.4, this amounts to

$$\text{Repair costs} = 1(8 \times \$10) + 12\frac{10 + 30 + 50}{60 \times 6}\$24$$

$$= \$80 + \$72$$

$$= \$152$$

The same simulation procedure is followed for each alternative waiting-time policy. For the example situation we would next determine the total repair costs with a crew of two. Repair costs for each alternative are then compared to decide which policy is most suitable.

QUESTIONS

24A.1 Simulate a waiting-time situation by using two dice to represent an arrival distribution. The total number of points on a toss of the two dice represents the time between arrivals. Let a single die represent the service times. Each time an arrival occurs, throw a die to determine how long that arrival spends in the service station.

    (*a*) What is the average arrival rate?

    (*b*) What is the average service time?

    (*c*) What is the average length of the queue?

    (*d*) What is the average waiting time in the queue?

    (*e*) Compare the answers to *c* and *d* above with the answers you get by using a Poisson arrival distribution and an exponential service rate. Comment on your results.

(*f*) Reverse the distribution for arrival and service rates described in the problem statement. How many service stations are required to limit the average waiting time to that determined in *d*?

CHAPTER 25

DECISION MAKING UNDER CONDITIONS OF UNCERTAINTY

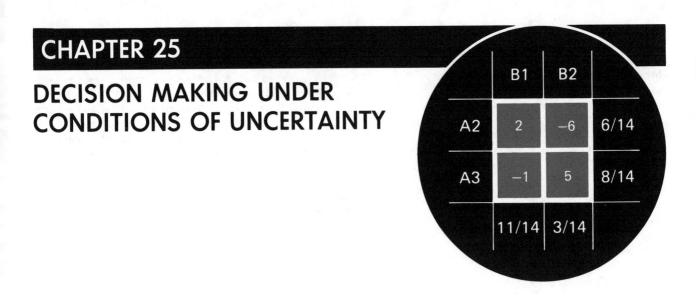

OVERVIEW

By definition a theory is a plausible or scientifically acceptable body of principles offered to explain phenomena. Two theories and a set of decision criteria are presented in this chapter. All are plausible and all contribute to an explanation of decision-making behavior in evaluating investment alternatives. Although they suggest quantitative frameworks for selection between opposing courses of action, none are noted for their practical application. Nonetheless, their acknowledged furtherance of comprehension for decisions afflicted with uncertainties makes them valuable to engineering economists.

Several *principles of choice* have been developed to systematize choices made in a neutral decision environment. Once the outcomes of alternatives are arranged in a payoff matrix, attention to the best or worst outcomes leads to the *maximin* or *maximax* criterion. An intermediate outlook is defined by a coefficient of optimism in the *Hurwicz* criterion. Opportunity costs assigned to less-favored alternatives guide decisions according to the *minimax-regret* criterion. The *equal-likelihood* criterion registers the belief that each state must have an equal chance to occur if there is insufficient reason to rate one more likely than another. Each criterion has certain unattractive properties, but each is still useful in directing attention to the attitudes and objectives of the decision makers.

Utility theory provides a structure for personalizing the value of money in terms of risk—its utility to the decision maker. When investors struggle to avoid risk (or, conversely, relish gambling), the attractiveness of a risky investment to them is not adequately measured by the expected monetary value of possible returns. For these cases, a more rational decision should result from comparing the *expected utility*, rather than expected monetary value, of the alternatives. It is supposedly possible to construct a *utility function*, or reference contract, that specifies on a scale of 0.0 to 1.0 the relative worth that a given individual places on gains and losses over a given range. The resulting numbers from the utility scale replace the dollar value of outcomes in a payoff matrix. The alternative with the highest expected utility is preferred.

Game theory is the basis for *competitive* decisions under uncertainty. In a *zero-sum*, *two-person* game, two opponents have equal knowledge of outcomes, and the winnings of one are the losses of the other. An optimal *pure* or *mixed* strategy, selecting a single course of action or randomly mixing actions within given proportions, can be calculated to ensure a minimum return, the value of the game. Strategies for multiplayer and nonzero-sum games are complicated to formulate, but future developments in these areas could increase the practicality and acceptance of game-theory analyses for competitive situations.

PERSPECTIVES FOR DECISION MAKING

This concluding chapter of *Engineering Economics* explores mental sets for decision making. It has been suggested in several preceding chapters that subjective assessments of the economic environment vary among individuals and result in contrary decisions based on identical data. Differences are caused by the way in which the data are perceived and personal perspectives are shaped by past experiences and current stresses.

Everyone knows how the appeal of a given type of food varies as a function of hunger and abstinence. The value placed on food corresponds to its appeal, and appeal depends on a variable appetite. Although this value-depends-on-want relation is credible, it is difficult to quantify. Not only are there inconsistencies between individuals' value-want scales, the scale for each individual vacillates over time. However, an appreciation of the transitional nature of economic values provides insights into the mechanics of decision making, which is the rationale for examining *utility theory*.

Still another perspective emerges when a decision is studied from a competitor's viewpoint. In our previous analyses we assumed a *neutral* environment characterized by future states in which the likelihood of outcomes is not influenced by the decision-maker's choice. A *competitive* environment presupposes intelligent opponents capable of exerting influence over our outcomes through their choice of action, while concurrently we choose a course of action that maximizes our returns with respect to the opponents' anticipated activities. This is the subject of *game theory*.

Yet another way to appreciate the complexities of decision making under conditions of uncertainty is to examine the *principles of choice* that a decision maker can employ to select an alternative that best suits his or her attitude about the situation. This is the subject of *decision criteria*.

All three perspectives postulate explanations for the behavior of decision makers. They explore the effects of risk taking and risk aversion, adventurism and conservative tactics, possibilities of collusion or mutual destruction, and differences in the perceived value of returns. But they do *not* provide definitive decision rules. Very few actual applications of the theories have been cited, yet they remain influential because they provide a logical structure for reasoning that helps us to better understand our own mental processes and the actions of others.

COMPARISON OF DECISIONS UNDER CONDITIONS OF CERTAINTY, RISK, AND UNCERTAINTY

When outcomes are presumed to be known, a decision is said to be made under a condition of certainty. There is always some doubt, of course, that a given cash flow will materialize, but an assumption of certainty narrows consideration of the future to a single state and one set of data for that state. The data set may include numbers to represent intangible factors.

"Uncertainty and expectation are the joys of life." (W. Congreve, *Love for Love*, 1695.)

Decisions under risk and uncertainty are initially treated in the same way. Possible future states are identified, and outcomes are estimated for each alternative by assuming that each state will indeed occur. The difference between risk and uncertainty depends on the capability of estimating the likelihood of future states. When state probabilities can legitimately be prescribed, it is a decision under risk and the expected-value criterion is appropriate. When probabilities are unattainable, other criteria must be applied to select the preferred alternative. Similarities and differences among decisions under certainty, risk, and uncertainty are illustrated by their associated matrices in Figure 25.1.

Payoff Matrix

As indicated in matrix (C) of Figure 25.1, no probabilities are available for states of nature $N1$, $N2$, . . . , Nn for the decision condition of *uncertainty*.

FIGURE 25.1
Decision matrices for conditions of certainty, risk, and uncertainty.

FUTURE STATES

STATE OF NATURE

	$P(1.0)$
A	O_i
B	O_i

(A) *CERTAINTY.* SOLUTION: PREFERRED OUTCOME

	$P(0.7)$	$P(0.3)$
A	O_{ij}	O_{ij}
B	O_{ij}	O_{ij}

(B) *RISK.* SOLUTION: HIGHEST EXPECTED VALUE

	$N1$	$N2$
A	O_{ij}	O_{ij}
B	O_{ij}	O_{ij}

(C) *UNCERTAINTY.* SOLUTION: PREFERRED STRATEGY

Each state is assumed to be independent of other states and immune to manipulation by the decision maker; *the decision environment is neutral.* Because there are no probabilities associated with the states of nature—only payoffs for each option in each state —selection of an alternative depends on the evaluation strategy employed by the decision maker.

It is important to identify all the states pertinent to the decision situation. Omitting states of nature is equivalent to neglecting states of risk, but it is easier to detect an omission under risk, because the associated probabilities would sum to less than 1. Sometimes related states of nature are lumped together because their effect on alternatives is the same. The check for using composite states comes when estimating outcomes O_{ij}. If one payoff for each alternative accurately measures the effect of every condition implied by the composite state, the combination is valid.

Dominance

After the payoff matrix has been developed, the next step is to check for dominance. The problem objective is to select the best-paying alternative. If one alternative produces a greater payoff than another alternative for *every* state of nature, then a rational decision maker would never select the lower-paying course of action. The higher-paying alternative is said to *dominate* the lower one. In Figure 25.2, alternative A dominates alternative C. Consequently, the dominated alternative C may be dropped from the matrix. An early check for dominance avoids unnecessary calculations by reducing the size of the problem matrix.

When *all* payoffs for one alternative are better than those of a second, a condition called *strict dominance* exists. Moreover, dominance can still exist when some payoffs are *equal to* and all others are *greater than* corresponding payoffs of another alternative.

CRITERIA FOR DECISIONS UNDER UNCERTAINTY

The most difficult aspect of noncompetitive problems under uncertainty is to decide what type of criteria to use for making a decision. In essence, we must determine the criteria for the criterion. The choice should be consistent with management philosophy. Is the current management outlook optimistic or pessimistic, conservative or adventurous? Certain criteria are compatible only with certain management views. Thus, it is necessary to understand both

STATE OF NATURE

	N1	N2	N3	N4
A	10	1	3	6
B	6	2	5	3
C	7	0	2	4

FIGURE 25.2
Dominance relationship.

management policy and the principles of choice before selecting a decision criterion.

Minimum-Maximum Criterion

A conservative approach to a decision is to look at the worst possible outcome for each alternative and select the course of action which ensures the best results for the worst conditions. The underlying viewpoint is that nature is malicious. If things can go wrong, they will. This pessimistic philosophy dictates that attention be focused only on the most damaging outcomes in order to limit the damage as much as possible.

The words "minimax" and "maximin" are derived from the measures taken to identify the limiting loss or the guaranteed gain. A *minimax* decision minimizes the maximum loss. The *maximin* principle is associated with positive payoffs, where it maximizes the minimum gain or profit. For either criterion the smallest payoff (or greatest loss) for each alternative is noted. Then the alternative having the most favorable of the collected worst payoffs is selected.

Maximax Criterion

A maximax philosophy is one of optimism and adventure. Nature is considered to be benevolent, so greatest gains are highlighted. The principle of choice is to identify the maximum gain possible for each alternative and then choose the course of action with the greatest maximum gain.

Hurwicz Criterion

A moderate outlook between the extremes of optimism and pessimism is allowed by the Hurwicz criterion. The degree of optimism is established by a coefficient called alpha (α), which may take any value between 0 and 1.0, with the following interpretation:

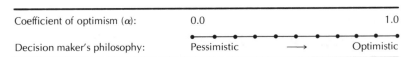

Coefficient of optimism (α):	0.0	1.0
Decision maker's philosophy:	Pessimistic $\longrightarrow$	Optimistic

After deciding the value of α which measures the decision maker's degree of optimism, maximum and minimum gains are identified for each alternative. Then the maximum payoffs are multiplied by α and the minimum payoffs by $1 - \alpha$. The two products for each alternative are added and the alternative with the largest sum is chosen.

The minimax-maximin and maximax criteria are special cases of the Hurwicz criterion. When $\alpha = 1$, only the maximum payoffs are included in the final alternative selection, because the minimum payoffs have been eliminated by zero multiplication. The opposite is true for $\alpha = 0$, a completely pessimistic outlook. Any value of α other than 1 or 0 is a compromise opinion about the hostility or benevolence of nature.

Example 25.1
Different Degrees of Optimism

Two sons and their mother own and operate an import shop in a medium-sized city. They have been successful enough to be in a position to expand their operations. Three courses of action are deemed most desirable: (1) expand their present operations by opening a store in a nearby city; (2) start a catalog business from their present location; or (3) invest their extra money in real estate and rentals. Each alternative will utilize about the same amount of capital and will require equivalent management. They recognize that the returns from each of these investments depend on the national economy (prosperity versus recession) and on the local economy (growth versus stagnation). However, they have no consensus of the probabilities of future conditions.

Outcomes for each of the alternatives have been developed for four possible levels of business activity: very high (VH), high (H), medium (M), and low (L). The payoffs shown are the estimated percentage returns on invested capital for expanding (E), starting a catalog service (C), and investing in property (P):

| | State of Nature | | | |
	VH	H	M	L
E	20	12	8	4
C	26	10	4	−4
P	10	8	7	5

The youngest son is an optimist and a risk taker; the mother is conservative; and the other son is midway between his mother and brother. Which alternative would probably appeal to each member of the family?

Solution 25.1

The youngest son, being an optimist, would use the maximax criterion. By doing so he would limit his selection area to the VH level of business activity, where the largest gains occur. From the possible gains of 20, 26, and 10 percent, he would naturally select the highest, which results from opening a catalog store.

The conservative mother would lean toward the maximin criterion, where she could be assured of a minimum gain of 5 percent (the maximum gain under the worst condition of low business activity) by choosing to invest in real estate and rentals.

The other son might use the Hurwicz criterion with a coefficient of optimism α of 0.5. The consequent calculations

reveal that his choice would be to expand the import business to the nearby town.

Alternative	Max $O_i \times \alpha$ + min $O_i (1 - \alpha)$ = total
E	$20 \times 0.5 + 4(1 - 0.5)$ = 12 ←Maximum
C	$26 \times 0.5 + (-4)(1 - 0.5)$ = 11
P	$10 \times 0.5 + 5(1 - 0.5)$ = 7.5

The relation of the three philosophies can be better interpreted by plotting the maximum and minimum gains from each as a function of α. As shown in Figure 25.3, the topmost lines indicate the alternative that would be selected for different

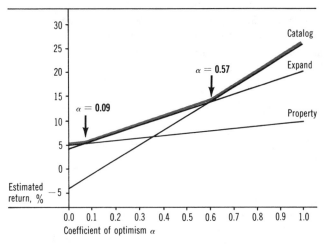

FIGURE 25.3
Sensitivity of alternatives to the decision maker's degree of optimism.

levels of optimism. At $\alpha = 0$ the maximin criterion is in effect, and property appears to be the most attractive investment. Property continues to be favored until the less pessimistic attitude of $\alpha = 0.09$ is attained. The next switch point occurs at the intersection of the lines representing "expand" and "catalog." Using the equation for the return expected from these two alternatives, we calculate the value of α at which a decision maker is indifferent to the choice between the two:

$$20\alpha + 4 \times (1 - \alpha) = 26\alpha + (-4)(1 - \alpha)$$
$$16\alpha + 4 = 30\alpha - 4$$
$$\alpha = 0.571$$

Mapping the alternatives gives an indication of their sensitivity with respect to the degree of optimism of the decision maker. The property alternative is quite sensitive because its selection requires a very pessimistic attitude. The remaining alternatives are relatively insensitive because each would be chosen over a considerable range of α values. The apparent range of attitudes which favor each alternative could aid the mother and sons in a search for a compromise solution.

Minimax-Regret Criterion

Opportunity costs have been used in previous chapters to express the loss incurred by not selecting the best alternative. The minimax-regret criterion is based on similar costs. The opportunity costs are determined for each state of nature by subtracting the largest payoff in each column from all other payoffs in the column. The absolute value of each subtraction is the amount of "regret" that results from not selecting the best alternative for the occurrence of a given state. This procedure converts the original matrix to a regret matrix.

A rational decision maker attempts to minimize regret. By applying the minimax principle, we select the alternative with the minimum maximum (lowest value of the worst regret for each row). The minimax-regret procedure applied to the data from Example 25.1 is shown in Figure 25.4. The indicated preference is for the "expand" alternative. In general, the minimax-regret criterion tends toward a conservative viewpoint.

Equal-Likelihood Criterion

When it is possible to assign probabilities to future states, we use the expected-value criterion to select a preferred alternative. An extension of this approach is the basis of the *equal-likelihood* criterion. Under uncertainty we admit that we cannot reasonably estimate outcome probabilities. Therefore, since we have no excuse to believe otherwise, why not treat each outcome as the same? The rationale behind this theory is that there is insufficient reason to believe one state of nature more probable than another, so each should be assigned an equal probability of occurrence.

ORIGINAL PAYOFF MATRIX

	VH	H	M	L
E	20	12	8	4
C	26	10	4	-4
P	10	8	7	5

REGRET MATRIX

	VH	H	M	L	Worst Regret
E	6	0	0	1	6
C	0	2	4	9	9
P	16	4	1	0	16

FIGURE 25.4 Original and regret matrix.

The equal-likelihood criterion is certainly the simplest to apply. Under the assumption that each future is equally likely to occur, the expected value of an alternative becomes its average outcome. The alternative with the largest average payoff is preferred. As applied to the data from Example 25.1, we have

$$E(E) = 20 \times 1/4 + 12 \times 1/4 + 8 \times 1/4 + 4 \times 1/4 = 11$$
$$E(C) = 26 \times 1/4 + 10 \times 1/4 + 4 \times 1/4 - 4 \times 1/4 = 9$$
$$E(P) = 10 \times 1/4 + 8 \times 1/4 + 7 \times 1/4 + 5 \times 1/4 = 7.5$$

which again lead to the decision to expand.

EVALUATION OF DECISION CRITERIA

Several different criteria for making noncompetitive decisions under uncertainty have been offered because no one criterion is unanimously preferred. Each one has certain weaknesses. Often one criterion is more intuitively appealing than are the others. This appeal seems to vary among individuals and to vary with time or circumstances. Inasmuch as there is no universal preference, a sound recourse is to investigate the criteria limitations in order to select the principle which accommodates a given decision environment.

Partial-Optimist Principle

The key factor in applying the Hurwicz criterion is the choice of a value for α. An arbitrary choice defeats the intent of portraying an individual outlook. A deliberate choice is a forced judgment often based on slim evidence. However, the judgment does allow a measure of added knowledge, even though it is undefined, to be included in the decision. This additional knowledge could be described as "a feel for the problem" or an "educated guess." Until methods are developed to determine α objectively, its value will remain uniquely individual.

Pure optimism and pessimism are special cases of the Hurwicz criterion. Since these are extreme outlooks, they often leave a decision maker uncomfortable in applying them to special situations. For instance, the maximin criterion applied to

	N1	N2	Minimum
A	$0.01	$0.01	$0.01 ← Maximin
B	0	$100	0

would indicate that alternative A should be selected, yet most people would be inclined to choose B. In the same vein, the maximax criterion applied to the next matrix would lead to the choice of B; most decision makers would express a decided preference for alternative A.

	N1	N2	Maximum
A	$99	$99	$99
B	0	$100	$100 ←Maximax

By selecting a value for α other than 1 or 0, a more moderate outlook is achieved. However, the criterion still possesses some dissatisfying aspects. According to the criterion, the two alternatives in the following matrix are considered equivalent, because attention is given only to the best and worst outcomes:

	N1	N2	N3	N4	N5	N6
A	$100	$100	$100	$100	$100	0
B	0	0	0	0	0	$100

Many people faced with this choice would cast a strong vote in favor of A over B.

There is also another difficulty. Some critics object to a decision criterion that changes preferences when a constant is added to all the outcomes in one column. Such a criterion is said to lack the property of "column linearity." To illustrate, the maximin criterion would indicate a preference for alternative B when applied to Figure 25.5(a). The outcomes in both matrices are the same, with the exception of 100 added to both alternatives under N1 in Figure 25.5(b). Such an exception could be caused by a discovery that a bonus payoff would result from the occurrence of state N1 regardless of which alternative was selected. The disturbing feature is that although the bonus has the same effect on both alternatives, it changes the preference from B to A.

Opportunity-Loss Principle   The minimax-regret criterion is plagued by many of the same defects as is the Hurwicz criterion. Attention is focused on only the largest opportunity costs, with a resultant disregard of other payoffs.

A further argument against the minimax-regret criterion is that the addition of irrelevant information to the matrix can switch the alternative preferences.

FIGURE 25.5
Column linearity.

	N1	N2
A	0	100
B	50	50

(a)

	N1	N2
A	100	100
B	150	50

(b)

From the payoff matrix

	N1	N2	N3
A	100	125	25
B	25	125	75

the resulting regret matrix is

	N1	N2	N3	Worst Regret
A	0	0	50	50 ← Minimax
B	75	0	0	75

which leads to the choice of alternative A. Now an additional, rather unattractive, course of action is included in the matrix as

	N1	N2	N3
A	100	125	25
B	25	125	75
C	25	25	125

The addition, alternative C, changes the regret matrix to

	N1	N2	N3	Worst regret
A	0	0	100	100
B	75	0	50	75 ←Minimax
C	75	100	0	100

which indicates that alternative B rather than A should be selected. Although the switch caused by irrelevant information may seem disturbing, it can also be argued that the additional alternative is not necessarily "irrelevant," because it can reveal more information about the states of nature, such as the change in the maximum possible payoff for $N3$.

Average-Outcome Principle

Most of the doubts raised for the other criteria are not applicable to the equal-likelihood criterion. It has column linearity and includes all the outcomes in an evaluation, and the addition of an unattractive alternative cannot switch an earlier preference. However, it also has one serious defect: The selection is very sensitive to the number of future states identified for a problem.

Consider the problem faced by a company that is bidding on a contract. They can build special low-operating-cost equipment which will be ready if they get the contract, or they can buy higher-operating-cost equipment after they know that they have won the bidding. The payoffs, in thousands of dollars, are shown in the accompanying matrix.

	No Contract	Win Contract	Average Outcome
Build	−50	125	37.5
Buy	0	100	50.0 ←Maximum

When the future states are limited to winning or losing the one contract, the equal-likelihood criterion shows a preference for buying the equipment. If other future states, such as the possibilities of winning two more similar contracts, are included, the same criterion indicates a preference for building the equipment.

	No Contract	Win Contract 1	Win Contract 2	Win Contract 3	Average Outcome
Build	−50	125	125	125	81.25 ←Maximum
Buy	0	100	100	100	75.0

Beginning with an exhaustive list of futures will eliminate the chance of a preference change. If one state is subdivided into substates such that the sum of the probabilities for the substates is equal to the probability of the original state, the problem is also avoided. However, each of these methods presumes some knowledge of the future that contradicts the assumption of uncertainty upon which the criterion is based.

Application of Principles None of the criteria is perfect. None can take the place of an accurate forecast. They should be considered guidelines which will help in the interpretation and consideration of possible choices.

It should be noted that most of the reservations about the criteria were more intuitive than deductive. Perhaps the adoption of a certain criterion must also rely to some degree upon intuition, because such insight is often a function of knowledge not yet formalized into distinct views. Nevertheless, the decision maker must understand the characteristics of each principle to be able to select the one which corresponds most closely to the uncertainties of a situation.

Example 25.2
One Set of Data with Many Interpretations

That different people make different decisions based on the same data is an accepted, albeit confusing, fact. An explanation of this phenomenon is inferred from applying the given decision criteria to a single payoff matrix.

	N1	N2	N3	N4
A	2	2	2	2
B	1	5	1	0
C	1	4	1	1
D	1	3	1	4
E	3	4	3	0

For the data shown, which alternative would be selected under each of the described decision criteria? Assume that the coefficient of optimism is 0.375 ($\alpha = 3/8$) for the Hurwicz criterion.

Solution 25.2

The procedures followed for each criterion reduce the original matrix to a single column of outcomes from which the desired value is selected. The results and preferences indicated by applying all the criteria are as shown.

It should not be too surprising that each criterion indicates a different preferred alternative. Each criterion has a slightly distinctive underlying principle. The key is to decide which criterion best fits the decision environment for each specific application.

Alternative	Maximin (pessimist)	Maximax (optimist)	Hurwicz ($\alpha = 3/8$)	Minimax Regret	Equal Likelihood
A	**2**	2	2	3	2.0
B	0	**5**	15/8	4	1.75
C	1	4	**17/8**	3	1.75
D	1	4	17/8	**2**	2.25
E	0	4	12/8	4	**2.5**

UTILITY THEORY

The value attached to an outcome for any alternative and a given future can vary among individuals or from one business enterprise to another. Response to risk is a very personal matter. It permeates our daily activities without causing particular awareness. Most pedestrians crossing a street probably know that their chances of being hit by an automobile are significantly greater than zero. Some apparently choose to ignore or consider insignificant the probability of an accident as they blithely jaywalk through a busy intersection. Others, by virtue of training or awareness of the value of safety, wait for the proper time to cross. A minority, because of an extreme sensitivity to the possibility of an accident, wait until no cars at all are in sight before attempting to cross. With such divergent reactions to danger, it is logical that each of the three groups would place a different value on an insurance policy for pedestrians. The *utility* of the policy would be unique to each.

Not only does the utility vary among parties, it also varies with time. In the pedestrian example, either the jaywalker or the normally cautious street crosser might be converted to the hypercautious category after being an eyewitness to a serious accident. Putting the situation in monetary terms, we would anticipate that the witness would put a higher value on the insurance policy after viewing the accident. If such eyewitnesses could buy protection, they would probably be willing to pay more for it after the accident, even though the probability that an accident will occur to them has not changed.

An individual or corporation places a value on an alternative which achieves a balance between the potential return and the probability of receiving those returns. Under common conditions a gamble may be attractive to one party and a conservative approach desirable to another. The positions could later be reversed. Each party rates the anticipated returns of an alternative in relation to its chance of successful culmination in a manner that reflects the current economic status of each.

Concepts

The idea of assigning a number to human feelings, such as pain and pleasure, has been around a long time. Blaise Pascal, the seventeenth century French mathematician, and Jeremy Bentham, the eighteenth century English philosopher, explored mathematical proofs for the value of being virtuous. Alfred Marshall expounded the concept of *cardinal* utility in which psychic properties are believed to be measurable and quantifiable; a person is thus assumed to get *x* units of satisfaction from a hedonistic activity, and each additional unit of that activity provides successively fewer units of satisfaction.* A theory of *ordinal* utility was later proposed that assumed that consumers could rank the desirability of goods without measuring the exact utility of each good.†

The most celebrated concept of utility theory is owed to John von Neumann and Oskar Morgenstern.‡ Their dealings with decision making under conditions of uncertainty led to a numerical method of evaluating the riskiness of outcomes.

Mechanics

A von Neumann–Morgenstern utility index is the certainty equivalent of a risky transaction. It is obtained by measuring the relative worth of monetary returns to the decision maker by varying the probabilities of getting those returns. Depending on the attitude toward risk, a utility rating is assigned to a certain-to-be-received sum that is considered equivalent to a gamble at given probabilities of a certain gain or loss. Under plausible conditions an individual is assumed to have a utility function with the following key properties:

Principles of Economics, London 1920.
†R. Hicks and R. G. D. Allen, "A Reconsideration of the Theory of Value," *Econometrica*, February 1934.
‡*Theory of Games and Economic Behavior,* Princeton University Press, Princeton, NJ, 1947.

1   If outcome A is preferred to outcome B, the utility of A is greater than is the utility of B, and the converse is true.
2   If an individual has a contract that carries a payoff of A with a probability of p and a payoff of B with a probability of $1 - p$, the utility of the contract is the expected value of the utilities of the payoffs.

The first statement confirms that utility values move in the same direction as payoffs, rising and falling as outcomes become more or less favorable. The second statement allows the utility of a contract, $u(C)$, to be calculated from the utilities of the payoffs, $u(A)$ and $u(B)$, and their respective probabilities, p and $1 - p$:

$$u(C) = u(A)p + u(B)(1 - p)$$

This relation underlies the *utility scaling method* of assigning numbers to intangibles, as discussed in Chapter 17. It is also the basis for *reference contracts* as an approach to the development of a utility scale, and *expected utility*, as a measure of merit for evaluating proposals.

Example 25.3
Reaction to Risk

Suppose that each of two university students was offered a proposition of putting up $100 for a double-or-nothing return on the throw of a single die. Each would win $100 if a 3, 4, 5, or 6 showed or would lose the ante if a 1 or a 2 turned up. Assume that both students have the same gambling instincts but one is working for school fees and the other is the recent recipient of a large inheritance. The alternatives for acceptance or rejection of the proposition are expressed in the payoff table:

	Win (2/3)	Lose (1/3)
Accept	$100	− $100
Reject	0	0

The two students would undoubtedly consider the proposition with different feelings. The lucky student with ample resources would probably consider it a "good" bet. This student might question the reason behind the odds or the fairness of the die but if the probabilities appeared true would likely accept the proposition because of the adventure aspects and because the consequence of a $100 loss would not be disastrous. On the other hand, the working student would quite possibly reject the proposition because the loss of $100 would be a painful experience. The working student would appreciate the gain, but even with favorable odds, the potential reverse in fortunes would outweigh the opportunity of winning. This student might accept the proposition if the odds of winning were upped to 10 to 1 or if the original odds could be applied to a smaller sum, such as $10. The reaction to risk is a function of the degree of that risk relative to the importance of the gain or loss.

REFERENCE CONTRACT

The concept of a reference contract is a means of taking into account the unique value that individual parties place on different alternatives. To develop this reference contract, we must determine the amount of money a party would

FIGURE 25.6
Payoff table for three alternatives.

	S1 (0.5)	S2 (0.3)	S3 (0.2)
C	$80	0	−$40
T	$100	$40	−$100
N	0	0	0

demand or be willing to pay to be relieved of the obligation stated in a proposition. The fact that individuals are often inconsistent in stating their preferences for specified combinations of risk and the associated consequences necessitates a measure of caution in the analysis.

With respect to the two students described in Example 25.3, consider their reactions to the opportunity of investing the money instead in campus services. The three alternatives are to invest in a food catering service C, a tutoring service T, or nothing N. Figure 25.6 shows the returns to be expected. The probability of success S in the ventures is based on enrollment in the university and the general state of the economy. We assume that the stated probabilities and the expected rewards are realistic. The outcomes are the profits from an investment of $100 over a year's period.

Let us look first at the preferences of the wealthy student, whom we will call Mr. Loaded. At the time of the proposition, he is inclined to be a risk taker. He feels that he will not be hurt by the consequences of a poor investment of this magnitude but will receive disproportionately high satisfaction from a success.

"To win without risk is to triumph without glory." (Corneille, *The Cid*, 1636.)

By asking him to relate the value he attaches to wins or losses associated with various probabilities, we hope to establish the cash equivalence of his preferences. We will consider a range of values from a gain of $100 to a loss of $100, although we could use any scale. The starting question could be, "What amount of cash would you be willing to accept in lieu of a contract that ensures you (probability of 1.0) a gain of $100?" Any sensible person would ask for at least $100, even though hoping to receive more for it. Next we could ask what amount he would take in place of a contract that gave him an 80 percent chance of winning $100 and a 20 percent chance of losing $100. Since he is a risk taker, he would probably ask for at least $80. This means that he places an $80 value on the opportunity to win $100 at a probability of 0.8 with the associated 20 percent chance of losing $100. We could say that he is indifferent to the two alternatives. By continuing this procedure through a selected range of discrete probabilities, we would obtain his reference contract, or *utility function*, as depicted in Figure 25.7.

The preferences for the working student, denoted as Ms. Broke in the table, could be determined by the same procedure. Because of her tight financial condition, we would expect her to try to avoid risk. Her cautious nature is evident in her response to the offer of an 80 percent chance of a $100 gain versus a 20 percent chance of a $100 loss. She would be willing to accept $40 in place of the alternative. That is, she is indifferent to a sure gain of $40 in

| P($100 loss) | P($100 gain) | REFERENCE CONTRACT | |
		Mr. Loaded	Ms. Broke
0.0	1.0	$100	$100
0.2	0.8	80	40
0.4	0.6	40	0
0.6	0.4	0	−40
0.8	0.2	−40	−70
1.0	0.0	−100	−100

FIGURE 25.7
Reference contracts for a risk taker and a risk avoider.

place of winning $100 with a probability of 0.8 or losing $100 with a probability of 0.2. The table continues to reflect her reluctance to incur a large debt. She would pay $40 to be relieved of a proposition that gave her only a 40 percent chance of winning and a 60 percent chance of losing $100.

EXPECTED UTILITY

Now that we have a utility index for each of the two students, we can personalize the original contract to indicate individual preferences. In the first statement of the payoffs the expected values of the alternatives were

$$EV(C) = \$80 \times 0.5 + 0 \times 0.3 + (-\$40) \times 0.2 = \$32$$

$$EV(T) = \$100 \times 0.5 + \$40 \times 0.3 + (-\$100) \times 0.2 = \$42$$

$$EV(N) = 0 + 0 + 0 = \$0$$

An immediate assumption is that a reasonable person would always choose the tutoring investment. However, by expressing the outcome in terms of the utility of each investment as expressed in the reference contract, we can see how conclusions would vary according to individual preferences. This is done by substituting the equivalent probability, or utility rating, of each return for the monetary outcome. Thus, the dollar payoff table is translated to a utility payoff table.

Mr. Loaded, for instance, is indifferent to a sure reward of $80 and the probability of 0.8 of a $100 gain. We can therefore substitute 0.8 for the $80 in his payoff table. This procedure is repeated for all the outcomes for each student, as is shown in Figure 25.8 along with the original payoff table.

Then the expectation for the utility EU of each alternative is

Mr. Loaded

$$EU(C) = 0.5 \times 0.8 + 0.3 \times 0.4 + 0.2 \times 0.2 = 0.56$$
$$EU(T) = 0.5 \times 1.0 + 0.3 \times 0.6 + 0.2 \times 0.0 = 0.68$$
$$EU(N) = 0.40$$

FIGURE 25.8
Dollar payoff table with comparable utility payoff tables.

	S1 (0.5)	S2 (0.3)	S3 (0.2)
C	$80	0	−$40
T	$100	$40	−$100
N	0	0	0

(a) ORIGINAL PAYOFF

	S1	S2	S3
C	0.8	0.4	0.2
T	1.0	0.6	0.0
N	0.4	0.4	0.4

(b) MR. LOADED'S UTILITY PAYOFF

	S1	S2	S3
C	0.96	0.6	0.4
T	1.0	0.8	0.0
N	0.6	0.6	0.6

(c) MS. BROKE'S UTILITY PAYOFF

Ms. Broke

$$EU(C) = 0.5 \times 0.96 + 0.3 \times 0.6 + 0.2 \times 0.4 = 0.74$$
$$EU(T) = 0.5 \times 1.0 + 0.3 \times 0.8 + 0.2 \times 0.0 = 0.74$$
$$EU(N) = 0.60$$

If the goal is to maximize the expected utility, Mr. Loaded would reasonably select an investment in the tutoring service, but Ms. Broke would be highly uncertain. Both investments appear to be equal if we assume that Ms. Broke has been consistent in her stated preferences and that neither investment is too much larger than the utility she places on doing nothing. She should obviously investigate the investment potentials further and assess the value of any other influencing factors.

Example 25.4
Using Utility Functions to Assist Decision Making

A small company is faced with the prospect that one of its products will shortly be outmoded by new developments achieved by a competitor. It would be possible to launch a crash program to develop a new version of the product that would be equal to or better than the competition. However, there is a better than even chance that neither the company's own new version nor the competitor's will receive significantly greater acceptance than will the current model. A suggested compromise between doing nothing and developing a new model is to retain the present model but increase the sales budget. From a market survey on the likelihood of acceptance of the new product and the plant engineers' estimates of anticipated developmental costs, the following payoff table was obtained:

	New Model		
Alternative	Acceptance (0.4)	Rejection (0.6)	Expected Value
Develop new model (A)	$300,000	−$200,000	$ 0
Increase sales budget (B)	−50,000	100,000	40,000
Retain present model (C)	−100,000	0	−40,000

Regardless of the reaction to risk, it is apparent that alternative C can be eliminated, because B is a more attractive choice in both possible futures. If the new product fails, B earns a reward of $100,000, compared with no gain for C; and if the new version succeeds, the expected loss from B will be $50,000 less than that from C. We can say that B *dominates* C. An alternative that is dominated by another can be eliminated from the decision process.

The next step is to consider the utility function of the company. It is a small but ambitious firm. It has made rapid growth and its products are well diversified. A loss of $200,000 would certainly be serious, but other successful products could carry the company without too much hardship. The management feels that it needs a major accomplishment to build its reputation. A significant success would improve prestige and enlarge the horizons of the whole organization. The company has a proven engineering staff and is willing to gamble. The accompanying utility index describes its current attitude.

Utility	Dollars
0.00	− 400,000
0.40	− 200,000
0.58	− 100,000
0.65	− 50,000
0.71	0

Utility	Dollars
0.82	100,000
0.90	200,000
0.95	300,000
0.98	400,000

Judging only from the described attitude, we might guess that the company would decide to choose alternative A, even though the expected dollar value is zero. This is confirmed by converting the payoffs to utilities and calculating the expected value.

	New Model		
Alternative	P(A) (0.4)	P(R) (0.6)	Expected Value (Utility)
A	0.65	0.06	0.296
B	0.16	0.30	0.244
C	0.12	0.20	0.168

We can achieve additional insights about utility functions by plotting the utilities of the money values involved in the decision. The shape of the curve in Figure 25.9 typifies a risk taker; it shows an increasing *marginal utility* for money. This means that as the rewards become higher, more value is placed on each additional dollar.

The alternatives can be shown by lines connecting the points on the utility curve which represent the outcomes. The dotted line to each chord shows the expected dollar value of each alternative. For A, the expected dollar value is zero and the expected utility is about 0.3. Alternative C is shown in a similar manner and can be seen to have an expected loss of $40,000, with the corresponding utility rating of 0.17. The dollar expectation of B is greatest, but its utility is exceeded by that for A.

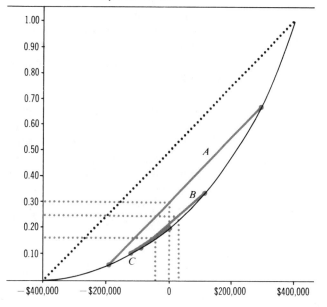

FIGURE 25.9
Utility function.

If the utility function for the company had been linear over its entire range, it would have appeared as the straight dotted line in the figure. Such a function shows the utility in proportion to the monetary value. In this case the expected value calculated from the dollar-payoff table would directly indicate the utility of each alternative.

PRACTICAL CONSIDERATIONS IN UTILITY THEORY

It is easier to manipulate utilities than it is to obtain them. Determining an individual's utility index is often a tedious and time-consuming task, but it can be even more difficult to obtain one that represents an organization's policy. While asking the series of questions necessary to establish a utility scale, side issues arising from the wording of the questions may cause distractions that show up as inconsistent responses. There is always the possibility that different responses would have been obtained on another day because of a change in mood or the temporary outlook of the person being questioned.

It should also be apparent that a utility function for a particular set of alternatives is not necessarily valid for another set of alternatives. Many intangible considerations fringe the choice of any specific rating. A manager might indicate a utility function which clearly shows conservative attitudes toward industrial actions, but he or she might have an entirely different set of attitudes for gambling at a gaming table or in the stock market.*

Under some conditions it is expedient to employ methods that retain the concepts of utility functions without the requirement of actually enumerating the full range of utilities.

An alternative can be deemed acceptable if it fulfills a minimum level of accomplishment. When applied to management, this is called an *aspiration level*. it is essentially the same concept as that applied to operating systems, where it is called a *standard of performance*. In money decisions it commonly takes the form of a minimum level of profit that is acceptable or a level of cost that is not to be exceeded. Perhaps its most useful application is in decisions where the outcomes vary over periods of time. In a sales situation in which bids on merchandise are received or made intermittently, as in offers made to a used-car dealer, an aspiration level is almost indispensable.

When the range of outcomes is relatively small, the expected dollar values can be used directly without introducing significant errors in preferences. A graph of a utility function is closely approximated by a line over a small range between maximum and minimum dollar returns. The question, "How small is

*R. O. Swalm, "Utility Theory—Insights into Risk Taking," *Harvard Business Review*, 6, 1966.

small?'' is a matter of judgment. A range of $500 may be trivial to a large corporation, but it could be a meaningful sum to a struggling business.

COMPETITIVE DECISIONS

In competitive decision making, two or more decision makers are pitted against each other. Both are considered to be equally informed and intelligent. They are referred to as *players*, and their conflicts are called *games*. The rationale of their competition is the basis of *game theory*.

Some of the terms associated with game theory need a special explanation. The generic term ''game'' in no way implies that game theory is limited to parlor games or similar entertainment contests. The players could be nations maneuvering on the brink of war, labor and management negotiating to ward off a strike on their own terms, or department store owners seeking a strategy to end a price war with an acceptable compromise.

In a game context, alternative courses of action are *strategies*. There is a subtle difference between alternatives in games of skill (akin to conditions of certainty), games of chance (risk), and games of strategy. In the last category the best course of action for a player depends on what that player's adversaries can do. Thus, an optimum strategy in a competitive environment may be always to use one alternative or to mix alternatives—that is, to use different alternatives for successive plays. The considerations inherent in determining strategies make game theory a rich source of fundamental ideas for decision making.

The 1947 publication of *Theory of Games and Economic Behavior* by von Neumann and Morgenstern laid the framework for and stirred subsequent interest in game theory.

GAME THEORY

In our treatment of game theory we will limit attention to games involving two players. Allowing more than two players greatly increases the mathematical complication without a corresponding accrual of basic logic. Multiplayer games rely on essentially the same type of inductive and deductive reasoning as do two-person games. The two-player limitation is actually not as restrictive as a first impression might indicate. Strategy for one side in a conflict is often designed to counter the aims of the most dangerous member of an opposing team. In other cases, a player may view the entire array of business competitors as a single opponent (a viewpoint similar to treating nature as hostile) and thereby reduce the conflict to a two-person contest.

Payoff Matrix

In two-person games the rows of the payoff matrix contain the outcomes for one player, and the columns show the outcomes for the other player. The

FIGURE 25.10
Two-person zero-sum game.

PLAYER B

		B1	B2	B3	B4
PLAYER A	A1	−1	−2	4	5
	A2	−1	4	−3	−2
	A3	0	3	1	2
	A4	−2	−3	3	1

outcomes are written in terms of the player on the left, player A in Figure 25.10. The alternatives for player A are A1, A2, A3, and A4. Player B also has four possible courses of action, B1, B2, B3, and B4. If player A selects alternative A1 and player B uses B1, the outcome is a loss of 1 for A and the gain of 1 for B. At A1, B4 the outcome is a gain of 5 for A and a corresponding loss of 5 for B. Thus, a positive number is always a gain for A, and a negative number is a gain for B. Because the sum of the payoffs for any choice of alternatives is zero (when A wins, B loses), the game described by Figure 25.10 is called a *zero-sum* game.

Several assumptions are implied for a conflict situation represented by a two person zero-sum payoff matrix:

1 The conflict is between only two opponents.
2 Each opponent has a finite number of alternatives.
3 Both players know all the alternatives available.
4 Each gain or loss can be quantified to a single number.
5 Both players know all the outcomes (numbers).
6 The sum of the payoffs for each outcome is zero.

Dominance

The first step in analyzing a game matrix is to check for dominance. In previous matrices we had to compare only the rows for dominance. In competitive decisions the columns as well as the rows represent alternatives. Therefore it is necessary to check both horizontally and vertically for dominance. Furthermore, the discovery of one dominant relation may reveal another dominant condition that was previously indistinguishable.

A sequential dominance relation can be observed in Figure 25.10. A check of the columns, the alternatives available to player B, shows that no one alternative is better for every outcome than is any of the other alternatives. For instance, B3 is preferred to B4 for all outcomes except those in the bottom row, A4.

Turning to the alternatives available to player A, we see that A1 is always preferred to A4. Eliminating A4 from the matrix, we have

	B1	B2	B3	B4
A1	−1	−2	4	5
A2	−1	4	−3	−2
A3	0	3	1	2

Now, from the viewpoint of player *B*, *B*3 is always preferable to *B*4. A further check reveals no additional dominant relations.

The assumption that allows two-way dominance checks is that both players are intelligent. Thus *B* would recognize that *A* would never use *A*4, which means that *B* would never have a use for *B*4. Viewed from either side, the end result would be the 3 × 3 matrix (three alternatives for each player) shown in Figure 25.11.

Saddle Point

When a player uses the same alternative at every play, we say that the player is following a *pure strategy*. If it is advantageous for both players to use pure strategy, a saddle point is present. A *saddle point* is identified by an outcome which is *both* the smallest number in its row *and* the largest number in its column. In Figure 25.11 the smallest numerical values for each row are listed to the right of the matrix and the largest numerical values for the columns are shown below the matrix. The outcome 0 at *A*3, *B*1 satisfies the requirements for a saddle point.

The significance of a saddle point develops from an investigation of the player's motives. Alternatives *A*1 and *A*2 are attractive to *A* because they allow a potential gain of four units. However, *B* can be assured that *A* will lose one unit (−1) by using *B*1 whenever *A* chooses to utilize *A*1 or *A*2. *A* would also be attracted to *A*3 because no negative outcomes can occur from this alternative. Again *B* can thwart gains by *A* through the use of *B*1. Since *A* is astute enough to observe the advantage *B*1 affords *B*, *A* would select the alternative that minimizes *B*'s gain from following the pure *B*1 strategy. This

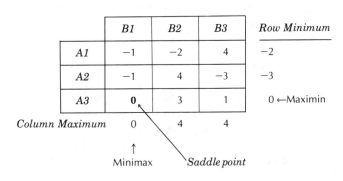

FIGURE 25.11
Saddle point for the reduced matrix.

alternative is $A3$. Therefore both players would use pure strategy, with A always employing alternative $A3$ and B always using $B1$. The result is a standoff, where neither side gains an advantage.

The underlying principle of choice for the players is maximin-minimax. B seeks the least of the maximum losses, minimax; A identifies the greatest of the minimum gains, maximin. This conservative viewpoint acknowledges that each player is capable of selecting the strategy which will satisfy the objective of maximizing gain or minimizing loss.

Value of a Game

The return from playing one game is the amount each player nets from the ensuing outcome. For instance, if one player gets a return of $+5$ after each has followed a certain course of action, the other player gets -5 in a zero-sum game. The average gain or loss per play, taken over an extended series of plays, is called the *value* of the game.

When the optimal strategy for both players is a pure strategy, the value of the game is the outcome at the saddle point. In Figure 25.11, the value of the game is zero; neither player wins or loses.

When no saddle point exists, the players turn to a policy called *mixed strategy*. This means that different alternatives are used for a fixed proportion of the plays, but the alternative employed for each play is a random choice from those available. The value of the game is the average return resulting from each player's having followed an optimal mixed strategy.

SOLUTION METHODS FOR MIXED-STRATEGY ZERO-SUM GAMES

Several different methods have been developed to solve zero-sum games. Every two-person zero-sum game with a finite number of alternatives can be transformed into a linear-programming problem and solved accordingly. In addition, there are special methods appropriate to certain conditions. Often these special methods are much less demanding than is the general approach. A different solution method is offered in each of the following sections for games of different sizes. Other methods are available from the references at the end of the book.

2 × 2 Games

A game with only two alternatives for each player readily reveals the wisdom of mixed strategy. Referring to Figure 25.12, we see that all the payoffs are positive. Therefore player A is bound to be a winner. The objective of B is to limit losses (A's gains) as much as possible. A pure strategy will not do this.

A quick check shows that there is no dominance among the alternatives

	B1	*B2*
A1	1	5
A2	3	2

FIGURE 25.12
2 × 2 matrix.

and no saddle point. The absence of a saddle point means that the players should rely on mixed strategy. Sometimes player A will use $A1$ and other times $A2$. Similarly, B switches randomly between $B1$ and $B2$. The problem is to determine the proportion of time each player will use each alternative.

There is a very simple method for solving 2×2 games. The procedural steps are listed in Figure 25.13 and illustrated by application to the matrix of Figure 25.12.

The indicated strategy is that A should use $A1$ 20 percent of the time, and $A2$ for 80 percent of the plays. B should play $B1$ 60 percent of the time, and $B2$ the other 40 percent.

The strategies appear intuitively sound when we remember the conservative viewpoint inherent in game theory. A would use $A2$ more often because of its higher minimum gain (two units instead of one). B would rely on $B1$

Step 1   Obtain the absolute value of the difference in payoff for each row and column.

	B1	B2	Payoff Difference
A1	1	5	**4** = 5 − 1
A2	3	2	**1** = 3 − 2
Payoff Difference	**2**	**3**	

Step 2   Add the values obtained in step 1 for the rows and columns. The sum of the column differences should equal the sum of the row differences.

	B1	B2	
A1	1	5	4
A2	3	2	1
	2 +	3	= **5**

Step 3   Form a fraction associated with each row and column by using the values obtained in step 1 as the numerator and the number obtained in step 2 as the denominator.

	B1	B2	
A1	1	5	4/5
A2	3	2	1/5
	2/5	3/5	

Step 4   Interchange each pair (row and column) of fractions obtained in step 3. The fraction now associated with each alternative is the proportion of plays that that alternative should be used for an optimal mixed strategy.

	B1	B2	
A1	1	5	**1/5**
A2	3	2	**4/5**
	3/5	**2/5**	

FIGURE 25.13   Solution steps for 2×2 games.

FIGURE 25.14
Probability of outcomes.

	B1 (3/5)	B2 (2/5)
A1 (1/5)	3/25	2/25
A2 (4/5)	12/25	8/25

more than B2 because of the lower minimum and maximum losses. The validity of the strategies becomes even more apparent when the value of the game is considered.

The long-run game return is the expected value of all the outcomes to one player. In order to calculate the probability of each outcome we must know how often each player plans to follow a strategy which includes the outcome. This information is provided by the optimal mixed strategy. Figure 25.13 shows the strategies determined for the sample-problem data. The cells of the matrix in Figure 25.14 contain the probable occurrence of each outcome in the original payoff matrix (Figure 25.12). Each cell value is the product of the probabilities for its row and column. The sum of these joint probabilities must equal 1.0. From Figure 25.14,

$$3/25 + 2/25 + 12/25 + 8/25 = 25/25 = 1.0$$

By multiplying each outcome by the probability of its occurrence and adding them together, we have the expected value of the game. In terms of player A, the value of the game is

$$3/25 \times 1 = 3/25 \qquad 2/25 \times 5 = 10/25$$
$$12/25 \times 3 = 36/25 \qquad 8/25 \times 2 = 16/25$$

$$EV(A) = 3/25 + 10/25 + 36/25 + 16/25 = 65/25 = 2.6 \text{ units}$$

which means that A can expect an average payoff of 2.6 units per play over a large number of plays. Since we are dealing with a zero-sum game, we would expect B's loss to be equal to A's gain. This is confirmed by

$$EV(B) = 3/25 \times (-1) + 2/25 \times (-5) + 12/25 \times (-3) + 8/25 \times (-2)$$
$$= (-3/25) + (-10/25) + (-36/25) + (-16/25)$$
$$= -2.6 \text{ units}$$

The best B can do is an average loss of 2.6 units.

If B tried to improve this position by changing strategy, say to a pure B1 strategy, A would soon recognize the pattern and switch accordingly to a pure A2 strategy. The value of the game would then change to 3, the outcome of A2, B1. If A continued with the optimal mixed strategy while B switched to a pure $B1$ strategy, the value of the game for A would remain unchanged:

$$EV(A) = 1/5 \times 1 + 4/5 \times 3 = 13/5 = 2.6 \text{ units}$$

	B1 (1.0)
A1 (1/5)	1/5
A2 (4/5)	4/5

From this result, we can observe that in 2×2 games one player can ensure the value of the game by using the optimal strategy, regardless of what the other player does.

Example 25.5
2 × 2 Matching Contest

Two prudent but impoverished gamblers were matching dimes. One player, Alice, won both dimes when two heads or two tails occurred. The other player, Bob, won both dimes when the coins did not match. After a long series of flips, both players, tired and bored, were nonwinners. Bob then offered a suggestion: "Let's stop flipping coins and just turn up whichever face we want. To make it interesting, you win the dime if we both show heads, I win a nickel if we don't match, and neither of us wins if we both show tails. It's a fair game, because you have one chance of winning a dime while I have two chances of winning a nickel."

Alice stopped to think. She visualized the game in matrix form and recognized it as a two-person zero-sum game. With no saddle point, it had to be mixed strategy. Calculating her optimal strategy, she found that she should turn up a head one-fourth of the time and a tail three-fourths of the time. After calculating Bob's strategy, she found her value of the game to be

$$EV(\text{Alice}) = 1/16 \times \$0.10 + 3/16(-\$0.05)$$
$$+ 3/16(-\$0.05) + 9/16 \times 0$$
$$= -\$0.0125$$

		Bob	
		H (1/4)	T (3/4)
Alice	H (1/4)	$0.10	−$0.05
	T (3/4)	−$0.05	0

After cogitating a bit, she said to Bob: "Your game sounds like fun, but we don't have any nickels to use as payoffs, so let's use only dimes. I'll give you four dimes if we both show

heads and one dime for two tails. You give me two dimes if I show a head while you turn up a tail, and three dimes if I have a tail and you have a head. Since both our total payoffs come to 50 cents, the game is just as fair as the one you suggested."

		Bob	
		H	T
Alice	H	−4	2
	T	3	−1

Now Bob stopped to think. He mentally pictured the payoff matrix, determined the strategies, and calculated his expected return:

		Bob	
		H (0.3)	T (0.7)
Alice	H (0.4)	−$0.40	$0.20
	T (0.6)	$0.30	−$0.10

$$EV(\text{Bob})$$
$$= 0.3 \times 0.4 \times \$0.40 + 0.3 \times 0.6(-\$0.30)$$
$$+ 0.7 \times 0.4(-\$0.20) + 0.7 \times 0.6 \times \$0.10$$
$$= \$0.048 - \$0.054 - \$0.056 + \$0.042$$
$$= -\$0.02$$

Noting his negative average payoff, he replied: "This takes too much thinking. Let's go back to flipping coins."

FIGURE 25.15
2 × 2 game solution

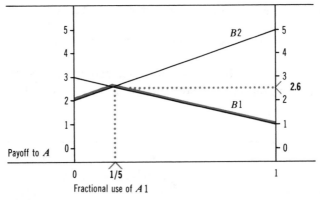

Payoff to *A*

Fractional use of *A* 1

(a) Graphical solution

	*B*1	*B*2
*A*1	1	5
*A*2	3	2

(b) Payoff matrix from Fig. 15.8

2 × *n* and *n* × 2 Games

When the number of alternatives increases beyond two, the method used to solve 2 × 2 games is not directly applicable. However, through the use of a graphical technique we can convert a 2 × *n* or *n* × 2 game to a 2 × 2 game. The obvious limitation is that one of the players still has to have only two alternatives.

To introduce the graphical method, we can apply it to the 2 × 2 game already solved in the previous section and shown again in Figure 25.15(b). The data from the payoff matrix are entered on a graph which shows payoffs on the ordinates and mixed strategy on the abscissa. The vertical scale must accommodate the largest payoff, and the horizontal scale runs from 0 to 1.0.

Figure 25.15 represents player *A*'s interests. Alternative *A*1 is depicted by the abscissa, and the ordinates are *A*'s payoffs. If *A* used pure *A*2 (left vertical scale), *A* could expect to win 2 or 3 units, depending on *B*'s strategy. The right vertical scale shows the possible payoffs from a pure *A*1 strategy. The lines connecting points on the two vertical scales represent *B*'s alternatives. They serve to constrain the payoffs to *A* as indicated by the heavy line, the minimum possible gains by *A*. Using the maximin criterion, *A* selects the strategy which gives the best minimum payoff. This payoff is the value of the game (2.6) and is indicated by the intersection of lines *B*1 and *B*2. The point on the horizontal scale directly below the intersection designates the fraction of the time that *A* should use *A*1.

The graphical method contributes little to the solution of a 2 × 2 game; it was applied above only to illustrate the technique. Its value lies in identifying the two limiting alternatives among several possessed by one player. As will be observed later, when one opponent has only two alternatives, the other player can advantageously play only two alternatives, even though several others are available. This condition allows the game to be reduced to 2 × 2 and solved accordingly, once the limiting alternatives are known.

An exception occurs when one player has an optimal pure strategy. Problem 25.19 illustrates this condition.

FIGURE 25.16
2 × 4 game solution.

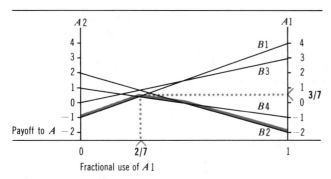

(a) Graphical solution

	B1	B2	B3	B4
A1	4	−2	3	−1
A2	−1	2	0	1
A3	−2	1	−2	0

(b) Payoff matrix

A 3 × 4 payoff matrix is shown in Figure 25.16(b). The first step in its solution is to check for dominance and a saddle point. The dominance of $A2$ over $A3$ reduces the matrix to a 2 × 4, which allows a graphical approach. The scales of the graph always represent the player with only two alternatives. Letting the abscissa denote $A1$, we plot each of B's alternatives with $A2$ payoffs (0 use of $A1$) on the left and $A1$ payoffs on the right.

After the graph has been completed, there are two ways to determine the mixed strategies and value of the game. The first is to read the value of the game and the strategy for A directly from the graph. If the scale were sufficiently large for accurate readings, the value of the game could be taken from Figure 25.16(a) as 3/7, and the strategy for A as 2/7 $A1$ and 5/7 $A2$. The value of the game is the maximum amount A can gain if B uses the optimal strategy. This strategy is defined by the intersection of $B1$ and $B4$ at the highest point of A's minimum gains. Knowing the value of the game and the payoffs for each of B's limiting alternatives, we can set up the relations

$$x \times 4 + (1 - x)(-1) = 3/7$$
$$x(-1) + (1 - x)1 = 3/7$$

where x = fractional use of $B1$
$1 - x$ = fractional use of $B4$

Then, solving for x,

$$4x + x - 1 = -x - x + 1$$
$$7x = 2$$
$$x = 2/7$$

We find that B should use $B1$ two-sevenths of the time and $B4$ five-sevenths of the time. $B2$ and $B3$ are never used. The validity of this strategy can be interpreted from the graph and confirmed by calculating the value of the game for any other strategy.

	B1*	B4*
A1	4	−1
A2	−1	1

*Limiting alternatives are selected graphically.

	B1	B4	
A1	4	−1	2/7
A2	−1	1	5/7
	2/7	5/7	

FIGURE 25.17
Solution of a graphically reduced payoff matrix.

The other way to calculate the optimal mix of strategies is to let the graph simply point out the two limiting alternatives and solve the evident 2 × 2 game. This method of solution for the sample problem is shown in Figure 25.17. A short-cut method to calculate the value of the game by applying strategy fractions to payoffs in a single column is given below.

$$EV(A) = 2/7 \times 4 + 5/7 \times (-1) = 2/7 \times (-1) + 5/7 \times 1 = 3/7 \text{ units}$$

The solution of an $n \times 2$ game is analogous to that used for $2 \times n$ games. The payoff scales on the graph represent the player with two alternatives (by our previous notation, this would be player B). Then the alternatives are plotted for the other player. Since the payoffs are written in terms of A, the player with two alternatives (B) seeks to minimize the maximum losses. Therefore, *the solution plane lies along the topmost set of lines.* The minimax solution is the lowest point on the plane. The lines intersecting at this point indicate the limiting alternatives for A. An $n \times 2$ game is illustrated in the following example.

Example 25.6
Competitive Selling Efforts

A new restaurant, Anton's, has opened in a city where the main competition is from Blue's restaurant. Anton's management has observed that Blue's promotional effort varies between reduced prices (B1) on certain nights and special menus (B2) on other nights. To attract customers from Blue's, Anton's can also reduce prices (A1), serve special meals (A2), or emphasize a takeout service (A3).

Both restaurants use the same advertising media to promote one special attraction each week. Because of differences in physical facilities and experience, each restaurant excels in certain specialties. Payoffs, the percentage of gain or loss in net revenue, have been estimated by Anton's management for the different outcomes shown.

		Blue's	
		B1	B2
Anton's	A1	−3%	6%
	A2	2%	−6%
	A3	−1%	5%

What strategy should Anton's follow in scheduling its weekly specials?

Solution 25.6

Anton's can expect a slight gain of 0.286 percent in revenue by serving special menus 44 percent of the time and featuring takeout meals 56 percent of the time, as indicated in Figure 25.18.

n × n Games

When a game matrix has more than two alternatives per player, the first effort should be a check for dominance and a saddle point. If these steps fail to produce a solution or a matrix reduced to two alternatives, an exact solution by linear programming or other methods can be undertaken. These measures

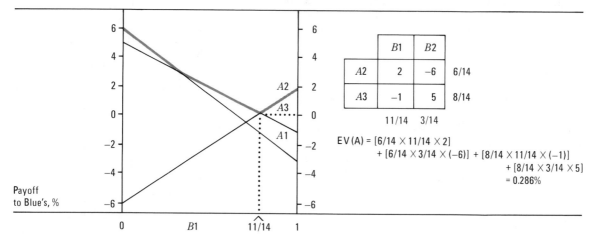

FIGURE 25.18
Solution of an $n \times 2$ game.

can be very tedious. There is also an approximation method which avoids mathematical complications and provides good estimates of strategy and the value of the game.

The approximation method is based on a series of fictitious plays of the game. In this respect it resembles the simulation methods we applied to queueing and inventory problems. The difference between the two methods is that the game returns are not a function of chance. When one player makes a play, the other player reacts by selecting the course of action which minimizes the gain of the opponent. In this way both players act under the assumption that past plays offer the best guide to the future. At each play an optimal pure strategy is selected according to the mixture of outcomes which represent all the opponent's past plays.

Detailed procedures for fictitious play can best be described by a sample application. A normal payoff matrix represents the contest. There can be any number of alternatives for either player. The sample problem shown in Figure 25.19 is a two-person zero-sum 3×3 game.

The value of the game is estimated by dividing the boldface payoff at each play by the number of that play. In Figure 25.20, the tenth play for A shows a payoff of 22 and the tenth play for B shows 18. The average payoff at the tenth play is therefore $22/10 = 2.2$ for A and $18/10 = 1.8$ for B. The value of the game lies between the *highest* average payoff to B (maximum loss or minimum gain) and the *lowest* average payoff to A. From the sample problem, the value of the game lies between 1.9 (occurring at play 20 for B) and 2.0, which occurs several times for A.

The optimal strategy is approximated by the number of times each alternative is used during fictitious play. In our illustration $A2$ and $A3$ were each used half the time, while $B2$ and $B3$ were used, respectively, for 7/20 and

Step 1 One of player A's alternatives is arbitrarily selected for use in the first play. We shall choose $A1$. The payoffs for $A1$ are then entered directly below the bottom row of the matrix.

	B1	B2	B3
A1	4	2	0
A2	2	0	3
A3	2	4	1
	4	2	0

Step 2 The response of player B to $A1$ is to select and play an alternative which minimizes the gain to A. Therefore the smallest payoff, 0, is selected (boldface), to indicate that $B3$ is B's best counter to A's use of $A1$.

	B1	B2	B3
A1	4	2	0
A2	2	0	3
A3	2	4	1
	4	2	**0**

Step 3 Now B's first play, the use of $B3$, is entered to the right of the matrix. The best pure strategy for A against the use of $B3$ is $A2$. The largest number in the column, 3, indicates this choice.

	B1	B2	B3	
A1	4	2	0	0
A2	2	0	3	**3**
A3	2	4	1	1
	4	2	**0**	

Step 4 Add the payoff from the alternative selected in step 3 ($A2$) to the last play by A as shown below:

4	2	0	first play
2	0	3	second play
6	2	3	cumulative payoff

Enter the cumulative payoff as the next row. Again select the smallest number, 2.

	B1	B2	B3	
A1	4	2	0	0
A2	2	0	3	**3**
A3	2	4	1	1
	4	2	**0**	
	6	**2**	3	

Step 5 Add the payoff from the indicated column ($B2$ from step 4) to the last column on the right of the matrix. Enter the payoff sums as a new column, and indicate the largest number.

Step 6 For each successive play the boldface number designates the row or column alternative to be added to the cumulative payoff. When a tie for high or low payoffs exists, any consistently applied rule may be used to select the next number. In the sample problem, to break a tie we designate an alternative different from the one last chosen. The number of plays is continued until a desired accuracy is obtained. Figure 25.15 shows 20 iterations for the sample game.

	B1	B2	B3		
A1	4	2	0	0	2
A2	2	0	3	3	3
A3	2	4	1	1	5
	4	2	**0**		
	6	**2**	3		

FIGURE 25.19

A 3 × 3 game and initial solution steps by the approximation method.

	B1	B2	B3	3.0	2.5	2.0	2.5	2.2	2.0	2.14	2.13	2.0	2.2	2.09	2.0	2.08	2.07	2.0	2.12	2.06	2.0	2.05	2.05	Av payoff
A1	4	2	0	0	2	2	4	4	4	6	6	8	8	8	8	10	10	12	12	12	12	14		0
A2	2	0	3	3	3	6*	6	9	12*	15	15	18*	18	21	24*	27	27	30*	30	33	36*	39	39	10
A3	2	4	1	1	5	6	10	11	12	13	17	18	22	23	24	25	29	30	34	35	36	34	41	10

Response by A

Av payoff	B1	B2	B3	Play
0.0	4	2	0	1
1.0	6	2	3	2
1.33	8	6	4	3
1.5	10	6	7	4
1.6	12	10	8	5
1.5	14	14	9	6
1.72	16	14	12	7
1.75	18	14	15	8
1.78	20	18	16	9
1.8	22	18	19	10
1.82	24	22	20	11
1.75	26	26	21	12
1.85	28	26	24	13
1.86	30	26	27	14
1.87	32	30	28	15
1.88	34	30	31	16
1.88	36	34	32	17
1.83	38	38	33	18
1.89	40	38	36	19
1.90	42	38	39	20
	0	7	13	

Av payoff / Response by B / Play

Results of fictitious play

Value of game between 1.9 and 2.0

Strategy for A:

A1	A2	A3
0	10/20	10/20

Strategy for B:

B1	B2	B3
0	7/20	13/20

Exact solution

Value of game 2.0

Mixed strategies:

A1	A2	A3	B1	B2	B3
0	1/2	1/2	0	1/3	2/3

*Ties broken by avoiding the last used alternative

FIGURE 25.20
$n \times n$ approximation method carried out for 20 fictitious plays.

13/20 of the plays. The proximity of the exact values to the approximate values is indicated in Figure 25.20.

The approximation method is a powerful tool for very large games. Its iterative nature makes it a natural candidate for machine computations. If there are several optimal solutions to a game, fictitious play will approximate one of them. It can be shown that the approximate solutions will converge on the actual solution with continued playing.

NONZERO-SUM GAMES

Many, if not most, real-life games are nonzero-sum games. In a conflict between nations neither side completely wins; if anything, both sides in a war are losers, regardless of which side surrenders. On the other hand, in labor-management negotiations both sides may be winners; labor can win an increase in wages, while the company wins increased productivity. Perhaps there is collusion in the game by the two players, labor and management, against an unrepresented but respected third party, the consumers.

The famous "prisoner's dilemma" game is discussed in Extension 25A.

Bargaining positions occur in nonzero-sum games because different players place different values on the same outcomes. A conciliation offer by a food processor to farmers might be less than either side wanted, but it could be accepted to avoid consequences that both sides fear. In a compromise each side gives up certain alternatives on the basis that the other side will not take advantage of the reduced game.

Unfortunately, there is no generally acceptable solution method for nonzero-sum games. However, the concepts of pure and mixed strategies, decision criteria, matrix representation, and game values provide considerable assistance in analyzing any conflict situation.

Example 25.7
Analysis of a Two-Person Nonzero-Sum Game

Two large shopping centers are competing for the same market. The Alpha company attracts customers by staging contests (A1) and by special sales or loss leaders (A2). The other firm, Beta, also uses contests (B1) and special bargains (B2). In addition, Beta has the exclusive franchise in the area for trading stamps (B3). Both firms periodically change advertising emphasis in their promotions. During these periods the entire advertising budget is spent exclusively on only one of the promotional alternatives.

The estimated gains and losses for each possible outcome are shown in the accompanying payoff matrix. The payoffs represent the net daily change in revenue resulting from competing promotional activities.

management philosophy. The utility each firm places on changes in daily revenue is indicated by the following utility functions.

| | Net Change, Dollars | | | | | |
	500	400	300	200	100	0
Alpha (U)	1.0	0.9	0.8	0.7	0.6	0.5
Beta (U)	1.0	0.95	0.9	0.85	0.8	0.7

	−100	−200	−300	−400	−500
Alpha (U)	0.4	0.3	0.2	0.1	0.0
Beta (U)	0.6	0.5	0.3	0.2	0.0

Beta

Alpha		B1	B2	B3
	A1	− $100	0	$100
	A2	$400	$200	− $300

Alpha is a new locally owned firm. The company's utility for money is directly proportional to the amount gained or lost. Beta is one store of a large chain which has a conservative

If the monetary payoffs are converted to utility payoffs, the game matrix takes the values shown, where the first payoff in each cell is the utility of that outcome to Alpha and the second number is Beta's utility payoff for the same outcome. Analyze the positions of the two competitors.

	B1	B2	B3
A1	4,8	5,7	6,6
A2	9,2	7,5	2,9

Solution 25.7

The usual check for dominance and a saddle point is unrewarding. Considerable insight can be gained by plotting graphs for both the utility payoffs. The resulting 3 × 2 games are then solved in terms of each player's utility function. From the first graph in Figure 25.21 it appears that Alpha can expect an average payoff of 5.11 utiles by using A1 seven-ninths of the time. The second graph indicates that this strategy for Alpha will allow a value of the game for Beta of 6.667 if Beta uses 1/3 B1 and 2/3 B3. If Beta decides to use maximum strategy for its own utility payoffs, the value of the game for Alpha would not be changed:

$$
\begin{aligned}
EV(A) &= 7/27 \times 4 \\
&\quad + 14/27 \times 6 \\
&\quad + 2/27 \times 9 \\
&\quad + 4/27 \times 2 \\
&= 5.11
\end{aligned}
$$

	B1 (1/3)	B3 (2/3)
A1 (7/9)	4	6
A2 (2/9)	9	2

Both firms should therefore be quite pleased with their mixed strategy: Alpha receives a slight cash gain ($11.11 per play); and Beta's concern for maintaining the status quo (as implied by its utility function) is satisfied by the high utility payoff.

The competition might include other factors that could lead to collusion, conciliation, or compromise. Perhaps both firms consider contests too expensive and difficult to administer. Alpha might avoid using contests if Beta would agree to use B2 70 percent of the time and B3 30 percent of the time. This strategy would allow an average utility for Alpha of $0.7 \times 7 + 0.3 \times 2 = 5.5$, and Beta's utility would be $0.7 \times 5 + 0.3 \times 9 = 6.2$.

If Beta wanted to stop staging contests and special bargains (B1 and B2), it might get Alpha to agree to use only A1. This compromise would allow both firms an equal utility payoff, 6. Beta would be giving Alpha an increase in average payoff for the privilege of administering only trading stamps. There are many such bargaining positions that can be established in nonzero-sum games.

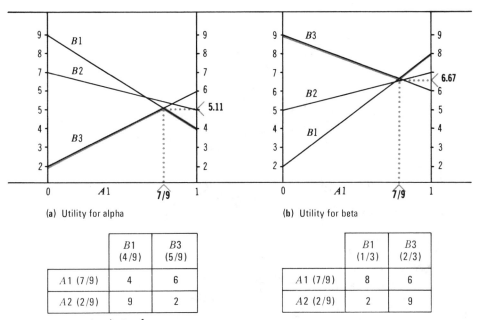

(a) Utility for alpha

(b) Utility for beta

	B1 (4/9)	B3 (5/9)
A1 (7/9)	4	6
A2 (2/9)	9	2

	B1 (1/3)	B3 (2/3)
A1 (7/9)	8	6
A2 (2/9)	2	9

FIGURE 25.21   Analysis of a nonzero-sum game.

PRACTICALITY OF GAME THEORY

Game theory is not a panacea for decision makers; nor is it a purely theoretical exercise in logic. Even though few practical industrial applications have been realized, it must be remembered that game theory is still a relatively new approach to decision making. Proponents of game theory believe continued research and development will make it more applicable to realistic problems. Some of the main difficulties preventing more widespread usage are:

1 Assigning meaningful payoffs
2 Solving very large matrices
3 Handling nonzero-sum and multiplayer games
4 Taking into consideration the possibility of collusion, conciliation, irrational players, and conditions which do not conform to the conventional game assumptions

Despite the difficulties, there is much to be gained from simply an awareness of the reasoning behind game theory. It forces attention to an opponent's strategy as well as our own. Payoffs must be quantified, ranked, or otherwise evaluated. The potential of mixed strategy is stressed. It suggests a philosophy for dealing with human conflict, an area in which decision makers have long needed assistance (see Figure 25.22).

FIGURE 25.22
Postscript to Figure 17.16.

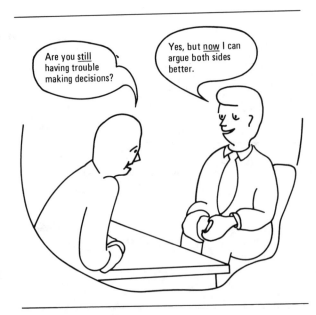

PROBLEMS

25.1 The profit expected from four alternative courses of action A, B, C, and D, under four states of nature, are given in the matrix below.

	1	2	3	4
A	7	9	5	2
B	8	1	10	4
C	6	6	6	6
D	5	7	9	8

(a) Which alternative would be selected by applying each of the following criteria: maximax, maximin, regret, and equal likelihood?

(b) Subtract 12 from each number in the matrix and multiply the resulting difference by 3. Apply the criteria from Problem 25.1a to the modified payoffs. What do the results indicate?

(c) Suppose that the states of nature are hair colors: brown, black, blonde, and red. A payoff will be made according to the color of hair of the next person that passes. You do not know which state applies to each hair color. Which alternative would you choose if the numbers in the matrix represented $100 bills? If the numbers in the matrix represented $1 bills? Which criterion do the above choices resemble?

(d) Add 10 to each number in the first column of the original matrix. Apply the same criteria as in Problem 25.1a. What do the results indicate?

25.2 Apply the Hurwicz criterion to the matrix below and show how different degrees of optimism will affect the selection of a preferred alternative:

	V	W	X	Y	Z
A	−8	6	2	−5	4
B	4	−2	−3	3	2
C	0	1	2	1	0
D	6	11	−10	4	−7

If each payoff represented years added to or subtracted from your life expectancy, which alternative would you choose?

25.3 An enterprising young man reads about a national Nature Club meeting to be held in a remote desert setting. He visualizes a great potential for the sale of ice-cream products during their meeting in the hot desert. He can carry up to 3000 ice-cream bars in a rented refrigeration truck. The bars will be sold for $1.50 each and cost only $0.15. Any bars he fails to sell will have no salvage value because of melting and refreezing. His only problem is that he does not know how many will attend the meeting or how many of those who do attend will buy his products. A rough guess is that 750 people will attend, and his expenses, exclusive of the cost of ice-cream bars, would be $375 to make the trip. What criterion would you suggest he use to help decide whether or not to go and how many bars to take? Why? What preference is indicated? Assume that his alternatives are increments of 500 bars.

25.4   The expected rates of return for investment in securities and investment in expanded plant facilities are estimated for two levels of future business activity:

	Recession	Inflation
Securities	5	7
Expansion	1	15

The company is undecided about the likelihood of each of the future conditions. It has been suggested that each future should be considered equally likely. Then the probability of each future at which the two alternatives are equivalent can be calculated. These are "indifference probabilities." The decision rule would be to select the alternative which has the highest return for the future state that has the greatest difference between the equal-likelihood probability and the indifference probability. Apply this decision rule and comment on the results.

25.5   Fancyfree Products has received a low-interest industrial loan to enable it to build a manufacturing plant in an economically depressed area. Effects of future economic and political conditions influence the site selection. As a preliminary study of the situation, ratings have been assigned as outcomes for operating the plant at each site under each of the more likely future conditions.

Site	Possible Conditions			
	1	2	3	4
NW	0	2	8	20
NE	4	14	12	10
SW	16	6	6	8
SE	4	4	18	6
C	4	14	10	9

(a) Which site is dominated?
(b) Which site is preferred according to the maximin principle?
(c) Which site is preferred according to the maximax principle?
(d) Which site gets a preference when the equal-likelihood principle is applied?
(e) When $\alpha = 0.75$, which site has the highest rating by the partial-optimist principle?
(f) Which site is preferred according to the regret principle?

25.6   Venture Capitalists, Inc., has $500,000 to invest in any one of the mutually exclusive projects shown below. Each of the projects has a life of 4 years, and the outcomes shown are annual returns in thousands of dollars of after-tax cash flow. Since each of the proposals depends upon public acceptance of an untested new product, there is considerable doubt about the state of each return. Apply the five principles of choice listed in Problem 25.5 and comment on the diversification of preference. (Let $\alpha = 0.6$.)

	Potential Market Conditions			
Project	W	X	Y	Z
A	200	200	200	200
B	50	150	500	0
C	100	300	200	100
D	400	350	200	50
E	300	200	100	0

25.7 Based on the payoffs in the matrix answer the following.

	S1	S2	S3
A1	20	10	0
A2	40	0	−20
A3	5	25	−5

(a) If S1, S2, and S3 have respective probabilities of occurrence of 0.3, 0.4, and 0.3, which alternative is preferred according to the expected-value criterion?

(b) What is the value of perfect information?

(c) If the probabilities of S1, S2, and S3 are unknown, at what values of α in the Hurwicz criterion would different alternatives be preferred?

(d) For unknown future states, what would be the preferred alternative using a regret matrix?

25.8 Suppose that you were notified that you were to receive a relatively small inheritance from a distant relative but the exact amount had not yet been determined. Then someone offered to bet you double or nothing for the unknown sum on the flip of a fair coin. Make a payoff table to represent your alternatives, letting I be the value of your inheritance. At what value of I would you be willing to accept the bet? What is the significance of this indifference point to your utility index?

25.9 What could be said about the individual with the utility function shown?

Utility

Cost

25.10 An individual decides that her utility rating for a loss of $20 is 0.10, and for a gain of $100 is 0.60. In addition, she finds herself indifferent to the two alternatives in the payoff table below.

	P (0.3)	P (0.7)
A1	$50	$50
A2	−20	100

What is her utility rating for $50?

25.11 The company described in Example 25.9 has determined a new utility index since its decision to develop a refined version of its product.

(a) If this index had been applied to the decision to develop a refined product, what would the decision have been?

(b) What are some possibilities that could have influenced the altered utility index?

Utility	Dollars
0.00	− 400,000
0.40	− 200,000
0.58	− 100,000
0.65	− 50,000
0.71	0
0.82	100,000
0.90	200,000
0.95	300,000
0.98	400,000

(c) At what probability would the expected utility of alternatives A and B be equal under the new utility scale?

25.12 An investor is deciding whether to buy bonds or stocks. She estimates that the probabilities of inflation and recession are, respectively, 0.7 and 0.3. For her anticipated investment of $10,000, she believes that her choice of stocks would gain 10 percent per year during a period of inflation and would lose 5 percent of their value in a recession. Her selection of bonds would show a gain of 4 percent regardless of inflation or recession.

(a) Which alternative would she choose if her utility scale were $-\$1000 = 0.0$; $-\$500 = 0.2$; $\$0 = 0.4$; $\$400 = 0.6$; $\$800 = 0.8$; and $\$1000 = 1.0$? The scale shows the dollar values equated to an equivalent utility, and intermediate values can be obtained by straight-line interpolation.

(b) Which alternative would the investor choose if her utility scale were directly proportional to dollar values?

(c) At what probability of recession would she be indifferent to the two alternatives?

25.13 Another way of approaching a decision is temporarily to bypass specific assignments of the probabilities of possible futures by determining at what probability the decision maker would be indifferent to each occurrence. General observations can then be made about the future which can narrow the area of consideration. This is accomplished by finding the probabilities at which the utilities of the strategies are equal; that is

$$EV(A1) = EV(A2) = EV(A3)$$

	R	N	I
A1	0.9	0.5	0.2
A2	0.7	1.0	0.1
A3	0.3	0.6	0.8

The given payoff table shows the utility a firm attaches to three alternative means of replacing obsolete equipment under the possible futures of a recession R, normal activity N, and inflation I.

(a) At what probabilities for R, N, and I would the firm be indifferent to alternatives A1, A2, or A3?

(b) What is the expected value of the utility?

(c) How could the firm use this information in narrowing the choice of alternatives?

25.14 Would it be accurate to say that one outcome is preferred to another because the utility of one is greater than that of the other?

25.15 Determine the optimal strategies and values of the following 2×2 two-person zero-sum games:

-1	6
7	2

(a)

4	6
-1	3

(b)

-11	0
-1	-4

(c)

8	2
6	1

(d)

25.16   Determine the optimal strategies and values of the following $2 \times n$ and $n \times 2$ two-person zero-sum games:

−3	4	1	−3
5	0	6	−5

(a)

−8	−1	−7
−2	6	9

(b)

4	0	1	7
2	8	4	2

(c)

7	1
3	0
4	8

(d)

−6	9
−10	−10
8	−4

(e)

7	2
8	0
4	7

(f)

6	−4
−1	3
4	−2

(g)

25.17   Determine the optimal strategies and values of the following $n \times n$ two-person zero-sum games:

8	4	2	−1
9	2	4	3
4	−5	3	0
3	−1	5	2
7	3	0	−3

(a)

2	−4	3
5	0	−2
4	1	0
−2	3	1

(b)

1	6	3	−4	0
−2	7	−1	1	2
3	2	5	−2	4

(c)

4	−1	−7	2
−2	0	3	4
1	2	−1	−6

(d)

400	100	200	0	−500
0	−100	300	−200	−400
500	0	300	0	−100
400	200	−100	100	100
200	100	300	100	−100

(e)

1	−2	3
−2	3	−4
3	−4	5

(f)

25.18 Abner devised a new card game to play with Betty during the long winter months. The new game uses only eight playing cards: the ace, king, queen, and jack of spades and hearts. Each player has four different face cards in one suit. At each play a particular card is selected by each player from the four available. Each player shows the other the selected card at the same time. They then turn to the matrix below to see how many points were scored.

		Betty			
		A	K	Q	J
Abner	A	-2	3	4	-1
	K	4	1	-1	0
	Q	-3	-2	3	-2
	J	3	0	-3	-4

What strategy should each player use? Did Abner set up the game for his benefit?

25.19 What strategy should A and B use for the following game? What is the value of the game?

	B1	B2	B3
A1	$-\$1000$	$\$3000$	$\$1000$
A2	$\$4000$	0	$\$1000$

25.20 Two ruthless but modern underworld gangs are competing in the same city. Once a month they must make a decision about where to set up their illegal gambling operations to minimize police interference. Some locations are more profitable than are others. The outcomes of the competing alternatives are indicated in the matrix, where the payoffs are in thousands of dollars:

		Bandits		
		B1	B2	B3
Angels	A1	-5	8	6
	A2	4	-3	2
	A3	-10	7	-2

Suppose that the Angel gang were able to place a spy in the Bandit gang. What amount could the Angels afford to pay the spy for perfect information about where the Bandits plan to stage each new operation?

25.21 A hotel has 100 rooms that rent for an average of $66 per day. The fixed cost per room is $27 per day and the variable cost is $12 per day. The hotel policy is to accept more reservations than there are rooms available during rush periods. They have found that between 0 and 8 percent of those reserving rooms will fail to claim them. A disgruntled customer who has reserved a room but finds none available is estimated to cost the hotel $150 in loss of future business. Analyze the situation as a competitive and noncompetitive problem. How many rooms should be rented under each condition? Assume that no deposit is required with a reservation.

25.22   Two very large firms with big government contracts have a serious shortage of technical people. Both firms attempt to fill their personnel needs by recruiting in colleges and by pirating from other firms. Company B has a very desirable location and pays excellent extra benefits. Company A is not so well situated but is known for its enlightened management and superb working conditions. Because of their respective reputations, B appeals to recent college graduates and A is more attractive to people with industrial experience.

   Both companies have several teams scouring the country for workers. In any given locality a team can attempt to (1) recruit or (2) pirate, but not both at one time because of time and personnel limitations. Teams from both companies often meet in one locality, which sets up the conflict situation shown in the following matrix:

	B1	B2
A1	$-4, -1$	$0, 0$
A2	$-1, 0$	$-1, -3$

The payoffs denote extra thousands of dollars beyond an average annual salary required to hire an employee under the conditions of each outcome. The first number in each cell refers to A's payoff and the second to B's.

   (a) Analyze the problem and suggest independent solutions for each company.
   (b) Comment on the possibility of collusion. Which company has the strongest bargaining position? Suggest a possible compromise solution.

25.23   Two companies compete in a relatively small market with very similar products. Ajax is the larger of the two and sells approximately twice as many units as does Besta. Since the products sold by Ajax and Besta serve the same purpose with about equal efficiency, each company attempts to elevate its product through consumer-oriented promotional plans. Three basic promotions are currently in vogue: (1) television and magazine advertising, (2) special packaging with bonus "gifts," and (3) short-term price reductions advertised in stores and local newspapers. Alternatives 2 and 3 tend to differentiate the two brands of product, but alternative 1 tends to increase the general awareness of the buying public to the product type without regard to the manufacturer. A strong advertising policy by Ajax would thus also increase the sales of Besta. Since Ajax is larger and works with a bigger budget, it benefits proportionately less from Besta's advertising than does Besta from Ajax's. Other alternatives also show a carry-over effect between the gains registered by each company.

   Ajax has developed the payoff estimates for the most commonly used alternatives of the two companies. The values are in percent of net gain or increased increment of profit expected from the following various alternatives during one quarter. The payoffs for Ajax from advertising by television and magazines (A1) depend on the extent of advertising conducted; the values for Besta are based on an extrapolation of historical data:

Besta

Ajax		B1	B2	B3
	A1	$x,12$	$y,21$	$z,18$
	A2	$-6,4$	$26,-12$	$-12,-20$
	A3	$4,0$	$9,9$	$-6,0$

(a) What strategy should Ajax follow to maximize gains if x, y, and z are, respectively, 14, 8, and 12? Discuss the pros and cons of the alternatives open to Ajax.

(b) Based on the given values of x, y, and z, what is the best strategy for Besta?

(c) Ajax recognizes that Besta benefits from an advertising campaign conducted by Ajax, but the reciprocal is not true. Ajax believes that its advertising budget can be manipulated from the level given in Problem 25.23a; so that for every 2 percent reduction in profit for Ajax, Besta's profit will decrease 3 percent. However, if advertising is increased by Ajax, Besta will gain 3 percent for every 1 percent increase. The maximum and minimum payoffs possible for Ajax are 20 percent and 0. What strategy combined with what advertising policy will be most advantageous for Ajax?

EXTENSION

25A The Engineering Economist's Dilemma

The classic game-theory fable known as the *prisoner's dilemma* illustrates a decision situation that is not strictly competitive. It highlights considerations of trust and accommodation that are factors in many decisions faced by engineering economists.

The two prisoners in the fable were caught committing a crime and were immediately jailed in separate cells where they could not communicate. A police officer interviews each individually to deliver the same message: "I have enough evidence now to send you to jail for 18 months. If you confess to the crime that usually deserves a 10-year jail term, I'll let you off with a 6-month sentence, providing you will testify against your partner. If you both confess, you will both be sentenced to 3 years."

The payoff matrix, written in terms of lost years for prisoners 1 and 2, is shown in Figure 25.23. Prisoner 1's loss is shown in the lower left of each cell, and the loss for prisoner 2 is at the upper right. Neither prisoner knows what the other will do. Prisoner 1 wonders what prisoner 2 will do and prisoner 2 thinks about what prisoner 1 is thinking: "Should I confess to play it safe with the possibility of getting a really short term, or should I hold out in hope that my partner will also hold out so we'll both get only 18 months? But if I hold out and my partner doesn't, I'll be stuck for 10 years." The equilibrium outcomes for confessions from both prisoners suggest the generally preferred strategy: confess.

It is hoped that no engineering economist will be faced with a decision about confessing to a crime, but most likely somewhat similar circumstances will be faced in advising which actions will be most compatible with an adversary's probable strategy. For example, replace the prisoners with two competing firms supplying the same product. For simplicity, say that the options are to price the product either high or low. Each firm would profit most by undercutting the other's price and least by being undercut. Assume too that both would profit more when both charge high than when both charge low. The resulting payoff matrix has

PRISONER 2

		Don't Confess	Confess
PRISONER 1	Don't Confess	−18 months / −18 months	−6 months / −10 years
	Confess	−10 years / −6 months	−3 years / −3 years

exactly the same pattern as that for the prisoner's dilemma. The most reasonable independent course of action would appear to be to set low prices.

But how should the analyst explain this advice to set low prices when there is a possibility of much higher profits from setting a high price? Should a suggestion be made to initiate communication between the competitors to seek a mutually advantageous price? Such counsel could lead the analyst to a practical experience of the prisoner's dilemma if price-fixing charges were proved. Should the company state its strategy ahead of an actual price increase and hope that its competitor will "rationally" agree to raise its price too? However, a unilateral price increase might provoke consumer wrath and hurt sales regardless of what the competitor did.

Competitive situations are not exclusively restricted to industry. In the public sector it is often necessary to select from alternatives that will solicit differing degrees of support or opposition. Future states may be sensitive to "opposition" activities such as law suits, budget curtailments, new legislation, adverse publicity, and, in extreme cases, voter approval or rejection. Consider the following situations, and recommend a strategy for player *A*. Discuss the influencing factors that led to your recommendation.

QUESTIONS

25A.1 A government agency must provide increased capacity for handling nuclear wastes. The agency can expand and modernize its current noncontroversial facility or it can build a more efficient facility at a new site. Constructing at a new site would likely cause considerable public discussion and opposition which would increase costs if prolonged. The present worths of net benefits (in millions of dollars) from the two plans, in relation to delays and modifications that might arise from public agitation, are shown in the lower left of each cell.

The strategies may be opposed by local groups backed by national environmentalist organizations. Payoffs for the opposition are rated on a scale of 1 to 10, with 10 being the

Public Response

	Accept	Oppose
A1: Expand at present site	5 / $4	3 / $3
A2: Develop at a new site	1 / $10	10 / $2

most rewarding. The groups would gain most from stopping or delaying a new site development. Their credibility would suffer most from accepting the new site without a protest.

What should the agency do? What factors are implied in the agency outcomes that might be negotiable?

25A.2 Two companies have nearly identical pulpwood plants. Both have been accused by local-citizen groups of being too slow in adopting air-quality-improvement equipment to reduce discharges of noxious odors and pollutants. Resentment is growing, even though it is realized that the local economy would suffer from a plant closure; excusing the odor as the "smell of money" is no longer a pervasive argument.

The discharge-control equipment is expensive. Both firms have been reluctant to make the investment until forced to do so. Now they face the prisoner's dilemma from legal suits brought against them. If only one plant admits it should have cleaned up its discharges earlier and will do so now, it will gain public goodwill and avoid stiff fines that will be levied against the plant that continues to resist. If both plants continue to claim that they cannot afford the equipment and still remain competitive with plants in other locations, the eventual total outlay will probably be smaller than if both agree to install the needed equipment immediately.

Assume that the proportions of gains and losses are the same as are the outcomes in the payoff matrix for the prisoner's dilemma (Figure 25.23). What would you advise? Why? What other factors not expressly noted in the payoff matrix might influence the decision?

SOLUTIONS TO SELECTED PROBLEMS

CHAPTER 2

2.2	2.25 years.
2.4	20 years
2.6	$76.64.
2.8	$1010.10.
2.10	$2.62
2.12	(*a*) 12.36 percent. (*b*) 12.55 percent. (*c*) 12.68 percent. (*d*) 12.75 percent.

2.13	(*a*) 19.56 percent. (*b*) $307.30.
2.16	(*a*) $1242. (*b*) $771. (*c*) $297.
2.18	$179.95.
2.20	5.6 percent.
2.22	17 percent, 16 percent.
2.24	$4330.
2.26	$20,435.
2.27	7.75 years.

2.30	(*a*) 2,526,900. (*b*) 33,064,000.
2.32	$10,611.
2.34	(*a*) 6 months. (*b*) 12 months. (*c*) 24 months. (*d*) 93 months. (*e*) Forever.

CHAPTER 3

3.2	9.15 percent.
3.4	$1498.
3.6	11.49 percent.
3.8	(*a*) $14,572. (*b*) $2362. (*c*) $964. (*d*) $6498. (*e*) $2270.
3.10	6.86 years.
3.12	$5641.
3.14	$522.
3.16	$19,926.
3.18	$754.

3.20	$7647.
3.22	(*a*) 52 percent. (*b*) 67.8 percent.
3.24	$599.
3.26	$690.
3.28	$5002.
3.30	58,406.
3.32	$660,800.
3.34	$2164.
3.36	Now.
3.39	(*a*) $3554. (*b*) $1004.

3.40	−$849.
3.42	$12,455.
3.44	5.4 percent.
3.46	4.8 percent.
3.48	$647.
3.50	$300.
3.52	$58.32.
3.54	$3177.

CHAPTER 4

4.1	$199,855.
4.4	$64,358.
4.6	−$15,182.
4.8	$22,274.
4.10	PW(diesel) = $3622.
4.12	$96,444.

4.14	$20,804 vs. $19,935.
4.16	$533,333 vs. $471,600.
4.18	$1,673,892.
4.20	$4350.
4.22	(*a*) $1167. (*b*) $875.
4.24	$67.35.

4.26	6.7 percent.
4.28	$665.239.
4.30	$83,218 vs. $80,773.
4.32	$2000.
4.34	6.76 years.
4.36	$19,666 vs $24,779.

CHAPTER 5

5.2 1,460,667 tons/year.
5.4 $4166 vs. $3450.
5.6 $1,183,070.
5.8 14.5 months.
5.10 (*a*) $2,071. (*b*) 1.4 years.

5.12 $555 vs. $555.
5.14 $1661.
5.16 $26,389.
5.18 $129,746.
5.20 $62,637.

5.22 $6072.
5.24 (*a*) $5844 vs. $5921.
(*b*) $6610.
5.26 Alternative *B* $332,900.

CHAPTER 6

6.2 5.9 percent.
6.4 4.2 percent.
6.6 12.36 percent.
6.8 (*a*) 19.0 percent. (*b*) 0 percent. (*c*) 13.1 percent.

6.10 13.89 percent.
6.13 18 percent, 19.56 percent.
6.14 2-year subscription.
6.16 (*a*) Program *A*. (*b*) Program *B*.

6.18 8.9 percent.
6.20 14.4 percent.
6.22 7.1 percent.
6.25 21.3 percent.

CHAPTER 7

7.2 Plan *I*, IRR = 9.5 percent.
7.4 (*a*) Alternative *X*, 11.4 percent. (*b*) Incremental *Z* to *X*, 15 percent.
7.6 Alternative 2.
7.8 Alternative 2.
7.10 Project *E*.

7.12 (*a*) Projects *C*, *D*, and *B*. (*b*) Projects *C, D*, and *B*; overall IRR = 16.9 percent. (*c*) Project *D*.
7.14 (*a*) Up quality. (*b*) All proposals. (*c*) 32.8 percent. (*d*) Up quality plus rearrange. (*e*) Neither are optimum.

7.16 (*a*) Alternatives *A* and *C*. (*b*) Alternatives A + B + C.
7.18 (*a*) Alternative *C*, EAC = $651,770. (*b*) Alternative *C*, cost = $9,311,000.
7.20 Alternative *D*, EAC = $40,577.

CHAPTER 8

8.2 $6067 vs. $5616.
8.4 $5905 vs $5522.
8.6 (*a*) $3122. (*b*) $2705 for 1 year; $3128 for 2 years.
8.8 $24,431 vs $25,581.

8.10 PW for 10 years; $35,114 vs. $38,588.
8.12 Retain old van for 3 more years.
8.14 $56,673 vs. $52,942.

8.16 N = 3 years.
8.18 Defender should be replaced each year.
8.20 5 years.
8.22 5 years.

CHAPTER 9

9.2   13.3 percent.
9.3   (a) 6.5 percent. (b) $24,870.
9.6   2.43 years.
9.8   (a) 1A, 1B, 2A, 2B, 4B. (b) 1B, 2A, 4B. (c) 13 percent.
9.10   (a) $180,000. (b) 19 percent. (c) 17.8 percent. (d) $180,000. (e) 17 percent.

9.12   (a) 1C, 2A, 2C, 3A, 3B; 15.1 percent. (b) 3A, 1C, 15.6 percent. (c) 3A, 1C, 2A. (d) 1C, 2A, 3A, 15.9 percent.
9.14   (a) C, D, A, B, E. (b) C, A, D, B, E. (c) A, C, D. (d) No change.

CHAPTER 10

10.2   (a) Projects B, E, D, A. (b) Project D, B/C = 2.65. (c) Project D, ROR = 19.2 percent.
10.4   B/C = 2.03.

10.6   (a) A2, B2, C2, D2. (b) A1, B1, C1, D1.
10.8   (a) 0.97; 1.04. (b) 1.10; 1.11.
10.10   (a) 2.59. (b) 0.56. (c) 2.00

10.12   The hill route with B/C = 13.13.
10.14   B/C = 156.

CHAPTER 11

11.1   (a) $D(7) = \$3000$; $BV(7) = \$9000$. (b) $D(7) = \$1333$; $BV(7) = \$6667$. (c) $D(7) = \$0$; $BV(7) = \$6000$.
11.4   (a) $BV(4) = \$200$; $D(5) = \$50$. (b) $BV(4) = \$125$; $D(5) = \$37.50$. (c) $BV(4) = \$104$; $D(5) = \$29.68$.
11.6   (a) $30,000. (b) $10,456. (c) $6538.
11.8   (a) $D(1) = \$40,000$; $BV(4) = \$16,000$, (b) $D(1) = \$20,000$; $BV(4) = \$60,000$.
11.10   6 years.
11.11   Change at end of fourth year.

11.13   (a) −$24,963. (b) −$7208.
11.14   ATCF(1) = $28,400.
11.15 $D(86) = \$17,820$; $D(87) = \$24,300$; $D(88) = \$7920$.
11.16   Tax credit = $3240.
11.19   (a) Macro = $1,129,750; Meso = $209,750; Micro = $8250. (b) Macro = 47.9 percent Meso = 44.7 percent; Micro = 20.6 percent.
11.20 $368,000.
11.22   ROR = 18.6 percent.
11.24   ROR = 20.8 percent.
11.28   Purchase new lot if interest rates exceed 7.7 percent.

11.30   PW = $3594; PW = $4330.
11.32   (a) 26.6 percent. (b) 15.4 percent. (c) 13.2 percent.
11.34   EUAW = −$18,147.

CHAPTER 12

12.2   3 percent per period.
12.4   (a) $392,000. (b) $180,000.
12.6   ROE = 9.07 percent.
12.8   Purchase: $12,654; lease: $18,249 (assuming FOY payments).

12.10   (a) −$28,301. (b) $15,554. (c) −$23,856.
12.12   (a) ROR = 30.8 percent, payback = 2.0 years. (b) Indicates turnover rate of capital. (d) ROR = 24.1 per-

cent, payback = 2.3 years. (e)   PWs = $104,280, $43,548, −$388.
12.14   (a) $7896 (assuming EOY payments). (b) $9013 (assuming EOY payments).

CHAPTER 13

13.1 (*a*) 4.61 percent. (*c*) 7.75 percent.
13.2 (*a*) $-20.2¢$. (*c*) $-37.0¢$.
13.4 PW = 4.1509 base amount.

13.6 $1730.
13.8 (*a*) 15.5 percent. (*b*) $121
13.10 $137,762.
13.12 $-$263.

13.14 (*a*) 14.2 percent. (*b*) 12.9 percent.

CHAPTER 14

14.2 $R = \$3019$.
14.3 (*a*) $B/C = 0.868$. (*b*) $P = \$1,604,400$.
14.5 Annual receipts.

14.6 3006 hours.
14.8 (*a*) Expected PW = $2238. (*b*) PW's is highly sensitive to variability in all input parameters.

14.9 EAC(*P*) = $2,155,713; EAC(*M*) = $1,007,238; EAC(*O*) = $781,593.
14.11 PW(*P*) = $-$245,652; PW(*M*) = $39,736; PW(*O*) = $213,959.

CHAPTER 15

15.2 226,000 barrels/year.
15.4 (*a*) 100 campers. (*b*) $9600. (*c*) $3000.
15.6 (*a*) 65 percent. (*b*) 46 units.

15.8 $12.17.
15.10 (*a*) $14. (*b*) $14.
15.12 (*a*) $130,000. (*b*) $130,000.
15.14 Reduced production.

15.16 (*a*) $51. (*b*) $1,750,000. (*c*) $1,080,000.
15.18 (*a*) $n_1 = 3.26$; $n_2 = 6.23$. (*b*) 4910 units.

CHAPTER 16

16.1 (*a*) 0.875. (*b*) 0.125. (*c*) 0.875.
16.3 (*a*) $345,000; $260,000.
16.5 4 months; $115,500.
16.6 Method one at $0.80.

16.8 (*a*) Alternative *A*. (*b*) Alternative *C*. (*c*) Alternative *C*. (*d*) Alternative B.
16.10 (*a*) 20 percent.

CHAPTER 17

17.2 (*a*) 0.76. (*b*) 0.99.
17.4 560.
17.6 Pull drills; $E(V) = \$75,000$.
17.8 0.98.
17.10 (*a*) EV(*A*1) = $4438. (*b*) EV(*B*) = $1835. (*c*) EVPI = $0.

17.12 (*b*) EVSI = $-$7750.
17.14 Plan *B*, $802,075.
17.16 Purchase.
17.19 (*a*) Design *B*.
17.23 Select site 1, $R = 258$.

CHAPTER 18

18.3   Net worth = \$150,000
18.4   \$240,000
18.5   \$40,000
18.6   (a) \$1,266,207; (b) \$1,467,310
18.8 $N = 1$, alpha = 1
18.10 \$424 difference

18.12 (a) 114.5
        67.9
        76.9
        58.2
    (b) 110.6
        69.6
        73.6
        54.0

18.14 3368 units
18.16 (a) 677,712; (b) $r = 0.983$; (c) 713,763

CHAPTER 19

19.3   Item 4
19.4   Mimo = \$0.268/unit, Print = \$0.216/unit
19.5   (a) One machine: \$1.90/unit; Two machines: \$2.26/unit. (b) \$0.81/acceptable unit.
19.8   Two machines: \$1.53/unit; Three machines: \$1.41/unit

19.9   \$46.00/unit
19.12 Total savings equal \$26,023
19.14 Minimum crew plus one finisher and one laborer; total cost = \$184.80

19.15 Seven spans

19.16 $A = \sqrt{\dfrac{I^2 C_I}{C_A(C_C + C_T)}}$

19.19 Six inspectors

CHAPTER 20

20.3   (b) CP = 16; (c) LF = 39
20.4   CP = 12
20.6   CP = 36
20.8   Duration = 16; minimum cost = \$5410
20.10 Cost overrun: 20; time overrun: 1.5 months
20.12 Order every 10 weeks

20.13 EOQ = 2981 pounds
20.15 (a) EOQ = 14,852 units; (b) total cost = \$162,020; (c) time between orders: 18.6 days
20.16 EPQ = 936 units
20.18 (a) EOQ = 11 packages; (b) order interval = 69 days

20.20 Discount saves \$330/year
20.22 Order 600 units

CHAPTER 22

22.2   (*a*) Fabrication PPI = 1.06
Assembly PPI = 1.01
(*b*) Labor PI = 1.03

22.4   (*b*) Labor productivity:
+2.90 percent
Factor productivity:
−3.16 percent
Total productivity:
−2.10 percent
(*b*) No, productivity delined

22.6   (*a*) Labor productivity =
2.931; (*b*) Total output =
2.625 million tons

22.8   (*a*) EPI = 1.254
(*b*) EPRI = 0.674
(*c*) ECEI = 0.944
(*e*) MPI = 1.097
(*f*) CFPI = 1.097
(*g*) TPI = 1.060

23.10   (*a*) Labor productivity:
+87.45 percent
Factor productivity:
+70.39 percent
Total productivity:
+22.69 percent

CHAPTER 23

23.2   Place order when stock equals 90 units

23.4   Place order when stock equals 30 gallons

23.6   (*a*) $p(x; 5) = 0.1755, 0.1755,$ 0.1462; (*b*) Mean = 25, variance = 25; (*c*) 0.7290; (*d*) 0.7286

23.8   Mean = 6.0, variance = 1.0

23.10   Replace every fourth day; minimum total cost = $122.50

23.12   Use two standby machines

23.14   (*a*) $15.19/unit/week; (*b*) Stay with 4-week interval; (*c*) new facility not cost effective; total cost = $16.77/unit/week

23.16 $98.02/unit

23.17 $132.85/unit

CHAPTER 24

24.2 $E(PW) = \$80450,$
$SD(PW) = \$7485$

24.4   (*a*) Rank: *B, D, E, C, F, A*;
(*b*) rank: *D, F, B, E, A, C*

24.6   (*a*) $E(PW) = \$9859,$
$SD(PW) = \$4262$;
(*b*) about 1 percent

24.8   (*a*) $E(PW) = \$-21532,$
$SD(PW) = \$2125$;
(*b*) $E(PW) = \$-21532,$
$SD(PW) = \$2014$

24.10 $E(AW) = \$7403,$
$V(AW) = 46,493,183$

24.12 $P = 0.25$

24.14   (*a*) $E(AP) = \$49,500,$
$V(AP) = 236,400,000$;
(*b*) $E(PW) = \$40,577,$
$SD(PW) = \$17,826$

24.16   (*a*) $E(PW) = \$143,114,$
$V(PW) = 13,127,031$ (*b*)
$8885/ton

CHAPTER 25

25.1

	(*a*)		(*b*)		(*d*)	
Maximax:		B	Maximax:	B	Maximax:	B
Maximin:		C	Maximin:	C	Maximin:	D
Regret:		D	Regret:	D	Regret:	D
Equal:		D	Equal:	D	Equal:	D
Worst regret:		D	Worst regret:	D	Worst regret:	D

25.2   Alternative *C* when alpha < 0.526; alternative *D* when alpha > 0.526

25.5   (*a*) Site *C* is dominated; (*b*) SW; (*c*) NW; (*d*) NE; (*e*) SE; (*f*) NE or SW

25.6   Maximax: *B*
      Maximin: *A*
      Regret: *A*, *C*, or *D*
      Hurwicz: *B*
      Equal: *D*

25.10 $U = 0.45$

25.11   (*a*) Option *B*; (*b*) a more conservative attitude; (*c*) $P(A) = 0.584$, $P(R) = 0.416$

25.12   (*a*) $U = 0.76$ indicates a preference for stocks; (*b*) stocks would be selected; (*c*) $P(R) = 0.4$, $P(I) = 0.6$

25.13   (*a*) $P(R) = 0.401$, $P(N) = 0.243$, $P(I) = 0.364$; (*b*) $E(A1) = E(A2) = E(A3) = 0.555$

25.15   (*a*) $EV = 3.67$; (*b*) $EV = 4.00$; (*c*) $EV = -3.14$; (*d*) $EV = 2.0$

25.16   (*a*) $EV = -3.0$; (*b*) $EV = -2.0$; (*c*) $EV = 2.8$; (*d*) $EV = 5.2$; (*e*) $EV = 1.78$; (*f*) $EV = 5.125$; (*g*) $EV = 1.0$

25.19 $EV = \$1000$

25.20   Without spy: Angels, $EV = \$850$; with spy: Angels, $EV = \$5,800$

APPENDIX B

GLOSSARY OF TERMS COMMONLY USED IN ENGINEERING ECONOMICS

ABC inventory system Application of the Pareto principle by classifying inventory into three categories, with the A class having the highest value.

Accelerated Cost Recovery System Specific schedules for the depreciated recovery of investments in various property classes.

Accounting life The period of time over which asset cost will be allocated by the accountant.

Amortization 1. (*a*) As applied to a capitalized asset, the distribution of the initial cost by periodic charges to operations as in depreciation; (*b*) the reduction of a debt by either periodic or irregular payments. 2. A plan to pay off a financial obligation according to some prearranged program.

Annual equivalent In time value of money, one of a sequence of equal end-of-year payments which would have the same financial effect when interest is considered as another payment or sequence of payments.

Annuity 1. An amount of money payable to a beneficiary at regular intervals for a prescribed period of time out of a fund reserved for that purpose. 2. A series of equal payments occurring at equal periods of time.

Balance sheet A display of the monetary values of a firm at a given time that equates assets to liabilities plus net worth.

Benefits Consequences of an investment either measured in or converted to economic terms.

Benefit-cost analysis An analysis in which all consequences of the investment are measured in or converted to economic terms.

Book value Original cost of an asset less the accumulated depreciation.

Breakeven chart A graphic representation of the relation between total income and total costs for various levels of production and sales, indicating areas of profit and loss.

Breakeven point The rate of operations, output, or sales at which income will barely cover expenses.

Capital 1. The financial resources involved in establishing and sustaining an enterprise or project. 2. Wealth that may be utilized to economic advantage.

Capital budgeting The allocation of funds to projects.

Capital recovery 1. The act of charging periodically to operations amounts that will ultimately equal the amount of capital expenditure. 2. Replacement of the original cost of an asset plus interest. 3. The process of regaining the new investment in a project by means of revenue in excess of the costs from the project.

Capital recovery factor A factor used to calculate the sum of money required at the end of each of a series of periods to regain the net investment plus the compounded interest on the unrecovered balance.

Capitalized cost The present worth of a uniform series of periodic costs that continue for an indefinitely long time (hypothetically infinite).

Cash flow 1. The real dollars passing into and out of the treasury of a financial venture. 2. The flowback of profit plus depreciation from a given project.

Cause-and-effect (C&E) diagram Pattern of arrows that display the causes of a situation and arrows that show resulting effects.

Challenger The proposed property or equipment which is being considered as a replacement for the presently owned property or equipment (defender).

Common stocks Part of the capital of a business, which furnishes variable dividends and partial control through voting rights.

Compound amount factor 1. The function of interest rate and time that determines the compound amount from a stated initial sum. 2. A factor which when multiplied by the single sum or uniform series of payments will give the future worth at compound interest of that single sum or series.

Compound interest 1. The type of interest that is periodically added to the amount of investment (or loan) so that subsequent interest is based on the cumulative amount. 2. The interest charges under the condition that interest is charged on any previous interest earned in any time period, as well as on the principal.

Compounding, continuous 1. A compound interest situation in which the compounding period is zero and the number of periods is infinitely great. 2. A mathematical procedure for evaluating compound amount factors based on a continuous interest function rather than discrete interest periods.

Consumer's surplus The difference between what consumers actually pay for a public good or service and the maximum they would be willing to pay.

Correlation analysis A study of the degree of relationship existing between variables.

Cost-effectiveness analysis A diagnosis that includes all prospective consequences of a proposed investment, economic as well as noneconomic.

Cost of capital The expense involved in borrowing capital, usually expressed as a percentage.

Costs Consequences of an investment either measured in or converted to economic terms.

Cost index A dimensionless number that indicates how cost for a class of items changes over time.

Crash cost The expense of completing an activity at its minimum duration.

Critical path The longest chain of activities through a network.

Critical-path method (CPM) A deterministic network scheduling method.

Cutoff rate of return The rate of return after taxes determined by management that is based on the supply and demand for funds and will be used as a criterion for approving projects or investments; may or may not be equal to the minimum attractive rate of return.

Decision tree A graphical representation of the anatomy of a decision showing the interplay between a present decision, change events, possible future decisions, and their results or payoffs.

Decisions under certainty Decisions that assume complete information and no uncertainty connected with the analysis of the decisions.

Decisions under risk Decisions for which the analyst elects to consider several possible futures, the probabilities of which can be estimated.

Decisions under uncertainty Decisions for which the analyst elects to consider several possible futures, the probabilities of which *cannot* be estimated.

Declining balance depreciation (*Also known as percent of diminishing value.*) A method of computing depreciation in which the annual charge is a fixed percentage of the depreciated book value at the beginning of the year to which the depreciation applies.

Defender Presently owned property or equipment that is being considered for replacement.

Dependent Events or alternatives that are related to each other in any way which influences the selection process.

Depletion 1. A form of capital recovery applicable to extractive property (e.g., mines). Can be on a unit-of-output basis the same as straight-line depreciation related to original or current appraisal of extent and value of deposit. 2. A lessening of the value of an asset because of a decrease in the quantity available. Similar to depreciation except that it refers to such natural resources as coal, oil, and timber in forests.

Deferred annuity   A series of payments in which the first payment does not occur until later than the end of the first period.

Depreciation   1. (*a*) Decline in value of a capitalized asset; (*b*) a form of capital recovery applicable to a property with two or more years lifespan, in which an appropriate portion of the asset's value is periodically charged to current operations. 2. The loss of value because of obsolescence or attrition. In accounting, the allocation of this loss of value according to some plan.

Direct cost   The traceable cost that can be segregated and allocated against specific products, operations, or services.

Discounted cash flow   1. The present worth of a sequence in time of sums of money when the sequence is considered as a flow of cash into and/or out of an economic unit. 2. An investment analysis which compares the present worth of projected receipts and disbursements occurring at designated future times in order to estimate the rate of return from the investment or project.

Dummy arrow   A dashed arrow used to show a precedence relationship among activities in an arrow network.

Economic life   The period of time, extending from the date of installation to the date of retirement from the intended service, over which a prudent owner expects to retain the property, in order to obtain a minimum cost.

Effectiveness   1. The degree to which a strategic or tactical plan achieves economic targets. 2. Consequences of an investment not measured in economic terms, e.g., reliability, safety, customer satisfaction.

Exponential smoothing   A forecasting method that predicts demand for the next period by applying a weighting factor to preceding demands.

Face value   A stated amount for which a bond may be redeemed at maturity.

Factor productivity   Ratio of output minus intermediate goods to the sum of labor and capital inputs.

First cost   The initial cost of a capitalized property, including transportation, installation, preparation for service, and other related initial expenditures.

Fixed cost   A cost that tends to be unaffected by changes in the number of units produced.

Float   Amount of surplus time or leeway allowed in scheduling activities.

Future worth   1. The equivalent value at a designated future date based on the time value of money. 2. The monetary sum at a given future time which is equivalent to one or more sums at given earlier times when interest is compounded at a given rate.

Goodwill   Value inherent in the fixed and favorable consideration of customers arising from an established well-known and well-conducted business.

Holding cost   Total annual cost of holding an item in storage and interest foregone on money invested in the item.

Independence   A condition where acceptance or rejection of an alternative has no effect on the acceptability of any other alternative.

Indirect cost   The traceable or common cost which is not allocated against specific products, operations, or services but is allocated against all products, operations, or services by a predetermined formula.

Inflation   A persistent rise in the general price level, usually resulting in the decline of purchasing power.

Interest   1. (*a*) Financial share in a project or enterprise; (*b*) periodic compensation for the lending of money; (*c*) in engineering economic study, synonymous with required return, expected profit, or charge for the use of capital. 2. The cost for the use of capital. (*Sometimes referred to as the time value of money.*)

Interest rate, effective   The true value of interest rate computed by equations for compound interest for a 1-year period.

Interest rate, nominal   The customary type of interest rate designation on an annual basis without consideration of compounding periods. The usual basis for computing periodic interest payments.

Internal rate of return   The rate of return which takes into account only what a project or investment earns and disregards earnings on cash flows reinvested in other projects.

Investment tax credit   A reduction in income tax liability allowed for specific types of interest.

Kanban   An inventory system designed to have material arrive precisely when it is needed in the production process.

Least-squares method   A procedure for calculating the formula of a best-fit line for a set of data points.

Learning curve   Pattern of declining cost over time as a function of more experience or learning.

MAPI method   1. A procedure for replacement analysis sponsored by the Machinery and Allied Products Institute. 2. A method of capital investment analysis formulated by the Machinery and Allied Products Institute, a prominent feature of which is that it explicitly includes obsolescence; uses a fixed format and provides charts and graphs to facilitate calculations.

Macroeconomics   The study of entire economic systems in terms of income, flow of money, consumption, and general prices.

Marginal cost   1. The cost of one additional unit of production, activity, or service. 2. The rate of change of cost with production or output.

Microeconomics The study of activity in very small segments of the economy, such as a particular firm or household.

Minimum attractive rate of return (MARR) The effective annual rate of return on investment which must meet the investor's threshold of acceptability.

Moving average A forecasting method that predicts demand for the next period from the average of demands experienced during several preceding periods.

Mutually exclusive alternative A choice such that the selection of it rules out the selection of all other alternatives in the set.

Net worth Current worth of investments in an organization.

Opportunity cost The cost of not being able to invest in an alternative because of limited resources being applied to another"approved" alternative, and thus not being available for investment in other income-producing alternatives.

Overhead The portion of operating cost that cannot be clearly associated with particular operations or products.

Pareto principal The point at which 10 to 20 percent of a total population possesses 60 to 80 percent of a given value.

Payback period 1. With respect to an investment, the number of years (or months) required for the related profit or savings in operating cost to equal the amount of said investment. 2. The period of time that a machine, facility, or other investment has produced sufficient net revenue to recover its investment costs.

PERT An acronym standing for Program Evaluation and Review Technique—a probabilistic network scheduling method.

Preferred stocks Part of the capital of a business, usually having fixed dividends and no voting rights.

Present worth 1. The equivalent value at the present, based on time value of money. 2. The monetary sum equivalent to a future sum(s) when interest is compounded at a given rate. 3. The discounted value of future sums.

Productivity Quality or state of being productive in the creation of goods and services.

Productivity index Rate of change of productivity from one time period to the next.

Rate of return 1. The interest rate at which the present worth of the cash flows of a project is zero. 2. The interest rate earned by an investment.

Repeated life An analysis based on the assumptions that an existing asset will be replaced with an asset having identical cash flows and that there is a continuing requirement for its services.

Replacement   1. A broad concept embracing the selection of similar but new assets to replace existing assets; 2. The evaluation of entirely different ways of performing an asset's function.

Retained earnings   Profits reinvested in an organization.

Salvage value   1. The cost recovered or which could be recovered from a used property when removed, sold, or scrapped. A factor in appraisal of property value and in computing depreciation. 2. The market value of a machine.

Safety stock   A reserve supply beyond the amount consumed by average usage.

Sensitivity   The relative magnitude of the change in one or more elements of an engineering economics problem that will reverse a decision among alternatives.

Simple interest   1. Interest that is not compounded; it is not added to the income-producing investment or loan. 2. The interest charges under the condition that interest in any time period is charged only on the principal.

Sinking fund   1. A fund accumulated by periodic deposits and reserved exclusively for a specific purpose, such as retirement of a debt or replacement of a property. 2. A fund created by making periodic deposits (usually equal) at compound interest in order to accumulate a given sum at a given future time for some specific purpose.

Social discount rate   The composite rate of interest used in evaluating public projects.

Spillover benefit or cost   The financial effect of an activity on third parties not directly involved in a project or program.

Straight-line depreciation   A method of depreciation whereby the amount to be recovered (written off) is spread uniformly over the estimated life of the asset in terms of time periods; or units of output. May be designated as percent of initial value.

Strategy   The setting of ultimate objectives.

Study period   In engineering economic study, the length of time that is presumed to be covered in the schedule of events and appraisal of results. Often the anticipated life of the project under consideration, but a shorter time may be more appropriate for decision making.

Sum-of-the-years-digits method   A method of computing depreciation in which the amount for any year is based on the ratio (Years of remaining life)/(1 + 2 + 3 $\cdots$ + n), n being the total anticipated life.

Sunk cost   A cost, already paid, that is not relevant to the decision that is being made concerning the future. Capital already invested that for some reason cannot be retrieved.

Tactical suboptimization   A solution which optimizes tactical efficiency with little or no regard for strategic effectiveness.

Tangibles   Things that can be *quantitatively* measured or valued, such as items of cost and physical assets.

Time value of money   1. The cumulative effect of elapsed time on the money value of an event, based on the earning power of equivalent invested funds. (See *future worth* and *present worth*.) 2. The expected interest rate that capital should or will earn.

Uniform gradient series   A pattern of receipts or disbursements for a series, that has a uniform linear (arithmetic) or geometric increase in each time period.

Unit cost   Cost divided by number of units of output.

Variable cost   A cost that tends to fluctuate according to the changes in the number of units produced.

Weighted cost of capital   A composite interest rate which represents the average cost of all acquired funding used by an organization.

Working capital   1. The portion of investment represented by current assets (assets that are not capitalized) less the current liabilities. The capital necessary to sustain operations. 2. Funds that are required to make the enterprise or project a going concern.

Yield   The ratio of return or profit over the associated investment, expressed as a percentage or decimal usually on an annual basis.

APPENDIX C
SELECTED REFERENCES

BOOKS

American Telephone and Telegraph: *Engineering Economy,* 3d ed., McGraw-Hill, New York, 1977.

Au, T., and T. P. Au: *Engineering Economics and Capital Investment Analysis,* Allyn and Bacon, Boston, MA, 1983.

Barish, N. N., and S. Kaplan: *Economic Analysis for Engineering and Managerial Decision Making,* 2d ed., McGraw-Hill, New York, 1978.

Bierman, H., and S. Smidt: *The Capital Budgeting Decision,* 4th ed., Macmillan, New York, 1975.

Blanchard, B. S.: *Life Cycle Cost*, M/A Press, Portland, OR, 1978.

Blank, L. T., and A. J. Tarquin: *Engineering Economy*, 2d ed., McGraw-Hill, New York, 1983.

Bussey, L. E.: *The Economic Analysis of Industrial Projects*, Prentice-Hall, Englewood Cliffs, NJ, 1978.

Byrd, J., Jr., and L. T. Moore: *Decision Models for Management,* McGraw-Hill, New York, 1982.

Canada, J. R., and J. A. White, Jr.: *Capital Investment Decision Analysis for Management and Engineering*, Prentice-Hall, Englewood Cliffs, NJ, 1980.

Clifton, D. S., and L. E. Fyffe: *Project Feasibility Analysis,* Wiley, New York, 1977.

Cornell, A. H.; *The Decision-Maker's Handbook*, Prentice-Hall, Englewood Cliffs, NJ, 1980.

DeGarmo, E. P., J. R. Canada, and W. G. Sullivan: *Engineering Economy*, 7th ed., Macmillan, New York, 1984.

English, J. M.: *Cost Effectiveness*, Wiley, New York, 1968.

Fabrycky, W. J., and G. J. Thuesen: *Economic Decision Analysis,* 2d ed., Prentice-Hall, Englewood Cliffs, NJ, 1980.

Fleischer, G. A. (ed.): *Risk and Uncertainty: Non-deterministic Decision Making in Engineering Economy*, American Institute of Industrial Engineers, Norcross, GA, 1975.

————: *Engineering Economy—Capital Allocation Theory*, Brooks-Cole, Monterey, CA, 1984.

Fukuda, R.: *Managerial Engineering,* Productivity, Inc., Stamford, CN, 1983.

Goicoecha, A., D. R. Hasen, and L. Duckstein: *Multiobjective Decision Analysis with Engineering and Business Applications*, Wiley, New York, 1983.

Grant, E. L., W. G. Ireson, and R. S. Leavenworth: *Principles of Engineering Economy*, 7th ed., Wiley, New York, 1982.

Hertz, D. B., and H. Thomas: *Risk Analysis and Its Applications*, Wiley, New York, 1983.

Holloway, C. A.: *Decision Making Under Uncertainty*, Prentice-Hall, Englewood Cliffs, NJ, 1979.

Jelen, F. C., and J. H. Black: *Cost and Optimization Engineering*, 2d ed., McGraw-Hill, New York, 1983.

Kenney, R. L., and H. Raiffa: *Decisions with Multiple Objectives: Preferences and Value Tradeoffs*, Wiley, New York, 1976.

Lin, S. A. Y.: *Theory and Measurement of Economic Externalities*, Academic, New York, 1976.

Luce, R. D., and H. Raiffa: *Games and Decisions: Introduction and Critical Survey*, Wiley, New York, 1958.

Mallik, A. K.: *Engineering Economy with Computer Applications*, Engineering Technology, Mahomet, IL, 1979.

Mao, J. C. T.: *Quantitative Analysis of Financial Decisions*, Macmillan, New York, 1969.

Mayer, R. R.: *Capital Expenditure Analysis*, Waveland, Prospect Heights, IL, 1978.

Morris, W. T.: *Engineering Economic Analysis*, Reston, Reston, VA, 1976.

Newnan, D. G.: *Engineering Economic Analysis*, 2d ed., Engineering Press, San Jose, CA, 1980.

Oakford, R. V.: *Capital Budgeting*, Ronald, New York, 1970.

Ostwald, P. F.: *Cost Estimating*, 2d ed., Prentice-Hall, Englewood Cliffs, NJ, 1984.

Palm, T., and A. Qayum: *Private and Public Investment Analysis*, South-Western, Cincinnati, 1985.

Radford, K. J.: *Managerial Decision Making*, Reston, Reston, VA, 1975.

Riggs, J. L.: *Production Systems: Planning, Analysis, and Control*, 3d ed., Wiley, New York, 1981.

———: *Productive Supervision*, Prentice-Hall, Englewood Cliffs, NJ, 1985.

———, and G. H. Felix: *Productivity by Objectives*, Prentice-Hall, Englewood Cliffs, NJ, 1983.

Schall, L. D., and C. W. Haley: *Financial Management*, McGraw-Hill, New York, 1977.

Shamblin, J. E., and G. T. Stevens, Jr.: *Operations Research*, McGraw-Hill, New York, 1974.

Shizuo, S. T., T. Fushimi, and S. Fujita: *Profitability Analysis for Managerial and Engineering Decisions*, Asian Productivity Organization, Tokyo, 1980.

Stevens, G. T., Jr.: *Economics and Financial Analysis of Capital Investments*, Wiley, New York, 1979.

Stuart, R. D.: *Cost Estimating*, Wiley, New York, 1982.

Sumanth, D. J.: *Productivity Engineering and Management*, McGraw-Hill, New York, 1984.

Taylor, G. A.: *Managerial and Engineering Economy,* Van Nostrand, New York, 1980.

Theusen, H. G., W. J. Fabrycky, and G. J. Theusen: *Engineering Economy*, 6th ed., Prentice-Hall, Englewood Cliffs, NJ, 1984.

Weist, J. D., and F. K. Levy: *Management Guide to PERT/CPM,* 2d ed., Prentice-Hall, Englewood Cliffs, NJ, 1977.

Wellington, A. M.: *The Economic Theory of Railway Location,* Wiley, New York, 1887.

White, J. A., M. H. Agee, and K. E. Case: *Principles of Engineering Economic Analysis*, 2d ed., Wiley, New York, 1984.

Wilkes, F. M.: *Capital Budgeting Techniques*, Wiley, New York, 1977.

ARTICLES

Pertinent articles are named throughout the text. The most frequently cited sources for subjects associated with engineering economics are listed below.

AIIE Transactions

American Institute of Industrial Engineers *Conference Proceedings*

Harvard Business Review

Industrial Engineering (called *The Journal of Industrial Engineering* prior to January, 1969)

Journal of Business

Journal of Finance

Journal of Financial and Quantitative Analysis

Management Science

The Economic Journal

The Engineering Economist

APPENDIX D

DISCRETE-COMPOUNDING INTEREST FACTORS

½% Interest Factors for Discrete Compounding Periods

	SINGLE PAYMENT		UNIFORM SERIES					
	Compound Amount Factor	Present Worth Factor	Capital Recovery Factor	Present Worth Factor	Sinking Fund Factor	Compound Amount Factor	Gradient Factor	
N	$(F/P, \frac{1}{2}, N)$	$(P/F, \frac{1}{2}, N)$	$(A/P, \frac{1}{2}, N)$	$(P/A, \frac{1}{2}, N)$	$(A/F, \frac{1}{2}, N)$	$(F/A, \frac{1}{2}, N)$	$(A/G, \frac{1}{2}, N)$	N
1	1.0050	.99503	1.0051	.9949	1.0001	.9998	.0000	1
2	1.0100	.99008	.50385	1.9847	.49885	2.0046	.4613	2
3	1.0150	.98515	.33674	2.9696	.33174	3.0143	.9537	3
4	1.0201	.98025	.25318	3.9497	.24818	4.0292	1.4531	4
5	1.0252	.97538	.20305	4.9248	.19805	5.0491	1.9462	5
6	1.0303	.97052	.16963	5.8951	.16463	6.0741	2.4413	6
7	1.0355	.96570	.14576	6.8606	.14076	7.1043	2.9364	7
8	1.0407	.96089	.12786	7.8213	.12286	8.1396	3.4304	8
9	1.0459	.95611	.11393	8.7772	.10893	9.1800	3.9231	9
10	1.0511	.95136	.10279	9.7282	.09779	10.225	4.4140	10
11	1.0563	.94663	.09368	10.674	.08868	11.276	4.9063	11
12	1.0616	.94192	.08609	11.616	.08109	12.332	5.3959	12
13	1.0669	.93723	.07966	12.553	.07466	13.394	5.8857	13
14	1.0723	.93257	.07415	13.485	.06915	14.460	6.3752	14
15	1.0776	.92793	.06938	14.413	.06438	15.532	6.8614	15
16	1.0830	.92332	.06520	15.336	.06020	16.610	7.3489	16
17	1.0884	.91872	.06152	16.255	.05652	17.693	7.8351	17
18	1.0939	.91415	.05824	17.168	.05324	18.781	8.3198	18
19	1.0993	.90961	.05531	18.078	.05031	19.874	8.8046	19
20	1.1048	.90508	.05268	18.983	.04768	20.974	9.2892	20
21	1.1103	.90058	.05029	19.883	.04529	22.078	9.7715	21
22	1.1159	.89610	.04812	20.779	.04312	23.183	10.253	22
23	1.1215	.89164	.04614	21.671	.04114	24.304	10.735	23
24	1.1271	.88721	.04433	22.558	.03933	25.425	11.216	24
25	1.1327	.88280	.04266	23.440	.03766	26.552	11.695	25
26	1.1384	.87841	.04112	24.318	.03612	27.685	12.173	26
27	1.1441	.87404	.03969	25.192	.03469	28.823	12.652	27
28	1.1498	.86969	.03837	26.062	.03337	29.967	13.129	28
29	1.1555	.86536	.03714	26.927	.03214	31.116	13.605	29
30	1.1613	.86106	.03599	27.788	.03099	32.272	14.081	30
31	1.1671	.85678	.03491	28.644	.02991	33.433	14.555	31
32	1.1730	.85251	.03390	29.497	.02890	34.600	15.029	32
33	1.1788	.84827	.03295	30.345	.02795	35.772	15.501	33
34	1.1847	.84405	.03206	31.189	.02706	36.951	15.974	34
35	1.1906	.83986	.03122	32.028	.02622	38.135	16.446	35
40	1.2207	.81918	.02765	36.164	.02265	44.147	18.790	40
45	1.2515	.79901	.02488	40.198	.01988	50.311	21.113	45
50	1.2831	.77933	.02266	44.133	.01766	56.630	23.416	50
55	1.3155	.76014	.02085	47.971	.01585	63.109	25.699	55
60	1.3487	.74142	.01934	51.715	.01434	69.751	27.960	60
65	1.3828	.72317	.01806	55.366	.01306	76.561	30.201	65
70	1.4177	.70536	.01697	58.928	.01197	83.543	32.422	70
75	1.4535	.68799	.01603	62.401	.01103	90.701	34.622	75
80	1.4902	.67105	.01520	65.790	.01020	98.040	36.802	80
85	1.5278	.65453	.01447	69.094	.00947	105.56	38.961	85
90	1.5663	.63841	.01383	72.318	.00883	113.27	41.099	90
95	1.6059	.62269	.01325	75.462	.00825	121.18	43.218	95
100	1.6464	.60736	.01273	78.528	.00773	129.29	45.316	100

1% Interest Factors for Discrete Compounding Periods

	SINGLE PAYMENT		UNIFORM SERIES					
	Compound Amount Factor	Present Worth Factor	Capital Recovery Factor	Present Worth Factor	Sinking Fund Factor	Compound Amount Factor	Gradient Factor	
N	(F/P, 1, N)	(P/F, 1, N)	(A/P, 1, N)	(P/A, 1, N)	(A/F, 1, N)	(F/A, 1, N)	(A/G, 1, N)	N
1	1.0100	.99010	1.0100	.9900	1.0000	.9999	.0000	1
2	1.0201	.98030	.50757	1.9701	.49757	2.0097	.4864	2
3	1.0303	.97059	.34006	2.9406	.33006	3.0297	.9813	3
4	1.0406	.96099	.25631	3.9014	.24631	4.0598	1.4751	4
5	1.0510	.95147	.20606	4.8528	.19607	5.1003	1.9675	5
6	1.0615	.94205	.17257	5.7947	.16257	6.1512	2.4581	6
7	1.0721	.93273	.14865	6.7273	.13865	7.2125	2.9469	7
8	1.0828	.92349	.13071	7.6507	.12071	8.2845	3.4349	8
9	1.0936	.91435	.11675	8.5649	.10675	9.3672	3.9209	9
10	1.1046	.90530	.10560	9.4701	.09560	10.460	4.4047	10
11	1.1156	.89634	.09647	10.366	.08647	11.565	4.8872	11
12	1.1268	.88746	.08886	11.253	.07886	12.680	5.3682	12
13	1.1380	.87868	.08242	12.132	.07242	13.807	5.8476	13
14	1.1494	.86998	.07691	13.002	.06691	14.945	6.3253	14
15	1.1609	.86137	.07213	13.863	.06213	16.094	6.8010	15
16	1.1725	.85284	.06795	14.716	.05795	17.255	7.2754	16
17	1.1842	.84440	.06427	15.560	.05427	18.427	7.7483	17
18	1.1961	.83604	.06099	16.396	.05099	19.611	8.2192	18
19	1.2080	.82776	.05806	17.223	.04806	20.807	8.6883	19
20	1.2201	.81957	.05542	18.043	.04542	22.015	9.1560	20
21	1.2323	.81145	.05304	18.854	.04304	23.235	9.6222	21
22	1.2446	.80342	.05087	19.658	.04087	24.467	10.086	22
23	1.2571	.79547	.04889	20.453	.03889	25.712	10.549	23
24	1.2696	.78759	.04708	21.240	.03708	26.969	11.010	24
25	1.2823	.77979	.04541	22.020	.03541	28.238	11.469	25
26	1.2952	.77207	.04387	22.792	.03387	29.521	11.927	26
27	1.3081	.76443	.04245	23.556	.03245	30.816	12.383	27
28	1.3212	.75686	.04113	24.313	.03113	32.124	12.838	28
29	1.3344	.74937	.03990	25.062	.02990	33.445	13.291	29
30	1.3478	.74195	.03875	25.804	.02875	34.779	13.742	30
31	1.3612	.73461	.03768	26.539	.02768	36.127	14.191	31
32	1.3748	.72733	.03667	27.266	.02667	37.488	14.640	32
33	1.3886	.72013	.03573	27.986	.02573	38.863	15.086	33
34	1.4025	.71301	.03484	28.699	.02484	40.251	15.531	34
35	1.4165	.70595	.03401	29.405	.02401	41.653	15.973	35
40	1.4887	.67169	.03046	32.831	.02046	48.878	18.164	40
45	1.5647	.63909	.02771	36.090	.01771	56.471	20.314	45
50	1.6445	.60808	.02552	39.192	.01552	64.452	22.423	50
55	1.7284	.57857	.02373	42.142	.01373	72.839	24.491	55
60	1.8165	.55049	.02225	44.950	.01225	81.655	26.520	60
65	1.9092	.52378	.02100	47.622	.01100	90.920	28.508	65
70	2.0065	.49836	.01993	50.163	.00993	100.65	30.457	70
75	2.1089	.47418	.01902	52.582	.00902	110.89	32.366	75
80	2.2164	.45117	.01822	54.883	.00822	121.64	34.236	80
85	2.3295	.42927	.01752	57.072	.00752	132.95	36.067	85
90	2.4483	.40844	.01690	59.156	.00690	144.83	37.859	90
95	2.5732	.38862	.01636	61.138	.00636	157.32	39.614	95
100	2.7044	.36976	.01587	63.024	.00587	170.44	41.330	100

1½% Interest Factors for Discrete Compounding Periods

	SINGLE PAYMENT		UNIFORM SERIES					
	Compound Amount Factor	Present Worth Factor	Capital Recovery Factor	Present Worth Factor	Sinking Fund Factor	Compound Amount Factor	Gradient Factor	
N	$(F/P, 1\frac{1}{2}, N)$	$(P/F, 1\frac{1}{2}, N)$	$(A/P, 1\frac{1}{2}, N)$	$(P/A, 1\frac{1}{2}, N)$	$(A/F, 1\frac{1}{2}, N)$	$(F/A, 1\frac{1}{2}, N)$	$(A/G, 1\frac{1}{2}, N)$	N
1	1.0150	.98522	1.0150	.9852	1.0000	1.0000	.0000	1
2	1.0302	.97066	.51131	1.9557	.49631	2.0148	.4917	2
3	1.0456	.95632	.34340	2.9120	.32840	3.0450	.9857	3
4	1.0613	.94219	.25946	3.8540	.24446	4.0905	1.4760	4
5	1.0772	.92827	.20910	4.7823	.19410	5.1518	1.9653	5
6	1.0934	.91455	.17554	5.6967	.16054	6.2290	2.4511	6
7	1.1098	.90103	.15157	6.5977	.13657	7.3223	2.9351	7
8	1.1264	.88772	.13359	7.4853	.11859	8.4320	3.4161	8
9	1.1433	.87460	.11962	8.3598	.10462	9.5585	3.8952	9
10	1.1605	.86168	.10844	9.2214	.09344	10.701	4.3716	10
11	1.1779	.84894	.09930	10.070	.08430	11.862	4.8456	11
12	1.1956	.83640	.09169	10.906	.07669	13.039	5.3169	12
13	1.2135	.82404	.08525	11.730	.07025	14.235	5.7863	13
14	1.2317	.81186	.07973	12.542	.06473	15.448	6.2524	14
15	1.2502	.79987	.07495	13.342	.05995	16.680	6.7165	15
16	1.2689	.78805	.07077	14.130	.05577	17.930	7.1781	16
17	1.2879	.77640	.06708	14.906	.05208	19.199	7.6374	17
18	1.3073	.76493	.06381	15.671	.04881	20.487	8.0939	18
19	1.3269	.75363	.06088	16.424	.04588	21.794	8.5482	19
20	1.3468	.74249	.05825	17.167	.04325	23.121	8.9998	20
21	1.3670	.73152	.05587	17.898	.04087	24.468	9.4493	21
22	1.3875	.72071	.05371	18.619	.03871	25.834	9.8959	22
23	1.4083	.71006	.05173	19.329	.03673	27.222	10.340	23
24	1.4294	.69957	.04993	20.028	.03493	28.630	10.782	24
25	1.4509	.68923	.04827	20.718	.03327	30.059	11.221	25
26	1.4726	.67904	.04674	21.397	.03174	31.510	11.658	26
27	1.4947	.66901	.04532	22.066	.03032	32.983	12.093	27
28	1.5171	.65912	.04400	22.725	.02900	34.477	12.525	28
29	1.5399	.64938	.04278	23.374	.02778	35.994	12.955	29
30	1.5630	.63979	.04164	24.014	.02664	37.534	13.382	30
31	1.5864	.63033	.04058	24.644	.02558	39.097	13.807	31
32	1.6102	.62102	.03958	25.265	.02458	40.683	14.229	32
33	1.6344	.61184	.03864	25.877	.02364	42.293	14.649	33
34	1.6589	.60280	.03776	26.479	.02276	43.928	15.067	34
35	1.6838	.59389	.03694	27.073	.02194	45.586	15.482	35
40	1.8139	.55129	.03343	29.913	.01843	54.261	17.522	40
45	1.9541	.51174	.03072	32.550	.01572	63.606	19.501	45
50	2.1051	.47504	.02857	34.997	.01357	73.673	21.422	50
55	2.2677	.44096	.02683	37.269	.01183	84.518	23.283	55
60	2.4430	.40933	.02539	39.378	.01039	96.201	25.087	60
65	2.6318	.37997	.02419	41.335	.00919	108.78	26.833	65
70	2.8351	.35271	.02317	43.152	.00817	122.34	28.523	70
75	3.0542	.32741	.02230	44.839	.00730	136.95	30.157	75
80	3.2903	.30392	.02155	46.405	.00655	152.68	31.737	80
85	3.5445	.28212	.02089	47.858	.00589	169.63	33.262	85
90	3.8185	.26188	.02032	49.207	.00532	187.89	34.734	90
95	4.1135	.24310	.01982	50.460	.00482	207.57	36.155	95
100	4.4314	.22566	.01937	51.622	.00437	228.76	37.524	100

2% Interest Factors for Discrete Compounding Periods

	SINGLE PAYMENT		UNIFORM SERIES					
	Compound Amount Factor	Present Worth Factor	Capital Recovery Factor	Present Worth Factor	Sinking Fund Factor	Compound Amount Factor	Gradient Factor	
N	(F/P, 2, N)	(P/F, 2, N)	(A/P, 2, N)	(P/A, 2, N)	(A/F, 2, N)	(F/A, 2, N)	(A/G, 2, N)	N
1	1.0200	.98039	1.0200	.9804	1.0000	1.0000	.0000	1
2	1.0404	.96117	.51507	1.9415	.49507	2.0199	.4934	2
3	1.0612	.94232	.34677	2.8837	.32677	3.0603	.9851	3
4	1.0824	.92385	.26263	3.8075	.24263	4.1214	1.4733	4
5	1.1040	.90573	.21217	4.7132	.19217	5.2038	1.9584	5
6	1.1261	.88798	.17853	5.6012	.15853	6.3078	2.4401	6
7	1.1486	.87056	.15452	6.4717	.13452	7.4339	2.9189	7
8	1.1716	.85350	.13651	7.3252	.11651	8.5826	3.3940	8
9	1.1950	.83676	.12252	8.1619	.10252	9.7541	3.8659	9
10	1.2189	.82035	.11133	8.9822	.09133	10.949	4.3347	10
11	1.2433	.80427	.10218	9.7865	.08218	12.168	4.8001	11
12	1.2682	.78850	.09456	10.574	.07456	13.411	5.2622	12
13	1.2935	.77304	.08812	11.347	.06812	14.679	5.7209	13
14	1.3194	.75788	.08261	12.105	.06261	15.973	6.1764	14
15	1.3458	.74302	.07783	12.848	.05783	17.292	6.6288	15
16	1.3727	.72846	.07365	13.577	.05365	18.638	7.0778	16
17	1.4002	.71417	.06997	14.291	.04997	20.011	7.5236	17
18	1.4282	.70017	.06670	14.991	.04670	21.411	7.9660	18
19	1.4567	.68644	.06378	15.677	.04378	22.839	8.4052	19
20	1.4859	.67298	.06116	16.350	.04116	24.296	8.8412	20
21	1.5156	.65979	.05879	17.010	.03879	25.781	9.2739	21
22	1.5459	.64685	.05663	17.657	.03663	27.297	9.7033	22
23	1.5768	.63417	.05467	18.291	.03467	28.843	10.129	23
24	1.6084	.62173	.05287	18.913	.03287	30.420	10.552	24
25	1.6405	.60954	.05122	19.522	.03122	32.028	10.972	25
26	1.6733	.59759	.04970	20.120	.02970	33.669	11.388	26
27	1.7068	.58588	.04829	20.706	.02829	35.342	11.802	27
28	1.7409	.57439	.04699	21.280	.02699	37.049	12.212	28
29	1.7758	.56313	.04578	21.843	.02578	38.790	12.619	29
30	1.8113	.55208	.04465	22.395	.02465	40.565	13.023	30
31	1.8475	.54126	.04360	22.937	.02360	42.377	13.423	31
32	1.8844	.53065	.04261	23.467	.02261	44.224	13.821	32
33	1.9221	.52024	.04169	23.987	.02169	46.108	14.215	33
34	1.9606	.51004	.04082	24.497	.02082	48.031	14.606	34
35	1.9998	.50004	.04000	24.997	.02000	49.991	14.994	35
40	2.2079	.45291	.03656	27.354	.01656	60.398	16.886	40
45	2.4377	.41021	.03391	29.489	.01391	71.888	18.701	45
50	2.6914	.37154	.03182	31.422	.01182	84.573	20.440	50
55	2.9715	.33652	.03014	33.174	.01014	98.579	22.103	55
60	3.2808	.30480	.02877	34.760	.00877	114.04	23.694	60
65	3.6223	.27607	.02763	36.196	.00763	131.11	25.212	65
70	3.9993	.25004	.02667	37.497	.00667	149.96	26.661	70
75	4.4155	.22647	.02586	38.676	.00586	170.77	28.041	75
80	4.8751	.20512	.02516	39.743	.00516	193.75	29.355	80
85	5.3824	.18579	.02456	40.710	.00456	219.12	30.604	85
90	5.9426	.16827	.02405	41.586	.00405	247.13	31.791	90
95	6.5611	.15241	.02360	42.379	.00360	278.05	32.917	95
100	7.2440	.13804	.02320	43.097	.00320	312.20	33.984	100

2½% Interest Factors for Discrete Compounding Periods

	SINGLE PAYMENT		UNIFORM SERIES					
	Compound Amount Factor	Present Worth Factor	Capital Recovery Factor	Present Worth Factor	Sinking Fund Factor	Compound Amount Factor	Gradient Factor	
N	$(F/P, 2\frac{1}{2}, N)$	$(P/F, 2\frac{1}{2}, N)$	$(A/P, 2\frac{1}{2}, N)$	$(P/A, 2\frac{1}{2}, N)$	$(A/F, 2\frac{1}{2}, N)$	$(F/A, 2\frac{1}{2}, N)$	$(A/G, 2\frac{1}{2}, N)$	N
1	1.0250	.97561	1.0250	.9756	1.0000	1.0000	.0000	1
2	1.0506	.95182	.51884	1.9273	.49384	2.0243	.4930	2
3	1.0768	.92860	.35014	2.8559	.32514	3.0755	.9827	3
4	1.1038	.90595	.26582	3.7618	.24082	4.1524	1.4681	4
5	1.1314	.88386	.21525	4.6457	.19025	5.2562	1.9496	5
6	1.1596	.86230	.18155	5.5079	.15655	6.3875	2.4269	6
7	1.1886	.84127	.15750	6.3492	.13250	7.5472	2.9002	7
8	1.2184	.82075	.13947	7.1699	.11447	8.7358	3.3695	8
9	1.2488	.80073	.12546	7.9707	.10046	9.9542	3.8346	9
10	1.2800	.78120	.11426	8.7518	.08926	11.203	4.2955	10
11	1.3120	.76215	.10511	9.5140	.08011	12.483	4.7524	11
12	1.3448	.74356	.09749	10.257	.07249	13.795	5.2052	12
13	1.3785	.72543	.09105	10.982	.06605	15.140	5.6539	13
14	1.4129	.70773	.08554	11.690	.06054	16.518	6.0985	14
15	1.4482	.69047	.08077	12.381	.05577	17.931	6.5391	15
16	1.4844	.67363	.07660	13.054	.05160	19.379	6.9756	16
17	1.5216	.65720	.07293	13.711	.04793	20.864	7.4081	17
18	1.5596	.64117	.06967	14.353	.04467	22.385	7.8365	18
19	1.5986	.62553	.06676	14.978	.04176	23.945	8.2609	19
20	1.6386	.61028	.06415	15.588	.03915	25.543	8.6813	20
21	1.6795	.59539	.06179	16.184	.03679	27.182	9.0976	21
22	1.7215	.58087	.05965	16.765	.03465	28.861	9.5100	22
23	1.7645	.56671	.05770	17.331	.03270	30.583	9.9183	23
24	1.8087	.55288	.05591	17.884	.03091	32.347	10.322	24
25	1.8539	.53940	.05428	18.424	.02928	34.156	10.723	25
26	1.9002	.52624	.05277	18.950	.02777	36.010	11.119	26
27	1.9477	.51341	.05138	19.463	.02638	37.910	11.512	27
28	1.9964	.50089	.05009	19.964	.02509	39.858	11.900	28
29	2.0463	.48867	.04889	20.453	.02389	41.854	12.285	29
30	2.0975	.47675	.04778	20.929	.02278	43.901	12.665	30
31	2.1499	.46512	.04674	21.395	.02174	45.998	13.042	31
32	2.2037	.45378	.04577	21.848	.02077	48.148	13.415	32
33	2.2588	.44271	.04486	22.291	.01986	50.352	13.784	33
34	2.3152	.43191	.04401	22.723	.01901	52.610	14.149	34
35	2.3731	.42138	.04321	23.144	.01821	54.926	14.511	35
40	2.6850	.37244	.03984	25.102	.01484	67.399	16.261	40
45	3.0378	.32918	.03727	26.832	.01227	81.512	17.917	45
50	3.4370	.29095	.03526	28.361	.01026	97.480	19.483	50
55	3.8886	.25716	.03365	29.713	.00865	115.54	20.959	55
60	4.3996	.22729	.03235	30.908	.00735	135.98	22.351	60
65	4.9777	.20089	.03128	31.964	.00628	159.11	23.659	65
70	5.6318	.17756	.03040	32.897	.00540	185.27	24.887	70
75	6.3719	.15694	.02965	33.722	.00465	214.87	26.038	75
80	7.2092	.13871	.02903	34.451	.00403	248.36	27.115	80
85	8.1565	.12260	.02849	35.095	.00349	286.26	28.122	85
90	9.2283	.10836	.02804	35.665	.00304	329.13	29.062	90
95	10.441	.09578	.02765	36.168	.00265	377.63	29.937	95
100	11.813	.08465	.02731	36.613	.00231	432.51	30.751	100

3% Interest Factors for Discrete Compounding Periods

	SINGLE PAYMENT		UNIFORM SERIES					
	Compound Amount Factor	Present Worth Factor	Capital Recovery Factor	Present Worth Factor	Sinking Fund Factor	Compound Amount Factor	Gradient Factor	
N	$(F/P, 3, N)$	$(P/F, 3, N)$	$(A/P, 3, N)$	$(P/A, 3, N)$	$(A/F, 3, N)$	$(F/A, 3, N)$	$(A/G, 3, N)$	N
1	1.0300	.97087	1.0300	.9709	1.0000	1.0000	.0000	1
2	1.0609	.94260	.52262	1.9134	.49262	2.0299	.4920	2
3	1.0927	.91514	.35354	2.8285	.32354	3.0908	.9795	3
4	1.1255	.88849	.26903	3.7170	.23903	4.1835	1.4622	4
5	1.1592	.86261	.21836	4.5796	.18836	5.3090	1.9401	5
6	1.1940	.83749	.18460	5.4170	.15460	6.4682	2.4129	6
7	1.2298	.81310	.16051	6.2301	.13051	7.6622	2.8809	7
8	1.2667	.78941	.14246	7.0195	.11246	8.8920	3.3440	8
9	1.3047	.76642	.12844	7.7859	.09844	10.158	3.8022	9
10	1.3439	.74410	.11723	8.5300	.08723	11.463	4.2555	10
11	1.3842	.72243	.10808	9.2524	.07808	12.807	4.7040	11
12	1.4257	.70139	.10046	9.9537	.07046	14.191	5.1475	12
13	1.4685	.68096	.09403	10.634	.06403	15.617	5.5863	13
14	1.5125	.66113	.08853	11.295	.05853	17.085	6.0201	14
15	1.5579	.64187	.08377	11.937	.05377	18.598	6.4491	15
16	1.6046	.62318	.07961	12.560	.04961	20.156	6.8732	16
17	1.6528	.60502	.07595	13.165	.04595	21.760	7.2926	17
18	1.7024	.58740	.07271	13.753	.04271	23.413	7.7072	18
19	1.7534	.57030	.06982	14.323	.03982	25.115	8.1169	19
20	1.8060	.55369	.06722	14.877	.03722	26.869	8.5219	20
21	1.8602	.53756	.06487	15.414	.03487	28.675	8.9221	21
22	1.9160	.52190	.06275	15.936	.03275	30.535	9.3176	22
23	1.9735	.50670	.06082	16.443	.03082	32.451	9.7084	23
24	2.0327	.49194	.05905	16.935	.02905	34.425	10.094	24
25	2.0937	.47762	.05743	17.412	.02743	36.457	10.475	25
26	2.1565	.46370	.05594	17.876	.02594	38.551	10.852	26
27	2.2212	.45020	.05457	18.326	.02457	40.707	11.224	27
28	2.2878	.43709	.05329	18.763	.02329	42.929	11.592	28
29	2.3565	.42436	.05212	19.188	.02212	45.217	11.954	29
30	2.4272	.41200	.05102	19.600	.02102	47.573	12.313	30
31	2.5000	.40000	.05000	20.000	.02000	50.000	12.666	31
32	2.5750	.38835	.04905	20.388	.01905	52.500	13.016	32
33	2.6522	.37704	.04816	20.765	.01816	55.075	13.360	33
34	2.7318	.36606	.04732	21.131	.01732	57.727	13.700	34
35	2.8137	.35539	.04654	21.486	.01654	60.459	14.036	35
40	3.2619	.30657	.04326	23.114	.01326	75.397	15.649	40
45	3.7814	.26445	.04079	24.518	.01079	92.715	17.154	45
50	4.3837	.22812	.03887	25.729	.00887	112.79	18.556	50
55	5.0819	.19678	.03735	26.774	.00735	136.06	19.859	55
60	5.8913	.16974	.03613	27.675	.00613	163.04	21.066	60
65	6.8296	.14642	.03515	28.452	.00515	194.32	22.183	65
70	7.9173	.12630	.03434	29.123	.00434	230.57	23.213	70
75	9.1783	.10895	.03367	29.701	.00367	272.61	24.162	75
80	10.640	.09398	.03311	30.200	.00311	321.33	25.034	80
85	12.334	.08107	.03265	30.630	.00265	377.82	25.834	85
90	14.299	.06993	.03226	31.002	.00226	443.31	26.566	90
95	16.576	.06033	.03193	31.322	.00193	519.22	27.234	95
100	19.217	.05204	.03165	31.598	.00165	607.23	27.843	100

4% Interest Factors for Discrete Compounding Periods

	SINGLE PAYMENT		UNIFORM SERIES					
	Compound Amount Factor	Present Worth Factor	Capital Recovery Factor	Present Worth Factor	Sinking Fund Factor	Compound Amount Factor	Gradient Factor	
N	$(F/P, 4, N)$	$(P/F, 4, N)$	$(A/P, 4, N)$	$(P/A, 4, N)$	$(A/F, 4, N)$	$(F/A, 4, N)$	$(A/G, 4, N)$	N
1	1.0400	.96154	1.0400	.9615	1.0000	1.0000	.0000	1
2	1.0816	.92456	.53020	1.8860	.49020	2.0399	.4900	2
3	1.1248	.88900	.36035	2.7750	.32035	3.1215	.9736	3
4	1.1698	.85481	.27549	3.6298	.23549	4.2464	1.4506	4
5	1.2166	.82193	.22463	4.4517	.18463	5.4162	1.9213	5
6	1.2653	.79032	.19076	5.2420	.15076	6.6328	2.3853	6
7	1.3159	.75992	.16661	6.0019	.12661	7.8981	2.8429	7
8	1.3685	.73069	.14853	6.7326	.10853	9.2140	3.2940	8
9	1.4233	.70259	.13449	7.4352	.09449	10.582	3.7387	9
10	1.4802	.67557	.12329	8.1108	.08329	12.005	4.1769	10
11	1.5394	.64958	.11415	8.7603	.07415	13.486	4.6086	11
12	1.6010	.62460	.10655	9.3849	.06655	15.025	5.0339	12
13	1.6650	.60058	.10014	9.9855	.06014	16.626	5.4529	13
14	1.7316	.57748	.09467	10.563	.05467	18.291	5.8655	14
15	1.8009	.55527	.08994	11.118	.04994	20.023	6.2717	15
16	1.8729	.53391	.08582	11.652	.04582	21.824	6.6716	16
17	1.9478	.51338	.08220	12.165	.04220	23.697	7.0652	17
18	2.0257	.49363	.07899	12.659	.03899	25.644	7.4526	18
19	2.1068	.47465	.07614	13.133	.03614	27.670	7.8338	19
20	2.1911	.45639	.07358	13.590	.03358	29.777	8.2087	20
21	2.2787	.43884	.07128	14.029	.03128	31.968	8.5775	21
22	2.3698	.42196	.06920	14.450	.02920	34.247	8.9402	22
23	2.4646	.40573	.06731	14.856	.02731	36.617	9.2969	23
24	2.5632	.39013	.06559	15.246	.02559	39.081	9.6475	24
25	2.6658	.37512	.06401	15.621	.02401	41.644	9.9921	25
26	2.7724	.36069	.06257	15.982	.02257	44.310	10.330	26
27	2.8833	.34682	.06124	16.329	.02124	47.083	10.663	27
28	2.9986	.33348	.06001	16.662	.02001	49.966	10.990	28
29	3.1186	.32066	.05888	16.983	.01888	52.964	11.311	29
30	3.2433	.30832	.05783	17.291	.01783	56.083	11.627	30
31	3.3730	.29647	.05686	17.588	.01686	59.326	11.936	31
32	3.5079	.28506	.05595	17.873	.01595	62.699	12.240	32
33	3.6483	.27410	.05510	18.147	.01510	66.207	12.539	33
34	3.7942	.26356	.05432	18.411	.01432	69.855	12.832	34
35	3.9460	.25342	.05358	18.664	.01358	73.650	13.119	35
40	4.8009	.20829	.05052	19.792	.01052	95.022	14.476	40
45	5.8410	.17120	.04826	20.719	.00826	121.02	15.704	45
50	7.1064	.14072	.04655	21.482	.00655	152.66	16.811	50
55	8.6460	.11566	.04523	22.108	.00523	191.15	17.806	55
60	10.519	.09506	.04420	22.623	.00420	237.98	18.696	60
65	12.798	.07814	.04339	23.046	.00339	294.95	19.490	65
70	15.570	.06422	.04275	23.394	.00275	364.27	20.195	70
75	18.944	.05279	.04223	23.680	.00223	448.60	20.820	75
80	23.048	.04339	.04181	23.915	.00181	551.21	21.371	80
85	28.042	.03566	.04148	24.108	.00148	676.05	21.856	85
90	34.117	.02931	.04121	24.267	.00121	827.93	22.282	90
95	41.508	.02409	.04099	24.397	.00099	1012.7	22.654	95
100	50.501	.01980	.04081	24.504	.00081	1237.5	22.979	100

5% Interest Factors for Discrete Compounding Periods

	SINGLE PAYMENT		UNIFORM SERIES					
N	Compound Amount Factor	Present Worth Factor	Capital Recovery Factor	Present Worth Factor	Sinking Fund Factor	Compound Amount Factor	Gradient Factor	N
	(F/P, 5, N)	(P/F, 5, N)	(A/P, 5, N)	(P/A, 5, N)	(A/F, 5, N)	(F/A, 5, N)	(A/G, 5, N)	
1	1.0500	.95238	1.0500	.9524	1.0000	1.0000	.0000	1
2	1.1025	.90703	.53781	1.8593	.48781	2.0499	.4874	2
3	1.1576	.86384	.36722	2.7231	.31722	3.1524	.9671	3
4	1.2155	.82271	.28202	3.5458	.23202	4.3100	1.4386	4
5	1.2762	.78353	.23098	4.3294	.18098	5.5255	1.9021	5
6	1.3400	.74622	.19702	5.0756	.14702	6.8017	2.3575	6
7	1.4070	.71069	.17282	5.7862	.12282	8.1418	2.8048	7
8	1.4774	.67684	.15472	6.4631	.10472	9.5488	3.2441	8
9	1.5513	.64461	.14069	7.1077	.09069	11.026	3.6753	9
10	1.6288	.61392	.12951	7.7216	.07951	12.577	4.0986	10
11	1.7103	.58469	.12039	8.3062	.07039	14.206	4.5140	11
12	1.7958	.55684	.11283	8.8631	.06283	15.916	4.9214	12
13	1.8856	.53033	.10646	9.3934	.05646	17.712	5.3211	13
14	1.9799	.50507	.10103	9.8985	.05103	19.598	5.7128	14
15	2.0789	.48102	.09634	10.379	.04634	21.577	6.0969	15
16	2.1828	.45812	.09227	10.837	.04227	23.656	6.4732	16
17	2.2919	.43630	.08870	11.273	.03870	25.839	6.8418	17
18	2.4065	.41553	.08555	11.689	.03555	28.131	7.2029	18
19	2.5269	.39574	.08275	12.085	.03275	30.538	7.5565	19
20	2.6532	.37690	.08024	12.462	.03024	33.064	7.9025	20
21	2.7859	.35895	.07800	12.821	.02800	35.718	8.2412	21
22	2.9252	.34186	.07597	13.162	.02597	38.503	8.5725	22
23	3.0714	.32558	.07414	13.488	.02414	41.429	8.8966	23
24	3.2250	.31008	.07247	13.798	.02247	44.500	9.2135	24
25	3.3862	.29531	.07095	14.093	.02095	47.725	9.5234	25
26	3.5555	.28125	.06956	14.375	.01957	51.111	9.8261	26
27	3.7333	.26786	.06829	14.642	.01829	54.667	10.122	27
28	3.9200	.25510	.06712	14.898	.01712	58.400	10.411	28
29	4.1160	.24295	.06605	15.140	.01605	62.320	10.693	29
30	4.3218	.23138	.06505	15.372	.01505	66.436	10.968	30
31	4.5379	.22037	.06413	15.592	.01413	70.757	11.237	31
32	4.7647	.20987	.06328	15.802	.01328	75.295	11.500	32
33	5.0030	.19988	.06249	16.002	.01249	80.060	11.756	33
34	5.2531	.19036	.06176	16.192	.01176	85.063	12.005	34
35	5.5158	.18130	.06107	16.374	.01107	90.316	12.249	35
40	7.0397	.14205	.05828	17.158	.00828	120.79	13.277	40
45	8.9846	.11130	.05626	17.773	.00626	159.69	14.364	45
50	11.466	.08721	.05478	18.255	.00478	209.33	15.223	50
55	14.634	.06833	.05367	18.633	.00367	272.69	15.966	55
60	18.678	.05354	.05283	18.929	.00283	353.56	16.605	60
65	23.838	.04195	.05219	19.161	.00219	456.76	17.153	65
70	30.424	.03287	.05170	19.342	.00170	588.48	17.621	70
75	38.829	.02575	.05132	19.484	.00132	756.59	18.017	75
80	49.557	.02018	.05103	19.596	.00103	971.14	18.352	80
85	63.248	.01581	.05080	19.683	.00080	1244.9	18.634	85
90	80.723	.01239	.05063	19.752	.00063	1594.4	18.871	90
95	103.02	.00971	.05049	19.805	.00049	2040.4	19.068	95
100	131.48	.00761	.05038	19.847	.00038	2609.7	19.233	100

6% Interest Factors for Discrete Compounding Factors

	SINGLE PAYMENT		UNIFORM SERIES					
N	Compound Amount Factor	Present Worth Factor	Capital Recovery Factor	Present Worth Factor	Sinking Fund Factor	Compound Amount Factor	Gradient Factor	N
	$(F/P, 6, N)$	$(P/F, 6, N)$	$(A/P, 6, N)$	$(P/A, 6, N)$	$(A/F, 6, N)$	$(F/A, 6, N)$	$(A/G, 6, N)$	N
1	1.0600	.94340	1.0600	.9434	1.0000	1.0000	.0000	1
2	1.1236	.89000	.54544	1.8333	.48544	2.0599	.4852	2
3	1.1910	.83962	.37411	2.6729	.31411	3.1835	.9610	3
4	1.2624	.79210	.28860	3.4650	.22860	4.3745	1.4269	4
5	1.3382	.74726	.23740	4.2123	.17740	5.6370	1.8833	5
6	1.4185	.70496	.20337	4.9172	.14337	6.9751	2.3301	6
7	1.5036	.66506	.17914	5.5823	.11914	8.3936	2.7673	7
8	1.5938	.62742	.16104	6.2097	.10104	9.8972	3.1949	8
9	1.6894	.59190	.14702	6.8016	.08702	11.491	3.6130	9
10	1.7908	.55840	.13587	7.3600	.07587	13.180	4.0217	10
11	1.8982	.52679	.12679	7.8867	.06679	14.971	4.4210	11
12	2.0121	.49698	.11928	8.3837	.05928	16.869	4.8109	12
13	2.1329	.46884	.11296	8.8525	.05296	18.881	5.1917	13
14	2.2608	.44231	.10759	9.2948	.04759	21.014	5.5632	14
15	2.3965	.41727	.10296	9.7121	.04296	23.275	5.9257	15
16	2.5403	.39365	.09895	10.105	.03895	25.671	6.2791	16
17	2.6927	.37137	.09545	10.477	.03545	28.212	6.6237	17
18	2.8542	.35035	.09236	10.827	.03236	30.904	6.9594	18
19	3.0255	.33052	.08962	11.158	.02962	33.759	7.2864	19
20	3.2070	.31181	.08719	11.469	.02719	36.784	7.6048	20
21	3.3995	.29416	.08501	11.763	.02501	39.991	7.9148	21
22	3.6034	.27751	.08305	12.041	.02305	43.390	8.2163	22
23	3.8196	.26180	.08128	12.303	.02128	46.994	8.5096	23
24	4.0488	.24698	.07968	12.550	.01968	50.814	8.7948	24
25	4.2917	.23300	.07823	12.783	.01823	54.862	9.0719	25
26	4.5492	.21982	.07690	13.003	.01690	59.154	9.3412	26
27	4.8222	.20737	.07570	13.210	.01570	63.703	9.6027	27
28	5.1115	.19564	.07459	13.406	.01459	68.525	9.8565	28
29	5.4182	.18456	.07358	13.590	.01358	73.637	10.102	29
30	5.7433	.17412	.07265	13.764	.01265	79.055	10.341	30
31	6.0879	.16426	.07179	13.929	.01179	84.798	10.573	31
32	6.4531	.15496	.07100	14.083	.01100	90.886	10.798	32
33	6.8403	.14619	.07027	14.230	.01027	97.339	11.016	33
34	7.2507	.13792	.06960	14.368	.00960	104.17	11.227	34
35	7.6858	.13011	.06897	14.498	.00897	111.43	11.431	35
40	10.285	.09723	.06646	15.046	.00646	154.75	12.358	40
45	13.764	.07265	.06470	15.455	.00470	212.73	13.141	45
50	18.419	.05429	.06344	15.761	.00344	290.32	13.796	50
55	24.649	.04057	.06254	15.990	.00254	394.14	14.340	55
60	32.985	.03032	.06188	16.161	.00188	533.09	14.790	60
65	44.142	.02265	.06139	16.289	.00139	719.03	15.160	65
70	59.071	.01693	.06103	16.384	.00103	967.86	15.461	70
75	79.051	.01265	.06077	16.455	.00077	1300.8	15.705	75
80	105.78	.00945	.06057	16.509	.00057	1746.4	15.903	80
85	141.56	.00706	.06043	16.548	.00043	2342.7	16.061	85
90	189.44	.00528	.06032	16.578	.00032	3140.7	16.189	90
95	253.52	.00394	.06024	16.600	.00024	4208.7	16.290	95
100	339.26	.00295	.06018	16.617	.00018	5637.8	16.371	100

7% Interest Factors for Discrete Compounding Periods

	SINGLE PAYMENT		UNIFORM SERIES					
	Compound Amount Factor	Present Worth Factor	Capital Recovery Factor	Present Worth Factor	Sinking Fund Factor	Compound Amount Factor	Gradient Factor	
N	$(F/P, 7, N)$	$(P/F, 7, N)$	$(A/P, 7, N)$	$(P/A, 7, N)$	$(A/F, 7, N)$	$(F/A, 7, N)$	$(A/G, 7, N)$	N
1	1.0700	.93458	1.0700	.9346	1.0000	1.000	.0000	1
2	1.1449	.87344	.55310	1.8080	.48310	2.0699	.4830	2
3	1.2250	.81630	.38105	2.6242	.31105	3.2148	.9548	3
4	1.3107	.76290	.29523	3.3871	.22523	4.4398	1.4153	4
5	1.4025	.71299	.24389	4.1001	.17389	5.7506	1.8648	5
6	1.5007	.66635	.20980	4.7665	.13980	7.1531	2.3030	6
7	1.6057	.62275	.18555	5.3892	.11555	8.6539	2.7302	7
8	1.7181	.58201	.16747	5.9712	.09747	10.259	3.1463	8
9	1.8384	.54394	.15349	6.5151	.08349	11.977	3.5515	9
10	1.9671	.50835	.14238	7.0235	.07238	13.816	3.9459	10
11	2.1048	.47510	.13336	7.4986	.06336	15.783	4.3294	11
12	2.2521	.44402	.12590	7.9426	.05590	17.888	4.7023	12
13	2.4098	.41497	.11965	8.3576	.04965	20.140	5.0647	13
14	2.5785	.38782	.11435	8.7454	.04435	22.550	5.4165	14
15	2.7590	.36245	.10980	9.1078	.03980	25.128	5.7581	15
16	2.9521	.33874	.10586	9.4466	.03586	27.887	6.0895	16
17	3.1587	.31658	.10243	9.7631	.03243	30.839	6.4108	17
18	3.3798	.29587	.09941	10.059	.02941	33.998	6.7223	18
19	3.6164	.27651	.09675	10.335	.02675	37.378	7.0240	19
20	3.8696	.25842	.09439	10.593	.02439	40.994	7.3161	20
21	4.1404	.24152	.09229	10.835	.02229	44.864	7.5988	21
22	4.4303	.22572	.09041	11.061	.02041	49.004	7.8723	22
23	4.7404	.21095	.08871	11.272	.01871	53.434	8.1367	23
24	5.0722	.19715	.08719	11.469	.01719	58.175	8.3922	24
25	5.4273	.18425	.08581	11.653	.01581	63.247	8.6389	25
26	5.8072	.17220	.08456	11.825	.01456	68.674	8.8772	26
27	6.2137	.16093	.08343	11.986	.01343	74.481	9.1070	27
28	6.6486	.15041	.08239	12.137	.01239	80.695	9.3288	28
29	7.1140	.14057	.08145	12.277	.01145	87.344	9.5425	29
30	7.6120	.13137	.08059	12.409	.01059	94.458	9.7485	30
31	8.1449	.12278	.07980	12.531	.00980	102.07	9.9469	31
32	8.7150	.11474	.07907	12.646	.00907	110.21	10.137	32
33	9.3250	.10724	.07841	12.753	.00841	118.92	10.321	33
34	9.9778	.10022	.07780	12.853	.00780	128.25	10.498	34
35	10.676	.09367	.07723	12.947	.00723	138.23	10.668	35
40	14.973	.06678	.07501	13.331	.00501	199.62	11.423	40
45	21.001	.04762	.07350	13.605	.00350	285.73	12.035	45
50	29.455	.03395	.07246	13.800	.00246	406.51	12.528	50
55	41.313	.02421	.07174	13.939	.00174	575.90	12.921	55
60	57.943	.01726	.07123	14.039	.00123	813.47	13.232	60
65	81.268	.01230	.07087	14.109	.00087	1146.6	13.475	65
70	113.98	.00877	.07062	14.160	.00062	1614.0	13.666	70
75	159.86	.00626	.07044	14.196	.00044	2269.5	13.813	75
80	224.21	.00446	.07031	14.222	.00031	3188.8	13.927	80
85	314.47	.00318	.07022	14.240	.00022	4478.2	14.014	85
90	441.06	.00227	.07016	14.253	.00016	6286.7	14.081	90
95	618.62	.00162	.07011	14.262	.00011	8823.1	14.131	95
100	867.64	.00115	.07008	14.269	.00008	12381.7	14.170	100

8% Interest Factors for Discrete Compounding Periods

N	SINGLE PAYMENT		UNIFORM SERIES					N
	Compound Amount Factor	Present Worth Factor	Capital Recovery Factor	Present Worth Factor	Sinking Fund Factor	Compound Amount Factor	Gradient Factor	
	$(F/P, 8, N)$	$(P/F, 8, N)$	$(A/P, 8, N)$	$(P/A, 8, N)$	$(A/F, 8, N)$	$(F/A, 8, N)$	$(A/G, 8, N)$	
1	1.0800	.92593	1.0800	.9259	1.0000	1.0000	.0000	1
2	1.1664	.85734	.56077	1.7832	.48077	2.0799	.4807	2
3	1.2597	.79383	.38803	2.5770	.30804	3.2463	.9487	3
4	1.3604	.73503	.30192	3.3121	.22192	4.5060	1.4038	4
5	1.4693	.68059	.25046	3.9926	.17046	5.8665	1.8463	5
6	1.5868	.63017	.21632	4.6228	.13632	7.3358	2.2762	6
7	1.7138	.58349	.19207	5.2063	.11207	8.9227	2.6935	7
8	1.8509	.54027	.17402	5.7466	.09402	10.636	3.0984	8
9	1.9989	.50025	.16008	6.2468	.08008	12.487	3.4909	9
10	2.1589	.46320	.14903	6.7100	.06903	14.486	3.8712	10
11	2.3316	.42889	.14008	7.1389	.06008	16.645	4.2394	11
12	2.5181	.39712	.13270	7.5360	.05270	18.976	4.5956	12
13	2.7196	.36770	.12642	7.9037	.04652	21.495	4.9401	13
14	2.9371	.34046	.12130	8.2442	.04130	24.214	5.2729	14
15	3.1721	.31524	.11683	8.5594	.03683	27.151	5.5943	15
16	3.4259	.29189	.11298	8.8513	.03298	30.323	5.9045	16
17	3.6999	.27027	.10963	9.1216	.02963	33.749	6.2036	17
18	3.9959	.25025	.10670	9.3718	.02670	37.449	6.4919	18
19	4.3156	.23171	.10413	9.6035	.02413	41.445	6.7696	19
20	4.6609	.21455	.10185	9.8181	.02185	45.761	7.0368	20
21	5.0337	.19866	.09983	10.016	.01983	50.422	7.2939	21
22	5.4364	.18394	.09803	10.200	.01803	55.455	7.5411	22
23	5.8713	.17032	.09642	10.371	.01642	60.892	7.7785	23
24	6.3410	.15770	.09498	10.528	.01498	66.763	8.0065	24
25	6.8483	.14602	.09368	10.674	.01368	73.104	8.2253	25
26	7.3962	.13520	.09251	10.809	.01251	79.953	8.4351	26
27	7.9879	.12519	.09145	10.935	.01145	87.349	8.6362	27
28	8.6269	.11592	.09049	11.051	.01049	95.337	8.8288	28
29	9.3171	.10733	.08962	11.158	.00962	103.96	9.0132	29
30	10.062	.09938	.08883	11.257	.00883	113.28	9.1896	30
31	10.867	.09202	.08811	11.349	.00811	123.34	9.3583	31
32	11.736	.08520	.08745	11.434	.00745	134.21	9.5196	32
33	12.675	.07889	.08685	11.513	.00685	145.94	9.6736	33
34	13.689	.07305	.08630	11.586	.00630	158.62	9.8207	34
35	14.785	.06764	.08580	11.654	.00580	172.31	9.9610	35
40	21.724	.04603	.08386	11.924	.00386	259.05	10.569	40
45	31.919	.03133	.08259	12.108	.00259	386.49	11.044	45
50	46.900	.02132	.08174	12.233	.00174	573.75	11.410	50
55	68.911	.01451	.08118	12.318	.00118	848.89	11.690	55
60	101.25	.00988	.08080	12.376	.00080	1253.1	11.901	60
65	148.77	.00672	.08054	12.416	.00054	1847.1	12.060	65
70	218.59	.00457	.08037	12.442	.00037	2719.9	12.178	70
75	321.19	.00311	.08025	12.461	.00025	4002.3	12.265	75
80	471.93	.00212	.08017	12.473	.00017	5886.6	12.330	80
85	693.42	.00144	.08012	12.481	.00012	8655.2	12.377	85
90	1018.8	.00098	.08008	12.487	.00008	12723.9	12.411	90
95	1497.0	.00067	.08005	12.491	.00005	18701.5	12.436	95
100	2199.6	.00045	.08004	12.494	.00004	27484.5	12.454	100

9% Interest Factors for Discrete Compounding Periods

	SINGLE PAYMENT		UNIFORM SERIES					
	Compound Amount Factor	Present Worth Factor	Capital Recovery Factor	Present Worth Factor	Sinking Fund Factor	Compound Amount Factor	Gradient Factor	
N	$(F/P, 9, N)$	$(P/F, 9, N)$	$(A/P, 9, N)$	$(P/A, 9, N)$	$(A/F, 9, N)$	$(F/A, 9, N)$	$(A/G, 9, N)$	N
1	1.0900	.91743	1.0900	.9174	1.0000	1.0000	.0000	1
2	1.1881	.84168	.56847	1.7591	.47847	2.0899	.4784	2
3	1.2950	.77219	.39506	2.5312	.30506	3.2780	.9425	3
4	1.4115	.70843	.30867	3.2396	.21867	4.5730	1.3923	4
5	1.5386	.64993	.25709	3.8896	.16709	5.9846	1.8280	5
6	1.6770	.59627	.22292	4.4858	.13292	7.5232	2.2496	6
7	1.8280	.54704	.19869	5.0329	.10869	9.2002	2.6572	7
8	1.9925	.50187	.18068	5.5347	.09068	11.028	3.0510	8
9	2.1718	.46043	.16680	5.9952	.07680	13.020	3.4311	9
10	2.3673	.42241	.15582	6.4176	.06582	15.192	3.7976	10
11	2.5804	.38754	.14695	6.8051	.05695	17.559	4.1508	11
12	2.8126	.35554	.13965	7.1606	.04965	20.140	4.4909	12
13	3.0657	.32618	.13357	7.4868	.04357	22.952	4.8180	13
14	3.3416	.29925	.12843	7.7861	.03843	26.018	5.1325	14
15	3.6424	.27454	.12406	8.0606	.03406	29.360	5.4345	15
16	3.9702	.25187	.12030	8.3125	.03030	33.002	5.7243	16
17	4.3275	.23108	.11705	8.5435	.02705	36.972	6.0022	17
18	4.7170	.21200	.11421	8.7555	.02421	41.300	6.2685	18
19	5.1415	.19449	.11173	8.9500	.02173	46.017	6.5234	19
20	5.6043	.17843	.10955	9.1285	.01955	51.158	6.7673	20
21	6.1086	.16370	.10762	9.2922	.01762	56.763	7.0004	21
22	6.6584	.15018	.10591	9.4423	.01591	62.871	7.2231	22
23	7.2577	.13778	.10438	9.5801	.01438	69.530	7.4356	23
24	7.9109	.12641	.10302	9.7065	.01302	76.787	7.6383	24
25	8.6228	.11597	.10181	9.8225	.01181	84.698	7.8315	25
26	9.3989	.10640	.10072	9.9289	.01072	93.321	8.0154	26
27	10.244	.09761	.09974	10.026	.00974	102.72	8.1905	27
28	11.166	.08955	.09885	10.116	.00885	112.96	8.3570	28
29	12.171	.08216	.09806	10.198	.00806	124.13	8.5153	29
30	13.267	.07537	.09734	10.273	.00734	136.30	8.6655	30
31	14.461	.06915	.09669	10.342	.00669	149.57	8.8082	31
32	15.762	.06344	.09610	10.406	.00610	164.03	8.9435	32
33	17.181	.05820	.09556	10.464	.00556	179.79	9.0717	33
34	18.727	.05340	.09508	10.517	.00508	196.97	9.1932	34
35	20.413	.04899	.09464	10.566	.00464	215.70	9.3082	35
40	31.408	.03184	.09296	10.757	.00296	337.86	9.7956	40
45	48.325	.02069	.09190	10.881	.00190	525.83	10.160	45
50	74.353	.01345	.09123	10.961	.00123	815.04	10.429	50
55	114.40	.00874	.09079	11.014	.00079	1260.0	10.626	55
60	176.02	.00568	.09051	11.047	.00051	1944.6	10.768	60
65	270.82	.00369	.09033	11.070	.00033	2998.0	10.870	65
70	416.70	.00240	.09022	11.084	.00022	4618.9	10.942	70
75	641.14	.00156	.09014	11.093	.00014	7112.7	10.993	75
80	986.47	.00101	.09009	11.099	.00009	10950.6	11.029	80
85	1517.8	.00066	.09006	11.103	.00006	16854.8	11.055	85
90	2335.3	.00043	.09004	11.106	.00004	25939.2	11.072	90
95	3593.1	.00028	.09002	11.108	.00003	39916.6	11.084	95
100	5528.4	.00018	.09002	11.109	.00002	61422.7	11.093	100

10% Interest Factors for Discrete Compounding Periods

N	SINGLE PAYMENT		UNIFORM SERIES					N
	Compound Amount Factor	Present Worth Factor	Capital Recovery Factor	Present Worth Factor	Sinking Fund Factor	Compound Amount Factor	Gradient Factor	
	$(F/P, 10, N)$	$(P/F, 10, N)$	$(A/P, 10, N)$	$(P/A, 10, N)$	$(A/F, 10, N)$	$(F/A, 10, N)$	$(A/G, 10, N)$	
1	1.1000	.90909	1.1000	.9091	1.0000	1.000	.0000	1
2	1.2100	.82645	.57619	1.7355	.47619	2.0999	.4761	2
3	1.3310	.75132	.40212	2.4868	.30212	3.3099	.9365	3
4	1.4641	.68302	.31547	3.1698	.21547	4.6409	1.3810	4
5	1.6105	.62092	.26380	3.7907	.16380	6.1050	1.8100	5
6	1.7715	.56448	.22961	4.3552	.12961	7.7155	2.2234	6
7	1.9487	.51316	.20541	4.8683	.10541	9.4870	2.6215	7
8	2.1435	.46651	.18745	5.3349	.08745	11.435	3.0043	8
9	2.3579	.42410	.17364	5.7589	.07364	13.579	3.3722	9
10	2.5937	.38555	.16275	6.1445	.06275	15.937	3.7253	10
11	2.8530	.35050	.15396	6.4950	.05396	18.530	4.0639	11
12	3.1384	.31863	.14676	6.8136	.04676	21.383	4.3883	12
13	3.4522	.28967	.14078	7.1033	.04078	24.522	4.6987	13
14	3.7974	.26333	.13575	7.3666	.03575	27.974	4.9954	14
15	4.1771	.23940	.13147	7.6060	.03147	31.771	5.2788	15
16	4.5949	.21763	.12782	7.8236	.02782	35.949	5.5492	16
17	5.0544	.19785	.12466	8.0215	.02466	40.543	5.8070	17
18	5.5598	.17986	.12193	8.2013	.02193	45.598	6.0524	18
19	6.1158	.16351	.11955	8.3649	.01955	51.158	6.2860	19
20	6.7273	.14865	.11746	8.5135	.01746	57.273	6.5080	20
21	7.4001	.13513	.11562	8.6486	.01562	64.001	6.7188	21
22	8.1401	.12285	.11401	8.7715	.01401	71.401	6.9188	22
23	8.9541	.11168	.11257	8.8832	.01257	79.541	7.1084	23
24	9.8495	.10153	.11130	8.9847	.01130	88.495	7.2879	24
25	10.834	.09230	.11017	9.0770	.01017	98.344	7.4579	25
26	11.917	.08391	.10916	9.1609	.00916	109.17	7.6185	26
27	13.109	.07628	.10826	9.2372	.00826	121.09	7.7703	27
28	14.420	.06935	.10745	9.3065	.00745	134.20	7.9136	28
29	15.862	.06304	.10673	9.3696	.00673	148.62	8.0488	29
30	17.448	.05731	.10608	9.4269	.00608	164.48	8.1761	30
31	19.193	.05210	.10550	9.4790	.00550	181.93	8.2961	31
32	21.113	.04736	.10497	9.5263	.00497	201.13	8.4090	32
33	23.224	.04306	.10450	9.5694	.00450	222.24	8.5151	33
34	25.546	.03914	.10407	9.6085	.00407	245.46	8.6149	34
35	28.101	.03559	.10369	9.6441	.00369	271.01	8.7085	35
40	45.257	.02210	.10226	9.7790	.00226	442.57	9.0962	40
45	72.887	.01372	.10139	9.8628	.00139	718.87	9.3740	45
50	117.38	.00852	.10086	9.9148	.00086	1163.8	9.5704	50
55	189.04	.00529	.10053	9.9471	.00053	1880.4	9.7075	55
60	304.46	.00328	.10033	9.9671	.00033	3034.6	9.8022	60
65	490.34	.00204	.10020	9.9796	.00020	4893.4	9.8671	65
70	789.69	.00127	.10013	9.9873	.00013	7886.9	9.9112	70
75	1271.8	.00079	.10008	9.9921	.00008	12709.0	9.9409	75
80	2048.2	.00049	.10005	9.9951	.00005	20474.0	9.9609	80
85	3298.7	.00030	.10003	9.9969	.00003	32979.7	9.9742	85
90	5312.5	.00019	.10002	9.9981	.00002	53120.2	9.9830	90
95	8555.9	.00012	.10001	9.9988	.00001	85556.8	9.9889	95
100	13780.6	.00007	.10001	9.9992	.00001	137796.1	9.9927	100

11% Interest Factors for Discrete Compounding Periods

	SINGLE PAYMENT		UNIFORM SERIES					
	Compound Amount Factor	Present Worth Factor	Capital Recovery Factor	Present Worth Factor	Sinking Fund Factor	Compound Amount Factor	Gradient Factor	
N	$(F/P, 11, N)$	$(P/F, 11, N)$	$(A/P, 11, N)$	$(P/A, 11, N)$	$(A/F, 11, N)$	$(F/A, 11, N)$	$(A/G, 11, N)$	N
1	1.1100	.90090	1.1100	.9009	1.0000	1.000	.0000	1
2	1.2321	.81162	.58394	1.7125	.47394	2.1099	.4739	2
3	1.3676	.73119	.40922	2.4437	.29922	3.3420	.9305	3
4	1.5180	.65873	.32233	3.1024	.21233	4.7097	1.3698	4
5	1.6850	.59345	.27057	3.6958	.16057	6.2277	1.7922	5
6	1.8704	.53464	.23638	4.2305	.12638	7.9128	2.1975	6
7	2.0761	.48166	.21222	4.7121	.10222	9.7831	2.5862	7
8	2.3045	.43393	.19432	5.1461	.08432	11.859	2.9584	8
9	2.5580	.39093	.18060	5.5370	.07060	14.163	3.3143	9
10	2.8394	.35219	.16980	5.8892	.05980	16.721	3.6543	10
11	3.1517	.31729	.16112	6.2065	.05112	19.561	3.9787	11
12	3.4984	.28584	.15403	6.4923	.04403	22.712	4.2878	12
13	3.8832	.25752	.14815	6.7498	.03815	26.211	4.5821	13
14	4.3104	.23200	.14323	6.9818	.03323	30.094	4.8618	14
15	4.7845	.20901	.13907	7.1908	.02907	34.404	5.1274	15
16	5.3108	.18829	.13552	7.3791	.02552	39.189	5.3793	16
17	5.8950	.16963	.13247	7.5487	.02247	44.500	5.6180	17
18	6.5434	.15282	.12984	7.7016	.01984	50.395	5.8438	18
19	7.2632	.13768	.12756	7.8392	.01756	56.938	6.0573	19
20	8.0622	.12404	.12558	7.9633	.01558	64.201	6.2589	20
21	8.9490	.11174	.12384	8.0750	.01384	72.264	6.4490	21
22	9.9334	.10067	.12231	8.1757	.01231	81.213	6.6282	22
23	11.026	.09069	.12097	8.2664	.01097	91.146	6.7969	23
24	12.238	.08171	.11979	8.3481	.00979	102.17	6.9554	24
25	13.585	.07361	.11874	8.4217	.00874	114.41	7.1044	25
26	15.079	.06631	.11781	8.4880	.00781	127.99	7.2442	26
27	16.738	.05974	.11699	8.5478	.00699	143.07	7.3753	27
28	18.579	.05382	.11626	8.6016	.00626	159.81	7.4981	28
29	20.623	.04849	.11561	8.6501	.00561	178.39	7.6130	29
30	22.891	.04368	.11502	8.6937	.00502	199.01	7.7205	30
31	25.409	.03935	.11451	8.7331	.00451	221.90	7.8209	31
32	28.204	.03545	.11404	8.7686	.00404	247.31	7.9146	32
33	31.307	.03194	.11363	8.8005	.00363	275.52	8.0020	33
34	34.751	.02878	.11326	8.8293	.00326	306.83	8.0835	34
35	38.573	.02592	.11293	8.8552	.00293	341.58	8.1594	35
40	64.999	.01538	.11172	8.9510	.00172	581.81	8.4659	40
45	109.52	.00913	.11101	9.0079	.00101	986.60	8.6762	45
50	184.55	.00542	.11060	9.0416	.00060	1668.7	8.8185	50

12% Interest Factors for Discrete Compounding Periods

	SINGLE PAYMENT		UNIFORM SERIES					
	Compound Amount Factor	Present Worth Factor	Capital Recovery Factor	Present Worth Factor	Sinking Fund Factor	Compound Amount Factor	Gradient Factor	
N	$(F/P, 12, N)$	$(P/F, 12, N)$	$(A/P, 12, N)$	$(P/A, 12, N)$	$(A/F, 12, N)$	$(F/A, 12, N)$	$(A/G, 12, N)$	N
1	1.1200	.89286	1.1200	.8929	1.0000	1.0000	.0000	1
2	1.2544	.79719	.59170	1.6900	.47170	2.1200	.4717	2
3	1.4049	.71178	.41635	2.4018	.29635	3.3743	.9246	3
4	1.5735	.63552	.32924	3.0373	.20924	4.7793	1.3588	4
5	1.7623	.56743	.27741	3.6047	.15741	6.3528	1.7745	5
6	1.9738	.50663	.24323	4.1114	.12323	8.115	2.1720	6
7	2.2106	.45235	.21912	4.5637	.09912	10.088	2.5514	7
8	2.4759	.40388	.20130	4.9676	.08130	12.299	2.9131	8
9	2.7730	.36061	.18768	5.3282	.06768	14.775	3.2573	9
10	3.1058	.32197	.17698	5.6502	.05698	17.548	3.5846	10
11	3.4785	.28748	.16842	5.9376	.04842	20.654	3.8952	11
12	3.8959	.25668	.16144	6.1943	.04144	24.132	4.1896	12
13	4.3634	.22918	.15568	6.4235	.03568	28.028	4.4682	13
14	4.8870	.20462	.15087	6.6281	.03087	32.392	4.7316	14
15	5.4735	.18270	.14682	6.8108	.02682	37.279	4.9802	15
16	6.1303	.16312	.14339	6.9739	.02339	42.752	5.2146	16
17	6.8659	.14565	.14046	7.1196	.02046	48.883	5.4352	17
18	7.6899	.13004	.13794	7.2496	.01794	55.749	5.6427	18
19	8.6126	.11611	.13576	7.3657	.01576	63.439	5.8375	19
20	9.6462	.10367	.13388	7.4694	.01388	72.051	6.0201	20
21	10.803	.09256	.13224	7.5620	.01224	81.698	6.1913	21
22	12.100	.08264	.13081	7.6446	.01081	92.501	6.3513	22
23	13.552	.07379	.12956	7.7184	.00956	104.60	6.5009	23
24	15.178	.06588	.12846	7.7843	.00846	118.15	6.6406	24
25	16.999	.05882	.12750	7.8431	.00750	133.33	6.7708	25
26	19.039	.05252	.12665	7.8956	.00665	150.33	6.8920	26
27	21.324	.04689	.12590	7.9425	.00590	169.37	7.0049	27
28	23.883	.04187	.12524	7.9844	.00524	190.69	7.1097	28
29	26.749	.03738	.12466	8.0218	.00466	214.58	7.2071	29
30	29.959	.03338	.12414	8.0551	.00414	241.32	7.2974	30
31	33.554	.02980	.12369	8.0849	.00369	271.28	7.3810	31
32	37.581	.02661	.12328	8.1116	.00328	304.84	7.4585	32
33	42.090	.02376	.12292	8.1353	.00292	342.42	7.5302	33
34	47.141	.02121	.12260	8.1565	.00260	384.51	7.5964	34
35	52.798	.01894	.12232	8.1755	.00232	431.65	7.6576	35
40	93.049	.01075	.12130	8.2437	.00130	767.07	7.8987	40
45	163.98	.00610	.12074	8.2825	.00074	1358.2	8.0572	45
50	288.99	.00346	.12042	8.3045	.00042	2399.9	8.1597	50

13% Interest Factors for Discrete Compounding Periods

	SINGLE PAYMENT		UNIFORM SERIES					
	Compound Amount Factor	Present Worth Factor	Capital Recovery Factor	Present Worth Factor	Sinking Fund Factor	Compound Amount Factor	Gradient Factor	
N	$(F/P, 13, N)$	$(P/F, 13, N)$	$(A/P, 13, N)$	$(P/A, 13, N)$	$(A/F, 13, N)$	$(F/A, 13, N)$	$(A/G, 13, N)$	N
1	1.1300	.88496	1.1300	.8850	1.0000	1.0000	.0000	1
2	1.2769	.78315	.59949	1.6680	.46949	2.1299	.4694	2
3	1.4428	.69305	.42352	2.3611	.29353	3.4068	.9187	3
4	1.6304	.61332	.33620	2.9744	.20620	4.8497	1.3478	4
5	1.8424	.54276	.28432	3.5172	.15432	6.4802	1.7570	5
6	2.0819	.48032	.25015	3.9975	.12015	8.3226	2.1467	6
7	2.3525	.42506	.22611	4.4225	.09611	10.404	2.5170	7
8	2.6584	.37616	.20839	4.7987	.07839	12.757	2.8684	8
9	3.0040	.33289	.19487	5.1316	.06487	15.415	3.2013	9
10	3.3945	.29459	.18429	5.4262	.05429	18.419	3.5161	10
11	3.8358	.26070	.17584	5.6869	.04584	21.813	3.8133	11
12	4.3344	.23071	.16899	5.9176	.03899	25.649	4.0935	12
13	4.8979	.20417	.16335	6.1217	.03335	29.984	4.3572	13
14	5.5346	.18068	.15867	6.3024	.02867	34.882	4.6049	14
15	6.2541	.15989	.15474	6.4623	.02474	40.416	4.8374	15
16	7.0672	.14150	.15143	6.6038	.02143	46.670	5.0551	16
17	7.9859	.12522	.14861	6.7290	.01861	53.737	5.2588	17
18	9.0240	.11081	.14620	6.8399	.01620	61.723	5.4490	18
19	10.197	.09807	.14413	6.9379	.01413	70.747	5.6264	19
20	11.522	.08678	.14235	7.0247	.01235	80.944	5.7916	20
21	13.020	.07680	.14081	7.1015	.01081	92.467	5.9453	21
22	14.713	.06796	.13948	7.1695	.00948	105.48	6.0880	22
23	16.626	.06015	.13832	7.2296	.00832	120.20	6.2204	23
24	18.787	.05323	.13731	7.2828	.00731	136.82	6.3430	24
25	21.229	.04710	.13643	7.3299	.00643	155.61	6.4565	25
26	23.989	.04168	.13565	7.3716	.00565	176.84	6.5613	26
27	27.108	.03689	.13498	7.4085	.00498	200.83	6.6581	27
28	30.632	.03265	.13439	7.4412	.00439	227.94	6.7474	28
29	34.614	.02889	.13387	7.4700	.00387	258.57	6.8295	29
30	39.114	.02557	.13341	7.4956	.00341	293.18	6.9052	30
31	44.199	.02262	.13301	7.5182	.00301	332.30	6.9747	31
32	49.945	.02002	.13266	7.5383	.00266	376.50	7.0385	32
33	56.438	.01772	.13234	7.5560	.00234	426.44	7.0970	33
34	63.775	.01568	.13207	7.5717	.00207	482.88	7.1506	34
35	72.065	.01388	.13183	7.5855	.00183	546.65	7.1998	35
40	132.77	.00753	.13099	7.6343	.00099	1013.6	7.3887	40
45	244.62	.00409	.13053	7.6608	.00053	1874.0	7.5076	45
50	450.71	.00222	.13029	7.6752	.00029	3459.3	7.5811	50

14% Interest Factors for Discrete Compounding Periods

	SINGLE PAYMENT		UNIFORM SERIES					
	Compound Amount Factor	Present Worth Factor	Capital Recovery Factor	Present Worth Factor	Sinking Fund Factor	Compound Amount Factor	Gradient Factor	
N	$(F/P, 14, N)$	$(P/F, 14, N)$	$(A/P, 14, N)$	$(P/A, 14, N)$	$(A/F, 14, N)$	$(F/A, 14, N)$	$(A/G, 14, N)$	N
1	1.1400	.87719	1.1400	.8772	1.0000	1.000	.0000	1
2	1.2996	.76947	.60729	1.6466	.46729	2.1399	.4672	2
3	1.4815	.67497	.43073	2.3216	.29073	3.4395	.9129	3
4	1.6889	.59208	.34321	2.9137	.20321	4.9211	1.3369	4
5	1.9254	.51937	.29128	3.4330	.15128	6.6100	1.7398	5
6	2.1949	.45559	.25716	3.8886	.11716	8.535	2.1217	6
7	2.5022	.39964	.23319	4.2882	.09319	10.730	2.4831	7
8	2.8525	.35056	.21557	4.6388	.07557	13.232	2.8245	8
9	3.2519	.30751	.20217	4.9463	.06217	16.085	3.1462	9
10	3.7071	.26975	.19171	5.2161	.05171	19.337	3.4489	10
11	4.2261	.23662	.18339	5.4527	.04339	23.044	3.7332	11
12	4.8178	.20756	.17667	5.6602	.03667	27.270	3.9997	12
13	5.4923	.18207	.17116	5.8423	.03116	32.088	4.2490	13
14	6.2612	.15971	.16661	6.0020	.02661	37.580	4.4819	14
15	7.1378	.14010	.16281	6.1421	.02281	43.841	4.6990	15
16	8.1371	.12289	.15962	6.2650	.01962	50.979	4.9010	16
17	9.2763	.10780	.15692	6.3728	.01692	59.116	5.0888	17
18	10.574	.09456	.15462	6.4674	.01462	68.392	5.2629	18
19	12.055	.08295	.15266	6.5503	.01266	78.967	5.4242	19
20	13.743	.07276	.15099	6.6231	.01099	91.022	5.5734	20
21	15.667	.06383	.14954	6.6869	.00955	104.76	5.7111	21
22	17.860	.05599	.14830	6.7429	.00830	120.43	5.8380	22
23	20.361	.04911	.14723	6.7920	.00723	138.29	5.9549	23
24	23.211	.04308	.14630	6.8351	.00630	158.65	6.0623	24
25	26.461	.03779	.14550	6.8729	.00550	181.86	6.1609	25
26	30.165	.03315	.14480	6.9060	.00480	208.32	6.2514	26
27	34.388	.02908	.14419	6.9351	.00419	238.49	6.3342	27
28	39.203	.02551	.14366	6.9606	.00366	272.88	6.4039	28
29	44.691	.02238	.14320	6.9830	.00320	312.08	6.4791	29
30	50.948	.01963	.14280	7.0026	.00280	356.77	6.5422	30
31	58.081	.01722	.14245	7.0198	.00245	407.72	6.5997	31
32	66.212	.01510	.14215	7.0349	.00215	465.80	6.6521	32
33	75.482	.01325	.14188	7.0482	.00188	532.01	6.6998	33
34	86.049	.01162	.14165	7.0598	.00165	607.49	6.7430	34
35	98.096	.01019	.14144	7.0700	.00144	693.54	6.7824	35
40	188.87	.00529	.14075	7.1050	.00075	1341.9	6.9299	40
45	363.66	.00275	.14039	7.1232	.00039	2590.4	7.0187	45
50	700.19	.00143	.14020	7.1326	.00020	4994.2	7.0713	50

15% Interest Factors for Discrete Compounding Periods

N	SINGLE PAYMENT		UNIFORM SERIES					N
	Compound Amount Factor	Present Worth Factor	Capital Recovery Factor	Present Worth Factor	Sinking Fund Factor	Compound Amount Factor	Gradient Factor	
	$(F/P, 15, N)$	$(P/F, 15, N)$	$(A/P, 15, N)$	$(P/A, 15, N)$	$(A/F, 15, N)$	$(F/A, 15, N)$	$(A/G, 15, N)$	
1	1.1500	.86957	1.1500	.8696	1.0000	1.000	.0000	1
2	1.3225	.75614	.61512	1.6257	.46512	2.1499	.4651	2
3	1.5208	.65752	.43798	2.2832	.28798	3.4724	.9071	3
4	1.7490	.57175	.35027	2.8549	.20027	4.9933	1.3262	4
5	2.0113	.49718	.29832	3.3521	.14832	6.7423	1.7227	5
6	2.3130	.43233	.26424	3.7844	.11424	8.7536	2.0971	6
7	2.6600	.37594	.24036	4.1604	.09036	11.066	2.4498	7
8	3.0590	.32690	.22285	4.4873	.07285	13.726	2.7813	8
9	3.5178	.28426	.20957	4.7715	.05957	16.785	3.0922	9
10	4.0455	.24719	.19925	5.0187	.04925	20.303	3.3831	10
11	4.6523	.21494	.19107	5.2337	.04107	24.349	3.6549	11
12	5.3502	.18691	.18448	5.4206	.03448	29.001	3.9081	12
13	6.1527	.16253	.17911	5.5831	.02911	34.351	4.1437	13
14	7.0756	.14133	.17469	5.7244	.02469	40.504	4.3623	14
15	8.1369	.12290	.17102	5.8473	.02102	47.579	4.5649	15
16	9.3575	.10687	.16795	5.9542	.01795	55.716	4.7522	16
17	10.761	.09293	.16537	6.0471	.01537	65.074	4.9250	17
18	12.375	.08081	.16319	6.1279	.01319	75.835	5.0842	18
19	14.231	.07027	.16134	6.1982	.01134	88.210	5.2307	19
20	16.366	.06110	.15976	6.2593	.00976	102.44	5.3651	20
21	18.821	.05313	.15842	6.3124	.00842	118.80	5.4883	21
22	21.644	.04620	.15727	6.3586	.00727	137.62	5.6010	22
23	24.891	.04018	.15628	6.3988	.00628	159.27	5.7039	23
24	28.624	.03493	.15543	6.4337	.00543	184.16	5.7978	24
25	32.918	.03038	.15470	6.4641	.00470	212.78	5.8834	25
26	37.856	.02642	.15407	6.4905	.00407	245.70	5.9612	26
27	43.534	.02297	.15353	6.5135	.00353	283.56	6.0318	27
28	50.064	.01997	.15306	6.5335	.00306	327.09	6.0959	28
29	57.574	.01737	.15265	6.5508	.00265	377.16	6.1540	29
30	66.210	.01510	.15230	6.5659	.00230	434.73	6.2066	30
31	76.141	.01313	.15200	6.5791	.00200	500.94	6.2541	31
32	87.563	.01142	.15173	6.5905	.00173	577.08	6.2970	32
33	100.69	.00993	.15150	6.6004	.00150	664.65	6.3356	33
34	115.80	.00864	.15131	6.6091	.00131	765.34	6.3705	34
35	133.17	.00751	.15113	6.6166	.00113	881.14	6.4018	35
40	267.85	.00373	.15056	6.6417	.00056	1779.0	6.5167	40
45	538.75	.00186	.15028	6.6543	.00028	3585.0	6.5829	45
50	1083.6	.00092	.15014	6.6605	.00014	7217.4	6.8204	50

20% Interest Factors for Discrete Compounding Periods

	SINGLE PAYMENT		UNIFORM SERIES					
	Compound Amount Factor	Present Worth Factor	Capital Recovery Factor	Present Worth Factor	Sinking Fund Factor	Compound Amount Factor	Gradient Factor	
N	$(F/P, 20, N)$	$(P/F, 20, N)$	$(A/P, 20, N)$	$(P/A, 20, N)$	$(A/F, 20, N)$	$(F/A, 20, N)$	$(A/G, 20, N)$	N
1	1.2000	.83333	1.2000	.8333	1.0000	1.0000	.0000	1
2	1.4400	.69445	.65455	1.5277	.45455	2.1999	.4545	2
3	1.7280	.57870	.47473	2.1064	.27473	3.6399	.8791	3
4	2.0736	.48225	.38629	2.5887	.18629	5.3679	1.2742	4
5	2.4883	.40188	.33438	2.9906	.13438	7.4415	1.6405	5
6	2.9859	.33490	.30071	3.3255	.10071	9.9298	1.9788	6
7	3.5831	.27908	.27742	3.6045	.07742	12.915	2.2901	7
8	4.2998	.23257	.26061	3.8371	.06061	16.498	2.5756	8
9	5.1597	.19381	.24808	4.0309	.04808	20.798	2.8364	9
10	6.1917	.16151	.23852	4.1924	.03852	25.958	3.0738	10
11	7.4300	.13459	.23110	4.3270	.03110	32.150	3.2892	11
12	8.9160	.11216	.22527	4.4392	.02527	39.580	3.4840	12
13	10.699	.09346	.22062	4.5326	.02062	48.496	3.6596	13
14	12.839	.07789	.21689	4.6105	.01689	59.195	3.8174	14
15	15.406	.06491	.21388	4.6754	.01388	72.034	3.9588	15
16	18.488	.05409	.21144	4.7295	.01144	87.441	4.0851	16
17	22.185	.04507	.20944	4.7746	.00944	105.92	4.1975	17
18	26.623	.03756	.20781	4.8121	.00781	128.11	4.2975	18
19	31.947	.03130	.20646	4.8435	.00646	154.73	4.3860	19
20	38.337	.02608	.20536	4.8695	.00536	186.68	4.4643	20
21	46.004	.02174	.20444	4.8913	.00444	225.02	4.5333	21
22	55.205	.01811	.20369	4.9094	.00369	271.02	4.5941	22
23	66.246	.01510	.20307	4.9245	.00307	326.23	4.6474	23
24	79.495	.01258	.20255	4.9371	.00255	392.47	4.6942	24
25	95.394	.01048	.20212	4.9475	.00212	471.97	4.7351	25
26	114.47	.00874	.20176	4.9563	.00176	567.36	4.7708	26
27	137.36	.00728	.20147	4.9636	.00147	681.84	4.8020	27
28	164.84	.00607	.20122	4.9696	.00122	819.21	4.8291	28
29	197.81	.00506	.20102	4.9747	.00102	984.05	4.8526	29
30	237.37	.00421	.20085	4.9789	.00085	1181.8	4.8730	30
31	284.84	.00351	.20070	4.9824	.00070	1419.2	4.8907	31
32	341.81	.00293	.20059	4.9853	.00059	1704.0	4.9061	32
33	410.17	.00244	.20049	4.9878	.00049	2045.8	4.9193	33
34	492.21	.00203	.20041	4.9898	.00041	2456.0	4.9307	34
35	590.65	.00169	.20034	4.9915	.00034	2948.2	4.9406	35
40	1469.7	.00068	.20014	4.9966	.00014	7343.6	4.9727	40
45	3657.1	.00027	.20005	4.9986	.00005	18281.3	4.9876	45
50	9100.1	.00011	.20002	4.9994	.00002	45497.2	4.9945	50

25% Interest Factors for Discrete Compounding Periods

	SINGLE PAYMENT		UNIFORM SERIES					
	Compound Amount Factor	Present Worth Factor	Capital Recovery Factor	Present Worth Factor	Sinking Fund Factor	Compound Amount Factor	Gradient Factor	
N	$(F/P, 25, N)$	$(P/F, 25, N)$	$(A/P, 25, N)$	$(P/A, 25, N)$	$(A/F, 25, N)$	$(F/A, 25, N)$	$(A/G, 25, N)$	N
1	1.2500	.80000	1.2500	.8000	1.0000	1.0000	.00000	1
2	1.5625	.64000	.69444	1.4400	.44444	2.2500	.44444	2
3	1.9531	.51200	.51230	1.9520	.26230	3.8125	.85246	3
4	2.4414	.40960	.42344	2.3616	.17344	5.7656	1.2249	4
5	3.0518	.32768	.37185	2.6893	.12185	8.2070	1.5631	5
6	3.8147	.26214	.33882	2.9514	.08882	11.259	1.8683	6
7	4.7684	.20972	.31634	3.1661	.06634	15.073	2.1424	7
8	5.9605	.16777	.30040	3.3289	.05040	19.842	2.3872	8
9	7.4506	.13422	.28876	3.4631	.03876	25.802	2.6048	9
10	9.3132	.10737	.28007	3.5705	.03007	33.253	2.7971	10
11	11.642	.08590	.27349	3.6564	.02349	42.566	2.9663	11
12	14.552	.06872	.26845	3.7251	.01845	54.208	3.1145	12
13	18.190	.05498	.26454	3.7801	.01454	68.760	3.2437	13
14	22.737	.04398	.26150	3.8241	.01150	86.949	3.3559	14
15	28.422	.03518	.25912	3.8593	.00912	109.687	3.4530	15
16	35.527	.02815	.25724	3.8874	.00724	138.109	3.5366	16
17	44.409	.02252	.25576	3.9099	.00576	173.636	3.6084	17
18	55.511	.01801	.25459	3.9279	.00459	218.045	3.6698	18
19	69.389	.01441	.25366	3.9424	.00366	273.556	3.7222	19
20	86.736	.01153	.25292	3.9539	.00292	342.945	3.7667	20
21	108.420	.00922	.25233	3.9631	.00233	429.681	3.8045	21
22	135.525	.00738	.25186	3.9705	.00186	538.101	3.8365	22
23	169.407	.00590	.25148	3.9764	.00148	673.626	3.8634	23
24	211.758	.00472	.25119	3.9811	.00119	843.033	3.8861	24
25	264.698	.00378	.25095	3.9849	.00095	1054.791	3.9052	25
26	330.872	.00302	.25076	3.9879	.00076	1319.489	3.9212	26
27	413.590	.00242	.25061	3.9903	.00061	1650.361	3.9346	27
28	516.988	.00193	.25048	3.9923	.00048	2063.952	3.9457	28
29	646.235	.00155	.25039	3.9938	.00039	2580.939	3.9551	29
30	807.794	.00124	.25031	3.9950	.00031	3227.174	3.9628	30
31	1009.742	.00099	.25025	3.9960	.00025	4034.968	3.9693	31
32	1262.177	.00079	.25020	3.9968	.00020	5044.710	3.9746	32
33	1577.722	.00063	.25016	3.9975	.00016	6306.887	3.9791	33
34	1972.152	.00051	.25013	3.9980	.00012	7884.609	3.9828	34
35	2465.190	.00041	.25010	3.9984	.00010	9856.761	3.9858	35

30% Interest Factors for Discrete Compounding Periods

	SINGLE PAYMENT		UNIFORM SERIES					
	Compound Amount Factor	Present Worth Factor	Capital Recovery Factor	Present Worth Factor	Sinking Fund Factor	Compound Amount Factor	Gradient Factor	
N	$(F/P, 30, N)$	$(P/F, 30, N)$	$(A/P, 30, N)$	$(P/A, 30, N)$	$(A/F, 30, N)$	$(F/A, 30, N)$	$(A/G, 30, N)$	N
1	1.3000	.76923	1.3000	.7692	1.0000	1.000	.0000	1
2	1.6900	.59172	.73478	1.3609	.43478	2.2999	.4348	2
3	2.1969	.45517	.55063	1.8161	.25063	3.9899	.8277	3
4	2.8560	.35013	.46163	2.1662	.16163	6.1869	1.1782	4
5	3.7129	.26933	.41058	2.4355	.11058	9.0430	1.4903	5
6	4.8267	.20718	.37839	2.6427	.07839	12.755	1.7654	6
7	6.2748	.15937	.35687	2.8021	.05687	17.582	2.0062	7
8	8.1572	.12259	.34192	2.9247	.04192	23.857	2.2155	8
9	10.604	.09430	.33124	3.0190	.03124	32.014	2.3962	9
10	13.785	.07254	.32346	3.0915	.02346	42.619	2.5512	10
11	17.921	.05580	.31773	3.1473	.01773	56.404	2.6832	11
12	23.297	.04292	.31345	3.1902	.01345	74.326	2.7951	12
13	30.287	.03302	.31024	3.2232	.01024	97.624	2.8894	13
14	39.373	.02540	.30782	3.2486	.00782	127.91	2.9685	14
15	51.185	.01954	.30598	3.2682	.00598	167.28	3.0344	15
16	66.540	.01503	.30458	3.2832	.00458	218.46	3.0892	16
17	86.503	.01156	.30351	3.2948	.00351	285.01	3.1345	17
18	112.45	.00889	.30269	3.3036	.00269	371.51	3.1718	18
19	146.18	.00684	.30207	3.3105	.00207	483.96	3.2024	19
20	190.04	.00526	.30159	3.3157	.00159	630.15	3.2275	20
21	247.06	.00405	.30122	3.3198	.00122	820.20	3.2479	21
22	321.17	.00311	.30094	3.3229	.00094	1067.2	3.2646	22
23	417.53	.00240	.30072	3.3253	.00072	1388.4	3.2781	23
24	542.79	.00184	.30055	3.3271	.00055	1805.9	3.2890	24
25	705.62	.00142	.30043	3.3286	.00043	2348.7	3.2978	25
26	917.31	.00109	.30033	3.3297	.00033	3054.3	3.3049	26
27	1192.5	.00084	.30025	3.3305	.00025	3971.6	3.3106	27
28	1550.2	.00065	.30019	3.3311	.00019	5164.1	3.3152	28
29	2015.3	.00050	.30015	3.3316	.00015	6714.4	3.3189	29
30	2619.9	.00038	.30011	3.3320	.00011	8729.7	3.3218	30
31	3405.9	.00029	.30009	3.3323	.00009	11350.0	3.3242	31
32	4427.6	.00023	.30007	3.3325	.00007	14756.0	3.3261	32
33	5755.9	.00017	.30005	3.3327	.00005	19184.0	3.3276	33
34	7482.7	.00013	.30004	3.3328	.00004	24940.0	3.3287	34
35	9727.5	.00010	.30003	3.3329	.00003	32423.0	3.3297	35

40% Interest Factors for Discrete Compounding Periods

	SINGLE PAYMENT		UNIFORM SERIES					
	Compound Amount Factor	Present Worth Factor	Capital Recovery Factor	Present Worth Factor	Sinking Fund Factor	Compound Amount Factor	Gradient Factor	
N	$(F/P, 40, N)$	$(P/F, 40, N)$	$(A/P, 40, N)$	$(P/A, 40, N)$	$(A/F, 40, N)$	$(F/A, 40, N)$	$(A/G, 40, N)$	N
1	1.4000	.71429	1.40000	.7143	1.00000	1.0000	.0000	1
2	1.9600	.51020	.81667	1.2244	.41667	2.3999	.4167	2
3	2.7440	.36443	.62936	1.5889	.22936	4.3599	.7798	3
4	3.8415	.26031	.54077	1.8492	.14077	7.1039	1.0923	4
5	5.3782	.18593	.49136	2.0351	.09136	10.945	1.3579	5
6	7.5295	.13281	.46126	2.1679	.06126	16.323	1.5810	6
7	10.541	.09486	.44192	2.2628	.04192	23.853	1.7663	7
8	14.757	.06776	.42907	2.3306	.02907	34.394	1.9185	8
9	20.660	.04840	.42034	2.3790	.02034	49.152	2.0422	9
10	28.925	.03457	.41432	2.4135	.01432	69.813	2.1419	10
11	40.495	.02469	.41013	2.4382	.01013	98.738	2.2214	11
12	56.693	.01764	.40718	2.4559	.00718	139.23	2.2845	12
13	79.370	.01260	.40510	2.4685	.00510	195.92	2.3341	13
14	111.11	.00900	.40363	2.4775	.00363	275.29	2.3728	14
15	155.56	.00643	.40259	2.4839	.00259	386.41	2.4029	15
16	217.79	.00459	.40184	2.4885	.00185	541.98	2.4262	16
17	304.91	.00328	.40132	2.4918	.00132	759.77	2.4440	17
18	426.87	.00234	.40094	2.4941	.00094	1064.6	2.4577	18
19	597.62	.00167	.40067	2.4958	.00067	1491.5	2.4681	19
20	836.67	.00120	.40048	2.4970	.00048	2089.1	2.4760	20
21	1171.3	.00085	.40034	2.4978	.00034	2925.8	2.4820	21
22	1639.8	.00061	.40024	2.4984	.00024	4097.1	2.4865	22
23	2295.8	.00044	.40017	2.4989	.00017	5737.0	2.4899	23
24	3214.1	.00031	.40012	2.4992	.00012	8032.8	2.4925	24
25	4499.8	.00022	.40009	2.4994	.00009	11247.2	2.4944	25

50% Interest Factors for Discrete Compounding Periods

N	SINGLE PAYMENT		UNIFORM SERIES					N
	Compound Amount Factor	Present Worth Factor	Capital Recovery Factor	Present Worth Factor	Sinking Fund Factor	Compound Amount Factor	Gradient Factor	
	$(F/P, 50, N)$	$(P/F, 50, N)$	$(A/P, 50, N)$	$(P/A, 50, N)$	$(A/F, 50, N)$	$(F/A, 50, N)$	$(A/G, 50, N)$	
1	1.5000	.66667	1.5000	.6667	1.00000	1.000	.0000	1
2	2.2500	.44444	.90000	1.1111	.40000	2.500	.4000	2
3	3.3750	.29630	.71053	1.4074	.21053	4.750	.7368	3
4	5.0625	.19753	.62308	1.6049	.12308	8.125	1.0153	4
5	7.5937	.13169	.57583	1.7366	.07583	13.187	1.2417	5
6	11.390	.08779	.54812	1.8244	.04812	20.781	1.4225	6
7	17.085	.05853	.53108	1.8829	.03108	32.171	1.5648	7
8	25.628	.03902	.52030	1.9219	.02030	49.257	1.6751	8
9	38.443	.02601	.51335	1.9479	.01335	74.886	1.7596	9
10	57.665	.01734	.50882	1.9653	.00882	113.33	1.8235	10
11	86.497	.01156	.50585	1.9768	.00585	170.99	1.8713	11
12	129.74	.00771	.50388	1.9845	.00388	257.49	1.9067	12
13	194.61	.00514	.50258	1.9897	.00258	387.23	1.9328	13
14	291.92	.00343	.50172	1.9931	.00172	581.85	1.9518	14
15	437.89	.00228	.50114	1.9954	.00114	873.78	1.9656	15
16	656.84	.00152	.50076	1.9969	.00076	1311.6	1.9756	16
17	985.26	.00101	.50051	1.9979	.00051	1968.5	1.9827	17
18	1477.8	.00068	.50034	1.9986	.00034	2953.7	1.9878	18
19	2216.8	.00045	.50023	1.9991	.00023	4431.6	1.9914	19
20	3325.2	.00030	.50015	1.9994	.00015	6648.5	1.9939	20
21	4987.8	.00020	.50010	1.9996	.00010	9973.7	1.9957	21
22	7481.8	.00013	.50007	1.9997	.00007	14961.7	1.9970	22
23	11222.7	.00009	.50004	1.9998	.00004	22443.5	1.9979	23
24	16834.1	.00006	.50003	1.9998	.00003	33666.2	1.9985	24
25	25251.2	.00004	.50002	1.9999	.00002	50500.3	1.9990	25

60% Interest Factors for Discrete Compounding Periods

	SINGLE PAYMENT		UNIFORM SERIES					
	Compound Amount Factor	Present Worth Factor	Capital Recovery Factor	Present Worth Factor	Sinking Fund Factor	Compound Amount Factor	Gradient Factor	
N	$(F/P, 60, N)$	$(P/F, 60, N)$	$(A/P, 60, N)$	$(P/A, 60, N)$	$(A/F, 60, N)$	$(F/A, 60, N)$	$(A/G, 60, N)$	N
1	1.6000	.62500	1.6000	.6250	1.0000	1.000	.0000	1
2	2.5600	.39063	.98462	1.0156	.38462	2.6000	.3846	2
3	4.0959	.24414	.79380	1.2597	.19380	5.1599	.6977	3
4	6.5535	.15259	.70804	1.4123	.10804	9.2559	.9464	4
5	10.485	.09537	.66325	1.5077	.06325	15.809	1.1395	5
6	16.777	.05960	.63803	1.5673	.03803	26.295	1.2863	6
7	26.843	.03725	.62322	1.6045	.02322	43.072	1.3958	7
8	42.949	.02328	.61430	1.6278	.01430	69.915	1.4759	8
9	68.719	.01455	.60886	1.6424	.00886	112.86	1.5337	9
10	109.95	.00909	.60551	1.6515	.00551	181.58	1.5748	10
11	175.92	.00568	.60343	1.6571	.00343	291.53	1.6037	11
12	281.47	.00355	.60214	1.6607	.00214	467.45	1.6238	12
13	450.35	.00222	.60134	1.6629	.00134	748.92	1.6377	13
14	720.57	.00139	.60083	1.6643	.00083	1199.2	1.6472	14
15	1152.9	.00087	.60052	1.6652	.00052	1919.8	1.6536	15
16	1844.6	.00054	.60033	1.6657	.00033	3072.7	1.6579	16
17	2951.4	.00034	.60020	1.6661	.00020	4917.4	1.6609	17
18	4722.3	.00021	.60013	1.6663	.00013	7868.8	1.6628	18
19	7555.7	.00013	.60008	1.6664	.00008	12591.0	1.6641	19
20	12089.0	.00008	.60005	1.6665	.00005	20147.0	1.6650	20

70% Interest Factors for Discrete Compounding Periods

	SINGLE PAYMENT		UNIFORM SERIES					
	Compound Amount Factor	Present Worth Factor	Capital Recovery Factor	Present Worth Factor	Sinking Fund Factor	Compound Amount Factor	Gradient Factor	
N	$(F/P, 70, N)$	$(P/F, 70, N)$	$(A/P, 70, N)$	$(P/A, 70, N)$	$(A/F, 70, N)$	$(F/A, 70, N)$	$(A/G, 70, N)$	N
1	1.7000	.58824	1.7000	.5882	1.0000	1.000	.0000	1
2	2.8900	.34602	1.0703	.9343	.37037	2.700	.3704	2
3	4.9130	.20354	.87889	1.1378	.17889	5.590	.6619	3
4	8.3520	.11973	.79521	1.2575	.09521	10.502	.8845	4
5	14.198	.07043	.75304	1.3279	.05304	18.855	1.0497	5
6	24.137	.04143	.73025	1.3693	.03025	33.053	1.1692	6
7	41.033	.02437	.71749	1.3937	.01749	57.191	1.2537	7
8	69.757	.01434	.71018	1.4080	.01018	98.224	1.3122	8
9	118.58	.00843	.70595	1.4165	.00595	167.98	1.3520	9
10	201.59	.00496	.70349	1.4214	.00349	286.56	1.3787	10
11	342.71	.00292	.70205	1.4244	.00205	488.16	1.3963	11
12	582.62	.00172	.70120	1.4261	.00120	830.88	1.4079	12
13	990.45	.00101	.70071	1.4271	.00071	1413.5	1.4154	13
14	1683.7	.00059	.70042	1.4277	.00042	2403.9	1.4202	14
15	2862.4	.00035	.70024	1.4280	.00024	4087.7	1.4233	15
16	4866.0	.00021	.70014	1.4282	.00014	6950.1	1.4252	16
17	8272.3	.00012	.70008	1.4284	.00008	11816.0	1.4265	17
18	14063.0	.00007	.70005	1.4284	.00005	20089.0	1.4272	18
19	23907.0	.00004	.70003	1.4285	.00003	34152.0	1.4277	19
20	40642.0	.00002	.70002	1.4285	.00002	58059.0	1.4280	20

80% Interest Factors for Discrete Compounding Periods

	SINGLE PAYMENT		UNIFORM SERIES					
	Compound Amount Factor	Present Worth Factor	Capital Recovery Factor	Present Worth Factor	Sinking Fund Factor	Compound Amount Factor	Gradient Factor	
N	$(F/P, 80, N)$	$(P/F, 80, N)$	$(A/P, 80, N)$	$(P/A, 80, N)$	$(A/F, 80, N)$	$(F/A, 80, N)$	$(A/G, 80, N)$	N
1	1.8000	.55556	1.8000	.5556	1.00000	1.0000	.0000	1
2	3.2400	.30864	1.1571	.8642	.35714	2.8000	.3571	2
3	5.8319	.17147	.96556	1.0356	.16556	6.0399	.6291	3
4	10.497	.09526	.88423	1.1309	.08423	11.871	.8288	4
5	18.895	.05292	.84470	1.1838	.04470	22.369	.9706	5
6	34.012	.02940	.82423	1.2132	.02423	41.265	1.0682	6
7	61.221	.01633	.81328	1.2295	.01328	75.277	1.1337	7
8	110.19	.00907	.80733	1.2386	.00733	136.49	1.1767	8
9	198.35	.00504	.80405	1.2437	.00405	246.69	1.2044	9
10	357.04	.00280	.80225	1.2465	.00225	445.05	1.2219	10
11	642.68	.00156	.80125	1.2480	.00125	802.10	1.2328	11
12	1156.8	.00086	.80069	1.2489	.00069	1444.7	1.2396	12
13	2082.2	.00048	.80038	1.2494	.00038	2601.6	1.2437	13
14	3748.1	.00027	.80021	1.2496	.00021	4683.8	1.2462	14
15	6746.5	.00015	.80012	1.2498	.00012	8431.9	1.2477	15

90% Interest Factors for Discrete Compounding Periods

	SINGLE PAYMENT		UNIFORM SERIES					
	Compound Amount Factor	Present Worth Factor	Capital Recovery Factor	Present Worth Factor	Sinking Fund Factor	Compound Amount Factor	Gradient Factor	
N	$(F/P, 90, N)$	$(P/F, 90, N)$	$(A/P, 90, N)$	$(P/A, 90, N)$	$(A/F, 90, N)$	$(F/A, 90, N)$	$(A/G, 90, N)$	N
1	1.9000	.52632	1.9000	.52632	1.00000	1.0000	.00000	1
2	3.6100	.27701	1.2448	.80332	.34483	2.9000	.34483	2
3	6.8589	.14579	1.0536	.94912	.15361	6.5099	.59908	3
4	13.032	.07673	.97480	1.0258	.07480	13.368	.77867	4
5	24.760	.04039	.93788	1.0662	.03788	26.401	.90068	5
6	47.045	.02126	.91955	1.0874	.01955	51.161	.98081	6
7	89.386	.01119	.91018	1.0986	.01018	98.207	1.0319	7
8	169.83	.00589	.90533	1.1045	.00533	187.59	1.0637	8
9	322.68	.00310	.90280	1.1076	.00280	357.42	1.0831	9
10	613.10	.00163	.90147	1.1093	.00147	680.11	1.0947	10

100% Interest Factors for Discrete Compounding Periods

	SINGLE PAYMENT		UNIFORM SERIES					
	Compound Amount Factor	Present Worth Factor	Capital Recovery Factor	Present Worth Factor	Sinking Fund Factor	Compound Amount Factor	Gradient Factor	
N	$(F/P, 100, N)$	$(P/F, 100, N)$	$(A/P, 100, N)$	$(P/A, 100, N)$	$(A/F, 100, N)$	$(F/A, 100, N)$	$(A/G, 100, N)$	N
1	2.000	.50000	2.0000	.50000	1.0000	1.000	.00000	1
2	4.000	.25000	1.3333	.75000	.33333	3.000	.33333	2
3	8.000	.12500	1.1428	.87500	.14286	7.000	.57143	3
4	16.000	.06250	1.0666	.93750	.06667	15.000	.73333	4
5	32.000	.03125	1.0322	.96875	.03226	31.000	.83871	5
6	64.00	.01562	1.0158	.98438	.01587	63.00	.90476	6
7	128.00	.00781	1.0078	.99219	.00787	127.00	.94488	7
8	256.00	.00391	1.0039	.99609	.00392	255.00	.96863	8
9	512.00	.00195	1.0019	.99805	.00196	511.00	.98239	9
10	1024.0	.00098	1.0009	.99902	.00098	1023.0	.99022	10

CONTINUOUS-COMPOUNDING INTEREST FACTORS

**Continuous-Compounding, Continuous-Flow Interest Factors
at an Effective Interest Rate of 8%; r = 7.696%**

N	SINGLE PAYMENT		UNIFORM SERIES				N
	$(F/\bar{P}, 7.696, N)$	$(P/\bar{F}, 7.696, N)$	$(\bar{A}/P, 7.696, N)$	$(P/\bar{A}, 7.696, N)$	$(\bar{A}/F, 7.696, N)$	$(F/\bar{A}, 7.696, N)$	
1	1.039	.9625	1.03897	962	.96201	1.039	1
2	1.123	.8912	.53947	1.854	.46251	2.162	2
3	1.212	.8252	.37329	2.679	.29633	3.375	3
4	1.309	.7641	.29045	3.443	.21349	4.684	4
5	1.414	.7075	.24094	4.150	.16398	6.098	5
6	1.527	.6551	.20810	4.805	.13114	7.626	6
7	1.650	.6065	.18478	5.412	.10782	9.275	7
8	1.781	.5616	.16740	5.974	.09044	11.057	8
9	1.924	.5200	.15400	6.494	.07704	12.981	9
10	2.078	.4815	.14337	6.975	.06641	15.059	10
11	2.244	.4458	.13476	7.421	.05779	17.303	11
12	2.424	.4128	.12765	7.834	.05069	19.726	12
13	2.618	.3822	.12172	8.216	.04475	22.344	13
14	2.827	.3539	.11669	8.570	.03973	25.171	14
15	3.053	.3277	.11239	8.897	.03543	28.224	15
16	3.297	.3034	.10869	9.201	.03172	31.522	16
17	3.561	.2809	.10546	9.482	.02850	35.083	17
18	3.846	.2601	.10265	9.742	.02569	38.929	18
19	4.154	.2409	.10017	9.983	.02321	43.083	19
20	4.486	.2230	.09798	10.206	.02102	47.569	20
21	4.845	.2065	.09604	10.412	.01908	52.414	21
22	5.233	.1912	.09431	10.604	.01735	57.647	22
23	5.651	.1770	.09276	10.781	.01580	63.298	23
24	6.103	.1639	.09137	10.945	.01441	69.401	24
25	6.592	.1518	.09012	11.096	.01316	75.993	25
26	7.119	.1405	.08899	11.237	.01203	83.112	26
27	7.688	.1301	.08797	11.367	.01101	90.800	27
28	8.303	.1205	.08705	11.487	.01009	99.103	28
29	8.968	.1116	.08621	11.599	.00925	108.071	29
30	9.685	.1033	.08545	11.702	.00849	117.756	30
31	10.460	.0956	.08476	11.798	.00780	128.216	31
32	11.297	.0886	.08413	11.887	.00717	139.513	32
33	12.201	.0820	.08355	11.969	.00659	151.714	33
34	13.177	.0759	.08303	12.044	.00606	164.890	34
35	14.231	.0703	.08254	12.115	.00558	179.121	35
40	20.910	.0478	.08067	12.395	.00371	269.286	40
45	30.723	.0326	.07945	12.587	.00249	401.768	45
50	45.142	.0222	.07864	12.717	.00168	596.427	50

**Continuous-Compounding, Continuous-Flow Interest Factors
at an Effective Interest Rate of 9%; r = 8.618%**

	SINGLE PAYMENT		UNIFORM SERIES				
N	$(F/\bar{P}, 8.618, N)$	$(P/\bar{F}, 8.618, N)$	$(\bar{A}/P, 8.618, N)$	$(P/\bar{A}, 8.618, N)$	$(\bar{A}/F, 8.618, N)$	$(F/\bar{A}, 8.618, N)$	N
1	1.044	.9581	1.04371	.958	.95753	1.044	1
2	1.138	.8790	.54433	1.837	.45815	2.183	2
3	1.241	.8064	.37828	2.644	.29210	3.423	3
4	1.352	.7398	.29556	3.383	.20938	4.776	4
5	1.474	.6788	.24617	4.062	.16000	6.250	5
6	1.607	.6227	.21345	4.685	.12727	7.857	6
7	1.751	.5713	.19025	5.256	.10407	9.609	7
8	1.909	.5241	.17300	5.780	.08682	11.518	8
9	2.081	.4808	.15971	6.261	.07354	13.599	9
10	2.268	.4411	.14920	6.702	.06302	15.867	10
11	2.472	.4047	.14071	7.107	.05453	18.339	11
12	2.695	.3713	.13372	7.478	.04754	21.034	12
13	2.937	.3406	.12789	7.819	.04172	23.971	13
14	3.202	.3125	.12298	8.131	.03680	27.173	14
15	3.490	.2867	.11879	8.418	.03261	30.663	15
16	3.804	.2630	.11519	8.681	.02901	34.467	16
17	4.146	.2413	.11208	8.923	.02590	38.614	17
18	4.520	.2214	.10936	9.144	.02318	43.133	18
19	4.926	.2031	.10699	9.347	.02081	48.060	19
20	5.370	.1863	.10489	9.533	.01872	53.429	20
21	5.853	.1710	.10305	9.704	.01687	59.282	21
22	6.380	.1568	.10141	9.861	.01523	65.662	22
23	6.954	.1439	.09995	10.005	.01377	72.616	23
24	7.580	.1320	.09865	10.137	.01247	80.196	24
25	8.262	.1211	.09748	10.258	.01130	88.458	25
26	9.006	.1111	.09644	10.369	.01026	97.463	26
27	9.816	.1019	.09550	10.471	.00932	107.279	27
28	10.699	.0935	.09465	10.565	.00848	117.979	28
29	11.662	.0858	.09389	10.651	.00771	129.641	29
30	12.712	.0787	.09320	10.729	.00702	142.353	30
31	13.856	.0722	.09258	10.802	.00640	156.209	31
32	15.103	.0663	.09201	10.868	.00584	171.313	32
33	16.462	.0608	.09150	10.929	.00533	187.775	33
34	17.944	.0558	.09104	10.984	.00486	205.719	34
35	19.559	.0512	.09062	11.035	.00444	225.278	35
40	30.094	.0332	.08901	11.234	.00283	352.869	40
45	46.303	.0216	.08800	11.364	.00182	549.183	45
50	71.244	.0140	.08735	11.448	.00117	851.236	50

Continuous-Compounding, Continuous-Flow Interest Factors
At an Effective Interest Rate of 10%; r = 9.531%

N	SINGLE PAYMENT		UNIFORM SERIES				N
	$(F/\bar{P}, 9.531, N)$	$(P/\bar{F}, 9.531, N)$	$(\bar{A}/P, 9.531, N)$	$(P/\bar{A}, 9.531, N)$	$(\bar{A}/F, 9.531, N)$	$(F/\bar{A}, 9.531, N)$	N
1	1.049	.9538	1.04841	.954	.95310	1.049	1
2	1.154	.8671	.54917	1.821	.45386	2.203	2
3	1.270	.7883	.38326	2.609	.28795	3.473	3
4	1.396	.7166	.30068	3.326	.20537	4.869	4
5	1.536	.6515	.25143	3.977	.15612	6.406	5
6	1.690	.5922	.21884	4.570	.12353	8.095	6
7	1.859	.5384	.19577	5.108	.10046	9.954	7
8	2.045	.4895	.17865	5.597	.08334	11.999	8
9	2.249	.4450	.16550	6.042	.07019	14.248	9
10	2.474	.4045	.15511	6.447	.05980	16.722	10
11	2.721	.3677	.14674	6.815	.05143	19.443	11
12	2.994	.3343	.13988	7.149	.04457	22.437	12
13	3.293	.3039	.13418	7.453	.03887	25.729	13
14	3.622	.2763	.12938	7.729	.03407	29.352	14
15	3.984	.2512	.12531	7.980	.03000	33.336	15
16	4.383	.2283	.12182	8.209	.02651	37.719	16
17	4.821	.2076	.11882	8.416	.02351	42.540	17
18	5.303	.1887	.11621	8.605	.02090	47.843	18
19	5.833	.1716	.11394	8.777	.01863	53.676	19
20	6.417	.1560	.11195	8.932	.01664	60.093	20
21	7.059	.1418	.11020	9.074	.01489	67.152	21
22	7.764	.1289	.10866	9.203	.01335	74.916	22
23	8.541	.1172	.10729	9.320	.01198	83.457	23
24	9.395	.1065	.10608	9.427	.01077	92.852	24
25	10.334	.0968	.10500	9.524	.00969	103.186	25
26	11.368	.0880	.10404	9.612	.00873	114.554	26
27	12.505	.0800	.10318	9.692	.00787	127.059	27
28	13.755	.0728	.10241	9.765	.00710	140.814	28
29	15.131	.0661	.10172	9.831	.00641	155.944	29
30	16.644	.0601	.10110	9.891	.00579	172.588	30
31	18.308	.0547	.10055	9.945	.00524	190.896	31
32	20.139	.0497	.10005	9.995	.00474	211.035	32
33	22.153	.0452	.09960	10.040	.00429	233.188	33
34	24.368	.0411	.09919	10.081	.00388	257.556	34
35	26.805	.0373	.09883	10.119	.00352	284.360	35
40	43.169	.0232	.09746	10.260	.00215	464.371	40
45	69.525	.0144	.09664	10.348	.00133	754.279	45
50	111.970	.0089	.09618	10.403	.00082	1221.180	50

**Continuous-Compounding, Continuous-Flow Interest Factors
at an Effective Interest Rate of 11%; r = 10.436%**

N	SINGLE PAYMENT		UNIFORM SERIES				N
	$(F/\bar{P}, 10.436, N)$	$(P/\bar{F}, 10.436, N)$	$(\bar{A}/P, 10.436, N)$	$(P/\bar{A}, 10.436, N)$	$(\bar{A}/F, 10.436, N)$	$(F/\bar{A}, 10.436, N)$	
1	1.054	.9496	1.05309	.950	.94873	1.054	1
2	1.170	.8555	.55399	1.805	.44963	2.224	2
3	1.299	.7707	.38823	2.576	.28387	3.523	3
4	1.442	.6943	.30580	3.270	.20144	4.964	4
5	1.600	.6255	.25670	3.896	.15234	6.564	5
6	1.776	.5635	.22426	4.459	.11990	8.340	6
7	1.971	.5077	.20133	4.967	.09697	10.312	7
8	2.188	.4574	.18436	5.424	.08000	12.500	8
9	2.429	.4121	.17134	5.836	.06698	14.929	9
10	2.696	.3712	.16110	6.208	.05674	17.626	10
11	2.993	.3344	.15286	6.542	.04850	20.619	11
12	3.322	.3013	.14613	6.843	.04177	23.941	12
13	3.688	.2714	.14055	7.115	.03619	27.628	13
14	4.093	.2445	.13588	7.359	.03152	31.721	14
15	4.543	.2203	.13194	7.579	.02757	36.265	15
16	5.043	.1985	.12857	7.778	.02421	41.308	16
17	5.598	.1788	.12568	7.957	.02132	46.906	17
18	6.214	.1611	.12319	8.118	.01883	53.120	18
19	6.897	.1451	.12102	8.263	.01666	60.017	19
20	7.656	.1307	.11914	8.394	.01478	67.673	20
21	8.498	.1178	.11749	8.511	.01313	76.171	21
22	9.433	.1061	.11604	8.618	.01168	85.603	22
23	10.470	.0956	.11477	8.713	.01041	96.074	23
24	11.622	.0861	.11365	8.799	.00929	107.696	24
25	12.901	.0776	.11265	8.877	.00829	120.597	25
26	14.320	.0699	.11177	8.947	.00741	134.916	26
27	15.895	.0630	.11099	9.010	.00663	150.811	27
28	17.643	.0567	.11030	9.066	.00594	168.454	28
29	19.584	.0511	.10968	9.118	.00532	188.038	29
30	21.738	.0460	.10913	9.164	.00477	209.777	30
31	24.129	.0415	.10864	9.205	.00428	233.906	31
32	26.784	.0374	.10820	9.242	.00384	260.690	32
33	29.730	.0337	.10780	9.276	.00344	290.420	33
34	33.000	.0303	.10745	9.306	.00309	323.420	34
35	36.630	.0273	.10714	9.334	.00278	360.050	35
40	61.724	.0162	.10599	9.435	.00163	613.270	40
45	104.009	.0096	.10532	9.495	.00096	1039.960	45
50	175.261	.0057	.10493	9.530	.00057	1758.958	50

Continuous-Compounding, Continuous-Flow Interest Factors
at an Effective Interest Rate of 12%; r = 11.333%

N	SINGLE PAYMENT		UNIFORM SERIES				N
	$(F/\bar{P}, 11.333, N)$	$(P/\bar{F}, 11.333, N)$	$(\bar{A}/P, 11.333, N)$	$(P/\bar{A}, 11.333, N)$	$(\bar{A}/F, 11.333, N)$	$(F/\bar{A}, 11.333, N)$	
1	1.059	.9454	1.05773	.945	.94441	1.059	1
2	1.186	.8441	.55880	1.790	.44547	2.245	2
3	1.328	.7537	.39320	2.543	.27987	3.573	3
4	1.488	.6729	.31093	3.216	.19760	5.061	4
5	1.666	.6008	.26199	3.817	.14866	6.727	5
6	1.866	.5365	.22970	4.353	.11638	8.593	6
7	2.090	.4790	.20694	4.832	.09361	10.683	7
8	2.341	.4277	.19011	5.260	.07678	13.024	8
9	2.622	.3818	.17725	5.642	.06392	15.645	9
10	2.936	.3409	.16714	5.983	.05382	18.582	10
11	3.289	.3044	.15905	6.287	.04572	21.870	11
12	3.683	.2718	.15246	6.559	.03913	25.554	12
13	4.125	.2427	.14702	6.802	.03369	29.679	13
14	4.620	.2167	.14248	7.018	.02915	34.299	14
15	5.175	.1935	.13866	7.212	.02533	39.474	15
16	5.796	.1727	.13542	7.385	.02209	45.270	16
17	6.491	.1542	.13265	7.539	.01932	51.761	17
18	7.270	.1377	.13027	7.676	.01694	59.032	18
19	8.143	.1229	.12822	7.799	.01489	67.174	19
20	9.120	.1098	.12644	7.909	.01311	76.294	20
21	10.214	.0980	.12489	8.007	.01156	86.508	21
22	11.440	.0875	.12354	8.095	.01021	97.948	22
23	12.813	.0781	.12236	8.173	.00903	110.761	23
24	14.350	.0698	.12132	8.243	.00799	125.111	24
25	16.072	.0623	.12041	8.305	.00708	141.183	25
26	18.001	.0556	.11961	8.360	.00628	159.184	26
27	20.161	.0497	.11890	8.410	.00558	179.345	27
28	22.580	.0443	.11828	8.454	.00495	201.925	28
29	25.290	.0396	.11773	8.494	.00440	227.215	29
30	28.325	.0353	.11724	8.529	.00391	255.539	30
31	31.724	.0316	.11681	8.561	.00348	287.263	31
32	35.530	.0282	.11643	8.589	.00310	322.793	32
33	39.794	.0252	.11609	8.614	.00276	362.587	33
34	44.569	.0225	.11578	8.637	.00246	407.157	34
35	49.918	.0201	.11552	8.657	.00219	457.074	35
40	87.972	.0114	.11456	8.729	.00123	812.248	40
45	155.037	.0065	.11402	8.770	.00070	1438.185	45
50	273.228	.0037	.11372	8.793	.00039	2541.300	50

Continuous-Compounding, Continuous-Flow Interest Factors
at an Effective Interest Rate of 13%; r = 12.222%

N	SINGLE PAYMENT		UNIFORM SERIES				N
	$(F/\bar{P}, 12.222, N)$	$(P/\bar{F}, 12.222, N)$	$(\bar{A}/P, 12.222, N)$	$(P/\bar{A}, 12.222, N)$	$(\bar{A}/F, 12.222, N)$	$(F/\bar{A}, 12.222, N)$	
1	1.064	.9413	1.06235	.941	.94014	1.064	1
2	1.202	.8330	.56360	1.774	.44138	2.266	2
3	1.358	.7372	.39817	2.512	.27595	3.624	3
4	1.535	.6524	.31607	3.164	.19385	5.159	4
5	1.734	.5773	.26729	3.741	.14508	6.893	5
6	1.960	.5109	.23518	4.252	.11296	8.853	6
7	2.215	.4521	.21257	4.704	.09036	11.067	7
8	2.502	.4001	.19591	5.104	.07369	13.570	8
9	2.828	.3541	.18320	5.458	.06099	16.397	9
10	3.195	.3133	.17326	5.772	.05104	19.593	10
11	3.611	.2773	.16531	6.049	.04310	23.203	11
12	4.080	.2454	.15887	6.294	.03665	27.283	12
13	4.611	.2172	.15357	6.512	.03135	31.894	13
14	5.210	.1922	.14917	6.704	.02695	37.104	14
15	5.887	.1701	.14548	6.874	.02326	42.991	15
16	6.653	.1505	.14236	7.024	.02014	49.644	16
17	7.517	.1332	.13971	7.158	.01749	57.161	17
18	8.495	.1179	.13745	7.275	.01523	65.656	18
19	9.599	.1043	.13551	7.380	.01329	75.254	19
20	10.847	.0923	.13383	7.472	.01161	86.101	20
21	12.257	.0817	.13238	7.554	.01017	98.358	21
22	13.850	.0723	.13113	7.626	.00891	112.208	22
23	15.651	.0640	.13004	7.690	.00782	127.859	23
24	17.685	.0566	.12909	7.747	.00687	145.544	24
25	19.984	.0501	.12826	7.797	.00604	165.529	25
26	22.582	.0443	.12753	7.841	.00532	188.111	26
27	25.518	.0392	.12690	7.880	.00468	213.629	27
28	28.835	.0347	.12634	7.915	.00412	242.465	28
29	32.584	.0307	.12585	7.946	.00364	275.049	29
30	36.820	.0272	.12542	7.973	.00321	311.869	30
31	41.607	.0241	.12505	7.997	.00283	353.476	31
32	47.016	.0213	.12471	8.018	.00250	400.491	32
33	53.128	.0188	.12442	8.037	.00220	453.619	33
34	60.034	.0167	.12416	8.054	.00195	513.653	34
35	67.839	.0148	.12394	8.069	.00172	581.491	35
40	124.988	.0080	.12315	8.121	.00093	1078.253	40
45	230.283	.0043	.12272	8.149	.00050	1993.505	45
50	424.281	.0024	.12249	8.164	.00027	3679.796	50

Continuous-Compounding, Continuous-Flow Interest Factors
at an Effective Interest Rate of 14%; r = 13.103%

N	SINGLE PAYMENT		UNIFORM SERIES				N
	$(F/\bar{P}, 13.103, N)$	$(P/\bar{F}, 13.103, N)$	$(\bar{A}/P, 13.103, N)$	$(P/\bar{A}, 13.103, N)$	$(\bar{A}/F, 13.103, N)$	$(F/\bar{A}, 13.103, N)$	N
1	1.068	.9373	1.06694	.937	.93592	1.068	1
2	1.218	.8222	.56837	1.759	.43734	2.287	2
3	1.389	.7212	.40313	2.481	.27210	3.675	3
4	1.583	.6326	.32121	3.113	.19018	5.258	4
5	1.805	.5549	.27262	3.668	.14159	7.063	5
6	2.057	.4868	.24068	4.155	.10965	9.120	6
7	2.345	.4270	.21825	4.582	.08722	11.465	7
8	2.674	.3746	.20176	4.956	.07073	14.139	8
9	3.048	.3286	.18921	5.285	.05818	17.187	9
10	3.475	.2882	.17943	5.573	.04840	20.661	10
11	3.961	.2528	.17164	5.826	.04061	24.622	11
12	4.516	.2218	.16535	6.048	.03432	29.138	12
13	5.148	.1945	.16019	6.242	.02917	34.286	13
14	5.868	.1706	.15593	6.413	.02490	40.154	14
15	6.690	.1497	.15238	6.563	.02135	46.844	15
16	7.627	.1313	.14939	6.694	.01836	54.471	16
17	8.694	.1152	.14686	6.809	.01583	63.165	17
18	9.912	.1010	.14471	6.910	.01368	73.077	18
19	11.299	.0886	.14288	6.999	.01185	84.376	19
20	12.881	.0777	.14131	7.077	.01028	97.258	20
21	14.685	.0682	.13996	7.145	.00893	111.942	21
22	16.740	.0598	.13880	7.205	.00777	128.682	22
23	19.084	.0525	.13780	7.257	.00677	147.766	23
24	21.756	.0460	.13693	7.303	.00590	169.522	24
25	24.802	.0404	.13617	7.344	.00515	194.324	25
26	28.274	.0354	.13552	7.379	.00449	222.598	26
27	32.232	.0311	.13495	7.410	.00392	254.830	27
28	36.745	.0273	.13446	7.437	.00343	291.574	28
29	41.889	.0239	.13403	7.461	.00300	333.463	29
30	47.753	.0210	.13365	7.482	.00262	381.217	30
31	54.439	.0184	.13332	7.501	.00230	435.655	31
32	62.060	.0161	.13304	7.517	.00201	497.716	32
33	70.749	.0142	.13279	7.531	.00176	568.464	33
34	80.653	.0124	.13257	7.543	.00154	649.118	34
35	91.945	.0109	.13238	7.554	.00135	741.063	35
40	177.032	.0057	.13173	7.592	.00070	1433.916	40
45	340.860	.0029	.13139	7.611	.00036	2767.945	45
50	656.298	.0015	.13122	7.621	.00019	5336.505	50

**Continuous-Compounding, Continuous-Flow Interest Factors
at an Effective Interest Rate of 15%; r = 13.976%**

N	SINGLE PAYMENT		UNIFORM SERIES				N
	$(F/\bar{P}, 13.976, N)$	$(P/\bar{F}, 13.976, N)$	$(\bar{A}/P, 13.976, N)$	$(P/\bar{A}, 13.976, N)$	$(\bar{A}/F, 13.976, N)$	$(F/\bar{A}, 13.976, N)$	
1	1.073	.9333	1.07151	.933	.93175	1.073	1
2	1.234	.8115	.57313	1.745	.43337	2.307	2
3	1.419	.7057	.40808	2.450	.26832	3.727	3
4	1.632	.6136	.32636	3.064	.18660	5.359	4
5	1.877	.5336	.27795	3.598	.13819	7.236	5
6	2.159	.4640	.24620	4.062	.10644	9.395	6
7	2.483	.4035	.22395	4.465	.08419	11.877	7
8	2.855	.3508	.20764	4.816	.06788	14.732	8
9	3.283	.3051	.19527	5.121	.05551	18.015	9
10	3.776	.2653	.18565	5.386	.04589	21.791	10
11	4.342	.2307	.17803	5.617	.03827	26.133	11
12	4.993	.2006	.17189	5.818	.03213	31.126	12
13	5.742	.1744	.16689	5.992	.02712	36.868	13
14	6.604	.1517	.16277	6.144	.02300	43.472	14
15	7.594	.1319	.15934	6.276	.01958	51.066	15
16	8.733	.1147	.15648	6.390	.01672	59.799	16
17	10.043	.0997	.15408	6.490	.01432	69.842	17
18	11.550	.0867	.15205	6.577	.01229	81.392	18
19	13.282	.0754	.15032	6.652	.01056	94.674	19
20	15.274	.0656	.14886	6.718	.00910	109.948	20
21	17.565	.0570	.14760	6.775	.00784	127.513	21
22	20.200	.0496	.14653	6.824	.00677	147.714	22
23	23.230	.0431	.14561	6.868	.00585	170.944	23
24	26.715	.0375	.14482	6.905	.00506	197.659	24
25	30.722	.0326	.14414	6.938	.00438	228.381	25
26	35.330	.0284	.14355	6.966	.00379	263.711	26
27	40.630	.0247	.14305	6.991	.00329	304.341	27
28	46.724	.0214	.14261	7.012	.00285	351.066	28
29	53.733	.0186	.14223	7.031	.00247	404.799	29
30	61.793	.0162	.14191	7.047	.00214	466.592	30
31	71.062	.0141	.14162	7.061	.00186	537.654	31
32	81.721	.0123	.14138	7.073	.00161	619.375	32
33	93.980	.0107	.14116	7.084	.00140	713.355	33
34	108.076	.0093	.14098	7.093	.00122	821.431	34
35	124.288	.0081	.14082	7.101	.00106	945.719	35
40	249.987	.0040	.14029	7.128	.00052	1909.415	40
45	502.814	.0020	.14002	7.142	.00026	3847.752	45
50	1011.339	.0010	.13989	7.148	.00013	7746.440	50

APPENDIX F

AREAS OF A STANDARD NORMAL DISTRIBUTION

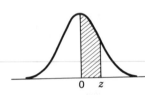

An entry in the table is the proportion under the entire curve which is between $Z = 0$ and a positive value of Z. Areas for negative values of Z are obtained by symmetry. The areas correspond to probabilities. Thus the area between $Z = 0$ and $Z = 2$ is 0.4772, which leads to a probability of $0.5000 - 0.4772 = 0.0228$ that Z will be larger than 2 in a normally distributed population. By symmetry, the probability that Z is less than -2 is $0.5000 - 0.4772 = 0.0228$.

Z	0.00	0.01	0.02	0.03	0.04	0.05	0.06	0.07	0.08	0.09
.0	.0000	.0040	.0080	.0120	.0160	.0199	.0239	.0279	.0319	.0359
.1	.0398	.0438	.0478	.0517	.0557	.0596	.0636	.0675	.0714	.0753
.2	.0793	.0832	.0871	.0910	.0948	.0987	.1026	.1064	.1103	.1141
.3	.1179	.1217	.1255	.1293	.1331	.1368	.1406	.1443	.1480	.1517
.4	.1554	.1591	.1628	.1664	.1700	.1736	.1772	.1808	.1844	.1879
.5	.1915	.1950	.1985	.2019	.2054	.2088	.2123	.2157	.2190	.2234
.6	.2257	.2291	.2324	.2357	.2389	.2422	.2454	.2486	.2517	.2549
.7	.2580	.2614	.2642	.2673	.2703	.2734	.2764	.2794	.2823	.2852
.8	.2881	.2910	.2939	.2967	.2995	.3023	.3051	.3078	.3106	.3133
.9	.3159	.3186	.3212	.3238	.3264	.3289	.3315	.3340	.3365	.3389
1.0	.3413	.3438	.3461	.3485	.3508	.3531	.3554	.3577	.3599	.3621
1.1	.3643	.3665	.3686	.3708	.3729	.3749	.3770	.3790	.3810	.3830
1.2	.3849	.3869	.3888	.3907	.3925	.3944	.3962	.3980	.3997	.4015
1.3	.4032	.4049	.4066	.4082	.4099	.4115	.4131	.4147	.4162	.4177
1.4	.4192	.4207	.4222	.4236	.4251	.4265	.4279	.4292	.4306	.4319
1.5	.4332	.4345	.4357	.4370	.4382	.4394	.4406	.4418	.4429	.4441
1.6	.4452	.4463	.4474	.4484	.4495	.4505	.4515	.4525	.4535	.4545
1.7	.4554	.4564	.4573	.4582	.4591	.4599	.4608	.4616	.4625	.4633
1.8	.4641	.4649	.4656	.4664	.4671	.4678	.4686	.4693	.4699	.4706
1.9	.4713	.4719	.4726	.4732	.4738	.4744	.4750	.4756	.4761	.4767
2.0	.4772	.4778	.4783	.4788	.4793	.4798	.4803	.4808	.4812	.4817
2.1	.4821	.4826	.4830	.4834	.4838	.4842	.4846	.4850	.4854	.4857
2.2	.4861	.4864	.4868	.4871	.4875	.4878	.4881	.4884	.4887	.4890
2.3	.4893	.4896	.4898	.4901	.4904	.4906	.4909	.4911	.4913	.4916
2.4	.4918	.4920	.4922	.4925	.4927	.4929	.4931	.4932	.4934	.4936
2.5	.4938	.4940	.4941	.4943	.4945	.4946	.4948	.4949	.4951	.4952
2.6	.4953	.4955	.4956	.4957	.4959	.4960	.4961	.4962	.4963	.4964
2.7	.4965	.4966	.4967	.4968	.4969	.4970	.4971	.4972	.4973	.4974
2.8	.4974	.4975	.4976	.4977	.4977	.4978	.4979	.4979	.4980	.4981
2.9	.4981	.4982	.4982	.4983	.4984	.4984	.4985	.4985	.4986	.4986
3.0	.4987	.4987	.4987	.4988	.4988	.4989	.4989	.4989	.4990	.4990

INDEX

INDEX